ENZYMOLOGY PRIMER FOR
Recombinant DNA Technology

ENZYMOLOGY PRIMER FOR
Recombinant DNA Technology

Hyone-Myong Eun
Laboratoires Virbac
France

Academic Press
San Diego   New York   Boston   London   Sydney   Tokyo   Toronto

This book is printed on acid-free paper. ∞

Copyright © 1996 by ACADEMIC PRESS, INC.

All Rights Reserved.
No part of this publication may be reproduced or transmitted in any form or by any means, electronic or mechanical, including photocopy, recording, or any information storage and retrieval system, without permission in writing from the publisher.

Academic Press, Inc.
A Division of Harcourt Brace & Company
525 B Street, Suite 1900, San Diego, California 92101-4495

United Kingdom Edition published by
Academic Press Limited
24-28 Oval Road, London NW1 7DX

Library of Congress Cataloging-in-Publication Data

Eun, Hyone-Myong.
    Enzymology primer for recombinant DNA technology / Hyone-Myong Eun.
        p.   cm.
    Includes index.
    ISBN 0-12-243740-3 (alk. paper)
    1. Enzymes. 2. Geentic engineering. 3. Recombinant DNA
I. Title.
QP601.E84   1996
54.19'25--dc20
                                                  95-30009
                                                  CIP

PRINTED IN THE UNITED STATES OF AMERICA
96 97 98 99 00 01 MM 9 8 7 6 5 4 3 2 1

*To the memory of
Sidney A. Bernhard, Professor of Biochemistry,
Institute of Molecular Biology, University of Oregon,
and my mentor.*

Contents

Preface xi
Acknowledgments xv
List of Abbreviations and Symbols xvii
List of Journal Abbreviations xxiii

1 Enzymes and Nucleic Acids: General Principles

I. Structure and Function of Enzymes 1
   A. Biosynthesis of Proteins 2
   B. Structure 18
   C. Function 33
      References 48
II. Structure and Function of Nucleic Acids 51
   A. General Description 51
   B. Denaturation and Renaturation 59
   C. Structure of DNA: Secondary and Tertiary Structures 68
   D. Structure and Function of RNA 83
      References 103

2 Ligases

I. DNA Ligases 109
   A. General Description 109
      References 119
   B. *Escherichia coli* DNA Ligase 120
      References 124
   C. T4 DNA Ligase 125
      References 132
II. RNA Ligase: T4 RNA Ligase 133
      References 143

3 Nucleases

I. DNA Endonuclease: Deoxyribonuclease I   146
    References   158
II. DNA/RNA Endonuclease: Staphylococcal Nuclease   160
    References   167
III. RNA Endonucleases   169
    References   170
    A. Ribonuclease A   170
       References   182
    B. Ribonuclease H   184
       References   194
    C. Ribonuclease T1   196
       References   203
IV. Single-Strand-Specific Endonucleases   204
    A. Nuclease S1   205
       References   211
    B. Mung Bean Nuclease   212
       References   215
V. DNA Exonucleases   215
    A. 3'→5'-Exonuclease: *E. coli* Exonuclease III   215
       References   225
    B. General Exonuclease: Nuclease Bal31   226
       References   231

4 Restriction Endonucleases and Modification Methylases

I. General Description   233
    References   238
II. Restriction Endonucleases   238
    A. Type II Restriction Endonucleases: General Properties   238
       References   249
    B. Type II Restriction Endonuclease: *Eco*RI   251
       References   265
    C. Type IIS Restriction Endonuclease: *Fok*I Restriction–Modification System   267
       References   274
    D. Type I Restriction Enzymes   275
       References   280
    E. Type III Restriction Enzymes   281
       References   284

III. Methyltransferases (Methylases) 284
   A. General Description 284
      References 293
   B. Type II Methylase: *Eco*RI Methyltransferase 294
      References 299
   C. Nontype II Methylases: Dam and Dcm Methyltransferases 300
      References 305

5 Phosphatases and Polynucleotide Kinase

I. Phosphatases (Alkaline) 307
   A. Bacterial Alkaline Phosphatase 308
      References 324
   B. Calf Intestinal Alkaline Phosphatase 326
      References 332
II. Polynucleotide Kinase: T4 Polynucleotide Kinase 333
     References 343

6 DNA Polymerases

I. General Description 345
     References 350
II. DNA-Directed DNA Polymerases 351
   A. *E. coli* DNA Polymerase I 351
      References 374
   B. T4 DNA Polymerase 377
      References 393
   C. T7 DNA Polymerase 394
      References 406
   D. *Taq* DNA Polymerase 407
      References 425
III. RNA-Directed DNA Polymerases (Reverse Transcriptases) 427
   A. General Description 427
      References 447
   B. AMV Reverse Transcriptase 450
      References 463
   C. MoLV Reverse Transcriptase 465
      References 475
IV. Template-Independent DNA Polymerase: Terminal Deoxynucleotidyltransferase 477
     References 488

7 RNA Polymerases

I. DNA-Directed RNA Polymerases 491
   A. General Description 491
      References 518
   B. T7 RNA Polymerase 520
      References 546
   C. SP6 RNA Polymerase 549
      References 552
   D. T3 RNA Polymerase 553
      References 555
II. DNA-Independent RNA Polymerase: Poly(A) Polymerase 555
      References 564

8 Marker/Reporter Enzymes

   A. β-Galactosidase 568
      References 593
   B. β-Lactamase 595
      References 611
   C. Chloramphenicol Acetyltransferase 613
      References 626
   D. Luciferases 627
      References 643

Appendix A: Important Molecular Biological Methods

I. DNA Labeling 647
      References 650
II. Nucleotide Sequencing 651
      References 657
III. Polymerase Chain Reaction 657
      References 658
IV. Site-Specific *in Vitro* Mutagenesis 659
      References 671

Appendix B: Genotypes of *Escherichia coli* Strains

I. Genotypes of Some Frequently Used *E. coli* Strains 674
II. Useful Genetic Markers and Phenotypic Traits 677

Appendix C: Practical Guide for Enzyme Handling 683

Index 687

Preface

A vast array of molecular biological techniques currently used in gene analyses and nucleic acid manipulations depend on the catalytic power of enzymes. With the discovery of site-specific restriction endonucleases and their use in combination with other enzymes that synthesize, cut, link, degrade, and modify DNA and RNA, the burgeoning era of recombinant DNA technology has begun. Thanks to recombinant DNA technology, the structure and function of a gene or a protein can now be dissected and characterized at a molecular level with greater ease, and valuable proteins can be produced in a safe, convenient, controlled, and economic manner. In the past decade, we have witnessed explosive advances in and diversification of recombinant DNA technology with far-reaching implications. These advances have been drastically changing the pace of progress in bioscience and reshaping bioindustries. Their applications are limited only by our imaginations. They have given rise to new disciplines such as "genetic engineering," "protein engineering," and "biotechnology." At the foundation of these new technologies is recombinant DNA technology, which in turn is based on the specificity and efficiency of the catalytic enzyme molecules. Past progress of these technologies is due to a large extent to enzymes, and further advances will undoubtedly require greater roles for enzymes.

By classical definition, an enzyme is a protein, but this definition should now be expanded to include nucleic acid enzymes as well. As a biological macromolecule endowed with catalytic function(s), each enzyme has its reaction specificity, substrate specificity, and optimal reaction conditions. It can be inhibited, degraded, or denatured by physical interactions with nonenzymic molecules during reaction and storage conditions. Enzymes are sophisticated, ingenious molecular machines, "operating at the boundary where chemistry just becomes biology" [Kraut, J. (1988). *Science* **242**, 533]. Only judicious use of enzymes leads to the best results in recombinant DNA research. Yet enzymes are more often regarded as simple reagents, the use of which is frequently guided by insufficient description in technical manuals and catalogs.

Enzymology has made tremendous progress in recent years, and our knowledge of enzyme catalysis is rapidly increasing. This is due in large part to the application of recombinant DNA technology to the study of enzymes themselves. For most nonenzymologists involved in recombinant DNA research, however, it is difficult, if not impossible, to keep pace with the rapid advances in enzymology. The primary goal of this book is, therefore, to help molecular biologists and recombinant DNA technologists to better understand the inner workings of these biocatalysts as they are utilized in recombinant DNA techniques. Thus it provides information for more judicious, appropriate, and creative use of enzymes as tools.

The book is organized as a quick reference source to the biochemical, biophysical, and catalytic properties of selected enzymes. It is designed to be a companion to technical manuals which focus on step-by-step procedures for certain molecular biological techniques. Since every book must be selective in its content, I have chosen to include only those enzymes that are most commonly used in present-day recombinant DNA technology. A conscious effort has been made to present only information relevant to and practical for recombinant DNA technology, largely omitting historical and theoretical aspects of enzymology.

The topics appear in eight chapters. Chapter 1 describes the functional and structural principles of enzymes and nucleic acids in general. It provides the reader with the basic concepts of enzymology, nucleic acid chemistry, and molecular biology on which recombinant DNA technology is based. It presents terminology relevant to new concepts or techniques. Chapters 2 through 8 describe the biochemical and biophysical properties and molecular genetics of selected enzymes, which are arranged in alphabetical order. Chapter 8 is followed by *Appendixes A–C:* (A) a glossary of methods treating fundamental concepts of some landmark molecular biological techniques; (B) genotypes of *E. coli* strains commonly used in recombinant DNA technology; and (C) a general guide for enzyme handling.

Each chapter is subdivided into sections beginning with a short introduction or general description, which gives a brief overview of the subject. Then functional, structural, and genetic aspects of each enzyme are described, with examples of typical applications. Functional aspects include (1) reaction conditions (optimal or recommended reaction conditions, kinetic parameters, cofactor requirements, and inhibitors/inactivators); (2) activity assay and unit definition; (3) substrate specificity; and (4) catalytic mechanism. The section on enzyme *functions* should enable the reader to predict the possible outcome of a reaction under a given condition and provide the basis for which the activity of an enzyme can be optimized or fine tuned to obtain the desired results.

The *applications* section presents some typical examples illustrating the basic ideas and strategies behind a recombinant DNA method. It shows how intricate enzyme specificities (both reaction and substrate specificities) can be exploited and/or integrated into recombinant DNA research.

The *structure* section provides basic information on primary, secondary, tertiary, and, where appropriate, quaternary structures. The enzyme active site is described, with emphasis on structure–function relationships.

The *genetics* section presents available genetic information on enzymes with regard to gene structure, organization, cloning status, and physiological role. As more and more cloned enzymes are introduced in recombinant DNA technology, the genetic information should be valuable for designing improved enzymes with more specific or expanded utility. Chapter 8 (marker/reporter enzymes) provides typical examples of how some enzymes are used more on the genetic level, that is, as an integral part of recombinant DNA strategy, than as exogenous proteins.

The *sources* section gives a relatively general description of the current methodologies used to purify and/or prepare the specific enzyme in question. This section should not discourage the use of commercially available enzymes.

The *reference* section guides the more inquisitive reader to direct sources for further information. Although no attempts have been made to be exhaustive, much of the earlier or additional literature can be readily traced back from the publications cited.

The organization of the book allows each section to be consulted rapidly and independently. I will have achieved my major goal if this work helps readers make an educated choice in the proper use of enzymes in their recombinant DNA research.

Hyone-Myong Eun

Acknowledgments

This book would not have materialized without the generous help of numerous people. It is my pleasure to acknowledge some of those who gladly sacrificed their precious time to review part of the manuscript: H.-C. Chen, L. Coulter, H.-M. Kang, A. Kim, D. Kruger, J. Diver, A. Maclean, R. Pon, M. Sexton, and P. Wishart. I also express my gratitude to J.-F. Petit, J. Yon, P. von Hippel, E. Miles, J. W. Yoon, and A. Flamand for their encouragement.

I owe so much to my family, especially to my spouse Michelle Perennec whose understanding, endurance, and assistance have been indispensable throughout this work. I hope my children, Valerie and Nathalie, will forgive me for so often depriving them of the tender moments of being together.

Finally, my sincere thanks go to Dolores Wright and Shirley Light of Academic Press for their valuable comments and assistance in bringing this manuscript to fruition.

List of Abbreviations and Symbols

A	adenine
A_{260}	absorbance at 260 nm
aa	amino acids
Ab	antibody (antibodies)
Ac	acetyl (group)
Ado	adenosyl (group)
AdoHcy	S-adenosylhomocysteine
AdoMet	S-adenosylmethionine
AMV	avian myeloblastosis virus
Ap (Ap^r)	ampicillin (ampicillin resistance)
AP	apurinic/apyrimidinic
6-APA	6-aminopenicillanic acid
APase	alkaline phosphatase
BAP	bacterial alkaline phosphatase
B. cereus	*Bacillus cereus*
BCIP	5-bromo-4-chloro-3-indolyl phosphate
βGal	β-galactosidase
B. licheniformis	*Bacillus licheniformis*
Bluo-Gal	5-bromoindolyl-β-O-galactopyranoside
bp	base pair

BSA	bovine serum albumin
B. subtilis	*Bacillus subtilis*
C	cytosine
CAT	chloramphenicol acetyltransferase
ccc	covalently closed circular (form of plasmid)
CD	circular dichroism
c^7dGTP	7-deaza-2′-deoxyguanosine 5′-triphosphate
CHES	2-(cyclohexylamino)ethanesulfonic acid (buffer pK_a = 10.4 at 25°C)
Ci	Curie (= 3.7×10^{10} Bq, Bq for Becquerel)
CIAP	calf intestinal alkaline phosphatase
Cm (Cm^r)	chloramphenicol (chloramphenicol resistance)
CMV	cytomegalovirus
3-D	Three-dimension(al)
Da	daltons(s) [molecular mass unit corresponding to one-twelfth mass of carbon 12 or 1.66×10^{-24} g, kDa (kilodalton) = 10^3 Da]
dAPTP	deoxyaminopurine 5′-triphosphate
ddNTP	2′,3′-dideoxyribonucleoside 5′-triphosphate
DEA	diethylamine (buffer pK_a = 11.0 at 25°C)
DEPC	diethyl pyrocarbonate
DFP	diisopropyl fluorophosphate
DMF	dimethylformamide
DNA	deoxyribonucleic acid
DNase	deoxyribonuclease
dNDP	2′-deoxyribonucleoside 5′-diphosphate (e.g., dADP, dCDP)
dNMP	2′-deoxyribonucleoside 5′-monophosphate (e.g., dAMP, dCMP)
DNP (-P)	dinitrophenol (dinitrophenylphosphate)
dNTP	2′-deoxyribonucleoside 5′-triphosphate (e.g., dATP, dCTP)
dNTPαS	2′-deoxynucleoside 5′-O-(1-thiotriphosphate)
dpm	disintegrations per minute
ds	double-strand(ed)
DTE	dithioerythritol
DTNB	5,5′-dithiobis(2-nitrobenzoate)
DTT	dithiothreitol
ε	molar absorption coefficient (or molar absorptivity, M^{-1} cm^{-1})
E. coli	*Escherichia coli*
EDTA	ethylenediaminetetraacetic acid
EGTA	ethyleneglycol bis(β-aminoethyl ether)-N,N,N′,N′-tetraacetic acid
ELISA	enzyme-linked immunosorbent assay
EMC	encephalomyocarditis (virus)
ENase	endonuclease
FAD	flavin-adenine dinucleotide

FMN	flavin mononucleotide (riboflavin 5′-monophosphate)
FSH	follicle-stimulating hormone
g	gram: m (milli-), μ (micro-), n (nano-), p (pico-), and f (femto-) gram
G	guanine
GuHCl	guanidine hydrochloride
GuSCN	guanidinium thiocyanate
hr	hour
H-bond(ing)	hydrogen bond(ing)
HBV	hepatitis B virus
HEPES	N-2-hydroxyethylpiperazine-N′-2-ethanesulfonic acid (buffer pK_a = 7.55 at 20°C, 7.31 at 37°C)
HIAP	human intestinal alkaline phosphatase
HIV	human immunodeficiency virus
HLH	helix–loop–helix
hm^{5}C	5-hydroxymethylcytosine
HPLC	high-performance liquid chromatography
HSV	herpes simplex virus
HTH	helix–turn–helix
I	ionic strength
IAA	iodoacetic acid
IC$_{50}$ (ID$_{50}$)	50% inhibiting concentration (dose)
Ig	immunoglobulin
IPTG	isopropyl-1-thio-β-D-galactoside
kb (kbp)	kilobase (kilobase pair)
K-P	potassium phosphate (as buffer component)
LH	luteinizing hormone
LTR	long terminal repeat (or redundancy)
M	molar concentration (mol/liter) (mM = 10^{-3} M, μM = 10^{-6} M, nM = 10^{-9} M)
m^{6}A (^{m}A)	N^6-methyl-A
m^{5}C (^{m}C)	5-methylcytosine (cf. m^{4}C for N^4-methyl-C)
MCS	multiple cloning site
Me	methyl (group)
Me$_2$SO	dimethyl sulfoxide (=DMSO)
MES	2-(N-morpholino)ethanesulfonic acid (buffer pK_a = 6.15)
min	minute(s)
M. luteus	*Micrococcus luteus*
MLV	murine leukemia virus
MMS	methylmethane sulfonate
MOPS	3-(N-morpholino)propanesulfonic acid (buffer pK_a = 7.2)
MoLV	Moloney MLV
M_r	relative molecular mass (without unit)
2-MSH	2-mercaptoethanol
MTase	methyltransferase (methylase)

MUG	methylumbelliferyl-β-D-galactopyranoside
MUP	4-methylumbelliferyl phosphate
MW	molecular weight (used for molecular mass)
NAD^+	nicotinamide adenine dinucleotide (oxidized, NADH for reduced)
$NADP^+$	NAD phosphate (oxidized, NADPH for reduced)
Na-P	sodium phosphate (as buffer component)
NBS	N-bromosuccinimide
NBT	nitroblue tetrazolium
NC	nitrocellulose
NEM	N-ethylmaleimide
NMN	nicotinamide mononucleotide
NMR	nuclear magnetic resonance (spectroscopy)
NP-40	Nonidet (nonionic detergent) P-40
nt	nucleotide(s): dnt, deoxyribonucleotide(s); rnt, ribonucleotide(s)
NTP (rNTP)	ribonucleoside 5'-triphosphate (e.g., ATP, CTP)
oc	open circular (form of plasmid)
ORD	optical rotatory dispersion (spectroscopy)
ORF	open reading frame
ONPG	o-nitrophenyl-β-D-galactoside
P. aeruginosa	*Pseudomonas aeruginosa*
PAGE	polyacrylamide gel electrophoresis
PALP	pyridoxal 5'-phosphate
PBS	phosphate-buffered saline (70 mM Na_2HPO_4 + 30 mM KH_2PO_4 + 150 mM NaCl)
PCA	perchloric acid
PCMB	p-chloromercuribenzoic acid
PCR	polymerase chain reaction
PEG	polyethylene glycol
PFU	plaque-forming unit
PHMB	p-hydroxymercuribenzoic acid
pI	isoelectric point
P_i	inorganic orthophosphate
PMS (C, F)	phenylmethane sulfonyl (chloride, fluoride)
PNK	polynucleotide kinase
PNP	p-nitrophenol
PNPP	p-nitrophenyl phosphate
Pol	polymerase
PP_i	inorganic pyrophosphate
Pu (or R)	purine base(s)
PVDF	polyvinylidene difluoride (membrane material)
Py (or Y)	pyrimidine base(s)
R	molar gas constant (= 1.987 cal/°C · mol)
RBS	ribosome-binding site

RDS	rate-determining step
RF	replicative form (double-stranded)
RNA	ribonucleic acid
RNase	ribonuclease
(sn)RNP	(small nuclear) ribonucleoprotein (particle)
RSV	Rous sarcoma virus
RTase	reverse transcriptase
sec	second(s)
$s^{\circ}_{20,w}$	sedimentation coefficient in Svedberg (S) units (in water at 20°C) 1S = 1 × 10^{-13} sec
S. aureus	*Staphylococcus aureus*
S. cerevisiae	*Saccharomyces cerevisiae* (baker's yeast)
SDS	sodium dodecyl (=lauryl) sulfate
SNase	staphylococcal nuclease
sp.act.	specific activity
ss	single-strand(ed)
SSC	a buffer system used in nucleic acid hybridization (1× = 150 m*M* NaCl + 15 m*M* trisodium citrate, pH 7.0)
S. typhimurium	*Salmonella typhimurium*
T	thymine
T_a	annealing temperature
T. aquaticus	*Thermus aquaticus*
Tc (Tc^r)	tetracycline (tetracycline resistance)
TCA	trichloroacetic acid
TdT (TDTase)	terminal deoxynucleotidyltransferase
TLC	thin-layer chromatography
T_m	melting temperature
Tris	2-amino-2-hydroxymethylpropane-1,3-diol or tris(hydroxymethyl)aminomethane (buffer pK_a = 8.1)
Tris–Cl	Tris base–HCl salt
ts	temperature sensitive
U	uracil
U	unit of enzyme activity
UPS	upstream promoter sequence(s)
UTR	untranslated region (in mRNA)
UV	ultraviolet light
v	rate (or velocity in enzyme kinetics)
V	maximum rate (or maximal velocity, V_{max})
VRC	vanadyl ribonucleoside complexes
X-Gal	5-bromo-4-chloro-3-indolyl-β-D-galactoside

List of Journal Abbreviations

ABB	Archives of Biochemistry and Biophysics
Adv. Enzy. (Mol. Biol.)	Advances in Enzymology (and Related Areas of Molecular Biology)
Acc. Chem. Res.	Accounts of Chemical Research
Am. J. Clin. Path.	American Journal of Clinical Pathology
Anal. Biochem.	Analytical Biochemistry
Annals N.Y.A.S.	Annals of the New York Academy of Sciences
Ann. Rev. Biochem.	Annual Review of Biochemistry
Ann. Rev. B. BC.	Annual Review of Biophysics and Biophysical Chemistry
Ann. Rev. Genet.	Annual Review of Genetics
Ann. Rev. Microbiol.	Annual Review of Microbiology
BBA	Biochimica and Biophysica Acta
BBRC	Biochemical and Biophysical Research Communications
Biochem.	Biochemistry
BJ	Biochemical Journal
Carbohyd. Res.	Carbohydrate Research
Cell. Mol. Biol. Res.	Cellular and Molecular Biology Research
Clin. Biochem.	Clinical Biochemistry
Clin. Chem.	Clinical Chemistry

Crit. Rev. Biochem.	CRC Critical Reviews in Biochemistry
Crit. Rev. B. MB.	CRC Critical Reviews in Biochemistry and Molecular Biology
CSHSQB	Cold Spring Harbor Symposium on Quantitative Biology
EJB	European Journal of Biochemistry
FEBS Lett.	Federation of European Biochemical Societies Letters
Fed. Proc.	Federation Proceedings
Gen. Eng.	Genetic Engineering. Principles and Methods
Genes Dev.	Genes and Development
JACS	Journal of American Chemical Society
J. Bacteriol.	Journal of Bacteriology
J. Biochem.	Journal of Biochemistry (Japan)
J. Bioenerg. Biomemb.	Journal of Bioenergetics and Biomembranes
J. Gen. Virol.	Journal of General Virology
JBC	Journal of Biological Chemistry
JMB	Journal of Molecular Biology
J. Virol.	Journal of Virology
Meth. Enzy.	Methods in Enzymology
Mol. Cell. Biochem.	Molecular and Cellular Biochemistry
Mol. Cell. Biol.	Molecular and Cellular Biology
MGG	Molecular and General Genetics
Mut. Res.	Mutation Research
NAR	Nucleic Acids Research
Nature NB	Nature New Biology
Proteins (S.F.G)	Proteins: Structure, Function, and Genetics
Prot. Eng.	Protein Engineering
PNARMB	Progress in Nucleic Acid Research and Molecular Biology
PNAS	Proceedings of the National Academy of Sciences (USA)
Q. Rev. Biol.	Quarterly Review of Biology
Q. Rev. Biophys.	Quarterly Review of Biophysics
Sci. Am.	Scientific American
TIBS	Trends in Biochemical Sciences
TIG	Trends in Genetics
TIPS	Trends in Pharmacological Sciences
Virol.	Virology
Z. Physiol. Chem.	Zeitschrift für Physiologische Chemie (Hoppe-Seyler's)

1
Enzymes and Nucleic Acids
General Principles

I. STRUCTURE AND FUNCTION OF ENZYMES

Enzymes are bioreactors that run all biochemical reactions of a living cell in a steady, controlled manner with extraordinary efficiency and specificity. *Efficiency* means high and widely ranging levels of rate acceleration; *specificity* means a specific molecular recognition between enzymes and their substrates. The catalytic function is harnessed in a three-dimensional (3-D) protein structure that is simple yet sophisticated and stable yet dynamic. Physiologically, an enzyme is an essential, life-sustaining biocatalyst. Many chemical and biochemical reactions *in vitro* require enzymes to proceed at reasonable rates under mild conditions of temperature, pH, and in aqueous solutions. Enzymes are thus an indispensable tool not only in recombinant DNA technology, but also in many other areas related to biochemical conversions. As a result, enzymology continues to expand and enzymological techniques are used by workers in a number of disciplines.

By a classical definition, an enzyme is a protein endowed with catalytic functions. As a catalyst, it can only promote a reaction in a thermodynamically favorable direction—it does not change the direction of a reaction. Certain RNA species have been found to possess catalytic functions either alone or in combination with tightly binding protein components. Such catalytic RNAs are referred to as *ribozymes*. Ribozymes are an exciting new category of enzymes (see Section II,D,4 in this chapter) making an impact on recombinant DNA technology (see Section I,B Chapter 7).

Despite the time-honored recognition of enzymes and their catalytic abilities, the detailed mechanism of enzymatic catalysis has been studied in only a few cases. However, our understanding of how an enzyme works is now entering a new era ushered in by burgeoning recombinant DNA technology. Protein and genetic engineering techniques are now applied to modify and redesign enzymes not only to decipher their structure–function relationships, but also to create novel substrate and reaction specificities and to improve stability and catalytic efficiency. The application of enzymatic principles to the production of antibody has given rise to enzyme-mimicking *catalytic antibodies* called *abzymes*. The abzymes, which for the time being "stand at the crossroads of chemistry and immunology" (1), are potentially valuable enzymological tools having tailor-made reaction and substrate specificities. Understanding the intricate inner workings of enzymes offers opportunities to discover novel-concept enzymes and to find innovative and broader uses for biocatalysts.

This volume focuses on the "classical" enzymes only, simply because they are the present-day workhorses in recombinant DNA technology. Starting from some molecular biological and genetic aspects of protein biosynthesis, we will briefly review the fundamental concepts of enzymology—the physicochemical bases of protein structure and catalytic function. This chapter also provides a background for relating the fundamental knowledge of protein biosynthesis, structure, and function to recombinant DNA technology.

A. Biosynthesis of Proteins

1. FLOW OF GENETIC INFORMATION

Proteins are translational products of the genetic information encoded in nucleotide sequences of RNA and/or DNA. The genetic code specifying each amino acid consists of three nucleotide units called *codons*. Among the possible 64 codons arising from the combination of four natural nucleotides, 61 codons specify amino acids, and the remaining 3 codons code for translation "stop" (Table 1.1). Since there are only 20 natural amino acids used as building blocks for proteins, many amino acids are specified by more than one codon, a phenomenon known as *codon degeneracy*.

Conventionally, the DNA sequences that code for functional RNA chains or proteins are termed *genes*. The RNA that carries the message for *translation* into a protein is called a *messenger RNA* (mRNA) and is derived from DNA by a process called *transcription*. This flow of information from DNA to RNA to protein was once the "central dogma" of molecular biology, but it is now known that the information can also flow in the reverse direction from RNA to DNA, a process called *reverse transcription*. Among various biological systems, viruses constitute the simplest one which displays diverse modes of genetic information

TABLE 1.1 The Genetic Code[a]

					SECOND					
FIRST	5′	U		C		A		G		3′ THIRD
	U	UUU	Phe	UCU	Ser	UAU	Tyr	UGU	Cys	U
		UUC		UCC		UAC		UGC		C
		UUA	Leu	UCA		UAA	Stop	UGA	Stop	A
		UUG		UCG		UAG	Stop	UGG	Trp	G
	C	CUU	Leu	CCU	Pro	CAU	His	CGU	Arg	U
		CUC		CCC		CAC		CGC		C
		CUA		CCA		CAA	Gln	CGA		A
		CUG		CCG		CAG		CGG		G
	A	AUU	Ile	ACU	Thr	AAU	Asn	AGU	Ser	U
		AUC		ACC		AAC		AGC		C
		AUA		ACA		AAA	Lys	AGA	Arg	A
		AUG	Met	ACG		AAG		AGG		G
	G	GUU	Val	GCU	Ala	GAU	Asp	GGU	Gly	U
		GUC		GCC		GAC		GGC		C
		GUA		GCA		GAA	Glu	GGA		A
		GUG[b]		GCG		GAG		GGG		G

[a] Notable exceptions to the "universal" genetic code include UGA (stop) for Trp in vertebrate mitochondria, UAA (stop) and UAG (stop) for Gln in ciliated protozoa, CUN (leu) for Thr in yeast mitochondria, and AUA (Ile) for Met in vertebrate mitochondria. In some bacterial and mammalian genes, the UGA (stop) codon is used in part for selenocysteine (SeCys).

[b] Codes for fMet if in the initiator position.

flow. For example, DNA viruses usually express their genes according to the DNA–RNA–protein pathway, whereas some RNA viruses (retroviruses) first reverse-transcribe their genomic RNA into DNA and then follow the "central dogma." Certain other RNA viruses replicate through an RNA intermediate of opposite sense and do not involve a message conversion to or from DNA at all. Within this group of RNA viruses, some have *positive-sense* genomic RNAs that can be directly translated into proteins, whereas others have *negative-sense* genomic RNAs that cannot be directly translated.

2. *Protein Biosynthesis*

Protein biosynthesis occurs in *ribosomes,* the cellular assembly factory where amino acids are covalently linked into a polypeptide chain along the template mRNA. The amino acids are transported into the ribosome in an activated state via coupling to specific carriers known as *transfer RNAs* (tRNA). The selection of specific amino acids from the cellular pool of amino acids (Tables 1.2 and 1.3) and the task of "charging" to cognate tRNAs are performed by specific aminoacyl-

TABLE 1.2 Common Amino Acids

Amino acid (symbol, MW)	Side chain (R) in R—CH(NH$_3^+$)CO$_2^-$ (at pH 7.0)	pK$_a$ —COOH	pK$_a$ —NH$_3^+$	R in aa	R in protein
Alanine (Ala, 89)	CH$_3$—	2.3	9.8	—	
Arginine (Arg, 174)	H$_2$N$^+$=C—NH—(CH$_2$)$_3$— | NH$_2$	1.8	9.0	12.5	≥12
Asparagine (Asn, 132)	H$_2$N—C—CH$_2$— ‖ O	2.0	8.8	—	
Aspartic acid (Asp, 133)	$^-$O—C—CH$_2$— ‖ O	2.0	9.9	3.9	4.4–4.6
Cysteine (Cys, 121)	HS—CH$_2$—	1.8	10.8	8.3	8.5–8.8
Glutamic acid (Glu, 147)	$^-$O—C—(CH$_2$)$_2$— ‖ O	2.2	9.6	4.2	4.4–4.6
Glutamine (Gln, 146)	H$_2$N—C—(CH$_2$)$_2$— ‖ O	2.2	9.1	—	
Glycine (Gly, 75)	H—	2.3	9.7	—	
Histidine (His, 155)	imidazole—CH$_2$—	1.8	9.2	6.0	6.5–7.0
Isoleucine (Ile, 131)	CH$_3$CH$_2$—CH— | CH$_3$	2.3	9.7	—	
Leucine (Leu, 131)	CH$_3$CH—CH$_2$— | CH$_3$	2.3	9.6	—	
Lysine (Lys, 146)	H$_3$N$^+$—(CH$_2$)$_4$—	2.2	9.2	10.8	10.0–10.2
Methionine (Met, 149)	CH$_3$S—(CH$_2$)$_2$—	2.2	9.2	—	
Phenylalanine (Phe, 165)	C$_6$H$_5$—CH$_2$—	1.8	9.1	—	
Proline (Pro, 115)	pyrrolidine ring with $^+$NH$_2$ and COO$^-$	2.0	10.6	—	
Serine (Ser, 105)	HO—CH$_2$—	2.2	9.2	—	
Threonine (Thr, 119)	HO—CH— | CH$_3$	2.6	10.4	—	
Tryptophan (Trp, 204)	indole—CH$_2$—	2.4	9.4	—	
Tyrosine (Tyr, 181)	HO—C$_6$H$_4$—CH$_2$—	2.2	9.1	10.1	9.6–10.0
Valine (Val, 117)	CH$_3$—CH— | CH$_3$	2.3	9.6	—	

TABLE 1.3 One-Letter Notation of Amino Acids[a]

1-Letter	3-Letter	1-Letter	3-Letter
A	Ala	M	Met
B	Asx	N	Asn
C	Cys	P	Pro
D	Asp	Q	Gln
E	Glu	R	Arg
F	Phe	S	Ser
G	Gly	T	Thr
H	His	V	Val
I	Ile	W	Trp
K	Lys	Y	Tyr
L	Leu	Z	Glx

[a] As recommended by the IUPAC-IUB Commission on Biochemical Nomenclature [*BJ* (1969) **113**, 1–4].

tRNA synthetases. Transfer RNAs contain *anticodons*, the nucleotide triplets serving as the counterpart of codons. The repertoire of anticodons is substantially smaller than that of codons. This apparent discrepancy is resolved by the fact that anticodons often contain, especially at the first (5' end) base known as *wobble* base, noncomplementary bases or inosine which allow nonstandard (or wobble) base pairing with codons. Thus codon degeneracy in mRNA is adequately matched by anticodon wobbling in tRNA. The general process of protein biosynthesis can be divided into three phases of translation: initiation, elongation, and termination.

a. Initiation of translation

i. **Translation in prokaryotes.** The start of a prokaryotic mRNA translation is usually signaled by the binding of a ribosome at the *ribosome binding site* (RBS) adjacent to a translation *initiator codon* (Fig. 1.1). The initiation involves at least three *initiation factors* (IF_{1-3}) and GTP binding (2). The translation initiation is one of the major regulatory steps in gene expression. The direction of mRNA reading is from the 5' end to the 3' end, corresponding to the protein sequence from the amino (NH_2 or N) terminus to the carboxyl (COOH or C) terminus.

The translation start of *Escherichia coli* mRNAs usually occurs within a special sequence context such as AGGA or (G)GAGG centered some 4 to 15 nucleotides (nt) upstream of the initiator codon AUG (with A conventionally numbered +1). These purine-rich sequences, known as the Shine–Dalgarno (S–D) sequence, function as the RBS by forming base pairs with the complementary pyrimidine-rich (-CCUCCUUA-3') sequence (anti-S–D). The anti-S–D sequence is located at the 3' end of 16S rRNA, a component of the 30S ribosomal subunit (3). In this "standard" mode of translation initiation, the efficiency of initiation is dependent on (i) the length of complementarity at the RBS, (ii) the nucleotides at positions −1 and +4, (iii) the distance between the RBS and the AUG codon, and (iv) the degree of secondary structure around the RBS.

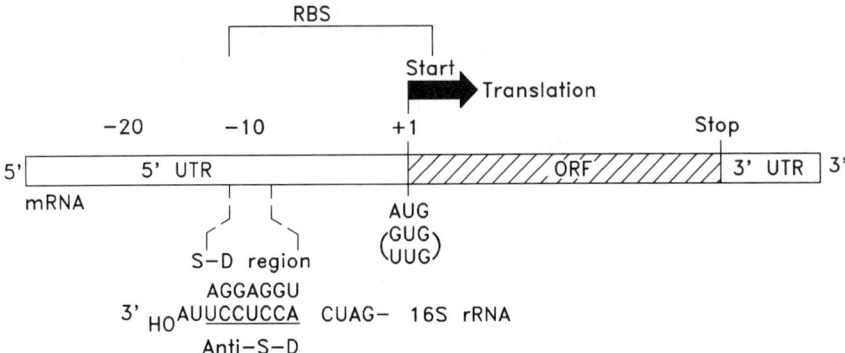

FIGURE 1.1 Prokaryotic translation initiation region. A mRNA can be functionally divided into three regions: 5'UTR (untranslated region), ORF (open reading frame), and 3'UTR. The first nucleotide A of the initiator codon AUG is conventionally numbered as +1. The region comprising approximately −15 to +5 positions is called the initiation region. This region features a ribosome-binding site (RBS) which is typically represented by the Shine–Dalgarno (S–D) sequence. The S–D sequence is recognized by the 3'-terminal sequence (anti-S–D) of the 16S rRNA.

The synthesis of virtually all bacterial polypeptides starts with N-formyl-Met (fMet). The N-terminal Met and other residues may be removed from certain proteins by *posttranslational processings* (see below). Two different species of tRNAs, $tRNA_i^{Met}$ (or $tRNA^{fMet}$) and $tRNA_m^{Met}$, carry the initiating Met and internal Met, respectively, in bacteria, chloroplasts, and mitochondria. Eukaryotic cells also contain two distinct $tRNA^{Met}$ having respective functions in initiation and elongation; however, the Met on the eukaryotic initiator tRNA is not formylated.

ii. Translation in eukaryotes. The translation initiation of eukaryotic mRNAs is significantly different from that of prokaryotes. Eukaryotic mRNAs possess a characteristic 5' cap structure (m^7GpppN) and their RBS are not the same type as those of prokaryotes. Furthermore, the anti-S–D sequence of the prokaryotic type is absent in 18S rRNA which is the counterpart of prokaryotic 16S rRNA. The process by which ribosomes bind to eukaryotic mRNAs and find the appropriate initiation codon is thought to occur via a *bind-and-scan* mechanism (4,5). According to the scanning mechanism, the 40S ribosome preinitiation complex, which includes associated initiator tRNA and eukaryotic initiation factors (eIFs), binds at or near the 5' cap of the mRNA and subsequently slides along the leader region until the first "appropriate" AUG is encountered, whereupon a 60S (large) ribosomal subunit binds to form a functional 80S initiation complex.

Based on the analysis of 5' noncoding (5'NC) sequences from 699 vertebrate mRNAs, a consensus initiation sequence (also called *Kozak sequence*) has been identified as 5'–CC(A/G)CCAUGG–3', in which the nucleotides at −3 (Pu) and +4 seem to be particularly important (5). Database analysis of approximately 2600 vertebrate mRNAs provided further evidence for strong but not exactly identical trinucleotide biases at the −3, −2, and −1 positions as (A/G)NCAUG

(6). In general, the translation initiation sequences exhibit a considerable degree of intertaxon variations: (A/Y)A(A/U)AAUGU in yeast (7), AACAAUGGC in plants (8), (C/A)AA(A/C)AUG in *Drosophila* (9), and AAAAAUGA in protozoa (10). If the 5′ proximal AUG codon is in an "unfavorable" context, scanning may continue until the initiation complex encounters an AUG in a better context or a non-AUG initiator codon may be used for the initiation. The discriminating information is presumed to be located within a sequence of ~35 nt (-20 to $+15$ nt positions) (11). Based on the new set of information on the -3, -2, -1 trinucleotide bias in the initiation site of mammalian mRNAs, the -321 of mRNA has been proposed to base pair with one of three triplet frames found in a five nucleotide segment (3′–UUUGG–5′) of the 18S rRNA (6).

b. Elongation of translation

The stepwise growth of a polypeptide chain begins with the free amino group (N terminus). The growing carboxyl end during the elongation mode is always terminated by a covalently attached tRNA molecule. Each ribosome has three tRNA-binding sites, i.e., *P-site* for peptidyl-tRNA binding, *A-site* for aminoacyl-tRNA binding, and *E-site* for the binding of discharged tRNA before exit from the ribosomal complex. As a new peptide bond is formed, the aminoacyl-tRNA translocates from A-site to P-site and the original peptidyl-tRNA from P-site to E-site, completing a cycle of protein synthesis. The peptide bond formation is catalyzed by a peptidyltransferase; the enzymatic activity of this transferase is presumed to reside in an RNA component of the ribosome. The elongation of a peptide requires the participation of at least three *elongation factors* (EFs) and GTP hydrolysis. Under optimal conditions, the elongation in *E. coli* proceeds at a rate of ~12 amino acids/sec, while the ribosomes load onto the same mRNA at ~3.2-sec intervals (2).

As the ribosome moves along the mRNA, occupying ~30 nt, several ribosomes can attach to the mRNA molecule forming a *polysome* and translating simultaneously. As the synthesis continues, the nascent polypeptide chain folds into a precursor or mature 3-D form, depending on the nature and destination of the protein.

The elongation of a polypeptide chain on a given mRNA is a continuous process for most of the proteins until the elongation reaches a translation stop signal. During the elongation, ribosomes may slip by or skip one or more bases either upward or downward from the triplet codons, resulting in fusion proteins due to -1 or $+1$ frameshifting or deletion of a segment of mRNA. In addition to the random as well as programmed "errors," alternative readings of the "universal" genetic code may occur. In bacteria and retroviruses, stop-codon readthrough and ribosomal frameshifts are an essential step in gene expression. The alternative modes of decoding genetic information, termed *recoding*, imply that the protein sequence cannot always be simply deduced from the sequence of mature mRNA, not to mention the sequence of genomic DNA. The nonstandard modes of decoding are extremely context dependent, and *recoding signals* are only beginning to be understood (see Section I,A,6, in this chapter).

c. Termination of translation

Protein synthesis stops when a protein *release factor* (RF) recognizes specific termination signals contained in the RNA sequence. Three *stop* or *nonsense* codons have been identified: UAG (amber), UAA (ochre), and UGA (opal). [All ciliated protozoa deviate from the universal genetic code by translating either one or two termination codons into Gln or Cys (13).] Chain termination involves the nucleophilic attack by water on the ester bond between the 3'-adenosine of tRNA and the amino acid.

Although the translation of mRNA is generally terminated at the trinucleotide stop codons, efficient stop signals are consistently associated, in a context-dependent manner, with a fourth base both before and after the trinucleotide codon, presumably functioning as *tetranucleotide stop signals*, e.g., UAA(A/G) and UGA(A/G) (14). Prokaryotic stop signals show a strong bias for U immediately 3' to stop codons, whereas eukaryotes prefer a purine or a purine and U. Both groups share a strong bias against C at this position and especially against the CG dinucleotide (6).

3. CO- AND POSTTRANSLATIONAL MODIFICATIONS, PROCESSING, AND MATURATION

Following the termination of or even during the course of translation, nascent proteins fold into a 3-D structure and undergo various chemical and physical modifications. The structural alterations accompanying these co- and posttranslational modifications determine biological activities, intra- or extracellular localizations, stability, and turnover (or degradation) of the proteins.

Some of the modifications are almost universal. The formyl group of the N-terminal N-formyl-Met in bacterial polypeptides is removed by the action of *deformylase*. The resulting N-terminal Met may or may not be removed by the action of Met *aminopeptidase*, as may several other amino acid residues near the N terminus in a process likely to be general in all organisms. The processing of N-terminal Met is particularly important in recombinant proteins that are expressed in *E. coli*. Often the additional Met comes from the AUG initiator codon that has been engineered into the protein. The N-terminal Met of such recombinant proteins can be cleaved *in vitro* by aminopeptidase M. The new N-terminal amino acids may be subject to N^α-acetylation as on most eukaryotic cellular proteins. Cys residues in a protein often form intramolecular as well as intermolecular disulfide linkages.

Many additional modifications are protein class dependent and specific for certain amino acid residues (Table 1.4), critically affecting structural and functional properties of the proteins. Some proteins, e.g., most viral proteins, that are translated from a polycistronic message into a polyprotein need to be proteolytically processed to become mature proteins (see below).

Nascent polypeptides may be either retained in the cytoplasm or transferred to various locations inside or outside the cell. This process is known as cellular

TABLE 1.4 Co- and Posttranslational Modifications of Proteins

Reaction type	Target amino acid
Acylation	
N-Acetylation	N-terminal amino acids
Myristoylation	N-terminal Gly
Palmitoylation	Cys, Ser, Thr
ADP-ribosylation	Arg (ε-NH$_2$)
Carboxylation	Glu [to γ-carboxy-Glu (Gla)]
Deamidation	Asn (to Asp), Gln (to Glu)
Decarboxylation	
Glutamylation	Glu (γ-COOH)
Glycosylation	N-glycosylation: Asn
	O-glycosylation: Ser and Thr
Hydroxylation	Asp and Asn (on β-carbon)
	Lys (to 5-hydroxy-Lys)
	Pro (to 3- or 4-hydroxy-Pro)
Iodination	Tyr
Methylation	O-methylation: Glu (γ-COOH)
	N-methylation: Lys, Arg and His
Phosphorylation	O-phosphorylation: Ser, Thr and Tyr
	N-phosphorylation: His and Lys
Prenylation	Cys (—SH on C-terminal —CXXX$_{OH}$ motif)
Farnesyl (C$_{15}$) or Geranylgeranyl (C$_{20}$) isoprenoids	
Ubiquitination	Lys

translocation or *compartmentalization*. Regardless of their locations, proteins should fold into a "right" shape to be active and/or to be transported. The folding of proteins *in vivo* appears to be actively modulated by posttranslational modifications as well as by interactions with various other molecules. In fact, the maturation of a protein is not a single process but a multiple process often interlinked with translocation. Protein translocation in turn requires specific interactions between certain sequence elements and translocation machinery, and involves posttranslational processing.

a. Protein folding

The folding of a polypeptide into its native state requires the selection of one conformation from an enormous ensemble of sterically available but incorrect conformations. *In vitro,* a correct folding may be attained spontaneously via an array of competing, incorrectly folded molecules. It may thus accompany varying degrees of irreversible aggregate states. How a protein sorts out a correct conformation from this complex "protein folding problem" has yet to be determined (15). Based on data from a small protein BPTI (bovine pancreatic trypsin inhibitor) which contains three disulfide bonds, incorrect folding and nonnative intermediates have been presumed to play a key role in the correct folding of proteins (16). Recent reinvestigation of the same model protein has shown, however, that the folding intermediates very close to the native state are the predominant species within the time scale of the techniques used to trap the folding intermediates (17). These short-lived intermediates, which are nevertheless stable enough to be trapped, may be qualitatively similar to a "molten globule," another term designating the folding intermediates that are relatively stable and nearly as compact as the native state.

Recombinant proteins, which are usually produced in heterologous hosts, face the same, if not more severe, folding problems. In fact, overexpression of the recombinant proteins often leads to their precipitation and aggregation, making it necessary for purification to solubilize the proteins using denaturing agents such as urea and GuHCl. In these cases, the recovery yield of biologically active molecules is directly tied to the efficiency of refolding, and conventional methods of renaturation relying on gradual removal of the denaturing agents by dialysis have been only moderately successful due largely to aggregation of the proteins during the refolding process. One method that substantially increases the yield is based on a rapid dilution of the protein plus urea (or GuCHl) solution in the presence of PEG (126). PEG (MW 3350), when added at a stoichiometric molar ratio to the protein of interest, binds only to the hydrophobic molten globule intermediate, reduces aggregation, and enhances correct refolding.

How does a protein fold *in vivo*? Evidence suggests that, for many proteins, the folding and the assembly of oligomeric proteins do not simply follow a spontaneous process, but require the active participation of other proteins, such as protein disulfide-isomerase (EC 5.3.4.1) and peptidyl prolyl isomerase. The protein disulfide-isomerase, a multifunctional enzyme, catalyzes the formation of intra- or interchain disulfide bonds (S–S), while peptidyl prolyl isomerase catalyzes the cis–trans isomerization at Pro residues. The folding and assembly of oligomeric proteins are also assisted by a group of auxiliary molecules known as *molecular chaperones*. Chaperone proteins, e.g., eukaryotic heat-shock proteins Hsp90, Hsp70 (or prokaryotic DnaK), and Hsp60 (or prokaryotic GroE), bind to unfolded or partially denatured (molten globule) proteins, prevent the formation of incorrect aggregates, and promote the formation of α-helix and acquisition of ordered tertiary structure. Chaperone molecules, which are usually oligomeric proteins, act without covalently modifying their substrates and without being part of the finished products. Release of the folded molecules may be either an ATP-driven process (with Hsp70 and Hsp60) or an ATP-independent process (with Hsp90). By modulating and/or maintaining a certain folded state, molecular chaperones such as *E. coli* SecB play an important role in protein translocations as well (18).

b. Signal (leader) peptides

Certain proteins are synthesized in "precursor" forms that contain a leader sequence at the amino and/or carboxyl terminus. The leader sequences (or peptides) are also called *signal* sequences (or peptides) or *pre*-sequences, implying their major functional roles as the signals for intracellular translocation or export across the membranes. The precursor protein with a signal sequence is generally designated the *pre*-protein. Leader sequences are presumed to play an active role in maintaining the proteins in a conformation suitable for translocation and/or final processing. Following the translocation or export, leader peptides are cleaved from the functional proteins. Despite their seemingly common roles, leader peptides display remarkable sequence and functional diversities. Indeed, some signal sequences, which characteristically occur at or near the C terminus, are used in anchoring the proteins to cellular membranes. The cleavage of signal peptides ushers in the maturation of proteins in their appropriate cellular or extracellular locations. Many proteins further undergo proteolytic cleavages to generate smaller, biologically active proteins. The proteins before such further processing are generally called *pro*-proteins, and the sequences which do not appear in the mature protein are generally designated the *pro*-sequence. In a nascent protein, a signal sequence may be attached to a proprotein, and such a protein is called *prepro*-protein (e.g., preproinsulin). Deciphering the structure–function relationship of precursor molecules in general and signal sequences in particular remains a challenge in molecular biology. Understanding the subject is also important in recombinant DNA technology, especially when foreign genes are expressed as fusion proteins with certain leader sequences.

i. Export signals. The signal sequences mainly used for membrane transport of proteins are 15–30 residues long. They are usually located at or near the N terminus of nascent protein, although they do not have to be N-terminal (19). In eukaryotes, the first interaction of signal peptides is not with the membrane but with an 11S *signal recognition particle* (SRP). The SRP is a ribonucleoprotein complex and, while functioning as an antifolding device by segregating the bound signal sequence from the rest of the polypeptide chain, targets the nascent protein to transport across the membrane of the endoplasmic reticulum. In prokaryotes (*E. coli*), a series of Sec proteins (Sec A, B, D, E, F, and Y) function as the mediator of secretion (20,21). Although all other Sec proteins are membrane-associated, SecB is a cytoplasmic molecular chaperone that maintains the target proteins in a loosely folded, export-competent conformation.

One notable feature of signal sequences is the absence of any consensus sequence. Their function in protein export is therefore associated with other properties such as hydrophobicity and conformation. Most of the signal sequences of, e.g., β-lactamase, *E. coli* alkaline phosphatase, and OmpA, retain their membrane translocation function when they are fused to the N terminus of a variety of heterologous proteins. Despite the heterogeneous appearance, signal sequences present three distinct domains (22): (a) a charged residue within the first five amino acids, (b) a central hydrophobic core of at least nine amino acids, of which Leu is the most abundant, and (c) a more polar carboxyl terminus where -3 and -1 positions (counting from the leader peptide cleavage site) are typically small, aliphatic residues such as Ala and are rarely aromatic or charged residues but never Pro. The hydrophobic core generally adopts an α-helical conformation. When efficiently functioning signal sequences are compared with nonfunctional or less efficiently functional signal sequences, the helical structure of the C-terminal region turns out to be less stable in functional signal sequences. The reduced stability is large attributed to a Pro residue located at the -4 to -6 position (23). Nevertheless, the Pro is not essential for export (24). These observations suggest that the signal function may depend on the ability of the sequence to adopt a β structure in a highly hydrophobic environment, most likely by a dynamic conformational shift of the α helix which is normally found in a hydrophilic environment (25).

The presence of a signal sequence is not always sufficient to ensure efficient membrane translocation. One feature of the nascent chain that adversely affects secretion, at least in *E. coli,* is the presence of positively charged amino acids immediately downstream of the signal sequence. The downstream sequences form a sharply delimited domain of about 30 residues and, together with the signal sequence, comprise an "export initiation domain" (26).

ii. **Membrane attachment signal.** Certain enzymes, e.g., mammalian alkaline phosphatases, that function as membrane-bound proteins commonly carry a strongly hydrophobic C-terminal signal of 15–30 amino acid residues. Following the cleavage of the C-terminal signal sequence, the newly formed C terminus (e.g., Asp in human placental alkaline phosphatase) is covalently linked to glycosylphosphatidylinositol of the plasma membrane (27).

c. *Processing of leader or prosequences*

The signal sequences of secretory proteins are cleaved following translocation across the membrane, a timely feature for activating the protein. With notable exceptions, the precursor form of a polypeptide is usually inactive. Cleavage of the signal sequence converts the precursor into an active, "mature" tertiary structure (28). Among the best known examples are the maturation of *zymogens* such as trypsinogen, chymotrypsinogen, and pepsinogen, which are converted to the respective active proteases, trypsin, chymotrypsin, and pepsin. A separated prosequence has been shown to actively guide the folding of the processed, but inactive, form to the active conformation in an intermolecular process (29). Combined with the fact that leader peptides initially retard the folding of precursor polypeptides (30,31), the kinetic effect of slow or retarded folding afforded by the signal sequence is believed to play an active and crucial role in protein transport and maturation. The delayed folding allows nonnative precursor forms to interact with cytosolic chaperone molecules and to enter the export pathway.

Analyses of a number of signal sequence cleavage sites have shown that the amino acids at positions -3 and -1 play an important role in the selection of the sites and efficiency of cleavages (32,33). However, efficient transport does not depend on the structural preference of the cleavage region (24).

The precursor form of insulin, preproinsulin (H_2N–signal–B–C–A–COOH), is converted to a biologically active insulin molecule by successive cleavages from preproinsulin to proinsulin (H_2N–B–C–A–COOH) to insulin (A + B chains).

The proteases involved in the processing of prohormones and many other cellular precursor molecules are called *converting enzymes* or *convertases*, e.g., KEX[1,2] proteins of yeast (34), human furin, and mPC[1,2] proteins (or mouse prohormone convertases) (35). These proprotein convertases commonly recognize and cut within a basic amino acid sequence motif –Lys/Arg–Arg.

d. Polyprotein processing

Apart from the processing of prosequences, the maturation of some proteins involves systematic cleavages of a large precursor molecule. These cleavages, called *polyprotein processing*, are essential in the maturation of many eukaryotic viral proteins and peptide hormones.

The polyprotein processing of picornaviruses (e.g., polio-, rhino-, and EMC viruses) is carried out by two virus-encoded proteinases: 2A and 3C (36,37). The cleavage of precursor by 2A protein occurs cotranslationally at a single site. All 10 (or 11 depending on the virus) remaining sites are cleaved posttranslationally by 3C proteinase with remarkable specificities for the Gln/Gly sequence.

4. PROTEIN TURNOVER

Proteins newly synthesized *in vivo* are eventually subject to degradation. Although detailed degradative pathways for most proteins remain unknown, multiple pathways have been identified for intracellular protein degradation (38). Some protein degradations are selectively mediated by ubiquitin conjugation and/or other posttranslational modifications. Other proteins are selectively degraded within mitochondria, endoplasmic reticulum, and the Golgi apparatus, while other proteins follow either a selective or a nonselective lysosomal pathway (39).

What are the molecular determinants of protein degradations and how do they enter a particular degradation pathway? The stability of a protein is thought to be encoded in its primary structure and so is the susceptibility to degradation. For instance, an N-terminal peptide sequence motif, KFERQ, has been identified to target cytosolic proteins for specific lysosomal proteolysis. Studies on protein stability and degradation have established a number of other systematic patterns, notably *N-end rule* and *C-terminal rule*. The specific posttranslational modifications may well be considered as programmed events that direct the eventual location of proteins, their functions, and fates.

a. N-End rule

The N-end rule states that the stability of a protein is largely determined by the presence of "destabilizing" or "stabilizing" N-terminal amino acids; the exact form of the rule is dependent on the physiological state of the cell (40). In mammalian reticulocytes, the primary destabilizing residues include Arg, Phe, and Ala, whereas Asp and Glu are secondary. Met, Val, and Gly are classified as stabilizing residues. In yeast, the N-end rule is essentially similar but not identical to that of mammalian reticulocytes: Ala, Ser, and Thr are stabilizing residues and Ile is a destablizing residue.

b. C-Terminal rule

Some proteins are subject to the C-terminal rule which predicts that proteins having a C-terminal stretch comprised predominantly of a PEST (Pro, Glu, Ser, and Thr) sequence motif are rapidly degraded (41). For example, ornithine decarboxylase containing the C-terminal PEST sequence has a half-life at least four times shorter than the same protein without the PEST motif (42). The PEST rule is subject to the polarity of the terminal residues; nonpolar amino acids are destabilizing when placed at the five amino acid positions of the C terminus, whereas charged and polar residues are stabilizing (43).

5. NORMAL AND MUTANT PROTEINS

a. Origins of mutation

Accurate incorporation of amino acids into proteins depends on correct aminoacylation of tRNAs by their cognate aminoacyl-tRNA synthetases. However, errors can occur during translation if the protein synthetic machinery suffers certain alterations or the synthesis occurs under limiting conditions. Errors can also be introduced naturally or artificially at the DNA level due to misincorporation of a nucleotide(s) during DNA polymerase-catalyzed replication and/or to various chemical and physical modifications. Errors occur at the RNA level as well due to incorporation of wrong nucleotides during transcription catalyzed by RNA polymerase(s). In fact, errors arising from transfer of mutated genetic information are more frequent than errors introduced during translation.

Spontaneous mutations occur at a frequency of $10^{-5}-10^{-6}$ per locus per generation due to misincorporation by DNA (or RNA) polymerases. Random and sometimes programmed amino acid substitutions and unusual elongations occur during translation at frequencies ranging from a few percent to as low as 10^{-5} per event (44). Mutations can also be induced, and at substantially higher frequencies, by chemical mutagens and by other mutagenesis techniques. Mutations can be *random* or partially random as is the case with most chemical mutagens. In contrast, mutations can be concentrated on "hot spots" due largely to some structural characteristics of the DNA at the region. Mutations can also be introduced in a *site-specific* manner. Indeed, *site-specific in vitro mutagenesis* is a powerful technique to generate predetermined mutations (see Appendix A, at end of volume).

b. Catalog of mutations

A mutation may be associated with a change in a biological trait(s) called *phenotype*. [The genetic trait giving rise to the phenotype is called *genotype*. Genotypes are described using italic type, e.g., *lac,* whereas the phenotypes, which are described by the same terms, are not italicized but the first letter is capitalized, e.g., Lac.] The usual active form of a gene or a phenotypic feature is referred to as *wild-type*. A mutation that results in no phenotypic changes is called a *silent mutation*.

When a codon specific for one amino acid is changed to a codon specific for another by single-base replacements or *point mutations*, it is called a *missense mutation*. When a mutation involves substitution of one Py by another Py, it is called a *transition*. The substitution between Pu and Py is called a *transversion*. Transversion mutation is less frequent than transition mutation.

Missense mutant proteins/enzymes may have full, partially impaired, or null (biological) activities depending on the location of the mutation. When the functional impairment manifests itself only at a higher or lower than "normal" temperature that is considered optimal for the wild-type enzyme, the mutation is called *temperature sensitive* (ts). Although most ts mutants are sensitive to higher than normal temperatures, some ts mutants are *cold sensitive*. When an amino acid codon is changed to a termination codon, it is called a *nonsense mutation*. Nonsense mutations result in incomplete or truncated polypeptides of varying sizes and functional profiles.

With respect to cell viability, a mutation can be *lethal, conditionally lethal,* or *nonlethal*. Conditionally lethal mutations (e.g., ts mutations) are lethal under one set of conditions (*nonpermissive*) whereas they exhibit no critical deficiency for cell viability under alternative or *permissive* conditions.

Mutations also occur from gene rearrangements and errors in gene recombinations that result in either *insertions* or *deletions* of one or more nucleotides. Insertions or deletions may lead to a change in the codon reading frame. From a biological perspective, insertions and deletions are part of the normal cellular processes that have evolved as a control mechanism of gene expression (refer to Section II,D,2, in this chapter). A mutation arising from the lack of an entire gene is termed a *null mutation*.

Viruses often present a special class of mutants called *escape mutants*. The mutation(s) occurs at neutralizing immunogenic sites such that the alteration of antigenicity allows the virus to escape from the neutralizing antibodies of the host. Influenza viruses are notorious for their propensity to undergo such mutations, which are specifically called *antigenic drift*.

c. Secondary and/or suppressor mutations

Lethal or ts mutations under nonpermissive conditions often generate *revertants*. Revertants arise from point mutations and insertions, but not from deletions. Reversion in point mutations may involve the reversal of the original mutation (*true reversion*) or a compensatory mutation elsewhere in the gene (*second-site* or more generally *suppressor mutation*). The reversion from insertional mutations usually occurs by appropriate deletion of the inserted part. The *suppression* restores a wild- or pseudo-wild-type phenotype to a mutant organism in which the primary mutation is still maintained.

The quantitative effect of a second mutation on a mutant enzyme may be antagonistic, absent, partially additive, synergistic, or a combination of these with respect to the first mutation. A double-mutation site specifically introduced in staphylococcal nuclease has been shown to demonstrate a combination of effects depending on the kinetic or thermodynamic criteria applied (45).

Suppressions occur *in vivo* through various mechanisms, the most frequent being the mutations in tRNA genes (46,47). The suppressor mutations generate new tRNAs with the ability to recognize termination codons, missense codons, and frameshifts. Mutant proteins can also occur from normal stop codons by *nonsense suppression* or alternate reading programs such as stop-codon read-through and ribosomal frameshifting. Such translation results in elongated (fusion or poly-)proteins. Nonsense suppressors, e.g., *supD*, *supE*, and *supF*, introduce Ser, Gly, and Tyr, respectively, to the UAG (amber) site. Similarly, other suppressors introduce Gln, Leu, Lys, Ser, and Tyr to the UAA (ochre) site, and Arg, Gly, and other amino acids to the UGA (opal) site.

d. Significance of mutant studies

From an enzymological point of view, a mutation can have various effects on the function of an enzyme. It can affect the substrate or cofactor binding by direct interference or by indirect conformational changes. It can affect catalysis by direct substitution of a catalytically essential group or by an indirect effect on the function of the essential group(s). Mutation affecting catalysis can have far-reaching biological consequences leading to, for example, cell death and the induction of tumor, cancer, drug resistance, and inherited diseases. In contrast, some mutations can be structural and affect (thermo)stabilities, but have no effect on catalysis. Mutations, both random and site specific, thus provide a powerful tool for the investigation of the functional and structural roles of particular amino acids in a given protein.

If an appropriate selection can be made from the assorted mutations that block each step in enzymatic catalysis and lead to the accumulation of intermediates, such mutant enzymes can serve as direct tools to delineate a complex catalytic pathway. From both theoretical and practical standpoints, mutations and mutagenesis have significant impacts on genetic engineering. Generating mutant enzymes (or proteins) with either improved catalytic (or functional) and/or structural features or *de novo* specificities is a lofty goal of protein engineering.

6. NONSTANDARD MODES OF TRANSLATION INITIATION AND TRANSLATIONAL CONTROLS OF GENE EXPRESSION

A gene is expressed through a variety of mechanisms which not only create diversity in encoded proteins but also provide a means to regulate the expression. The control of gene expression can be divided into four functional levels: transcriptional, posttranscriptional, translational, and posttranslational. The foremost regulation occurs at the DNA-to-RNA transcriptional level, in particular at the initiation of transcription (see Section I,A in Chapter 7). Posttranscriptional control comprises mRNA modifications, notably splicing and editing (see Section II,D,2, this chapter). Posttranslational controls involve various modifications and processings of proteins as described earlier. The following discussion is concerned with translational controls. Of the different mechanisms used for translation initiation and elongation, two translational controls, namely translation of overlapping reading frames and alternative codon reading, will be emphasized.

a. Nonstandard modes of translation initiation

The information on where and how to start translating a message is all contained in mRNA. Specific sequences along the mRNA chain direct where the ribosome can start binding, which initiator codon is to be used, and where to stop translation (see Fig. 1.1 for "standard" mode of translation initiation). The segment of nucleotide sequences that can be translated into a protein is called an *open reading frame* (ORF). An ORF is thus a long stretch of triplet codons demarcated by an initiator codon and a terminator codon. When searching for an ORF, an apparent amino acid codon can be read in three different frames. In certain *polycistronic* mRNAs (an mRNA encoding more than one protein is called polycistronic), either the proteins encoded in different reading frames are somehow expressed or the proteins encoded in the same reading frame are expressed differently. The secondary (or hidden) reading frames are apparently used for coordinated expressions of proteins by such mechanisms as ribosomal frameshifting, initiation on alternative non-AUG initiation codons, or reinitiation on internal AUG codons.

i. **Ribosomal slipping and hopping.** Ribosomal slipping, which results in frameshifts, is often the mechanism of alternate gene expression when the first ORF overlaps with a second one. In this programmed genetic recoding, the translation begins at the first AUG codon and continues until the frameshift signal is encountered, whereupon a ribosomal frameshifting occurs in either direction creating a fusion protein. The frameshift signals (or recoding signals) consist of at least two components within the RNA. The first is a 7-nt stretch at the frameshift site, and the second is a stem–loop or pseudoknot structure which occurs 6–7 nt downstream from the first signal. A possible third component is the presence of a subset of particular tRNAs with nonmethylated guanosine at the nt position 37 of tRNAs (next to the 3' end of the anticodon). The unmodified nucleoside would allow a 4-nt base pairing to occur and may occasionally cause the tandem slippage of tRNAs one nucleotide leftward (-1 frame). The presence of a methyl group, e.g., 1-methylguanosine, has been shown to prevent frameshifting in *Salmonella typhimurium* (48). Regardless of the mechanism, the frameshifting appears to involve ribosomes pausing on the mRNA due to either low availability of the next tRNA or some physical hindrance caused by an RNA secondary structure(s).

Because only a fraction of the ribosomes change frame at the frameshift signal, fusion proteins are produced in addition to, rather than instead of, the "normal" protein. Frameshifting in the -1 frame is used to access the polymerase reading frames of most retroviruses (49) and of an yeast dsRNA virus (50). The $+1$ frameshift occurs in the yeast retrotransposon Ty (51). Various other forms of frameshifting have also been observed, for example, with *E. coli* ribosomes (52).

Ribosomal shifts can occur over a considerable distance without intermediate steps, a phenomenon known as *ribosomal hopping* (124). An extreme case of hopping is the skip of 50 nt that separate codon 46 from codon 47 in the mature message of T4 topoisomerase subunit gene 60 (12). The elements contributing to the skip are presumed to be located both at the coding gap which contains a pseudoknot and at the nascent, 46-aa peptide which precedes the interruption.

ii. **Internal initiation.** The translation initiations taking place at the first AUG codon in "favorable" contexts of the translation initiation region comprise ~90% of all translations. The remainders are classified as "alternative" initiations. For example, ACG in the same reading frame as AUG can also serve as an initiator codon, allowing synthesis of multiple proteins from a single mRNA (53). The sequence contexts preferentially used for internal initiation are often not favorable compared with the "consensus" initiation sequences (54), consensus meaning the most frequent among a set of similar sequences. The use of ACG in a favorable context is as efficient as AUG in a less favorable context, but only 10–20% as efficient as AUG in a more favorable context. Other codons such as GUG, CUG, UUG, and AUU also function in certain contexts as alternative but less efficient initiation codons. The translation initiation at GUG is ~5% as efficient as AUG.

In a special class of internal initiations, the ribosomes bind at a specific internal AUG codon to start translation. The internal AUG may be located as far as ~800 bases from the 5′ end as in picornaviruses with uncapped, (+)-sense genomic RNAs. According to the model of *cap-independent translation* or *internal ribosome entry* (55), the internal RBS or *ribosome landing pad* is a 5′ UTR sequence that forms a stable stem–loop structure. Hepatitis B virus (HBV) and some retroviruses also use an internal AUG codon, rather than ribosomal frameshifting, to express their reverse transcriptase genes (56,57).

b. *Controls at translation elongation and termination*

In addition to the important characteristics of mRNA sequences at the 5′ UTR and the regions of translation initiation, the 3′ UTR also plays an important role in the control of translation efficiency. The presence of the 3′ end poly(A) not only enhances mRNA stability but also appears to affect the translation. In some mRNAs, UA-rich structural elements, e.g., interspersed repeats of UUAUUUAU, are located downstream from the stop codon and function as a signal for selective mRNA degradation (58) and/or translation inhibition (59).

B. Structure

1. STRUCTURAL UNITS

a. *Primary structure*

Proteins (and enzymes) consist of amino acids that are linked together by peptide (or amide) bonds (–CO–NH–) between the α-carboxyl (COOH) group of one residue and the α-amino (NH$_2$) group of the next. The 20 amino acids usually found in proteins are listed in Tables 1.2 and 1.3. They differ only in their side

TABLE 1.5 Physicochemical Groups of Amino Acid Side Chain (R)

Groups	Amino acid	Hydropathy scale[a]
Aliphatic	Ala	1.8
	Ile	4.5
	Leu	3.8
	Val	4.2
Nonpolar	Cys	2.5
	Gly	−0.4
	Pro	−1.6
	Met	1.9
Aromatic	His	−3.2
	Phe	2.8
	Trp	−0.9
	Tyr	−1.3
Polar	Asn	−3.5
	Gln	−3.5
	Ser	−0.8
	Thr	−0.7
Charged	Arg (+)	−4.5
	Asp (−)	−3.5
	Glu (−)	−3.5
	Lys (+)	−3.9

[a] Hydropathy scale according to Kyte and Doolittle (1982) [*JMB* **157**, 105–132].

chains (R), the combination of which imparts to the protein (enzyme) its unique physicochemical and functional properties (Table 1.5). The sequential order of amino acid residues, which is written by convention from N terminus to C terminus, is called the *primary* structure.

The amino acid sequence gives an identity to the protein and provides fundamental information on the nature of the protein, such as structure, mechanism of action, evolutionary origin, and classification into enzyme families. This primary information also serves as a guide for gene search, molecular cloning, and manipulation of genes in recombinant DNA technology.

The primary sequence of a protein is determined by a series of "standard" procedures: (*i*) chemical and/or enzymatic cleavages of the protein into small fragments called *peptides*, (*ii*) separation of the free or derivatized peptides, (*iii*) sequential degradation of residues from either the N or the C terminus, and (*iv*) identification of amino acids by chromatography and/or mass spectrometry (60, 61). Another elegant and often less ambiguous approach is to deduce the amino acid sequence from the nucleotide sequence of the gene encoding the protein.

b. Secondary structure

The polypeptide chain of a protein assumes distinctive ordered or disordered local structures known as *secondary* structures. The α *helix* (right-handed or left-handed) is an ordered structure in which the backbone amide group (NH and

CO) makes a H-bond with a third amide group away from it. There are 3.6 residues in each turn of the right-handed helix (thus 3.6-fold), and the rise of the helix per turn, the *pitch*, is ~5.4 Å. The diameter of the helix backbone is ~6 Å. Certain amino acids have helix-forming propensities (measured by s values, the helix propagation parameter). They are in the order Ala > Leu > Phe > Ile > Val > Thr > Gly at internal positions, where Pro is a helix destabilizer (62,63). At N- and C-terminal positions of an α helix, however, Gly shows a substantially increased helix-stabilizing tendency which may be greater than that of Ala (125).

Another form of the ordered structure is the *β sheet*. In this structure, an extended chain makes interstrand rather than intrastrand H-bondings with one or two of the neighboring *parallel* or *antiparallel* strands. The peptide backbone in the β sheets has a geometry that approaches the most extended chain conformation allowed by normal bond lengths and angles. The displacement of each amino acid residue in the pleated β sheet is 3.47 Å.

In addition to the two ordered secondary structures, a polypeptide chain may also contain nonregular structures often classified as *random coil*. The random coils, which include *turns* and *loops*, also play a significant role in the function of a protein. One notable substructure is a *β bend* or turn in which 3–5 successive residues are involved as a unit to sharply change the direction of the main chain. A *loop* is a continuous segment of a polypeptide chain that connects other defined structures. A special class of loops, called omega-loop, consists of 6 to 16 residues that fold into a compact substructure resembling the Greek letter ω (omega) (64). The omega-loop has been shown to be associated with prohormonal cleavage sites (65).

The information on local secondary structures (α helix and β sheet) is usually obtained by using spectroscopic methods such as NMR, ORD, and/or CD. However, secondary structures can be best derived from the tertiary structural information.

c. Tertiary structure

To be biologically active, a protein must adopt a specific, folded structure known as a *tertiary* (or 3-D) structure. The formation of a 3-D structure requires various posttranslational modifications and/or processings as discussed earlier (see Section I,A,3, this chapter). Although our knowledge on this subject has significantly increased during the past decade, the general principle of folding and the process by which patches of distinct secondary structures interact to arrive at a stable 3-D structure remain to be elucidated.

Single polypeptide enzymes of large molecular size ($\geq$20 kDa) are often formed from two or more structural *domains*. The term *domain* refers to compact, autonomously folding units with a minimum surface-to-volume ratio. The domains are linked by an oligopeptide having an average of 6.5 residues with a noticeable preference for the STG sequence: Ser (10.1%), Thr (9.0%), and Gly (9.0%) (66). In many instances, structural domains are stable and enzymatically active on their own if each domain happens to have different catalytic activities. Genetically engineered proteins are often produced as fusion proteins having distinct bi- or multifunctional activities. In other instances where a single peptide has a single catalytic activity, e.g., eukaryotic protein kinase catalytic subunit (67) and *E. coli* tryptophan synthase β subunit (68), the active site is often located in the interdomain cleft.

The 3-D structure of an enzyme (or protein), either alone or as a complex with substrates, inhibitors, or cofactors, has been traditionally determined by X-ray diffraction of crystals. X-ray crystallography can now achieve a near atomic resolution of 1.5 Å. Combined with the vastly improved display techniques supported by 3-D computer graphics, crystallography has spawned an exciting subfield known as *rational drug design*. The term describes the process by which pharmacologically important compounds (e.g., enzyme inhibitors) are designed to fit the active site with high specificity based on 3-D structural information.

Because functional properties of an enzyme are usually studied in solution, a nagging question has been whether the crystal and solution structures are identical. The evidence that the two structures are identical, at least globally, comes from the observations that crystalline enzymes are capable of performing chemical catalysis when substrates are diffused into the crystals.

Multidimensional NMR spectroscopy provides an alternative method for 3-D structure determination. In fact, this NMR technique is unique in the sense that protein structures can be determined in a solution and noncrystalline state (69, 70). Although still limited in capability, 4-D NMR techniques, in which a 3-D measurement based on ^{15}N spectra is extended into a fourth dimension by ^{13}C spectra, have been successfully applied to the 153-aa interleukin 1β (71), providing further insights into the dynamics and folding of the protein molecule. Probing crystalline enzymes by NMR spectroscopy seems to support the claim that crystal structures are representative of the solution structures for most functional purposes. Nevertheless, some differences have been noted when detailed microenvironments are studied. For instance, the pK_a of His-57 in the *catalytic triad* of α-lytic protease in crystal is 7.9, nearly one unit higher than that in solution (72). Certain regional sequences may be more flexible and disordered in solution than in the crystal, as has been shown by staphylococcal nuclease (73).

In the absence of alternative 3-D structural information, fluorescence quenching can be used to study the structural and functional dynamics of a protein. This method is based on the reduction of fluorescence efficiency of Trp, Tyr, or an artificially introduced fluorophor in the presence of substrates (analogs), inhibitors, or small molecules (e.g., iodide ion and acrylamide).

d. Quaternary structure

When two or more independent tertiary structural units, called *subunits*, interact to form a noncovalent multisubunit complex, the closely packed, spatial arrangement of such an ensemble is referred to as the *quaternary* structure. The mode of interactions between subunits and the behavior of subunits are remarkably diverse. The subunits of a quaternary structure may be of identical size and/or species, giving rise to *homo*(di)mers or *hetero*(di)mers. For some multisubunit enzymes, e.g., T7 DNA polymerase (see Section I,D, Chapter 6) and Qβ replicase, the subunits are not of a single genetic origin but of both phage and bacterial origins. The subunits may have distinct catalytic activities on their own. When one subunit apparently lacks a catalytic function, it may play a regulatory role. The arrangement of subunits in the quaternary structure may be symmetric or asymmetric. The dissociation into monomers often requires rigorous disruptive conditions (see Section I,B,1,f).

A change in quaternary structure means that the subunits move relative to each other. In the native quaternary structure, each subunit somehow alters the properties of the other and enhances its intrinsic catalytic rate either through the induction of a catalytically more active conformation or by more efficient substrate/product channeling. Multisubunit enzymes may have evolved as a means to achieve a concerted control of activity in efficient response to complex metabolite balance. Among the well-characterized multisubunit enzymes are tryptophan synthase which has an $\alpha_2\beta_2$ form (68) and many allosteric enzymes including tetrameric glyceraldehyde-3-phosphate dehydrogenase.

e. Higher ordered structure

Certain enzymes involved in a common metabolic pathway naturally form either covalent or noncovalent *supra*structures termed *multienzyme complexes*. For example, the eukaryotic fatty acid synthase system is composed of seven enzymatic entities which perform in concert in the synthesis of saturated, long-chain fatty acids (74). Other enzymes, such as those involved in glycolytic pathways, appear to form a rather specific interactive system at high *in vivo* concentrations of proteins. The higher ordered enzyme structures may increase the catalytic efficiency of the group of enzymes by facilitated transfer of the substrate metabolites via enzyme–enzyme complexes (75). The subunit channeling offers several advantages: (i) it minimizes the diffusion time of an intermediate from one enzyme to the next, (ii) it alleviates the competition between enzymes for the substrate, and (iii) it shields reactive intermediates from aqueous solution.

f. Denaturation and stability

Under physiological conditions the 3-D native structure of an enzyme is presumed to be biologically most active. The specific active conformation is the result of a large number of interactions. When the environment is changed, some of those interactions are weakened or broken which may cause the breakup of a large portion of the structure, usually but not necessarily leading to the inactivation of an enzyme. The transition from an *ordered* conformation to an essentially disordered *random* or *unfolded* conformation is referred to as *denaturation* (76). Concomitant with denaturation, large changes in a number of physical properties ensue: typically in optical rotation, viscosity, UV, and Raman spectra. In most cases, the denaturation process is reversible, and denatured enzymes can be *renatured* by slowly removing the denaturing conditions. Susceptibility to proteases is often a sensitive indicator of protein stability/denaturation.

Numerous factors and/or reagents induce denaturation of proteins by various mechanisms (76,77). These include (a) breaking of hydrophobic interactions (GuHCl, urea), (b) strong organic solvents (urea solution, alcohols), (c) high or low temperatures (e.g., RNase T1 is most stable near $-5°C$ at pH 7), (d) detergents (SDS, Triton X-100, NP-40) causing dissociation of subunits or aggregation, (e) changes in ionic strengths, (f) changes in pH causing excessive positive or negative charges, (g) chelating agents causing dissociation of coenzymes, and (h) modifications of structurally and/or functionally essential residues.

Protein stability may be regarded as the opposite of denaturation. The stability of enzymes (and proteins) can be increased in many ways, e.g., by microenvironmental changes, immobilization, and protein engineering (78). Enzymes are more stable in the presence of polyols (ethylene glycol, glycerol, erythritol, and sorbitol), polymers (PEG, dextrans), and carbohydrates (sucrose, lactose, and trehalose). Hydrophilic enzymes are stabilized by the presence of salts (LiCl, NaCl, and KCl), whereas hydrophobic enzymes are hardly affected by salts. Proteins are also stabilized by compounds that bind specifically to the folded conformation. Most of the metalloenzymes and the enzymes that have an anion-binding site fall into this category.

2. PHYSICAL PARAMETERS

In addition to the amino acid sequence, the identity of a protein can be distinguished by certain physical parameters such as molecular weight, size, and shape. The average density of a soluble protein is 1.33 g/cm^3.

a. Molecular weight

The term *molecular weight* denotes the "relative" molecular mass (M_r), which is the ratio of the mass of a molecule to one-twelfth of the mass of carbon 12. The M_r is dimensionless; for example, the M_r of ribonuclease A is 13,680. The M_r of a protein can be calculated as the sum of the "molecular weight" of its constituent amino acids when the primary sequence or the minimal amino acid composition is known.

If the molecular weight (M_r) is not available, a molecule can be expressed by *molecular mass* (using MW as its abbreviation). The molecular mass is not a ratio and hence is expressed by a unit called a *dalton* (Da) which corresponds to one-twelfth of the mass of carbon 12. The MW of a protein is usually estimated by comparing its physicochemical behaviors with those of other well-characterized proteins, *protein standards,* such as BSA and ovalbumin. The number of amino acid residues constituting a given protein can be estimated by dividing the molecular weight of the protein by 120, the average M_r of amino acid residues. The analytical methods frequently used to estimate the MW of a protein (and other macromolecules in general) include sedimentation velocity (Svedberg units) in ultracentrifugation, electrophoretic mobility, and elution order from gel-filtration chromatography. The MW may be more or less different from one method to another and from M_r as well since each analytical method relies on a specific set of physicochemical and hydrodynamic properties of the proteins.

b. Protein concentration

The concentration of a protein can be determined by various colorimetric methods including the "old standard" Folin–phenol reagent assay of Lowry (79) or the "new standard" Coomassie blue R-250 method of Bradford (80) or bicinchoninic acid (BCA) assay. *Silver stain* (108) and *gold stain* are also widely used to identify and quantitate proteins with a 10–50 times greater sensitivity than Coomassie blue staining. Once a reliable estimate of the protein concentration is obtained, the UV absorptivity of the known protein can be used to determine unknown concentrations of the same protein. From the *Beer–Lambert law*, $A_{(\lambda)} = \log(I_o/I) = C\varepsilon_{(\lambda)}l$, where $A_{(\lambda)}$ is the *absorbance* (or *optical density*) at the wavelength λ (lambda); I_o and I are the intensities of incident and exiting light; C is the molar concentration; $\varepsilon_{(\lambda)}$ is the *molar absorption* (or *extinction*) *coefficient* in $(M^{-1}cm^{-1})$; and l is the length of the light path (in cm).

An average protein with no prosthetic groups has an UV absorption at around 280 nm. (The peptide bond has an absorption at 210 nm.) The absorption at 280 nm is mostly due to the presence of Trp and Tyr. Consequently the protein displays a characteristic extinction coefficient at or near 280 nm. Trp and Tyr as free amino acids have an ε_{280} of 5700 and 1300 M^{-1} cm^{-1}, respectively. As a rough approximation, the absorption properties of free amino acids can be taken to be the same as those of a protein. Then the UV absorption of a protein (1 mg/ml) is a function of the total number N of Trp and Tyr residues per molecule: $A_{280} = (5700 \times N_{Trp} + 1300 \times N_{Tyr})/MW$. This relationship enables one to estimate the MW of a protein when the number of aromatic amino acid residues (Trp and Tyr) is known.

3. STRUCTURAL PRINCIPLES

How a sequence of DNA bases is translated into a sequence of peptides, the so-called first half of the genetic code, is now fairly well understood. The second half of the code—what rules determine the folding and packing of a linear amino acid sequence into a 3-D protein—remains to be decoded.

Once the covalent backbone of peptide bonds is formed, it is the physical principles of noncovalent or sometimes covalent interactions among the amino acid side chains and the peptide backbone that govern the pathway of protein folding and packing. *In vitro*, the process can be spontaneous, suggesting that the basic rule of folding is inscribed in the primary structure. The process also involves cooperativity. During the (un)folding transitions of small proteins, the molecules present at equilibrium are usually in two states: either fully folded (N) or fully unfolded (U). Partially folded intermediates with distinguishable physical or thermodynamic properties are rare and account for less than 5% of the total. Most likely the folding process involves relatively short-lived transition state intermediates.

Compared with the denatured forms, the active folded states are only marginally stable, favored by 5 to 15 kcal/mol (21–63 kJ/mol) for globular proteins. Energetically, the final, native conformation of a protein is not a rigid structure in its lowest energy state, but rather in one of its multiple energy minima. The small degree of conformational flexibility associated with the multiple energy minima is presumed to be the powerhouse for catalytic functions.

Disulfide bonds stabilize proteins by constraining the unfolded conformation and thereby raising the free energy of the unfolded state. In addition to the covalent peptide backbone and occasional disulfide bonds, a number of weak and subtle forces cooperate to maintain the 3-D structure of an enzyme. These forces also play a key role in the interaction of an enzyme with its substrates. These "weak" forces include H-bonds, electrostatic forces between charged groups, and hydrophobic or van der Waals forces among nonpolar groups (see Section I,B,4,g in this chapter).

Polar groups are usually located at the surface of a protein and apolar groups preferentially in the interior. Charged groups are normally solvated and have nearby counterions. Globular proteins are most stable near their isoelectric points. Protein interiors are, however, more tightly packed than can be extrapolated from apolar liquids, and the additional forces are referred to as *packing effects*. The packing effects on protein stability are roughly additive among the constituent amino acids. A single mutation contributes 0.4–2.4 kcal/mol to destabilization primarily due to unfavorable contacts with the rest of the protein (81). In proteins where imperfect packing may be a cause of limited stability, an amino acid substitution that fills the internal cavity increases the stability of the protein. One such example is the Gly-316 to Ala mutation in glyceraldehyde-3-P dehydrogenase that increases the thermostability of the enzyme by 7°C (82).

Although all constituent amino acids of a protein may be considered important for the maintenance of an optimally folded native structure, some residues have been found to be structurally and functionally more important than others. Residues that are directly involved in essential structure and/or function of a protein are among the most conserved, and their modifications or replacements usually result in a drastic decrease in the activity of the protein. In contrast, nonessential residues can be replaced by other amino acids and sections of polypeptides can even be removed without large negative effects on the stability of a protein. Proteins tolerate amino acid substitutions by various mechanisms (83): (a) some substitutions preserve critical interactions, (b) some interactions apparently do not make large contributions to stability, and (c) conformational adjustments occur to compensate for the changes in primary structure.

4. Structural Contributions to Enzymatic Function

a. Active site

The catalytic function of an enzyme is ordained in its characteristic tertiary structure. Within this structural integrity, the catalytic function of an enzyme is largely determined by a few particular functional groups of amino acids that are strategically positioned in the *active site* or *active center*. The physical dimensions, shape, and nature of the functional groups that compose the active site determine, by the principle of complementarity, which substance can be used as a substrate (i.e., *substrate specificity*) and, perhaps more importantly, what type of chemical reaction is to occur (i.e., *reaction specificity*).

Active-site residues act as an ensemble to create a unique microenvironment not found elsewhere on an enzyme molecule. Particular functional groups (e.g., the OH group of Ser in the catalytic triad of serine proteases) may thus become "abnormally" reactive due to significant perturbations in chemical properties. The ε-NH_2 group of the Lys residue in the active site of aspartate aminotransferase participates in the proton transfer by acting as the endogenous general base with an "effective concentration" equal to 10^6 M ethylamine (84). Although the microenvironment is unique, an active site can be constructed by combinations of different amino acids in different enzymes. Active sites generally provide specific polar rather than nonpolar environments and enhance electrostatic stabilization of ionic transition states by solvation which is stronger than by using water (85).

Certain enzymes such as DNA polymerases require the binding of multiple substrates. In these multisubstrate enzymes, each substrate often occupies respective *subsites* within the active center. Although the residues in an active site (or center) are important for catalysis, certain residues are used for the binding of substrates, whereas others are used almost exclusively for chemical catalysis. The noncovalent enzyme–substrate complex that is enzymatically competent is called the Michaelis–Menten complex (see Section I,C,3).

b. Substrate specificities

i. Substrates and analogs. An enzyme may bind one substrate tightly and another relatively loosely. The strength of binding (or binding affinity) can be compared by the magnitude of binding (or association) constants. The binding constant is the reverse of the dissociation constant which is the parameter usually measured in enzyme kinetics.

The compounds that structurally resemble true substrates but cannot be "turned over" by the enzyme are called *pseudo-substrates, substrate analogs,* or *inhibitors*. The relative magnitude of binding affinities for a series of substrates and/or analogs often reflects the physical dimension and physicochemical nature of the active site and enzyme–substrate interactions. Although most of the substrate analogs bind the enzyme relatively poorly, some analogs (e.g., *transition-state analogs*) bind more tightly than the substrates, resulting in "specific" inhibition of the enzyme. The vanadyl ribonucleoside complex (VRC) is a good example of a "transition-state" inhibitor for RNase A.

ii. Stereospecificity. With a few notable exceptions (e.g., racemases), enzymes exhibit an exquisite *stereospecificity* for the binding of a substrate. By stereospecificity, enzymes discriminate *chiral* substrates, also called *optical isomers, enantiomers,* or *stereoisomers*. In all common amino acids, except Gly in which the side chain is simply a hydrogen atom, the α carbon atom is asymmetric and the atoms or groups are disposed around it in L-configuration (left-handed), i.e., with respect to the H–C–R plane, the NH$_2$ lies to the left of the COOH. Therefore an enzyme itself is inherently a chiral molecule. The mirror image of the L-form is the D-form (right-handed). Chiral molecules have the ability to rotate a plane of polarized light, and thus have *optical activities*: *dextro*rotatory (*d* or +) when the rotation is clockwise and *levo*rotatory (*l* or −) when the rotation is counterclockwise. The notation for D or L should not be confused with *d* or *l*, and an amino acid having a D configuration may exhibit a (+) or (−) rotation. Asymmetric centers having equal amounts of (+) and (−) rotation neutralize the optical rotatory power of each other. If this happens within a single molecule, the molecule is called a *meso*compound. If two different molecules in a mixture neutralize the optical activities of each other, the mixture is called *racemic*. All enzymes except the racemases bind only one stereomeric form. If one substitution (but not two) can make a molecule chiral, the molecule (e.g., ethanol) is designated as *prochiral*.

iii. **Alteration of substrate specificities.** The substrate specificity of an enzyme can be altered by introducing some secondary modifications on or near the active site residues. It can also be changed by amino acid substitutions at the genetic level. Protein engineering, which is largely based on recombinant DNA technology, is a powerful process through which one can generate the enzymes (or more generally proteins) with new or altered reaction and/or substrate specificities and higher stability.

c. Enzyme families

i. **Functional and/or structural families.** Despite individual differences in optimum reaction conditions and catalytic turnover rates, certain groups of enzymes carry out catalysis through a similar if not identical pathway. Such groups of enzymes (or proteins) are called *enzyme (protein) families*. The members of a family possess a sequence motif(s) or essential functional groups distinct from those of other families.

Comparison (or alignment) of amino acid sequences, also called *homology search,* often provides first-hand information on such conserved structural features and enables one to classify enzymes into families and predict the possible function of a new enzyme (86). A family of enzymes usually folds into similar 3-D structures, at least at the active site region. A typical example is the serine protease family whose members—trypsin, chymotrypsin, elastase, and subtilisin—commonly contain three active-site residues, Asp/His/Ser, which are known as the *catalytic triad* or *charge relay system*. Another example is the conserved features of catalytic domains of the highly diverse protein kinase family. In this kinase family, the ATP-binding (or phosphate-anchoring) sites present a consensus sequence motif of Gly–X–Gly–X–X–Gly (67,87).

It should be noted, however, that many proteins with apparently unrelated sequences fold into similar 3-D structures, providing families of proteins based on folding motifs (88). For example, remarkable structural similarities have been found in apparently disparate proteins such as ubiquitin, ferredoxin, and the IgG-binding domain of protein G. Whether a particular structural motif indicates a possible evolutionary relationship among the proteins or simply an energetically favorable folding unit remains debatable.

ii. Isoenzymes. Isoenzymes (or *isozymes*) are a group of enzymes that catalyze the same reaction but have different enzyme forms and catalytic efficiencies. Isozymes are usually distinguished by their electrophoretic mobilities. All living systems apparently require multiple molecular forms of certain enzymes in order to maximize biological capacity.

Isozymes arise from gene duplications and/or different epigenetic modifications of a gene product(s). In this sense, most of the recombinant enzymes with deletion, insertion, and/or other mutations at the genetic level fall into the category of isozymes (89). In a restricted definition, "isozymes" are different in genetic origins. An example is the human alkaline phosphatases which have at least three different genetic origins, i.e., for placental, intestinal, and liver/bone/kidney enzymes (see Section I,B, Chapter 5). The enzymes that have epigenetic differences due to differential precursor processings, covalent modifications, and/or tissue distributions are then called *isoforms*. Examples of isoforms are the liver/bone/kidney alkaline phosphatases which are encoded by the same gene but differentially modified in a tissue-specific manner.

The five "classical" isozymes of lactate dehydrogenase (LDH) arise from combinations of the two restricted definitions described earlier. LDH isozymes consist of two genetically distinct polypeptide chains, A (or M for muscle type) and B (or H for heart type), which form varying combinations of tetrameric structures (90).

d. Reaction specificities: enzyme classification and nomenclature

The ever-increasing number of enzymes requires a systematic grouping of and coherent nomenclature for the enzymes. The current classification and nomenclature scheme recommended by IUBMB (International Union of Biochemistry and Molecular Biology) is based on the reaction specificity of enzymes (91). It should be noted that certain enzymes are capable of catalyzing more than one type of reactions. Those multifunctional enzymes are classified according to their "principal" reaction specificity under presumed physiological conditions. Each classified enzyme is given a code number prefixed by EC (which stands for Enzyme Commission). The code numbers contain four numbers separated by decimal points, with the following meaning. The first number designates one of the six classes of enzymes (or major divisions). The second, third, and fourth figures give the subclass, sub-subclass, and the serial number of the enzyme in its sub-subclass, respectively. The six classes of enzymes are

1. *Oxidoreductases* such as dehydrogenases and oxidases that catalyze oxidation–reduction reactions
2. *Transferases* that carry out group-transfer reactions between donors and acceptors
3. *Hydrolases* that catalyze the hydrolytic cleavage of C–O, C–N, C–C, and other bonds

4. *Lyases* which cleave C–C, C–O, C–N, and other bonds by elimination, leaving double bonds or rings, or conversely adding groups to double bonds
5. *Isomerases* that catalyze geometric or structural changes within one molecule
6. *Ligases* that catalyze the joining of two molecules in a reaction coupled with the hydrolysis of a pyrophosphate bond in ATP or a similar triphosphate

e. Determination of active-site residues

The identification of essential amino acid residues (structural or catalytic) is sometimes a difficult but rewarding task in structural enzymology. This information is a prerequisite to the understanding of catalytic mechanisms. Among the methods used for these purposes are classical methods such as acid–base titrations of the enzyme monitoring the activity and/or specific NMR signals and, more frequently, group-specific modifications of essential residues. Perhaps the most satisfying source of information on the active site in general and the residues in particular is the 3-D structure determined in the presence and/or absence of a substrate (or inhibitor). Site-specific mutagenesis (see Appendix A at the end of this volume) is a novel, powerful method that can be applied to assess, and more appropriately to confirm, the essential nature of certain amino acid residues.

Inactivation of enzymes often follows covalent bond or noncovalent complex formation between active reagents and essential functional groups of an enzyme. The essential functional groups have thus been identified by monitoring the inactivation kinetics as a function of chemical modifications with "group-specific" reagents (92). In fact, the so-called group-specific reagents are not necessarily specific in all circumstances. Protection from modification of a certain residue(s) by substrates (or inhibitors) and the subsequent modification of the protected residue(s) in the absence of the substrates (or inhibitors) provide strong evidence that the residue(s) is located in the active site. This technique is called *differential modification* or *labeling*.

By relying on the unparalleled specificity of another enzyme (transferase), enzymatic modification of an essential residue provides an ideal alternative for group-specific modifications. For example, when a reactive Arg residue cannot be distinguished from the reactive active-site Cys residues by the apparently Arg-specific reagent phenylglyoxal or (*p*-hydroxyphenyl)glyoxal (93), the single Arg can be unambiguously modified (or labeled) by the enzymatic transfer of the ADP-ribosyl group from NAD in a reaction catalyzed by $NAD(P)^+$–arginine ADP-ribosyltransferase (EC 2.4.2.31). When multiple modifications occur, which is more often than not the case, the modification and inactivation data should be subject to kinetic and/or statistical analyses to estimate the number of "true" essential residues (94,95).

To localize the essential residue(s) by following a typical procedure of *protein chemistry*, the modified (or labeled) protein is first cleaved by site-specific proteases or chemical reagents (e.g., cyanogen bromide). The modified fragment(s) is then separated, and the position of the modified residue(s) is determined by amino acid sequencing.

The essential nature of selected amino acid residues may be deduced or confirmed by comparison with some different enzymes having similar structures at various levels. This evolutionary approach hinges on two basic premises: (a) if two proteins are similar in their amino acid sequences, then they would fold into similar structures, and (b) if two proteins differ in their amino acid sequences but have similar functions, the essential functional groups may be disposed in similar positions in a 3-D structure.

f. Nonprotein components of enzymes

Certain classes of enzymes require small, auxiliary, nonprotein molecules called *cofactors, coenzymes,* and *prosthetic groups.* Definitions for these three terms are somewhat arbitrary and, in fact, the term *cofactor* will be used in the following chapters to represent broadly the identity and functional roles of cocatalysts. The roles of cofactors are structural, functional, or both. They provide the enzyme with the chemical or photochemical capabilities lacking in the normal amino acid side chains. An enzyme devoid of a cofactor is called an *apoenzyme*. Apoenzymes are catalytically inactive. The active complex of the protein and the cofactor is termed a *holoenzyme*. The cocatalysts can be defined on the basis of the catalytic functions that are mediated (76).

i. Cofactor. A cofactor forms part of an active site and is regenerated on each turnover of substrate. The cofactors can be reversibly dissociated from the enzyme. A typical example is pyridoxal phosphate which is a cofactor in a variety of group-transferring enzymes (e.g., aminotransferases).

ii. Coenzyme. A coenzyme is a cocatalyst that is converted to a new product concomitant with each turnover of substrate. The coenzyme product is regenerated via other enzyme-catalyzed reactions. Examples of coenzymes are pyridine nucleotides (e.g., NAD^+ and $NADP^+$), flavins (e.g., FMN, FAD), and heme compounds, which participate in various oxidation–reduction reactions.

iii. Prosthetic group. In contrast to cofactors and coenzymes which undergo chemical conversions as part of the catalysis, a prosthetic group is a stable constituent that remains unchanged. Prosthetic groups tend to bind the enzyme (or protein) either covalently or by noncovalent but strong forces such that their bindings are relatively tighter than those of cofactors or coenzymes. Typical examples of prosthetic groups are metal ions (e.g., Zn^{2+} in dehydrogenases and alkaline phosphatases and Fe^{2+} in oxidoreductases) or a small organic compound TTQ (tryptophan tryptophylquinone) in methylamine dehydrogenase (EC 1.4.99.3) (96).

g. Forces involved in substrate binding

The specific binding between enzymes and substrates is thought to be based on structural and electrostatic *complementarity*. A substrate is firmly held in the active site by an assortment of weak forces contributed by hydrophobic, hydrophilic, H-bonding, and/or electrostatic interactions. The bound substrate is poised next to an acidic, basic, or nucleophilic group(s) of amino acid(s) that usually serve as catalytic residues.

i. Hydrophobic forces. The hydrophobic interactions between nonpolar compounds are measured by van der Waals forces which range from 20 to 25 cal/mol Å^2 of surface area. Hydrophobic residues contribute 1.5–5 kcal/mol to ligand binding and catalysis (97, 98).

ii. Electrostatic forces. The Coulombic forces involved in electrostatic interactions between two point charges, e_1 and e_2, separated by a distance of r can be calculated by using the formula $E = e_1e_2/Dr$, where D is the dielectric constant of the medium. In the active-site environment which is "unusual," the electrostatic interaction energy becomes substantially greater due to the lowering of the dielectric constant. Despite the fact that the definition and calculation of the dielectric constant(s) in proteins remain ill-defined, case studies show that the electrostatic (or salt bridge) interaction between the α-NH$_3^+$ of a substrate and the —CO$_2^-$ of an enzyme contributes 2–4 kcal/mol to the binding energy. In superoxide dismutase, the electric field of the enzyme is estimated to enhance the association rate of the substrate anion by a factor of 30 or more (99).

iii. H-bonding. The energetics of H-bonding in proteins and their complexes (protein–ligand) involves the energy gained by the H-bonding interactions and that lost on displacement of H-bonds with the solvent. As a rule, (a) an uncharged H-bond between a protein (enzyme) and its ligand (substrate) contributes 0.5–1.5 kcal/mol, (b) a H-bond with a charged group on the substrate contributes 3.5–6.0 kcal/mol, and (c) the shortening of a "normal" H-bonding distance improves binding energy by 1.2 kcal/mol (100). A single specific H-bonding interaction has been estimated to contribute ~4.1 kcal/mol to the instrinsic binding energy of peptide inhibitors to thermolysin (101,102).

iv. Steric repulsion. Steric repulsion contributes to the decrease in catalytic efficiency by a factor of up to 5000 times or ~2.6 kcal/mol in transition-state binding energy per 100 Å^3 of excess volume (97). (Note that 100 Å^3 is approximately the volume of a leucyl side chain or 5 times the volume of a water molecule.)

5. ENZYMATIC CATALYSIS: TRANSITION-STATE THEORY

Among a number of proposals, the *transition-state* theory best accounts for how an enzyme achieves catalytic efficiency in excess of 10^8 times compared with the rate of uncatalyzed reactions. The essence of this theory is that an enzyme increases the reaction rate by stabilizing the transition state of a substrate and thus lowering the highest energy barrier on its reaction pathway.

The initial interaction of the substrate with the enzyme is deemed thermodynamically "unfavorable" due to unfavorable entropy, geometry, solvation, or charges. The changes in electron density and geometry developed in the substrate on going from the initial unfavorable ground state to the transition state are presumed to be stabilized by complementary charges and the shape of the active site. The transition-state binding energy thus compensates for the unfavorable ground-state interactions. For example, a single OH group appropriately placed in a molecule can enhance the affinity of the molecule for an enzyme by a factor greater than 10^8 (or -9.8 kcal/mol of free energy of binding) (103). The optimum binding of the transition state is attained by cooperative interactions of individual binding forces, an effective strategy for amplifying enzyme specificity (104).

The *Gibbs free energy* change (ΔG^*) between the transition state and the ground state of the substrate is related to the acceleration of the reaction rate by Eq. (1.1)

$$k_1 = (kT/h)K^* = (kT/h)\exp(-\Delta G^*/RT), \qquad (1.1)$$

where k_1 is the first-order rate constant ($= k_{cat}$); k is the Boltzmann constant ($= 1.380 \times 10^{-16}$ erg/deg/molecule); h is the Planck constant ($= 6.625 \times 10^{-27}$ erg sec); K^* is the equilibrium constant; R is the universal gas constant ($= 1.987$ cal/deg/mol); and T is absolute (or K) temperature (degrees Kelvin $= 273 +$ degrees Celsius).

In enzymatic reactions that do not occur via a covalent intermediate(s), chemical catalysis involves general or specific acids or bases played by an active-site residue(s) or sometimes by a water molecule within the particular active-site environment. In these noncovalent catalyses, one of the most important factors contributing to the high efficiency of the enzymatic reaction is the *entropy* (ΔS), the degree of freedom. The Gibbs free energy (ΔG) is a function of entropy by $\Delta G = \Delta H - T\Delta S$, where ΔH is the enthalpy or heat content of the molecule. The binding of a substrate to a specific position in an enzyme results in the loss of entropies both translational (~ 30 cal/deg/mol) and rotational (10–30 cal/deg/mol). Therefore the initial interaction of the substrate is "unfavorable" for the catalytic reaction. However, the binding affords an intramolecular type of reaction to occur at a high "effective concentration" of the catalytic groups, leading to a substantial enhancement of the reaction rate.

In covalent catalysis where an enzyme forms a covalent intermediate with a substrate, the chemical bonding activates the substrate and enhances the electrophilic or nucleophilic catalysis. Electron-rich groups (or elements) are called *nucleophiles*, and electron-poor groups (or elements) are called *electrophiles*. The nucleophilic groups of enzymes that participate in covalent catalysis include —OH (Ser in serine proteases and alkaline phosphatases), —OH [Tyr in glutamine synthetase (glutamate–ammonia ligase)], OH^- (Zn-bound, in DNase I and alcohol dehydrogenases), —SH (Cys in thiol proteases), —CO_2 (Asp in Na^+, K^+-ATPase), —NH_2 (Lys in pyridoxal enzymes and DNA ligases), and imidazole (His in phosphoglycerate mutase). Cofactors such as pyridoxal phosphate and, in particular, metal ions typically participate in electrophilic catalysis.

C. Function

1. SPECIFIC ACTIVITIES

The term *specific activity*, which is defined as the enzyme units per milligram protein, is a practical expression of the functional state of an enzyme. It gives a measure of not only enzyme purity but also the concentration. Specific activity is particularly useful when the actual molar concentration of an enzyme is not known.

The comparison of specific activities of different enzyme preparations requires the consideration of the following variables: (i) the assay conditions may not be identical, (ii) the unit definition may be arbitrary, and (iii) the active conformation of an enzyme, which often depends on purification procedures and storage conditions, is usually unknown.

2. CONCEPT OF EQUILIBRIUM: EQUILIBRIUM CONSTANTS AND GIBBS FREE ENERGY

A general reversible reaction involving two reactants (A, B) and two products (P, Q) with forward rate constant k_f and reverse rate constant k_r can be written as Scheme 1.1:

$$A + B \underset{k_r}{\overset{k_f}{\rightleftharpoons}} P + Q$$

SCHEME 1.1

In this reversible system, the reaction starting with A and B proceeds to the forward direction until enough products P and Q are formed. The state in which no further concentration changes occur is said to be "equilibrium." [Note: All concentration terms will be written in brackets.] At equilibrium, the ratio of products and reactants is a constant, K_{eq}, given by Eq. (1.2):

$$K_{eq} = [P][Q]/[A][B] = k_f/k_r. \quad (1.2)$$

In chemical kinetics, the equilibrium is still a dynamic state in which the rate of product formation (v_f) is equal to the rate of reactant formation (v_r), i.e., $v_f = k_f[A][B] = v_r = k_r[P][Q]$. At preequilibrium, the ratio of product/reactant is then defined by Eq. (1.3):

$$K' = [P]'[Q]'/[A]'[B]'. \quad (1.3)$$

When reactants are added to the initial reaction mixture or to the mixture at equilibrium, the reaction proceeds forward, generating more products until a new equilibrium is attained. Thermodynamically, the forward reaction has been made "favorable" because the energy content of the reactants has been made higher than the energy content of the products. The net energy change under this situation is a negative value: ΔG (Gibbs free energy change) = $G_{product} - G_{reactant}$. Similarly, a positive ΔG indicates that the reverse reaction is favorable and can occur spontaneously. The Gibbs free energy is related to the constant K by Eq. (1.4):

$$\Delta G = -RT \ln K. \tag{1.4}$$

The ΔG is synonymous to *reaction potential*. It indicates the direction and measures the displacement of the actual reaction system ($\Delta G'$) from the ideal equilibrium ($\Delta G°$). The free energy concept applies to any circumstances in which there is a change whether chemical, physical, or biological. For example, the conformations of proteins and DNA are driven by the sign and magnitude of ΔG. Equation (1.4) can then be rewritten as Eqs. (1.5) and (1.6):

$$\Delta G° = -RT \ln K° \tag{1.5}$$

$$\Delta G = \Delta G° - \Delta G' = \Delta G° + RT \ln K'. \tag{1.6}$$

The $\Delta G°$ is the "standard" free energy of reaction, which is a constant for a given reaction at a given temperature. In fact, the $\Delta G°$ is defined as the ΔG under the "standard-state" conditions in which all reactants and products are considered to be maintained at steady-state concentrations of 1 M and thus ln K' = ln ([P]'[Q]'/[A]'[B]') = 0. For a reaction in which the forward reaction is 10 times more favored than the reverse reaction, i.e., K_{eq} = 10, the $\Delta G°$ is -1.36 kcal/mol at 25°C.

The K_{eq} is a constant for a specific chemical reaction, as is the $\Delta G°$. An enzyme does not change the equilibrium. It only expedites, often by a factor of over 10^{10}, the process through which the chemical reaction occurs.

In a living cell, a true reaction equilibrium does not exist because either the product(s) of a reaction is the reactant(s) of the next reaction or the product(s) is actively transported out of the reaction domain. The reaction components in a living cell are said to be in dynamic equilibrium or a metabolic steady state in which the ΔG of the whole system is maintained at a negative value.

3. ENZYME KINETICS

The catalytic function of an enzyme is described by enzyme kinetics usually determined under steady-state conditions. A *steady state* refers to a complete balance of a particular quantity between its rate of formation and its rate of disappearance. In steady-state enzyme kinetics, the concentrations of enzyme-bound intermediates are meant to be in a steady state. On mixing an enzyme with a large excess of substrates, there is an initial period, known as a *presteady state*, during which the concentrations of the intermediates build up to a maximal level under the reaction conditions. Then the reaction rate changes relatively slowly with time and the intermediates are considered to be at steady-state concentrations. Note that the steady state is an approximation because the substrate is gradually depleted during the course of reaction. Therefore, steady-state kinetic measurements should be performed in a relatively short time interval over which the concentration of substrates does not greatly change. The enzyme concentration should be within the range in which the enzyme activity is proportional to the enzyme concentration.

a. Types of reactions

Enzyme-catalyzed reactions can be classified according to *reactancy*, the number of kinetically significant substrates and products: A ⇌ P (Uni–Uni), A ⇌ P + Q (Uni–Bi), A + B ⇌ P (Bi–Uni), A + B ⇌ P + Q (Bi–Bi), A + B + C ⇌ P + Q (Ter–Bi), and so on.

The term *kinetic mechanism* refers to the binding order of the substrates (or products), which can be *ordered* or *random*. [The term *catalytic mechanism* refers to the process whereby enzymes and substrates undergo transformation during catalysis.] The term *sequential* denotes the system in which all substrates bind to the enzyme before any product is released.

In any multistep reaction, the rate of conversion of reactants to products is critically dependent on a single step called *rate-determining (or limiting) step* (RDS). The RDS is not the step with the smallest rate constant, but the step that is the slowest in rate. Thus a RDS is a function of both the rate constant and the concentration of species undergoing that step. The power of enzyme catalysis is that the rate of the RDS can be 10^6 to 10^{15} times greater than that of the uncatalyzed reaction at ambient temperatures.

b. Michaelis–Menten equation

i. Michaelis–Menten mechanism: a rapid equilibrium assumption. Michaelis–Menten kinetics is perhaps the best known mathematical treatment of the course of an enzyme-catalyzed reaction (105). The Michaelis–Menten kinetic equation provides the framework from which the current kinetic parameters are defined. The Michaelis–Menten kinetic scheme describes the enzymatic conversion of a substrate (S) to a product (P) (Scheme 1.2):

$$E + S \underset{k_{-1}}{\overset{k_1}{\rightleftharpoons}} ES \overset{k_2}{\longrightarrow} E + P$$

SCHEME 1.2

This mechanism is based on the assumption that the enzyme (E) binds the substrate in a rapid and reversible step to give a noncovalent enzyme–substrate complex (ES) known as the *Michaelis–Menten complex*. The ES slowly "turns over" to the product with a first-order rate constant k_2. The free enzyme can resume the catalytic cycle. When $k_{-1} \gg k_2$, the rapid equilibrium assumption holds and the ES is in equilibrium with E and S. Under the rapid equilibrium assumption, the rate expression is given by Eq. (1.7)

$$v = k_2[E]_o[S]/(K_m + [S]) = V[S]/(K_m + [S]), \quad (1.7)$$

where $[E]_o$ is the total enzyme concentration, v is the initial rate, and K_m is the

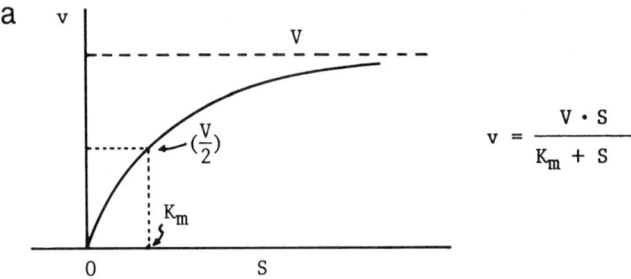

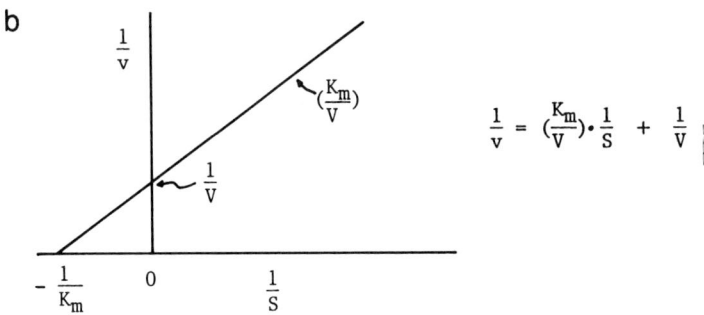

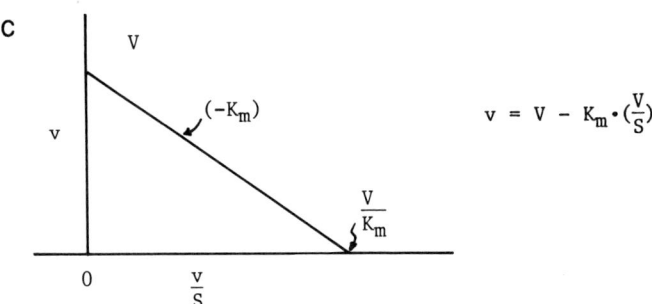

FIGURE 1.2 Graphic representations of Michaelis–Menten equation. (a) Michaelis–Menten equation, (b) Lineweaver–Burk plot, and (c) Eadie–Hofstee plot.

Michaelis constant. The v is usually measured by monitoring the rate of appearance of a product or disappearance of a substrate.

Equation (1.7) indicates that v increases as a function of [S] in a hyperbolic manner (Fig. 1.2). When [S] approaches infinity, v becomes the maximal rate V (or V_{max}). The V ($= k_{cat}[E]_o$) is given by $k_2[E]_o$ and thus $k_{cat} = k_2$. The K_m is equivalent to the apparent dissociation constant of the noncovalent enzyme–substrate complex, i.e., $K_m = K_s = [E][S]/[ES] = k_{-1}/k_1$. The value of K_m is the substrate concentration at which $v = V/2$. At an [S] of 10 K_m, the concentration often employed to saturate the enzyme, the v is ~90% of V. Enzyme kinetic experiments are sometimes carried out, for various reasons, under subsaturating conditions for one or more substrates. The kinetic parameters estimated from such "imperfect" experiments are still useful as approximate value and are expressed with the word "apparent," e.g., apparent K_m or $K_{m,app}$.

ii. Briggs–Haldane mechanism: a steady-state assumption. For many enzymes, the magnitude of k_{-1} is approximately similar to that of k_2, implying that the ES is not in equilibrium with E and S. For such systems, the ES is considered to be in a steady state, at least in the initial phase of the reaction. Under this steady-state assumption, which was put forward by Briggs and Haldane (106), the rate expression is the same as Eq. (1.7) except that K_m is now $K_s + k_2/k_1$. This also indicates that the equilibrium treatment in the Michaelis–Menten mechanism can be regarded as a special case of the more general steady-state theory.

c. Kinetic parameters

i. Catalysis constant (k_{cat}). The catalysis constant (k_{cat}) refers to the maximum number of substrate molecules converted to products per active site per unit time, or the number of times the enzyme "turns over" per unit time. Therefore the k_{cat} is also called *turnover number*. For the majority of enzymes, the k_{cat} is $\leq 10^3$ sec^{-1}, although it can range from 1 to 10^7 sec^{-1}.

In the Michaelis–Menten mechanism, k_{cat} is equal to k_2, the first-order rate constant. In more general situations to which steady-state assumptions adequately apply, k_{cat} is a combination of rate constants from several microscopic steps.

ii. Michaelis constant (K_m). The K_m is an apparent dissociation constant of all enzyme-bound species. In its reciprocal form ($1/K_m$), K_m can be regarded as the binding affinity of an enzyme for its substrate. The lower the constant, the higher the affinity. In the Michaelis–Menten mechanism, K_m is equal to the substrate constant K_s ($= k_{-1}/k_1$). In general, however, K_m is not identical to K_s and its expression involves the rate constants of several microscopic steps. The bimolecular rate constant k_1 usually ranges between 10^6 and 10^8 M^{-1} sec^{-1} (104), with the upper limit being the diffusion-controlled rate of 10^9 M^{-1} sec^{-1}.

iii. Specificity constant (k_{cat}/K_m). This parameter is not a microscopic rate constant but rather a measure of enzymatic efficiency for a specific substrate. This parameter is derived from Eq. (1.7) when the reaction occurs at very low substrate concentrations so that $[E]_o \approx [E]$ and $v = (k_{cat}/K_m)[E][S]$. The significance of this composite parameter is that the reaction rate can be related to the concentration of free, rather than total, enzyme. Indeed this relationship holds at any substrate concentration as long as the enzyme–substrate complex is in thermodynamic equilibrium with the free enzyme and the substrates (107). In the Briggs–Haldane mechanism in which $k_2 \gg k_{-1}$ requiring steady-state assumptions, the relationship of k_{cat}/K_m to free enzyme does not hold. If k_{cat}/K_m is equal to k_1 and of the order of $10^8\ M^{-1}\ sec^{-1}$, it is indicative of the equilibrium system fitting the Michaelis–Menten mechanism.

The k_{cat}/K_m is the kinetic parameter (an apparent second-order rate constant) that is most appropriate for distinguishing substrate specificity. Kinetically, *specificity* means the capability of an enzyme to discriminate between competing substrates and is a function of both K_m and k_{cat}.

d. Graphic representations of Michaelis–Menten equation

One of the major goals of enzyme kinetics is to determine the enzyme parameters, V and K_m. Although many methods have been devised (109), including computer-aided curve fittings based on least-squares regression analysis (110, 111), enzyme parameters have been traditionally and conveniently estimated from graphical representations of the kinetic data using transformations of the Michaelis–Menten equation into linear forms. One of the best known methods is the double-reciprocal or *Lineweaver–Burk* (112) plot based on Eq. (1.8):

$$1/v = (K_m/V)(1/[S]) + 1/V. \tag{1.8}$$

The plot of $1/v$ versus $1/[S]$ gives an intercept of $1/V$ on the y axis (ordinate) and $-1/K_m$ on the x axis (abscissa) (Fig. 1.2).

Another commonly used plot is that of *Eadie and Hofstee* (113,114) based on Eq. (1.9):

$$v = V - K_m(v/[S]). \tag{1.9}$$

The plot of v versus $v/[S]$ gives a straight line where V can be directly obtained from the y intercept and K_m from the value of V/K_m, the x intercept (Fig. 1.2).

The Lineweaver–Burk plot has the advantage that values of v can be easily read for a given value of [S]. Its disadvantage is that it compresses the data points at high substrate concentrations into a small region while emphasizing the points at lower concentrations. The Eadie–Hofstee plot does not compress the data points at high [S], but it is more difficult to determine the values of v against [S]. Taken together, the Eadie–Hofstee plot is considered more accurate and generally superior (115,116).

e. Inhibition

The activity of an enzyme can be modulated *reversibly* or *irreversibly* by inhibitors and inactivators. The kinetics of inhibition and/or inactivation provide valuable insights into the nature of essential and/or catalytic residues as well as the mechanism of enzyme catalysis.

Reversible inhibition is a phenomenon whereby the noncovalent binding of inhibitors competes either directly or indirectly with the substrate binding and modulates the enzyme activity. The strength of inhibition is measured by K_i (= [E][I]/[EI]). Three main types of reversible inhibitions are usually recognizable and graphically distinguishable by their characteristic effects on K_m and/or V (Fig. 1.3).

i. Competitive inhibition. Competitive inhibition is usually caused by substances that are structurally related to the substrate, and thus combine at the same binding site as the substrate. The bindings are exclusive to each other, forming either an enzyme–substrate (ES) or an enzyme–inhibitor (EI) complex but not a ternary complex (EIS) (Scheme 1.3, Fig. 1.3). This type of inhibition

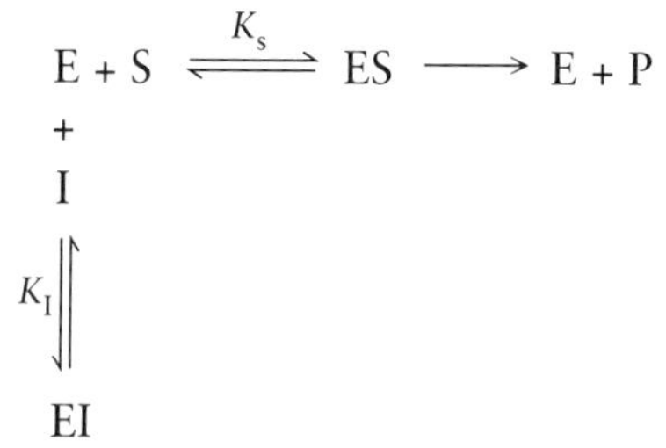

$$\begin{array}{c} E + S \xrightleftharpoons{K_s} ES \longrightarrow E + P \\ + \\ I \\ K_I \updownarrow \\ EI \end{array}$$

SCHEME 1.3

can be completely overcome by high substrate concentrations and thus does not affect the V. The K_m is increased by a factor of $(1 + [I]/K_i)$.

ii. Noncompetitive inhibition. In noncompetitive inhibition, the inhibitor binding site on the enzyme is separate from that for the substrate. Therefore the noncompetitive inhibitor can bind equally well with the enzyme giving EI or with its enzyme–substrate complex giving ESI (Scheme 1.4, Fig. 1.3). This type of

$$\begin{array}{c} E + S \xrightleftharpoons{K_s} ES \longrightarrow E + P \\ + \quad\quad + \\ I \quad\quad I \\ K_I \updownarrow \quad\quad \updownarrow \\ EI + S \rightleftharpoons ESI \end{array}$$

SCHEME 1.4

inhibition, which is very common with multisubstrate enzymes, cannot be completely overcome by high substrate concentrations. Therefore it does not affect the K_m but decreases the V by a factor of $(1 + [I]/K_i)$.

iii. Uncompetitive inhibition. Certain substances bind only with the enzyme–substrate complex but not with free enzyme, causing uncompetitive inhibition

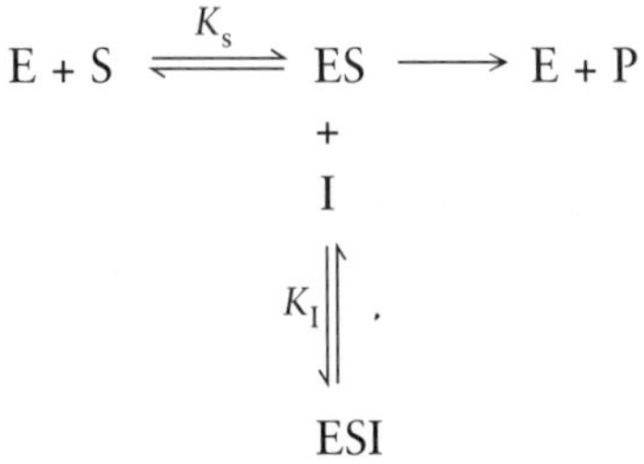

$$E + S \underset{}{\overset{K_s}{\rightleftharpoons}} ES \longrightarrow E + P$$
$$+$$
$$I$$
$$K_I \updownarrow$$
$$ESI$$

SCHEME 1.5

(Scheme 1.5, Fig. 1.3). This type of inhibition is relatively rare with single-substrate enzymes. It is not completely overcome by high substrate concentrations and lowers both K_m and V by a factor of $(1 + [I]/K_i)$. No change occurs for k_{cat}/K_m.

Although kinetically indistinguishable, the uncompetitive mode of inhibition is different from the case of inhibition due to the *nonproductive binding* of substrates (Scheme 1.6). Here a substrate binds in an alternative unreactive mode

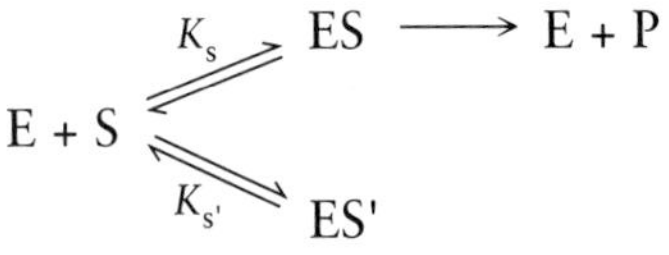

$$E + S \underset{K_{s'}}{\overset{K_s}{\rightleftharpoons}} \begin{matrix} ES \longrightarrow E + P \\ ES' \end{matrix}$$

SCHEME 1.6

(K'_s) at the active site in competition with the productive mode of binding (K_s). Nonproductive binding lowers both K_m and V by a factor of $(1 + K_s/K'_s)$.

iv. Mixed-type inhibition. In many instances, an inhibition cannot be simply classified as competitive, noncompetitive, or uncompetitive. Instead it may appear as a combination of all these types (Fig. 1.3). One example is the case in which the dissociation constant of the substrate from ESI (K_{si}) is different than that from ES (K_s) due to the binding of the inhibitor. Such inhibition in which both K_m and V are altered is known as a mixed-type inhibition.

v. Irreversible inhibition. The irreversible inhibition of enzymes is the result of either extremely tight noncovalent binding of inhibitors or, more usually, the formation of stable covalent bonds between small molecules and essential enzymatic functional groups. When the irreversible inhibitors are substrate analogs, they are relatively specific for target enzymes and are called *affinity labels* or *active site-directed* inhibitors. Typical examples are alkylating reagents of the halomethyl ketone group such as TPCK (tosyl-L-phenylalanyl chloromethyl ketone) and TLCK (tosyl-L-lysyl chloromethyl ketone) for chymotrypsin and trypsin, respectively.

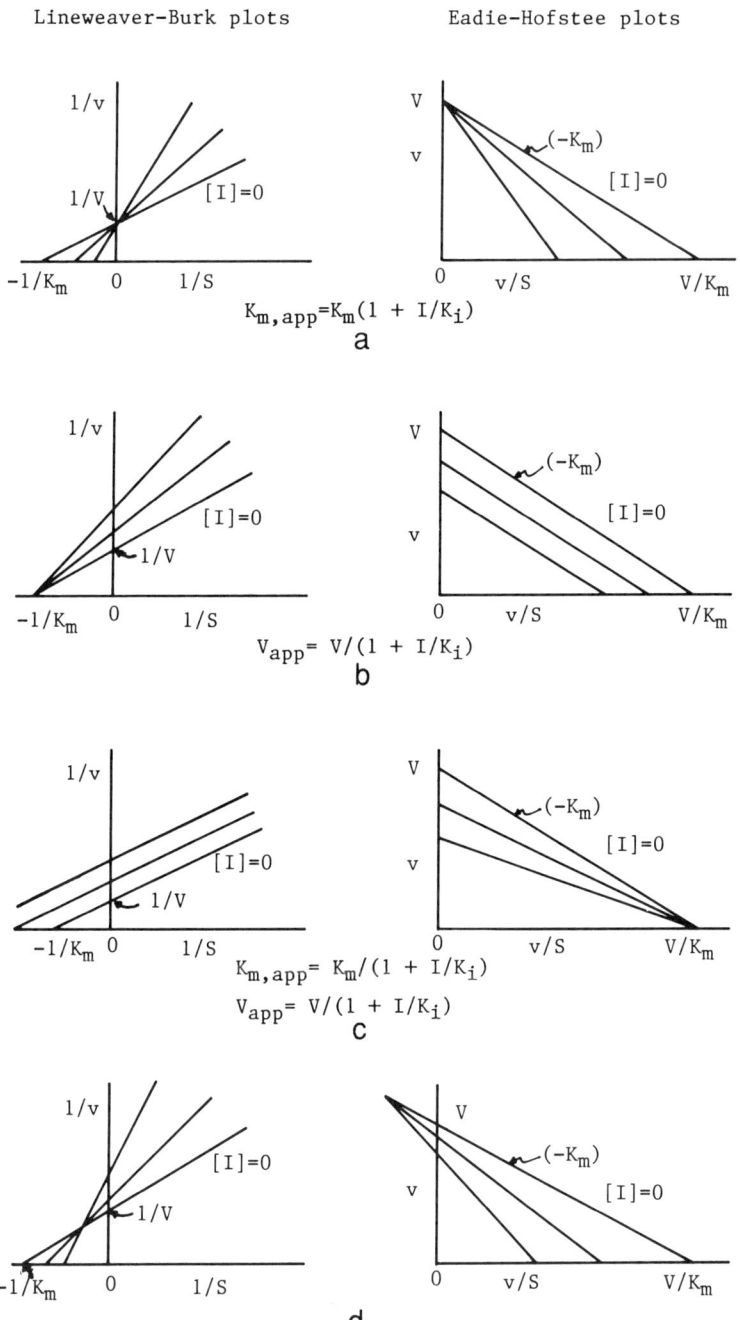

FIGURE 1.3 Types of enzyme inhibition represented in both Lineweaver–Burk and Eadie–Hofstee plots: (a) competitive, (b) noncompetitive, (c) uncompetitive, and (d) mixed type.

The inactivation presents distinct kinetics as a function of modifying reagents and the modified groups involved (117). For the simple cases in which preequilibrium binding is followed by a slow chemical step, the inactivation scheme (Scheme 1.7) resembles that of the Michaelis–Menten mechanism.

$$E + I \xrightleftharpoons{K_I} E \cdot I \xrightarrow{k_1} E - I$$

SCHEME 1.7

The rate of inactivation is then given by Eq. (1.10):

$$-d[E]/dt = k_1[E][I]/(K_I + [I]). \qquad (1.10)$$

The inactivation kinetics of this type can indeed be treated as a special case of the Michaelis–Menten mechanism in which the turnover of the substrate is too slow compared with the rate (k_1) of enzyme inactivation. Penicillins are typical examples of such substrate inhibitors for β-lactamases. Certain classes of irreversible inhibitors, called *suicide inhibitors,* are chemically unreactive in the absence of target enzymes. When an enzyme binds the innocuous inhibitor with the same specificity as the substrate, however, the inhibitor is activated into a powerful irreversible inhibitor.

f. Generalization and limits of the Michaelis–Menten equation

The Michaelis–Menten kinetic scheme, which involves a single substrate and a single product, is obviously the simplest type of enzyme catalysis. Equation (1.7) may hold for many mechanisms, but the mechanisms can be different from each other and the expression of kinetic parameters may also differ. When there is a substrate inhibition or activation due to the binding of a second substrate molecule, the Michaelis–Menten equation does not hold.

The steady-state and rapid equilibrium kinetics do not give detailed information on the existence of multiple intermediates or on their lifetimes. Such information is provided by fast (or transient) kinetics. The methods can be divided in two categories: *rapid-mixing* techniques (stopped-flow, rapid-scanning stopped-flow, quenched flow) which operate in a millisecond time scale and *relaxation* techniques (temperature jump, pressure jump) which monitor a transient reaction in a microsecond time scale. Most of the transient kinetic methods rely on spectrophotometrically observable substrate changes during the course of enzyme catalysis.

The opposite of fast kinetics is low-temperature kinetics in which the catalytic rate is lowered by applying subzero temperatures. Using this technique in a field known as *cryoenzymology,* one can trap and/or observe the intermediates which would have been otherwise elusive.

The kinetic parameters characterizing the enzymes in the following chapters are mostly those from steady-state kinetics, which are usually obtained within a period of 30 sec to 1 hr.

g. Multisubstrate enzyme systems

The kinetics of many enzymes such as kinases and polymerases involve two or more substrates and products. Such multisubstrate enzymes present far more complex kinetics than the Michaelis–Menten type because of the order of substrate and/or product interactions with the enzyme. For the formulation of detailed rate equations for multisubstrate enzyme systems, interested readers should consult the treatises on enzyme kinetics (110,118).

When the concentration of one substrate is made variable, while all the others are held constant or at saturation ($>10\ K_m$), such a multisubstrate system usually reduces to a practical single-substrate system that obeys Michaelis–Menten kinetics. It is through such a reduced, single-variable system that most steady-state kinetics are used to evaluate kinetic parameters and to distinguish between various reaction mechanisms.

The most common, important type of kinetic mechanism to be briefly mentioned here is the Bi–Bi mechanism. The letters A and B represent the substrates in the order that they bind the enzyme, whereas P and Q represent products in the order that they leave the enzyme.

i. Ordered Bi–Bi mechanism.

In this sequential mechanism, the binding of two substrates (A, B) and the release of the two products (P, Q) are "ordered." According to the diagrammatic notation of Cleland (119), this mechanism is shown by Scheme 1.8. The rate equation for this mechanism involves not only

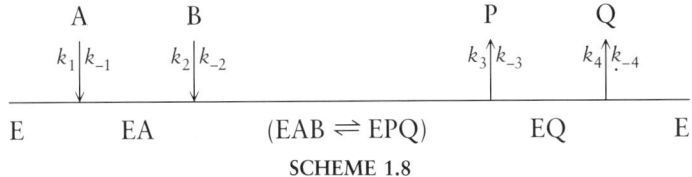

SCHEME 1.8

the concentration terms of the substrates A and B but also the products P and Q. When the concentrations of the products are not significant, the expression reduces to a pseudo-Michaelis–Menten equation (Fig. 1.4). The addition of P or Q results in a *product inhibition*, which is characteristic of the kinetic mechanism involved (Table 1.6).

ii. Random Bi–Bi mechanism.

Regardless of the term "random," this mechanism is a sequential mechanism in which product release occurs only after the binding of all substrates. As to the order of binding, however, multiple situations can be envisaged: (a) ordered addition of substrates and random release of products, (b) random addition of substrates and ordered release of products, and (c) random addition of substrates and random release of products.

The rate equations for these mechanisms are extremely complex. Only under circumstances in which all steps (except for the central conversion of EAB ⇌

EPQ) are in rapid equilibrium (Scheme 1.9) do the rate equations reduce to

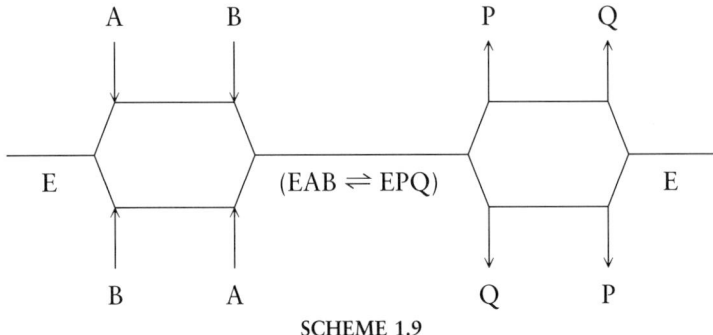

SCHEME 1.9

relatively simple forms similar to that of an ordered Bi–Bi mechanism (Fig. 1.4). The patterns of product inhibition depend on whether dead-end complexes EBQ and/or EAP are formed (Table 1.6).

ii. Ping-Pong Bi–Bi mechanism. The Ping-Pong reaction mechanism is usually observed in group transfer reactions such as transamination between an amino acid and a keto acid. In this mechanism (Scheme 1.10), the enzyme oscillates

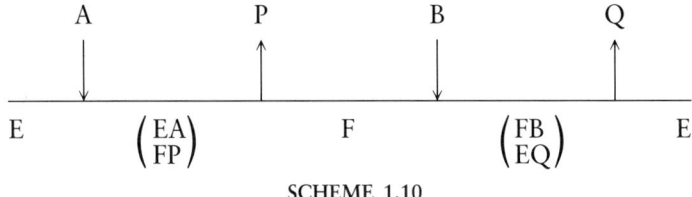

SCHEME 1.10

between two stable forms, i.e., the free enzyme and an intermediate form. A product is released between each substrate addition. Therefore the Ping-Pong system is nonsequential. The rate equation and graphic representation (e.g., reciprocal plots) of the Ping-Pong Bi–Bi mechanism are characterized by a series of parallel lines depending on fixed concentrations of A or B. The patterns of product inhibition are summarized in Table 1.6.

h. Allosteric enzymes

For oligomeric enzymes in which each monomer is catalytically active and independent, the catalytic rate is hyperbolic with [S]. Such oligomeric enzymes behave as one-site enzymes, regardless of the distribution of site occupation by substrates.

For other oligomeric enzymes, e.g., hexameric aspartate carbamoyltransferase (ATCase; EC 2.1.3.2), the binding of a substrate (more generally a ligand) molecule on one site exerts either a stimulatory or an inhibitory effect on the binding of

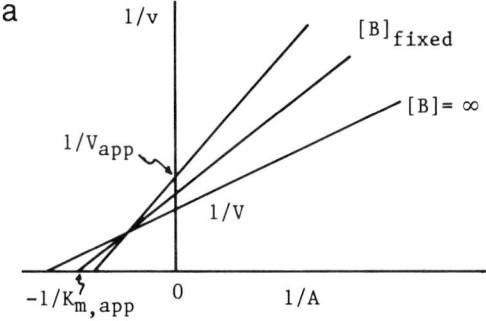

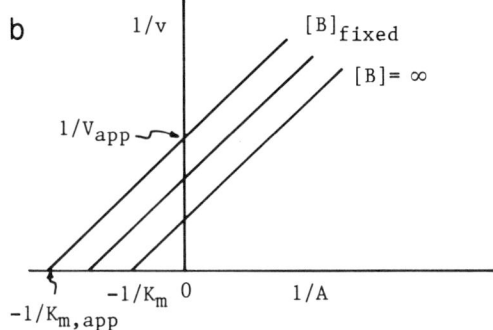

FIGURE 1.4 Graphic representations of Bi–Bi enzyme kinetics: (a) ordered Bi–Bi mechanism and rapid equilibrium random Bi–Bi mechanism and (b) Ping-Pong Bi–Bi mechanism.

the next molecule. Such regulatory enzymes are called *allosteric* enzymes, and the ligand is a *modifier* or an *allosteric effector*. The allosteric effectors can be substrates, products (in case of feedback inhibition), or other small molecules. With allosteric enzymes, the plot of v versus [S] is *sigmoidal*. The shapes of these curves vary depending on the type of ligand interactions with the enzymes. When the interaction involves identical ligands, the allosteric effect is *homotropic* and usually positively *cooperative* (120). When the binding of one kind of ligand affects the binding of a second kind of ligand, the allostery is called *heterotropic* whereas cooperativity can be either positive or negative. The allostery is a complex function of dissociation constants involving each conformational state and the number of ligands. The allostery serves as an effective control mechanism of enzyme functions *in vivo*. It is believed to occur when the interaction of a ligand (allosteric activator or inhibitor) causes an alteration in the tertiary structure of a protomer. [A protomer is a structural unit that bears one site for each of the various ligands.]

The allosteric effect may be propagated either sequentially from one subunit to another [*sequential* or *induced-fit* model (121)] or by a concerted transition between "R" (relaxed) state and "T" (tense) state [*concerted* or *symmetry transition* model (120)]. The ATCase from *E. coli* is a typical example that shows a

TABLE 1.6 Patterns of Product Inhibition in Bi–Bi Kinetic Mechanisms[a]

	Inhibition by Q P = 0		Inhibition by P Q = 0	
Mechanism	A varies B fixed	B varies A fixed	A varies B fixed	B varies A fixed
Ordered Bi–Bi	comp	mixed or noncomp (A < sat) No inhib (A = sat)	mixed (B < sat) uncomp (B = sat)	mixed
Random Bi–Bi (rapid equilibrium)				
No dead-end complex	comp (A < sat) No inhib (B = sat)	comp (A < sat) No inhib (A = sat)	comp (B < sat) No inhib (B = sat)	comp (A < sat) No inhib (A = sat)
+ EBQ dead-end complex	comp	mixed or noncomp (A < sat) No inhib (A = sat)	N.A.	N.A.
+ EAP dead-end complex	N.A.	N.A.	mixed or noncomp (B < sat) No inhib (B = sat)	comp
Ping-Pong Bi–Bi	comp	mixed (A < sat) No inhib (A = sat)	mixed (B < sat) No inhib (B = sat)	comp

[a] Inhibition types are *comp* (for competitive), *noncomp* (for noncompetitive), *mixed* (for mixed-type) or *No inhib* (for no inhibition). Substrate concentrations are divided into =*sat* (for saturation) and <*unsat* (for unsaturation), N.A., not applicable.

concerted transition from the low activity T to the high activity R state in response to the binding of a single ligand molecule to one of its six active sites (122). The heterotetrameric complex of anthranilate synthase (EC 4.1.3.27) and anthranilate phosphoribosyltransferase (EC 2.4.2.18) of *S. typhimurium* undergoes multiple conformational transitions commensurate with the sequential model (123).

i. Factors affecting enzyme catalysis

i. pH. The activities of many enzymes vary with pH, sometimes in a way similar to the simple ionizations of acids and bases. The pH dependence indicates that the active sites of enzymes generally contain important acids and bases. If the catalysis is dependent on the concentration of one active form (acidic or basic), kinetic parameters are affected by the ionization of the enzyme and the enzyme–substrate complex(es). In fact, more than one acid or base group or combination of them often participate in enzyme catalysis in a concerted manner. In such cases, the pH–activity relationship is more complex, and the enzyme exhibits the optimal activity within a defined pH range.

To illustrate the complexity, we will take a simple case in which the enzymes typically follow the Michaelis–Menten kinetic mechanism. We will also make the following assumptions: (a) only one ionic form of the enzyme is active (monoprotic), (b) no change occurs in RDS due to ionization, and (c) the enzyme maintains active conformation within the experimental pH range. For this type of enzyme, an observable quantity, e.g., rate constant (k), depends on the pH of the system as follows: (i) the plot of k against pH resembles a titration curve, yielding at the inflection point the pK_a of the acid involved, (ii) the k_{cat} follows the ionization of the enzyme–substrate complex, (iii) the K_m follows the ionizations of both the free enzyme and the enzyme–substrate complex, and (iv) the k_{cat}/K_m follows the ionization of the free enzyme only, even when there are multiple intermediates on the reaction pathway.

ii. **Temperature.** In any chemical reactions, including those catalyzed by enzymes, the rate (k) of the reaction is affected by the temperature. The temperature effect is most conveniently explained on the basis of the collision theory: the frequency of molecular encounter depends on the concentration and the kinetic energy of the molecules. The minimum energy required for a reaction is called the *activation energy* (E_a), which is related to temperature by the *Arrhenius equation* [Eq. (1.11)]

$$k = A \exp(-E_a/RT), \qquad (1.11)$$

where E_a is the activation energy and A is a *frequency factor* corresponding to the term (kT/h) used in the transition-state theory [Eq. (1.1)]. When the E_a is ~12.6 kcal/mol, a 10°C rise in temperature results in a twofold increase in the reaction rate. Temperature is also a critical factor for the stability of enzymes. The thermostability of a protein is a net result of interactions among various physical forces affecting the tertiary and quaternary structures.

iii. **Ionic strength.** Ionic strength (I) is a measure of the concentration of charges in a solution. It is defined as $\Sigma C_i z_i^2 / 2$, where C_i is the molar concentration of the ion species i and z_i is the net charge of the ion i. As the ionic strength of a solution increases, the activity coefficient (γ) of an ion decreases according to the *Debye–Hückel* limiting law, $\log \gamma = -0.509 z^2 I^{1/2}$. The modulation of enzyme activity by ionic strength (or salts) is largely due to two effects: (a) the neutralizing effect by the counterion of the salts on the electrostatic interactions required for substrate binding and (b) the influence on the pK_a values of the ionizable residues on enzyme and substrates, thus affecting their electrostatic interactions. The ionic strength may also affect the activity of an enzyme by changing the stability and solubilities of the enzyme as well as those of the substrates. The effects of salts on stability becomes more important with more hydrophilic enzymes.

iv. Solvent polarity. Ions are stabilized by polar solvents. The electrostatic dipoles of the solvent directly interact with the electrical charges of the ions. Solvents with strong dielectric constants decrease the tendency of the ions to reassociate and thus affect the pK_a values of the ionizable groups on enzymes and substrates. The ionization of a neutral acid (HA $\rightleftharpoons$ A$^-$ + H$^+$) in an aqueous solution is depressed by the addition of a solvent having lower polarity than water, e.g., methanol. In contrast, the ionization of a cationic ion (BH$^+$ $\rightleftharpoons$ B + H$^+$) is insensitive to solvent polarity because there is no change in charge in the equilibrium.

References

1. Lerner, R. A., Benkovic, S. J., and Schultz, P. G. (1991). *Science* **252**, 659–667.
2. Gualerzi, C. O., and Pon, C. L. (1990). *Biochem.* **29**, 5881–5889.
3. Shine, J., and Dalgarno, L. (1975). *Nature* **254**, 34–38.
4. Kozak, M. (1983). *Microbiol. Rev.* **47**, 1–45.
5. Kozak, M. (1987). *NAR* **15**, 8125–8148.
6. Cavener, D. R., and Ray, S. C. (1991). *NAR* **19**, 3185–3192.
7. Cigan, A. M., and Donahue, T. F. (1987). *Gene* **59**, 1–18.
8. Heidecker, G., and Messing, J. (1986). *Ann. Rev. Plant Physiol.* **37**, 439–466.
9. Cavener, D. R. (1987). *NAR* **15**, 1353–1361.
10. Yamauchi, K. (1991). *NAR* **19**, 2715–2720.
11. Dreyfus, M. (1988). *JMB* **204**, 79–94.
12. Huang, W. M., Ao, S. Z., Casjens, S., Orlandi, R., Zeikus, R., Weiss, R., Winge, D., and Fang, M. (1988). *Science* **239**, 1005–1012.
13. Meyer, F., Schmidt, H. J., Plumper, E., Hasilik, A., Mersmann, G., Meyer, H. E., Engstrom, A., and Heckmann, K. (1991). *PNAS* **88**, 3758–3761.
14. Brown, C. M., Stockwell, P. A., Trotman, C. N. A., and Tate, W. P. (1990). *NAR* **18**, 6339–6345.
15. Richards, F. M. (1991). *Sci. Am.* **264**, 54–63.
16. Creighton, T. E., and Goldenberg, D. P. (1984). *JMB* **179**, 497–526.
17. Weissman, J. S. and Kim, P. S. (1991). *Science* **253**, 1386–1393.
18. Wickner, W., Driessen, A. J. M., and Hartl, F.-U. (1991). *Ann. Rev. Biochem.* **60**, 101–124.
19. Talmadge, K., Brosius, J., and Gilbert, W. (1981). *Nature* **294**, 176–178.
20. Schatz, P. J., and Beckwith, J. (1990). *Ann. Rev. Genet.* **24**, 215–248.
21. Bieker-Brady, K., and Silhavy, T. J. (1992). *EMBO J.* **11**, 3165–3174.
22. Von Heijne, G. (1985). *JMB* **184**, 99–105.
23. Yamamoto, Y., Ohkubo, T., Kohara, A., Tanaka, T., Tanaka, T., and Kikuchi, M. (1990). *Biochem.* **29**, 8998–9006.
24. Laforet, G. A., and Kendall, D. A. (1991). *JBC* **266**, 1326–1334.
25. Reddy, G. L., and Nagaraj, R. (1989). *JBC* **264**, 16591–16597.
26. Andersson, H., and von Heijne, G. (1991). *PNAS* **88**, 9751–9754.
27. Micanovic, R., Gerber, L. D., Berger, J., Kodukula, K., and Udenfriend, S. (1990). *PNAS* **87**, 157–161.
28. Neurath, H. (1989). *TIBS* **14**, 268–271.
29. Zhu, X., Ohta, Y., Jordan, F., and Inouye, M. (1989). *Nature* **339**, 483–484.
30. Park, S., Liu, G., Topping, T. B., Cover, W. H., and Randall, L. L. (1988). *Science* **239**, 1033–1035.
31. Laminet, A. A., and Pluckthun, A. (1989). *EMBO J.* **8**, 1469–1477.
32. Von Heijne, G. (1986). *NAR* **14**, 4683–4690.
33. Folz, R. J., and Gordon, J. I. (1987). *BBRC* **146**, 870–877.
34. Thomas, L., Cooper, A., Bussey, H., and Thomas, G. (1990). *JBC* **265**, 10821–10824.

35. Benjannet, S., Rondeau, N., Day, R., Chretien, M., and Seidah, N. G. (1991). *PNAS* **88**, 3564–3568.
36. Krausslich, H.-G., and Wimmer, E. (1988). *Ann. Rev. Biochem.* **57**, 701–754.
37. Palmenberg, A. C. (1990). *Ann. Rev. Microbiol.* **44**, 603–623.
38. Dice, J. F. (1987). *FASEB J.* **1**, 349–357.
39. Hershko, A. (1988). *JBC* **263**, 15237–15240.
40. Gonda, D. K., Bachmair, A., Wuenning, I., Tobias, J. W., Lane, W. S., and Varshavsky, A. (1989). *JBC* **264**, 16700–16712.
41. Rogers, S., Wells, R., and Rechsteiner, M. (1986). *Science* **234**, 364–368.
42. Ghoda, L., van Daalen Wetters, T., Macrae, M., Ascherman, D., and Coffino, P. (1989). *Science* **243**, 1493–1495.
43. Parsell, D. A., Silber, K. R., and Sauer, R. T. (1990). *Genes Dev.* **4**, 277–286.
44. Parker, J. (1989). *Microbiol. Rev.* **53**, 273–298.
45. Weber, D. J., Serpersu, E. H., Shortle, D., and Mildvan, A. S. (1990). *Biochem.* **29**, 8632–8642.
46. Morgola, E. J. (1985). *Ann. Rev. Genet.* **19**, 57–80.
47. Eggertsson, G., and Soell, D. (1988). *Microbiol. Rev.* **52**, 354–374.
48. Bjork, G. R., Wikstrom, P. M., and Bystrom, A. S. (1989). *Science* **244**, 986–989.
49. Jacks, T., Madhani, H. D., Masiarz, F. R., and Varmus, H. E. (1988). *Cell* **55**, 447–458.
50. Icho, T., and Wickner, R. B. (1989). *JBC* **264**, 6716–6723.
51. Belcourt, M. F., and Farabaugh, P. J. (1990). *Cell* **62**, 339–352.
52. Weiss, R. B., Dunn, D. M., Atkins, J. F., and Gesteland, R. F. (1987). *CSHSQB* **52**, 687–693.
53. Curran, J., and Kolakofsky, D. (1988). *EMBO J.* **7**, 245–251.
54. Herman, R. C. (1989). *TIBS* **14**, 219–222.
55. Sonenberg, N., and Pelletier, J. (1989). *BioAssays* **11**, 128–132.
56. Schlicht, H. J., Radziwill, G., and Schaller, H. (1989). *Cell* **56**, 85–92.
57. Chang, L. J., Pryciak, P., Ganem, D., and Varmus, H. E. (1989). *Nature* **337**, 364–368.
58. Shaw, G., and Kamen, R. (1986). *Cell* **46**, 659–667.
59. Kruys, V., Marinx, O., Shaw, G., Deschamps, J., and Huez, G. (1989). *Science* **245**, 852–855.
60. Kent, S., Hood, L., Aebersold, R., Teplow, D., Smith, L., Farnsworth, V., Cartier, P., Hines, W., Hughes, P., and Dodd, C. (1987). *BioTechniques* **5**, 314–321.
61. Findlay, J. B. C., and Geisow, M. J., Eds. (1989). "Protein Sequencing: A Practical Approach." IRL Press, Oxford.
62. O'Neil, K. T., and DeGrado, W. F. (1990). *Science* **250**, 646–651.
63. Lyu, P. C., Liff, M. I., Marky, L. A., and Kallenbach, N. R. (1990). *Science* **250**, 669–673.
64. Leszczynski, J. F., and Rose, G. D. (1986). *Science* **234**, 849–855.
65. Bek, E., and Berry, R. (1990). *Biochem.* **29**, 178–183.
66. Argos, P. (1990). *JMB* **211**, 943–958.
67. Knighton, D. R., Zheng, J., TenEyck, L. F., Ashford, V. A., Xuong, N.-H., Taylor, S. S., and Sowadski, J. M. (1991). *Science* **253**, 407–414.
68. Miles, E. W. (1991). *Adv. Enzy. (Mol. Biol.)* **64**, 93–172.
69. Wuthrich, K. (1989). *Science* **243**, 45–50.
70. Clore, G. M., and Gronenborn, A. M. (1991). *Ann. Rev. B. BC.* **20**, 29–63.
71. Kay, L. E., Clore, G. M., Bax, A., and Gronenborn, A. M. (1990). *Science* **249**, 411–414.
72. Smith, S. O., Farr-Jones, S., Griffin, R. G., and Bachovchin, W. W. (1989). *Science* **244**, 961–964.
73. Torchia, D. A., Sparks, S. W., and Bax, A. (1989). *Biochem.* **28**, 5509–5524.
74. Wakil, S. J., Stoops, J. K., and Joshi, V. C. (1983). *Ann. Rev. Biochem.* **52**, 537–579.
75. Srivastava, D. K., and Bernhard, S. A. (1986). *Science* **234**, 1081–1086.
76. Bernhard, S. A. (1968). "The Structure and Function of Enzymes." Benjamin, Inc., New York.
77. Mozhaev, V. V., Berezin, I. V., and Martinek, K. (1988). *Crit. Rev. Biochem.* **23**, 235–281.
78. Gianfreda, L., and Scarfi, M. R. (1991). *Mol. Cell. Biochem.* **100**, 97–128.
79. Lowry, O. H., Rosebrough, N. J., Farr, A. L., and Randall, R. J. (1951). *JBC* **193**, 265–275.
80. Bradford, M. (1976). *Anal. Biochem.* **72**, 248–254.
81. Sandberg, W. S., and Terwilliger, T. C. (1989). *Science* **245**, 54–57.
82. Ganter, C., and Pluckthun, A. (1990). *Biochem.* **29**, 9395–9402.

83. Alber, T. (1989). *Ann. Rev. Biochem.* **58**, 765–798.
84. Toney, M. D., and Kirsch, J. F. (1989). *Science* **243**, 1485–1488.
85. Warshel, A., Aqvist, J., and Creighton, S. (1989). *PNAS* **86**, 5820–5824.
86. Argos, P., Vingron, M., and Vogt, G. (1991). *Prot. Eng.* **4**, 375–383.
87. Hanks, S. K., Quinn, A. M., and Hunter, T. (1988). *Science* **241**, 42–52.
88. Vriend, G., and Sander, C. (1991). *Proteins (S.F.G.)* **11**, 52–58.
89. Markert, C. L. (1990). *In* "Isozymes: Structure, Function, and Use in Biology and Medicine," pp. xvii–xxiv. Wiley-Liss, New York.
90. Li, S. S.-L. (1990). *In* "Isozymes: Structure, Function, and Use in Biology and Medicine," pp. 75–99. Wiley-Liss, New York.
91. Enzyme Nomenclature 1992. (IUB publication) Academic Press, San Diego.
92. Glazer, A. N., DeLange, R. J., and Sigman, D. S. (1975). Chemical Modification of Proteins: Selected Methods and Analytical Procedures. A volume in the series of Laboratory Techniques in Biochemistry and Molecular Biology (T. S. Work and E. Work, Eds.). Elsevier Biomedical Press, Amsterdam.
93. Eun, H. M., and Miles, E. W. (1984). *Biochem.* **23**, 6484–6491.
94. Ray, W. J., Jr., and Koshland, D. E., Jr. (1961). *JBC* **236**, 1973–1979.
95. Tsou, C. L. (1969). *Scientia Sinica* (English ed.) **11**, 1535–1558.
96. McIntire, W. S., Wemmer, D. E., Chistoserdov, A., and Lidstrom, M. E. (1991). *Science* **252**, 817–824.
97. Estell, D. A., Graycar, T. P., Miller, J. V., Powers, D. B., Burnier, J. P., Ng, P. G., and Wells, J. A. (1986). *Science* **233**, 659–663.
98. Benkovic, S. J., Fierke, C. A., and Naylor, A. M. (1988). *Science* **239**, 1105–1110.
99. Sharp, K., Fine, R., and Honig, B. (1987). *Science* **236**, 1460–1463.
100. Leatherbarrow, R., and Fersht, A. R. (1987). *In* "Enzyme Mechanisms" (M. I. Page and A. Williams, Eds.), pp. 78–96. Royal Society of Chemistry, London.
101. Bartlett, P. A., and Marlowe, C. K. (1987). *Science* **235**, 569–571.
102. Tronrud, D. E., Holden, H. M., and Matthews, B. W. (1987). *Science* **235**, 571–574.
103. Kati, W. M., and Wolfenden, R. (1989). *Science* **243**, 1591–1593.
104. Kraut, J. (1988). *Science* **242**, 533–540.
105. Michaelis, L., and Menten, M. L. (1913). *Biochem. Z.* **49**, 333.
106. Briggs, G. E., and Haldane, J. B. S. (1925). *Biochem. J.* **19**, 338–339.
107. Fersht, A. (1985). "Enzyme Structure and Mechanism," 2nd ed. Freeman and Co., New York.
108. Morrissey, J. H. (1981). *Anal. Biochem.* **117**, 307–310.
109. Dixon, M., and Webb, E. C. (1979). "Enzymes," 3rd ed. Academic Press, New York.
110. Roberts, D. V. (1977). "Enzyme Kinetics." Cambridge University Press, Cambridge.
111. Atkins, G. L., and Nimmo, I. A. (1975). *BJ* **149**, 775–777.
112. Lineweaver, H., and Burk, D. (1934). *JACS* **56**, 658–666.
113. Eadie, G. S. (1942). *JBC* **146**, 85–93.
114. Hofstee, B. H. J. (1952). *Science* **116**, 329–331.
115. Dowd, J. E., and Riggs, D. S. (1965). *JBC* **240**, 863–869.
116. Atkins, G. L., and Nimmo, I. A. (1975). *BJ* **149**, 775–777.
117. Tsou, C. L. (1988). *Adv. Enzymol.* **61**, 381–436.
118. Segel, I. H. (1975). "Enzyme Kinetics: Behavior and Analysis of Rapid Equilibrium and Steady-State Enzyme Systems." Wiley, New York.
119. Cleland, W. W. (1963). *BBA* **67**, 104–137.
120. Monod, J., Wyman, J., and Changeux, J.-P. (1965). *JMB* **12**, 88–118.
121. Koshland, D. E., Jr., Nemethy, G., and Filmer, D. (1966). *Biochem.* **5**, 365–385.
122. Foote, J., and Schachman, H. K. (1985). *JMB* **186**, 175–184.
123. Caligiuri, M. G., and Bauerle, R. (1991). *Science* **252**, 1845–1848.
124. Atkins, J. F., Weiss, R. B., and Gesteland, R. F. (1990). *Cell* **62**, 413–423.
125. Serrano, L., Niera, J.-L., Sancho, J., and Fersht, A. R. (1992). *Nature* **356**, 453–455.
126. Cleland, J. L., Builder, S. E., Swartz, J. R., Winkler, M., Chang, J. Y. and Wang, D. I. C. (1992). *BioTechnology* **10**, 1013–1019.

II. STRUCTURE AND FUNCTION OF NUCLEIC ACIDS

Nucleic acids, the storehouses of genetic information, are present in all organisms from replicating units smaller than viruses to higher eukaryotes. They constitute the major component of cells (5–15% of cellular dry weight). Nucleic acids are chain-linked macromolecules made up of *nucleotides,* just as proteins are composed of amino acids. DNA consists of *deoxyribo*nucleotides, while RNA consists of *ribo*nucleotides. Like proteins, the structural complexity of nucleic acids can be divided into a (i) primary structure (the nucleotide sequence), (ii) secondary structure (2-D arrangement), (iii) tertiary structure (3-D conformation), and (iv) quaternary structure (supercoils).

The vital functions of nucleic acids are decoded through interactions with proteins. DNA and RNA provide the script, while proteins perform supportive roles in the storage, inheritance, and expression of genetic information. Understanding the physicochemical basis of nucleic acid structure and function is therefore not just an important goal of molecular biology but also an indispensable part of recombinant DNA technology. This section presents an overview of the present knowledge regarding the basic structural features and relevant functions of nucleic acids and relates this fundamental knowledge to practical concepts applicable to recombinant DNA technology by exploring the ways to efficiently synthesize, restrict, ligate, (de)stabilize, modify, and ultimately express recombinant genes.

A. General Description

1. COMPONENTS OF NUCLEIC ACIDS

The building blocks of nucleic acids are called *nucleotides* (nt). Each nucleotide consists of three parts (Fig. 1.5): (i) a nitrogen-containing heterocyclic base (pyrimidine or purine base), (ii) a pentose (C_5 sugar), and (iii) a molecule of phosphoric acid.

DNA contains four major bases: two *pyrimidines* (cytosine and thymine) and two *purines* (adenine and guanine). Native RNA also contains four major bases identical to DNA except that the thymine is replaced by uracil. In addition to the four major bases, nucleic acids may contain rare or minor bases. Most of them are derivatives of major bases, for example, they may be modified by methylation (methyladenine, methylcytosine, etc.). Modified bases are particularly abundant in tRNA molecules.

The purine (Pu or R for short) and pyrimidine (Py or Y for short) bases are connected by a N-glycosidic linkage to the C-1' of the pentose sugar, either a ribose in RNA or a 2'-deoxyribose in DNA. The base–(2'-deoxy)ribose unit is called a *nucleoside* (ribonucleoside or 2'-deoxyribonucleoside). When a nucleoside is phosphorylated either at the 3'-OH or 5'-OH, it is called *3'-nucleotide* (Np) or *5'-nucleotide* (pN), respectively. The number of phosphate groups, for example, at the 5'-OH of adenosine, determines whether the nucleotides are designated as adenosine 5'-*mono*phosphate (AMP), adenosine 5'-*di*phosphate (ADP), and

52 Chapter 1 Enzymes and Nucleic Acids

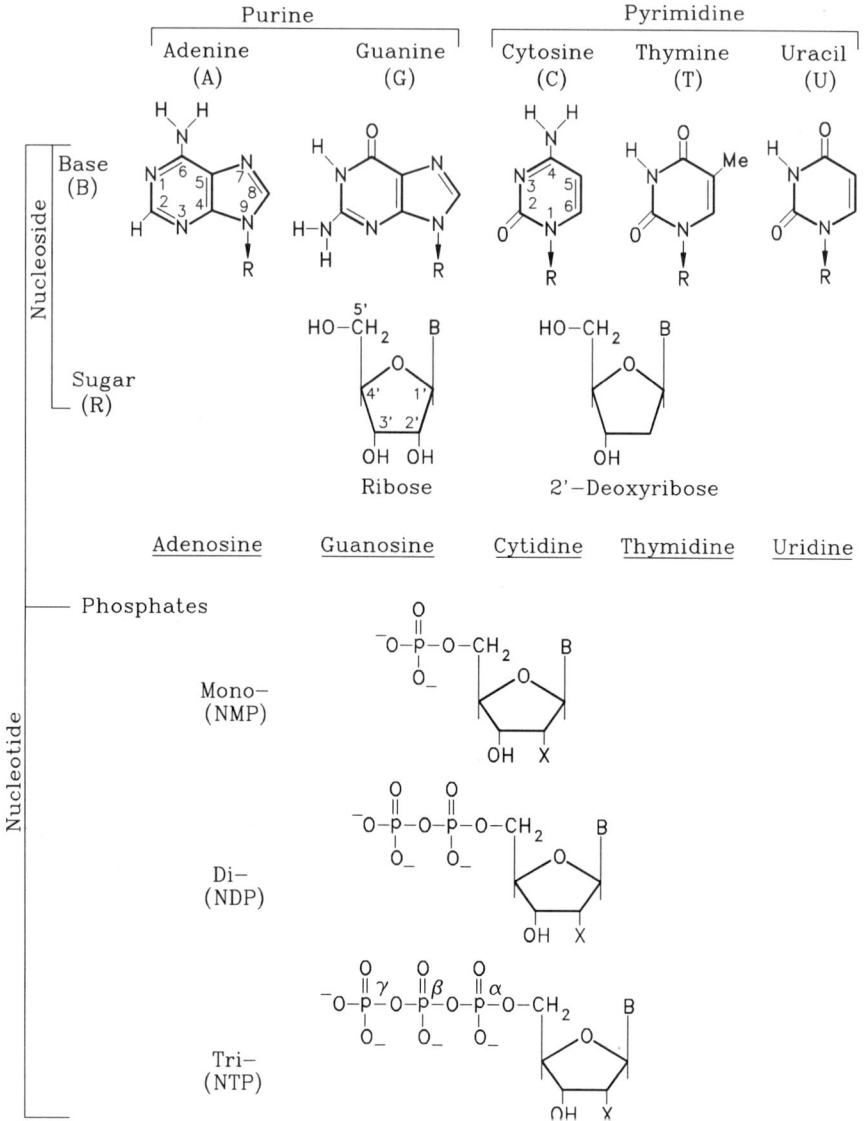

FIGURE 1.5 Molecular constituents of nucleic acids. B denotes a purine or pyrimidine base and R denotes a ribose (X = OH) or a deoxyribose (X = H).

adenosine 5'-*tri*phosphate (ATP) or more generally NMP, NDP, and NTP. The phosphoryl residues are designated by α, β, and γ starting from the 5'-OH position of the pentose moiety.

When the phosphoryl residue forms a phosphodiester bond within a single nucleotide, it gives rise to cyclic nucleotides, e.g., 3',5'-cyclic adenosine monophos-

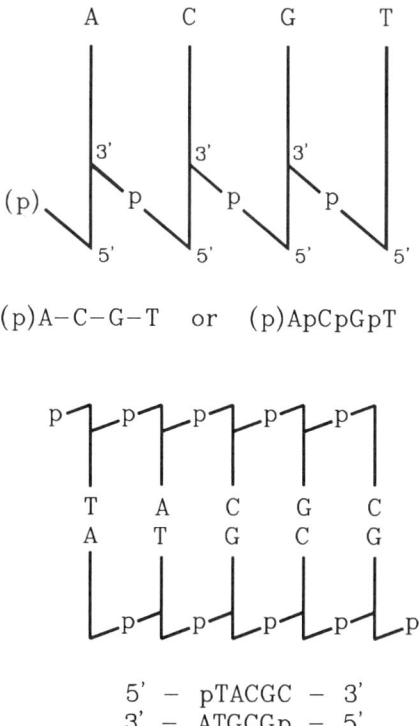

FIGURE 1.6 Schematic representation of polynucleotide structures and shorthand notations. In the shorthand notation for double strands, the top strand is written by convention in 5' → 3' polarity.

phate (3',5'-cyclic AMP or cAMP) and 2',3'-cyclic adenosine monophosphate (2',3'-cyclic AMP or A > p).

In nucleic acids, the 5'-phosphoryl residue of NMP is covalently linked to the 3'-OH of the next NMP, forming an interunit phosphodiester bond (Fig. 1.6). The number of nucleotide units in the chain determines whether the nucleic acid is a *mono*nucleotide, *di-*, *tri-*, *oligo-*, or *poly*nucleotide.

2. Base Pairing

a. Watson–Crick base pairing

Most DNAs are double stranded. The two strands are generally "opposite" in polarity. For a given strand written in the conventional 5' → 3' orientation, the other strand is 3' → 5' or *antiparallel* (Fig. 1.6). Between the two ribose–phosphate backbones, the purine base A pairs with the pyrimidine base T (or U) via H-bonding and so does G with C. These H-bonded base pairs (bp), i.e., A·T (or A·U in RNA) and G·C, are known as Watson–Crick base pairs (Fig. 1.7). In this base pairing scheme, the base A is the *complement* of T (or U), while G is the complement of C. The antiparallel strand is called the *complementary strand*.

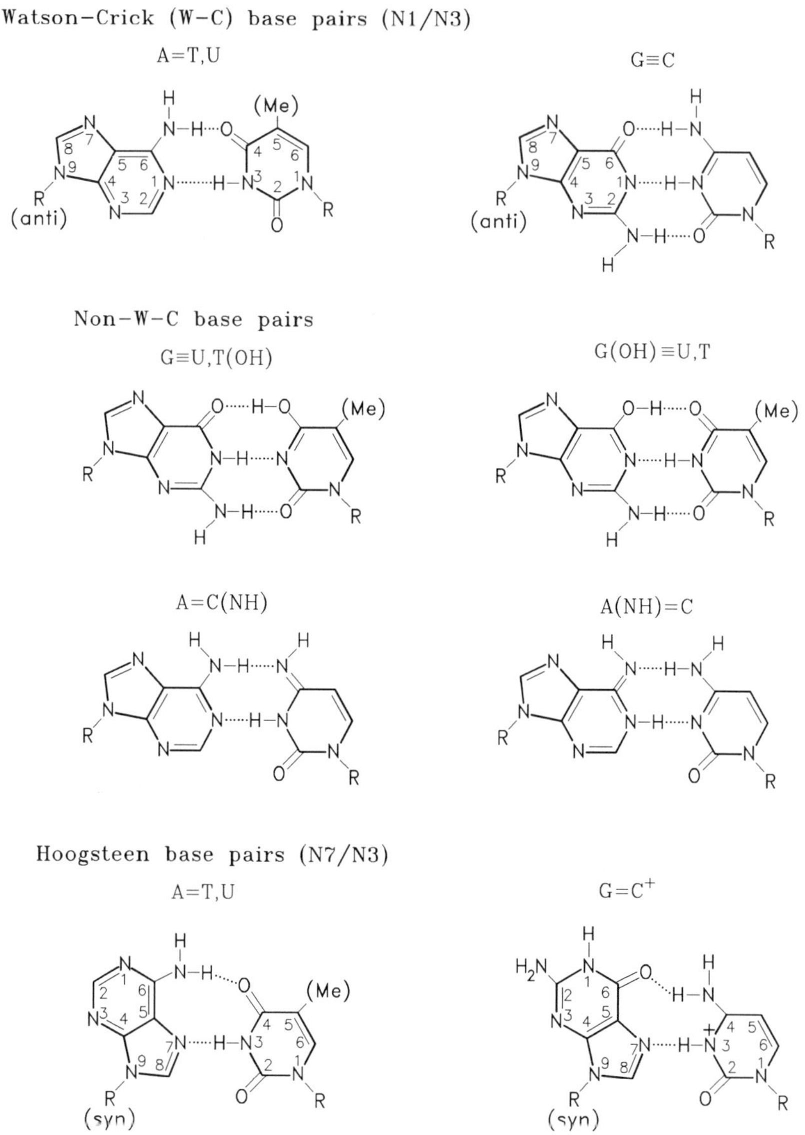

FIGURE 1.7 Base pairing schemes of Watson–Crick and non-Watson–Crick types. Among the non-Watson–Crick base pairs, wobble pairs are formed by H-bonds between tautomeric forms of hydroxyl (OH) and imino (NH) groups. Non-Watson–Crick base pairs also comprise Hoogsteen base pairs and reverse Watson–Crick base pairs which form the basal structural units in triplex DNA and in parallel-stranded DNA (psDNA), respectively.

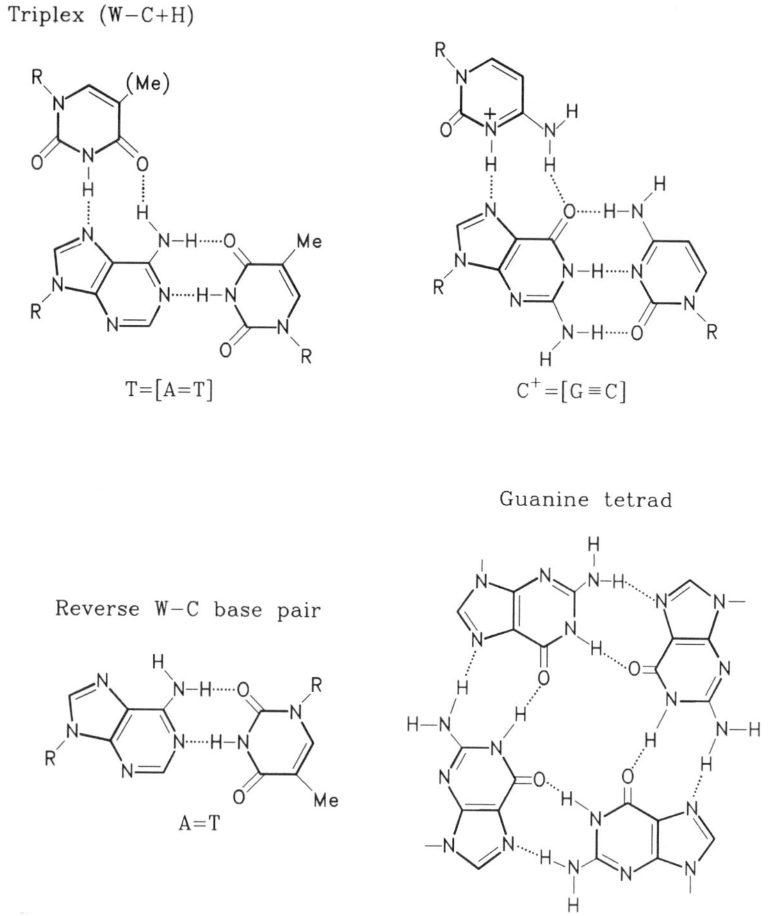

Triplex (W–C+H)

T=[A=T]

C⁺=[G≡C]

Reverse W–C base pair

A=T

Guanine tetrad

FIGURE 1.7 *Continued.*

The G·C base pair contains three H-bonds and is more stable by a factor of 100 than the A·T(U) base pair which contains two H-bonds. When two bases opposite each other are not complementary, they are termed *mismatches*.

As confirmed by the Watson–Crick base pairing rule, the base composition of a nucleic acid follows *Chargaff's rule:* for double-stranded DNAs, the total number of pyrimidine bases is equal to that of purine bases. In terms of mole fraction (x), $x_A = x_T$, $x_G = c_C$, and consequently $x_{G+C} = 1 - (x_{A+T})$.

b. Non-Watson–Crick base pairing

Bases do not always pair according to the Watson–Crick base pairing rule. There are a variety of alternative H-bonded base pairing arrangements called *non-Watson–Crick* or *wobble* base pairs. Wobble base pairs occur at a high frequency in tRNAs, but are relatively rare in other nucleic acids. Wobble base pairs play an important role in codon–anticodon interactions. Among the most frequent of wobble base pairs are G·T(U) base pairing which arises as a result of keto–enol tautomerism and A·C base pairing which is based on amino–imino tautomerism (Fig. 1.7). These non-Watson–Crick base pairs, including the rare type C·T(U) base pair, are stabilized in the helix structure by one or more water molecules which in one part bridge between the ring nitrogens and in other part link the bases and backbone phosphates (1).

The non-Watson–Crick type of base pairing also comprises the base pairs formed by inosine (I) with C, T, or A, purine base pairs A·G(or I), A·A, and G·C, and the *reverse Watson–Crick* base pair A·T. In the reverse Watson–Crick A·T pairing, the T ring is rotated 180° around the N-3–C-6 axis from the normal Watson–Crick pair. One peculiarity of the reverse Watson–Crick base pairing is that the two pairing strands can form an as yet unnatural *parallel*-stranded, right-handed double helix (2, 3).

Another important base pairing scheme is the *Hoogsteen* base pairing, A·T and G·C$^+$ (C residue is protonated at the N-3 position). In Hoogsteen base pairing, the purine rotates 180° with respect to the helix axis and adopts a *syn* conformation. Hoogsteen base pairing permits the formation of triple-helix DNA, also called *triplex* or *H-DNA* (refer to Section II,C,1). Hoogsteen base pairs are less stable than Watson–Crick base pairs by 0.22 to 0.64 kcal/mol depending on ionic conditions and base compositions (4). The Hoogsteen base pair A·T is as stable as the G·C$^+$ base pair at low pH in the presence of Mg^{2+} ions.

3. Physicochemical Properties

a. Nucleotides

The characteristic physicochemical properties of nucleotides, such as molecular size, acid ionization constant (pK_a), and spectroscopic parameters, are summarized in Table 1.7. In mononucleotides, the phosphate group has two ionizing protons, pK_{a1} = 0.9 and pK_{a2} = 6.1.

b. Nucleotide distribution in DNA: base bias

In natural DNAs, the four major bases, A, T, G, and C, are present in accordance with Chargaff's rule. However, the abundance of G·C base pairs relative to A·T base pairs is rather characteristic of an organism, ranging from 19% in *Plasmodium falciparum* to 73% in *Mycobacterium phlei* (5). Mammalian DNAs have a G + C (or GC) content of 40–45%. The *base bias* or *preference* of an organism is most often represented in the third position of "degenerate" amino acid codons. (Note: Two or more codons which specify the same amino acid are called *degenerate* codons. Most of the degenerate codons have variations in the third base of the nucleotide triplets.) For example, the gene for *Taq* DNA polymerase of the thermophilic bacteria, *T. aquaticus*, shows a 68% GC content compared to the *E. coli* Pol I gene which has a 52% GC content. The *Taq* DNA Pol gene has over 90% GC bias in the third position of the amino acid codons (6). A higher GC content is equivalent to a higher melting temperature (T_m) (see Section

TABLE 1.7 Physicochemical Properties of Nucleotide 5'-Phosphates

Compounds	MW	λ_{max} (pH 7.0)	ε (10^3) (pH 7.0)	A_{280}/A_{260}	pK_a (base)		Remark
ATP	507.2	259	15.4	0.15	4.0	(N1)	
CTP	483.2	271	9.1	0.97	4.5	(N3)	
		280	12.8	2.12			At pH 2
GTP	523.2	252	13.7	0.66	2.4	(N7)	
					9.4	(N1)	
UTP	484.2	262	10.0	0.38	9.5	(N3)	
dTTP	482.2	267	9.6	0.72	9.5	(N3)	
dATP	491.2	259	15.2	0.15	—		
dCTP	467.2	271	9.1	0.97	—		
dGTP	507.2	253	13.7	0.67	—		
		(pH 2)	(pH 2)				
AMP	347.2	257	15.0	0.16			
CMP	323.2	280	13.2	0.98			
GMP	363.2	256	12.2	0.66			
UMP	324.2	262	10.0	0.38			
dTMP	322.2	267	10.2	0.73			
dAMP	331.2	258	14.3	—			
dCMP	307.2	280	13.5	—			
dGMP	347.2	255	11.8	—			

II,B,1 in this chapter), suggesting that the observed base bias is a form of adaptation to the environment. Base bias is also linked to the level of gene expression, i.e., highly expressed genes show a greater base bias in *E. coli* (7).

Base bias also appears in codon usages in a manner dependent on the organism. Among the combinations of eight least-used codons, the bias for a dinucleotide sequence CG (including CGA and CGG for Arg) has been found to be most remarkable and is shared by *E. coli*, yeast, fruit fly, and primates (8). Low codon usage is relatively insensitive to gross base composition.

Understanding the pattern of base bias for a given organism helps to design "degenerate" oligonucleotide probes used for gene cloning or hybridization.

c. Density and sedimentation properties

A higher GC content increases the buoyant density (d_o) of dsDNA according to the linear relationship: $d_o = 1.660 + 0.098 (x_{GC})$ (9). The buoyant density is generally in the order: RNA (~1.9 g/ml) > ssDNA (~1.72 g/ml) > dsDNA (~1.70 g/ml) > protein (~1.3 g/ml).

The differences in buoyant density as well as molecular size and shape provide a means for separating the nucleic acids by *isopycnic ultracentrifugation*, a centrifugation technique in which particles sediment through a gradient medium (CsCl or sucrose) until they reach a zone of equal densities and band at that position.

Even when the GC content is identical, isopycnic ultracentrifugation separates linear dsDNA from circular DNA. Covalently closed circular (ccc) plasmid DNA appears heavier in CsCl density gradient than linear molecules such as chromosomal DNA (generally sheared) or open circular and linearized plasmid. In the presence of ethidium bromide (EtBr), the differences are accentuated between linear and circular molecules, and the *ccc* plasmids are well separated as the lower band from other linear molecules. In *rate-zonal* centrifugation using a sucrose gradient, particles move at different rates depending on their mass. Therefore, the size of a molecule is the main factor for separation. (Note: The density range of sucrose medium does not extend to that of RNA species.) RNA molecules can be separated according to their size in preformed gradient zones by their differences in sedimentation velocity.

For linear dsDNA, the determination of sedimentation coefficient allows an estimate of the molecular size up to 10^8 Da (or 150 kbp) by an empirical relationship (10); $s^o_{20,w} = 2.8 + 0.00834 M^{0.479}$.

d. Counterions and solvents

Cations are an important factor in stabilizing DNA. They coordinate with electron-rich elements or groups such as the base N, O, and sugar OH group of nucleotides and, most importantly, counter the anionic internucleotide phosphate–phosphate repulsion. Therefore high concentrations of cations stabilize the duplex structure, whereas low or zero concentrations of cations destabilize it. Divalent cations are 15- to 20-fold more effective than monovalent cations at stabilizing duplex structures. The neutralizing effect of divalent cations decreases in the following order: Mg > Ca > Mn > Co > Ni > Zn > Cu. The most important natural cations *in vivo* are polyamines, e.g., spermine, spermidine, and putrecine. DNA is destabilized in apolar or organic solvents in the order: *tert*-butanol > ethanol > methanol > *n*-butanol > ethylene glycol > formamide > water. DNA is destabilized in the presence of anions such as perchlorate (ClO_4^-), iodide (I^-), trifluoroacetate (CF_3COO^-), and thiocyanate (CNS^-). These anions are classified as "chaotropic" (chaos-loving) anions.

e. Acids and bases

RNA is readily hydrolyzed to 2'- and 3'-mononucleotides in alkali (0.3 M NaOH at 37°C) due to the presence of the 2'-OH *cis* to the 3'-phosphodiester linkage (Fig. 1.8a). In contrast to RNA, the backbone of DNA is extremely resistant to alkali, although duplex DNA can be readily denatured to single strands in alkaline solutions. DNA can be hydrolyzed under extreme conditions with uncharged bases like ammonia and piperidines, producing nucleosides as the final product. Under acidic conditions, nucleic acids undergo depurination in an amine-catalyzed reaction. The purine N-glycosidic bond is much less stable than its pyrimidine counterpart. The purine N-glycosidic bond cleavage occurs much more readily with DNA than with RNA, and concomitant phosphodiester bond cleavage leads to the hydrolysis of the nucleic acid to low molecular weight compounds.

f. Chemicals

Purine and pyrimidine bases have distinct reactivities with various chemicals (Fig. 1.8). In fact, the chemicals are often employed as mutagens: both nitric and nitrous acids for deamination of A to hypoxanthine, G to xanthine, and C to U; bisulfite for deamination of pyrimidine bases; and hydrazine or hydroxylamine for depyrimidination (5). The following reagents methylate DNA and RNA: dimethyl sulfate (Me_2SO_4), methylmethane sulfonate, and *N*-methyl-*N'*-nitro-*N*-nitrosoguanidine (MNNG). The alkylation occurs at nucleophilic centers, with particular preference at the N-7 of G and in the decreasing order: N^7-G $\gg$ N^3-A $>$ N^1-A $=$ N^3-G $=$ O^6-G. The following chemicals are known to bind DNA covalently: *N*-methyl-*N*-nitrosourea, β-propionolactone, propylene oxide, streptozotocin, and nitrogen mustard. The differences in chemical reactivity of the nucleosides toward acid, Me_2SO_4, and hydrazine form the basis of the Maxam–Gilbert chemical nucleotide sequencing method (see Appendix A at the end of this volume).

B. Denaturation and Renaturation

Under extreme physical or chemical conditions, the H-bond network which normally keeps the bases paired in a double helix breaks down. Consequently base unstacking occurs and the double helix dissociates into randomly coiled single strands in a process known as *denaturation*. Denaturation conditions include high pH, high concentrations of urea (8 *M*) or GuHCl (6 *M*), and temperatures higher than the *melting* (or *transition*) temperature (T_m). For example, most DNA molecules are readily denatured by brief heating at 100°C or in 0.4 *M* NaOH at room temperature. Under these denaturing conditions, the resilient phosphate–sugar backbone of DNA remains almost intact, unlike that of RNA. Unless the DNA (or RNA) suffers chemical damage, the denaturation is largely *reversible*. On lowering the temperature or eliminating the *denaturant(s)*, the single strands usually regain intermolecular complementarity and (re)*anneal* to form duplexes indistinguishable from the native form. This process is called *renaturation*. When the annealing occurs between two complementary strands from different origins, it is called *hybridization*.

DNAs are constantly subjected to denaturing forces *in vivo* due to physical interactions with various dsDNA-binding or destabilizing proteins and ssDNA-binding proteins. The local reversible unwinding (or melting) is indeed of central biological importance because it is the primary step in DNA replication performed by DNA polymerases (Chapter 6) and in transcription performed by RNA polymerases (Chapter 7).

With the advent of *nucleic acid hybridization* as a pivotal technique in recombinant DNA technology, the theoretical and practical aspects of nucleic acid denaturation–renaturation have received renewed, more rigorous attention (11). How does a small piece of DNA (or RNA) find its complementary sequence from among often astronomical numbers of similar sequences? Finding a 1000 base stretch in the human genome would be like finding a needle in a 2-ton haystack.

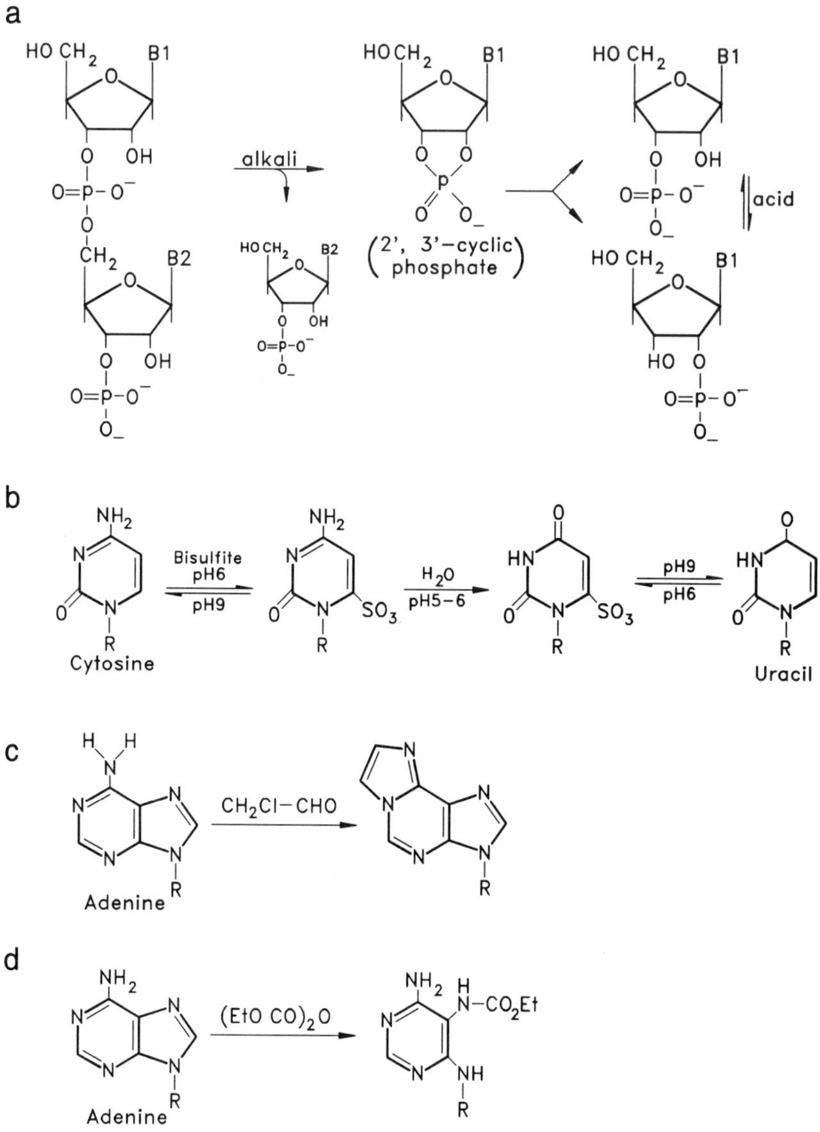

FIGURE 1.8 Some chemical reactions of bases and nucleotides. (a) Hydrolysis of a dinucleotide by alkali. Bases are denoted by B1 and B2. R denotes ribose or deoxyribose. (b to k) Chemical reactions: (b) bisulfite, (c) chloroacetaldehyde, (d) diethyl pyrocarbonate, (e) dimethyl sulfate, (f) formaldehyde, (g) glyoxal, (h) hydrazine, (i) nitric acid, (j) nitrous acid, and (k) osmium tetroxide. Partly from R. L. P. Adams, J. P. Knowler, and D. P. Leader (1986). *In* "The Biochemistry of the Nucleic Acids," 10th ed., pp. 5–34. Chapman & Hall, with permission.

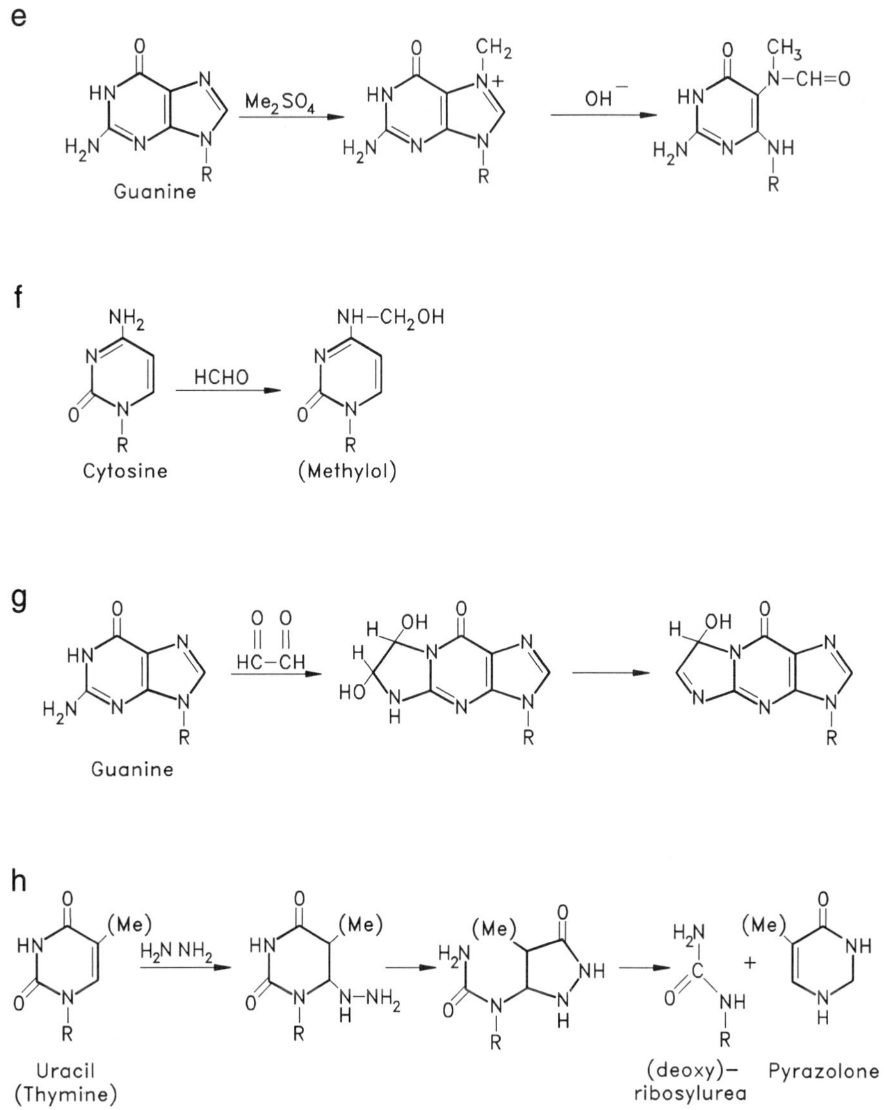

II. Structure and Function of Nucleic Acids

FIGURE 1.8 *Continued.*

Yet nucleic acids do it relatively quickly and with amazing precision. Indeed this reannealing (or hybridization) capability between a polynucleotide and a complementary piece of oligonucleotide (commonly called a *primer* or a *probe*) forms the basis of such diverse and critical techniques as dideoxynucleotide sequencing, site-specific mutagenesis, cDNA cloning, and hybridization screening of recombinant DNA clones. Nucleic acid hybridization also serves as the basis

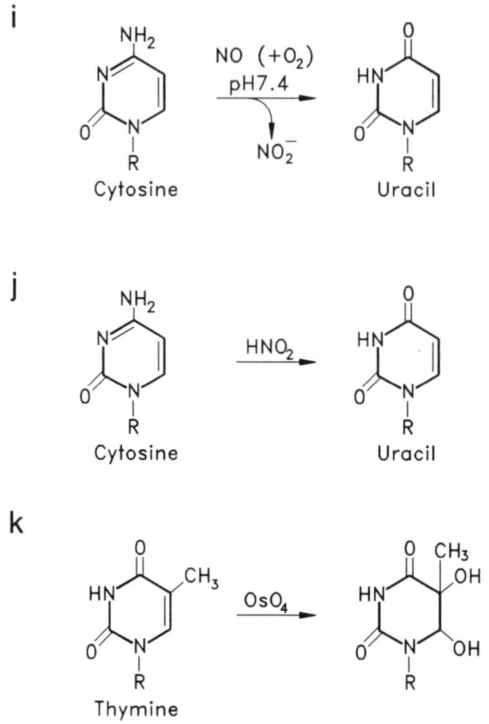

FIGURE 1.8 *Continued.*

for polymerase chain reaction (PCR) techniques and antisense oligonucleotide technology.

The denaturation/renaturation process can be quantitatively monitored by a number of methods (12,13). The most commonly used ones include (i) binding to hydroxyapatite (only double strands bind in 0.12–0.14 M Na–P buffer), (ii) resistance (or sensitivity) to single strand-specific nucleases, and (iii) optical hyperchromicity (the breakage of H-bonds and base unstacking result in a 30% increase in UV absorbance at 260 nm).

As considered in the following section, several factors influence the kinetics of denaturation and renaturation. A thorough understanding of the denaturation/renaturation phenomenon as well as the factors influencing the process should prove valuable in performing and refining the many important techniques mentioned earlier.

1. Factors Affecting Denaturation and Melting Temperature

The phenomenon of denaturation and renaturation can ideally be treated using a two-state model involving the transition between "ordered" double helices and "disordered" or "random" coils. When the transition monitored by A_{260}, for example, is plotted as a function of temperature, the midpoint of the curve resembling phase transition gives the T_m. Frequently, however, the "melting" (and reannealing as well) of natural DNA may not be so straightforward since the phenomenon is more appropriately described by a cooperative process involving a collection of discrete subtransitions, also called *thermalites*. Thermalites are Gaussian in form and have a relatively narrow distribution width of 0.33°C (14). They have an average sequence length of 900 bp but with a broad length distribution. Thermalites are sequence dependent and are usually centered in AT-rich regions.

The T_m is thus dependent on base composition, GC content, and length of dsDNA as well as various other factors, e.g., the ionic strength of the medium, according to the general equation (15) [Eq. (1.12)],

$$T_m (°C) = 81.5 + 0.41 (\% \ GC) + 16.6 \log M - 675/L, \quad (1.12)$$

where M is the molar concentration of the cation (Na^+ or K^+) and L is the length (in bp) of the base-paired region. For DNAs longer than 1 kb, the T_m may be considered independent of length.

Equation (1.12) holds for salts within the concentration range of 0.01 to 0.2 M at pH 7. With certain salts, for example, tetraethylammonium chloride (2.4 M), the effect of GC content becomes negligible. Divalent cations have a much greater effect on T_m than do monovalent cations, whereas chaotropic anions depress T_m in the order of $CCl_3COO^- > CNS^- > ClO_4^- > I^- > Br^- > CH_3COO^-$.

The T_m of long dsDNA in SSC buffers can be calculated from Eq. (1.13) (16), which is in fact a special case of the more general Eq. (1.12)

$$T_m (°C) = Tc + 0.41 (\% \ GC), \quad (1.13)$$

where Tc is a basal constant, 81.5°C in 6× SSC, 69.3°C in 1× SSC, and 53.9°C in 0.1× SSC.

For oligonucleotides of unit lengths less than 100 nt, the T_m can be most reasonably predicted by Eq. (1.14)

$$T_m (°C) = \frac{\Delta H°}{\Delta S° + R \ln(C/4)} - 273 + 16.6 \log M, \quad (1.14)$$

where C is the molar concentration of non-self-complementary oligonucleotides. The thermodynamic parameters can be calculated from Table 1.8. The values in Table 1.8 are derived from the nearest-neighbor model in which the stability of a duplex DNA or RNA is considered as the sum of its nearest-neighbor interactions (17–19). The T_m for oligonucleotides used in hybridization on membrane supports (or filters) is ~7°C lower than the calculated value of T_m in solution.

2. Factors Influencing Renaturation (Annealing, Hybridization)

a. Kinetics of reassociation

i. *Second-order reassociation kinetics.* Renaturation is believed to proceed, like denaturation, in two microscopic steps, i.e., an initial rate-determining nucleation step followed by rapid "zippering" (20). For practical purposes, however,

TABLE 1.8 Thermodynamic Parameters for Nearest-Neighbor Pairs of Bases[a]

	DNA				RNA		
Sequence	$\Delta H°$	$\Delta S°$	$\Delta G°$	Sequence	$\Delta H°$	$\Delta S°$	$\Delta G°$
AA/TT	9.1	24.0	1.9	AA/UU	6.6	18.4	0.9
AT/TA	8.6	23.9	1.5	AU/UA	5.7	15.5	0.9
TA/AT	6.0	16.9	0.9	UA/AU	8.1	22.6	1.1
CA/GT	5.8	12.9	1.0	CA/GU	10.5	27.8	1.8
GT/CA	6.5	17.3	1.3	GU/CA	10.2	26.2	2.1
CT/GA	7.8	20.8	1.6	CU/GA	7.6	19.2	1.7
GA/CT	5.6	13.5	1.6	GA/CU	13.3	35.5	2.3
CG/GC	11.9	27.8	3.6	CG/GC	8.0	19.4	2.0
GC/CG	11.1	26.7	3.1	GC/CG	14.2	34.9	3.4
GG/CC	11.0	26.6	3.1	GG/CC	12.2	29.7	2.9

[a] The parameters for DNA are for 1 M NaCl and 25°C [Breslauer, K. J., Frank, R., Blocker, H., and Marky, L. A. (1986). *PNAS* **83**, 3746–3750.] The parameters for RNA are for 1 M NaCl and 37°C [Freier, S. M., Kierzek, R., Jaeger, J. A., Sugimoto, N., Caruthers, M. H., Neilson, T., and Turner, D.H. (1986). *PNAS* **83**, 9373–9377.] The units for $\Delta H°$ and $\Delta G°$ are kcal/mol. $\Delta S°$ is in entropy unit (cal/°K mol).

the kinetics of reassociation can be described by a simple second-order reaction (21) (Scheme 1.11)

$$R + D \rightarrow H$$
SCHEME 1.11

where R stands for unhybridized RNA, D for unhybridized DNA, and H for heteroduplex. The formation of the product H as a function of time t(sec) is given by Eq. (1.15)

$$H = \frac{R_o D_o [1 - e^{(D_o - R_o)kt}]}{R_o - D_o e^{(D_o - R_o)kt}}, \quad (1.15)$$

where k is a bimolecular rate constant ($M^{-1}\text{sec}^{-1}$) and R_o and D_o are the initial (time zero) concentrations (mol nt/liter) of R and D, respectively. Equation (1.15) can be simplified to a more practical form under two conditions:

(a) When the concentration of one hybridization component greatly exceeds the other (>20:1), a pseudo first-order condition ($R_o \gg D_o$) is established and Eq. (1.15) reduces to Eq. (1.16):

$$H = D_o(1 - e^{-kR_o t}). \quad (1.16)$$

When the reassociation (or annealing) reaches 50% completion, i.e., $H/D_o = 0.5$, then

$$R_o t_{1/2} = \ln 2/k \text{ or } k = \ln 2/R_o t_{1/2} \quad (1.16a)$$

(b) When the initial concentrations of the two components are equal ($R_o =$

D_o) as in DNA–DNA reannealing, the kinetics of the bimolecular reaction can be simplified as in Eq. (1.17)

$$C/C_o = (1 + kC_o t)^{-1}, \qquad (1.17)$$

where C is the concentration of single-stranded molecules, C_o is the total concentration of DNA (mol/liter) at time zero, and C/C_o is the fraction of unpaired strands in hydroxyapatite-binding assays. At 50% reannealing,

$$C_o t_{1/2} = 1/k \text{ or } k = 1/C_o t_{1/2}. \qquad (1.17a)$$

When the single-stranded fraction is analyzed by other methods such as nuclease S1 or optical hyperchromicity, Eq. (1.17) is slightly modified by a method-dependent factor to give Eq. (1.18):

$$C/C_o = (1 + kC_o t)^{-0.44}. \qquad (1.18)$$

ii. **Significance of reassociation kinetics.** The bimolecular constant k is a factor that reflects the nature of the molecule and its environment. From Eq. (1.19)

$$t_{1/2} = N \ln 2/(3.5 \times 10^5 \times L^{0.5} R_o) \qquad (1.19)$$

thus,

$$k = (3.5 \times 10^5) L^{0.5}/N, \qquad (1.20)$$

where N is the total number of base pairs in a nonrepeating sequence (originally termed *base complexity*). Therefore the determination of $t_{1/2}$ and consequently k from Eq. (1.16a) or Eq. (1.17a) provides the information on the base complexity (N) of DNA, its concentration (R_o), or other factors (e.g., ionic strength and metal ion) that influence the stability of a duplex molecule by a comparison with similar data from the reassociation kinetics of "standard" molecules or under "standard" conditions.

iii. **Practical applications of reassociation kinetics.** The principle of second-order reassociation kinetics has been used to develop a method for the construction of a uniform abundance (normalized) cDNA library. The approach is based on the assumption that rarer cDNA species anneal less rapidly than abundant cDNA species, and therefore the duplex form of the abundant species can be selectively removed by passage through a hydroxyapatite column (22,23). When the resulting cDNA is amplified by PCR and put through repeated cycles of normalization (i.e., denaturation, reannealing, separation, and PCR amplification), the rare cDNA can be progressively enriched, yielding a normalized cDNA library. The same principle can be applied to a reverse situation in which unhybridized DNAs are removed, thus enriching a certain class of DNAs (e.g., repetitive DNA sequences) that would form double strands faster than others (e.g., single or nonrepetitive sequences).

b. Factors affecting reassociation

The reassociation process is dependent on various factors. We will emphasize here those aspects related to more practical situations such as short oligonucleotide annealing and/or hybridization, especially in solution at neutral pH. The factors involved in other types of hybridization, e.g., filter (or membrane) hybridization and *in situ* hybridization, are similar, at least qualitatively, to those for hybridization in solution.

i. Length and base composition. The rate of annealing for DNA–DNA and RNA–DNA is proportional to the square root ($L^{0.5}$) of the shorter strand (24). In contrast to the strong effect of base composition on denaturation, the base composition has little effect on the rate of DNA–DNA or RNA–DNA annealing (24,25). In the hybridization of an oligonucleotide (primer or probe) to larger DNA, the probe size, composition, and especially the sequence are of critical importance.

For optimal hybridization with fully complementary oligonucleotides, three main factors have to be considered in choosing the right oligonucleotide: (a) stability of the duplex to be formed, (b) specificity for the target sequence, and (c) self-complementarity. The thermodynamic parameters from nearest-neighbor interaction models (Table 1.8) provide a framework for the computerized calculation of T_m (26). Based on the predicted T_m, one can choose an oligonucleotide most appropriate for a given hybridization or annealing temperature. The critical length for oligonucleotides used as primers and hybridization probes is estimated to be 17–18 nt (27). The specificity for a target sequence increases as a function of probe length. In general, a probe length of 20–30 bases gives an acceptable specificity. With longer probes, the higher specificity may be compromised due to the increase in tolerant mismatches. For the same reason, primers and hybridization probes should be made longer if stable mismatching is desired.

ii. Mismatching bases. In general, a 1% base mismatch reduces the T_m by 1.0–1.4°C (28). For an oligonucleotide, single base pair mismatches can affect the T_m by as much as 10°C (29). When the T_m is decreased by ~15°C due to mismatches, the annealing rate of the DNA is reduced by a factor of 2 (30). Mismatching hybrids are more stable at high salt than at low salt concentrations, approximately 66% being the minimum match. The *stringency* of hybridization can be adjusted by several factors such as temperature, ionic strength, and chaotropic agents (see below). For small oligonucleotide probes, the stringency can be controlled up to the point of distinguishing single base pair mismatches. Discriminating single base pair mismatches in RNA·DNA heteroduplexes or RNA·RNA duplexes is more difficult than in DNA·DNA, especially when the mismatches involve a G·U(T) base pair. To enhance the mismatch discrimination, one technique, called *competition hybridization,* employs an excess of unlabeled noncomplementary oligonucleotide probes for nonspecific masking and for enhancing the specificity of the labeled complementary oligonucleotide probes during the hybridization (31).

iii. **Ionic strength and pH.** The rate of DNA–DNA annealing is strongly dependent on Na$^+$ concentrations up to 2.5 M. At concentrations of up to 0.2 M, the rate increases proportionally to the cube of the ionic strength (32). Between 0.4 and 1.0 M NaCl (the ionic strengths usually employed for hybridization, e.g., with 6× SSC), the increase in the annealing rate (k) is minimal (1.5-fold). When the NaCl concentration is above 0.4 M, the effect of pH (5.0–9.0) on the annealing rate is small (less than 1.3-fold).

iv. **Viscosity.** The rate of hybridization in solutions containing organic substances such as sucrose, ethylene glycol, and glycerol is inversely proportional to the *microscopic* viscosity. The microscopic viscosity affects the microenvironment around the DNA bases (33).

Polymers that affect *macroscopic* viscosity by changing the hydration at the DNA surface substantially increase the rate of annealing (30): Ficoll [neutral polymer, at 5.7% (w/v)] and sodium-dextran sulfate [anionic polymer, at 2% (w/v)] increase the rate by 1.5 and 4 times, respectively.

v. **Denaturing agents.** The base pairing interactions of nucleic acids are also affected by the presence of denaturing agents, such as formamide, urea, and guanidinium thiocyanate, or are temporarily masked by chemical reactions with formaldehyde which reversibly blocks amino groups. When appropriately used, these compounds can provide selective advantages in nucleic acid hybridization.

(a) *Formamide.* At concentrations of up to 50%, formamide lowers the T_m of DNA by 0.6°C for each percent increase of formamide in the presence of NaCl (0.035–0.88 M) (34). Formamide also decreases the DNA–DNA renaturation rate by 1.1% for each percent increase in concentration. In RNA–DNA hybridization, thermal stability does not show a simple linear relationship with formamide concentration (35). At 50% formamide, the rate of RNA–DNA hybridization is reduced at least fourfold compared with the rate at 68°C in 2× SSC.

Despite the decrease in annealing rate, formamide presents unique advantages in nucleic acid hybridization. It is used to lower the hybridization temperature and to reduce the risk of thermal strand scissions or to control the stringency of annealing (36). A 30–50% formamide concentration allows optimal hybridization to be performed at 30–42°C rather than at 68°C without formamide.

The T_m of RNA · DNA hybrids in aqueous solutions containing 70% formamide is ~20°C higher than the corresponding DNA · DNA hybrid under similar conditions (37). Therefore, with 70% formamide, the annealing of DNA strands is completely inhibited at 41–50°C, whereas RNA–DNA hybridization proceeds to its full extent (37,38).

(b) *Urea.* Urea at concentrations of up to 8 M decreases the T_m of DNA by 2.25°C for each increase of unit molar concentration (34). It also decreases the renaturation rate by ~8% per unit molar concentration increase.

(c) *Guanidinium thiocyanate.* A sensitive RNA–RNA hybridization can be performed in the presence of the chaotropic agent GuSCN (3–6 M) at 23–37°C

(39). The use of GuSCN allows detection of a specific DNA or RNA sequence directly from the GuSCN-dissolved cell lysate. The rate of hybridization in GuSCN at room temperature is greater than that in 50% formamide at 42°C.

(d) Formaldehyde. Formaldehyde reacts with free amino groups of nucleoside bases, forming methylol or Schiff base derivatives and consequently denaturing dsDNA. Formaldehyde-treated, immobilized DNA can be hybridized with oligonucleotide probes with efficiencies 5–10 times greater than untreated DNA. Formaldehyde reacts initially with DNA at the single-stranded regions that occur during the "breathing" of the double helix. As the chemical modification progresses, optimally at 6× SSC, 10% HCHO, and 60°C for 20–30 min with free DNAs, the double helix eventually collapses, further accelerating the DNA–HCHO reaction. Formaldehyde adducts are stable in neutral buffers in the presence of formaldehyde and prevent DNA from reassociation. Removal of the reagent from the buffer during the (pre)hybridization steps leads to regeneration of the amino groups, thereby allowing unimpeded, efficient hybridization. Formaldehyde can be used on RNA in a similar manner to break temporarily and/or to prevent secondary structures.

vi. Temperature. The rate of reassociation (or hybridization) of DNA is maximal at ~25°C below the T_m (28,40). A further reduction of temperature reduces the annealing rate; the dependence of rate on temperature can be described by a bell-shaped curve. In contrast to DNA–DNA reassociation, RNA–DNA hybridization is optimal at about 10–15°C below the T_m of the hybrid (41).

The approximate T_m for an oligonucleotides (≤100 bases) in up to 0.5 M NaCl can be calculated from the following empirical formula (42) [Eq. (1.21)]:

$$T_m(°C) = 81.5 + 0.41\,(\%\ GC) + 16.6\log M - 675/L \\ - 1.0\,(\%\ mismatch) - 0.65\,(\%\ formamide) \quad (1.21)$$

For perfectly matching DNAs, hybridization is usually carried out at 10–25°C below the T_m in aqueous solution or at lower temperatures (e.g., 42°C) in 50% formamide solution. In PCR amplification, which is performed at 72°C, the choice of optimal temperature for primer annealing can be critical. The T_m calculation should take into account both the T_m for the primer and the T_m for the PCR product (43).

C. Structure of DNA: Secondary and Tertiary Structures

1. HELICES

a. Structural basis of double helix

i. Major forces contributing to DNA conformations. The spontaneous formation of double-stranded DNA or RNA is attributed to two major forces (20,40). One is the H-bonding between complementary bases according to the Watson–Crick and non-Watson–Crick base pairing rules (see Section II,A,2, this chapter).

TABLE 1.9 Structural Parameters for Representative Double Helices

Form	Helix sense	Helical pitch (Å)	Helix diameter (Å)	Residues per turn	Base tilt (°)	Examples
A	Right	28–32	23	11	16–19	Variant of B form in Na$^+$ salts and at 75% humidity. RNA·DNA hybrid, RNA duplex
B	Right	34	19.3	10.6	−6	Most natural and synthetic DNAs
C	Right	31	19.2	9.3	−8	Variant of B form in Li$^+$ salts at 66% humidity
D	Right or left	24	17.5	8.0	−16	Poly[d(A-T)] RNA·DNA hybrid at low humidity
Z	Left	45	18	12	−7	Poly[d(G-C)]

The other is *base stacking*, the vertical arrangement of bases such that one base plane is at the van der Waals distance (~3.4 Å) and parallel to the adjacent one. Base stacking drives the relatively hydrophobic bases to the interior. It is also the major contributing factor to H-bonding interactions between bases in aqueous solution. Phosphate–phosphate anionic repulsions are also minimized when the sugar–phosphate backbone forms the hydrophilic exterior, making contact with counterions such as Mg^{2+} and Na^+. The stacking of bases on top of each other occurs in decreasing order from Pu/Pu > Pu/Py > Py/Py.

Although built of common structural elements according to the rules of base complementarity, double-stranded DNAs can form double helices that are quite different from each other. Even a single DNA double helix can adopt highly variable conformations depending on nucleotide sequence, salt concentration, hydration, molecular packing, and temperature (45,46).

The conformations termed A, B, and Z DNA are derived from idealized DNA models based on X-ray diffraction data (Table 1.9). This implies that only broad and relaxed definitions should be applied to the DNA structures in solution. Perhaps more appropriately, local structures of DNA in solution should not be considered static.

ii. **Structural parameters.** Some of the structural parameters used to differentiate one helix from another include (a) helix *pitch* or *repeat* (increment of helix axis for a 360° turn of the helix), (b) residues per turn, (c) rise per residue, (d) helix *diameter* or *radius*, (e) dimensions (width and depth) of *major* and *minor grooves*, (f) base *translations* (slide, shift) (Fig. 1.9) and *rotations* (tilt, twist, propeller twist, and roll) (Fig. 1.10), and (g) curvature (or bending).

The conformation of a double helix is also defined by the stereoscopic nature of glycosidic bonds and the sugar conformation (Fig. 1.11). The bases adopt two main orientations relative to the sugar moiety: *anti* and *syn* with respect to the

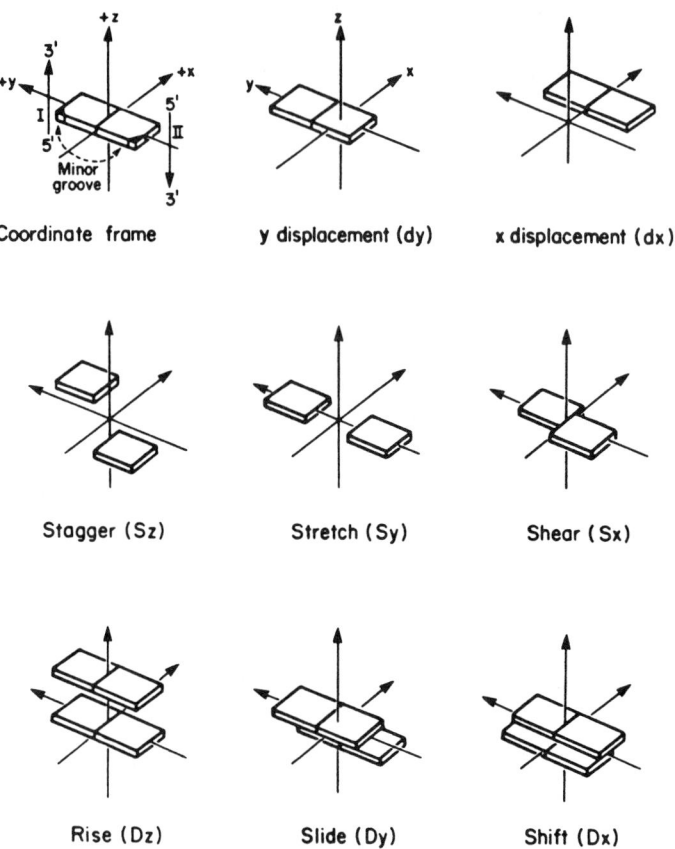

FIGURE 1.9 Definitions of various base translations. The translations involve two bases of a pair (upper two rows) or two successive base pairs (bottom row). In the top row the motions of the two bases are coordinated, and in the middle row their motions are opposed. Columns at left, center, and right describe translations along the z, y, and x axes, respectively. The standard coordinate frame is defined at the upper left. From EMBO Workshop on DNA Curvature and Bending (1989). *EMBO J.* **8**, 1–4, with permission of Oxford University Press.

C-1′–N glycosidic bond. A base is in *anti* conformation when the bulk of the heterocycle (N-1–C-2 side of Py, N-9–C-4 side of Pu) is pointing away from the sugar, and in *syn* conformation when it is over or toward the sugar. The ranges of *syn* and *anti* are defined by *torsion angle,* chi (χ), with respect to the C-1′–N-1 glycosidic bond (20). Pyrimidines tend to adopt the *anti* conformations, whereas purines adopt both conformations.

Sugar (β-D-furanose) puckers can be either *endo* or *exo* with respect to the plane of C-1′–O–C-4′. A sugar has an *endo* pucker when C-2′ or C-3′ is most remarkably in a position above the plane on the same side as C-5′. The opposite position is the *exo* pucker. The base conformation and sugar puckers have preferred combinations, e.g., *C3′-endo* for *syn* and *C2′-endo* for *anti*. [The term

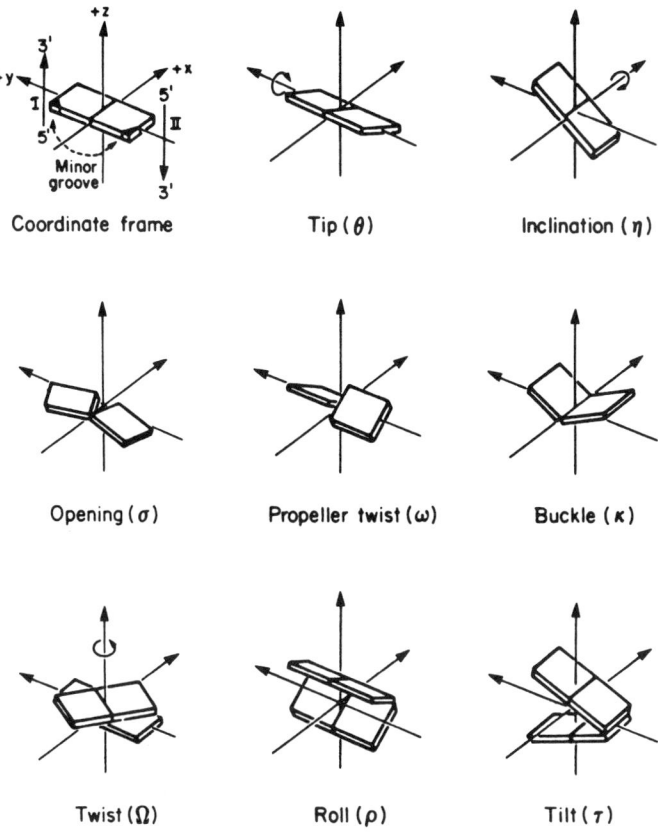

FIGURE 1.10 Definitions of various base rotations. The rotations involve two bases of a pair (upper two rows) or two successive base pairs (bottom row). In the top row the motions of the two bases are coordinated, and in the middle row their motions are opposed. Columns at left, center, and right describe translations along the z, y, and x axes, respectively. The standard coordinate frame is defined at the upper left. From EMBO Workshop on DNA Curvature and Bending (1989). *EMBO J.* 8, 1–4, with permission of Oxford University Press.

conformation is used here to refer to the 3-D shape of a molecule, i.e., orientations of groups that are, in principle, capable of movement by internal rotation. The conformation depends on the environment of the molecule and represents an average over energetically possible atomic structures. The *configuration* is used to refer to the atomic structure of a molecule. The configuration is constant for a particular molecule, e.g., the stereochemistry of asymmetric centers.]

iii. **Conformational heterogeneity.** DNA conformations can be interconverted by relatively mild changes in conditions or can coexist in a single, continuous double helix. Sequence-dependent polymorphism is exhibited at the level of single base pairs and at dinucleotide steps. The upstream sequence in the

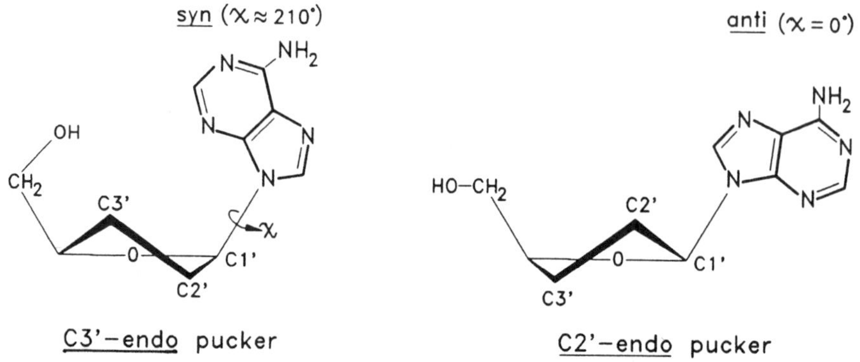

FIGURE 1.11 Idealized examples of sugar puckering modes and base (adenosine) orientations about the glycosidic bond. Pyrimidines and purines have different ranges of *syn* and *anti* domains defined by the torsion angle (χ) about the C-1′–N bond.

transcription–initiation region is particularly rich in "sequence motifs" formed by specific sets of nucleotide sequences (see Section I,A, Chapter 7). The structural variations associated with the sequence motifs are presumed to play a key role in specific interactions with transcription factors regulating the transcription and gene expression. The conformational *microheterogeneity* of the double helix is also believed to play a vital role in other cellular processes involving interactions with DNA-binding enzymes and/or proteins. The heterogeneity of DNA (and RNA) conformation is becoming more diverse and biologically more important in light of our deeper understanding regarding the nature of various "unusual" DNA forms, such as parallel-stranded double helices (ps-DNA), triple helices (H-DNA), quadruple helices (G-DNA), DNA curvature, and supercoils (see below). Binding of certain proteins to DNA introduces further local structural changes such as DNA melting or bending.

iv. Interaction with proteins. How a protein (or enzyme) interacts at the molecular level with DNA (or RNA) remains a central unresolved issue in molecular biology. For a small number of well-studied cases, it has been shown that sequence-specific interactions of a protein with DNA are based on the recognition of features of the major groove; nonspecific interaction or less sequence-specific interactions are based mostly on the recognition of a minor groove and phosphate backbone. The recognition involves essentially two components, the relative contributions of which may be highly variable depending on particular proteins and DNA or RNA: (a) direct interactions with individual base pairs and (b) interactions with global and local structural features.

A brief account of some of the better known nucleic acid–protein interactions is presented in later chapters: e.g., the *Eco*RI enzyme which recognizes not only specific base sequences but also an unusual structure of the B-DNA helix (Section II,B, Chapter 4) and DNase I whose interaction is predominently structural in the minor groove of B-DNA (Section I, Chapter 3). DNA looping generated by

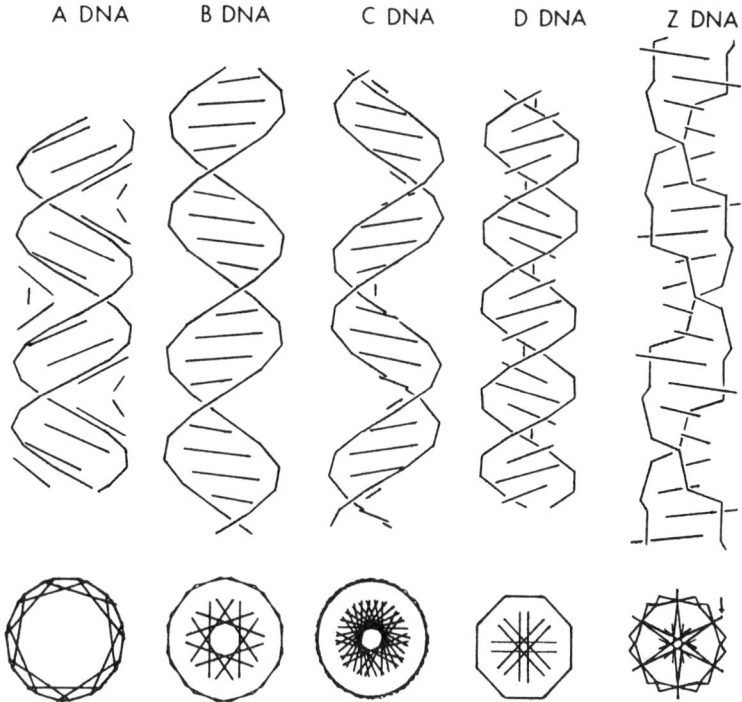

FIGURE 1.12 Models for various helical conformations of double-stranded DNA. Segments containing 20 bp are shown for right-handed models of A-, B-, C-, or D-DNA and for the left-handed Z-DNA. The upper views are perpendicular to the helical axis and the lower views look along the helical axis. The continuous helical lines are the sugar–phosphate backbone. The line segments indicate the positions of the base pairs and are formed by joining the C-1' atoms of each base pair. This simplified mode of representation emphasizes the differences in helical parameters and in positions of the base pairs among these models. The arrow in the lower view of Z-DNA indicates the C-1' of the deoxyguanosine residue, the base of which is relatively exposed in this structure; the deoxycytidine residue is more centrally located. From S. B. Zimmerman (1982). *Ann. Rev. Biochem.* **51**, 395–427.

specific interactions between DNA elements and DNA-bending or DNA-binding proteins (e.g., transcription factors) is described in the section on RNA polymerase (Section I,A, Chapter 7).

b. B-DNA and A-DNA double helices

Watson–Crick base pairing has two important features: (a) the formation of linear H-bonds, two in A·T pairing and three in G·C pairing, and (b) virtually identical geometry of ~10.7 Å between the pair of bases. The right-handed DNA double helix modeled after the Watson–Crick base pairing is typically close to the X-ray crystallographic structure known as a B helix (or B-DNA) (Fig. 1.12).

In B-DNA, all bases have *anti* conformation and sugars have a C2'-*endo* pucker. The two strands are antiparallel. The B helix is considered the standard form for most natural DNAs, although they can adopt alternative conformations (A, C, and D) depending on hydration and ionic environment. Low hydration, low ionic strength, and high GC content favor an A conformation (*C3'-endo*) (Table 1.9). Double-stranded RNA is always in A conformation and RNA·DNA heteroduplexes readily adopt the A conformation. [The DNA chain should transform to an A-DNA for the hybridization to occur.] In chimeric DNA–RNA–DNA oligonucleotide duplexes, the internal RNA segment (4 nt) adopts a characteristic A conformation while the two flanking DNA segments (4 nt each) adopt a fairly normal B conformation (47). Only the DNA–RNA junction residues display intermediate characteristics.

A high degree of hydration, high ionic strength, and high AT content stabilize the B conformation (*C2'-endo*). Although the global conformation may fall into one or another of the main conformational types, a DNA may harbor numerous local irregularities, such as bending and *kink* (deviation of the helix axis from a straight line), due to sequence-dependent peculiarities or interactions with DNA-binding proteins.

Many proteins bind and distinguish B-DNA via base-specific H-bond donor and acceptor groups that lie in the major groove. The major groove is sufficiently wide to accommodate an α helix or other structural motifs and contains nucleophilic groups or elements such as N-7 of G and A, O-6 of G, 6-amino group of A, O-4 of T, and 4-amino group of C. The groups on proteins that provide complementary H-bonding include amino acid side chains and the peptide backbone.

In contrast to B-DNA, the contacts of proteins that bind RNA commonly involve base-specific H-bond interactions between protein groups and minor groove bases. The A-form structure of RNA helices provides minor grooves that are wide and shallow (rather than deep and narrow major grooves as in B-DNA) and are readily accessible to protein side chains, although there are fewer H-bonding possibilities to allow discrimination among the base pairs in duplex RNA.

c. Z-DNA double helix

Z-DNA is a Watson–Crick base-paired, left-handed helix that is distinct from the Watson–Crick right-handed B-DNA (48,49). Z-DNA is thinner (by ~10%), more extended (by 29%), and has more base pairs per turn than B-DNA (see Fig. 1.12 and Table 1.9). The purine bases of Z-DNA, formed for example with d(CG)$_3$, are rotated into the *syn* (*C3'-endo*) conformation. The pyrimidine bases maintain an *anti* conformation. Z-DNA is thus characterized by a dinucleotide repeat in which *anti* and *syn* conformations of the bases alternate in succession along the chain. The deoxyribose–phosphate backbone follows a zigzag left-handed course rather than the smoothly spiraling right-handed path found in B-DNA. The interphosphate distances also differ from those of B-DNA. There is one deep groove that corresponds to the minor groove in B-DNA. The major groove of B-DNA is replaced in Z-DNA by a convex surface on which purine N-7 and C-8 and the pyrimidine C-5 positions are exposed.

Alternating Pu–Py tracts such as the hexanucleotide d(CGCGCG) or poly [d(CG)] have the propensity to adopt the Z-helix (48,50). In addition to the intrinsic sequence dependence, the Z-form is stabilized by a variety of environmental factors such as a high degree of hydration, high ionic strengths, transition–metal complexes, spermine, cytosine methylation, and/or negative supercoiling (49,51,52). The $\Delta G°$ of B- to Z-DNA transitions are estimated to be -0.5 kcal/mol for $d(m^5CG)$, 0.69–1.30 kcal/mol for unmethylated d(CG), 1.32–1.48 kcal/mol for d(CA)-d(TG), and 2.3–2.4 kcal/mol for d(TA). Longer sequences require less energy than shorter sequences to adopt a Z-helix. The junction region between contiguous B-DNA and Z-DNA can be as small as four nonalternating bases. The junction is conformationally flexible and usually shows a hypersensitivity to single-strand cutting nucleases. Under high salt conditions, $d(CG)_2$ tetranucleotides adopt a variant form of Z-helix (50).

The Z-form DNAs are presumed to play a role in various cellular functions such as gene expression and chromosomal recombination (49). Several lines of evidence point to the presence of at least transient left-handed Z-DNA conformations *in vivo*. For example, a segment of DNA inserted into a plasmid adopts a left-handed Z-helix in *E. coli* (53). Segments of CG tracts as short as 12 bp can readily adopt the Z-form *in vivo* when they are located in a region of high negative supercoiling (54). Such sites are often found in the upstream region of promoter sites. In contrast, no Z-DNA is detected even with a 74-bp CG tract when this sequence is located in a region of low negative supercoiling. A DNA target (recognition) site is not methylated by its specific methylase or cleaved by its specific restriction endonuclease (*Bam*HI, *Eco*RI, or *Hha*I) when the site is in or near a left-handed Z-DNA tract (55–57).

d. Parallel-stranded double helices

The Watson–Crick rule of base pairing dictates that DNA double helices (e.g., B, A, and Z forms) be composed of two antiparallel strands (aps-DNA). Accumulating evidence suggests that DNA conformations cannot always be fitted into "standard" molds, and diverse secondary structural forms of DNA can be constructed on the basis of variant H-bonding schemes collectively known as non-Watson–Crick base pairing.

One of the unconventional forms of DNA double helices is the *parallel-stranded DNA* (ps-DNA) in which two H-bonded strands have the same $5' \rightarrow 3'$ orientation (2,3). The ps-DNA is based on reverse Watson–Crick base pairing between A·T (see Fig. 1.7). A model of ps-DNA consisting of $d(A)_n \cdot d(T)_n$ has the following features. (i) Like aps-helices, ps-helices have a $C2'$-*endo* sugar puckering in both strands and a helical twist of about 42°. The rise per base pair and propeller twist are ~3.2 Å and 20° in the ps-helix, as compared to ~3.0 Å and 22° in the aps-helix. (ii) ps-DNA lacks a clear distinction between a major and a minor groove. Consequently, the *bis*benzimidazole drug Hoechst-33258, which binds in the minor groove of B-DNA, exhibits very little fluorescence in the presence of the ps-DNA. EtBr intercalates equally well into ps- and aps-helices. (iii) The H-bonding potential of a reverse Watson–Crick A·T base pair is essentially the same as that of the conventional Watson–Crick structure. However, thermal transitions of oligonucleotide hairpins (10-bp stem and 4-nt loop) have shown that the ps hairpins denature 10°C lower than the corresponding aps hairpins. (iv) The ps hairpins are substrates for T4 polynucleotide kinase, T4 DNA ligase, and *E. coli* Exonuclease III, and have the same electrophoretic mobilities as aps oligomers in polyacrylamide gels.

In addition to the $d(A)_n \cdot d(T)_n$, which can form a stable, right-handed ps-DNA, alternating d(G-A) sequences have also been shown to form, in the pH range 4–9, a parallel-stranded, right-handed homopurine duplex. The duplex consists of alternating, symmetrical $G_{syn} \cdot G_{syn}$ and $A_{anti} \cdot A_{anti}$ base pairs with N-1H–O-6 and N-6H–N-7 H-bonds, respectively (203). The ps-DNA formed by $d(G-A)_{25}$ has a T_m of 40°C and shows very little fluorescence upon staining with EtBr. In contrast to the $d(A)_n \cdot d(T)_n$ ps-DNA, d(G-A) ps-DNA displays *C3'-endo* sugar puckering and apparently shares some essential structural elements with left-handed Z-DNA.

e. DNA triple helices

Homopurine–homopyrimidine sequences, $d(G-A)_n \cdot d(T-C)_n$ and $d(G)_n \cdot d(C)_n$, have a propensity for adopting triple helices in which the third pyrimidine strand binds via Hoogsteen H-bonding to the regular Watson–Crick base-paired double helix. A triple helix (triplex) can also be formed when a third purine strand binds to the regular Watson–Crick double helix. Sequence-specific third-strand recognition is a general phenomenon for all four combinations of Watson–Crick base pairs. Triple helices are presumed to play important biological functions, and intensive studies are directed toward triplex-based strategies in the design of sequence-specific gene regulators.

i. Py·Pu·Py(C^+) triplex: H-DNA. A non-Watson–Crick base pairing known as Hoogsteen base pairing occurs under physiological and certain nonphysiological conditions: at acidic pH values in the absence of Mg^{2+} or in a high degree of negative supercoiling (58,59). The Hoogsteen base pairing has been recognized as the basis for triple helices in which a Watson–Crick homopurine·homopyrimidine duplex is associated with a third pyrimidine strand in DNA (and RNA) based on standard base triplets, T·AT and C^+·GC. The formation of the standard base triplets requires that the cytosine in the pyrimidine strand be protonated (pK_a of 4.5 in the free state); the N-3 imino protons have been directly observed by NMR (60). Polyamines promote DNA triplex formation based on standard base triplets at neutral pH (61). In fact, the conditions of neutral pH with Mg^{2+} favor other nonstandard types of "pyrimidine" triplets such as C·GC, T·CG, C·CG (at pH 4.2), and G·TA (204). RNA can also form triplexes based on pyrimidine triplets which include r(U·AU), r(U·CG), and r(C^+·GC).

Triple helices based on standard triplex motifs are observed to a greater degree in regions containing Py·Pu tracts. A triplex model based on fiber diffraction data with poly[d(T·A·T)] shows that the third pyrimidine strand is base paired according to the Hoogsteen pairing with an underlying A-type Watson–Crick duplex along the major groove (62,63). The third strand is "parallel" to the purine

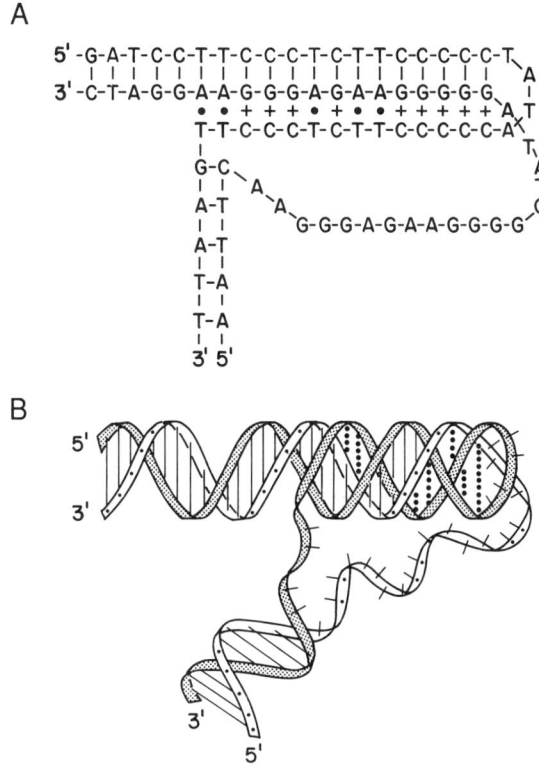

FIGURE 1.13 Structural models of H-DNA. (A) A two-dimensional model of H-DNA consisting of the Watson–Crick duplex associated with an intramolecular homopyrimidine stretch via Hoogsteen base pairing (·, +). From S. M. Mirkin, V. I. Lyamichev, K. N. Drushlyak, V. N. Dobrynin, S. A. Filippov, and M. D. Frank-Kamenetskii (1987). *Nature* **330**, 495–497. (B) A three-dimensional model of H-DNA. The strand containing the pyrimidine-rich sequences is shaded whereas the strand containing the purine-rich sequences is unshaded. The Hoogsteen base-paired pyrimidine tract occupies the major groove of the Watson–Crick duplex region. After J. C. Harvey, M. Shimizu, and R. D. Wells (1988). *PNAS* **85**, 6292–6296.

strand (64). Double helices containing Py·Pu tracts can undergo intramolecular transition to a triplex known as *H-DNA* (65–67). Recent 3-D models for H-DNA feature a triplex structure with a "kink" consisting of a hinge-like, flexible single-stranded region (Fig. 1.13). The binding affinity of the third strand for the duplex has been estimated to be 2–5 orders of magnitude lower than that of the two strands forming the duplex (4). The third strands that bind with either single-base bulges or single mismatches are destabilized by 2.5–2.9 or 3.2–4.0 kcal/mol, respectively, relative to the perfect triplex. This is essentially equivalent to the corresponding values determined for duplex DNA and RNA (68).

ii. **Pu·Pu·Py triplex.** Purine-rich oligonucleotides can also bind in the major groove of DNA double helix in an orientation "antiparallel" to the Watson–Crick purine strand and form a Pu·Pu·Py triplex (69). The "purine" triplets (G·GC and A·AT) which, in addition to the pyrimidine triplets, underlie this type of triple helix are pH independent, stable in basic pH, and occur in neutral solutions (pH 5–8) in the presence of $MgCl_2$ (70,71). Evidence suggests that the third strand pairs through Hoogsteen alignment with the Pu strand to generate G·GC and A·AT triples (72). The stability of the third strand in the Pu·Pu·Py triplex at pH 7.3 is approximately twice that of the third strand in the corresponding Py·Pu·Py(C^+) triplex at pH 5.5. Purine triplets such as r(A·AU), r(G·GC), and r(A·GC) also form the basis of RNA triplex.

iii. Detection of triplex. Triplex DNA can be readily distinguished from DNA duplexes by various physicochemical techniques such as CD, NMR, UV absorption spectroscopy, immunoblotting, PAGE, and electron microscopy (70,73). The formation of intramolecular H-DNA results in a ssDNA region that is hypersensitive to nuclease S1 and chemical probes such as OsO_4 (for T), DEPC (for A), and methoxylamine (for C). Triplex formation can also be detected by the resistance of certain sequences (e.g., TTTAAA) to the cleavage by restriction enzymes (e.g., *Dra*I). Triplex formation can protect the DNA from UV-induced pyrimidine dimerization, making the photofootprinting assay a promising method for studying triplexes (75).

The stability of a triplex can be modulated by small ligands which intercalate the helices in a sequence-specific manner. For example, the intercalation of a benzo[*e*]pyridoindole derivative into a stretch of six T·AT triplets stabilizes a 14-mer triplex by 20°C, whereas the same oligonucleotide triplex in which two interior T·AT triplets are replaced with C^+·GC triplets is stabilized by the compound by merely 2°C (205). The intercalation of EtBr results in moderate destabilization of certain triplexes (206), whereas EtBr can intercalate the poly [d(T·AT)] triplex more strongly than the duplex poly[d(T·A)] and in a noncooperative manner (74).

iv. Biological significance of triplex. Triplex structures (H-DNA and Pu·Pu·Py) are presumed to play a regulatory role in the expression of eukaryotic genes (76). Some of the observations that support this *in vivo* role are as follows. (a) A triplex can occur under physiological conditions of ionic strength, pH, temperature, and the presence of polyamines. (b) The H-DNA formation relaxes negative supercoiling and can provide a buffer region for a proximal site involved in facilitated DNA–protein interactions (77). (c) Natural and recombinant plasmid DNAs containing alternating Pu/Py sequences can adopt the H-DNA conformation *in vitro*. (d) Py·Pu tracts, which are hypersensitive to nuclease S1, frequently occur in DNA regions presumed to be involved in gene promoter function and/ or initiation of replication. For example, a cis-acting upstream element of the c-*myc* proto-oncogene has a large Pu/Py symmetry and can assume a H-DNA conformation (78). The origins of DNA replication in eukaryotes, as well as prokaryotes, characteristically contain an AT-rich region called the *initiator recog-*

nition element [e.g., (A/T)TTTATRTTT(A/T) in yeast (79)] 5' to the DNA unwinding elements.

The observation that a dsDNA can bind a third single strand to form a triplex has opened up a variety of triplex-based strategies applicable to recombinant DNA techniques as well as genetic therapies. For example, the site-specific DNA recognition by a third, synthetic oligonucleotide attached to a chemical probe (e.g., EDTA·Fe) or nuclease makes the oligonucleotide a novel tool for site-specific cleavages of DNA (80). The selection of cleavage sites can be further controlled by the action of DNA-binding proteins. In a different strategy, the nuclease action of restriction enzymes can be modulated or masked by a triplex-forming oligonucleotide. In therapeutic applications, the triplex strategy offers new ways of controlling gene expression. For example, introduction of an oligonucleotide capable of forming a triplex at the promoter region of a gene has been shown to result in the repression of transcription of c-*myc* oncogene *in vitro* (81) and of IL2Rα gene in lymphocytes (82).

2. UNUSUAL STRUCTURES

Not all natural DNAs are double-strand helices. The DNAs from many phages such as phage φX174 and M13 are single covalent strands. Even within the apparent double helices, transient as well as stable single-stranded regions coexist with other types of unusual structures such as hairpin (stem–loop), cruciform, and G-DNA. These unusual structures are presumed to play critical roles in DNA replication and gene expression. Experimental methods used to probe these and other unusual structures employ single-strand-specific nucleases, conformation-specific chemical reagents, and physical probes such as NMR (83,84). With a few notable exceptions, naturally occurring RNAs are single covalent strands that adopt characteristic secondary structures. Because the stem–loop is the most prominent feature of RNA secondary structures, it will be described in detail in Section II,D,1 on RNA structures. We now describe the structural and functional features of cruciform and G-DNA.

a. Cruciform

Certain DNA sequences have a twofold inverted symmetry and are called *self-complementary* or *palindromic* sequences. Palindromic sequences are the usual recognition sites for restriction enzymes and frequently occur as essential elements in regulatory regions. The inverted repeat sequences in mRNA also play an important role as critical sites of interaction with various protein factors that are involved in the initiation of translation, termination of transcription, and hence in the regulation of gene expression. Under certain conditions, such as high negative supercoiling (85), low ionic strength, and the presence of Mg^{2+}, palindromic or alternating Pu–Py sequences [e.g., $d(A-T)_n$] can fold back on themselves to form two opposed hairpins, called a *cruciform* (cross-like) structure. A cruciform by itself is not as stable as a linear duplex due to the energetically unfavorable single-stranded loop, and most likely exists as a minor species in equilibrium with duplex structures under "normal" or unstressed conditions. The cruciform DNA generated from alternating Pu–Py sequences is presumed to be a transient intermediate in genetic recombination. DNA superhelix-induced cruciform formation in the promoter region has been shown to repress transcription (86).

b. Four-stranded DNA: G-DNA

G-rich DNA sequences with a common motif of G_{3-4} can form stable four-stranded helices (tetraplex) called *G4-DNA* or *G-DNA* (87,88). G-DNA is based on Hoogsteen-paired quartets of G residues (see Fig. 1.7) and is stabilized in high salt (K^+ and/or Na^+) conditions. Similar stable tetramerizations occur in RNAs (e.g., 5S RNA of *E. coli* or synthetic oligonucleotides) containing G_4 sequences (89). The G-rich single strands in the tetraplex can be oriented parallel (low-mobility form when electrophoresed on polyacrylamide gel) or antiparallel (high-mobility form) to each other depending on the kinetics of assembly. The assembly process is anomalously dependent on monovalent cations (87,90) and involves cation-dependent folding intermediates. The K^+ is the strongest stabilizer of G-DNA, while Na^+ is less effective and Li^+ is ineffective. The K^+-driven G-DNA assumes a predominantly antiparallel (fold-back) structure, although all parallel tetraplex can be formed, for example, with $d(TG_3T)$ and $d(TG_3T)_2$ (230). G-DNA is stable even at 95°C in the presence of 125 mM KCl and 40 mM NaCl. The quadruplex is also extremely stable even in low salt buffers containing 8 M urea. G-DNA structures have been found at the ends of chromosomes, the part called *telomeres* (repetitive TTAGGG in humans and TTGGGG in *Tetrahymena thermophila*). G-DNA is believed to play a key role in meiosis and in maintaining integral chromosomal termini "out of harm's way" of exonucleases. Guanine tracts also occur in immunoglobulin switch regions and gene regulatory regions.

3. Circular Structures

DNA does not always assume helical *rod*- or random *coil*-like structures but can accommodate various other structures such as the bending of the helix axis (91). Curvature or bending in the absence of external forces is an inherent property of DNA. Sequence-directed DNA curvature is often associated with AA dinucleotide wedge and non-AA sequence wedges as well. Furthermore, DNA can exist in a stable circular form when its 5′ and 3′ termini are covalently joined by ligases. Typical examples of such circular molecules are *plasmids*, the self-replicating, small-sized (average 2–8 kbp) molecules of DNA (or RNA). The smallest dsDNA circle constructed out of two synthetic oligonucleotides and T4 DNA ligase is reported to be a 42-bp microcircle (92). Depending on the degree of built-in or induced strains caused by overwinding or partial unwinding, *covalently closed circles* (ccc) adopt superhelical structures (see below). When one or more nicks are introduced into the covalently closed circles, the supercoils relax into *open-circular* (oc) forms. The *ccc* DNA differs considerably from linear or *oc* DNA in several physical properties, notably electrophoretic mobility and sedimentation velocity. When heat denatured, the *ccc* DNA collapses into a compact, fast-sedimenting complex having two interlocked random coils and has an electrophoretic mobility ~20% faster than a comparable *oc* DNA.

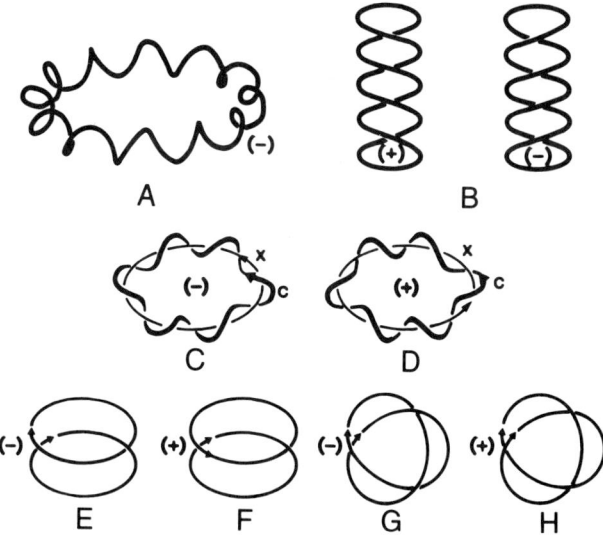

FIGURE 1.14 Topological elements of supercoiled DNA. (A and B) Two forms of supercoiled duplex DNA: solenoidal and plectonemic. (C and D) Left-handed and right-handed supercoils. The curves x and c represent the duplex axis (or spanning surface) and one of the phosphodiester backbone strands of DNA, respectively. The surface normal points up for the two supercoils. With respect to the spanning surface defined by the circular boundary curve x, the left-handed DNA strand (C) punctures the surface in the direction opposite to the surface normal, and the punctures are given a negative sign. Punctures made by the right-handed DNA strand (D) are in the same direction as the surface normal and are given a positive sign. The sum of all such signed punctures is the Lk for the molecule. The Lk of the supercoil (C) is -6 and that of D is $+6$. (E and F) Two forms of singly linked catenanes. When the overlying arrow indicating the orientation of the DNA primary sequence can be aligned with the underlying one by a *clockwise* rotation (of less than 180°), the node for duplex DNA has a value of -1. When the assignment is made by a counterclockwise rotation, the node has a value of $+1$. (G and H) Two forms of the simplest knot, the trefoil (a torus knot). After N. R. Cozzarelli, T. C. Boles, and J. H. White (1990). *In* "DNA Topology and Its Biological Effects," pp. 139–184. CSHL Press, New York.

4. QUATERNARY STRUCTURES: SUPERCOILS

A single double-helical DNA can be further coiled in space, forming a higher-ordered structure called a *supercoil*. A supercoil can be either *solenoidal* (or *toroidal*) when the DNA wraps helically about a circular superhelix axis or *plectonemic* (or *interwound*) when the DNA winds helically up and back down about a superhelix axis (Figs. 1.14A and 1.14B).

A plectonemic supercoil obtained by removing positive primary turns is a *negative supercoil* although the direction of the primary DNA helix is right handed. Conversely, a plectonemic supercoil obtained by the addition of positive turns or overwinding is a *positive supercoil* although the DNA helix is left handed.

Various forms of DNA supercoiling are involved in many biological processes such as DNA recombination, replication, and transcription (93,94). During the transcription process, positive supercoils are generated in the DNA template ahead of the advancing RNA polymerase by an RNA polymerase-anchoring mechanism, while negative supercoils are generated simultaneously in the DNA behind the polymerase (95).

Multiple chains of DNA can be oriented in space in different topological forms (or *topoisomers*) such as linked rings (chains, *catenanes*) and *knots* (Figs. 1.14E–1.14H). Reactions leading to topological changes are catalyzed by DNA topoisomerases which transiently break and then reseal the DNA helix. Catenanes are interlocked rings that are held together by "topological bonds" independent of direct interaction. Among the knots are *torus knots* which can be drawn on the surface of a doughnut-shaped object and *twist knots* which are formed by strand passage between two looped ends of the molecule twisted on itself.

The principal *topological parameters* used to describe higher-ordered structures of DNA include linking numbers, writhe, and twist (96). When the DNA strand of a supercoil punctures the imaginary surface spanning the superhelical axis in the same direction as the *surface normal* as in the case of right-handed (+) supercoil, each puncture is assigned an index number of +1 (see Fig. 1.14D). When the puncture occurs in the direction opposite the surface normal as in the case of left-handed (−) supercoil, it is given an index number of −1 (see Fig. 1.14C). Linking number (Lk or α) is the sum of the index numbers over the whole molecule. For closed DNA, Lk is an integer. Lk is a topological invariant and changes only when the continuity of the chain is interrupted. *Twist* (or turn, T, previously β or α_o) is the number of turns in the strand in its *relaxed* state or free of torsional strain. T is not a measure of how many times a backbone strand winds about the superhelix axis (x). *Writhe* (W) is the number of supertwists (supercoils or superturns) and involves the crossing of the axis (x), with the sign given by the conventional rule of handedness. The topology of a *ccc* DNA is described by Eq. (1.22):

$$Lk = W + T. \quad (1.22)$$

For a relaxed DNA, $W = 0$ and $Lk = T$. Most plasmid DNAs are isolated as negative supercoils. Intercalation of EtBr or other dyes unwinds the DNA to such a point that the *ccc* DNA becomes relaxed in a way similar to *oc* DNA. Further addition of EtBr results in continuous unwinding of the DNA, eventually leading to coil flipping (to positive supercoiling). In this process, the Lk remains constant throughout, but the T and W change according to Eq. (1.22). In other words, the continued decrease in turns (T) by the addition of EtBr is exactly compensated by the increase in W from a negative value (negative supercoil) to 0 (open-circle) to a positive value (positive supercoil).

Superhelical density (σ) is the number of superhelical turns per turn in the relaxed primary helix, i.e., $\sigma = W/T$. The superhelicity can be modulated by several means; the most important are DNA intercalating agents.

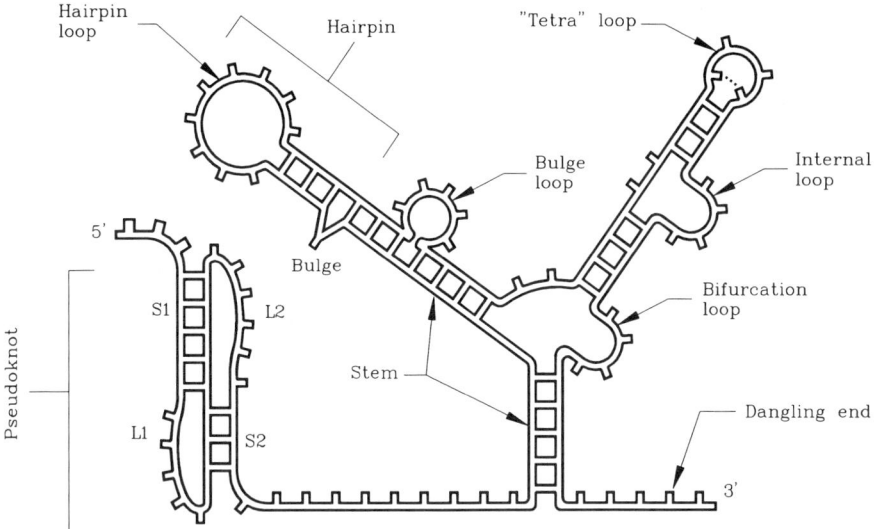

FIGURE 1.15 Secondary structural elements of RNA. In a typical tetraloop, the first and the last bases are H-bonded. In a pseudoknot (of type H), S and L denote stem and loop, respectively.

D. Structure and Function of RNA

In contrast to the usual duplex form of DNA, RNAs are usually single strands. However, regions of ssRNA may interact to form secondary structures which in turn become involved in tertiary interactions resulting in functional 3-D RNA molecules. Notable secondary structures of RNA include double-stranded stems and single-stranded loops (hairpin, internal, bulge) of varying lengths, pseudoknots, bulges due to single base pair mismatches, junctions (three-way or four-way), and unpaired termini (Fig. 1.15). Protruding hairpins are among the most prominent structural features of the RNA surface and are frequently involved in interactions with proteins, enzymes, receptors, and various gene regulatory factors.

RNAs can be functionally divided into messenger RNA (mRNA), ribosomal RNA (rRNA), transfer RNA (tRNA), and catalytic RNA (ribozyme). Other functional RNAs include protein-conjugated RNAs, such as small nuclear ribonucleoproteins (snRNPs).

Typical tertiary interactions of RNA include base triplets, H-bonds between base and sugar OH group, and between base and phosphate. As yet, tRNAs are the most abundant sources for the detailed 3-D structural information of RNA. Nevertheless, tertiary structures play an essential role in various catalytic functions of RNA whose expanding repertoire includes endonuclease/ligase (i.e., cleavage and ligation reactions of phosphodiester or phosphomonoester bonds in RNA and DNA), carbonyl esterase, and peptidyltransferase.

The extent and diversity of RNA-specific secondary and tertiary structures are reflected by the functional diversities of RNAs: mRNA, rRNA, tRNA, ribozyme, and snRNA.

1. SECONDARY STRUCTURES OF RNA

a. Analysis and prediction of secondary structures

Accurate prediction of the most plausible secondary structure of an RNA molecule based on its sequence can provide valuable insights into the tertiary structure and function of the RNA. Thermodynamically, an optimal structure is the structure with the minimum free energy. How can we find such a structure?

The most popular method is based on the thermodynamic interactions between two nearest-neighbor base pairs that have assigned free energy values (19,97). A typical set of free energy parameters for 10 nearest-neighbor sequences is listed in Table 1.8. The sum of free energy contributions by each base pair is taken as the measure of stability of the secondary structure. The precision and the accuracy of secondary structure prediction are only as good as those of the free energy parameter estimates (17,19). The computation and the resulting optimal structure(s) should accommodate various structural variations such as loops (hairpin, internal, bulge), mismatches (terminal, internal G·U), stacking regions, and dangling ends (unpaired terminal nucleotides). Therefore, the complexity involved in the determination of secondary structure increases approximately as the second power of the length of the RNA chain, indicating that the free energy minimization can best be accomplished by computer. Most of the currently available algorithms take into account the two most important determinants of RNA stability, namely base stacking and base pair destabilizing energies (98,99). One of the most popular RNA secondary structure prediction programs is the FOLD program (100), which can predict correct structures from the free energy estimate within 2% of that for the lowest free energy structure.

It is to be noted that thermodynamically predicted structures are only the most plausible of many similar or suboptimal possibilities. This is due to inherent uncertainties involved in the evaluation of thermodynamic parameters when tertiary interactions are not known. To be able to predict biologically relevant secondary structures, the thermodynamic approach has to be complemented with information from two other approaches: (a) phylogenetic identification of sequence constraints by comparing similarly functioning RNA molecules from different sources and (b) direct structural analyses using various chemical and/or enzymatic probes and site-specific mutagenesis. Among the experimental approaches are chemical and enzymatic *footprinting* techniques that employ a battery of "chemical nucleases" (101) and "enzymic" ribonucleases in conjunction with sequencing-grade PAGE. Among the useful RNases are single-strand-specific RNases (see Section IV, Chapter 3) such as RNase A (cutting at Py bases), RNase T1 (cutting at Gp residues), RNase T2 (cutting preferentially at Ap residues), PhyM (cutting at Ap and Up residues), and double-strand-specific RNases such as cobra venom RNases (giving 5'-P products).

b. Features of stem–loop structures

All naturally occurring RNAs, with a few notable exceptions (e.g., dsRNA of reovirus), consist of single covalent strands. However, some regions of the single strands form intramolecular double strands following Watson–Crick and sometimes non-Watson–Crick base pairing rules, giving rise to double-stranded *stem* and single-stranded *loop* structures (Fig. 1.15). The stem–loop structures, also called *hairpins,* play an important role in the tertiary folding of RNA molecules. They also serve as critical sites for the interaction with other nucleic acids and proteins. One extensively studied example of secondary structures is the cloverleaf model of tRNAs. Other RNAs such as mRNA and rRNA may have different secondary structures with varying structural motifs.

What are the structural features of a hairpin? A priori, a hairpin requires a self-complementary oligonucleotide to form the double-helical stem and a variable number of nucleotides to form the loop. The stability and dynamics of a hairpin thus depend on contributions from the stem as well as the loop. Thermodynamic studies have shown that four or five bases are optimal to form a loop in DNA as well as in RNA (102). Among all possible hairpin structures, a tetranucleotide loop or *tetraloop* is the predominant species in 16S and 23S rRNAs (103), transcription terminators (104), and the phage T4 genome (105). About 70% of these tetraloops have the consensus loop sequences UNCG or GNRA (where N is A, C, G, or U), while the closing base pairs tend to be C·G or less preferably G·C.

The predominance of tetraloop hairpins is somewhat paradoxical since the conventional estimates of the ΔG values are higher when the loop size is below or above 7–8 nt (106). It is now known that the unusual stability of the tetraloop is due to a number of specific interactions among the residues comprising the loop (107). For example, in the U^1NCG^4 sequences, the first (U^1) and the last (G^4) bases are H-bonded, effectively reducing the size of the loop to two unpaired bases. The third base (C^3) is partly stacked on the first and fourth bases, further constraining the loop. Similar structural constraints are observed with the synthetic DNA tetraloop carrying, for example, TTTA in the loop region (108). The attendant loss of configurational freedom is more than compensated for by other conformational changes. Hairpin loops containing only two bases (and a 5-bp stem) have been shown to be readily formed without any unusual torsional angles (109,110). Nevertheless, the contribution of H-bonds to the folding stability in the phylogenetically conserved tetraloop motifs is strongly context dependent and maximally 0.7 kcal/mol, suggesting that the conserved tetraloop motifs may have a functional rather than a thermodynamic basis (207).

For stem structures, a stable duplex requires at least 3 bp. Lower temperatures and high concentrations of oligomer and salt favor the formation of double-helical stems. In DNA hairpins with a 2-, 3-, and 4-nt loop, T_m values decrease in that order when the helical stem is in the B-form, whereas they increase when the stem is in the Z-form (208). This observation suggests that 2-nt loops can be formed in B-hairpins but not in Z-hairpins.

c. Pseudoknot

A *pseudoknot* is a double-hairpin structure with an extended quasi-continuous double-helical stem region (Fig. 1.15). It is formed when bases outside a hairpin structure pair with bases within the hairpin or internal loop (111,112). The pseudoknot is a potentially important tertiary structural motif of RNA and it has been identified in 16S rRNA, U2 snRNA, and some plant viral RNAs with tRNA-like structures (113). This type of compact folding is presumed to play a key role in ribosomal function, catalysis of RNase P ribozyme, RNA splicing, and/or recognition of tRNA-like structures. The pseudoknot structure is sensitive to salt concentrations, metal ions, and sequences as it can readily adopt alternative secondary structures such as a 5'- or 3'-hairpin (114).

2. MESSENGER RNA

Messenger RNAs, which constitute about 1–2% of the total cellular RNAs, have a protein-coding function, playing a central role in the transfer of genetic information from DNA to protein. Eukaryotic mRNAs are initially transcribed by RNA polymerase II as precursor mRNAs (pre-mRNAs) in the nucleus. The pre-mRNAs are processed to translatable templates before being transported across the nuclear membrane to the cytoplasm. The processing or maturation of eukaryotic mRNAs involves 5'-end capping, methylation, intron splicing, RNA editing, and polyadenylation. These posttranscriptional modifications are essential for mRNA function and stability. Prokaryotic messages are often *polycistronic* and contain several discrete coding regions in an mRNA unit. Their expression is controlled by the *operon* (see Section I,A, Chapter 7). In contrast to the prokaryotic mRNAs, mature eukaryotic mRNAs are mostly *monocistronic* and contain a single coding region with a general structure of 5'-Cap-(5')UTR-ORF-(3')UTR-poly(A)-3'.

a. 5'-Capping

All eukaryotic nascent mRNAs, except for some eukaryotic viral RNAs, contain 5'-end cap structures. The cap consists of an N-methylguanine nucleotide connected to the 5' end (nucleotide N) of the RNA by three phosphate groups, i.e., $m^7G(5')ppp(5')N^{(m)}$. The cap is added to the free 5' ends of mRNAs in the nucleus before the polymerase has transcribed more than ~20 nt. The biogenesis of the 5'-end cap involves a series of enzymatic reactions catalyzed by nucleotide phosphorylase (pppN → ppN + P_i), RNA guanylyltransferase (ppN + GTP → GpppN + PP_i), and RNA methyltransferase (GpppN + AdoMet → m^7GpppN + AdoHcy).

The 5'-end cap influences gene expression at several levels. Most importantly, the cap renders the native mRNA resistant to RNases. It also assists the ribosome in binding mRNA via a cap-binding translation initiation factor. The monomethylated G cap also serves as a signal for specifying nuclear export of RNA, while a trimethylated G cap ($m^{2,2,7}$GpppG) functions as one of the bipartite signals specifying the import of snRNP particles into the nucleus (115,116).

b. 3'-Adenylation

The majority of eukaryotic mRNAs and eukaryotic viral RNAs have a long (150–200 bases) poly(A) tail at the 3' end. The tail is attached posttranscriptionally in the nucleus by nuclear poly(A) polymerase (see Section II, Chapter 7). When mRNA is transported to cytoplasm, the long poly(A) tails are first recognized by a poly(A)-binding protein (PAB) and are subsequently shortened to a smaller, translation-competent size by a PAB-dependent poly(A) nuclease. The relationship between the initial poly(A) tail shortening which is slow and at least partially coupled to a translational capacity of the mature mRNA and the rapid shortening which is a part of mRNA degradation pathway is not clearly known. However, the presence and the length of the poly(A) tail affect the stability of most mRNAs, mRNA translational efficiency, and ultimately the level of gene expression or infectivity of viral RNAs. The circadian rhythm of vasopressin peptide levels has been related to the 240-nt difference in the poly(A) tail lengths between two species of vasopressin mRNA (117). A mRNA does not need a poly(A) tail to be translated. In fact some mRNAs, e.g., histone mRNAs, are not polyadenylated.

Polyadenylation of nascent mRNAs is signaled by a consensus sequence, *AAUAAA,* located ~20 nt upstream from the 3'-end polyadenylation site. Polyadenylation is an event coincident with transcription termination (118).

c. Splicing of introns

Eukaryotic genes (and some eukaryotic viral genes) are often discontinuous and composed of segments called *exons* (the RNA sequences carried into mature mRNA) and *introns* (the RNA sequences not carried into mature mRNA). In human genomes, the total length of exons is estimated to be only about 3% of the DNA. Unlike eukaryotic genes, prokaryotic genes do not contain introns except for a few notable cases, e.g., the coliphage T4 *td* gene encoding thymidylate synthase (119) and *B. subtilis* phage SP01 (120). [Note that some introns code for "intronic" proteins that have essential functions in self-splicing (maturase) and/or intron mobility (endonuclease) (121).]

The introns, also called *intervening sequences* (IVS), are cut out of the primary transcript (pre-mRNA), and the exons are spliced together as a single translation unit. The splice junctions of all eukaryotic protein-encoding genes have the global consensus sequences for exon–intron junctions of (..AG|gta..) and intron–exon junctions of (..ag|G..). In fact, the frequent occurrence in the protozoan *Tetrahymena* of the amber codons (UAG, used as a Gln codon) at both exon and intron termini invokes a tantalizing hypothesis that exons were once "microgenes," originally terminating with amber and encoding relatively short oligopeptides that assembled spontaneously into active proteins (209). The consensus splice–junction sequences do not occur in the nonprotein-encoding rRNA and tRNA genes which have quite different splicing mechanisms known as self-splicing (see Section II,D,4,b, this chapter).

The pre-mRNA splicing occurs by a two-step transesterification mechanism (Fig. 1.16). The first step is the cleavage at the 5' splice site and covalent joining of the 5'-terminal G of the intron to an internal A residue (at C-2'–OH) near the 3' splice site. The 2'–5' phosphodiester-linked intermediate structure is called a *lariat*. [The lariat is also produced by self-splicing group II introns.] The branching A residue is part of a conserved sequence called a *branch point sequence,* e.g., UACUAAC in yeast. The branch point sequence is normally located between 18 and 40 nt upstream from the 3' splice site. The second step involves cleavage at the 3' splice site and ligation of the two exons. The released intron is still in

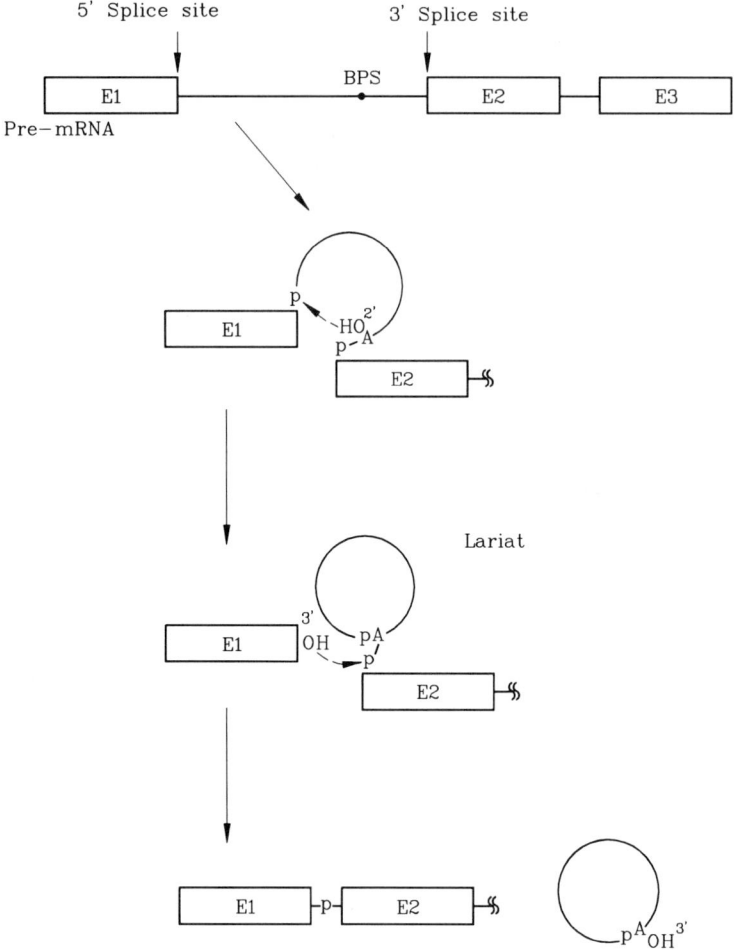

FIGURE 1.16 Schematic mechanism of nuclear mRNA splicing. The intron splicing of a pre-mRNA occurs in a spliceosome, a complex structure in which at least five snRNPs recognize, associate with the pre-mRNA, and assist in the correct 5' and 3' cleavages of the intron. The recognition of 5' splice sites occurs by the binding of U1 snRNP as a result of direct base pairing between nucleotides of the 5' splice site and the 5' end of U1 RNA. The recognition of 3' splice sites is more complex and involves the binding of initially U2 snRNP and subsequently U4, U5, and U6 RNAs. BPS, branch point sequence.

the form of a lariat. The introns appear to have a minimum length of 25–30 nt. In many genes, the combined length of introns greatly exceeds that of the exons.

The non-self-splicing reaction of pre-mRNAs takes place within the dynamic environment of spliceosome (122). The *spliceosome* is a multicomponent complex containing pre-mRNA, U-series snRNAs complexed with proteins (called snRNPs), and non-snRNP splicing factors. The major nucleoplasmic snRNAs (U1, U2, U4, U5, and U6) range in size from 106 nt (U6) to 189 nt (U2), have distinct stem–loop structures, and are essential for pre-mRNA splicing (123–125). The splicing components are present on the nascent RNA, suggesting that RNA splicing can take place during transcription and probably prior to polyadenylation.

The processing of pre-mRNA is not always a unique process but may involve the selection of different 5' or 3' splice sites. This regulated or alternative mode of splicing occurs in a variety of combinations, including alternative selections of translation initiation codons and/or polyadenylation sites (126). Alternative splicing is an important regulatory mechanism in tissue-specific gene expression as well as in generating multiple gene products from a single gene. Developmentally regulated alternative splicing of a transcription factor gene (the chorion transcription factor CF2 of *Drosophila*) has been shown to result in at least three zinc-finger forms which bind to distinct promoters and DNA target sequences (210).

What regulates the choice between alternative 5' or 3' splice sites? One important modulatory factor is the secondary structure of RNA. The alternative splicing of chicken β-tropomyosin pre-mRNA has been shown to depend on whether stable double-stranded regions are formed between parts of an intron and the neighboring exon (127,128). The formation of a double-stranded structure represses the splicing of the exon involved by inhibiting the formation of spliceosomes, whereas disruption of this structure leads to derepression and splicing of the exon. The driving force for the alternative mRNA secondary structures is presumed to be the cell-specific differences in the concentration or activity of trans-acting non-snRNA splicing factors (129).

d. mRNA editing

The term *mRNA editing* is used to define the various processes by which mature, translatable mRNA is generated posttranscriptionally from defective or *cryptogenes*. Whereas RNA splicing removes noncoding regions of RNA, RNA editing alters the nucleotide sequence of the coding region to provide initiator codon (AUG), termination codons, and/or extension of ORF by correcting internal frameshifts. The RNA editing may thus modulate mRNA stability and provide a means of regulating gene expression by dictating when and which proteins are to be made.

Two types of mRNA editing are *insertional* and *substitutional editing*. Insertional editing includes (a) insertion or deletion of one or more U residues [typically in mitochondrial mRNAs such as cytochrome oxidase subunits II (*coxII*) and III of trypanosomatids (130–132)], (b) insertion of nontemplated C residues [e.g., single C insertion at 54 sites within 528 codons of mitochondrial mRNA for ATP synthetase α subunit of a slime mold, *Physarum polycephalum* (133)], and (c) insertion of one or more nontemplated G's [in paramyxovirus P mRNAs (134–136)]. In *substitutional editing*, templated C is replaced by U (or vice versa) in the mitochondrial mRNAs [e.g., cytochrome *c* oxidase subunit 2 (*cox2*)] of plants (137–139), chloroplast mRNA (140), and apolipoprotein B mRNA of mammalian cells (141,142).

How does the base (U or C) insertion or deletion occur? A probable sequence of events is as follows: (a) the annealing to improper template sequences of a

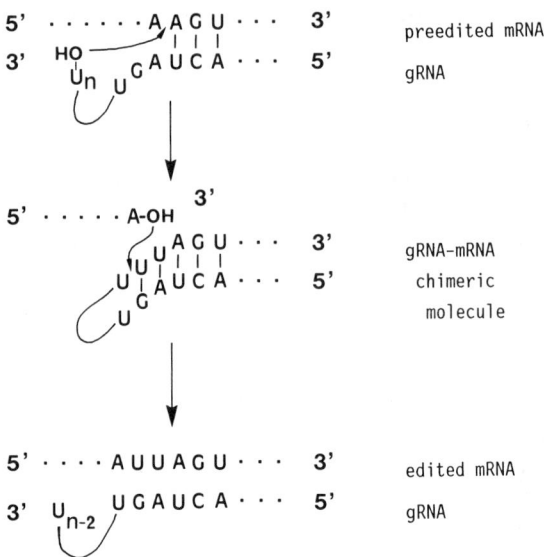

FIGURE 1.17 A RNA editing mechanism involving transesterification reactions under the direction of a guide RNA.

short (60–80 bases) guide RNA (gRNA) which contains a nonencoded 3' oligo(U) tail of 5–20 nt, (b) endonucleolytic cleavage of the preedited mRNA, (c) addition or deletion of a single or stretch of U residues, (d) religation, and (e) the next cycle of editing while the whole editing machinery (*editosome*) moves in the 3' → 5' direction (143–145). The extent and mode of editing may differ considerably among organisms.

The crucial feature of the mechanism is that the addition (or deletion) of U (or C) residues requires a nucleolytic activity which can be either an endonuclease or terminal uridylyltransferase (TUTase, EC 2.7.7.12). The inserted U residues could be derived directly from UTP which is added to the 3'-OH of the cleaved mRNA by the mitochondrial TUTase (143). Alternatively, the cleavage and insertion of the U residue(s) could be a concomitant event involving a transesterification reaction(s) analogous to that operating in RNA self-splicing and/or reverse splicing (146). Chimeric gRNA–mRNA molecules with oligo(U) tails covalently linked at sites of RNA editing provide evidence for the formation of covalent intermediates expected from the transesterification mechanism both *in vivo* (147) and *in vitro* (211,212). At least for the RNA editing in the mitochondria of kinetoplastid protozoa, it appears that successive transesterifications result in the transfer of U residues from the gRNA 3' oligo(U) tail into an editing site (Fig. 1.17). The number of U residues transferred are determined by the base pairing of U residue(s) with A and G "guide" nucleotides in the gRNA. The preedited mRNA may contain a single *editing block* (the editing mediated by a single gRNA) or an *editing domain* (a region involving multiple, overlapping gRNAs). The polarity of 3'-to-5' editing is attributed to the formation of anchor sequences for upstream gRNAs by downstream editing (213). The gRNAs are transcribed from the same mitochondrial DNA (maxicircles and minicircles) that contains the cryptogene(s). Considering the system as a whole, mRNA editing is not an independent process, but a process controlled by the instructions encoded in the genomic DNA.

The insertion of nontemplated G residues, which always takes place following a stretch of guanosines, is most likely a cotranscriptional event resulting from RNA polymerase "stuttering." The substitutional editing of C-to-U changes does not involve any catalytic or guide RNA component (148). The editing is apparently carried out by a site-specific cytidine deaminase which recognizes some structural features of the preedited mRNA. The C-to-U editing proceeds without a strong directional bias (149).

e. Functional roles of mRNA secondary structure

The secondary structure of mRNA, especially in the 5'UTR (or leader region), is an important determinant of the efficiency of translation initiation. For most of the known structural genes in prokaryotes, the initiator codon AUG is found either within or immediately downstream of a potential stem–loop structure. Well-translated RNAs have a relatively unstructured translation initiation region, whereas poorly translated RNAs contain a relatively high degree of stem–loop structures. For *lacZ* mRNAs in which various site-specific mutations have been introduced, the efficiency of translation varies up to 160-fold depending on the secondary structures within the initiation region (150). The translation efficiency reflects the binding affinity of 30S ribosomal subunit and is linearly proportional to the fraction of unfolded molecule. The efficiency of prokaryotic (and phage) RBS can be reduced by 10-fold when the stability of its secondary structure is increased by 2.3 kcal/mol (151). In certain cases in which the 5' cap is absent, the stem–loop structure of the 5'UTR can provide a critical shield for nascent RNAs from degradation, directly affecting the level of translation and gene expression (152).

Secondary structures of eukaryotic mRNAs also critically influence the initiation and termination of protein synthesis (153). For instance, a strong stem–loop structure ($\Delta G = -30$ kcal/mol) prevents a mRNA from engaging the 40S ribosomal subunit when the secondary structure occurs at 12 nt from the cap. Another stem–loop structure with a ΔG of -61 kcal/mol results in ribosome "stalling," probably because the structure is too stable to be unwound by the 40S subunit.

Certain eukaryotic RNA viruses (e.g., picornaviruses) contain a single-stranded genomic RNA (~7.5 kb) that functions as a polycistronic mRNA. The 5' ends of the picornaviral RNAs are not "capped," but are covalently linked to a small viral protein, VPg. In contrast to the average of 55-nt-long 5'UTR in yeast and higher eukaryotic mRNAs, the 5'UTRs of picornaviruses are over 700 bases long and contain extensive stem–loop structures. The particular 5'UTR structures of picornaviruses by no means fit the widely accepted *scanning mechanism* of translation, in which a ribosome binds first at the 5′-end cap and slides along the 5′UTR until it locates an initiator codon in a favorable context (see Section I,A,2, this chapter). How do picornaviruses have their proteins efficiently translated? In these RNAs, ribosomes are believed to enter the translation initiation site guided by certain specific secondary structural features and to start translation: a mechanism known as *direct (or internal) entry* of ribosome (154).

The secondary structures in the 5′UTR of mRNA may also be used for selective binding of regulatory proteins. The binding alters the mRNA structure and affects translational efficiency or simply blocks translation initiation, thus providing another mechanism for the translational regulation of gene expression (155).

3. RIBOSOMAL RNA (rRNA)

Ribosomal RNAs constitute 80–90% of total cellular RNAs and are the essential components (50–60%) of ribosome structure. Ribosomes (70S in prokaryotes, 80S in eukaryotes) provide the platform for translation of the genetic code and the link between genotype and phenotype. Ribosomal RNAs have the most extensive secondary structure of all RNAs and, by cooperative interactions with associated proteins, fold into complex tertiary structures within the ribosome (156,157). In contrast to the prokaryotic and eukaryotic rRNAs, protozoan rRNAs are derived by self-splicing (see Section II,D,4,b, this chapter).

a. Prokaryotic rRNAs

Prokaryotic ribosomes contain three rRNA molecules: 5S (~120 nt), 16S (~1.5 kb), and 23S (~2.9 kb). Together with about 21 different proteins, the 16S rRNA is a part of the small 30S ribosomal subunit which is the site of codon–anticodon interaction. The 3′ terminus of 16S rRNA is known to interact with the initiation region of mRNA via the Shine–Dalgarno sequence. Together with about 36 proteins, the 5S and 23S rRNAs constitute the large 50S ribosomal subunit that is thought to be involved in the regulation of translation accuracy. The 23S rRNA has been proposed to bind the 3′ terminus (—CCA) of tRNA in the ribosomal exit site (E-site) and to promote actively translocation of tRNA from the P-site (158). Indeed, there is evidence that the peptidyltransferase activity resides in the 23S rRNA which contains, in its presumed catalytic center, the sequences that are among the most highly conserved in biology (214). The catalytic role of 23S rRNA in protein synthesis is further supported by the demonstration that a *Tetrahymena* ribozyme can be engineered to express a "modest" aminoacyl esterase activity catalyzing the hydrolysis of N-formyl-Met from the substrate CAACCA-fMet (215).

All three species of *E. coli* rRNAs are derived from a single transcript which contains the rRNAs in the order of 16S–23S–5S. The transcript also contains one or more intermittent tRNAs. In *E. coli,* there are seven such rRNA transcription units dispersed throughout the genome and arranged in *rrn* operons. The processing of the spacer RNAs between each rRNA and tRNA is carried out by a series of RNases.

b. Eukaryotic rRNAs

Eukaryotic rRNAs include 18S (1.8 kb), 5.8S, and 28S (4.7 kb) molecules. They are derived from the nucleolytic processing of a single precursor molecule (pre-rRNA) which is transcribed by RNA polymerase I. The 5.8S RNA remains attached to the 28S RNA by H-bonds. In humans, the pre-rRNA is ~13 kb (45S) long and has a structure in which three exons are separated by both internal and external RNA spacers (total length of 31 kb). The fourth component of rRNA is 5S RNA which is transcribed by RNA polymerase III. The 5S rRNA gene is not normally linked to that of other rRNAs. Together with ribosomal proteins, the 18S rRNA constitutes the small 40S ribosomal subunit; 5S, 5.8S, and 28S rRNAs make up the large 60S subunit.

4. CATALYTIC RNA: RIBOZYMES

Ribozymes are RNA molecules with the ability to catalyze phosphodiester cleavage and/or ligation reactions on RNA substrates. Most naturally occurring ribozyme activities are intramolecular and have roles involved with RNA processing such as intron removal by self-splicing and cleavage of concatameric transcripts from replicating RNA viruses. The self-splicing introns fall into two groups: *group I* introns comprise, among others, nuclear rRNA genes of *Tetrahymena* and the majority of introns in fungal mitochondrial DNAs (mtDNAs), while *group II* introns comprise plant and fungal mtDNAs and the majority of introns in chloroplasts. The splicing mechanisms of group I and group II introns both involve two sequential transesterification reactions. However, for group I introns, splicing is initiated by the addition of a guanosine to the 5′ end of the intron (see details below), whereas the splicing of group II introns is initiated by formation of an intron lariat similar to that involved in splicing of nuclear mRNA introns. Although group I and group II introns catalyze their own splicing, for efficient splicing *in vivo*, they depend on protein factors such as "maturases" encoded by the intron themselves and other host proteins (e.g., aminoacyl-tRNA synthetases).

The interest pertinent to the application of ribozyme technology lies in the fact that, under appropriate *in vitro* conditions, the cleavage reactions of self-splicing introns conform to true enzymatic catalysis exhibiting successive cycles of substrate annealing, cleavage, and dissociation. Ribozymes fall into several structurally and mechanistically distinct classes. The following descriptions focus on three relatively well-known classes of ribozymes: (a) a hammerhead ribozyme, (b) the self-splicing intron of *Tetrahymena* (a ciliate protozoan), and (c) the ribonucleoprotein RNase P.

a. Hammerhead ribozyme

i. Activity and structure. The processing and maturation of several plant viroid and virusoid RNAs involve a self-cleavage reaction within a structural domain consisting of three helices and 13 phylogenetically conserved nucleotides known as a *hammerhead* (216). The *in vivo* reaction takes place in a unimolecular complex whose basic structure contains the following elements (Fig. 1.18): (i) three double helices (I, II, and III) which are held together by 11 nonhelical

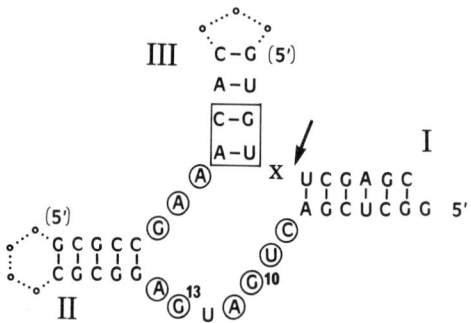

FIGURE 1.18 A consensus hammerhead structure. Three base-paired stems are labeled I, II, and III. Conserved sequences are either boxed (in helix) or circled (in nonhelical regions). The nucleotide numbering is based on a 19-mer hammerhead complex, but the whole structure can be a single molecule with helices II and III covalently closed via respective loops (broken lines). The cleavage or self-cleavage site is indicated by arrow, and X is any base except guanosine.

nucleotides with any base sequence; (ii) the cleavage site, located between helices I and III, has a sequence UXpN, and the cleavage occurs on the 3′ side of X which is preferably C but not G; and (iii) helix II connects two stretches of highly conserved nucleotides of sequence 5′-CUGANGA-3′ and 5′-GAAA.

The structural features can be separated *in vitro* into two (or even three) parts, i.e., endonuclease and substrate, as long as the conserved nucleotide residues and the hammerhead structure are maintained by the complex (159,217). The ribozyme component can be as small as 19 nt and have a strong catalytic activity (K_m = 0.6–0.9 μM, k_{cat} = 0.5–0.9 min^{-1}). Indeed, it can be further reduced in size by replacing the helix II with four or five nucleotides that cannot form Watson–Crick base pairs between themselves. Remarkably, the nucleotides replacing helix II and those forming the substrate-binding arms of helices I and III can be substituted with DNA and retain the full activity (218). In contrast, nine of the conserved nucleotides that are located in the nonhelical regions are critical for nucleolytic activity. In particular, the 2′-OH groups of the guanosines in the conserved stretch, 5′-CU$\underline{G}^{10}$AN$\underline{G}^{13}$A-3′ (numbering in a 19-mer ribozyme), cannot be replaced with hydrogen atoms (i.e., 2′-deoxyguanosine) without resulting in a drastic decrease (at least 150-fold) in catalytic efficiency (219,220). Deletion of the 2-NH$_2$ at G^{10} (replacement with inosine) also results in a similar reduction in catalytic efficiency. The phosphorothioate substitution of the bond between G^{13}–A^{14} results in a strong reduction of catalytic activity, which can be stimulated by Mn^{2+} (221,222).

The implication of the hammerhead structures is that, because the base sequences of helices I and III are not conserved, a hammerhead ribozyme can be constructed from completely unrelated RNAs and perform rapid, site-specific cleavage reactions *in trans* on virtually any target sequence. This versatility, combined with the structural simplicity and portability, makes the hammerhead ribozyme an increasingly important player whose potential remains to be fully explored in recombinant DNA technology (see Fig. 7.7 in Chapter 7).

ii. Catalytic mechanism. Hammerhead ribozymes require a divalent metal ion (preferably Mg^{2+}) as a cofactor. In the presence of Mg^{2+}, or Mn^{2+} in some cases, highly specific cleavages occur at the target phosphodiester bond. The cleavage reaction, which is optimal at a neutral or higher pH, proceeds as a transesterification of the 2'-OH, resulting in a free 5'-OH on one RNA fragment and a 2',3'-cyclic phosphodiester on the other. As in the catalysis of pancreatic RNase A (see Section III,A in Chapter 3), the stereochemistry of hydrolysis catalyzed by the hammerhead ribozyme involves an in-line mechanism with inversion of configuration at the phosphorus atom (223,224). Evidence suggests that the Mg^{2+} cofactor is bound to the pro-*R* oxygen in the active ternary complex (224,225).

b. Tetrahymena ribozyme

i. Activity and tertiary structure. In *Tetrahymena*, nuclear pre-rRNAs mature through autocatalytic cleavages (or self-splicing) of the introns. The IVS of *Tetrahymena thermophila* is a self-splicing group I intron consisting of 413 nt. Its secondary structure comprises a conserved catalytic core (about one-third of the molecule) and surrounding stem–loop elements that offer structural support but are not essential for catalytic activity. The catalytic activity requires a unique, high ordered folding much like that of proteinaceous enzymes (160,161). A stereochemical model based on the sequence alignment of 87 group I self-splicing introns suggests a 3-D structure which is extremely compact with all of the most evolutionarily conserved residues converging around the two helices that constitute the substrate of the core ribozyme and the binding site for guanosine (226). The folding of ribozyme is unique in the sense that, unlike protein folding, it involves the juxtaposition of negatively charged phosphodiester groups and thus requires neutralizing cations. In fact, cations not only play a structural role but also affect the specificity of ribozymes. For instance, the high ordered RNA foldings in the *Tetrahymena* ribozyme can be similarly produced by Mg(II), Ca(II), and Sr(II), but only the Mg(II)-stabilized RNA is catalytically active (162). Mn(II) can also fully activate the ribozyme in strand scission reactions under physiological conditions (163). The catalytic domain of the ribozyme is believed to contain some nonclassical H-bonded base triplets (164).

ii. Catalytic properties: transesterification. The active site (self-cleaving domain or catalytic core) of the intact (413 nt) or some shortened forms of *Tetrahymena* ribozyme contains an *internal guide sequence* (IGS, 5'-GGAGGG-3') at the 5' end of the RNA and a highly specific binding site for the substrate guanosine. (IGS is a feature commonly found in group I introns such as phage T4 *td* and *sun*Y.) The G-binding site consists of the G264-C311 bp with A207 in close proximity. Note that A207 is far from the G-binding site in the secondary structure.

96 Chapter 1 Enzymes and Nucleic Acids

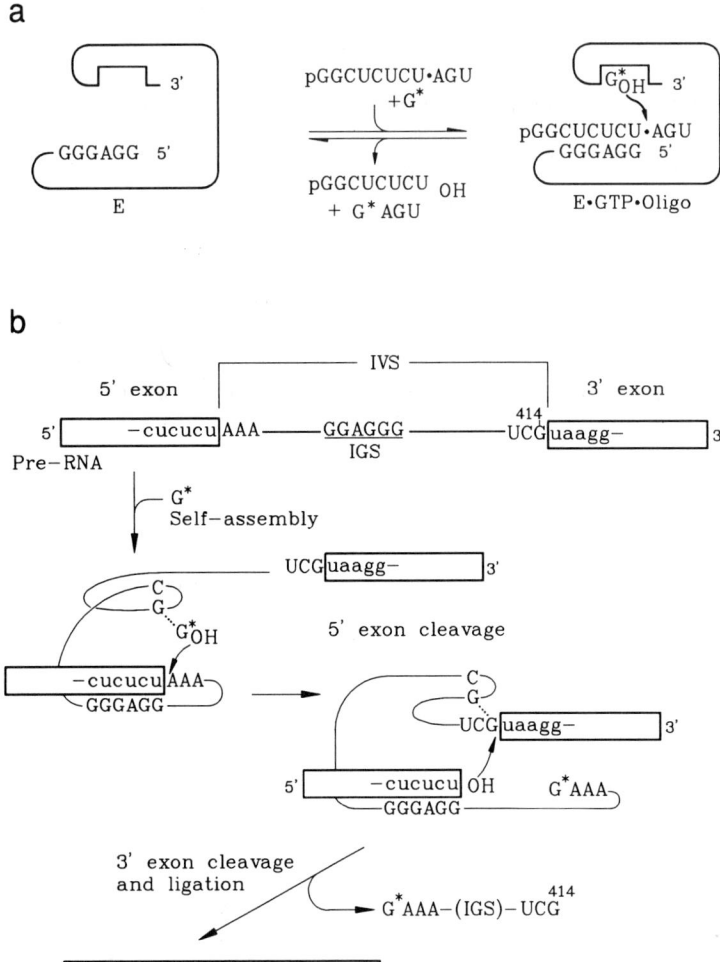

FIGURE 1.19 Putative base-pairing interactions during the splicing reactions catalyzed by *Tetrahymena* ribozyme. G* is the guanosine nucleotide added to the ribozyme to initiate the normal splicing reaction. (a) Cleavage of a synthetic oligonucleotide substrate (the dot denotes the scissile bond), and (b) sequential steps of IVS self-splicing. The nucleophilic attack on the phosphorus by the 3'-hydroxyl of guanosine occurs in-line with the labile phosphorus–oxygen bond.

With an RNA substrate (e.g., GGCUCUCU · AGU) that bears complementarity to the IGS and thus forms a double helix, the ribozyme catalyzes a guanosine-dependent endonuclease reaction to produce GGCUCUCU and G · AGU (Fig. 1.19a). The k_{cat} and K_m of the reaction with GGCCCUCU · AGU are 0.13 min^{-1} and 0.07 μM, respectively (165). The chemical cleavage (phosphodiester transfer) occurs with an inversion of configuration at phosphorus, suggesting that the nucleophilic attack by the 3'-OH of guanosine occurs in-line with the labile phosphorus–oxygen bond (165,166). The chemical step is estimated to be very fast (~350 min^{-1}) relative to the rate-limiting substrate binding or release of the cleaved oligonucleotide products. This *in vitro* reaction is analogous to the first step in the self-splicing of the IVS, i.e., the 5' exon cleavage initiated by transesterification of guanosine (from GTP, GDP, GMP, and guanosine) to the 5' end of the intron (Fig. 1.19b). In self-splicing reactions, the 3'-terminal CU(OH) of the 5' exon engages in a reverse transesterification at the 3' splice site (intron-G|pU-exon) concluding the exon ligation and removal of the intron.

RNA substrates and cleavage products are associated with the ribozyme with a binding energy greater than that of base pairing alone by ~3 kcal/mol (at 42°C). This is mostly attributed to an additional H-bonding between the N-1 of a particular base (A302) in the catalytic core and the 2'-OH group of the third nucleotide (U) from the cleavage site of the substrate (167). The tertiary interaction apparently contributes to the packing of the RNA helix into a folded structure, thereby bringing the scissile bond to a conformationally restricted position close to the guanosine nucleophile.

iii. Extended catalytic properties. The *Tetrahymena* ribozyme is capable of catalyzing limited (up to ~15 nt) polymerization of ribonucleotides onto a short primer annealed to a sequence within the intron (168). The ribozyme can also catalyze template-directed oligonucleotide ligation reactions (169–171). The template-directed primer extension involves nucleophilic attack by the 3'-OH group of an oligonucleotide primer on the phosphate of a GpN (where N is A, C, or U) dinucleotide, releasing a guanosine nucleoside (Scheme 1.12).

$$E \cdot \text{Primer} + \text{GpN} \rightleftharpoons E \cdot \text{Primer} \cdot \text{GpN} \rightarrow E \cdot \text{Primer-pN} + G$$
SCHEME 1.12

Each addition of mononucleotides is analogous to the second step of self-splicing, in which the primer represents the 5' exon and is extended by a single nucleotide 3' exon.

The *Tetrahymena* ribozyme can catalyze a sequence-specific cleavage of ssDNA under conditions of high temperature (50°C) or high MgCl$_2$ concentration (50 mM) or both (172,173). The ssDNA is an extremely poor substrate with a 10^4 times greater K_m and a 10 times lower k_{cat} compared with the analogous RNA. Under physiologic conditions (37°C, 10 mM MgCl$_2$), the DNA cleavage reaction is undetectable ($k_{cat}/K_m = 36\ M^{-1}\ \text{min}^{-1}$). This observation reveals the importance of the 2'-OH group of the substrate in binding and in catalysis of the ribozyme. Nevertheless, the ribozyme can be forced to directionally evolve *in vitro*, via application of repeated cycles of mutation selection and amplification steps, to exhibit a 100-fold improved DNA cleavage activity (227).

Particularly noteworthy is the Mg^{2+}-dependent, sequence-specific aminoacyl esterase activity demonstrated with a *Tetrahymena* ribozyme which was engineered at the IGS (GGAGGG to GGGUUG) to recognize the aminoacylated oligonucleotide CAACCA derived from the 3' end of N-formylmethionyl-tRNA^{fMet} (215). The catalytic rate enhancement of the N-formyl-Met hydrolysis was a mere fivefold, which is 6–7 orders of magnitude smaller than RNA-catalyzed reactions with natural substrates. Nevertheless, the ribozyme-mediated carboxylester hydrolysis suggests the catalytic versatility of RNA beyond the making and breaking of phosphodiester bonds, which is further supported by the finding that a peptidyltransferase activity can be assigned to 23S rRNA (214).

c. Ribonuclease P

i. Activity and molecular structure. Another well-known ribozyme is RNase P, a ubiquitous endonuclease that specifically cleaves pre-tRNAs *in vivo* to produce the mature 5′ terminus of tRNA molecules (174,175). RNase P also cleaves many tRNA-like structures *in vitro* (113). In *E. coli* (and *B. subtilis*), RNase P is a ribonucleoprotein composed of one catalytic RNA subunit (M1 RNA) and one protein cofactor (C5 protein). The C5 protein is a small, basic protein with 119 amino acids (13.8 kDa). The M1 RNA consists of 377 nt (125 kDa) with an extensive secondary structure. It contains two pseudoknots which strongly constrain the 3-D structure of the RNA near the active site (176).

ii. Catalytic properties. Under high ionic conditions, e.g., 0.3–1 M NH$_4$Cl or KCl and 20–200 mM MgCl$_2$, RNase P RNA can cleave tRNA precursors in the absence of the protein cofactor (177). The RNase P RNA thus contains features for substrate (pre-tRNA) binding and catalysis, as well as for interaction with RNase P protein. The main function of the RNase P protein is apparently the stabilization of the complex (pre-tRNA + RNase P RNA) by screening the electrostatic repulsions between the two RNAs. Depending on the substrate (e.g., pre-4.5S rRNA), however, the protein subunit plays a crucial role in the recognition of the substrate and contributes to the RNA cleavage reaction in both the K_m (2- to 85-fold increase) and the k_{cat} (1- to 30-fold increase) (178). Whereas the k_{cat} of RNase P RNA-alone reaction is nearly constant throughout the functional ionic strength, the holoenzyme and the apoenzyme (RNA-alone) have, at their respective ionic optima, significantly different k_{cat} values, i.e., 10 and 0.5 min^{-1}, respectively. The difference in the rate is attributed not to substrate cleavage but to product release, which is apparently facilitated by the protein cofactor.

iii. Active site. A comparison of several eubacterial RNase P RNAs has revealed some characteristic secondary structures, in which the degree of conservation correlates with the phylogenetic relationship of the organisms. No single residue in the RNase P RNA is absolutely essential for activity, and the nucleotides essential for forming the structure of active RNase P are distributed throughout the length of the RNA. Nevertheless, a 263-nt fragment consisting only of evolutionarily conserved features can fully replace natural RNase P (of 354–417 nt) (179).

Unlike the *Tetrahymena* self-splicing intron, RNase P is believed to recognize its substrates via highly conserved structural motifs. With pre-tRNA molecules, the primary site of interaction is believed to be the 3′ terminal (but not internal)

CCA sequence (or CCC in pre-4.5S rRNA) (180,181). Two other possible regions are the acceptor stem near the site of cleavage and the purine nucleotides of the T-stem/loop. The affinity of the enzyme and substrate RNAs is most likely determined by nonionic contacts, presumably through the H-bonds that do not involve the base complementarity of the canonical Watson–Crick type.

iv. *Noneubacterial RNases P.* In contrast to the eubacterial RNase P, eukaryotic (yeast, mammalian) RNase P activities are rather poorly defined. The common feature among the nuclear RNases P is that the activity resides in a large cylindrical complex (~15S) containing as the major component at least 14 polypeptides and an RNA as the minor component. In some eukaryotic species (e.g., yeast) but not others (e.g., *Xenopus* oocytes), the minor RNA component has been shown to be essential for RNase P activity, although its catalytic role has not yet been proven (182,183).

5. Transfer RNA (tRNA)

Transfer RNAs function as specific adapters for amino acids in the biosynthesis of proteins. The size of tRNAs is almost uniform, about 80 nt, which is fairly small compared with mRNAs or rRNAs. The tRNAs carry an extensive array of modified bases (~10%), featuring deamination, thioketo substitution, ring N-alkylation, and saturation of the pyrimidine C-5–C-6 double bonds. Transfer RNAs have been found to contain 43 different rare nucleosides (184). Transfer RNAs adopt characteristic secondary and L-shaped tertiary structures suitable for amino acid carrying and codon recognition functions. This bifunctional role of tRNAs suggests that tRNAs interact with two major partners: aminoacyl-tRNA synthetases and the protein synthesis compartment of ribosomes. The tRNAs composed of entirely unmodified nucleosides, as can be obtained by *in vitro* transcriptions, still function as substrates for aminoacyl-tRNA synthetases. This suggests that the structural features of tRNA recognized by the enzyme are not critically dependent on the presence of modified and/or rare nucleosides (185). In fact, the structure of the unmodified yeast tRNA^{Phe} molecule is presumed to be similar to the native tRNA^{Phe} (186). The natural initiator codon is predominantly AUG, which in bacteria calls for a near-universal role of initiation by N-formylmethionyl-tRNA (tRNA^{fMet}). The tRNA$_i^{Met}$ is a part of the *scanning ribosome* (40S) in eukaryotes and it mediates ribosomal recognition of the eukaryotic initiator region (187).

a. Pre-tRNAs

Mature tRNAs are derived from precursors (pre-tRNAs) by posttranscriptional processing, which consists of extensive base modifications, base transpositions, and the removal of both 5′ leader and 3′ trailer sequences.

Transfer RNAs are transcribed as a cluster in a precursor form. The pre-tRNA usually contains rRNAs as well. All (eukaryotic and prokaryotic) pre-tRNA transcripts are processed to mature 5′ termini by RNase P. [Note: Mito-

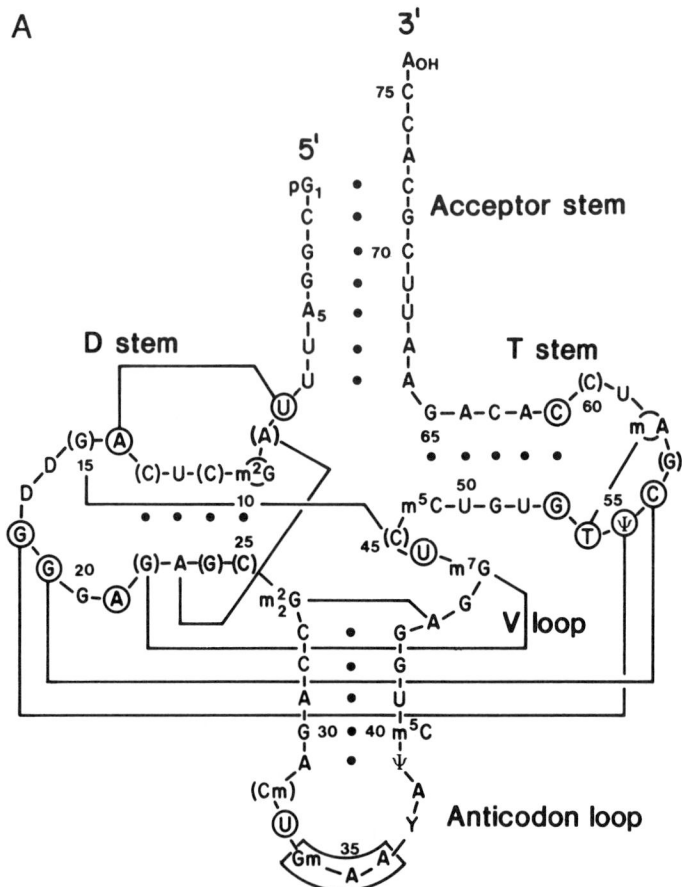

FIGURE 1.20 Structures of tRNA. Cloverleaf model of secondary structure and 3-D structure of yeast tRNA$^{Phe.}$ (A) In the cloverleaf model, the bases that form base–base tertiary H-bonds are connected by lines. Circles and parentheses indicate conserved and semiconserved bases, respectively. (B) In this schematic representation of a 3-D structure, the ribose–phosphate backbone is shown as a continuous tube. Base pairs are presented by long connected bars and single bases by short bars. Tertiary base–base H-bonds are indicated by filled bars between the two bases. Adapted from S. R. Holbrook, J. L. Sussman, R. W. Warrant, and S. H. Kim (1978). *JMB* **123**, 631, with permission.

chondrial RNA is processed mostly by a mitochondrial RNA processing enzyme, RNase MRP, and also by RNase P.]

b. *2-D and 3-D structures*

The 2-D structure of tRNA, which is generally described by a cloverleaf model, consists of five elements: a CCA acceptor stem, a D (dihydrouridine)-loop, an anticodon loop, a variable loop, and a T (ribothymidine)-loop. Tertiary interactions among the secondary structural motifs contribute to the folding that gives rise to the typical L-shaped molecule (Fig. 1.20). X-ray crystallographic structures

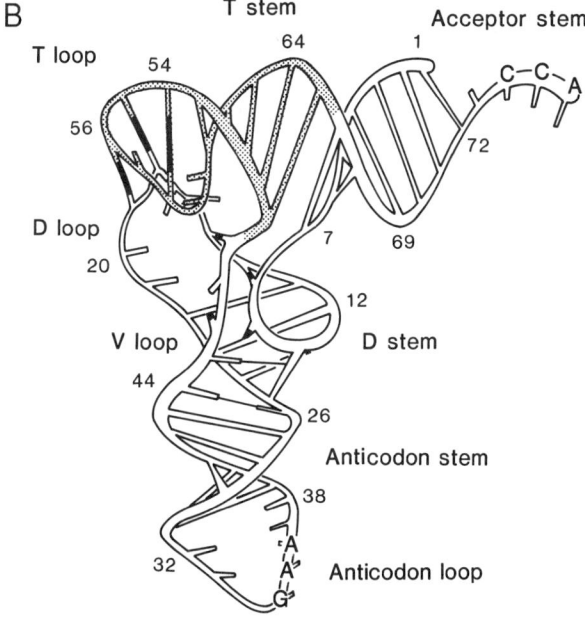

FIGURE 1.20 *Continued.*

have been determined for several tRNA molecules including yeast tRNA^{Phe} (188,189), yeast tRNA^{Asp} (190), yeast initiator tRNA^{Met} (191), *E. coli* initiator tRNA^{Met} (192), and *E. coli* tRNA^{Gln} (193).

Among the 76 bases in the tRNA^{Phe} molecule, only 4 bases (D16, D17, G20, and U47) are not involved in base stacking. Some of the bases are invariant in all tRNA sequences, and they are mainly involved in the tertiary interactions. The yeast tRNA^{Phe} cloverleaf contains 20 bp consisting of 52 H-bonds. The tertiary interactions contribute at least 40 additional H-bonds to the structure. All the tertiary H-bonds are of the non-Watson–Crick type and are highly twisted and bent. Due to extensive base pairing and stacking in tRNA, all the bases except the 5′ terminal CCA and the three anticodon bases are inaccessible to solvent.

All *E. coli* tRNAs have a genetically encoded 3′ CCA terminus. However, some tRNAs in gram-positive bacteria, e.g., *B. subtilis*, do not have the transcribed CCA terminus (194), and none of the eukaryotic tRNA genes encode the CCA terminus. In those tRNAs, the CCA terminus is added posttranscriptionally by tRNA nucleotidyltransferase. The 3′ CCA terminus of the acylated tRNA specifically interacts with a characteristic set of ~6 bases in 23S rRNA in the 50S ribosomal P-site (195).

c. Interaction of tRNA with aminoacyl-tRNA synthetase

Once an amino acid is charged onto the 3′-terminal A residue (either 2′- or 3′-OH) of a tRNA, the genetic relationship between a codon and an amino acid is fixed. It is the aminoacyl-tRNA synthetase (aa-RS) that catalyzes this reaction, whose specificity is the cornerstone for faithful transmission of genetic information. Aminoacyl-tRNA synthetases are a family of enzymes that come in various sizes and oligomeric structures. The synthetases fall into two distinct classes (I and II) based on an 11-aa element, the *signature sequence,* that is known to occur only in synthetases. The synthetases use ATP as their energy source for the aminoacylation reaction (Scheme 1.13).

$$tRNA(CCA)\text{-}3'\text{---}OH + H_3N^+\text{---}CH(R)\text{---}COO^- + ATP$$
$$\rightleftharpoons tRNA(CCA)\text{---}O\text{---}CO\text{---}CH(R)\text{---}NH_2 + AMP + PP_i$$

SCHEME 1.13

About 20 aminoacyl-tRNA synthetases discriminate their cognate tRNAs from other extremely similar tRNAs. What is the structural basis of this remarkable substrate specificity?

Depending on the tRNA species, the interaction of tRNAs with their cognate aminoacyl-tRNA synthetases involves varying degrees of contacts over the surface of the two macromolecules (196). The structural elements of tRNA responsible for defining the amino acid acceptor specificity are called *identity elements* or *discriminator bases.* Among the major identity elements are the amino acid acceptor stem and the anticodon.

The *E. coli* Gln-RS is a monomeric enzyme (553 amino acids, 63.5 kDa). It is an elongated molecule with an axial ratio of >3. The synthetase interacts with the tRNA^{Gln} molecule (~50 Å length) from the anticodon to the acceptor stem along the entire inside of its L-shaped structure (193). For tRNA^{Gln}, the identity elements include U35 in the anticodon (C—U^{35}—G) and G73 in the acceptor stem. The interaction of yeast tRNA^{Asp} with its cognate Asp-RS [an α_2 dimer binding two tRNAs, 557 amino acids (63 kDa) per monomer] occurs similarly through two major contacts: the single-stranded acceptor stem (in particular, G73) which assumes a regular helical conformation and the anticodon (G^{34}—U—C) loop that undergoes a large conformational change (197,198). For tRNA^{Phe}, the bases at positions A73, G34—A—A anticodon, and G20 in the D-loop have been identified as the discriminator bases for Phe-RS (186).

The ribose backbone of a tRNA is not an absolute requirement for the interaction between tRNAs and aminoacyl-tRNA synthetases. The corresponding tDNA (e.g., for *E. coli* tRNA^{Phe}) carrying the 3'-terminal rA can be specifically aminoacylated, albeit at a two- to threefold slower rate (202).

In contrast to the group of tRNAs mentioned earlier, tRNA^{fMet} is recognized by the fMet-RS uniquely through interactions with the anticodon. The alteration of Met anticodon (CAU) in tRNA^{fMet} to GAU (Ile), GAC (Val), or GAA (Phe) leads, albeit >10 times less efficiently, to the corresponding aminoacylation of the tRNA and insertion of Ile, Val, or Phe into the protein from complementary initiation codons (199,200).

For tRNA^{Ala} and tRNA^{His}, the anticodon and other parts of the tRNAs are dispensable for aminoacylation, while the acceptor stem sequences with the 7-bp helix containing G3·U70 (in tRNA^{Ala}) serve as the only sufficient discriminator (201). The conserved G3·U70 base pair in tRNA^{Ala} is a major determinant in prokaryotes as well as in eukaryotes. For this group of tRNAs, which also include tRNA^{Gly} (228) and yeast tRNA^{Val} (229), the full-size tRNA can be efficiently replaced by mini- or microhelices mimicking only the acceptor stem as far as the discriminator function is concerned. Nevertheless, the aminoacylation efficiency of the synthetase for the minihelices is over 4 orders of magnitude smaller than that for the corresponding full-size tRNAs. In tRNA^{Ala}, the exocyclic 2-NH$_2$ of G3 provides a critical contact with 2'—OH groups in the minor groove of the acceptor stem, making a highly specific, thermodynamically significant contribution to RNA recognition by the synthetase.

References

1. Holbrook, S. R., Cheong, C., Tinoco, I., Jr., and Kim, S.-H. (1991). *Nature* **353**, 579–581.
2. Pattabiraman, M. (1986). *Biopolymers* **25**, 1603–1606.
3. van de Sande, J. H., Ramsing, N. B., Germann, M., Elhorst, W., Kalisch, B. W., Kitzing, E. V., Pon, R. T., Clegg, R. C., and Jovin, T. M. (1988). *Science* **241**, 551–557.
4. Pilch, D. S., Brousseau, R., and Shafer, R. H. (1990). *NAR* **18**, 5743–5750.
5. Adams, R. L. P., Knowler, J. P., and Leader, D. P. (1986). "The Biochemistry of the Nucleic Acids," 10th ed., pp. 5–34. Chapman and Hall, London.
6. Lawyer, F. C., Stoffel, S., Saiki, R. K., Myambo, K., Drummond, R., and Gelfand, D. H. (1989). *JBC* **264**, 6427–6437.
7. Kalnins, A., Otto, K., Ruether, U., and Mueller-Hill, B. (1983). *EMBO J.* **2**, 593–597.
8. Zhang, S., Zubay, G., and Goldman, E. (1991). *Gene* **105**, 61–72.
9. Schildkraut, C. L., Marmur, J., and Doty, P. (1962). *JMB* **4**, 430–443.
10. Freifelder, D. (1982). "Physical Biochemistry," 2nd ed. Freeman, San Francisco.
11. Wetmur, J. G. (1991). *Crit. Rev. B. MB.* **26**, 227–259.
12. Britten, R. J., Graham, D. E., and Neufeld, B. R. (1974). *Meth. Enzy.* **29**, 363–418.
13. Young, B. D., and Anderson, M. L. M. (1985). *In* "Nucleic Acid Hybridisation: A Practical Approach" (B. D. Hames and S. J. Higgins, Eds.), pp. 47–71. IRL Press, Oxford.
14. Vizard, D. L., and Ansevin, A. T. (1976). *Biochem.* **15**, 741–750.
15. Schildkraut, C., and Lifson, S. (1965). *Biopolymers* **3**, 195–208.
16. Mandel, M., and Marmur, J. (1976). *Meth. Enzy.* **12(B)**, 195–206.
17. Freier, S. M., Kierzek, R., Jaeger, J. A., Sugimoto, N., Caruthers, M. H., Neilson, T., and Turner, D. H. (1986). *PNAS* **83**, 9373–9377.
18. Breslauer, K. J., Frank, R., Blocker, H., and Marky, L. A. (1986). *PNAS* **83**, 3746–3750.
19. Borer, P. N., Dengler, B., Tinoco, I., Jr., and Uhlenbeck, O. C. (1974). *JMB* **86**, 843–853.
20. Saenger, W. (1984). "Principles of Nucleic Acid Structure." Springer-Verlag, New York.
21. Anderson, M. L. M., and Young, B. D. (1985). *In* "Nucleic Acid Hybridisation: A Practical Approach" (B. D. Hames and S. J. Higgins, Eds.), pp. 73–111. IRL Press, Oxford.
22. Ko, M. S. H. (1990). *NAR* **18**, 5705–5711.
23. Patanjali, S. R., Parimoo, S., and Weissman, S. M. (1991). *PNAS* **88**, 943–947.
24. Hutton, J. R., and Wetmur, J. G. (1973). *JMB* **77**, 495–500.
25. Birnstiel, M. L., Sells, B. H., and Purdom, I. F. (1972). *JMB* **63**, 21–39.
26. Rychlik, W., and Rhoads, R. E. (1989). *NAR* **17**, 8543–8551.
27. Lathe, R. (1985). *JMB* **183**, 1–12.
28. Bonner, T. I., Brenner, D. J., Neufeld, B. R., and Britten, R. J. (1973). *JMB* **81**, 123–135.
29. Wallace, R. B., Shaffer, J., Murphy, R. F., Bonner, J., Hirose, T., and Itakura, K. (1979). *NAR* **6**, 3543–3556.
30. Chang, C.-T., Hain, T. C., Hutton, J. R., and Wetmur, J. G. (1974). *Biopolymers* **13**, 1847–1858.
31. Nozari, G., Rahbar, S., and Wallace, R. B. (1986). *Gene* **43**, 23–28.
32. Studier, F. W. (1969). *JMB* **41**, 199–209.

33. Wetmur, J. G., and Davidson, N. (1968). *JMB* **31**, 349–370.
34. Hutton, J. R. (1977). *NAR* **4**, 3537–3555.
35. Schmeckpeper, B. J., and Smith, K. D. (1972). *Biochem.* **11**, 1319–1326.
36. Howley, P. M., Israel, M. A., Law, M.-F., and Martin, M. A. (1979). *JBC* **254**, 4876–4883.
37. Casey, J., and Davidson, N. (1977). *NAR* **4**, 1539–1552.
38. Vogelstein, B., and Gillespie, D. (1977). *BBRC* **75**, 1127–1132.
39. Thompson, J., and Gillespie, D. (1987). *Anal. Biochem.* **163**, 281–291.
40. Marmur, J., and Doty, P. (1961). *JMB* **3**, 585–594.
41. Straus, N. A., and Bonner, T. I. (1972). *BBA* **277**, 87–95.
42. Baldino, F., Jr., Chesselet, M.-F., and Lewis, M. E. (1989). *Meth. Enzy.* **168**, 761–777.
43. Rychlik, W., Spencer, W. J., and Rhoads, R. E. (1990). *NAR* **18**, 6409–6412.
44. Cantor, C. R., and Schimmel, P. R. (1980). "Biophysical Chemistry. Part 1. The Conformation of Biological Macromolecules." Freeman, San Francisco.
45. Zimmerman, S. B. (1982). *Ann. Rev. Biochem.* **51**, 395–427.
46. Shakked, Z., Guerstein-Guzikevich, G., Eisenstein, M., Frowlow, F., and Rabinovich, D. (1989). *Nature* **342**, 456–460.
47. Chou, S.-H., Flynn, P., Wang, A., and Reid, B. (1991). *Biochem.* **30**, 5248–5257.
48. Wang, A. H.-J., Quigley, G. J., Kolpak, F. J., Crawford, J. L., van Boom, J. H., van der Marel, G., and Rich, A. (1979). *Nature* **282**, 680–686.
49. Rich, A., Nordheim, A., and Wang, A. H.-J. (1984). *Ann. Rev. Biochem.* **53**, 791–846.
50. Drew, H., Takano, T., Tanaka, S., Itakura, K., and Dickerson, R. E. (1980). *Nature* **286**, 567–573.
51. Klysik, J., Stirdivant, S. M., Singleton, C. K., Zacharias, W., and Wells, R. D. (1983). *JMB* **168**, 51–71.
52. Jovin, T. M., Soumpasis, D. M., and McIntosh, L. P. (1987). *Ann. Rev. Phys. Chem.* **38**, 521–560.
53. Jaworski, A., Hsieh, W.-T., Blaho, J. A., Larson, J. E., and Wells, R. D. (1987). *Science* **238**, 773–777.
54. Rahmouni, A. R., and Wells, R. D. (1989). *Science* **246**, 358–363.
55. Singleton, C. K., Klysik, J., and Wells, R. D. (1983). *PNAS* **80**, 2447–2451.
56. Zacharias, W., Larson, J. E., Kilpatrick, M. W., and Wells, R. D. (1984). *NAR* **12**, 7677–7692.
57. Vadimon, L., and Rich, A. (1984). *PNAS* **81**, 3268–3272.
58. Ughetto, G., Wang, A. H. J., Quigley, G. J., Van der Marel, G. A., Van Boom, J. H., and Rich, A. (1985). *NAR* **13**, 2305–2323.
59. Quigley, G. J., Ughetto, G., Van der Marel, G. A., Van Boom, J. H., Wang, A. H. J., and Rich, A. (1986). *Science* **232**, 1255–1258.
60. Rajagopal, P., and Feigon, J. (1989). *Nature* **339**, 637–640.
61. Hampel, K. J., Crosson, P., and Lee, J. S. (1991). *Biochem.* **30**, 4455–4459.
62. Arnott, S., and Selsing, E. (1974). *JMB* **88**, 509–521.
63. Arnott, S., Bond, P. J., Selsing, E., and Smith, P. J. C. (1976). *NAR* **3**, 2459–2470.
64. Voloshin, O. N., Mirkin, S. M., Lyamichev, V. I., Belotserkovskii, B. P., and Frank-Kamenetskii, M. D. (1988). *Nature* **333**, 475–476.
65. Mirkin, S. M., Lyamichev, V. I., Drushlyak, K. N., Dobrynin, V. N., Filippov, S. A., and Frank-Kamenetskii, M. D. (1987). *Nature* **330**, 495–497.
66. Hanvey, J. C., Shimizu, M., and Wells, R. D. (1988). *PNAS* **85**, 6292–6296.
67. Htun, H., and Dahlberg, J. E. (1988). *Science* **241**, 1791–1795.
68. Roberts, R. W., and Crothers, D. M. (1991). *PNAS* **88**, 9397–9401.
69. Beal, P. A., and Dervan, P. B. (1991). *Science* **251**, 1360–1363.
70. Chen, F.-M. (1991). *Biochem.* **30**, 4472–4479.
71. Pilch, D. S., Levenson, C., and Shafer, R. H. (1991). *Biochem.* **30**, 6081–6087.
72. Radhakrishnan, I., de los Santos, C., and Patel, D. J. (1991). *JMB* **221**, 1403–1418.
73. Lee, J. S., Latimer, L. J. P., Haug, B. L., Pulleyblank, D. E., Skinner, D. M., and Burkholder, G. D. (1989). *Gene* **82**, 191–199.
74. Scaria, P. V., and Shafer, R. H. (1991). *JBC* **266**, 5417–5423.
75. Lyamichev, V. I., Frank-Kamenetskii, M. D., and Soyfer, V. N. (1990). *Nature* **344**, 568–570.

76. Wells, R. D., Collier, D. A., Hanvey, J. C., Shimizu, M., and Wohlrab, F. (1988). *FASEB J.* **2**, 2939–2949.
77. Htun, H., and Dahlberg, J. E. (1989). *Science* **243**, 1571–1576.
78. Kinniburgh, A. J. (1989). *NAR* **19**, 7771–7778.
79. Umek, R. M., Linskens, M. H. K., Kowalski, D., and Huberman, J. A. (1989). *BBA* **1007**, 1–14.
80. Dervan, P. B. (1992). *Nature* **359**, 87–88.
81. Cooney, M., Czernuszewicz, G., Postel, E. H., Flint, S. J., and Hogan, M. E. (1988). *Science* **241**, 456–459.
82. Orson, F. M., Thomas, D. W., McShan, W. M., Kessler, D. J., and Hogan, M. E. (1991). *NAR* **19**, 3435–3441.
83. Yagil, G. (1991). *Crit. Rev. B. MB.* **26**, 475–559.
84. Palececk, E. (1991). *Crit. Rev. B. MB.* **26**, 151–226.
85. Dayn, A., Malkhosyan, S., Duzhy, D., Lyamichev, V., Panchenko, Y., and Mirkin, S. (1991). *J. Bacteriol.* **173**, 2658–2664.
86. Horwitz, M. S. Z., and Loeb, L. A. (1988). *Science* **241**, 703–705.
87. Sen, D., and Gilbert, W. (1990). *Nature* **344**, 410–414.
88. Williamson, J. R., Raghuraman, M. K., and Cech, T. R. (1989). *Cell* **59**, 871–880.
89. Kim, J., Cheong, C., and Moore, P. B. (1991). *Nature* **351**, 331–332.
90. Hardin, C. C., Henderson, E., Watson, T., and Prosser, J. K. (1991). *Biochem.* **30**, 4460–4472.
91. Hagerman, P. J. (1990). *Ann. Rev. Biochem.* **59**, 755–781.
92. Wolters, M., and Wittig, B. (1989). *NAR* **17**, 5163–5172.
93. Wasserman, S. A., and Cozzarelli, N. R. (1986). *Science* **232**, 951–960.
94. Pruss, G. J., and Drlica, K. (1989). *Cell* **56**, 521–523.
95. Ostrander, E. A., Benedetti, P., and Wang, J. C. (1990). *Science* **249**, 1261–1265.
96. Cozzarelli, N. R., Boles, T. C., and White, J. H. (1990). *In* "DNA Topology and Its Biological Effects" (N. R. Cozzarelli and J. C. Wang, Eds.), pp. 139–184. CSHL Press, New York.
97. Tinoco, I., Jr., Borer, P. N., Dengler, B., Levine, M. D., Uhlenbeck, O. C., Crothers, D. M., and Grallia, J. (1973). *Nature NB.* **246**, 40–41.
98. Zuker, M., and Stiegler, P. (1981). *NAR* **9**, 133–148.
99. Williams, A., Jr., and Tinoco, I., Jr. (1986). *NAR* **14**, 299–315.
100. Zuker, M., Jaeger, J. A., and Turner, D. H. (1991). *NAR* **19**, 2707–2714.
101. Sigman, D. S., and Chen, C. B. (1990). *Ann. Rev. Biochem.* **59**, 207–236.
102. Groebe, D. R, and Uhlenbeck, O. C. (1988). *NAR* **16**, 11725–11735.
103. Woese, C. R., Winker, S., and Gutell, R. R. (1990). *PNAS* **87**, 8467–8471.
104. d'Aubenton-Carafa, Y., Brody, E., and Thermes, C. (1990). *JMB* **216**, 835–858.
105. Tuerk, C., and 11 coauthors (1988). *PNAS* **85**, 1364–1368.
106. Turner, D. H., Sugimoto, N., and Freier, S. M. (1988). *Ann. Rev. B. BC.* **17**, 167–192.
107. Cheong, C., Varani, G., and Tinoco, I., Jr. (1990). *Nature* **346**, 680–682.
108. Blommers, M. J. J., van de Ven, F. J. M., van der Marel, G. A., van Boom, J. H., and Hilbers, C. W. (1991). *EJB* **201**, 33–51.
109. Howard, F. B., Chen, C., Ross, P. D., and Miles, H. T. (1991). *Biochem.* **30**, 779–782.
110. Raghunathan, G., Jernigan, R. L., Miles, H. T., and Sasisekharan, V. (1991). *Biochem.* **30**, 782–788.
111. Pleij, C. W. A., Reitveld, K., and Bosch, L. (1985). *NAR* **13**, 1717–1731.
112. Pleij, C. W. A., and Bosch, L. (1989). *Meth. Enzy.* **180**, 289–303.
113. Mans, R. M. W., Pleij, C. W. A., and Bosch, L. (1991). *EJB* **201**, 303–324.
114. Wyatt, J. R., Puglisi, J. D., and Tinoco, I., Jr. (1989). *BioEssays* **11**, 100–106.
115. Hamm, J., and Mattaj, I. M. (1990). *Cell* **63**, 109–118.
116. Hamm, J., Darzinkiewicz, E., Tahara, S. M., and Mattaj, I. W. (1990). *Cell* **62**, 569–577.
117. Robinson, B. G., Frim, D. M., Schwarz, W. J., and Majzoub, J. A. (1988). *Science* **241**, 342–344.
118. Levitt, N., Briggs, D., Gil, A., and Proudfoot, N. J. (1989). *Genes Dev.* **3**, 1019–1025.
119. Chu, F. K., Maley, G. F., Maley, F., and Belfort, M. (1984). *PNAS* **81**, 3049–3053.
120. Goodrich, H. A., Gott, J. M., Xu, M.-Q., Scarlato, V., and Shub, D. A. (1989). *In* "Molecular Biology of RNA" (T. R. Cech, Ed.), Vol. 94, pp. 59–66. A. R. Liss, New York.

121. Perlman, P. S., and Butow, R. A. (1989). *Science* **246**, 1106–1109.
122. Sharp, P. A. (1987). *Science* **235**, 766–771.
123. Luehrmann, R., Kastner, B., and Bach, M. (1990). *BBA* **1087**, 265–292.
124. Guthrie, C. (1991). *Science* **253**, 157–163.
125. Kuo, H., Nasim, F. H., and Grabowski, P. J. (1991). *Science* **251**, 1045–1050.
126. Breitbart, R. E., Andreadis, A., and Nadal-Ginard, B. (1987). *Ann. Rev. Biochem.* **56**, 467–495.
127. Clouet d'Orval, B., D'Aubenton-Carafa, Y., Sirand-Pugnet, P., Gallego, M., Brody, E., and Marie, J. (1991). *Science* **252**, 1823–1828.
128. Libri, D., Piseri, A., and Fiszman, M. Y. (1991). *Science* **252**, 1842–1845.
129. Maniatis, T. (1991). *Science* **251**, 33–34.
130. Benne, R., Van den Burg, J., Brakenhoff, J. P. J., Sloof, P., Van Boom, J. H., and Tromp, M. C. (1986). *Cell* **46**, 819–826.
131. Feagin, J. E., Abraham, J. M., and Stuart, K. (1988). *Cell* **53**, 413–422.
132. Simpson, L., and Shaw, J. (1989). *Cell* **57**, 355–366.
133. Mahendran, R., Spottswood, M. R., and Miller, D. L. (1991). *Nature* **349**, 434–438.
134. Thomas, S. M., Lamb, R. A., and Paterson, R. G. (1988). *Cell* **54**, 891–902.
135. Cattaneo, R., Kaelin, K., Baczko, K., and Billeter, M. A. (1989). *Cell* **56**, 759–764.
136. Vidal, S., Curran, J., and Kolakofsky, D. (1990). *J. Virol.* **64**, 239–246.
137. Gualberto, J. M., Lamattina, L., Bonnard, G., Weil, J.-H., and Grienenberger, J.-M. (1989). *Nature* **341**, 660–662.
138. Covello, P. S., and Gray, M. W. (1989). *Nature* **341**, 662–666.
139. Schuster, W., Wissinger, B., Unseld, M., and Brennicke, A. (1990). *EMBO J.* **9**, 263–269.
140. Hoch, B., Maier, R. M., Appel, K., Igloi, G. L., and Kossel, H. (1991). *Nature* **353**, 178–180.
141. Chen, S.-H., and 12 coauthors (1987). *Science* **238**, 363–366.
142. Powell, L. M., Wallis, S. C., Pease, R. J., Edwards, Y. H., Knott, T. J., and Scott, J. (1987). *Cell* **50**, 831–840.
143. Blum, B., Bakalara, N., and Simpson, L. (1990). *Cell* **60**, 189–198.
144. Stuart, K. (1991). *TIBS* **16**, 68–72.
145. Benne, R. (1990). *TIG* **6**, 177–181.
146. Cech, T. R. (1991). *Cell* **64**, 667–669.
147. Blum, B., Sturm, N. R., and Simpson, L. (1991). *Cell* **65**, 543–550.
148. Greeve, J., Navaratnam, N., and Scott, J. (1991). *NAR* **19**, 3569–3576.
149. Yang, A. J., and Muligan, R. M. (1991). *Mol. Cell. Biol.* **11**, 4278–4281.
150. Schulz, V. P., and Reznikoff, W. S. (1990). *JMB* **211**, 427–445.
151. de Smit, M. H., and van Duin, J. (1990). *PNARMB* **38**, 1–35.
152. Fuerst, T. R., and Moss, B. (1989). *JMB* **206**, 333–348.
153. Kozak, M. (1989). *Mol. Cell. Biol.* **9**, 5134–5142.
154. Sonenberg, N., and Pelletier, J. (1989). *BioAssays* **11**, 128–132.
155. Thomas, M. S., and Nomura, M. (1987). *NAR* **15**, 3085–3096.
156. Noller, H. F. (1984). *Ann. Rev. Biochem.* **53**, 119–162.
157. Stern, S., Powers, T., Changchien, L.-M., and Noller, H. F. (1989). *Science* **244**, 783–790.
158. Lill, R., Robertson, J. M., and Wintermeyer, W. (1989). *EMBO J.* **8**, 3933–3938.
159. Uhlenbeck, O. C. (1987). *Nature* **328**, 596–600.
160. Kim, S.-H., and Cech, T. R. (1987). *PNAS* **84**, 8788–8792.
161. Cech, T. R. (1990). *Ann. Rev. Biochem.* **59**, 543–568.
162. Celander, D. W., and Cech, T. R. (1991). *Science* **251**, 401–407.
163. Dange, V., Van Atta, R. B., and Hecht, S. M. (1990). *Science* **248**, 585–588.
164. Michel, F., Ellington, A. D., Couture, S., Szostak, J. W. (1990). *Nature* **347**, 578–580.
165. McSwiggen, J. A., and Cech, T. R. (1989). *Science* **244**, 679–683.
166. Rajagopal, J., Doudna, J. A., and Szostak, J. W. (1989). *Science* **244**, 692–694.
167. Pyle, A. M., Murphy, F. L., and Cech, T. R. (1992). *Nature* **358**, 123–128.
168. Been, M. D., and Cech, T. R. (1988). *Science* **239**, 1412–1416.
169. Doudna, J. A., and Szostak, J. W. (1989). *Nature* **339**, 519–522.
170. Zang, A. J., and Cech, T. R. (1986). *Science* **231**, 470–475.

171. Bartel, D. P., Doudna, J. A., Usman, N., and Szostak, J. W. (1991). *Mol. Cell. Biol.* **11**, 3390–3394.
172. Herschlag, D., and Cech, T. R. (1990). *Nature* **344**, 405–409.
173. Robertson, D. L., and Joyce, G. F. (1990). *Nature* **344**, 467–468.
174. Altman, S. (1989). *Adv. Enzymol.* **62**, 1–36.
175. Pace, N. R., and Smith, D. (1990). *JBC* **265**, 3587–3590.
176. Haas, E. S., Morse, D. P., Brown, J. W., Schmidt, F. J., and Pace, N. R. (1991). *Science* **254**, 853–856.
177. Reich, C., Olsen, G. J., Pace, B., and Pace, N. R. (1988). *Science* **239**, 178–181.
178. Peck-Miller, K. A., and Altman, S. (1991). *JMB* **221**, 1–5.
179. Waugh, D. S., Green, C. J., and Pace, N. R. (1989). *Science* **244**, 1569–1571.
180. Guerrier-Takada, C., Lumelsky, N., and Altman, S. (1989). *Science* **246**, 1578–1584.
181. Thurlow, D. L., Shilowski, D., and Marsh, T. L. (1991). *NAR* **19**, 885–891.
182. Castano, J. G., Ornberg, R., Koster, J. G., Tobian, J. A., and Zasloff, M. (1986). *Cell* **46**, 377–387.
183. Lee, J.-Y., and Engelke, D. R. (1989). *Mol. Cell. Biol.* **9**, 2536–2543.
184. Sprinzl, M., and Gauss, D. H. (1982). *NAR* **10**, r1–r115.
185. Sampson, J. R., DiRenzo, A. B., Behlen, L. S., and Uhlenbeck, O. C. (1989). *Science* **243**, 1363–1366.
186. Hall, K. B., Sampson, J. R., Uhlenbeck, O. C., and Redfield, A. G. (1989). *Biochem.* **28**, 5794–5801.
187. Cigan, A. M., Feng, L., and Donahue, T. F. (1988). *Science* **242**, 93–97.
188. Holbrook, S. R., Sussman, J. L., Warrant, R. W., and Kim, S. H. (1978). *JMB* **123**, 631–660.
189. Robertus, J. D., Ladner, J. E., Finch, J. T., Rhodes, D., Brown, R. S., Clark, B. F. C, and Klug, A. (1974). *Nature* **250**, 546–551.
190. Moras, D., Comarmond, M. B., Fischer, J., Weiss, R., Thierry, J. C., Ebel, J. P., and Giege, R. (1979). *Nature* **288**, 669–674.
191. Schevitz, R. W., Podjarny, A. D., Krishnamachari, N., Hughes, J. J., and Sigler, P. B. (1979). *Nature* **278**, 188–190.
192. Woo, N. H., Roe, B. A., and Rich, A. (1980). *Nature* **286**, 346–351.
193. Rould, M. A., Perona, J. J., Soell, D., and Steitz, T. A. (1989). *Science* **246**, 1135–1154.
194. Vold, B. S. (1985). *Microbiol. Rev.* **49**, 71–80.
195. Moazed, D., and Noller, H. F. (1991). *PNAS* **88**, 3725–3728.
196. Schimmel, P. (1991). *FASEB J.* **5**, 2180–2187.
197. Ruff, M., Krishnaswamy, S., Boeglin, M., Poterszman, A., Mitschler, A., Podjarny, A., Rees, B., Thierry, J. C., and Moras, D. (1991). *Science* **252**, 1682–1689.
198. Putz, J., Puglisi, J. D., Florentz, C., and Giege, R. (1991). *Science* **252**, 1696–1699.
199. Chattapadhyay, R., Pelka, H., and Schulman, L. H. (1990). *Biochem.* **29**, 4263–4268.
200. Pallanck, L., and Schulman, L. H. (1991). *PNAS* **88**, 3872–3876.
201. Musier-Forsyth, K., Usman, N., Scaringe, S., Doudna, J., Green, R., and Schimmel, P. (1991). *Science* **253**, 784–786.
202. Khan, A., and Roe, B. A. (1988). *Science* **241**, 74–79.
203. Rippe, K., Fritsch, V., Westhof, E., and Jovin, T. M. (1992). *EMBO J.* **11**, 3777–3786.
204. Yoon, K., Hobbs, C. A., Koch, J., Sardaro, M., Kutny, R., and Weis, A. L. (1992). *PNAS* **89**, 3840–3844.
205. Mergny, J.-L., Duval-Valentin, G., Nguyen, C. H., Perrouault, L., Faucon, B., Rougee, M., Montenay-Garestier, T., Bisagni, E., and Helene, C. (1992). *Science* **256**, 1681–1684.
206. Mergny, J.-L., Collier, D., Rougee, M., Montenay-Garestier, T., and Helene, C. (1991). *NAR* **19**, 1521–1526.
207. SantaLucia, J., Jr., Kierzek, R., and Turner, D. H. (1992). *Science* **256**, 217–219.
208. Xodo, L. X., Manzini, G., Quadrifoglio, F., van der Marel, G., and van Boom, J. (1991). *NAR* **19**, 1505–1511.
209. Seidel, H. M., Pompliano, D. L., and Knowles, J. R. (1992). *Science* **257**, 1489–1490.
210. Hsu, T., Gogos, J. A., Kirsh, S. A., and Kafatos, F. C. (1992). *Science* **257**, 1946–1950.

211. Koslowsky, D. J., Goringer, H. U., Morales, T. H., and Stuart, K. (1992). *Nature* **356**, 807–809.
212. Harris, M. E., and Hajduk, S. L. (1992). *Cell* **68**, 1091–1099.
213. Maslov, D. A., and Simpson, L. (1992). *Cell* **70**, 459–467.
214. Noller, N. F., Hoffarth, V., and Zimniak, L. (1992). *Science* **256**, 1416–1419.
215. Piccirilli, J. A., McConnell, T. S., Zaug, A. J., Noller, H. F., and Cech, T. R. (1992). *Science* **256**, 1420–1424.
216. Symons, R. H. (1989). *TIBS* **14**, 445–450.
217. Haseloff, J., and Gerlach, W. L. (1988). *Nature* **334**, 585–591.
218. McCall, M. J., Hendry, P., and Jennings, P. A. (1992). *PNAS* **89**, 5710–5714.
219. Fu, D.-J., and McLaughlin, L. W. (1992). *PNAS* **89**, 3985–3989.
220. Williams, D. M., Pieken, W. A., and Eckstein, F. (1992). *PNAS* **89**, 918–921.
221. Buzayan, J. M., van Tol, H., Feldstein, P. A., and Bruening, G. (1990). *NAR* **18**, 4447–4451.
222. Ruffner, D. E., and Uhlenbeck, O. C. (1990). *NAR* **18**, 6025–6029.
223. van Tol, H., Buzayan, J. M., Feldstein, P. A., Eckstein, F., and Bruening, G. (1990). *NAR* **18**, 1971–1975.
224. Koizumi, M., and Ohtsuka, E. (1991). *Biochem.* **30**, 5145–5150.
225. Slim, G., and Gait, M. J. (1991). *NAR* **19**, 1183–1188.
226. Michel, F., and Westhof, E. (1990). *JMB* **216**, 585–610.
227. Beaudry, A. A., and Joyce, G. F. (1992). *Science* **257**, 635–641.
228. Francklyn, C., Shi, J.-P., and Schimmel, P. (1992). *Science* **255**, 1121–1125.
229. Frugier, M., Florentz, C., and Giege, R. (1992). *PNAS* **89**, 3990–3994.
230. Jin, R., Gaffney, B. L., Wang, C., Jones, R. A., and Breslauer, K. J. (1992). *PNAS* **89**, 8832–8836.

2
Ligases

I. DNA LIGASES
[Polydeoxyribonucleotide synthase, EC 6.5.1]

A. General Description

1. INTRODUCTION

DNA ligases catalyze the formation of a phosphodiester bond between DNA single strands in the duplex form (Fig. 2.1). The covalent linkage of the 5'-P group of one chain with the adjacent 3'-OH group of another is coupled with the pyrophosphate hydrolysis of the cofactor ATP or NAD. Bacterial DNA ligases, e.g., from *E. coli, B. subtilis,* and *S. typhimurium,* use the hydrolysis of NAD as their energy source. In contrast, ATP is the cofactor for DNA ligases from bacteriophages (e.g., T4 and T7) and eukaryotic cells.

The covalent joining of polynucleotides catalyzed by the DNA ligase is a necessary event in DNA repair, recombination, and most notably DNA replication which requires the joining of "Okazaki" fragments (the small, nascent ssDNA fragments generated from the copying of the minus strand). Initially searched for

109

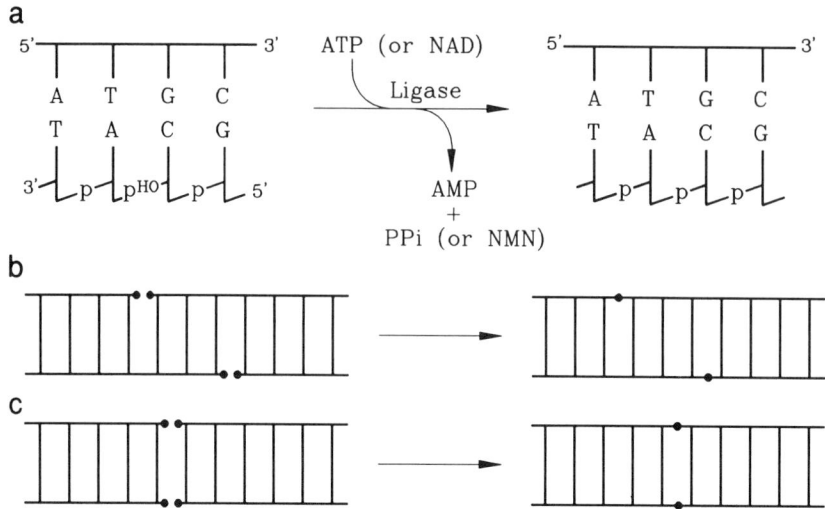

FIGURE 2.1 DNA ligase-catalyzed joining of duplex DNA. (a) Nick-sealing activity, (b) cohesive-end ligation, and (c) blunt-end ligation. In (b) and (c), the 5'-P and 3'-OH termini to be ligated or that have been ligated are denoted by small filled circles.

and found as a polynucleotide joining or sealing activity, DNA ligases are present ubiquitously in all living organisms.

This chapter focuses on two of the best characterized DNA ligases, the *E. coli DNA ligase* and the *T4 DNA ligase*. The two DNA ligases share many properties, but also have some distinct properties that make them uniquely useful depending on application. This section describes the general properties of the two enzymes.

2. Ligase Activity Assays and Unit Definitions

DNA ligase activities can be assayed by various methods involving either partial or whole ligation reactions. The principal assay methods and unit definitions are as follows.

i. Circularization assay. This assay measures the production of exonuclease III-resistant circular DNA from ^{3}H-labeled linear d(AT) copolymers (1). A typical mixture (0.1 ml) for *E. coli* DNA ligase assay contains 30 mM Tris–Cl (pH 8.0), 4 mM MgCl$_2$, 1.2 mM EDTA, 26 μM NAD, 50 μg/ml BSA, 0.16 mM [^{3}H](dAT)$_n$ (~2000 cpm/nmol). For T4 DNA ligase assay, NAD is replaced by 0.2 mM ATP and 10 mM DTT. After incubation at 30°C for 30 min, the reaction is terminated by boiling for 2 min. Ten microliters of 0.1 M 2-MSH and 150 U of Exo III are added to the mixture and the mixture is further incubated at 37°C for 30 min. The reaction mixture is chilled in ice, and 0.2 ml of 0.1 M Tris–Cl (pH 8), 50 μl of 0.25 mM calf thymus DNA, and 0.4 ml of cold 7% perchloric acid (PCA) are added. After 10 min at 0°C, the mixture is filtered on a glass filter, thoroughly washed with 1 N HCl and 95% (v/v) ethanol, and the amount of radioactivity is determined using a scintillation counter.

Unit definition: One *circularization* unit (or Modrich–Lehman unit) is the amount of enzyme which converts linear poly[d(AT)] (optimum length of 1000 nt) into 1 nmol of exonuclease III-resistant circular form in 30 min at 30°C.

A slightly different type of ligase assay, developed by Olivera and Lehman (2), involves the conversion of the $5'$-^{32}P termini in oligo(dT) into an alkaline phosphatase-resistant form in the presence of poly(dA).

Note that the ligase activity is actually dependent on the nature of the substrate DNA. Therefore a given enzyme unit may not be identical to others unless the assay conditions are identical.

ii. **ATP–PP$_i$ exchange assay.** This assay measures the amount of PP$_i$ exchange between the ATP and ^{32}PP$_i$ (3). The assay is based on the partial reaction in the ligase-catalyzed reaction and does not require DNA substrates (see Fig. 2.2). The reaction conditions originally described for T4 DNA ligase assay are 3.3 μM ^{32}PP$_i$ (sp. act. = 1 $\times$ 10^8 cpm/μmol), 66 mM Tris–Cl (pH 7.6), 6.6 mM MgCl$_2$, 10 mM DTT, and 66 μM ATP. Note that *E. coli* DNA ligase can be assayed similarly by an NAD–NMN exchange reaction.

Unit definition: One *phosphate exchange* unit (or Weiss unit) is the amount of enzyme which catalyzes the conversion of 1 nmol of ^{32}PP$_i$ into a charcoal (Norit)-adsorbable form in 20 min at 37°C in an ATP–PP$_i$ exchange reaction. One Weiss unit corresponds to ~0.17 Modrich–Lehman units.

iii. **λ *Hin*dIII DNA ligation assay.** This is probably the assay method most commonly used by enzyme suppliers. The standard ligase reaction mixture contains 50 mM Tris–Cl (pH 7.5), 10 mM MgCl$_2$, 1 mM ATP, and 1 mM DTT.

Unit definition: One ligase unit is that amount of enzyme which catalyzes 50% ligation of 1 μg of λ DNA *Hin*dIII fragments in 30 min at 16°C. One "cohesive-end ligation" unit is equivalent to 0.015 Weiss units or 0.0025 Modrich-Lehman units.

iv. **Oligonucleotide ligation assay.** The ligation of defined oligonucleotides has long been used as part of characterizing enzymatic properties of ligases. The ability to join two synthetic oligodeoxynucleotides (e.g., 10-mer) in the presence of their complement (e.g., 20-mer) in turn provides an unambiguous simple assay for detecting ligase activity (4). Only one of the two short oligonucleotides is $5'$-phosphorylated and ^{32}P-labeled so that only one ligation product is generated. The product is detected by PAGE (12–15%) in the presence of 8 M urea.

3. SUBSTRATE SPECIFICITIES

i. **Basic structure of DNA termini: nick sealing.** DNA ligases join single-strand breaks with juxtaposed $3'$-OH and $5'$-P groups in a duplex DNA. Such single-strand breaks, called *nicks*, can be created by the action of pancreatic DNase or by annealing two DNA fragments with cohesive ends.

The nicks in DNA containing 3'-P and 5'-OH, 3'-OH and 5'-OH, 3'-dideoxynucleotide and 5'-P, or 3'-OH and 5-triphosphoryl termini (5) are not substrates.

ii. Length and nucleotide composition of DNA substrates

(a) Minimum length of substrates. Oligonucleotides as short as 6 or 7 (but not 4) bases in length can be joined if annealed to long complementary DNA strands (6). Both E. coli and T4 DNA ligases give comparable results when used in equivalent amounts. 5'-P mononucleotides (dNMPs) are not substrates. The minimum chain length for the ligation of oligo(dT) on poly(dA) is 7 whereas that for oligo(dA) on poly(dT) is 10, and an appreciable reaction occurs only with the T4 enzyme. The joining of oligo(dA) on poly(dT) is slower than that of oligo(dT) on poly(dA) by approximately three orders of magnitude (7).

The optimum temperature for the joining of a hexanucleotide, $pd(TG)_3$, is between 10° and 20°C; the ligation efficiency drops to 25% at 0°C and 5% at 37°C. The optimum temperature for ligation by the T4 enzyme of oligonucleotides in a series from $p(dT)_7$ to $p(dT)_{12}$ increases from 10° to 32°C (8).

(b) Differences between T4 and E. coli DNA ligases. The extent of joining two octanucleotides, $pd(TAAG)_2$, into a hexadecanucleotide $pd(TAAG)_4$ on the complementary hexadecanucleotide $(dTTAC)_4$ is approximately threefold higher with the T4 enzyme than with the E. coli enzyme. Under similar conditions with $pd(AG)_4$ on $d(TC)_8$, an appreciable (6%) reaction occurs only with the T4 enzyme. The temperature optimum is ~10°C. Compared with the 10°C reaction, the efficiency of ligation is 42% at 0°C and 21% at 20°C.

iii. Heteroduplex substrates.

Short oligodeoxynucleotides (4–10 bases) cannot be ligated on long complementary polyribonucleotides (6). Some oligodeoxynucleotides (50–500 bases), when annealed to complementary polyribonucleotides, can be joined by the T4 DNA ligase but not by the E. coli enzyme (2, 7). The ligation of oligo(dT) in DNA·RNA heteroduplexes occurs at less than 5% of the rate observed with DNA·DNA homoduplexes.

Both E. coli and T4 DNA ligases join 5'-P DNA to 3'-OH RNA in the presence of the complementary DNA strand. The joining in reverse combination, i.e., 5'-P RNA to 3'-OH DNA, is catalyzed by the T4 enzyme only (9,10).

iv. RNA substrates.

The T4 DNA ligase, but not the E. coli enzyme, can join RNA to RNA (11,12). At ligase concentrations higher than those normally used for nick sealing, short oligoribonucleotides (5–20 bases) such as oligo(rU), oligo(rI), and oligo(rC) can be joined head to tail, respectively, in the presence of the corresponding complementary polyribonucleotide. Oligo(rA) can be joined, when annealed to complementary poly(dT), more efficiently than other RNA to RNA ligations in RNA·RNA homoduplexes. In contrast, oligo(rA) cannot be joined in the presence of poly(rU).

The ligation of oligoribonucleotides in RNA·DNA heteroduplexes occurs at a rate of only 1 to 2% of that observed in DNA duplexes (7).

v. Substrates with blunt ends. The T4 DNA ligase can catalyze blunt-end joining of DNA duplexes, regardless of whether they are synthetic oligonucleotides (13–15), blunt-end-cut phage DNA, or plasmid DNA (16,17). The blunt-end joining between two duplex oligonucleotides (16 and 22 bp) is optimal at 25°C (18). A 16-bp duplex is more efficiently utilized in ligation than the 8-bp parent duplex by a factor of almost 100 (13). Compared with the nick repair, the rate of blunt-end ligation is slow and is dependent on enzyme concentrations typically requiring two to five times more enzyme.

The *E. coli* DNA ligase is far less efficient (about 400-fold) in blunt-end ligation than the T4 DNA ligase under the conditions used for a conventional ligation reaction. However, the efficiency of blunt-end ligation can be significantly increased in the presence of volume-exclusion macromolecules such as PEG and Ficoll.

vi. Macromolecular stimulants. *Macromolecules* such as PEG, Ficoll, albumin, and glycogen stimulate DNA ligases in blunt-end ligation. An over 1000-fold stimulation of the T4 DNA ligase is observed at high concentrations of PEG 8000 (formerly PEG 6000, 13–23%) or Ficoll 70 (a branched polysaccharide). In the presence of these macromolecules, the *E. coli* DNA ligase becomes active in catalyzing the ligation of blunt-ended duplex DNAs (19). Low MW PEG (of size ≤400) or sucrose gives little or no stimulation even at very high concentrations.

The stimulating effects of macromolecules on DNA ligase activity are largely directed to intermolecular joining to yield mostly linear oligomers and some high MW circular products. This suggests that the stimulating activity of macromolecules is derived from a volume-exclusion mechanism. The joining of single-strand breaks is relatively unaffected, whereas the sealing of cohesive termini is stimulated to a similar extent as in blunt-end ligation.

Although PEG is useful for stimulating blunt-end ligations, high concentrations of PEG may interfere with subsequent procedures, e.g., transformations. However, PEG may be easily removed from DNA solutions by extraction with chloroform. A single extraction with 3 volumes of $CHCl_3$ will remove 95% of the PEG 8000 (20).

For other macromolecular stimulants, refer to Section I,C on T4 DNA ligase.

4. Mechanism of Ligation

All known DNA ligases use the hydrolysis energy of either NAD ($\rightarrow$ AMP + NMN) or ATP ($\rightarrow$ AMP + PP_i) to catalyze the formation of a phosphodiester bond between two polynucleotide chains. The basic substrate structure is a duplex DNA with a juxtaposed 5′-P "donor" group and a 3′-OH "acceptor" group.

The ligation reaction is carried out by a three-step mechanism which involves two relatively stable covalent intermediates (Fig. 2.2). The overall reaction follows a Ping-Pong kinetic mechanism. The first step is the covalent transfer of the adenylyl group of NAD or ATP to an ε-amino group of a Lys residue in the enzyme, forming a ligase–adenylate intermediate with the concomitant release of NMN or PP_i (1,2). This reaction occurs regardless of the absence or presence of

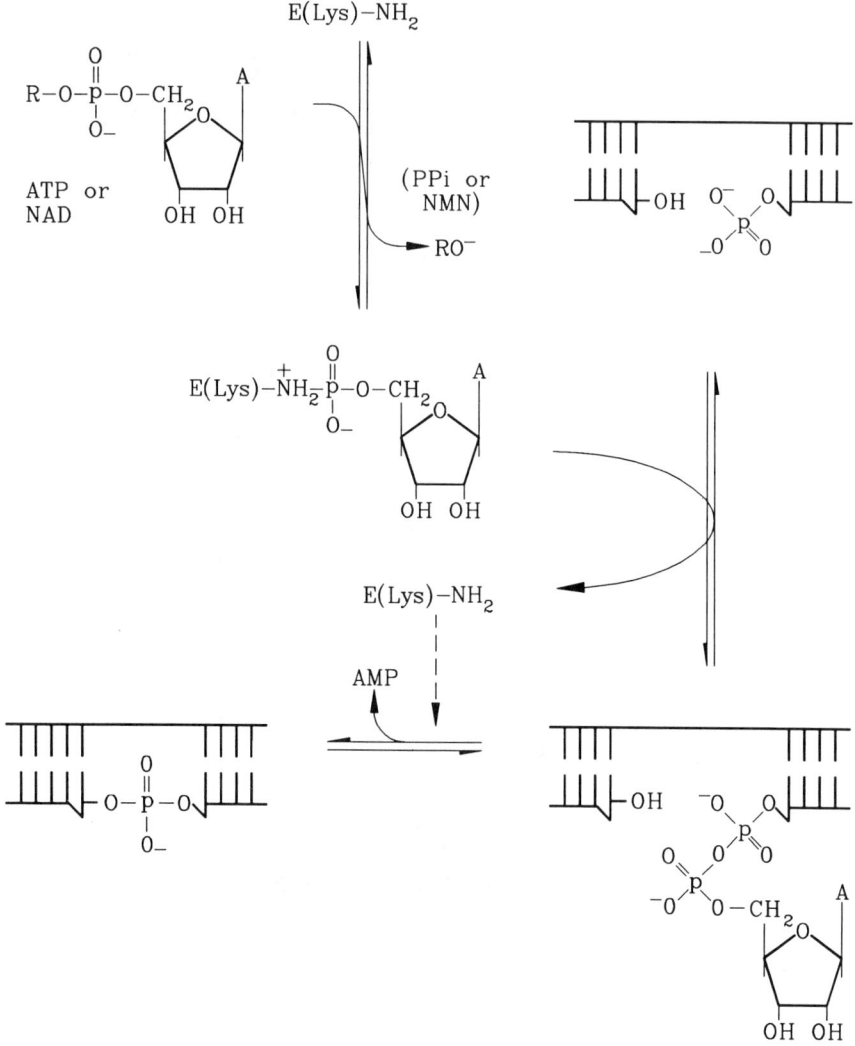

FIGURE 2.2 A general mechanism of DNA ligase-catalyzed reactions. The ligation reaction catalyzed by a DNA ligase involves three steps. The first step is the formation of "activated" ligase in which an adenylyl group (AMP) is covalently linked to the active site Lys with accompanying release of either PP_i (from ATP) or NMN (from NAD). In the presence of DNA substrates, the activated ligase transfers the AMP moiety to the 5'-P of the "donor" substrate, resulting in the formation of pyrophosphate-linked DNA-adenylate intermediate. The third and last step is the nucleophilic attack on the pyrophosphate linkage by the adjacent 3'-OH of the "acceptor" substrate, resulting in a covalently closed molecule and free AMP. All the steps involved in the ligation reaction are reversible. Therefore, the addition of exogenous $[^{32}P]PP_i$ or $[^{3}H]NMN$ in the absence of DNA results in PP_i or NMN exchange into $[\gamma\text{-}^{32}P]ATP$ or $[^{3}H]NAD$, respectively. The arrow with the broken line indicates that a DNA ligase may display a nicking activity in the presence of excess AMP.

DNA substrates. For the T4 DNA ligase, the self-adenylation occurs not only with ATP but also with dATP and their α-thio derivatives, albeit at lower rates (23). The isolated ligase–AMP complex is functionally competent and discharges the AMP moiety to a 5′-P donor substrate in a concentration-dependent manner, which is the second step of the reaction (24,25). Note that the enzyme complex is essentially inactive when dATP replaces ATP as the cofactor. The AMP transfer results in the activation of the 5′-P group by forming a new pyrophosphate linkage in the second intermediate, DNA–adenylate (13,24,26). The DNA–AMP intermediate, whether prepared synthetically (5,27) or isolated from a ligase-catalyzed reaction (26,27), is catalytically competent and, in the presence of an unadenylylated enzyme, releases a stoichiometric amount of AMP on formation of a phosphodiester bond with the 3′-OH group of acceptor molecules. This final step is the nucleophilic displacement of the AMP by the attack of 3′-OH at the activated 5′-P group. The 5′-P DNAs or polynucleotides activated by PP_i linkage with GMP or dAMP function poorly or not at all as donor substrates for both *E. coli* and T4 DNA ligases. Ligation of the oligonucleotides proceeds in a random mode, and a considerable accumulation of certain particular oligomers may occur during the course of the reaction.

The individual steps of a ligase reaction are essentially reversible and, depending on the presence of AMP and ATP or NAD, a ligase can show either closing (ligase) or nicking (endonuclease) activity. A covalently closed, superhelical plasmid can undergo a slow, but complete ATP-dependent conversion to a relaxed circle which may be covalently closed (resealed) or nicked.

In blunt-end ligation, low levels of nicked intermediates have been observed, suggesting a ligation mechanism involving two discrete steps: the formation of a nicked dimer, and then sealing of the nick (13). Although the rates for the two steps are presumed to be similar, the rate-determining step is apparently the initial bimolecular reaction where two duplexes meet to be joined as a nicked dimer.

5. THEORY OF LIGATION

Many different factors, including temperature, enzyme concentration, substrate DNA concentration, and the shape and flexibilities of the DNA molecules, affect the rate of ligation and the nature of the end products. The joining of two ends either of the same molecule resulting in circular products or of different fragments resulting in linear, concatemeric molecules is a matter of the probability of the ends finding each other. Therefore the molar ratio of vector and insert DNA as well as the lengths of the DNA fragments affect the outcome of the ligation products.

In theory, a ligation reaction depends on the following molecular factors (28,29): (i) the total concentration of DNA ends, i (ends/ml), and (ii) the "effective" concentration of one end of a DNA molecule in the immediate vicinity of the other end of the same molecule, j (ends/ml, J–S factor). The shorter the DNA molecule and the lower the extent of joining, the higher the effective concentration j is. The ratio of j/i generally determines whether more or less of the ligation products will be linear or circular.

In a single component system in which j and i values are identical ($j = i$), the linear and circular forms of products are equally likely to occur. The DNA concentration (mg/ml) fitting the condition $j = i$ is calculated as $24/(MW)^{0.5}$. When $i > j$ (or greater than $[DNA]_{j=i}$ concentration), linear products are favored. When $i < j$ (or less than $[DNA]_{j=i}$) and for DNA molecules longer than a minimum contour length (72 nm) which is necessary for circularization, circular products are more likely to occur.

Essentially similar but more elaborate theoretical treatments have been described for multiple component systems, e.g., the cloning of a DNA fragment in a linear λ-type cloning vectors or in circular plasmids linearized with two restriction enzymes (29). In such systems, the parameter i is the sum of individual components, whereas j is calculated in the same way as for a single component system.

Many ligation reactions, typically in cloning DNA fragments in plasmid vectors, involve two components in a more complex situation. An optimal ligation condition would be to favor the joining of DNA fragments to the vector initially, and then to lead to circularization of the chimeric molecule. This implies that the concentrations of plasmid and DNA fragments be such that the j/i value is minimal (<0.8) at time zero to favor the formation of linear products and, as the reaction proceeds, becomes higher to favor the circularization of the expected hybrid monomer (= plasmid + single insert).

Although such optimal ligation conditions may allow better control of the ligation reaction, finding the optimal conditions would be time- and effort-consuming. In practice, it is much easier to run a combination of preliminary ligation reactions and to measure the ligation efficiency by counting the number of transformants. Perhaps a more compelling reason for opting for this practical approach (see Section I,C in this chapter) is that the probability of finding two ligatable ends is in fact influenced by multiple factors such as ligase and cofactor concentrations, temperature, time of incubation, molar ratio of insert to vector, and other components that affect the fluid dynamics of the ligation milieu.

6. APPLICATIONS

DNA ligases are one of the most important tools in the construction of recombinant DNA molecules. The *E. coli* DNA ligase can in some cases substitute for the T4 DNA ligase and, in other cases, may be uniquely useful. However, the T4 DNA ligase is in general the enzyme of choice for blunt-end ligations. The T4 DNA ligase may also prove to be more useful for certain cohesive-end ligations because it requires a smaller overlapping sequence than does the *E. coli* DNA ligase.

i. Elongation or circularization of dsDNA. Natural or chemically synthesized DNA or oligonucleotides, e.g., linkers, adapters, and mutagenic sequences, can be ligated intra- or intermolecularly into circular or elongated molecules of variable lengths. This application is probably the major use of DNA ligase in constructing synthetic genes (41) and recombinant DNA molecules.

ii. As base of ligation amplification reactions. The *ligation amplification reaction* (LAR), also called *ligase chain reaction* (LCR), is a strategy to amplify specific DNA sequences using sequential rounds of template-dependent ligation (31). This strategy consists of annealing two oligonucleotides to a common target sequence such that the 5'-P of one of the oligonucleotides (donor) is adjacent to the 3'-OH of the second oligonucleotide (acceptor). The DNA ligase is then used to join the two oligonucleotides in a nick-closing reaction. When two pairs of oligonucleotides are utilized, one complementary to the upper strand and the other to the lower strand of a target sequence, an exponential amplification can be achieved by repeating cycles of annealing, ligation, and denaturation.

This strategy is critically dependent on maximizing the nick-sealing activity while minimizing the template-independent, blunt-end activity of the DNA ligase. This can be achieved by using two oligonucleotides whose termini structures are designed such that only a single ligation is possible (see Section I,A,2 on *ligase activity assay*). Alternatively, the amplification specificity can be improved by using the *E. coli* DNA ligase which has a strong preference for nick sealing (32). This ligase-based gene amplification technique, when applied to the detection of single-base mismatches, greatly increases the sensitivity of detection (see below).

iii. Detection of mismatched nucleotides in dsDNA. This technique is based on the fact that a DNA ligase can covalently join two oligonucleotides when they are hybridized immediately adjacent to each other on a complementary DNA. When the nucleotide(s) at the junction is incorrectly base paired, for example, due to a point mutation (as in human β-globin alleles responsible for sickle cell anemia), the efficiency of ligase-mediated phosphodiester bond formation decreases substantially. This distinction, especially when combined with the ligase-based gene amplification techniques described earlier, provides a sensitive mutant gene detection method which can be applied to the diagnosis of a variety of viral and genetic diseases (33–35).

A significant improvement to this ligation amplification and mismatch detection system is the introduction of thermostable DNA ligase, for example, from *T. aquaticus* (36). The use of thermostable DNA ligase simplifies the repetitive thermal cycles, as does the thermostable DNA polymerase in PCR, and also significantly improves the mismatch discrimination due to the high ligation temperature (65°C).

iv. Synthesis of long RNA molecules. The capacity of the T4 DNA ligase to ligate RNA junctions in double-stranded regions enables one to prepare long RNA molecules in a highly efficient and selective manner (43). Compared with the RNA ligation that can be performed using the T4 RNA ligase (see Section II, this chapter), the T4 DNA ligase offers the following advantages.

(a) The low K_m of the T4 DNA ligase for double-stranded polynucleotides (from 10^{-8} to $10^{-7} M$) permits efficient ligations to be performed at submicromolar to micromolar concentrations of RNA instead of the millimolar range of concentrations required for T4 RNA ligase.

(b) The T4 DNA ligase produces RNA ligation at a substantially higher specificity due to the requirement for double-stranded template structure. Thus side products such as circular and oligomeric RNA molecules are not generated even when the 3′-OH terminus of the phosphate donor RNA is not protected. In case RNA substrates are synthesized by *in vitro* transcription with T7, T3, or SP6 RNA polymerases (see Section I, Chapter 7), the by-products carrying an extra nontemplated nucleotide at the 3′ terminus (N + 1 products) are efficiently discriminated against the correct acceptor molecules (N products), reducing the possibility of mismatching nucleotide insertions.

v. Use of DNA ligase in DNA analyses

(a) *Identification of termini structure.* The favored substrate structure for DNA ligases is a duplex DNA containing a single-strand break with 3′-OH and 5′-P groups in juxtaposition. The ability to ligate provides a sensitive and specific test for identifying such juxtaposed termini structures (37).

(b) *Measurement of number of nicks.* The breakdown of ligase–AMP yielding the phosphodiester bond and AMP occurs only in the presence of nicks in duplex DNAs which display 3′-OH and 5′-P groups. Therefore the amount of AMP released from the radiolabeled AMP–ligase complex gives an estimate of the number of nicks in a given DNA (38).

(c) *Identification of 3′- and 5′-end groups.* The end groups of single-strand breaks in duplex DNA can be identified by nearest-neighbor analysis (3). In this analysis, the 5′ end of the nick is labeled with ^{32}P using T4 PNK and is then sealed using the T4 DNA ligase. The DNA is digested with a specific nuclease, e.g., spleen or micrococcal diesterase that releases 3′-mononucleotides or snake venom phosphodiesterase that releases 5′-mononucleotides.

(d) *Assessment of nicking activities of other enzymes.* The specific ability of DNA ligases to rejoin a nick is used to determine the substrate specificity of other enzymes with respect to nicks and gaps in duplex DNA. It also permits the determination of the parameters (e.g., temperature and salt concentrations) affecting the structure of substrate termini. For example, the nicking activity of an endonuclease can be detected and evaluated by the amount of ligatable nicks. The polymerase activity of an enzyme can be detected if the enzyme reaction on gapped DNA gives rise to a structure that can be covalently closed using DNA ligases.

(e) *Structural analyses of DNA.* Measurement of the rates of formation of a circular product and a linear dimer product in a T4 ligase-catalyzed reaction enables one to estimate the flexibilities (lateral and torsional) and/or curvature of a DNA molecule (39–41). The ratio of the two rate constants is termed the *molar cyclization factor,* j_M. For a 366-bp DNA with *Eco*RI termini, the j_M is 5.8×10^{-8} M (39,40). The j_M is independent of ligase concentration, but sensitive to very small changes in fractional twists and persistence length of presumed worm-like DNA coil. Using ligase-catalyzed ring-closure methodology, it has been shown that the DNA helix is essentially independent of NaCl concentrations below ~200 mM (+1 mM $MgCl_2$) and maintains a helical repeat of 10.44 bp per turn (40). In a different set of experiments, a dA_6 tract has been estimated to bend the DNA helix by 17–21° (42).

References

1. Modrich, P., and Lehman, I. R. (1970). *JBC* **245**, 3626–3631.
2. Olivera, B. M., and Lehman, I. R. (1968). *JMB* **36**, 261–274.
3. Weiss, B., Jacquemin-Sablon, A., Live, T. R., Fareed, G. C., and Richardson, C. C. (1968). *JBC* **243**, 4543–4555.
4. Goodchild, J., and Vishwanatha, J. K. (1991). *NAR* **19**, 3745.
5. Hall, Z. W., and Lehman, I. R. (1969). *JBC* **244**, 43–47.
6. Gupta, N. K., Ohtsuka, E., Weber, H., Chang, S. H., and Khorana, H. G. (1968). *PNAS* **60**, 285–292.
7. Fareed, G. C., Wilt, E. M., and Richardson, C. C. (1971). *JBC* **246**, 925–932.
8. Harvey, C. L., and Wright, R. (1972). *Biochem.* **11**, 2667–2671.
9. Nath, K., and Hurwitz, J. (1974). *JBC* **249**, 3680–3688.
10. Westergaard, O., Brutlag, D., and Kornberg, A. (1973). *JBC* **248**, 1361–1364.
11. Kleppe, K., van de Sande, J. H., and Khorana, H. G. (1970). *PNAS* **67**, 68–73.
12. Sano, H., and Feix, G. (1974). *Biochem.* **13**, 5110–5115.
13. Deugau, K. V., and van de Sande, J. H. (1978). *Biochem.* **17**, 723–729.
14. Sgaramella, V., van de Sande, J. H., and Khorana, H. G. (1970). *PNAS* **67**, 1468–1475.
15. Marians, K. J., Wu, R., Stawinski, J., Hozumi, T., and Narang, S. A. (1976). *Nature* **263**, 744–748.
16. Sgaramella, V. (1972). *PNAS* **69**, 3389–3393.
17. Sugino, A., Goodman, H. M., Heyneker, H. L., Shine, J., Boyer, H. W., and Cozzarelli, N. R. (1977). *JBC* **252**, 3987–3994.
18. Sgaramella, V., and Khorana, H. G. (1972). *JMB* **72**, 493–502.
19. Zimmerman, S. B., and Pheiffer, B. H. (1983). *PNAS* **80**, 5852–5856.
20. Pheiffer, B. H., and Zimmerman, S. B. (1983). *NAR* **11**, 7853–7871.
21. Gumport, R. I., and Lehman, I. R. (1971). *PNAS* **68**, 2559–2563.
22. Engler, M. J., and Richardson, C. C. (1982). *Enzymes* **15**(B), 3–29.
23. Montecucco, A., Lestingi, M., Pedrali-Noy, G., Spadari, S., and Ciarrochi, G. (1990). *BJ* **271**, 265–268.
24. Weiss, B., Thomson, A., and Richardson, C. C. (1968). *JBC* **243**, 4556–4563.
25. Little, J. W., Zimmerman, S. B., Oshinski, C. K., and Gellert, M. (1967). *PNAS* **58**, 2004–2011.
26. Harvey, C. L., Gabriel, T. F., Wilt, E. M., and Richardson, C. C. (1971). *JBC* **246**, 4523–4530.
27. Olivera, B. M., Hall, Z. W., and Lehman, I. R. (1968). *PNAS* **61**, 237–244.
28. Dugaiczyk, A., Boyer, H. W., and Goodman, H. M. (1975). *JMB* **96**, 174–184.
29. Legerski, R. J., and Robberson, D. L. (1985). *JMB* **181**, 297–312.
30. Chen, H.-B., Weng, J.-M., Jiang, K., and Bao, J.-S. (1990). *NAR* **18**, 871–878.
31. Wu, D. Y., and Wallace, R. B. (1989). *Genomics* **4**, 560–569.
32. Barringer, K. J., Orgel, L., Wahl, G., and Gingeras, T. R. (1990). *Gene* **89**, 117–122.
33. Landegren, U., Kaiser, R., Sanders, J., and Hood, L. (1988). *Science* **241**, 1077–1080.
34. Alves, A. M., and Carr, F. J. (1988). *NAR* **16**, 8723.
35. Wu, D. Y., and Wallace, R. B. (1989). *Gene* **76**, 245–254.
36. Barany, F. (1991). *PNAS* **88**, 189–193.
37. Richardson, C. C., Masamune, Y., Live, T. R., Jacquemin-Sablon, A., Weiss, B., and Fareed, G. C. (1968). *CSHSQB* **33**, 151–164.
38. Jacquemin-Sablon, A., and Richardson, C. C. (1970). *JMB* **47**, 477–493.
39. Shore, D., Langowski, J., and Baldwin, R. L. (1981). *PNAS* **78**, 4833–4837.
40. Taylor, W. H., and Hagerman, P. J. (1990). *JMB* **212**, 363–376.

41. McNamara, P. T., and Harrington, R. E. (1991). *JBC* **266**, 12548–12554.
42. Koo, H.-S., Drak, J., Rice, J. A., and Crothers, D. M. (1990). *Biochem.* **29**, 4227–4234.
43. Moore, M. J., and Sharp, P. A. (1992). *Science* **256**, 992–997.

B. *Escherichia coli* DNA Ligase
[Polydeoxyribonucleotide synthase (NAD$^+$), EC 6.5.1.2]

The *E. coli* DNA ligase was initially identified as the covalent-joining activity of H-bonded λ DNA circles (1). It is a monomeric enzyme (74 kDa) that catalyzes the formation of phosphodiester bonds at single-strand breaks with 3'-OH and 5'-P termini in a duplex DNA. The enzyme requires NAD as a cofactor, as do other bacterial DNA ligases. Under conventional conditions of ligation reaction, the *E. coli* DNA ligase does not show blunt-end joining activity. In the presence of macromolecules that generate molecular crowding effects, however, the ligase can efficiently catalyze the blunt-end ligation. Compared with the T4 DNA ligase (see Sections I,A and I,C, this chapter), the *E. coli* DNA ligase has narrower substrate specificities, making it a useful tool in some particular applications.

1. FUNCTIONS

a. Reaction conditions

i. Optimal reaction conditions. The ligation reaction with the *E. coli* DNA ligase is typically carried out in 40 m*M* Tris–Cl (pH 8.0), 10 m*M* MgCl$_2$, 5 m*M* DTT, and 0.1 m*M* NAD. The reaction mixture is incubated at 10° to 25°C for 2–16 hr.

It should be noted that the efficiency of a ligation reaction depends on multiple factors, in particular, the termini structure and size of the substrate. Therefore, each reaction should be optimized individually. Among the most important factors that can be readily controlled are pH and temperature.

(a) pH. The optimum pH for the joining reaction is pH 7.5–8.0 in 10 m*M* Tris–Cl buffer and pH 8.0 in Na–P buffer (2). The optimum pH is 6.5 for the NAD–NMN exchange reaction in 50 m*M* K–P buffer (3). In Tris–Cl (pH 8.0), which is the standard condition for the joining reaction, the rate of the exchange reaction is 20% of the maximum.

(b) Temperature. The ligation efficiency shows a sharp profile for temperature optimum as a function of size and nucleotide composition of DNA substrates. For cohesive-end ligation, the recommended reaction temperature is between 15° and 25°C. The ligation reaction can be performed at lower temperatures (4), but requires longer incubation times.

ii. Kinetic parameters. The overall ligation reaction follows a Ping-Pong kinetic mechanism (5). The K_m is 25–56 n*M* for 5'-P termini in poly[d(AT)]. The V for the ligation reaction is 20–25 min^{-1} at 30°C with poly[d(AT)]. The V for NAD–NMN exchange is 45–60 min^{-1}.

iii. Cofactor requirements

(a) **NAD.** The *E. coli* DNA ligase shows a highly specific cofactor requirement for NAD (2,6). The equilibrium for the reaction (ligase + ATP $\rightleftharpoons$ ligase-AMP + PP$_i$) lies 28 times in favor of the formation of the ligase–AMP. [In *E. coli* cells where the intracellular concentration of NAD is ~0.5 mM, virtually all the enzyme is in the adenylylated form.] NADH and (thionicotinamide)-NAD can replace NAD but with sevenfold greater K_m values. The *E. coli* DNA ligase also shows partial activity with 3-acetylpyridine–NAD. NADP and NADPH are inactive as cofactors. The K_m for NAD in ligation reactions is 0.1 μM (3,6) or 7 μM in the presence of stimulating NH_4^+ (5). The K_m for NAD under the NAD–NMN exchange conditions is 0.7 μM, whereas the K_m for NMN is 0.6–0.9 μM (3).

(b) **Divalent cations.** The *E. coli* DNA ligase requires Mg^{2+} (1–3 mM) for optimal activity. Mn^{2+} at low concentrations (0.2–1 mM) can substitute Mg^{2+}, but it can be inhibitory at higher concentrations (3,6). Zn^{2+}, Ca^{2+}, and Co^{2+} cannot replace Mg^{2+}.

iv. Activators and inhibitors

(a) **Monovalent cations.** NH_4^+ ions at saturating concentrations [cf. $K_m \approx 1$ mM] stimulate the V of the *E. coli* DNA ligase approximately 20-fold (5). Note that NH_4^+ has no effect on the T4 DNA ligase. It also has no effect on the rate of NMN exchange, suggesting that activation occurs at a step subsequent to the formation of the ligase–AMP complex. NH_4^+ markedly increases the dissociation of ligase–AMP from DNA. Other monovalent cations also stimulate *E. coli* ligase activity in the order of $NH_4Cl >$ CsCl $\approx$ RbCl $>$ KCl $>$ LiCl. NaCl has no effect at concentrations up to 10 mM.

(b) **PEG.** PEG 8000 greatly increases the rate of cohesive-end joining and enables the enzyme to carry out blunt-end ligation efficiently (7).

(c) **Sulfhydryl reagents.** The *E. coli* DNA ligase does not require the addition of sulfhydryl reagents (2-MSH or DTT).

b. Substrate specificities

The *E. coli* DNA ligase catalyzes the phosphodiester bond formation between the 3'-OH and 5'-P of duplex DNAs. Compared with the T4 DNA ligase, the *E. coli* DNA ligase has restrictive substrate specificities. The canonical activity of *E. coli* DNA ligase is the nick-sealing activity. Under nonconventional reaction conditions, however, the blunt-end joining activity can be substantially increased. Some of the notable properties pertaining to *E. coli* DNA ligase are as follows.

i. Ligation of blunt-end molecules.
Investigation of the factors affecting the blunt-end joining activity of *E. coli* DNA ligase has shown that the efficiency can be substantially increased, even without using volume-exclusion agents, under the following modified reaction conditions (8): (a) high substrate concentrations, e.g., ≥100 pmol each of oligonucleotides, (b) high ligase concentrations, e.g.,

≥50 U per 25 μl reaction volume, and (c) high concentration of Mg^{2+}, e.g., 8 mM. The efficiency of ligation (both blunt end and nick closing) is not significantly affected by β-NAD^+ concentrations over the range of 0 to 26 μM as there appears to be sufficient β-NAD carried with the ligase. Increasing the temperature from 16° to 42°C results in an approximate threefold reduction in blunt-end ligation.

These observations suggest that under situations where the nick-closing activity is maximally required, those conditions that favor blunt-end activity, such as high concentrations of ligase and substrates, should be avoided.

ii. **Single-strand ligation activity.** In certain cases, the *E. coli* DNA ligase has the ability to join two noncomplementary single-strand oligonucleotides (8).

2. APPLICATIONS

The *E. coli* DNA ligase can be used as an alternative to the T4 DNA ligase when blunt-end ligations are not required. The restrictive specificity for substrate termini makes the *E. coli* enzyme more suitable than the T4 DNA ligase for certain applications. For example, the *E. coli* DNA ligase can substitute the T4 DNA ligase in ligase-based *in vitro* DNA amplification systems that rely on specific nick-sealing activity (8). In addition, DNA to DNA ligation can be preferentially achieved in the presence of other possible combinations of ligation between RNA and DNA (9). Because of this high degree of selectivity in ligation, the transformation of *E. coli* using DNAs ligated with the *E. coli* DNA ligase generally gives a lower background. For other general applications, see Section I,A of this chapter.

3. STRUCTURE

i. **Primary and secondary structures.** The *E. coli* DNA ligase is a single polypeptide composed of 671 amino acids (M_r 73,690) (10). The N- and C-terminal residues are Met and Ser, respectively. Thus the N-terminal initiating Met is apparently retained in the mature ligase. The enzyme contains more acidic residues than basic ones. The enzyme has eight Cys residues, none of which seems to be essential for activity.

The secondary structure is predicted to be comprised of 44.4% α helix, 26.4% β sheet, and 5.2% turns (10). The primary and predicted secondary structures of the *E. coli* DNA ligase do not show any significant similarities with those of the T4 DNA ligase.

ii. **Active site.** The substrate-specificity profile suggests that the active site of the *E. coli* DNA ligase apparently contains three subsites, one for the cofactor NAD and the others for 5'-P and 3'-OH termini of DNA substrates. The NAD-binding site has an essential Lys residue that can be covalently modified by the NMN moiety of the cofactor.

iii. **Physical parameters.** The $s^o_{20,w}$ of the purified enzyme is 3.91–4.2. The A_{280} of 1 mg/ml solution of the unadenylylated form of the enzyme is 0.72 (11).

The A_{280}/A_{260} ratios for adenylylated and unadenylylated enzymes are 0.99 and 1.42, respectively.

4. GENETICS

i. Gene location and organization. The structural gene for the E. coli DNA ligase is *lig*. The gene was initially mapped at the 46-min locus (12) but is now repositioned to the 52-min locus on the E. coli genetic map (13). The *lig* gene (2552 bp) has been cloned and sequenced (10). The initiator AUG codon is preceded by a possible RBS (GATG) at the −11 region. The 671 codon ORF is terminated by a UGA codon.

ii. Regulatory regions. The upstream promoter region contains possible transcription regulatory sequences: TATGATG [−140 to −134] which is highly homologous to the Pribnow box (TATAATR) and TTGACG [−156 to −151] which is presumed to be an RNA Pol recognition site. (For general descriptions regarding regulatory elements, refer to Section I,A in Chapter 7.) No particular sequences characteristic of potential transcriptional terminator sites (e.g., stem–loop) have been found downstream of the *lig* gene ORF, leaving the possibility that the *lig* gene is part of a multicistronic operon.

iii. Mutation analysis of the ligase gene. A mutation at the operator of the *lig* gene, called the *lop* mutation, overproduces a normal enzyme (14). Studies with *lig*⁻ mutants defective in the DNA ligase show that the DNA ligase is essential for DNA replication, particularly in the sealing of Okazaki fragments. The *lig* mutants also show increased recombination (15) and mutation frequencies in all genes apparently through the DNA ligase deficiency-induced error-prone repair pathway (16). Some of the interesting DNA ligase mutants are described below.

(a) The *lop*8 and *lop*11 mutants produce a normal ligase but four to five times the amount of the parental strain (14). The growth of these mutants is normal under culture conditions of 30° to 42°C. These E. coli *lop* mutants permit the growth of the ligase-deficient T4 phage gene 30 mutants. This observation suggests that the E. coli ligase is able to act on the *in vivo* substrates of T4 (and also T7) DNA ligases when the latter enzymes are missing.

(b) The *lig*4 mutant is a *ts* mutant. The mutant ligase has ~35% at 30°C and 1% or less at 42°C of the wild-type ligase activity (12). The mutant grows normally under all culture conditions. It is as UV-resistant as wild-type and slightly more sensitive to MMS, suggesting that the DNA repair mechanism is not severely impaired. The recombination frequency is also normal (3).

(c) The *lig*ts7 mutant, originally isolated as a UV-sensitive TAU ts7 mutant (17), is a conditionally lethal *ts* mutant and shows severe ligase deficiency at all temperatures (18). It has 25% wild-type activity at 15°C and less than 1% at 42°C. The mutant grows normally at or below 34°C, but is unable to grow at temperatures above 37°C. The mutation causes an excessive accumulation of 10S DNA (Okazaki) fragments at permissive temperatures. The mutant is abnormally sensitive to UV and MMS compared with the wild-type controls.

The *E. coli lig*ts7 mutation can be "complemented" by a phage Mu mutation *lig*3 (*ts* or *amber*), which restores the viability of the *lig*ts7 strain to the wild-type level at 42°C (19). The suppression of *lig*ts7 by lysogenic Mu *lig*3 is due to the chromosomal (and plasmid as well) relaxation of supercoiling induced by the Mu *lig* gene product and to consequent transcriptional increase of the host *lig* gene. [Note: The phage Mu *lig* gene has been renamed *gem* (gene expression modulation) (20).]

5. SOURCES

The *E. coli* DNA ligase has been purified to an apparent homogeneity after a 1400-fold purification from *E. coli* LC81 (11), a mutant strain that produces four times the usual amount of normal ligase (14). Wild-type *E. coli* cells are estimated to contain about 300 DNA ligase molecules per cell (11).

Cloning of the *lig* gene in a λ phage vector (λgt4 *lop*11 *lig*$^+$) has significantly improved the production of the *E. coli* DNA ligase (21). The *lop*11 mutation itself leads to a 5-fold overproduction of the ligase, and the expression in *E. coli* of the cloned gene results in a 10-fold increase of DNA ligase over that in extracts of uninfected *lig*$^+$ cells (22). Further modification of the recombinant λ phage vector by introducing an amber mutation "S7" that prevents cell lysis (λgt4 *lop*11 *lig*$^+$ S7) results in a stable lysogen (*E. coli* strain 594) which, on induction, produces an over 500-fold increase in DNA ligase activity. This overexpression corresponds to ~5% of the total cellular protein. A simplified three-step purification procedure produces a ~40% yield of homogeneous enzyme at the level of 30 mg ligase per 120 g of cell paste (23).

References

1. Gellert, M. (1967). *PNAS* **57**, 148–155.
2. Zimmerman, S. B., Little, J. W., Oshinsky, C. K., and Gellert, M. (1967). *PNAS* **57**, 1841–1848.
3. Little, J. W., Zimmerman, S. B., Oshinski, C. K., and Gellert, M. (1967). *PNAS* **58**, 2004–2011.
4. Dugaiczyk, A., Boyer, H. W., and Goodman, H. M. (1975). *JMB* **96**, 174–184.
5. Modrich, P., and Lehman, I. R. (1973). *JBC* **248**, 7502–7511.
6. Olivera, B. M., and Lehman, I. R. (1967). *PNAS* **57**, 1700–1704.
7. Pheiffer, B. H., and Zimmerman, S. B. (1983). *NAR* **11**, 7853–7871.
8. Barringer, K. J., Orgel, L., Wahl, G., and Gingeras, T. R. (1990). *Gene* **89**, 117–122.
9. Okayama, H., and Berg, P. (1982). *Mol. Cell. Biol.* **2**, 161–170.
10. Ishino, Y., Shinagawa, H., Makino, K., Tsunasawa, S., Sakiyama, F., and Nakata, A. (1986). *MGG* **204**, 1–7.
11. Modrich, P., Anraku, Y., and Lehman, I. R. (1973). *JBC* **248**, 7495–7501.
12. Gottesman, M. M., Hicks, M. L., and Gellert, M. (1973). *JMB* **77**, 531–547.
13. Bachmann, B. (1987). *Escherichia coli* and *Salmonella typhimurium*. In "Cellular and Molecular Biology" (J. J. Ingraham, K. B. Low, B. Magasanik, M. Schaechter, and H. E. Umbargo, eds.), Vol. 2, pp. 807–876. American Society for Microbiology, Washington, D.C.
14. Gellert, M., and Bullock, M. L. (1970). *PNAS* **67**, 1580–1587.
15. Konrad, E. B. (1977). *J. Bacteriol.* **130**, 167–172.
16. Morse, L. S., and Pauling, C. (1975). *PNAS* **72**, 4645–4649.
17. Pauling, C., and Hamm, L. (1968). *PNAS* **60**, 1495–1502.
18. Konrad, E. B., Modrich, P., and Lehman, I. R. (1973). *JMB* **77**, 519–529.

19. Paolozzi, L., Ghelardini, P., Liebart, J. C., Capozzoni, A., and Marchelli, C. (1980). *NAR* **8**, 5859–5874.
20. Ghelardini, P., Liebart, J. C., Paolozzi, L., and Pedrini, A. M. (1989). *MGG* **216**, 31–36.
21. Panasenko, S. M., Cameron, J. R., Davis, R. W., and Lehman, I. R. (1977). *Science* **196**, 188–189.
22. Cameron, J. R., Panasenko, S. M., Lehman, I. R., and Davis, R. W. (1975). *PNAS* **72**, 3416–3420.
23. Panasenko, S. M., Alazard, R. J., and Lehman, I. R. (1978). *JBC* **253**, 4590–4592.

C. T4 DNA Ligase
[Polydeoxyribonucleotide synthase (ATP), EC 6.5.1.1]

The T4 DNA ligase is a single 55-kDa polypeptide. It uses ATP as the energy-yielding cofactor and requires Mg^{2+} ions for activity. The T4 DNA ligase has a broad specificity for substrate termini. It catalyzes the formation of a phosphodiester bond between adjacent 3'-OH and 5'-P termini of DNA and RNA in any combination of duplex structures, albeit at different rates. The enzyme can join, under normal conditions, DNA fragments having either cohesive or blunt ends. The T4 DNA ligase is thus the most widely used enzyme in DNA ligations.

1. FUNCTIONS

a. Reaction conditions

i. Optimal reaction conditions. Optimal conditions for ligation reactions can be best determined experimentally using a cloning system involving the construction of a recombinant DNA and bacterial transformation as a test of ligation efficiency (1). Because the conditions generating high MW ligation products are rather counterproductive due to the poor efficiency of transformation, the efficiency of ligation in experiments for cloning and amplification purposes is better represented by the yield of transformable products. Applying this rationale, the following conditions are recommended for standard blunt-end ligation. Cohesive end ligation can also be performed by a minor modification of the same conditions.

The *standard ligation* mixture (20 μl) contains 50 mM Tris–Cl (pH 7.6), 10 mM $MgCl_2$, 5% (w/v) PEG 8000, 1 mM ATP, 1 mM DTT, a vector-to-insert ratio of 3 [for example, 170 ng (60 fmol) of blunt-end, linear, dephosphorylated pBR322 DNA and 13 ng (20 fmol) of a 1-kb blunt-end DNA fragment], and 1 U (or 0.1 U for sticky-end ligation) of the T4 DNA ligase. The mixture is incubated at 14°C overnight.

The reaction is stopped by adding 1 μl of 0.5 M Na_2EDTA (pH 8.0). Transformation (100 μl) is performed with 2-μl sample aliquots diluted to 10 μl in buffer TE [10 mM Tris–Cl (pH 7.2) and 1 mM Na_2EDTA]. This dilution is important to avoid the inhibition of transformation which occurs at high DNA and PEG concentrations. For a small-scale (20 μl) transformation, 1 μl (1–2 ng) of the diluted ligase reaction mixture is used.

The ratio of vector to blunt-end insert does not influence greatly the efficiency of transformation. A less than threefold increase is obtained when the ratio is varied from 0.03 to 3.0.

ii. **Kinetic parameters.** The steady-state kinetics of the joining reaction of oligo(dT)$_{10}$ on poly(dA) follow a Ping-Pong mechanism with a K_m of 0.6 μM (5'-P ends) and a V of 5 min^{-1} at 20°C (2). Under the standard reaction conditions described earlier, the T4 DNA ligase catalyzes the ligation of a pair of oligonucleotides (8-mer and 14-mer) with a K_m of 50 nM at 30°C. The K_m for a duplex decamer in blunt-end ligation is 50 μM and is unaffected by the presence of the stimulating T4 RNA ligase (3). The K_m for nicked DNA is 1.5 nM expressed as the concentration of internal phosphomonoesters (4); V is 1.7 μmol per 20 min per mg protein.

iii. **Cofactor requirements**

(a) *ATP.* The T4 DNA ligase requires ATP as a cofactor (5). The K_m for ATP is 2 μM in an ATP–PP$_i$ exchange reaction (6) and 14–100 μM in a ligation reaction (4). Note that the enzyme does not catalyze an ATP–AMP exchange reaction. The ATP concentration can be varied from 10 μM to 1 mM without affecting the efficiency of blunt-end or sticky-end ligations. In contrast, the recircularization of the phosphorylated vector is maximal at 0.1 mM ATP.

The enzyme can use dATP but no other nucleotides as a cofactor substitute. The dATP (66 μM) results in a 60% decrease in ligation activity and is a competitive inhibitor against ATP with a K_i of 35 μM (4). NAD cannot serve as a cofactor.

(b) *Divalent cations.* The T4 DNA ligase shows no detectable activity without Mg^{2+}. The optimal concentration of Mg^{2+} is 10 mM. At 3 and 30 mM Mg^{2+}, the ligase activities are reduced to 35 and 80% of the maximum, respectively (4). Mn^{2+} can substitute Mg^{2+} and, depending on substrates, the rate can be either lower, e.g., 25% on nicked duplex DNA at 10 mM Mn^{2+} (4,6), or higher, e.g., two times on homopolydeoxynucleotides than that observed with Mg^{2+} (7).

iv. **Activators and inhibitors**

(a) *Macromolecular stimulants.* Macromolecules such as PEG, Ficoll, albumin, and glycogen stimulate T4 DNA ligase activity most remarkably in blunt-end ligation (see Section I,A,3, this chapter).

PEG: The addition of 5% (w/v) PEG 8000 increases the ligation efficiency three- to sixfold as measured by the number of transformants. Above 5%, PEG shows a dramatic inhibition of transformation, although the amount of high MW DNA is increased.

DNA-binding proteins: HMG 14, a *h*igh *m*obility *g*roup protein that preferentially associates with active chromatin through the electrostatic interaction with the phosphate backbone of DNA, binds DNA, stabilizes the DNA duplex, and enhances intermolecular ligation of linear DNA at low concentrations. At 50 mol of HMG 14 per mole of DNA, a 50% increase has been observed in the blunt-end ligation catalyzed by the T4 DNA ligase, whereas intramolecular ligation (or recircularization) of the DNA was completely inhibited (8). Phosphorylation of HMG 14 at Ser-20 specifically abolishes the ability of HMG 14 to stimulate intermolecular DNA ligation.

Protamine, up to a molar ratio of 40:1, gives a modest stimulation of the intermolecular (cohesive ends) ligation, predominantly generating dimers or trimers (8).

T4 RNA ligase: Blunt-end ligation at low concentrations of T4 DNA ligase is stimulated up to 20-fold in the presence of the T4 RNA ligase, approaching the efficiency of cohesive-end ligation (3). The joining of cohesive ends by the T4 DNA ligase is only slightly increased by the presence of the T4 RNA ligase.

(b) Monovalent cations. In contrast to the *E. coli* DNA ligase, the T4 DNA ligase is not stimulated by the presence of NH_4^+. At low concentrations (e.g., ~20 mM) of monovalent cations, however, some stimulation (~30%) is observed. At high concentrations above ~70 mM, monovalent cations such as Na^+, K^+, Cs^+, Li^+, and NH_4^+ show inhibitory effects on ligation reactions. The ligase activity is almost completely inhibited at 200 mM NaCl (2,9).

(c) Organic and/or inorganic compounds. Many organic or inorganic compounds interact with either the ligase or DNA substrates, leading to a stimulation or inhibition of the ligase activity.

Spermine and *spermidine* (but not some other polyamines) at 0.5–1 mM concentrations greatly stimulate the T4 DNA ligase-catalyzed ligation reaction (10,11). This stimulation is unique for intermolecular ligation, suggesting the role of specific polyamines in charge neutralization, condensation, and aggregation of DNA into high local concentrations. Spermine at high concentrations (>5 mM) inhibits the ligation reaction (9). The addition of polyamines results in a large increase in the apparent K_m for DNA substrates, whereas the V and the K_m for ATP remain unaffected.

Hexaaminecobalt chloride also stimulates the formation of high MW DNA but does not increase the number of transforming DNA molecules. The transformation is rather inhibitory at ≥3 mM.

Cibacron blue F3GA, which is often used as a chromophoric affinity adsorbent in the purification of the enzyme, is an inhibitor of the ligase, giving 50% inhibition at 10 μM (12). The inhibition is largely due to electrostatic interactions between the sulfogroups of the dye and the Lys residues of the enzyme.

Ethidium bromide inhibits the T4 DNA ligase-catalyzed ligation of poly[d(AT)] with an ID_{50} of ~4.3 μM (13). The inhibition is not due to the interaction of EtBr with the enzyme per se, but to the drug–DNA binding (or intercalation) so that formation of the active DNA–Mg^{2+}–AMP–enzyme complex is impaired.

(d) Anions. Phosphates at >25 mM concentrations and, more generally, salts at >50 mM concentrations are considerably inhibitory in blunt-end ligation.

(e) Sulfhydryl reagents. The T4 DNA ligase requires reducing agents such as 2-MSH or, more preferably, DTT (4). 2-MSH (10 mM) can replace 10 mM DTT but only with 40% of the optimal activity.

(f) Temperature. The T4 DNA ligase is relatively thermolabile, and an apparent thermal inactivation occurs at temperatures above 37°C. At 20°C, a tempera-

ture often used for DNA ligations, the enzyme loses most of its activity within 24 hr (14). A 5-min incubation at 60°C completely destroys the ligase activity (15).

b. Considerations on ligation reaction conditions

i. pH. The optimum pH for the joining of nicks is pH 7.2–7.6 in Tris–Cl (66 mM) (4). At pH 6.9 and 8.0, the enzyme has 46 and 65%, respectively, of the activity at pH 7.6. The ATP–PP_i exchange reaction exhibits a pH optimum similar to that for the joining reaction (6).

ii. Temperature. The rate of ligase reaction reflects in principle the dynamic interactions of the components in a ligation mixture: blunt-end versus cohesive-end termini, collision frequencies of the ligatable ends, and H-bonding strengths of cohesive termini on the dsDNA helix. Therefore, temperature is an important factor determining the ligation efficiency. Short oligomers are extremely sensitive to temperature. In general, the optimal joining occurs at temperatures at or slightly below the melting temperature (T_m) of the termini (but not of the whole duplex structure) of the substrate. For example, the T_m of the H-bonded EcoRI-cut cohesive ends is 5°–6°C, and 4°C has been shown to be the optimum temperature for the ligation (16). Although the temperature dependence of the joining of various substrates is generally described by sigmoid curves, the optimum temperature of each reaction can vary, and sometimes varies widely depending on substrate structures and reaction conditions. For example, the optimum temperature for ligation of nicks in circular DNA (pBR322) has been reported to be 37°C (17). The nick-closing activity on oligonucleotides has been shown to be biphasic with optima at 28° and 37°C (9). Blunt-end and cohesive-end joining of oligomeric (16 or more bases) duplex DNA can be carried out at ~25°C (18,19). However, smaller duplexes, e.g., self-complementary octanucleotides, require lower temperatures (12°–15°C) to be consistent with their T_m of 18°C (3,20).

Temperature also affects the maximum extent of the ligation reaction. The joining of oligo$(dT)_{10}$ on poly(dA) at three different temperatures (5°, 10°, and 25°C) shows that the extent of ligation decreases in the order of 5° > 10° > 25°C, although the initial reaction rates are in reverse order (2). The decrease in the level of maximal ligations at the higher temperatures is probably due to thermal denaturation of the duplex DNA as the T_m for oligo$(dT)_{10}$ on poly(dA) is 29°–32°C.

The ligation efficiency determined from the number of transformants provides the following insight on the ligation reaction temperature (1). (a) For sticky-end ligation, a 4-hr incubation at room temperature (26°C) gives approximately 25-fold more transformants than the 4-hr incubation at 4°C and reaches ~90% of the efficiency obtained during a 23-hr incubation. The highest efficiency was obtained with 0.1 U of ligase. (b) For the blunt-end ligation, incubation at 14°C overnight yields at least 4 times more transformants than incubation at room temperature for 4 hr. The optimal amount of ligase is approximately 1–2 U, at least 10 times greater than that for the sticky-end ligation.

c. Substrate specificities

As described in detail in Section I,A, this chapter, the T4 DNA ligase has rather broad substrate specificities compared with the *E. coli* DNA ligase. The following are additional substrate specificities pertinent to the T4 DNA ligase.

   i. **Partially double-stranded substrates.** When present at high concentrations (e.g., 60 Weiss units), the T4 DNA ligase carries out an intramolecular ligation of partially dsDNA having an extended single strand. The ligation results in a product with a stem–loop structure (21). One example of such partially dsDNA is that of two synthetic oligodeoxynucleotides of unequal length annealed so that the shorter oligonucleotide is completely base-paired to the 3' half of the longer oligonucleotide. Intramolecular ligation occurs between the extended, unpaired 5'-P and the paired 3'-OH ends. When the substrate has the alternative configuration in which the 5'-P is recessed instead of protruding, the yield of looped product is higher. For a given set of oligonucleotides, the efficiency of loop formation is dependent on the length of ssDNA and most probably on the composition of terminal nucleotides as well. At room temperature, the ligation products with 15- and 25-base loops are readily formed. The DNAs with potential 10- and 20-base loops are poor substrates.

   ii. **Gapped duplex DNA substrates and bulged/looped products.** Single-strand gaps of 1–5 nt in DNA molecules can be sealed by the T4 DNA ligase, generating bulged or looped products. The optimal reaction conditions are the same as those for the ligation of blunt-ended DNAs (10). Spermidine (2 mM) stimulates the ligation of gaps as well as general ligation activities. The ligation of single-strand gaps is a slow process, reaching a plateau after several hours at 25°C. Approximately 10% of the circular pBR322 DNA with a gap of 1–5 nt has been shown to be converted to a covalently closed form. The slow ligation is due not so much to the slow down in the formation of the activated intermediate but rather to the slow 3'-OH attack on the adenylyl-5'-P end (22).

   Transformation of *E. coli* cells with bulged or looped molecules results in mutant molecules with deletion at a high frequency. For unknown reasons, gaps formed by a 2-min incubation with exonuclease III do not produce mutants with deletions larger than 100 bp, although some of the molecules have considerably larger gaps.

   iii. **Substrates with nicks 3' or 5' to AP sites or to mispaired bases.** In the presence of complementary polynucleotides, an AP (apurinic/apyrimidinic) site at the 3' or the 5' end of juxtaposed oligodeoxynucleotides does not prevent ligation by the T4 DNA ligase. For example, a 5'-phosphorylated base-free oligonucleotide, pd(−)C(TC)$_4$, can be ligated when hybridized to poly[d(AG)], albeit at a slow rate (22). The rate decrease is more severe when the AP site is at the 3' end.

   A mismatch at the 3' or 5' end of oligodeoxynucleotides does not prevent their ligation, although the efficiency is substantially reduced. Mismatch ligation is fivefold more frequent if the single mismatch occurs at the 3' side rather than at the 5' side of the junction (9).

All of the ligations just mentioned require the presence of a continuous complementary strand. In situations of blunt-end ligation, i.e., with an AP site or a mispaired base at the 3' or 5' end of one strand of the duplexes, the T4 DNA ligase is unable to ligate these strands. A missing nucleotide (which is responsible for one unpaired nucleotide protruding at the 3' or 5' end of the complementary strand) does not stop ligation of the shorter oligodeoxynucleotide between independent duplexes (22).

iv. Intracellular ligation of blunt-ended DNA. Introduction of the T4 DNA ligase into eukaryotic cells by hypertonic shock has been shown to repair potentially lethal double-strand DNA breaks of the blunt-end type inflicted by agents, including the exogenous *Pvu*II restriction enzyme (23).

2. Applications

The T4 DNA ligase is usually the enzyme of choice for cohesive-end ligation and, in particular, for blunt-end ligation. (For general applications, see Section I,A, this chapter.)

3. Structure

The T4 DNA ligase is a single polypeptide with 487 deduced amino acids (M_r 55,230) (24). The enzyme contains six Cys and almost equal numbers of basic and acidic amino acid residues. The N- and C-terminal residues are Met and Leu, respectively, and have been confirmed by amino acid analyses. Like *E. coli* DNA ligase, the N-terminal initiating Met is maintained in the mature enzyme. Lys-221 is believed to be the covalent linkage site for AMP.

4. Genetics

i. Primary structure of T4 DNA ligase gene. The T4 DNA ligase is the product of the phage T4 gene 30. It is contained in the 1.9-kbp *Hin*dIII restriction fragment of the T4 genome (24). The ligase gene is flanked on the 3' and 5' sides by overlapping initiator and terminator codons of hitherto unidentified coding sequences. The terminator codon for the ligase is UGA. The ligase gene sequence is AT rich and has a composition of A (34.6%), T (30.5%), C (15.0%), and G (20.0%) in the coding strand, similar to the base composition of the whole T4 genome.

ii. Regulatory regions. Although a potential promoter site with appropriate Pribnow box (TATAAT) and −35 sequences is found in the 5' flanking "X" gene, available data suggest that the ligase gene may be transcribed *in vivo* from two different promoters, the second one being most likely located outside of the *Hin*dIII genomic region (24,25).

iii. Mutant analyses. Several *amber* and *ts* mutants of gene 30 have been isolated and studies with the mutants suggest that the T4 DNA ligase plays an important role in DNA replication. The T4 DNA ligase is also required for encapsidation of phage T4 DNA (26). The gene 30 mutations affect other processes of DNA metabolism such as increased recombination and high UV sensitivities (27,28). T4 DNA ligase-deficient mutants can grow on *E. coli lop* mutant hosts that overproduce the *E. coli* DNA ligase, suggesting that the host ligase is able to catalyze the ligations on nonnatural substrates. Some of the interesting gene 30 mutants are discussed below.

(a) T4*ts*B20, a gene 30 *ts* mutant, shows ~10% wild-type ligase activity at the permissive temperature (25°C) and less than 1% activity at 37°C (29). Another *ts* mutant, *ts*A80, shows ligase activity approximately fivefold higher than that of *ts*B20. Infection of gene 30 *ts* mutants (e.g., *ts*B20) under the nonpermissive conditions of 42°–44°C results in an accumulation in cells of low MW DNA fragments. The DNA fragments have an s_{20} of 5–15 S in alkali, similar to Okazaki fragments. The *ts*B20 mutation also increases recombination frequencies in *E. coli* (30).

(b) T4*am*H39, a gene 30 *amber* mutant, does not produce a functional ligase in a nonpermissive host *E. coli* B/5, but induces ligase activity comparable to that of the wild type in a permissive host *E. coli* CR63 (29). After infection in *E. coli*, gene 30 *amber* mutants (e.g., T4*am*H39X) initially show phage DNA synthesis at a normal rate, but the rate drops rapidly. DNA synthesis stops within 10 min with a yield of less than 10% of the DNA produced by the wild-type T4 (29,31).

5. Sources

The T4 DNA ligase has been purified from *E. coli* B cells infected with the wild-type T4 phage (4) or the replication-defective phage mutant *am*N82 (32). Ligase activity can be detected 5 min after infection and reaches a maximum at ~20 min (33). A purified enzyme is obtained after a 1000-fold purification.

The T4 DNA ligase is more conveniently prepared from an overexpression system using lysogens of λ T4 *lig* recombinant phages (15). Although the T4 *lig* gene can be transcribed from any of the three major λ promoters (p_L, p_R, and p_R'), transcription at the late promoter (p_R') is the most effective, with the amplification level reaching ~2% of the total cellular protein. When combined with a rapid purification procedure, the λ T4 *lig* lysogen yields a 300-fold increase in the amount of high purity T4 DNA ligase (17.5 ATP–PP_i exchange units/μg protein) (34).

Further improvement of the expression to ~20% of the total cellular protein has been achieved by subcloning the T4 *lig* gene in the λ NM540 T4*lig* phage into a vector (pCP3 *lig*14) provided with a thermoinducible expression (λ cI857, p_L) and temperature-regulated runaway replication (35).

T4 DNA ligase immobilized to Sepharose 4B using tresyl chloride activation is reported to be far more thermostable than the soluble form of the enzyme (14).

The enzyme can be stored in 50% glycerol at −20°C for 6 months with an activity loss of <10%.

References

1. King, P. V., and Blakesley, R. W. (1986). *Focus(BRL)* **8**(1), 1–3.
2. Raae, A. J., Kleppe, R. K., and Kleppe, K. (1975). *EJB* **60**, 437–443.
3. Sugino, A., Goodman, H. M., Heynecker, H. L., Shine, J., Boyer, H. W., and Cozzarelli, N. R. (1977). *JBC* **252**, 3987–3994.
4. Weiss, B., Jacquemin-Sablon, A., Live, T. R., Fareed, G. C., and Richardson, C. C. (1968). *JBC* **243**, 4543–4555.
5. Becker, A., Lyn, G., Gefter, M., and Hurwitz, J. (1967). *PNAS* **58**, 1996–2003.
6. Weiss, B., Thomson, A., and Richardson, C. C. (1968). *JBC* **243**, 4556–4563.
7. Fareed, G. C., Wilt, E. M., and Richardson, C. R. (1971). *JBC* **246**, 925–932.
8. Sheflin, L. G., Fucile, N. W., and Spaulding, S. W. (1991). *BBRC* **174**, 660–666.
9. Wu, D. Y., and Wallace, R. B. (1989). *Gene* **76**, 245–254.
10. Nilsson, S. V., and Magnusson, G. (1982). *NAR* **10**, 1425–1437.
11. Psoso, H., and Kuosmanen, M. (1983). *BBRC* **117**, 217–222.
12. Baronaite, Z. A., Martsishauskas, R. P., and Peslyakas, I.-G. I. (1984). *Biokhimiya* **49**, 142–148. [English translation: pp. 121–126]
13. Montecucco, A., Pedrali-Noy, G., Spadari, S., Lestingi, M., and Ciarrocchi, G. (1990). *BJ* **266**, 379–384.
14. Bulow, L., and Mosbach, K. (1982). *BBRC* **107**, 458–464.
15. Murray, N. E., Bruce, S. A., and Murray, K. (1979). *JMB* **132**, 493–505.
16. Ferretti, L., and Sgaramella, V. (1981). *NAR* **9**, 85–93.
17. Pohl, F. M., Thomae, R., and Karst, A. (1982). *EJB* **123**, 141–152.
18. Sgaramella, V., and Khorana, H. G. (1972). *JMB* **72**, 493–502.
19. Sgaramella, V., and Erlich, S. D. (1978). *EJB* **86**, 531–537.
20. Deugau, K. V., and van de Sande, J. H. (1978). *Biochem.* **17**, 723–729.
21. Western, L. M., and Rose, S. J. (1991). *NAR* **19**, 809–813.
22. Goffin, C., Bailly, V., and Verly, W. G. (1987). *NAR* **15**, 8755–8771.
23. Durante, M., Grossi, G. F., Napolitano, M., and Gialanella, G. (1991). *Int. J. Radiat. Biol.* **59**, 963–971.
24. Armstrong, J., Brown, R. S., and Tsugita, A. (1983). *NAR* **11**, 7145–7156.
25. Wilson, G. G., and Murray, N. E. (1979). *JMB* **132**, 471–491.
26. Zachary, A., and Black, L. W. (1981). *JMB* **149**, 641–658.
27. Kozinski, A. W., and Kozinski, P. B. (1969). *J. Virol.* **3**, 85–88.
28. Baldy, M. W. (1970). *Virol.* **40**, 272–287.
29. Richardson, C. C., Masamune, Y., Live, T. R., Jacquemin-Sablon, A., Weiss, B., and Fareed, G. C. (1968). *CSHSQB* **33**, 151–164.
30. Berstein, H. (1968). *CSHSQB* **33**, 325–331.
31. Warner, H. R., and Hobbs, M. D. (1967). *Virol.* **33**, 376–384.
32. Knopf, K. W. (1977). *EJB* **73**, 33–38.
33. Weiss, B., and Richardson, C. C. (1967). *PNAS* **57**, 1021–1028.
34. Tait, R. C., Rodriguez, R. L., and West, R. W., Jr. (1980). *JBC* **255**, 813–815.
35. Remaut, E., Tsao, H., and Fiers, W. (1983). *Gene* **22**, 103–113.

II. RNA LIGASE: T4 RNA LIGASE
[Polyribonucleotide synthase (ATP), EC 6.5.1.3]

RNA ligases catalyze the 3′ → 5′ phosphodiester bond formation of RNA molecules with concomitant hydrolysis of ATP to AMP and PP_i. Initially detected as an activity that catalyzes the circularization of polyribonucleotides with 3′-OH and 5′-P ends (1), the RNA ligase is found in *E. coli* after infection by T-even (but not T-odd) bacteriophages. The enzyme can carry out intermolecular as well as intramolecular ligation reactions on ssRNA, ssDNA, or a variety of oligonucleotides and polynucleotides having 5′-P and 3′-OH termini.

The T4 RNA ligase is a monomeric enzyme with 375 amino acid residues. It is the T4 gene 63 product. The *in vivo* role of RNA ligase activity is not known. The T4 RNA ligase demonstrates an unrelated activity that promotes tail fiber attachment to the phage base plate, thus playing an apparent phenotypic role in phage assembly.

The phage RNA ligase is distinct from other eukaryotic or fungal RNA ligase activities and self-splicing introns. This section focuses on the T4 RNA ligase, the only phage RNA ligase that has been extensively characterized and used in recombinant DNA technology.

1. FUNCTIONS

a. Reaction conditions

i. Optimal reaction conditions. The T4 RNA ligase has a pH optimum between 7.5 and 8.2 in Tris–Cl buffer (1). HEPES (50 mM, pH 7.9) and glycylglycine can also be used as buffers, although HEPES is preferred (2,3). Maximum adenylylation of the enzyme is achieved above pH 8.7 (4).

With properly blocked 5′-P donor oligodeoxynucleotides (25-mers), the following ligation conditions are optimal for ssDNA ligation (5): 50 mM Tris–Cl (pH 8.0), 10 mM $MgCl_2$, 10 μg/ml BSA, 25% PEG, 1 mM hexaaminecobalt chloride, and 20 μM ATP with a 5:1 acceptor:donor ratio at a ~1 mM donor DNA concentration. Incubation at 25°C with 0.1 U RNA ligase results in ~25% yield after 24 hr or with 1 U RNA ligase 40% yield at 6 hr, respectively. Incubation at 37°C gives predominantly adenylylated side products.

ii. Kinetic parameters. The kinetic parameters for T4 RNA ligase-catalyzed reactions have been obtained under unfulfilled steady-state conditions since the enzyme does not seem to be saturated with substrates. Moreover, the RNA ligase is subject to substrate and/or product inhibition. In general, the catalytic rate of the RNA ligase is very slow compared with DNA ligases.

The apparent kinetic constants for intermolecular ligations of mono- and oligonucleotide donor and acceptor substrates are listed in Table 2.1. Table 2.1 also shows some kinetic parameters for ssRNA cyclization and phosphodiester bond-forming half-reactions.

TABLE 2.1 Apparent Steady-State Kinetic Constants of T4 RNA Ligase

Substrates Donor–acceptor (pair)	$K_{m,app.}$	$V_{app.}$	pH (temperature)	Ref.
Intermolecular ligation				
pdTp	0.17 mM	0.07 hr^{-1}	8.3 (17°C)	7
d(A)$_5$	2.1 mM	0.2 hr^{-1}		
pdCp	1.1 mM	0.4 hr^{-1}	8.3 (17°C)	7
d(A)$_5$	1.8 mM	0.6 hr^{-1}		
pd(T)$_4$Cp	0.5 mM	0.08 hr^{-1}	8.0 (17°C)	8
d(A)$_6$	1.5 mM	0.2 hr^{-1}		
Cyclization				
A$_9$	10 μM	0.58 hr^{-1}	7.5 (37°C)	20
A$_{10}$	~1 μM	3.8 hr^{-1}		
A$_{100}$	~1 μM	1.2 hr^{-1}		
Half-reaction (phosphodiester bond formation)				
AMP–pGp	0.6 mM	180 min^{-1}	8.4 (37°C)	16
A$_3$	0.18 mM	165 min^{-1}		
AMP–pGp (+ U$_3$)	0.6 mM	96 min^{-1}		
AMP–pCp (+ A$_3$)	0.8 mM	192 min^{-1}		
AMP–pCp (+ U$_3$)	0.8 mM	48 min^{-1}		

iii. Cofactor requirements

(a) ATP. ATP is the specific cofactor for the T4 RNA ligase. The K_m for ATP is 0.2 μM in circularization reactions (1) or 12 μM in ATP–PP$_i$ exchange reactions (6). Of the common nucleoside triphosphates, only dATP can substitute for ATP but with only 10–20% efficacy.

Increasing the ATP concentrations up to 100 μM leads to an increase of ligation efficiency up to 50%. A higher concentration of ATP inhibits the reaction, resulting in excessive accumulation of the AMP–RNA (donor) intermediate. Eventually, however, this condition leads to a higher product yield after a long incubation time (7). At high enzyme and ATP concentrations, adenylylation at the 3'-P of both donor and product occurs as a side reaction. For practical purposes, an ATP concentration of 20 μM provides the best compromise between ligation and adenylylation (5).

(b) Divalent cations. The T4 RNA ligase requires divalent cations for full activity. The optimum concentration of Mg^{2+} is 5 mM (6); concentrations above 10 mM are inhibitory. In an ATP-regenerating reaction system, 10 mM MnCl$_2$ can substitute for 10 mM MgCl$_2$, offering a higher yield of ligation products (2). In other reaction systems, however, Mn^{2+} has been shown to be inhibitory, with a 50% inhibition of ligation at 5 mM Mn^{2+} (5).

iv. Activators and inhibitors

Spermine at 2–5 mM, in the copresence of 10 mM $MnCl_2$ but not $MgCl_2$, greatly enhances both the rate and yield of 3',5'-dNDP ligation (2).

RNase, as a ssDNA-binding protein, does not affect the initial rate but gives a modest increase in the ligation yield. The presence of RNase A enables the ligation between duplex forms of oligodeoxynucleotides which are not substrates under normal conditions (8).

PEG 8000 increases the efficiency of ssDNA ligation up to 30-fold, achieving a 67% yield at the optimum PEG concentration of 25% (5).

NaCl at high concentrations (e.g., 50 mM) inhibits the ligation reaction by more than 60% in the presence of 16% PEG.

Hexaaminecobalt chloride (1 mM) results in a threefold increase of ligation efficiency.

Dimethyl sulfoxide (Me_2SO) at 10–20% increases the yield by two- to threefold in some RNA ligation reactions (9,10). However, in DNA ligation reactions, e.g., with pdNp, Me_2SO has been shown to be inhibitory (2). The T4 RNA ligase by itself can tolerate up to 40% Me_2SO without any appreciable decrease in activity.

Sulfhydryl reagents: RNA ligase loses activity in the absence of DTT or 2-MSH. High concentrations of DTT protect the activity and increase the product yield. The enzyme is inactivated by SH group-modifying reagents.

Cibacron blue (F3GA), which is often used as an affinity adsorbent in the purification of RNA ligase (see *Sources* below), inhibits the enzyme, with 50% inhibition at 0.1 M (11). It is a noncompetitive inhibitor with respect to oligo$(A)_{20}$ and a competitive inhibitor with respect to ATP. The interaction of the dye with the RNA ligase is mainly mediated by electrostatic interactions between the sulfo groups of the dye and the Lys residues of the enzyme.

b. ACTIVITY ASSAYS AND UNIT DEFINITIONS

i. **5'-P ligation–phosphatase digestion assay.** This is probably the most common ligase assay. It is based on the conversion of 5'-^{32}P-labeled oligodeoxynucleotides or oligoribonucleotides to a form resistant to alkaline phosphatase digestion (1,6,12). The assay mixture (20 µl) consists of 50 mM Tris–Cl (or HEPES, pH 7.9), 10 mM $MgCl_2$, 20 mM DTT, 1 mM ATP, 10 µM 5'-^{32}P-labeled substrates [e.g., rA_{20} or poly(A)], and 0.1–0.6 U ligase. After 30 min at 37°C, the reaction mixture is heated at 100°C for 2 min and treated with 1 U of BAP at 65°C for 20 min. After termination of the BAP reaction by the addition of perchloric acid (1 M), Norit-adsorbable radioactivity is measured.

Unit definition: One ligase unit is the amount of enzyme that converts 1 nmol of [5'-^{32}P]rA_{20} to a phosphatase-resistant form in 30 min.

ii. **ATP–[^{32}P]PP$_i$ exchange assay.** This assay is based on a partial reaction up to the step in which the AMP–enzyme intermediate in formed (6). The assay mixture contains 50 mM Tris–Cl (pH 7.5), 10 mM $MgCl_2$, 0.1 mM ATP, 1 mM DTT, an appropriate amount of ligase, and 0.4 mM [^{32}P]PP$_i$ (5 cpm per pmol).

After the reaction, the radioactivity converted to a Norit-adsorbable form is quantitated. This activity is approximately 50-fold greater than that of the sealing reaction described earlier.

Unit definition: One enzyme unit converts 1 nmol of [^{32}P]PP$_i$ to a Norit-adsorbable form in 30 min at 37°C.

iii. **RNA ligase–AMP complex formation assay.** Although less commonly used for activity assays, the formation of a covalent RNA ligase–AMP complex is a useful method for estimating the active proportion of the enzyme within the total concentration of ligase (6).

c. Substrate specificities

i. **Minimal 5'-P donor substrates: 3',5'-diphosphates.** 5'-P mononucleotides are not substrates. Ribonucleoside 3',5'-diphosphates and a wide variety of base- and sugar-modified nucleoside 3',5'-diphosphates are active donors (13–15). The product is one nucleotide longer than the substrate and is terminated by a 3'-P (Scheme 2.1).

$$\text{RNA-OH} + \text{pNp} + \text{ATP} \rightarrow \text{RNA-pNp} + \text{PP}_i + \text{AMP}$$
SCHEME 2.1

The relative extent of adenylylation as well as ligation is pCp > pUp > pAp > pGp (10,16). The minimum length of the 5'-P oligoribonucleotides that can serve as donor substrates is a diribonucleotide (17).

The 3',5'-dNDP (pdNp) and DNA are also donor substrates although, with the exception of pdCp, they are significantly less efficient than the corresponding ribonucleotides (2,7).

In the ligation reactions with 3',5'-diphosphate donors, the T4 RNA ligase shows a strong preference for cytosine (see Table 2.1). However, once the 5'-P mononucleotides or a variety of mononucleotide or nonnucleotide derivatives are adenylylated to form a pyrophosphate-linked product, Ado-5'pp5'-X (AMP-pX), they become effective donor substrates in the presence of acceptors and enzyme, and no longer require the presence of the 3'-P group (13,18).

ii. **β-substituted ADP derivatives.** Various β-substituted ADP derivatives (Ado-5'pp5'-X) can be used as donor substrates in an ATP-independent reaction in which the P-X moiety is transferred to the 3'-OH of an acceptor (oligoribonucleotide) and the Ado-5'P is released (18,19). AMP, CMP, GMP, UMP, dTMP, NMN, FMN, and ribose-5P have all been added to (Ap)$_3$C from their corresponding AMP adducts. In marked contrast to the relative lack of specificity for the P-X group added, the RNA ligase shows a stringent specificity for the AMP portion of the substrates: for example, NADP$^+$, deamino-NAD$^+$, and CoA are not substrates. When the X moiety is a "good" leaving group, the initial, alternative addition product (RNA-3'-X) may undergo a nonenzymatic elimination of X, resulting in a 2',3'-cyclic phosphate.

iii. **Homopolynucleotide donors.** Among homopolyribonucleotides, polypurines (A and I) are effective substrates, whereas polypyrimidines (C and U) are 25% less effective than poly(A) (1). Compared with poly(A), poly(dA) is inactive as a substrate. Using pA_n, the best known substrate among homopolyribonucleotides, the majority of products occur as a result of intramolecular reactions. The minimum length for circularization is $n = 8$, and the optimum rate occurs at $n = 10-11$ (20). The rate of intramolecular ligation with pA_{34-40} is two- and ninefold greater than with pA_n of $n = 70-100$ and $n = 300$, respectively (1).

iv. **5′-P donor substrates: longer than minimal size.** The length of the donor polyribonucleotide has a slight effect on the reaction rate. When 5′-P hexaribonucleotide, $p(Up)_5U$, is used as a donor substrate and $(Ap)_5C$ as an acceptor, the intermolecular ligation product (dodecamer) is obtained at a near quantitative yield (21).

Oligodeoxynucleotides also behave similarly; for example, pdT_8 is the minimal length for effective circularization, although pdT_6 may form a detectable level of circular products (12). The rate increases with longer oligomers until it reaches the optimum chain length of ~20 nt. The rate of circularization with $n = 30$ is ~50% of that of the optimum length. In general, 5′-P DNA donors up to 6000 bases long and varying in structure from completely ssDNA to dsDNA with base-paired or 5′-protruding ends have similar activity (22).

The use of 5′-P donors with blocked 3′-OH termini limits the reaction to intermolecular ligation. The efficiency of ligation increases as the acceptor:donor ratio increases, with 5:1 being optimal (5).

v. **Modified donor substrates.** The T4 RNA ligase can tolerate certain modifications in the ribose, base, and 5′-P parts of donor molecules. For example, 2′-O-methylcytidine and 5′-thiophosphoryl-pNp derivatives are substrates (13,23). The 5′-P termini of some RNA (tRNA) molecules are not active donors presumably due to the H-bonded secondary structure (15).

When the *5′-phosphorothioate* donor, p(s)Ap, is incubated in the presence of ATP and a 3′-OH acceptor, the RNA ligase generates adenosine 2′,3′-cyclic phosphate 5′-phosphorothioate as the major product (24).

"*Cross-sectional*" *base pair analogs,* which comprise deoxyribonucleoside and ribonucleoside pairs that are held together by covalent bonds, can also be ligated by the T4 RNA ligase to an oligodeoxynucleotide acceptor with modest yields of ~20% (25).

Some *nucleoside 2′,5′-diphosphates* are neither substrates nor competitive inhibitors (14). The 2′,5′-ADP binds strongly to the enzyme in the presence (but not in the absence) of Mg^{2+} and inhibits the RNA ligase activity by ~90% (17).

vi. **Minimal structures for 3′-OH acceptor substrates.** The recognition of an acceptor involves the sugar, base, and chain length of the molecule. The minimal size for an acceptor is a trinucleotide (7,14,26). The optimum length of a homopolyribonucleotide is 6 nt (22). Homopolyribonucleotides of 10–20 nt are ligated at the half-maximal rate. RNAs as long as ribosomal and viral RNAs are acceptable substrates (9).

vii. **Nucleotide composition of acceptors.** The nucleotide composition has remarkable effects on the efficacy of acceptor substrates. In homopolydeoxynucleotides or homopolyribonucleotides, the base preference is A > C ≥ G (or I) > U (10,14,22). Similar discrimination against the U residue is also observed in heteropolymers: the trimers UAG and AUG are both poorer acceptors than AAG. DNA is a poor acceptor.

The addition of a single ribonucleotide to the 3′ terminus of an oligodeoxynucleotide increases its reactivity and, conversely, the addition of a deoxynucleotide to the 3′-end of an oligoribonucleotide acceptor renders it less reactive. In either case, molecules having mixed composition do not react like their pure DNA or RNA parent substrates (13).

viii. **Duplex structures.** Duplex structures of RNA may inhibit the ligase reaction. Circularization or intermolecular ligation of homopolymers in the presence of a complementary strand either has little effect on or inhibits the reaction (6,8). The 5′-terminal nucleotide in a single-strand structure but not in a duplex structure serves as the donor substrate (15). In contrast, duplex structures at the 3′-OH end of dsRNA (e.g., reovirus) are active as acceptors (9).

The ligation between duplex forms of oligodeoxynucleotides is not catalyzed by RNA ligase under normal conditions. However, the reaction can occur in the presence of ssDNA-binding protein, e.g., RNase A (8). Remarkably, the blunt-end DNA fragments (—GG/CC—) generated by *Hae*III restriction enzyme digestion are active donors (22).

ix. **Circularization reaction.** Oligonucleotides or RNAs with appropriate 3′-OH and 5′-P ends serve as a donor–acceptor for RNA ligase, resulting in circularization. Under normal intramolecular ligation conditions, the phosphodiester bond formation apparently occurs faster than the other steps in the ligation reaction. As a consequence, the adenylylated 5′-P RNA intermediate may not be observed unless the 3′-terminal side is made to be a poor acceptor.

When both intra- and intermolecular ligations can occur, circularization predominates by few orders of magnitude over intermolecular joining (15,27). Nucleotide composition and the length of the oligonucleotides affect the rate of circularization in a manner similar to that observed in intermolecular reactions.

d. Mechanism of ligation

The RNA ligase catalyzes the formation of 3′ → 5′ phosphodiester bonds between 3′-OH and 5′-P groups of RNA molecules. This reaction is coupled to the pyrophosphorolysis of ATP to AMP. The reaction mechanism, which is essentially identical to that of the DNA ligases described earlier (refer to Section I,A,4, this chapter), consists of three steps involving two covalent intermediates (see Fig. 2.2). The first step is the formation of a covalent ligase–AMP intermediate with concomitant pyrophosphorolysis of ATP (6,12,13,19). This step is reversible, allowing an active exchange reaction to occur between ATP and PP_i in the absence of an RNA substrate, but not between ATP and AMP. The second step is the transfer of AMP to 5'-P donor RNA and formation of an AMP–RNA complex (second intermediate) via a new pyrophosphate linkage (4). The adenylylation of the donor proceeds with a retention of configuration at the P^α of ATP, consistent with a double displacement mechanism involving a competent AMP–enzyme intermediate (28). The formation of an adenylylated donor requires the presence of acceptor molecules: in the absence of 3'-OH acceptors or in the presence of poor acceptors, very little activated intermediate is formed (12). The activated donor need not join to the acceptor that initially stimulated activation. It can also join to another acceptor, a process called *acceptor exchange*. The third and last step is the phosphodiester bond formation between the 5'-P donor and the 3'-OH acceptor by a direct nucleophilic displacement of AMP from the adenylylated donor. This nucleotidyl transfer occurs with an inversion of configuration at the phosphorus of AMP that is consistent with the "in-line" mechanism (24). The third step does not require active enzymatic catalysis, but the RNA ligase exerts both structural constraints for accepting appropriate substrates and stereochemical preferences at the phosphorus throughout the adenylylation and ligation steps of the reaction. As with DNA ligases, the three reaction steps of the RNA ligase-catalyzed reaction are reversible.

Despite the similarities in every aspect of the catalytic process, the RNA ligase differs from the DNA ligase in that the RNA ligase uses ssRNA molecules to align and join, whereas the DNA ligase requires a duplex structure. Furthermore, the RNA ligase prefers RNA as substrates, whereas the DNA ligase prefers DNA as substrates.

2. APPLICATIONS

Although the RNA ligase is infrequently used in general recombinant DNA work, the enzyme is a versatile tool for the construction of DNA as well as RNA molecules.

i. 3'-Labeling of RNA and DNA.
RNAs can be labeled at their 3'-OH ends by ligation with a nucleoside 3',5'-diphosphate, e.g., [5'-^{32}P]pCp (29), [5'-^{32}P]-pN$_3$Ap (30), [3'-β-^{32}P]pGpp (31), or A-5'pp5'-X (18). The 3'-end-labeled RNAs are useful for enzymatic or chemical sequencing (32) and for studies on RNA structure and RNA–protein interactions. The 3' termini of DNA can also be labeled similarly by the use of [5'-^{32}P]pdNp.

ii. 5'-Tailing of DNA with RNA.
The T4 RNA ligase is useful in joining defined oligoribonucleotides to the 5'-P terminus of DNA molecules (22). A variety of DNA donors up to 6 kbp long, single stranded or double stranded, and with base-paired or cohesive ends have similar affinity as substrates. The optimal length of a ribohomopolymer is 6 nt.

iii. Production of elongated molecules

(a) *Synthesis of long oligonucleotides.* This can be achieved by joining small defined single-strand oligodeoxynucleotide (8) or oligoribonucleotide fragments (33,34). Oligodeoxynucleotides with a defined sequence can be synthesized by the sequential ligation of pdNp (2,35).

(b) *Creation of ssDNA circles.* Single-strand DNA circles can be obtained in two ways: (1) intramolecular ligation on an oligodeoxynucleotide by RNA ligase (see Section II,1,c this chapter) and (2) a ssDNA circle held together with a terminal duplex by tailing the ssDNA at the 5' end with RNA ligase, adding the complementary sequence to the 3' end with terminal deoxynucleotidyltransferase (TdT), and by annealing the extensions (22).

(c) *Use in cDNA cloning.* The RNA ligase can be used to join a ssDNA to another ssDNA with a relatively high yield under optimized conditions. A cDNA cloning strategy is based on the ligation of a single-stranded oligodeoxynucleotide to the 3' end of single-stranded cDNA (36). This oligonucleotide is then used as a primer for PCR amplification of the cDNA. [Note that the return primer at the 5' end of cDNA is the first primer that has been used for reverse transcription. It can be a specific primer or an oligonucleotide tailored with a poly(dT) stretch.] The amplified cDNAs are then used for cDNA library construction.

(d) *Strategies for achieving a single ligation.* The use of a 5'-OH, 3'-OH terminated acceptor and a 5'-P, 3'-P terminated donor is one way to limit reaction to a single addition product. Donor cyclization or sequential additions can be prevented as well by blocking the 3'-OH terminus of the 5'-P donor with various reversible blocking groups, e.g., 3'-O-(α-methoxyethyl)ether (37) and 3'-O-isovaleryl ester (38). Attachment of a ddNTP to the 3'-OH terminus by the use of TdT enzyme is another efficient but irreversible blocking method (5).

iv. Modification of internal nucleotide(s). A RNA (tRNA) molecule with a modified internal nucleotide(s) is valuable for structure–function studies. The RNA chain is cleaved at the desired site by chemical or enzymatic methods, including partial digestion with a ribonuclease (e.g., RNase A) or a site-specific oligonucleotide-directed RNase H cleavage (see Section II,B, Chapter 3). Nucleotide modifications are made at the nick or gap, and the separated ribonucleotide fragments are joined using the T4 RNA ligase (39,40).

v. Stimulation of T4 DNA ligase. The blunt-end joining activity of the T4 DNA ligase is stimulated up to 20-fold in the presence of the T4 RNA ligase (41).

3. STRUCTURE

i. Primary structure. The T4 RNA ligase is a monomeric enzyme with 375 deduced amino acid residues (M_r 43,510) (42). Among the five Cys residues of the enzyme, three Cys are located within the C-terminal 74 amino acids (4). Of the two N-terminal ones (Cys-12 and Cys-53), Cys-12 appears to be structurally essential. A Cys-12 to Ser change gives rise to a highly unstable protein, reminiscent of the sensitive loss of enzyme activity on oxidation and/or SH group modifications. The enzyme remains fully active when Cys-53 is replaced with Ser.

ii. Active site. The active site of the T4 RNA ligase contains at least three subsites. The cofactor ATP-binding site is highly specific for the adenosine moiety of ATP. Lys-99 is the residue that binds AMP via a phosphonamide linkage at the ε-amino group (43). The 5'-P donor-binding site most strongly binds the 5'-terminal nucleoside residue with associated binding to its 3'-P in the form of free or phosphodiester linkage. Mutations at Asp-101 dramatically affect the reaction steps involving AMP–RNA complex formation and subsequent ligation, whereas the first adenylylation of the enzyme is unaffected. Asp-101 is thus likely to interact in some important way with the donor molecule rather than with the acceptor or the adenylyl moiety. The acceptor-binding site probably binds at least two or preferably three 3'-OH terminal nucleosides and two phosphates. This site strongly favors ribonucleosides to deoxyribonucleosides.

iii. Physical parameters. The T4 RNA ligase has a pI of 6.1. The molar absorptivity (ε_{280}) is 5.7×10^4 M^{-1} cm^{-1} and A_{280} (1 mg/ml) is 1.3. The A_{280}/A_{260} is 1.98.

Adenylylation of the RNA ligase markedly decreases the electrophoretic mobility of the enzyme in SDS–PAGE to a level corresponding to an increase in MW by ~4000 (26).

4. Genetics

i. Primary structure of RNA ligase gene. The T4 RNA ligase is the product of phage T4 gene 63. The gene 63 is located proximal to the *pseT* (or *pnk*) gene which encodes T4 5'-PNK-3'-phosphatase (refer to Section II, Chapter 5). The complete nucleotide sequence of the gene 63 has been determined (42). The cloned gene contains a strong T4 "early" promoter sequence immediately upstream of the 5' end of the gene 63 (42,43).

ii. RNA ligase-associated functions. The T4 RNA ligase possesses an activity that promotes phage tail fiber attachment (TFA) to the base plate, the final step in phage morphogenesis (44,45). The RNA ligase, which is a nonstructural protein, thus plays the role of a *molecular chaperone in vivo* in the self-assembly of the phage.

The dual activities of the gene 63 product (gp63) are apparently unrelated. Some gene 63 *amber* or missense mutants are defective only in TFA, whereas others are defective only in RNA ligase activity. The only known phenotype of gene 63 mutation is for TFA deficiency. Those mutants that are defective in TFA but active in ligase activity show normal protein synthesis, phage DNA replication, and packaging in *E. coli* B. The tail fiber attachment is a noncovalent interaction, and there is no phosphate present in any of the T4 structural proteins. Therefore the TFA activity is apparently distinct from the ligase activity within the same molecule.

RNA ligase activity is not essential for the phage growth in wild-type *E. coli*, or any putative function can be replaced by cellular enzymes, although no RNA ligase activity has been detected in *E. coli* extracts. Another beneficial function to the survival of phage T4 is that the T4 RNA ligase is involved, together with T4 PNK, in the repair of host endonuclease-cleaved tRNAs (46,47). The rli^- mutants, designating a class of T4 mutants that are RNA ligase defective but fully functional in TFA, are unable to grow in a hybrid *E. coli* strain, CTr5X (48), whereas they can grow normally on *E. coli* B or K. The restriction of *E. coli* CTr5X to the rli^- or pnk^- T4 phage mutants accompanies the accumulation of $tRNA^{Lys}$ cleaved near the anticodon. This *E. coli* mutation has been mapped to an apparent single locus, *prr*, which encodes a latent form of anticodon nuclease consisting of a core enzyme and a cognate masking factor (49). The T4-encoded proteins, which are yet to be identified, are thought to counteract the *prr*-encoded masking factor, thus activating the latent enzyme (50).

5. SOURCES

The T4 RNA ligase is conventionally purified from the *E. coli* cells infected with T4 phage. RNA ligase activity is detectable 3 min after infection, but the cells are usually harvested 2–5 hr postinfection to maximize the accumulation of early proteins. The infection with a T4 mutant, T4*am*H39X which is defective in the DNA ligase gene, enables preparation of RNA ligase free of DNA ligase (1). Use of T4 phage mutants (D0-*regA*) enables one to overproduce the ligase to as much as sevenfold over the amount of wild type or 0.5–1% of the total protein (26,51). The D0 indicates a "DNA-negative" mutation in gene 43 (DNA polymerase) or gene 45 (a DNA polymerase accessory protein) which allows the synthesis of enzyme without causing lysis of the cell, whereas the *regA* mutation is defective in the regulatory gene.

The T4 RNA ligase needs to be scrupulously purified by using various affinity columns to remove other contaminating nucleases; Cibacron blue-conjugated affinity matrix, Affi-Gel Blue (52,53), and dextran blue (54) have been shown to be useful in preparing a high purity RNA ligase (sp. act. ≈ 3000 U/mg) free of RNase and DNase at nearly 60-fold purification. A 2′,5′-ADP–Sepharose has also been used to purify the RNA ligase free of RNase activity (17).

The T4 gene 63 has been cloned in bacteriophage KR72 (a derivative of M13 phage) and plasmid vector pDR540 (55). A recombinant pDR540 clone (KR54) expresses efficiently the T4 RNA ligase under the control of the *tac* promoter. When induced by IPTG, the RNA ligase is produced at a level of 5–10% of the soluble proteins or 1.3–3.6 mg per g cell paste.

The purified enzyme is stored at ~1 mg/ml in 20 mM Tris–Cl or HEPES (pH 7.5), 1 mM DTT, 0.1 mM EDTA or 10 mM $MgCl_2$, and 50% glycerol at −20°C. RNA ligase activity in T4-infected cells stored at −70°C is stable for years.

References

1. Silber, R., Malathi, V. G., and Hurwitz, J. (1972), *PNAS* **69**, 3009–3013.
2. Hinton, D. M., and Gumport, R. I. (1979). *NAR* **7**, 453–464.
3. Uhlenbeck, O. C., and Cameron, V. (1977). *NAR* **4**, 85–98.
4. Heaphy, S., Singh, M., and Gait, M. J. (1987). *Biochem.* **26**, 1688–1696.
5. Tessier, D. C., Brousseau, R., and Vernet, T. (1986). *Anal. Biochem.* **158**, 171–178.
6. Cranston, J. W., Silber, R., Malathi, V. G., and Hurwitz, J. (1974). *JBC* **249**, 7447–7456.
7. Hinton, D. M., Baez, J. A., and Gumport, R. I. (1978). *Biochem.* **17**, 5091–5097.
8. McCoy, M. I. M., and Gumport, R. I. (1980). *Biochem.* **19**, 635–642.
9. England, T. E., and Uhlenbeck, O. C. (1978). *Nature* **275**, 560–561.
10. Romaniuk, E., McLaughlin, L. W., Neilson, T., and Romaniuk, P. J. (1982). *EJB* **125**, 639–643.
11. Baronaite, Z. A., Martsishauskas, R. P., and Peslyakas, I.-G. I. (1984). *Biokhimiya* **49**, 142–148. [English translation, pp. 121–126]
12. Sugino, A., Snopek, T. J., and Cozzarelli, N. R. (1977). *JBC* **252**, 1732–1738.
13. Uhlenbeck, O. C., and Gumport, R. I. (1982). *Enzymes* **15**(B), 31–58.
14. England, T. E., and Uhlenbeck, O. C. (1978). *Biochem.* **17**, 2069–2076.
15. Bruce, A. G., and Uhlenbeck, O. C. (1978). *NAR* **5**, 3665–3677.
16. McLaughlin, L. W., Piel, N., and Greaser, E. (1985). *Biochem.* **24**, 267–273.
17. Sugiura, M., Suzuki, M., Ohtsuka, E., Nishikawa, S., Uemura, H., and Ikehara, M. (1979). *FEBS Lett.* **97**, 73–76.
18. England, T. E., Gumport, R. I., and Uhlenbeck, O. C. (1977). *PNAS* **74**, 4839–4842.
19. Gumport, R. I., and Uhlenbeck, O. C. (1981). In "Gene Amplification and Analysis" (J. G. Chirikjian and T. S. Papas, eds.), Vol. II, pp. 313–345. Elsevier North Holland, New York.
20. Kaufmann, G., Klein, T., and Littauer, U. Z. (1974). *FEBS Lett.* **46**, 271–275.
21. Walker, G. C., Uhlenbeck, O. C., Bedows, E., and Gumport, R. I. (1975). *PNAS* **72**, 122–126.
22. Higgins, N. P., Geballe, A. P., and Cozzarelli, N. R. (1979). *NAR* **6**, 1013–1024.
23. Barrio, J. R., Barrio, M. C. G., Leonard, N. J., England, T. E., and Uhlenbeck, O. C. (1978). *Biochem.* **17**, 2077–2081.
24. Bryant, F. R., and Benkovic, S. J. (1982). *Biochem.* **21**, 5877–5885.
25. Petric, A., Bhat, B., Leonard, N. J., and Gumport, R. I. (1991). *NAR* **19**, 585–590.
26. Higgins, N. P., Geballe, A. P., Snopek, T.J., Sugino, A., and Cozzarelli, N. (1977). *NAR* **4**, 3175–3186.
27. Kaufmann, G., and Kallenbach, N. R. (1975). *Nature* **254**, 452–454.
28. Harnett, S. P., Lowe, G., and Tansley, G. (1985). *Biochem.* **24**, 7446–7449.
29. England, T. E., Bruce, A. G., and Uhlenbeck, O. C. (1980). *Metd. Enzy.* **65**, 65–74.
30. Sylvers, L. A., Wower, J., Hixson, S. S., and Zimmermann, R. A. (1989). *FEBS Lett.* **245**, 9–13.
31. Simoncsits, A. (1980). *NAR* **8**, 4111–4124.
32. Peattie, D. A. (1979). *PNAS* **76**, 1760–1764.
33. Ohtsuka, E., Nishikawa, S., Markham, A. F., Tanaka, S., Miyake, T., Wakabayashi, T., Ikehara, M., and Sugiura, M. (1980). *Biochem.* **17**, 4894–4899.
34. Middleton, T., Herlihy, W. C., Schimmel, P. R., and Munro, H. N. (1985). *Anal. Biochem.* **144**, 110–117.
35. Hinton, D. M., Brennan, C. A., and Gumport, R. I. (1982). *NAR* **10**, 1877–1894.
36. Edwards, J. B. D. M., Delort, J., and Mallet, J. (1991). *NAR* **19**, 5227–5232.
37. Sninsky, J. J., Last, J. A., and Gilham, P. T. (1976). *NAR* **3**, 3157–3166.
38. Kaufmann, G., Fridkin, M., Zutra, A., and Littauer, U. Z. (1971). *EJB* **24**, 4–11.
39. Bruce, A. G., and Uhlenbeck, O. C. (1982). *Biochem.* **21**, 855–861.
40. Grosjean, H., de Hanau, S., Doi, T., Yamane, A., Ohtsuka, E., Ikehara, M., Beauchemin, N., Nicoghosian, K., and Cedergren, R. (1987). *EJB* **166**, 325–332.
41. Sugino, A., Goodman, H. M., Heynecker, H. L., Shine, J., Boyer, H. W., and Cozzarelli, N. R. (1977). *JBC* **252**, 3987–3994.
42. Rand, K. N., and Gait, M. J. (1984). *EMBO J.* **3**, 397–402.
43. Liebig, H. D., and Ruger, W. (1989). *JMB* **208**, 517–536.

44. Wood, W. B., Conley, M. P., Lyle, H. L., and Dickson, R. C. (1978). *JBC* **253**, 2437–2445.
45. Snopek, T. J., Wood, W. B., Conley, M. P., Chen, P., and Cozzarelli, N. R. (1977). *PNAS* **74**, 3355–3359.
46. David, M., Borasio, G. D., and Kaufmann, G. (1982). *PNAS* **79**, 7097–7101.
47. Amitsur, M., Levitz, R., and Kaufmann, G. (1987). *EMBO J.* **6**, 2499–2503.
48. Runnels, J., Soltis, D., Hey, T., and Snyder, L. (1982). *JMB* **154**, 273–286.
49. Amitsur, M., Morad, I., and Kaufmann, G. (1989). *EMBO J.* **8**, 2411–2415.
50. Levitz, R., Chapman, D., Amitsur, M., Green, R., Snyder, L., and Kaufmann, G. (1990). *EMBO J.* **9**, 1383–1389.
51. Last, J. A., and Anderson, W. F. (1976). *ABB* **174**, 167–176.
52. McCoy, M. I. M., Lubben, T. H., and Gumport, R. I. (1979). *BBA* **562**, 149–161.
53. Brennan, C. A., Manthey, A. E., and Gumport, R. I. (1983). *Meth. Enzy.* **100**, 38–52.
54. Hu, M. H., Wang, A., and Hua, H. F. (1982). *Anal. Biochem.* **125**, 1–5.
55. Thogerson, H. C., Morris, H. R., Rand, K. N., and Gait, M. J. (1985). *EJB* **147**, 325–329.

3
Nucleases

Nucleases are broadly defined as the enzymes that degrade polynucleotides. They comprise a large family of enzymes that show diverse reaction and substrate specificities: DNA-specific (DNase) and/or RNA-specific (RNase), nucleotide sequence-specific or nonspecific, double-strand-specific and/or single-strand-specific, exonucleolytic ($5' \rightarrow 3'$ and $3' \rightarrow 5'$) and/or endonucleolytic, and generating $3'$-P or $5'$-P products. Many polymerases, including reverse transcriptases, also exhibit intrinsic nuclease activities with widely varying reaction and substrate specificities, which makes the rational division of nucleases even more complex.

Historically, nucleases have been divided into DNases and RNases, but many nucleases degrade both DNA and RNA, blurring the division based on sugar specificity. Nevertheless, this chapter adopts the conventional division scheme and another division scheme based on strand specificities. The nuclease activities associated with DNA polymerases will be described in Chapter 6.

The almost ubiquitous existence of a large variety of nucleases suggests that the functional roles for nucleases could be diverse; yet none of those suspected *in vivo* functions have been clearly identified. Nevertheless, our knowledge of the structural and functional characteristics of individual nucleases has dramatically increased. The accumulated knowledge has led to wide and important applications of the nucleases in the analysis of nucleic acid structure and nucleic acid–protein interaction. Nucleases are now an important tool in nucleotide sequencing and in various other recombinant DNA techniques.

The use of nucleases as "scissors" for cutting nucleic acids is continually expanding as new generations of nucleases are developed. One of the most prominent new generation nucleases is catalytic RNA (or ribozyme). Ribozymes are potentially important tools in site-specific cleavages of nucleic acids (see Section I,B, Chapter 7); some fundamental aspects of ribozymes are described in Section II of Chapter 1.

Another class of promising new generation nucleases includes "artificial" and "semiartificial" nucleases. *Chemical nucleases* are typical artificial nucleases that make use of redox-active compounds (e.g., phenanthroline–copper and ferrous–EDTA) as the cutter and an attached oligonucleotide as the site-specific recognition module. Although the importance of chemical nucleases and the potential for their applications are bound to grow, this chapter focuses on some of the "classical enzyme" nucleases that are most widely used in recombinant DNA technology.

Semiartificial nucleases include: (1) in concept, *chimeric nucleases* which are composed of heterogenous structural parts brought together by the use of recombinant DNA techniques and (2) *hybrid nucleases* which are derived from classical nucleases by incorporating a piece of synthetic oligonucleotide in the active site to harness a sequence specificity. An example of the hybrid nuclease is described in Section II (this chapter) on staphylococcal nuclease.

I. DNA ENDONUCLEASE: DEOXYRIBONUCLEASE I
[EC 3.1.21.1]

Bovine pancreatic deoxyribonuclease (DNase I) is an endonuclease that cleaves both ssDNA and dsDNA to produce primarily 5′-P dinucleotides and 5′-P oligonucleotides (Fig. 3.1). DNase I degrades dsDNA in a sequence-nonspecific manner. The enzyme requires divalent cations (both Ca^{2+} and Mg^{2+} or Mn^{2+}) as cofactors for full double-strand cutting activity. In the presence of Mg^{2+} alone as a cofactor, DNase I exhibits nicking activity and the single-strand nucleolytic activity is more discriminatory.

Unless otherwise specified, DNase I (30.5 kDa) is a mixture of four glycoprotein components of similar catalytic activity: DNase A (major), B, C, and D. For most purposes, a mixture of the four is a suitable catalyst and will be referred to as *DNase I*.

DNase I is distinct from DNase II, a lysosomal "acidic" DNase (38 kDa) found in various organs such as the thymus, liver, and spleen. DNase I prefers a duplex region for cleavage, whereas DNase II prefers a single-stranded region for its activity. Furthermore, DNase II differs from DNase I in its optimal pH and the requirement for Mg^{2+}. In contrast to DNase I, DNase II generates 3′-P oligonucleotides as the predominant products.

I. DNA Endonuclease: Deoxyribonuclease I

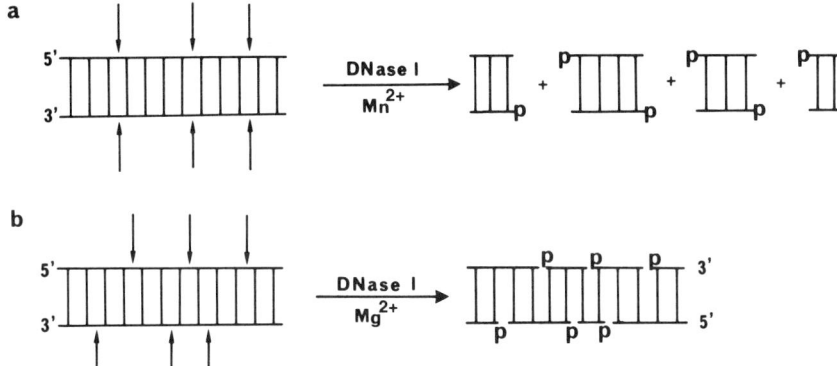

FIGURE 3.1 Reaction profile of DNase I. (a) Double-strand cutting activity and (b) nicking activity.

DNase I is one of the most thoroughly studied endonucleases. As a nonspecific nuclease whose activity can be easily modulated by metal ions to enhance specific double-strand scission or nicking activity, DNase I is an important and versatile tool in recombinant DNA technology.

1. FUNCTIONS

a. Reaction conditions

i. Optimal reaction conditions. The pH optimum with $MgCl_2$ (2.5 mM) and $CaCl_2$ (0.1 mM) is between pH 7.0 and 8.0 (1). With Mg^{2+} alone, the maximal activity is observed at pH 5.5. However, the activity is less than 3% of the optimal activity observed with both Mg^{2+} and Ca^{2+}. With Ca^{2+} alone, DNase I is optimally active at pH 8 and the maximal activity is only 1% of the value obtained with both Mg^{2+} and Ca^{2+}.

The recommended reaction conditions are 10–50 mM Tris–Cl (pH 7.5), 5 mM $MgCl_2$, 0.1 mM $CaCl_2$, 50 μg/ml BSA, DNA (1 μg/50 μl), and 1–5 U DNase I (or 1 μg of enzyme with a specific activity of 3000 U/mg protein). The mixture is incubated at 37°C for a time (e.g., 30 min) suitable for the degree of digestion desired. Nicking activity prevails under the reaction conditions or in the same buffer containing 10 mM $MgCl_2$ as the only metal cofactor. For double-strand cleavage activity, $MgCl_2$ is replaced by $MnCl_2$ in the reaction mixture described earlier. The reaction is stopped by adding EGTA or EDTA in excess of the Ca^{2+} and Mg^{2+} concentrations used.

ii. Kinetic parameters. Under optimal reaction conditions in the presence of 0.5 mM $CaCl_2$ and 5 mM $MgCl_2$, the V is 1300 H^+/min (protons released at pH 7 in a pH stat) (1). The K_m is 34–40 μg DNA/ml and increases as the concentration of Ca^{2+} is decreased.

iii. **Metal ion requirements.** DNase I requires Ca^{2+} for activity; 0.5 mM is the optimum concentration (1). Specific binding of a Ca^{2+} ion apparently stabilizes the active conformation of DNase I. Binding of Ca^{2+} ions protects the enzyme from tryptic digestion; a 50% protection is observed at 0.1 mM Ca^{2+}. Gel filtration experiments at pH 7.5 show that two Ca^{2+} are tightly bound ($K_d \approx 14$ μM) and that three Ca^{2+} are loosely bound ($K_d \approx 0.2$ mM) (2). One of the two strong Ca^{2+}-binding sites it not subject to competition from Mg^{2+} or Mn^{2+}. At pH 5.5, the enzyme binds only one Ca^{2+} ion strongly ($K_d \approx 0.23$ mM).

DNase I also requires Mg^{2+} for activity; the optimal concentration for Mg^{2+} is 5–10 mM. The activating effect of Mg^{2+} is largely ascribed to its binding to DNA substrates, although Mg^{2+} can bind the enzyme with a K_d of 0.23 mM.

In the presence of divalent cations (e.g., $MgCl_2$), DNase I turns out to be highly thermostable and in a Mg^{2+} concentration-dependent manner. DNase I remains active after heat treatment (30 min at 95°C) in 2 mM $MgCl_2$ buffer (10 mM Tris-Cl, pH 8.3, and 50 mM KCl), whereas it is fully inactivated by the same treatment in 6 mM $MgCl_2$ (62). The heat-inactivated DNase I regains its activity when it is incubated in 2 mM $MgCl_2$ buffer.

The nature of activating cations affects the substrate specificity of the enzyme, the optimum pH, and the maximum activity (3,4). DNase I introduces double-strand breaks when Mn^{2+} is used in place of Mg^{2+} (3,5). The combination of Ca^{2+} and Mg^{2+} may shift the Mg^{2+}-dependent nicking activity toward a double-strand cutting activity. In contrast to the stability of DNase I in the presence of Mg^{2+} (with which the optimum temperature is 37°C), the enzyme is considerably more stable in the presence of Mn^{2+}, losing no activity up to 60°C (6). The ability of transition metal ions such as Mn^{2+} and Co^{2+} to promote double-strand cleavage is believed to be due at least in part to the stabilization and/or activation of a different enzyme structure (6) and also in part to the effect on DNA structure consequent to the binding of these metal ions (but not of Mg^{2+} or Ca^{2+}) to the N7 of the G base and an adjacent phosphate (5).

iv. **Inhibitors and inactivators**

(a) *Metal ions and chelators.* DNase I can be completely inactivated by EGTA even in the presence of 2.5 mM $MgCl_2$. The EGTA chelates Ca^{2+} over 10^5 times more strongly than Mg^{2+} (7). EDTA, which chelates both Ca^{2+} and Mg^{2+}, also blocks DNase I activity. The presence of monovalent cations such as Na^+ and K^+ (5) or a high concentration of Ca^{2+} (1 mM) (8) is inhibitory.

(b) *Proteinaceous inhibitors.* Proteinaceous inhibitors isolated from calf spleen and thymus form tight 1:1 complexes with DNase I (9,10). The spleen inhibitor protein is probably a cytoplasmic, globular actin molecule (~42 kDa) (11,12). It is a major cellular component that usually constitutes 5–10% of the soluble protein. The monomeric actin (G-actin) from rabbit skeletal muscle is a competitive inhibitor of DNase I with a K_i of 2 nM (13). The binding of DNase I with filamentous actin (F-actin) causes depolymerization of the actin filament.

I. DNA Endonuclease: Deoxyribonuclease I

(c) **Disulfide-reducing agent.** In the presence of Ca^{2+}, 2-MSH can reduce one S–S bridge (C101/C104), but not the other (C173/C209), and the enzyme remains fully active (14). In the absence of Ca^{2+}, DNase I can be rapidly and fully reduced and inactivated by 2-MSH at pH 7.2 and 25°C. The addition of Ca^{2+} (4 mM), but not Mg^{2+} or Zn^{2+}, can fully restore the activity.

(d) **Denaturants.** DNase I is rapidly denatured with loss of activity by 0.005% (w/v) SDS, but not by neutral detergents such as Brij 35 and Triton X-100 at similar concentrations (15). The denatured enzyme can be fully renatured on dilution in the presence of Ca^{2+} via transient exposure to 6 M GuHCl.

b. Activity assays and unit definition

i. **Spectrophotometric assays.** The activity of DNase I is typically assayed in a mixture consisting of 0.1 M NaAc (pH 5.0), 5 mM $MgCl_2$, 50 μg/ml calf thymus DNA and enzyme in 1 ml for 10 min at 25°C.

For kinetic or analytical purposes, DNase I assays employ a PNP-ester as substrate (16).

Unit definition: One unit is the amount of enzyme that increases the absorbance of the DNA solution at a rate of 0.001 A_{260} units per min per ml of reaction mixture (17).

ii. **Product solubility assay.** DNase activity can also be assayed by monitoring the amount of acid-soluble digestion products. The activity is essentially linear as a function of enzyme concentration and incubation time except for a short initial lag period (18). Gradually, however, the activity falls into the pattern of autoretardation. This is caused by the continuous formation of products that are poorer substrates than those from which they are derived (19).

c. Substrate specificities

i. **Substrate preference.** Although ssDNAs are substrates, dsDNAs are 100 to 500 times better substrates (20). Single-stranded DNAs of ≤40 nt in length are cleaved faster in the central region beyond the 10 terminal nucleotides (21). This distinction disappears as the substrate length increases.

ii. **Product profile.** DNase I digestion gives primarily dinucleotides (60%), trinucleotides (25%), and other oligonucleotides as final products (19). The products have 5′-P and 3′-OH termini (Fig. 3.1). Double-strand cleavage of DNA, in the presence of Mn^{2+}, results in staggered ends with 2 to 4 nt in the 3′ direction. DNase I displays predominant nicking activity under certain suboptimal reaction conditions (see below).

iii. **Substrate size.** The smallest substrate for DNase I is a tetranucleotide d(NpN|pNpN) or a trinucleotide d(NpN|pNp) carrying a 3′-P group. For example, dAATp (a DNase II digestion product) can be further cleaved to dAA and pdTp, showing a tendency for preferential cleavage at Pu-Py linkages (22). Other trinucleotides without 3'-P or dinucleotides with 3'-P are not substrates. In general, oligonucleotides bearing a 3'-P (but not a 5'-P or 5'-OH) group are easily cleaved to give the terminal mononucleoside 3',5'-diphosphates (23). Otherwise, mononucleotide products are not obtained from DNase I digestion.

iv. Base specificity. Under or near optimal conditions, natural DNA substrates (e.g., calf thymus DNA, *E. coli* DNA) are cut at all four bases. The base specificity, estimated from the abundance of 5' and 3' terminal bases of the oligonucleotide products, is similar for all nucleotides within a factor of 3 (24). This relative nonspecificity for nucleotide sequences and the distribution of nearly uniform fragment sizes are the basis for many important applications of DNase I.

Under suboptimal conditions where the activity can be significantly modulated by various metal ions, however, the base preference for dsDNA cleavage noticeably increases (up to sixfold). Dinucleotide product analyses show a particular high abundance of Py(T)-Pu(N) pairs (25), suggesting a certain cleavage preference at Pu-Py linkages. Although detailed base specificities can vary depending on the DNA species, the preference for a given DNA is only marginally altered in the presence of Mg^{2+} or Mn^{2+} or during DNA digestions (18).

v. DNA conformation. Under nicking conditions, the cleavage of alternating copolymeric dsDNAs, e.g., $d(AT)_n \cdot d(AT)_n$ or $d(GC)_n \cdot d(GC)_n$, is influenced not only by the sequence but also by structural variations in the backbone of the copolymer (26). In such alternating Pu-Py copolymers, the cleavage at the 5' side of Py is favored by ~40-fold in $d(AT)_n$ and 6-fold in $d(GC)_n$. With homopolymers, the 5' termini patterns show that the cleavage preference is markedly reduced: the Py:Pu ratio of ~7 and ~1 in $(dA)_n \cdot (dT)_n$ and $(dG)_n \cdot (dC)_n$, respectively. Rapid DNase I cutting sites are correlated with high local helical twist.

vi. Synthetic substrates. Nitrophenyl esters (e.g., PNP-5'-pdTp-3'-PNP) are the simplest and most convenient synthetic substrates for assaying DNase I (16). With the substrate, PNP-5'-pdTp-3'-PNP, DNase I releases PNP mostly from the 3'-ester group rather than from the normally expected 5'-P. [Note that staphylococcal nuclease (27) and snake venom phosphodiesterase release PNP predominantly from the 5'-ester group.]

vii. DNAs containing ribonucleotides. DNAs such as the crab d(A-T) polymer that contain ribonucleotides can be digested by DNase I to give a mixture of dinucleotides, e.g., dC-rG, dT-rA, and dT-rG (28).

viii. Nicking activity in the presence of Mg^{2+}. Under certain suboptimal reaction conditions (e.g., with 10 m*M* $MgCl_2$ and with no other divalent metal ions present), DNase I produces nicks rather than the scission of both strands. The nicking activity is exclusive for some DNAs, e.g., *ccc* DNA (5), and for the dI strand in the dI·dC homopolymer. Nicking in dsDNA (e.g., λ or mitochondrial) occurs at random, multiple sites at an average of 2.7–4 nicks for every double-strand cut (24,28). The duplex cutting activity, e.g., on dI·dC homopolymer, is fully restored by adding Ca^{2+} to Mg^{2+} or by substituting Mn^{2+} for Mg^{2+} (29).

ix. Nicking activity in the presence of EtBr. In the presence of a high concentration of EtBr (20–80 μg/ml) and 2 mM $MgCl_2$ (or $CaCl_2$), DNase I introduces a single nick into ccc DNA (30,31). DNase I (0.2–1.0 μg/ml) will yield ~90% nicked molecules in 15 min at 30°C. In the presence of 89 μg/ml EtBr, the nicking activity is 180 times less than in the absence of this dye. The rate of DNase I cleavage of nicked ccc DNA molecules is much lower than the rate of the first nicking (31).

x. Hypersensitivity in the presence of actinomycin D. The intercalation of actinomycin D at a G·C site causes varying DNase I responses depending on the sequences flanking the drug-binding site. Binding of actinomycin D at a site with certain AT-rich flanking sequences induces DNase I hypersensitivity at sites distal to the drug-binding site. The intercalator-induced alteration of the DNA structure apparently propagates in a 5′ → 3′ direction, asymmetrically exposing the two strands, notably at the 5–6 nt in the 3′ direction and the 3–4 nt in the 5′ direction (32).

d. Mechanism of DNase I action

DNase I is an endonuclease that catalyzes the hydrolysis of phosphodiester bonds by nucleophilic attack on 3′O-P. The enzyme activity requires divalent metal ion cofactors such as Mg^{2+} and Ca^{2+} or Mn^{2+}. DNase I binds to the minor groove of dsDNA and cleaves each strand independently. The often observed sequence preference and different cleavage rates of DNase I are largely structural and are related to sequence-dependent variations of the double helix such as groove width, local rigidity to bending, radial asymmetry, and accessibility to backbone phosphates (26,29,33,34).

i. Role of metal ions. In the absence of Ca^{2+}, the binding of DNA to the enzyme is practically nil. The presence of Ca^{2+} not only renders the DNase I capable of binding DNA but also transforms the enzyme into a catalytically competent form. The rate acceleration by Ca^{2+} is estimated to be of the order of $10^{4.6}$ (29). The Ca^{2+}-bound enzyme is structurally more stable vis-à-vis denaturation and tryptic digestion.

Evidence indicates that Ca^{2+} induces a conformational change in the enzyme in a concentration-dependent manner, with half-transition occurring at 0.1 mM Ca^{2+}. At low concentrations of Ca^{2+} (i.e., a low active form of DNase 1), the cleavage is directed primarily toward the bonds for which the DNase has the greatest affinity, for example, single-strand cutting in the dI strand of the dI·dC homopolymer (1). At high concentrations of Ca^{2+} (i.e., a high active form of DNase 1), differential DNA-binding affinity becomes less critical than the relative rate of bond cleavage and double-strand scission occurs predominantly.

ii. Mode of DNA binding. The crystal structure of DNase I/DNA complexes shows that DNase I binds to one side of the B-type DNA double helix, forming contacts exclusively in the minor groove and with the phosphate backbones on both sides (29,35). Direct contact with the DNA (nicked octamer duplex) extends over a stretch of 6 bp plus a total of 6 (including 2 adjacent) phosphates. The phosphate of the scissile phosphodiester bond is anchored to the enzyme via the catalytic Ca^{2+} ion. The binding of DNase I to the minor groove is asymmetric, covering but making no contacts with the major groove on the 3' side of the cleavage site. Regarding the DNA-binding site, no specific secondary structure element is involved. Contacting residues of the protein all reside on loop regions and form an extensive network of interactions with the backbone phosphates and the bases: 14 H-bonds, one salt-bridge, several van der Waals contacts, and one stacking-type, hydrophobic interaction. Among those, the stacking interaction of Tyr-76 with a base (T), in conjunction with Arg-41 which forms H-bonds with two bases (T and C) on the opposite strand, apparently has a critical effect on the DNA conformation. The contacts induce a 3 Å widening of the minor groove to about 15 Å by bending the DNA toward the major groove by about 20°. (As a consequence, the major groove at the same position of the double helix is narrowed to about 15 Å.) The conformation of DNase I remains virtually invariable.

The observed low cleavage rates in the DNA regions with runs of adenines and thymines are attributed to (a) a narrower minor groove and (b) a conformational rigidity that resists bending. The slightly lower than average cutting rates in G·C-rich regions may be due to a wider than normal groove width, a conformational rigidity, and/or possible hindrance of the NH_2 group of G base (in G·C or C·G pair) to the stacking interaction with Tyr-76. DNase I cleaves A-form DNA more slowly than B-form DNA, presumably because the A-form DNA has a wider minor groove and more tilted base pairs than does the B-form DNA (36). In contrast, DNase I hypersensitivity is a sensitive indicator for certain unusual DNA structures whose function has been associated with gene promoters and enhancers.

iii. Patterns of double-strand cleavage. The pattern of major cutting sites in defined oligonucleotides provides the following insight into the mode of DNase I action (26). The rate of cleavage of favored bonds increases with distance from the 5' end, requiring a helix of ~4 bp inward from the 5' end. In contrast to the 5' end, DNase I cuts near to the 3' end of a duplex strand and shows a certain sequence preference. For example, when the 3' end is C in G-C copolymers, the 3' terminal C is hardly cleaved at all.

The nuclease treats loops as if they were duplex ends (20). As a result, hairpins are cut at the 3' end of duplex region which abuts the 5' beginning of a loop. In contrast, cuts occur three bonds inward on the duplex strand abutting the 3' end of the loop.

The asymmetric cleavages in double-strand helix result in a diagonally skewed cleavage pattern, reflecting the actions at helix grooves. DNase I recognizes the

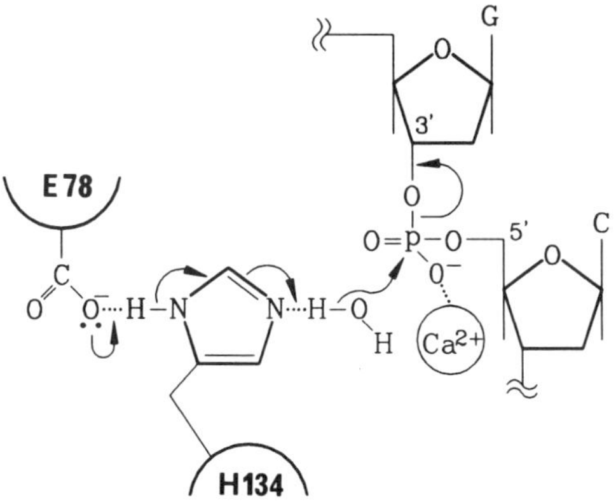

FIGURE 3.2 Mechanism of DNase I action. A nucleophilic attack of water hydroxyl is presumed to be mediated by a "charge relay system" consisting of Glu[78]–His[134]–H_2O.

disposition of the sugar–phosphate backbone rather than base pair arrangements. The most preferred cuts occur at the phosphates that lie directly across the minor groove and are in close proximity (20). This is consistent with the observation that cutting rates fall off dramatically when the groove becomes too wide or goes downward into an unpaired loop, or when phosphates cease to oppose one another near the 5′ end of the double helix.

iv. Cleavage by nucleophilic catalysis. Based on biochemical and crystallographic studies, the mechanism of DNase I action is thought to involve a water-mediated general acid–base catalysis (29,35). The hydrolysis by DNase I results in an inversion of configuration at the phosphorus, indicating that a single displacement reaction takes place at the reaction center (37). The present model of DNase I catalysis consists of three elements forming a proton acceptor–donor chain, Glu[78]–His[134]–H_2O (Fig. 3.2), similar to the better known "catalytic triad" (Asp–His–Ser) of serine proteases. The nucleophilic attack of the hydroxyl ion on the phosphorus results in the P-O3′ bond cleavage, which is facilitated by the positive charge on the Ca^{2+} ion that interacts with the phosphate group.

In DNase I, the carboxylate of Glu-78 accepts a proton most likely from His-134 (N1), which in turn accepts a proton from a water molecule. In the DNase I/DNA complex (29), His-134 is located in the immediate neighborhood of the scissile phosphate group. Note that His-252 is also placed in close proximity to the scissile phosphate group, and the possibility that His-252 plays a supportive catalytic role in combination with Glu-39 and a nearby water molecule has not been excluded.

Why the nucleophilic attack of the OH^- group preferentially cleaves the P-O3′ bond and not the P-O5′ in DNase I is not clearly known. Presumably the metal ion (Ca^{2+} or Mg^{2+}) is bound in such a way that the P-O5′ bond is more stabilized than the 3′O-P linkage, rendering the 3′O-R a better leaving group.

2. APPLICATIONS

As a nonspecific nuclease, DNase I is extensively used in molecular biology and recombinant DNA technology. The following examples are given to illustrate the contexts and strategies for using DNase I.

i. Nonspecific degradation of DNAs. DNase I is commonly employed as a nonspecific nuclease to eliminate contaminating cellular DNAs in routine RNA (or protein) preparations. DNase I is also used to remove the template DNAs after *in vitro* RNA transcription using SP6 or T7 RNA polymerase (see Section I,B in Chapter 7). For these applications, the use of "RNase-free" DNase I is recommended.

Immobilized DNase I can be particularly useful for the removal of contaminating genomic DNA from miniaturized RNA preparations. The RNA derived from a single cell or small numbers of cells has been shown to be suitable for reverse transcription-PCR when the RNA was treated with immobilized DNase I without subsequent phenol extraction (63).

ii. DNA labeling by nick translation. DNase I introduces nicks into dsDNA under certain suboptimal reaction conditions such as with 10 mM Mg^{2+} (5). The nick serves as the priming site for *E. coli* DNA Pol I which, by virtue of its $5' \rightarrow 3'$-exonuclease activity, removes nucleotides from the $5'$ side of the nick and at the same time synthesizes a new strand from the $3'$ side of the nick in a process called *nick translation* (see Appendix A: DNA labeling methods). In the presence of one or more radioactive dNTPs, labeled DNAs suitable for hybridization (approximate size of 500–1000 bp) can be generated by nick translation.

iii. Random deletion mutagenesis. DNase I introduces a single nick into supercoiled plasmids in the presence of a high concentration of EtBr (20–80 μg/ml) (30). Limited digestion of the nicked DNA with T4 DNA polymerase in the presence of a different single dNTP enables generation of gapped DNAs. The gapped DNAs can be converted to DNAs with various small deletions by treatment with mung bean nuclease and subsequently with T4 DNA ligase (31).

iv. Random insertion mutagenesis. In the presence of Mn^{2+}, DNase I can inflict a single, often random nick followed by double-strand cleavage on circular dsDNA molecules. The ends of the unit-length DNAs can be made blunt-ended and then ligated to a single synthetic oligonucleotide containing a unique restriction site. Recircularization and transfection of such engineered plasmid DNAs allow generation of a set of mutant plasmids into which a single new restriction site has been randomly inserted. This insertion mutagenesis is one of the methods used in gene mapping (38).

v. Generation of DNA fragments for dideoxy sequencing. DNase I digestion in the presence of Mn^{2+} produces double-strand breaks in DNA. The average size of the products from partial digestion is inversely proportional to the amount of enzyme used, and the distribution (e.g., in λ DNA) is smooth and unimodal. Therefore, DNase I can be used to generate random DNA fragments for both *shotgun* cloning and, more importantly, for *nonrandom* cloning.

(a) DNase I-based shotgun cloning and sequencing. The DNA fragments obtained from DNase I digestion are briefly treated with DNA polymerase to make the staggered ends flush and to repair the nicks and gaps. The DNA fragments are then subcloned into a suitable vector (M13 or one of its derivatives), and random clones are selected for dideoxy sequencing (39). Although the shotgun strategy rapidly provides sequence information at the beginning, the strategy requires progressively greater efforts to obtain a complete set of sequences, which is often the critical reason for resorting to nonrandom cloning and sequencing strategies.

(b) DNase I-based nonrandom cloning and sequencing. A number of strategies based on DNase I digestion have been developed that generate nested deletions. Typically a recombinant plasmid DNA is linearized by introducing a moderate number of random double-strand cuts using DNase I and is subjected to a suitable restriction enzyme digestion at a unique site proximal to the sequencing primer site. Progressively shortened DNAs are then selected and recircularized. The recircularization can be made by blunt-end ligation after treatment of the termini with an appropriate enzyme, e.g., *E. coli* Pol Ik or T4 DNA Pol (40). In an alternative strategy, restriction linkers (e.g., *Eco*RI linker) are attached to the DNase I-digested, blunt-ended DNA molecules, and then cut with the restriction enzyme to open the linker at one end. The restriction enzyme digestion simultaneously prepares a vector site for final cohesive-end ligation (41).

Another nonrandom strategy is to perform partial DNase I digestion on isolated restriction fragments (e.g., by the use of *Eco*RI) and to subclone them directly (42). The mixture of subfragments, having one cohesive end and the other end randomly produced by DNase I, is ligated first at the cohesive end with a vector. This linear, recombinant DNA molecule is then circularized by blunt-end ligation after polishing the termini. Subclones with progressively shortened insert sizes are selected to generate a nested set of subclone libraries for DNA sequencing. The use of linear dsDNA substrates with noncompatible staggered ends makes the strategy more efficient since the DNA subfragments carry one of two noncompatible staggered ends and one blunt end, thereby providing the possibilities for bidirectional deletion cloning into an appropriately restricted vector (66).

vi. **DNA-binding site analysis: footprinting.** *DNase I footprinting* is one of the powerful techniques used to study the interaction of ligands (proteins and drugs) with DNA in solution. When there is at least a 10-fold sequence specificity (differential binding constant), the technique provides direct and immediate information about the location of a protein- or drug-binding site(s) in the DNA sequence. The extent to which a footprint represents protein binding varies with the protein/DNA complex studied because the rates and thus the pattern of DNase

I cleavage are affected by both the width of minor groove and the flexibility of the DNA. Furthermore, these parameters can be affected by protein binding. DNase I footprinting data should thus be complemented by other footprinting techniques such as those based on micrococcal nuclease, hydroxyl radical (Fe^{2+}-EDTA and H_2O_2), and/or methylation by dimethyl sulfate (64).

Since the initial description and application of the DNase I footprinting techniques to the study of the *lac* repressor-binding site (43), various technical refinements have been made (44,45). Perhaps more importantly, a deeper understanding of the structural requirements for DNase I to cleave DNA has provided a solid basis for interpreting the footprint data.

DNase I footprinting requires singly end-labeled (usually 5'-end) DNA fragments. After incubation of DNA (preferably of a size between 30 and 200 bp) with a DNA-binding protein, the DNA is briefly digested with DNase I to create a nested set of labeled fragments. The recommended DNase I digestion conditions are 1 or 2 min at 20–21°C in 10–14 mM Tris–Cl (pH 8.0), 5 mM NaCl, 1 mM $MgCl_2$, 1 mM $CaCl_2$, 0.1 mM DTT, and 50 µg/ml BSA in the presence or absence of 6% glycerol. DNA fragments are then separated on a polyacrylamide gel to obtain a ladder of bands.

3. STRUCTURE

i. Chemical and physical compositions. DNase I is an enzyme mixture composed of four glycoprotein species: DNase A (major), B, C, and D (46). DNase A carries a neutral carbohydrate side chain of 8–10 residues which is attached to Asn-18 via an aspartamido-hexose linkage. DNase B is identical to DNase A except for the carbohydrate chain: DNase B but not DNase A contains a sialic acid and a galactose residue. DNase C is an isozyme of DNase A: it contains a Pro in place of His-121 in DNase A (47). DNase D is identical to DNase C except for the carbohydrate chain: DNase D but not DNase C contains a sialic acid and a galactose residue.

The primary structure of DNase A, which was renumbered and corrected (29), consists of 260 amino acid residues (M_r 30,500) (48,49). The N- and C-terminal residues are Leu and Thr, respectively. The enzyme contains 3 Trp residues and 2 disulfide bonds (C101/C104 and C173/C209).

C-terminal Thr is involved in many H-bonding contacts and is thus normally not available for attack by carboxypeptidase A. However, SDS denaturation renders the terminal residue(s) (–Leu–Thr) susceptible to digestion. Removal of the C-terminal residue(s) inactivates the enzyme and the activity cannot be restored under renaturing conditions.

ii. Physical parameters. The pI of DNase I is ~4.7 (50). The A_{280} of a 1% solution is 12.3 in 0.1 M potassium phosphate buffer (pH 7.6) (51). The $s^\circ_{20,w}$ is 2.78S.

iii. **Tertiary structure.** The 3-D structure of DNase I has been determined for the enzyme–inhibitor (Ca-pdTp) complex (35,52) and for the enzyme–DNA complex (29). Except for the protruding (~15 Å) carbohydrate side chain, DNase I has an overall compact structure with approximate dimensions of 45 Å × 40 Å × 35 Å. The core has a sandwich-type structure consisting of two six-stranded β sheets packed against each other. Approximately 34% of amino acid residues are located in the β sheets and 23% in α helices.

iv. **Cation-binding sites.** DNase I has two structural Ca^{2+}-binding sites. One is located near the flexible loop (D99 to D107) containing the small disulfide bond (C101/C104). The second Ca^{2+}-binding site is located in the loop formed by residues Asp-201 to Thr-207, close to the essential S-S bridge (C173/C209). The second Ca^{2+} is coordinated to the Asp-201.

v. **Active site.** The active site of the DNase I molecule is presumed to contain two catalytic subsites ~15 Å apart in shallow grooves on the surface between the central β-pleated sheets. Remarkably strong H-bonds exist between the carboxyl group of Glu-78 and the N-1 position of His-134 (2.6 Å) in the first catalytic subsite (35). The second subsite is presumed to have such residues as Glu-13, Arg-41, Asp-42, Ser-43, and His-44 (29). His-134 located close to the bound Ca-pdTp (47) is essential; the modification by iodoacetate (but not by iodoacetamide) in the presence of Mn^{2+} or Cu^{2+} inactivates the enzyme (53). Replacement of His-134 by Gln results in a mutant enzyme whose activity is only ~0.001% of the native enzyme (65).

Chemical modifications of Tyr-65 (54), Trp-158 (55), and a Ser (56) lead to inactivation of the enzyme. Modification of the COO^- side chains of Asp and Glu by carbodiimide-induced condensation with glycine ethyl ester causes inactivation of the enzyme (56). The presence of Ca^{2+} significantly slows down the rate of inactivation, apparently protecting six to eight critical carboxyl groups.

None of the primary amino groups are essential for catalysis because the enzyme remains active following modifications (guanidination, picolinimidylation) of $\alpha-$ and/or ε-amino groups (57).

4. GENETICS

Information on the native gene for bovine pancreatic DNase I is not yet available. However, a synthetic DNase I gene whose nucleotide sequence was deduced from the amino acid sequence has been cloned and expressed in *E. coli* (65). The cloning of the synthetic gene that produces an active DNase I and expression under the single control of *tac* or λ p_L promotor is lethal to the host. Successful expression (to a level of ~150 μg per liter of cell culture) has been achieved only when an additional promoter (*lac*) was provided in antisense to the normal sense message of the gene expressed from the λ p_L promoter.

5. Sources

The traditional purification of DNase I from bovine pancreatic extracts using chromatography on phosphocellulose or DEAE-cellulose results in the separation of four active components; DNase A is the principal component followed by DNases B, C, and D in the following ratio: 1:0.2:0.5:0.1 (46,58). DNase A is the last component to elute in a NaAc linear gradient (0.38 to 0.67 M, pH 4.7) in phosphocellulose chromatography. The four forms can also be resolved by DEAE-cellulose chromatography, using a Ca^{2+} (0 to 2 mM) gradient buffer (0.1 M Tris–Cl, pH 8.0) for elution (58). The specific activity of each form is 1158 U/mg for DNase A, 921 for DNase B, 1045 for DNase C, and 837 for DNase D.

Because DNase I is extremely sensitive to inactivation by proteases in the absence of Ca^{2+}, buffers routinely contain protease inhibitors such as DFP and PMSC. Otherwise DNase I is very stable, retaining full activity for more than 10 days at pH 8 and 37°C.

DNase I readily cocrystallizes with chymotrypsin B. Protease-free DNase I can be obtained by a combination of affinity chromatography and salting-out adsorption (59). Affinity chromatography on a column of lima bean protease inhibitor coupled to Sepharose eliminates tightly binding trypsin and chymotrypsin, whereas the salting-out adsorption using NaCl at 3 M helps to separate DNase I from the flow-through chymotrypsinogen. Traces of contaminating RNases can be efficiently removed by passage through 5′-(4-aminophenylphosphoryl)uridine 2′(3′)-phosphate-Sepharose (13,58,59).

DNase I is usually supplied either as a solution or as a lyophilized powder at a concentration of 2000–3000 U/mg protein. A common working solution is 1 mg/ml in 20 mM Tris–Cl (pH 7.5), 1 mM $MgCl_2$, and 50% (w/v) glycerol. It can be stored for at least 12 months at -20°C.

DNase I has been covalently coupled to porous glass (60) or to CNBr-activated Sepharose 4B (61). The immobilized DNase I facilitates the removal and/or repetitive use of the enzyme.

References

1. Price, P. A. (1975). *JBC* **250**, 1981–1986.
2. Price, P. A. (1972). *JBC* **247**, 2895–2899.
3. Bollum, F. J. (1965). *JBC* **240**, 2599–2601.
4. Melgar, E., and Goldthwait, D. A. (1968). *JBC* **243**, 4409–4416.
5. Campbell, V. W., and Jackson, D. A. (1980). *JBC* **255**, 3726–3735.
6. Pohl, F. M., Thomae, R., and Karst, A. (1982). *EJB* **123**, 141–152.
7. Blanchard, J. S. (1984). *Meth. Enzy.* **104**, 404–414.
8. Douvas, A., and Price, P. A. (1975). *BBA* **395**, 201–212.
9. Lindberg, U. (1967). *Biochem.* **6**, 343–347.
10. Lindberg, M. U., and Skoog, L. (1970). *EJB* **13**, 326–335.
11. Lazarides, E., and Lindberg, U. (1974). *PNAS* **71**, 4742–4746.
12. Hitchcock, S. E., Carlsson, L., and Lindberg, U. (1976). *Cell* **7**, 531–542.

13. Mannherz, H. G., Goody, R. S., Konrad, M., and Nowak, E. (1980). *EJB* **104**, 367–379.
14. Price, P. A., Stein, W. H., and Moore, S. (1969). *JBC* **244**, 929–932.
15. Liao, T. H. (1975). *JBC* **250**, 3831–3836.
16. Liao, T. H. (1975). *JBC* **250**, 3721–3724.
17. Brison, O., and Chambon, P. (1976). *Anal. Biochem.* **75**, 402–409.
18. Bernardi, G., Ehrlich, S. D., and Thiery, J. P. (1973). *Nature NB* **246**, 36–40.
19. Vanecko, S., and Laskowski, M., Sr. (1961). *JBC* **236**, 3312–3316.
20. Drew, H. R. (1984). *JMB* **176**, 535–557.
21. Hoard, D. E., and Goad, W. (1968). *JMB* **31**, 595–606.
22. Potter, J. L., Laurila, U.-R., and Laskowski, M., Sr. (1958). *JBC* **233**, 915–916.
23. Vanecko, S., and Laskowski, M., Sr. (1961). *JBC* **236**, 1135–1140.
24. Junowicz, E., and Spencer, J. H. (1973). *BBA* **312**, 85–102.
25. Bernardi, A., Gaillard, C., and Bernardi, G. (1975). *EJB* **52**, 451–457.
26. Lomonossoff, G. P., Butler, P. J. G., and Klug, A. (1981). *JMB* **149**, 745–760.
27. Cuatrecasas, P., Wilchek, M., and Anfinsen, C. B. (1969). *Biochem.* **8**, 2277–2284.
28. Pruch, J. M., and Laskowski, M., Sr. (1980). *JBC* **255**, 9409–9412.
29. Lahm, A., and Suck, D. (1991). *JMB* **221**, 645–667.
30. Greenfield, L., Simpson, L., and Kapan, D. (1975). *BBA* **407**, 365–375.
31. Kirkegaard, K., and Nelsen, B. (1990). *J. Virol.* **64**, 185–194.
32. Bishop, K. D., Borer, P. N., Huang, Y.-Q., and Lane, M. J. (1991). *NAR* **19**. 871–875.
33. Dickerson, R. E., and Drew, H. R. (1981). *JMB* **149**, 761–786.
34. Drew, H. R., and Travers, A. A. (1984). *Cell* **37**, 491–502.
35. Suck, D., and Oefner, C. (1986). *Nature* **321**, 620–625.
36. Rhodes, D., and Klug, A. (1986). *Cell* **46**, 123–132.
37. Mehdi, S., and Gerlt, J. A. (1984). *Biochem.* **23**, 4844–4852.
38. Heffron, F., So, M., and McCarthy, B. J. (1978). *PNAS* **75**, 6012–6016.
39. Anderson, S. (1981). *NAR* **9**, 3015–3027.
40. Hong, G. F. (1982). *JMB* **158**, 539–549.
41. Frischauf, A. M., Garoff, H., and Lehrach, H. (1980). *NAR* **8**, 5541–5549.
42. Li, Q., and Wu, G. (1987). *Gene* **56**, 245–252.
43. Galas, D. J., and Schmitz, A. (1978). *NAR* **5**, 3157–3170.
44. Dynan, W. S. (1987). *Gen. Eng.* **9**, 75–87.
45. Hudson, J. M., and Fried, M. G. (1989). *BioTechniques* **7**, 812–815.
46. Salnikow, J., Moore, S., and Stein, W. H. (1970). *JBC* **245**, 5685–5690.
47. Salnikow, J., and Murphy, D. (1973). *JBC* **248**, 1499–1501.
48. Salnikow, J., Liao, T. H., Moore, S., and Stein, W. H. (1973). *JBC* **248**, 1480–1488.
49. Liao, T. H., Salnikow, J., Moore, S., and Stein, W. H. (1973). *JBC* **248**, 1489–1495.
50. Kunitz, M. (1950). *J. Gen. Physiol.* **33**, 349–357.
51. Lindberg, U. (1967). *Biochem.* **6**, 335–342.
52. Suck, D., Oefner, C., and Kabsch, W. (1984). *EMBO J.* **3**, 2423–2430.
53. Price, P. A., Moore, S., and Stein, W. H. (1969). *JBC* **244**, 924–928.
54. Hugli, T. E., and Stein, W. H. (1971). *JBC* **246**, 7191–7200.
55. Sartin, J. L., Hugli, T. E., and Liao, T. H. (1980). *JBC* **255**, 8633–8637.
56. Poulos, T. L., and Price, P. A. (1974). *JBC* **249**, 1453–1457.
57. Plapp, B. V., Moore, S., and Stein, W. H. (1971). *JBC* **246**, 939–945.
58. Liao, T. H. (1974). *JBC* **249**, 2354–2356.
59. Wang, D., and Moore, S. (1978). *JBC* **253**, 7216–7219.
60. Neurath, A. R., and Weetall, H. H. (1970). *FEBS Lett.* **8**, 253–256.
61. Kastern, W. H., Eldridge, J. D., and Mullinix, K. P. (1979). *JBC* **254**, 7368–7376.
62. Bickler, S. W., Heinrich, M. C., and Bagby, G. C. (1992). *BioTechniques* **13**, 64–66.
63. Ziegler, B. L., Lamping, C., Thoma, S., and Thomas, C. A. (1992). *BioTechniques* **13**, 726–729.
64. Fairall, L., and Rhodes, D. (1992). *NAR* **20**, 4727–4731.
65. Worrall, A., and Connally, B. (1990). *JBC* **265**, 21889–21895.
66. Eberle, J. R. (1993). *BioTechniques* **14**, 408–411.

II. DNA/RNA ENDONUCLEASE: STAPHYLOCOCCAL NUCLEASE
[Micrococcal nuclease, EC 3.1.31.1]

Staphylococcal nuclease (SNase), also commonly known as *micrococcal nuclease*, is a Ca^{2+}-dependent phosphodiesterase which cleaves both DNA and RNA to yield 3'-phosphomononucleotide and 3'-phosphooligonucleotide end products (Scheme 3.1).

Mononucleotides $\quad$ R-pdN-R' $\rightarrow$ R-P + dN-R'

Polynucleotides $\quad$ $pN^1pN \ldots pN^k \rightarrow pN^1p + (Np)_2 + (Np)_3 + \ldots + N^k$

SCHEME 3.1

SNase is a monomeric enzyme with both endonucleolytic and exonucleolytic activities. It is a small globular protein of 149 residues containing no disulfide linkages, and is produced as an extracellular enzyme by certain strains of *S. aureus* (1).

Staphylococcal nuclease has been extensively used as a model protein for the study of enzyme mechanisms, stability, and the kinetics of (re)folding. In recombinant DNA technology, SNase is primarily used as a nonspecific endonuclease, as is DNase I which generates 5'-phosphonucleotides (Section I, this chapter). Staphylococcal nuclease has also served as the prototype model for a new generation of nucleases called *hybrid nucleases*, which can cleave DNA and RNA with tailored site specificities.

1. FUNCTIONS

a. Reaction conditions

i. Optimal reaction conditions. A typical reaction mixture consists of 40 mM Tris–Cl (pH 8.5), 10 mM $CaCl_2$, DNA (or RNA), and SNase. The mixture is incubated at 37°C, and the reaction is stopped by adding EDTA (20 mM).

SNase is optimally active between pH 9 and pH 10 (2). Above pH 10, the activity falls irreversibly. At pH values below 10, the optimal activity requires high levels of Ca^{2+} (e.g., 10 mM). The activity increases almost linearly with NaCl to a maximum of twofold at 150 mM, after which the activity decreases (3). For the purpose of nonspecific degradation, however, it is preferable to use no salt or low salt concentrations.

Above 20°C, the activity increases with temperature according to the Arrhenius equation [Eq. (1.11)] without any indication of heat inactivation up to 65°C (4). Below 20°C, a drastic decrease in activity is observed, signaling that the enzyme is inactivated by low temperatures.

ii. Kinetic parameters. The kinetics of PNP-phosphate release from the substrate PNP-pdT do not follow typical Michaelis–Menten kinetics, but show pronounced substrate inhibition at high substrate concentrations. The apparent ki-

TABLE 3.1 Apparent Kinetic Constants of Staphylococcal Nuclease for p-Nitrophenyl Ester Substrates

Substrate	K_m	k_{cat} (min^{-1})	Conditions
PNP-pdT	2.2 mM	9.1	a
	19.6 mM	10.2	b
PNP-pdTp	10 μM	10.6	a
	35 μM	13.9	b

[a] 50 mM borate (pH 8.8), 10 mM CaCl$_2$, and 24°C (18).
[b] 100 mM CHES (pH 9.5), 0.2 M KCl, I (ionic strength) = 0.3, and 26°C (17).

netic parameters for the two most commonly used synthetic substrates, PNP-pdT and PNP-pdTp, are listed in Table 3.1. The k_{cat} for the dinucleotide substrate dTdA is 0.23 min^{-1} (5). The K_m for ssDNA is 3.5–5.0 μg/ml (3,6).

iii. **Metal ion requirements.** SNase requires a high concentration of Ca^{2+} (10 mM) for full activity. The K_d for Ca^{2+} is 0.5 mM and K_m is 0.1 mM (6,7). Certain other metals such as Ba^{2+}, Co^{2+}, Mn^{2+}, and Zn^{2+} bind competitively at the Ca^{2+}-binding site and promote the binding of various inhibitors. However, SNase is virtually inactive with metal ions other than Ca^{2+} and is also inactive when Ca^{2+} is replaced with Sr^{2+} (8).

iv. **Inhibitors and inactivators**

(a) Mononucleotides. Deoxythymidine 3′,5′-diphosphate (pdTp), but not pdCp, is a potent inhibitor in the presence of Ca^{2+} (or other metal ions) with a K_i of 0.2 μM. 5′-Mononucleotides such as pA, pdA, and pdT are also inhibitors with K_i values ranging from 20 to 200 μM. Nucleosides and nucleoside 2′- or 3′-monophosphates are not inhibitory (2). However, vanadate complexes of dT (VO^{2+}-dT) and dTp (VO^{2+}-dTp) are, in the presence of Ca^{2+}, strong (possibly transition-state) inhibitors with K_i values of 80 and 170 nM, respectively (9).

(b) Group-modifying reagent. Phenylglyoxal, an Arg-modifying reagent, rapidly inactivates SNase, but the presence of pdTp and Ca^{2+} protects the enzyme from the inactivation (9).

(c) Calcium chelator. Because SNase absolutely requires Ca^{2+} for activity, it can be readily inactivated by a Ca^{2+} chelator, EGTA. The EGTA-inactivated SNase is still capable of forming a complex with free DNA or RNA which can be dissociated by the addition of synthetic polynucleotides, carrier RNA, or heparin. In the absence of Ca^{2+}, SNase binds to precursor tRNA with an apparent $K_d \approx 1.4$ μM, comparable to its reported affinity for DNA (10). The complex formation of the inactive SNase with pre-tRNA blocks tRNA processing.

b. Activity assays and unit definition

i. Spectrophotometric assays. Nuclease activity is conventionally assayed by observing the absorbance increase at 260 nm as DNA or RNA is hydrolyzed (2). A typical assay mixture contains 40 mM Tris–Cl (pH 8.5), DNA (salmon sperm, 20–90 µg/ml), and $CaCl_2$ (0.4–10 mM) in a volume of 1.0 ml at 24°C.

Kinetic assays are usually performed spectrophotometrically using PNP-pdT as a substrate. The initial product PNP-phosphate (+ dT) is converted to PNP in an alkaline phosphatase-coupled reaction. The rate of PNP production is monitored at 405 nm, where ε_{405} is 1.83×10^3 M^{-1} cm^{-1}.

Unit definition: One enzyme unit is defined as the amount of enzyme that causes a change of 1.0 unit of A_{260} per minute.

ii. Alternative assays. The following three methods provide useful alternatives for the assay of SNase under circumstances that require higher sensitivities.

(a) *Nicking activity assay on supercoiled plasmid DNA.* The nicked and linearized products are separated from the supercoiled DNA by gel electrophoresis and are quantitated by densitometric scanning (11).

(b) *Colorimetric assay.* This assay monitors the color change of a metachromatic dye (e.g., toluidine blue) from blue to pink on hydrolysis of DNA as the enzyme diffuses through a DNA-containing agar gel (11).

(c) *Radiometric assay.* This assay measures the release of acid-soluble nucleotides from the heat-denatured [^{3}H]thymidine-labeled DNA of *E. coli* (12).

c. Substrate specificities

i. Polynucleotides. Denatured DNA (ssDNA) is the preferred substrate over duplex DNA. Denatured DNA is digested by SNase sevenfold more effectively than RNA, and no inhibitory products are accumulated during the reaction. The endonucleolytic double-strand cutting activity is a result of nicking (primary cleavage) in one strand followed by secondary cleavage at a site two bases down in the 3' direction (13). The two-base staggered end is then rapidly trimmed by exonucleolytic activity.

ii. Cleavage preferences. With denatured DNA, the order of cleavage appears to be nearly random. In native dsDNA, however, Xp-dTp and Xp-dAp bonds are preferentially cleaved. The cleavage preference at the 5' side of A or T is estimated to be 30 times greater than that of G or C (14). Although AT-rich regions are cleaved preferentially, the transient single strandedness of such regions is not the only recognition feature. The most rapid cleavage sites tend to occur in stretches of more than three contiguous A and/or T residues. Nevertheless, long stretches of DNA containing only A and T have been observed not to be cleaved by the enzyme (15). Because of these known and unknown preferences, partial digestion of certain DNAs may produce a rather distinct pattern of fragments resembling sequence-specific cleavages at 5'-C|A|T|A-3' and 5'-C|T|A-3' (15).

iii. **Dinucleotides.** Dinucleotides (dNp$^\alpha$Np$^\beta$) are substrates with a distinct order of preference (dT ≤ dA ≫ dC ≫ dG) for the base in the β position but little base specificity in the α position (8,16). Corresponding dinucleoside monophosphates (dNpN) are hydrolyzed about 10-fold more slowly [e.g., k_{cat} for dTdA is 0.23 min^{-1} (5)], while the 5'-phosphodinucleotides (pdNpN) are even more resistant.

When ribo- and deoxyribodinucleotides are compared, the rate of hydrolysis is practically identical except that r(ApAp) is hydrolyzed about three times faster than d(ApAp) (8).

iv. **Mononucleotide esters.** Synthetic substrates with the basic structure of R-pdN-R' are generally cleaved by SNase with the release of R-P and dN-R' (1). The size and the nature of the R group are relatively unimportant as long as a diester bond is present (see exceptions below). The 5'-PNP esters of pdT and pdTp are the simplest known good substrates. Both of these esters are hydrolyzed exclusively to the PNP-phosphate and dT or dTp. Although the enzyme-catalyzed hydrolysis of the two substrates undergoes at almost the same rate, SNase has a 200–600 times smaller K_m for the pdTp ester (see Table 3.1). This difference in substrate-binding affinity highlights the importance of the 3'-P group and the nature (size and charge as well) of the R' substitution on the 3'-OH (1,17).

Both the 5'-methyl ester of pdT and thymidine 5'-fluorophosphate are poor substrates. The products are methylphosphate and fluorophosphate plus dT. When PNP-pdTp-PNP is used as a substrate, there is a rapid release of PNP-phosphate from the 5' position along with a slow release of PNP from the 3' position (18). Other substrates with a 3'-PNP consistently give rise to some PNP as a product. This secondary activity of SNase is poorly understood. 2',3'-Cyclic phosphate esters are not substrates for SNase (5).

d. *Mechanism of catalysis*

i. **Essential features of catalysis.** The phosphodiesterase reaction catalyzed by SNase is believed to be a concerted general acid–base catalysis involving Arg-87 and Glu-43 (refer to *Structure* below). This catalytic mechanism consists of the following three features: (a) hydrolysis proceeds via direct nucleophilic attack on the phosphate (O5'-PO$_2$-O3') with the formation of a five-coordinate, trigonal bipyramidal transition state intermediate followed by breakdown to a new phosphate species (O3'-PO$_3^-$); (b) the Ca^{2+} ion assists the catalysis through coordination with the 5'-P in a way that induces conformational changes both at the attacked phosphorus (activation) and at the leaving group (stabilization) of the bound substrate (5); and (c) attacking groups enter and leaving groups leave in a concerted manner at the apical positions of the bipyramid.

ii. **Interactions within active site.** The environment of Ca^{2+} within the enzyme–Ca^{2+}–inhibitor (pdTp) ternary complex comprises an octahedral geometry.

The metal is coordinated to the carboxylates of Asp-21 and Asp-40, the amide carbonyl group of Thr-41, the 5′-P of pdTp, and two water molecules. Asp-18 and Glu-43 are somewhat farther from the Ca^{2+} and lie in the second coordination sphere (19).

The three ternary complexes of SNase formed respectively with pdTp, pdGp, and PNP-pdTp display quite similar conformational features in NMR, suggesting that the conformation of the enzyme is not strongly dependent on whether the 5′-P is a mono- or diester (55). The conformational features of the bound nucleic acid determine the differences in catalytic rates between substrates. In the ternary complex, e.g., enzyme–Co^{2+}–dTdA, evidence suggests that the metal induces conformational changes both at the attacked phosphorus and at the leaving group of the enzyme-bound substrate (5).

Compared with the Mn^{2+}–enzyme binary complex, the ternary complex (enzyme–Mn^{2+}–pdTp) has one more water molecule in the second coordination sphere of the metal ion (20). The additional water molecule apparently replaces the Glu-43 carboxylate which is coordinated with the metal ion in the binary complex. This and other evidence suggests that, in the ternary complex, the side chain carboxylate of Glu-43 acts as a general base potentiating the attack of the water molecule on the phosphodiester bond. The guanidinium ion of Arg-87 is also appropriately positioned to both bind and catalytically activate the 5′-phosphate group of the substrate. Since all five Arg residues in the protein have pK_a values greater than 11.6, Arg-87 may be a candidate for the acidic catalyst that protonates the 5′-ribose alkoxide prior to product release (17).

The pH profile of k_{cat}/K_m is bell shaped and depends on two pK_a values, 8.4–8.9 and 9.2–9.7 (17,21). The upper pK_a has been assigned to Tyr-85 which is likely to be a H-bond donor to the 3′-phosphomonoester of the substrates poised for hydrolysis (17). The RDS in the SNase-catalyzed reaction is presumed to be the chemical step, a nucleophilic displacement at the phosphorus of the 5′-OH of the leaving polynucleotide.

In the presence of the strong electrostatic forces contributed by Ca^{2+}, the two residues (Glu-43 and Arg-87) exert various effects: anticooperativity in binding the substrate and cooperativity in binding the transition state. This combination of effects is presumed to lower the kinetic barrier to catalysis by 8.1 kcal/mol (7).

For oligonucleotides having a terminal 5′-phosphate, $(pdT)_n$, the maximum binding occurs when n is 3, suggesting the presence of a third, probably ionic, binding site in addition to those for the 5′- and 3′-phosphates of pdTp (1).

iii. Effect of site-specific amino acid substitutions.

According to site-specific mutagenesis studies, the substitution of Arg-87 by Cys or Gly reduces V by $>10^4$-fold (6,7). Perhaps surprisingly, the conservative substitution by Lys (R87K) results in a 10^5-fold reduction in catalytic efficiency (22). On the other hand, the substitution of Glu-43 by Asp (E43D) reduces the catalytic efficiency by 1400-fold, while the charge-neutral mutations of Glu-43 (by Gln, Asn, Ala, and Ser) result in about a 5000-fold reduction in the efficiency (23). In contrast, the double mutant (E43S + R87G) has only a 10^6-fold lower V than the wild-type enzyme.

The reason for this unpredictable sensitivity to amino acid changes is that the conservative or nonconservative substitutions seemingly entail minor but significant nonlocal structural changes affecting the active conformation (24). All of the Glu-43 mutant enzymes just mentioned show higher thermal stabilities than the wild-type SNase.

2. APPLICATIONS

i. **Preparation of 3'-phosphomononucleotides.** The relatively nonspecific activities of SNase can be used to generate 3'-P-mononucleotides from a wide variety of DNA and RNA substrates. The 5'-OH of the 3'-P mononucleotides can be efficiently phosphorylated by T4 PNK (see Section II, Chapter 5). As a 3'-P yielding endonuclease, SNase is useful in the analysis of the nearest neighboring base in synthetic DNA and enzyme-cleaved polynucleotide chains (25).

ii. **Removal of nonspecific DNA and RNA.** SNase is frequently used to prepare eukaryotic cell-free extracts (e.g., rabbit reticulocyte lysates, wheat germ lysates) which support the *in vitro* translation of a wide variety of viral, prokaryotic, and eukaryotic mRNAs into proteins. Brief digestion of endogenous mRNAs (e.g., globin mRNA) by SNase reduces the background of the *in vitro* translations to a minimum (26–28), leaving the rRNAs and tRNAs largely intact. In rabbit reticulocyte lysates, a variety of posttranslational processing activities remain functional such as proteolysis, phosphorylation, acetylation, and isoprenylation. SNase is conveniently inactivated by the addition of EGTA, a Ca^{2+} chelator.

iii. **Use as sequence-specific hybrid nucleases.** By fusing the relatively nonspecific SNase with specific oligonucleotides, the SNase is transformed to hybrid nucleases which can perform sequence-specific cleavages of ssDNA and RNA (29–31). A typical hybrid enzyme consists of an oligodeoxynucleotide (e.g., 22-mer, 3'-S-thiopyridyl-) fused to a unique site (e.g., Cys-116) on SNase (K116C mutant) through a disulfide linkage. The oligonucleotide has been shown to direct the nuclease activity of the enzyme to the nucleotides directly adjacent to the complementary target sequence on the substrate DNA or RNA. Hydrolysis occurs primarily at one phosphodiester bond two bases away from the 5' termini of the target site. A similar hybrid enzyme strategy has also been applied to cleave homopurine-containing dsDNA site-specifically via triple helix formation (32).

iv. **As probe for protein binding or distorted region of DNA.** SNase is a small protein (17 kDa) compared with DNase I (30 kDa) which is most widely used in DNA footprinting. SNase recognizes a single strand of the DNA double helix, whereas DNase I interacts with both strands in the minor groove. As an essentially single-strand-specific endonuclease, SNase relies on the breathing motion of the duplex DNA to make cleavages. Furthermore, the cleavage at G and C residues can be substantially slower than at A and T residues. These properties make the SNase an excellent complement for DNase I in footprinting analysis, by providing information in greater detail about the regions of protein binding and/or structural perturbations at the site of protein–DNA interactions (33,56).

3. STRUCTURE

i. Primary structure. The mature SNase is a single polypeptide having 149 residues (M_r 16,800) (1,34). The N- and C-terminal residues are Ala and Gln, respectively. The enzyme from the Foggi strain (SNase-Foggi) contains 4 His and 5 Arg residues but no Cys. [Note that the enzyme from the V8 strain (SNase-V8) is identical to SNase-Foggi except that it has Leu-124 in place of the His-124 found in SNase-Foggi.]

Limited tryptic digestion of SNase in the presence of pdTp and Ca^{2+} produces an enzymatically active (8% of the native enzyme) nuclease derivative, called *nuclease-T*, which contains two noncovalently bound peptide fragments nuclease-T-(6–48) and nuclease-T-(49–149) (35,36). *Nuclease-T'* refers to the complex formed on reconstitution of the isolated inactive nuclease-T fragments.

ii. Tertiary structure. The 3-D structures of SNase and its complex with a strong inhibitor pdTp have been determined by X-ray crystallography at 2.0 Å resolution (37) and at or near 1.5 Å resolution (19,38,39). Detailed solution structures of SNase have also been determined by NMR (40). SNase has an overall dimension of 30 Å × 30 Å × 40 Å. The formation of the ternary complex (SNase–pdTp–Ca^{2+}) induces only limited changes in backbone conformation but significant changes in side chain orientation primarily in the region of the Ca^{2+}-binding site. The ternary complex is more resistant than the free enzyme to urea, heat denaturation, high pH denaturation, and tryptic cleavages.

Binding studies in solution indicate that two Ca^{2+} ions bind to the nuclease along with the pdTp inhibitor (41), but only one site is observed in the crystal. The single Ca^{2+} ion is bound to the protein through the side chain carboxylates of Asp-21, Asp-40, and the backbone carbonyl oxygen of Thr-41. The side chain of Glu-43 is bridged by a water molecule to both the Ca^{2+} and the 5'-P of the nucleotide substrate, and is thought to play the role of a general base in the cleavage of 5'-P. This Ca^{2+} ion is at some 5 Å distance from the substrate phosphate.

The substrate-binding site contains Arg-35, Lys-84, Tyr-85, and Arg-87 residues. They assist in positioning the 3'-P and 5'-P of pdTp in the nuclease complex. The positive side chains of the two Arg residues are H-bonded to the oxygen atoms of the 5'-phosphate, activating it to nucleophilic attack. The 3'-phosphate oxygens interact with the side chain of Lys-84 and the phenolic OH of Tyr-85.

Two alternative monomeric forms (N, N*) of SNase arise from the cis/trans isomerism about the Lys-116/Pro-117 peptide bond (42). Certain mutations coupled to changes in the stability of the N protein alter the ratio of N'(trans)/N(cis) by as much as two orders of magnitude (43).

iii. Physical parameters. The A_{280} of a 0.1% solution is 0.93 at neutral pH (21). The pI of SNase is 10.2 (3).

iv. Immobilized enzyme. SNase immobilized on an agarose support has been shown to be remarkably stable, compared with the native nuclease, and active at 45°C during the prolonged incubation time (e.g., 24 hr) (44).

4. GENETICS

The gene (*nuc*) coding for SNase has been cloned from the Foggi strain of *S. aureus*, sequenced and expressed in *E. coli* (45). The potential coding region of the cloned gene contains 242 codons in which the C-terminal 149 amino acids correspond to the classical SNase, also known as *nuclease A*. Immediately preceding the codons for nuclease A (+1 to +149) are 19 codons (−19 to −1) which specify the additional amino acids present at the N terminus of *nuclease B*, a membrane-bound precursor of nuclease A found in exponentially growing cells (46). Further upstream, 74 codons contain five potential translation initiation triplets (three AUG and two GUG). The upstream sequence is presumed to encode a signal peptide by virtue of the fact that it contains some of the features typical of known prokaryotic signal peptides including a long hydrophobic core (−44 to −29) preceded by one or more positively charged residues.

Various mutant enzymes have been generated for both functional and structural studies by site-specific mutagenesis (6,7,17,22,23) or random mutagenesis (47,48).

5. SOURCES

Staphylococcal nuclease was initially isolated as an extracellular enzyme (nuclease A) of *S. aureus* (or *Micrococcus pyogenes*) (1,49). The extremely high affinity of the nuclease to phosphocellulose at pH 5.8 has provided the basis of convenient purification methods. SNase can also be efficiently purified by the use, as an affinity matrix, of 3′-(4-aminophenylphosphoryl)deoxythymidine 5′-P linked to the CNBr-activated Sepharose (50).

The *nuc* gene originally cloned into the *Bam*HI site of pBR322 (pFOG301) is poorly expressed in *E. coli* (HB101), at a level lower than that of the genomic expression in *S. aureus* (45). SNase can now be overproduced in *E. coli* as a fusion protein with the OmpA signal peptide and under the control of the *lac*UV5 promoter (51). The nuclease is fully and correctly processed at the signal sequence, it is secreted to the periplasmic space, and accumulates to ~10% of the total cellular protein. The SNase gene has also been overexpressed in yeast as fusion proteins with different signal peptides and under the GAL10 inducible promoter (52).

Various mutant enzymes have also been produced at high yields. For example, SNase-H124L, which is threefold less active but more thermostable than the wild-type enzyme (53), has been cloned and overexpressed under the control of phage T7 promoter in a T7 RNA polymerase expression vector pET3A (54).

References

1. Anfinsen, C. B., Cuatrecasas, P., and Taniuchi, H. (1971). "The Enzymes," 3rd ed., Vol. 4, pp. 177–204.
2. Cuatrecasas, P., Fuchs, S., and Anfinsen, C. B. (1967). *JBC* **242**, 1541–1547.

3. Shortle, D., and Meeker, A. K. (1989). *Biochem.* **28**, 936–944.
4. Pohl, F. M., Thomae, R., and Karst, A. (1982). *EJB* **123**, 141–152.
5. Weber, D. J., Mullen, G. P., and Mildvan, A. S. (1991). *Biochem.* **30**, 7425–7437.
6. Serpersu, E. H., Shortle, D., and Mildvan, A. S. (1987). *Biochem.* **26**, 1289–1300.
7. Weber, D. J., Meeker, A. K., and Mildvan, A. S. (1991). *Biochem.* **30**, 6103–6114.
8. Sulkowski, E., and Laskowski, M., Sr. (1970). *BBA* **217**, 538–540.
9. Tucker, P. W., Hazen, E. E., Jr., and Cotton, F. A. (1979). *Mol. Cell. Biochem.* **23**, 67–86.
10. Wang, M. J., and Gegenheimer, P. (1990). *NAR* **18**, 6625–6631.
11. Weber, D. J., Serpersu, E. H., Shortle, D., and Mildvan, A. S. (1990). *Biochem.* **29**, 8632–8642.
12. Kohli, J. K., Vakil, B. V., Patil, M. S., Pandey, V. N., and Pradhan, D. S., (1989). *Indian J. Biochem. Biophys.* **26**, 296–300.
13. Cockell, M., Rhodes, D., and Klug, A. (1983). *JMB* **170**, 423–446.
14. Dingwall, C., Lomonossoff, G. P., and Laskey, R. A. (1981). *NAR* **9**, 2659–2673.
15. Horz, W., and Altenburger, W. (1981). *NAR* **9**, 2643–2658.
16. Mikulski, A. J., Sulkowski, E., Stasiuk, L., and Laskowski, M., Sr. (1969). *JBC* **244**, 6559–6565.
17. Grissom, C. B., and Markley, J. L. (1989). *Biochem.* **28**, 2116–2124.
18. Cuatrecasas, P., Wilchek, M., and Anfinsen, C. B. (1969). *Biochem.* **8**, 2277–2284.
19. Cotton, F. A., Hazen, E. E., Jr., and Legg, M. J. (1979). *PNAS* **76**, 2551–2555.
20. Serpersu, E. H., McCracken, J., Peisach, J., and Mildvan, A. S. (1988). *Biochem.* **27**, 8034–8044.
21. Dunn, B. M., DiBello, C., and Anfinsen, C. B. (1973). *JBC* **248**, 4769–4774.
22. Pourmotabbed, T., Dell'Acqua, M., Gerlt, J. A., Stanczyk, S. M., and Bolton, P. H. (1990). *Biochem.* **29**, 3677–3683.
23. Hibler, D. W., Stolowich, N. J., Reynolds, M. A., Gerlt, J. A., Wilde, J. A., and Bolton, P. H. (1987). *Biochem.* **26**, 6278–6286.
24. Loll, P. J., and Lattman, E. E. (1990). *Biochem.* **29**, 6866–6873.
25. Josse, J., Kaiser, A. D., and Kornberg, A. (1961). *JBC* **263**, 864–875.
26. Pelham, H. R. B., and Jackson, R. J. (1976). *EJB* **67**, 247–256.
27. Merrick, W. C. (1983). *Meth. Enzy.* **101**, 606–615.
28. Henshaw, E. C., and Paniers, R. (1983). *Meth. Enzy.* **101**, 616–629.
29. Corey, D. R., and Schultz, P. G. (1987). *Science* **238**, 1401–1403.
30. Zuckerman, R. N., and Schultz, P. G. (1989). *PNAS* **86**, 1766–1770.
31. Corey, D. R., Pei, D., and Schultz, P. G. (1989). *Biochem.* **28**, 8277–8286.
32. Pei, D., Corey, D. R., and Schultz, P. G. (1990). *PNAS* **87**, 9858–9862.
33. Zhang, L., and Gralla, J. D. (1989). *NAR* **17**, 5017–5028.
34. Cotton, F. A., and Hazen, E. E., Jr. (1971). "The Enzymes," 3rd ed., Vol. 4, pp. 153–176.
35. Taniuchi, H., and Anfinsen, C. B. (1968). *JBC* **243**, 4778–4786.
36. Taniuchi, H., Anfinsen, C. B., and Sodja, A. (1967). *PNAS* **58**, 1235–1242.
37. Arnone, A., Bier, C. J., Cotton, F. A., Day, V. W., Hazen, E. E., Jr., Richardson, D. C., Richardson, J. S., and Yonath, A. (1971). *JBC* **246**, 2302–2316.
38. Loll, P. J., and Lattman, E. E. (1989). *Proteins (S. F. G)* **5**, 183–201.
39. Hynes, T. R., and Fox, R. O. (1991). *Proteins (S. F. G)* **10**, 92–105.
40. Torchia, D. A., Sparks, S. W., and Bax, A. (1989). *Biochem.* **28**, 5509–5524.
41. Tucker, P. W., Hazen, E. E., Jr., and Cotton, F. A. (1979). *Mol. Cell. Biochem.* **23**, 3–16.
42. Evans, P. A., Kautz, R. A., Fox, R. O., and Dobson, C. M. (1989). *Biochem.* **28**, 362–370.
43. Alexandrescu, A. T., Hinck, A. P., and Markley, J. L. (1990). *Biochem.* **29**, 4516–4525.
44. Alcantara, A., Ballesteros, A., Heras, A. M., Montero, J. M. S., and Sinisterra, J. V. (1988). *Annals N.Y.A.S.* **542**, 1–6.
45. Shortle, D. (1983). *Gene* **22**, 181–189.
46. Davis, A., Moore, I. B., Parker, D. S., and Taniuchi, H. (1977). *JBC* **252**, 6544–6553.
47. Shortle, D., and Lin, B. (1985). *Genetics* **110**, 539–555.
48. Shortle, D., and Meeker, A. K. (1986). *Proteins (S. F. G)* **1**, 81–89.
49. Moravek, L., Anfinsen, C. B., Cone, J. L., and Taniuchi, H. (1969). *JBC* **244**, 497–499.
50. Wilcheck, M., and Gorecki, M. (1974). *Meth. Enzy.* **34**, 492–496.
51. Takahara, M., Hibler, D. W., Barr, P. J., Gerlt, J. A., and Inouye, M. (1985). *JBC* **260**, 2670–2674.

52. Pines, O., and London, A. (1991). *J. Gen. Microbiol.* **137**, 771–778.
53. Alexandrescu, A. T., Ulrich, E. L., and Markley, J. L. (1989). *Biochem.* **28**, 204–211.
54. Wang, J., LeMaster, D. M., and Markley, J. L. (1990). *Biochem.* **29**, 88–101.
55. Stanczyk, S. M., and Bolton, P. H. (1992). *Biochem.* **31**, 6396–6401.
56. Fairall, L., Rhodes, D., and Klug, A. (1986). *JMB* **192**, 577–591.

III. RNA ENDONUCLEASES

Ribonucleases (RNases) constitute a superfamily of nucleases that catalyze the hydrolysis of 3′,5′-phosphodiester linkages in RNA. For the majority of RNases, the catalysis involves the formation of nucleoside 2′,3′-cyclic phosphate intermediates and generation of the products with a 3′-phosphate group. RNases are present in all organisms, including bacteria, yeast, plants, and animals, and in almost all tissues and body fluids of mammals. Some RNases are secreted (extracellular), probably suggesting their major roles in digestion. Other RNases, so-called cytoplasmic or intracellular RNases, are usually found in latent forms complexed with a protein inhibitor, and yet they share some structural characteristics typical of the secreted RNases. Within a single cell, a number of functionally distinct RNases can be found; for example, 15 RNases have been identified in *E. coli* (1). These RNases play essential roles in diverse aspects of RNA metabolism, including nonspecific turnover and degradation of RNAs in regulated environments and high specific processing of precursor forms of mRNA, tRNA, and rRNA. In fact, most of the enzymes involved in specific RNA processing belong to the group known as ribozymes (see Section II,D,4, Chapter 1). Along with the developments in the characterization of ribozymes, the classical proteinaceous RNases, which were once believed to be ordinary enzymes catalyzing nonspecific RNA turnover and degradation, have gained renewed respect as one of the central regulators in specific cellular recognition (2).

The history of the RNases began in the 1960s with the study of bovine pancreatic RNase (*RNase A*) (3,4). This enzyme continues to serve as a prime model in the study of protein structure and enzyme functions. The resurgence of interest in various aspects of RNases is bound to bring additional new information that will most likely change the way in which the *in vivo* and *in vitro* functions of RNases are described

Based on the work with RNase A, a number of other base-specific RNases have been characterized that enlarge the scope of RNase functions. These base-specific endoribonucleases are powerful and essential tools in the analysis of RNA structure and function as well as in RNA sequencing and various other manipulations of RNA (5). Their importance in molecular biology and recombinant DNA technology will continue to grow.

This section is devoted to three representative endoribonucleases: (a) bovine pancreatic RNase (RNase A) as the prime model of RNases, (b) *RNase H* as the model of a special class of RNases that catalyze specific cleavages of the RNA strand in RNA·DNA heteroduplexes, and (c) a guanine-specific nuclease, *RNase T1*.

References

1. Deutscher, M. P. (1988). *TIBS* **13**, 136–139.
2. Strydom, D. J., Fett, J. W., Lobb, R. R., Alderman, E. M., Bethune, J. L., Riordan, J. F., and Vallee, B. L. (1985). *Biochem.* **24**, 5486–5494.
3. Richards, F. M., and Wyckoff, H. W. (1971). "The Enzymes," 3rd ed., Vol. 4, pp. 647–806.
4. Blackburn, P., and Moore, S. (1982). "The Enzymes," 3rd ed., Vol. 15(B), pp. 317–433.
5. *Meth. Enzy.* (1989). **180**, whole volume.

A. Ribonuclease A

[Pancreatic ribonuclease, EC 3.1.27.5]

The ribonuclease from bovine pancreas, known as *RNase A,* is the most thoroughly characterized of all RNases (1,2). It is the first enzyme and only the second protein (the first being insulin) whose amino acid sequence has been determined. The proteolytic cleavage of RNase A by subtilisin results in two fragments, S-peptide and S-protein. When the two fragments are held together by noncovalent interactions, full RNase A-like activity is restored; this enzyme is called *RNase S*. Other species of RNases, e.g., *RNase B, C,* and *D,* have also been isolated from pancreatic secretions. They are now known to be the same as RNase A with the exception that they are glycosylated at Asn-34 with different sugar components.

RNase A is a small, monomeric enzyme of 124 amino acid residues. The catalytic core is unusually stable to heat and denaturants. RNase A is an endoribonuclease specific for pyrimidine bases. The enzyme cleaves phosphodiester bonds of RNA at the 3' side of C and U residues, resulting in mono- and oligoribonucleotide products with 3'-P and 5'-OH termini (Scheme 3.2):

$$5'\,p-G-p-\overset{\downarrow}{C}-p-\overset{\downarrow}{C}-p-A-p-A-p-\overset{\downarrow}{U}-p-G\ 3' \longrightarrow pGpCp + Cp + ApApUp + G$$

SCHEME 3.2

1. Functions

a. Reaction conditions

i. Optimal reaction conditions. The reaction conditions for RNase A are rather flexible and usually depend on the purpose of the experiment. Although various buffers can be used over a wide range of pH and temperature, the reactions in 10 mM Tris–Cl (pH 7.0–7.5) at 37°C are generally considered to be adequate. The digestion of RNA in the presence of DNA samples can be carried out under conditions most suitable for the structural integrity of the particular DNA, such as high salt (0.1–0.2 M NaCl and/or KCl or LiCl) in the presence or absence of Mg^{2+} or EDTA.

TABLE 3.2 Steady-State Kinetic Parameters of RNase A[a]

Substrate	K_m (mM)	k_{cat} (sec^{-1})	k_{cat}/K_m[b] ($10^{-5} \times M^{-1}$ sec^{-1})
Dinucleotides[c]			
CpA	1.0	3000	60
CpG	3.0	500	5.1
CpC	4.0	240	—
CpU	3.7	27	—
UpA	1.9	1200	41
UpG	2.0	69	1.8
UpC	3.0	40	—
UpU	3.7	11	—
Cyclic nucleotides[d]			
C > p	3.0	5.5	0.008[e]
U > p	5.0	2.2	—

[a] Reaction conditions are, unless specified otherwise, 0.1 M imidazole, pH 7.0, NaCl to give I = 0.2, 27°C. For more extensive listing of substrates, refer to Ref. (1).

[b] Data for k_{cat}/K_m are not the values calculated from the individual parameters in the first two columns but are from Ref. (79). Reaction conditions are 0.1 M Tris-Cl, pH 7.5, 100 mM NaCl, 25°C.

[c] The kinetics with dinucleotide substrates are from the transesterification reaction (the first step of catalysis) (10).

[d] The kinetics with 2′,3′-cyclic nucleotides are based on the hydrolysis reaction (the second step of catalysis) (10). *Note:* Do not confuse with cyclic 3′,5′-AMP (or -GMP) which are conventionally denoted by cAMP (or cGMP).

[e] Data from Ref. (63).

ii. Kinetic parameters. The catalytic parameters as well as binding constants are strongly dependent on pH and, to some extent, on ionic strength and buffer components. The k_{cat} is usually maximal at pH 7.0–7.5, although the catalytic efficiency (k_{cat}/K_m) is maximal at a lower pH of 6.0–6.5. The steady-state kinetic parameters for some typical dinucleotide and cyclic nucleotide substrates are listed in Table 3.2.

The k_{cat}/K_m for uridine 3′-P aryl esters in cyclization reactions ranges from 2.4 mM^{-1} sec^{-1} for the phenyl group to 70 mM^{-1} sec^{-1} for the 2-chloro-4-nitrophenyl group (3).

b. Activity assays and unit definition.

RNase activity can be assayed by measuring the rate of transphosphorylation, hydrolysis, or both (refer to *catalytic mechanism* below and Fig. 3.3). The following assay methods are among the most commonly used.

i. Kunitz spectrophotometric assay. This assay measures the first step in the cleavage of RNA. The assay is carried out by convention at pH 5.0 (4), although the optimum pH of RNase A is near 7. The RNA substrate (yeast RNA) is dissolved in 0.1 M NaAc (pH 5.0) at a concentration sufficient to yield a A_{300} of

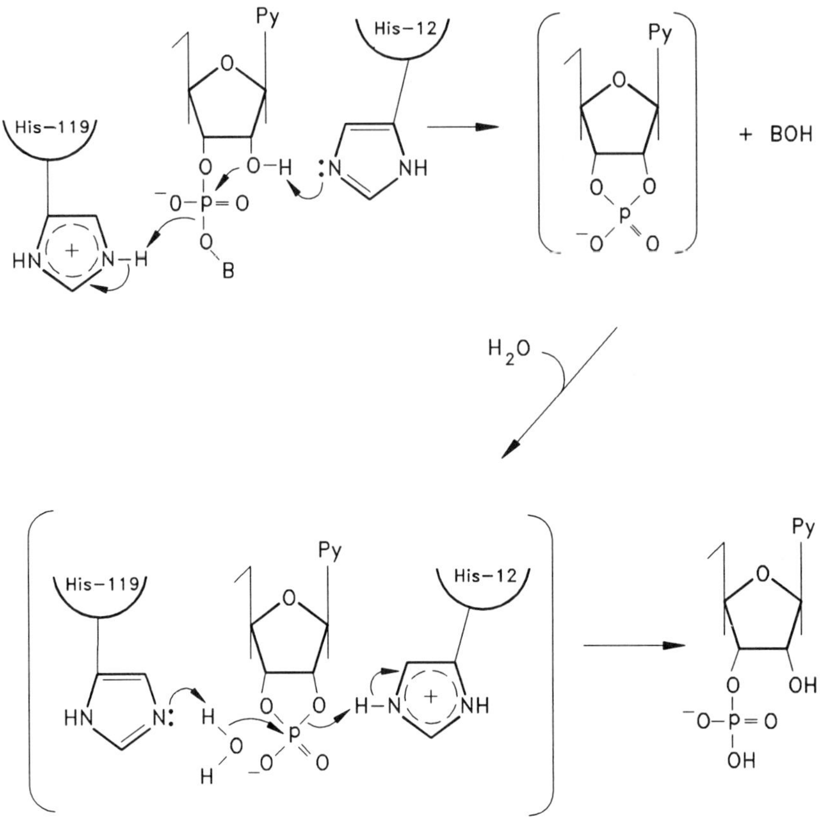

FIGURE 3.3 Reaction mechanism of RNase A-catalyzed hydrolysis of RNA. His-12 and His-119 are the catalytic residues of RNase A that perform concerted acid–base catalysis in the first (transesterification) and the second (hydrolysis) steps. B denotes the nucleotide 3' to the cleavage site (Py).

~1.0. The reaction is monitored spectrophotometrically at A_{300} as a function of time, usually over the course of 5 min at 25°C. The A_{300} decrease, which is linear at least during the initial phase of the reaction, results from the fact that cyclic phosphate intermediates have absorption maxima at shorter wavelengths than their corresponding mono- or oligonucleotides. This assay is rapid but only moderately accurate.

Unit definition: One "Kunitz" unit is defined as the amount of enzyme that produces a decrease in A_{300} from A_0 (initial) to A_f (final) in 1 min under the assay conditions. The A_0 to A_f decrease corresponds to the total conversion. The rate is obtained from the slope of a linear plot, $\log(A - A_f)$ versus t(min).

ii. Precipitation assay. This assay, which is often employed for crude enzyme preparations, measures the amount of acid-soluble nucleotides produced (5,6). The RNA substrate (0.5% in 0.1 M NaAc, pH 5) is digested with RNase for

15–25 min at 25°C. The reaction is stopped by the addition of an equal volume of precipitant composed of 0.25% uranyl acetate in 2.5% TCA (or PCA). The undegraded RNAs and proteins are removed by filtration or centrifugation at 1500 rpm for 5 min at 4°C, and the supernatant is used to measure A_{260}. The amount of released nucleotides is determined by comparison with a standard curve which is established with pure RNase for each batch of the substrate solution. This assay gives a linear response between 1 and 10 µg RNase and is well suited for the assay of a large number of samples.

iii. **Cyclic nucleotide spectrophotometric assay.** This assay measures the rate of 2′,3′-cyclic CMP [C > p, cCMP, or 2′(3′)-CMP] hydrolysis by monitoring the A_{286} increase as the cyclic nucleotide is hydrolyzed to 3′-CMP (7). The maximum increase in difference spectra occurs at 286 nm, and the ratio of molar absorption (3′-CMP/C > p) is 1.495. The assay is carried out in 0.2 M Tris–Cl (pH 7.0) with the substrate at 0.1 mg/ml. The initial rate is calculated from the linear slope of the plot, $\log(A_f - A)$ versus t(min). The dependence on enzyme concentration is linear up to 35 µg/ml.

iv. **Titrimetric assay of cyclic phosphate hydrolysis.** A freshly prepared substrate solution (3 ml of cCMP·Ba salt, 3 mg/ml, pH 7.0) is introduced into a pH-stat cell at 25°C. The pH is maintained at 7.0 by the automatic addition of 0.02 N NaOH. The enzyme solution to be assayed is added and the amount of base necessary to maintain pH 7 is recorded as a function of time for 10 min (8). The initial slope of this curve is proportional to the enzyme concentration within the range from 1 to 10 µg. This assay is the most precise and suitable for kinetic studies. However, it is relatively slow, expensive, and generally not adapted for routine assays.

c. *Substrate specificities*

i. **Polynucleotide substrates.** Natural or synthetic ssRNA is the preferred substrate. The RNA in RNA·DNA heteroduplexes is not cleaved, unless there is a mismatch preferably at a pyrimidine nucleotide in the RNA strand (9). DNA and dsRNA are not substrates.

ii. **Dinucleotides.** In dinucleotide substrates, the second (or leaving) base can be replaced by a variety of bases such as deoxynucleotides, modified ribonucleotides, and nucleotides with attached alkyl groups. The leaving groups have a varying influence on catalytic efficiencies (1). For example, the leaving groups 3′ to the pyrimidine bases (C and U) affect the k_{cat} in the decreasing order of A > G > C > U, or by a factor of >100 between CpA and CpU (10).

iii. **Mononucleotides and derivatives.** Pyrimidine 2′,3′-cyclic mononucleotides (C > p or U > p) are substrates. The two stereoisomeric derivatives of uridine 2′,3′-cyclophosphorothioate are also substrates but the V for both is 20% of U > p (11). Uridine 3′-P aryl esters are also substrates (3).

TABLE 3.3 Mononucleotide inhibitors of RNase A

Compounds	K_d (mM)	K_i (mM)	pH	Temperature (°C)	Ref.
2'-CMP	0.15		7.0	25	80
3'-CMP	0.62		7.0	25	80
2'-UMP	0.007		5.5	25	80
3'-UMP	0.44		7.0	25	80
U > p		2.4	7.6	37	26
A > p		35	7.6	37	26
3'-dUMP	0.12		7.0	25	82
Arabinose-Cyt-3'-P	0.10		7.0	25	81
Cytidine		25	7.6	37	26
Folic acid	80 (μM)		5.0	23	83
P_i	4.2		5.5	25	80
	14.5		7.0	25	80
PP_i	0.15		5.5	25	80
	1.33		7.0	25	80

iv. RNase activity in alcohols. RNase A is active in very high concentrations of alcohols. In aqueous methanol, C > p is hydrolyzed to 3'-CMP and 3'-CMP methyl ester, the product of alcoholysis (12). The alcoholysis is the reverse of the cyclic phosphate-forming step and is most effective with glycerol, followed by ethylene glycol, methanol, and ethanol. Alcohol inhibits the hydrolysis in a noncompetitive manner (e.g., K_i for glycerol is 60 mM) with respect to the nucleotides, but in a competitive manner with respect to water.

d. Inhibitors and inactivators

i. Mononucleotides and nucleotide analogs. Pyrimidine 2'-, 3'-, and 5'-P bind RNase A with a stoichiometry of 1:1 and are competitive inhibitors of the enzyme (Table 3.3). The binding constants are highly pH dependent with maxima at pH 5.5–6.0 (1). Table 3.3 lists some other nucleotides and analogs as well that are inhibitory to RNase A.

ii. Ribonucleoside–vanadyl complexes. Ribonucleoside–vanadyl complexes, more commonly called *VRC* for vanadyl–ribonucleoside complexes or simply vanadyl complexes, are stable complexes formed between ribonucleosides (generally an equimolar mixture of A, U, G, and C) and oxovanadium ions (VO^{2+}). The K_d of vanadyl complexes is estimated to be 10 μM. VRC (at 10 mM) is a very effective inhibitor for a variety of ribonucleolytic enzymes, including RNase A, T1, and U2 (13). VRC is believed to be a transition state analog of the activated 2',3'-cyclic phosphates that occur during the RNase catalysis. Nuclease S1 and DNase I are not inhibited by VRC. VRC is particularly useful in the preparation and manipulation of RNA (14). EDTA and SDS are known destabilizing agents for VRC. VRC apparently remains as an active complex in the presence of 10 mM Mg^{2+}. VRC coprecipitates with RNA when ethanol is added and is difficult to remove completely by phenol/chloroform extraction alone. Residual VRC may compete with low concentrations of dNTPs and may interfere with cDNA synthesis or other reactions. At a high concentration (1.5 mM each) of dNTPs, VRC (2 mM) does not inhibit the AMV reverse transcriptase activity (13). Potential problems due to high concentrations of VRC can be alleviated by various methods such as gel filtration, affinity column purification of mRNAs, and the addition of high concentrations of EDTA prior to the ethanol precipitation of RNA.

iii. **Aurintricarboxylic acid (ATA).** ATA is a heterogeneous polymer of the phenol–formaldehyde type. It has been recognized as a potent inhibitor of RNases. ATA binds to the active site of RNase A and displaces 2'- and 3'-CMP (15). The use of ATA (0.5 mM ammonium aurintricarboxylate, Sigma) enables one to isolate intact RNAs from RNase-rich sources at a yield higher than that obtainable with "standard" guanidinium isothiocyanate methods (16). ATA is deemed to be as good or slightly better as a RNase inhibitor than VRC or RNasin (see below).

ATA forms a tight complex with RNA and DNA (ethanol precipitation of nucleic acids with ATA colors the pellet red), and apparently dissociates from the complex by electrophoresis. Excess ATA can be quickly removed by passage through a Sephadex G-50 spin column. ATA does not interfere with hybridizations of RNA samples nor does it affect spectrophotometry at 260 or 280 nm. However, ATA interferes with some restriction enzymes (e.g., *Hin*dIII) and may also inhibit enzymes used in reverse transcription or *in vitro* translation.

iv. **RNase inhibitor proteins: placental RNase inhibitor (PRI).** RNase inhibitor proteins are present in various mammalian tissues, virtually wherever RNases (intracellular type) have been found. Normal human placenta has been the most abundant source of the RNase inhibitor protein (17), commercially known as *RNasin* or *RNAguard*, and now available in a recombinant form from Promega. PRI strongly binds to the active site of the RNase molecule within the S-protein part and forms a tight 1:1 complex at pH 7–8 (18,19). PRI is a competitive inhibitor with a K_i of 4×10^{-14} M. The K_i increases with increasing NaCl concentrations, suggesting the importance of ionic interactions. The half-time of dissociation of the PRI–RNase A complex is 13 hr at 25°C.

PRI is also an inhibitor ($K_i < 0.1$ nM) of both the angiogenic and ribonucleolytic activities of angiogenin, a blood vessel-inducing protein (20). PRI may therefore be called a ribonuclease/angiogenin inhibitor (RAI). Note that PRI does not inhibit some other RNases such as RNase T1, nuclease S1, and RNase H.

One *unit of PRI* inhibits the activity of 5 ng of RNase A by 50% in a spectrophotometric assay of the enzyme for 2',3'-cCMP (21).

Partial amino acid sequencing, as well as the nucleotide sequencing of the cloned gene, indicates that the PRI is composed of 460 amino acids (M_r 49,847) (22,23). Both the N- and the C-terminal residues of the mature protein are Ser. The N-terminal Ser is apparently blocked. PRI contains 32 Cys residues, at least

30 of which are in the reduced form. Leu is the most abundant component of PRI, comprising ~18% of the total residues. The amino acid sequence features seven tandem repeats of Leu-rich 28 amino acid domains.

PRI is an acidic protein with a pI of 4.7, in contrast to basic RNase A (pI 9.4) (17). PRI is denatured in 7 M urea or by heating for 10 min at 65°C; under these conditions RNase A is fully active. PRI is rapidly inactivated by SH group-modifying reagents such as PCMB, PHMB, and NEM. At DTT concentrations of less than 1 mM, PRI is irreversibly inactivated. The inactivation of PRI causes the RNase–inhibitor complex to dissociate into active RNase and inactive PRI. Excess free thiols cannot reverse the inactivation. Divalent cations have no effect on the stability of PRI.

Because of the ease of its elimination by phenol/chloroform extraction, PRI is often preferred to VRC for protecting RNA. In experiments where RNA is either a substrate or a product, e.g., in the isolation of RNA, cDNA synthesis, *in vitro* transcription and translation, PRI is used at concentrations of 1 U/μl.

Stratagene has introduced an RNase inhibitor protein, called *RNase Block II*, that is claimed to be at least 10 times more effective than PRI in protecting RNAs.

v. **Inactivators.** RNase A is reversibly inactivated by an anionic detergent SDS (2 mM) (24). The enzyme is fully active in the presence of 8 M urea (25), although there is evidence to the contrary (26). RNase A is also active in 6 M GuHCl, even though the protein molecule is in a largely denatured state. GuSCN (4–5 M) is often used in combination with reducing agents such as 2-MSH or DTT to inactivate RNase A during the purification of RNAs (27,28). The conformational stability of RNase A toward urea and GuHCl is maximal between pH 7 and pH 9 (29).

RNase A is extremely thermostable, especially at pH 5 ± 1; it remains virtually intact after heating at 100°C for 20 min (6). At pH 7.5, the same heat treatment results in a ~50% inactivation. However, a common procedure for preparing DNase-free RNase is to heat the RNase A stock (10 mg/ml in Tris–Cl, pH 7.5, and 15 mM NaCl) at 100°C for 15 min (30).

RNase A is known to be strongly adsorbed on *Macaloid*, a trade name for a negatively charged, purified clay mineral, hectorite (Na, Mg, Li-fluorosilicate). Macaloid has been used to block RNases during the preparation of RNA (31) and to eliminate contaminating RNases in preparing RNase-free DNases.

e. Mechanism of RNase action

i. **Binding of substrates or inhibitors.** RNase A specifically cleaves RNA chains at the 3' side of pyrimidine bases. RNase A binds up to 11 nt along an extended ssRNA or ssDNA chain (32,33). The complex between RNase and nucleic acid is principally a multisite (eight to nine) electrostatic interaction between cationic Lys and Arg residues of the enzyme and anionic phosphate groups of the substrate. The base and ribose moieties play only a minor role in binding, as evidenced by the ready binding of DNAs, d(pA)$_4$, and d(pT)$_4$ (33), but these moieties play a key role in determining cleavage specificity. As a result of similar modes of binding to RNase A, ssDNA acts as a competitive inhibitor against RNA substrates.

The binding at the active site of cyclic mono- and dinucleotide substrates or mono- and oligonucleotide inhibitors may be divided into at least four subsite interactions:

(a) The base-binding subsite (Thr-45 or perhaps Ser-123) distinguishes U from C probably through a charge or H-bonding interaction at the enolizable -CO-NH- group of the pyrimidine ring; tight binding affects the catalysis negatively. Purine bases are discriminated against due to their bulky size.

(b) The sugar-binding subsite binds ribose in the β-D-ribofuranoside ring structure with a preference for the 2'-OH in the *cis* configuration with respect to the 3'-OH. This subsite contains the presumed catalytic residues, His-12 and His-119.

(c) The phosphate-binding subsite (most likely Lys-41) binds the 2'-P, 3'-P, 5'-P, or inorganic phosphate somewhat nonspecifically.

(d) The purine-binding subsite binds the leaving group, regardless of whether it is a purine or pyrimidine.

ii. Catalytic mechanism. Catalysis by RNase A proceeds in two steps. The first step is the formation of a cyclic 2',3'-nucleoside phosphate intermediate via cyclization (also referred to as transesterification or transphosphorylation). The second step is the hydrolysis of the cyclic intermediate to 3'-P (Fig. 3.3). The first cyclization step, which is reversible, is usually much faster than the subsequent hydrolysis step so that the intermediate can be readily isolated. The enzyme also converts monomeric U > p and C > p to their corresponding 3'-phosphates. RNase A cannot cleave DNA due to the absence of 2'-OH that is essential for the formation of cyclic phophodiester bonds.

The catalytic mechanism is thought to be a concerted general acid–base catalysis involving two His residues (His-12 and His-119) that sandwich the phosphate moiety of the substrate. The two His residues reverse their roles as general acid and base in each catalytic step. Initially His-12 acts as a base on the 2'-OH group of the substrate, while His-119 acts as an acid to protonate the 5'-oxygen of the leaving group. Later His-12 reprotonates the 2'-O and His-119 deprotonates a water molecule which attacks the cyclic phosphate intermediate. Either by H-bonding or by charge–charge interaction, Lys-41 is presumed to stabilize the pentacoordinate transition state of phosphorus when an OH group (from the ribose 2'-OH in the transesterification step or from water in the ring-opening step) attacks the phosphate.

Despite general acceptance of the catalytic mechanism, the order of the microscopic steps through which the acid–base catalysis occurs remains arguable. An acid–base buffer system, with or without zinc ion, that promotes the chemical catalysis of cyclization and cleavage of phosphodiester bonds provides some insight into the mechanism. In this nonenzymatic system, acidic imidazolium (Im^+) ion initiates the conversion to a pentacoordinate phosphorane intermediate; then the basic imidazole (Im) catalyzes the decomposition of this intermediate to a cyclic phosphate (34). The active site environment of RNase A is obviously not identical to this chemical system. However, there is no evidence to exclude the possibility that similar microscopic steps might be operational with the two His residues (Im$^+$-119 and Im-12) in RNase A as well.

The pH–activity profile of RNase A supports a mechanism that involves a diprotonic enzyme acting in a monoprotonated form. The geometry for the two-step catalysis, probed by the use of phosporothioate analogs, is consistent with an "in-line" rather than an "adjacent" mechanism (11,35–37). Kinetic studies on the cyclization of uridine 3′-P aryl esters and the effect of leaving groups suggest that electrophilicity is also involved in the catalysis (3).

2. APPLICATIONS

i. **RNA sequencing.** RNase A cleaves RNA at pyrimidine bases (C and U); therefore, partial digestion of an RNA sample results in fragments terminating with C or U. The fragments are then separated by PAGE to form a ladder. When the pattern of the ladder is juxtaposed to those of other base-specific RNases such as RNase T1 (G specific), RNase U2 (A specific), and Phy M (U + A specific), the RNA sequence can be read from the sequence ladders.

ii. **Preparation of DNA materials.** RNase A is frequently used to degrade RNA in plasmid and other DNA preparations. Following digestion, RNase A is removed by phenol/chloroform extraction, and the DNA is purified by ethanol precipitation. It should be noted that it is nearly impossible to remove or denature RNase A completely. Therefore whenever RNase contamination is suspected or must be avoided, an RNase inhibitor(s) should be used.

iii. **Scission and removal of unhybridized regions of RNA.** The fine substrate specificity of RNase A can be utilized to cleave the single-strand region(s) of RNA in DNA · RNA or RNA · RNA hybrids. RNase A can cleave a significant percentage of single-base mismatches present in RNA:RNA and DNA:RNA hybrids (9,38). This mismatch cleavage can be used to detect point mutations (single-base substitutions, insertions, and deletions), as well as other genetic alterations involving larger deletions or duplications. Although the usefulness of the method is limited to the spots containing pyrimidine bases, RNase A is still the primary nuclease used as a molecular diagnostic tool (39).

iv. **Biochemical applications.** In addition to the applications in recombinant DNA technology, RNase A has various biochemical and genetic uses as noted below.

(a) *As a helix destabilizer.* Because of its ability to bind both DNA and RNA at the sugar–phosphate backbone (32,40), RNase A also functions as a "melting" or helix-destabilizing protein. RNase A binds noncovalently to both ssDNA ($K_d \approx 170\ \mu M$) and dsDNA ($K_d \approx 10$ mM) and lowers the melting temperature of dsDNA (41).

(b) As a model for protein structural studies. RNase A is a small, structurally and functionally well-characterized enzyme. Therefore it is an ideal model for the study of not only folding pathways (42) but also relationships between protein structure and function by means of various molecular biological techniques including site-specific mutagenesis.

(c) Gene expression and protein degradation. RNase A is also an excellent model for studies of gene expression, protein stability, and/or degradation (43).

3. STRUCTURE

i. General features. RNase A is a single polypeptide composed of 124 amino acid residues (M_r 13,680) (44). The N- and C-terminal residues are Lys and Val, respectively. The enzyme has no Trp, but has 4 His and 6 Tyr residues. The enzyme contains ~13% α helix and 33% β sheet.

Crystallographic structures have been determined for RNase A (1,45–48), RNase S (49), and RNase S with substrate analogs such as 2′,5′-CpA (50) and 2′-fluoro-dUpA (51). The enzyme is roughly kidney-shaped (approximate dimensions 20 Å × 25 Å × 45 Å) and has a deep cleft on one side which binds the nucleotide substrate or analogs.

ii. Disulfide bonds. All 8 Cys residues (numbered I to VIII in sequence) of a RNase A molecule are involved in intrachain disulfide (S–S) bonds: 26(I)–84(VI), 40(II)–95(VII), 58(III)–110(VIII), and 65(IV)–72(V) (52). The reaction of native RNase A with DTT or DTE proceeds in two distinct steps. The half-time for overall reduction by 10 mM DTT is ~10 hr under nondenaturing conditions (100 mM Tris–Cl, pH 8.7, 25°C) (42). In the presence of urea or other denaturants, the reduction proceeds much faster. Fully denatured RNase A can spontaneously refold, on removal of the denaturants, into active enzyme along with the formation of all four S–S bonds.

Reduction of RNase A with phosphorothioate at pH 9.0 in the absence of urea results in the opening of only two S–S bonds: IV–V and III–VIII (53). This partially reduced enzyme is fully active toward RNA substrates and is more active than the native enzyme toward C > p. The IV–V bond is the only disulfide bond that is reduced by 20 mM DTT or DTE (reaction time of ~1 hr at 23°C) (54). The IV–V bridge is structurally and functionally nonessential although, under certain reduced activity conditions, the region (residues 59–73) containing the IV–V bond contributes to an increase of over 200-fold in RNase activity toward dinucleotide substrates (55). The unusual stability of RNase A is largely attributed to the two S–S bonds: I–VI and II–VII.

Reduced and denatured RNase A can be reactivated in reactions catalyzed by thiol redox enzymes. For example, thioredoxin catalyzes the renaturation, probably via disulfide interchange, about 1000-fold more efficiently than DTT (56). Pituitary hormones such as lutropin (LH) and follitropin (FSH) also catalyze similar redox reactions resulting in reactivation of the denatured RNase A (57).

iii. Proteolysis and RNase S. Limited digestion of RNase A with subtilisin at pH 8.0 and 3°C results in proteolytic cleavages mostly at the peptide bond of Ala^{20}–Ser^{21}. The two fragments, 1–20 (S-peptide) and 21–124 (S-protein), each contain an essential His residue, His-12 and His-119. The separate fragments are inactive, but when mixed in equimolar proportions they combine to generate a fully active complex known as *RNase S* (58). Every one of the first eight N-terminal residues of the S-peptide positively contributes to the molecular recognition between the two fragments (59). The K_d for the S-peptide/S-protein complex is 7 n*M* at pH 5.0 and 27°C. The crystal structure of RNase S refined at 1.6 Å resolution is identical to that of RNase A, particularly in the active-site region (84). Residues 3–13 of the S-peptide form an α helix, just as they do in the RNase A. Despite the apparent structural similarities, RNase A and RNase S show substantial differences in stability to both acid and temperature denaturation and in susceptibility to proteolysis at neutral pH. RNase A is not degraded by trypsin or chymotrypsin in aqueous solution, although it can be readily degraded by trypsin in the presence of denaturants (e.g., 2 *M* GuHCl) or by pepsin in aqueous solution at pH 2. In contrast, RNase S is extremely susceptible to degradation by trypsin at neutral pH.

iv. Surface residues. Lys-41 and the highly conserved Tyr-92 are essential for interaction of RNase A with PRI (60). Among 10 Asn residues of RNase A, Asn-67 is the preferred site of deamidation. The deamidation rate is at least 30-fold lower in the native structure than in the unfolded state (61). Thirteen surface residues, including Lys-61 which is uniquely involved in crystal packing interactions, have discrete multiple conformations in crystal structure (62). His-48 lies far removed from the active site, and yet the NMR spectra of the imidazole ring is sensitive to the binding of 2′-CMP or 3′-CMP, probably mediated by a conformational change on inhibitor binding.

v. Active site. The active site of RNase A is formed principally by the side chains of His-12, Arg-39, Cys-40, Lys-41, Asn-71, Glu-111, His-119, and Phe-120.

(a) His residues. The substrate-binding cleft of RNase A contains two catalytically essential His residues, His-12 and His-119. The pH dependence of k_{cat}/K_m shows that the rate depends on the ionization of a base (pK_a 5.2–5.8) and an acid (pK_a 6.8–7.3) in the free enzyme; on substrate binding, the two groups which are assigned to imidazole groups demonstrate higher pK_a values of 6.3–6.6 and 8.1–8.4 (3,63,64). Direct titration of the His residues monitoring the C(2)–H NMR chemical shifts at 40°C gives pK_a values of 5.8 and 6.2 for His-12 and His-119, respectively (65).

The binding of 2′-CMP or 3′-CMP perturbs the C-2–proton resonances of His-12 (upfield shift) and His-119 (strong downfield shift) due to a protonation of the Im ring between pH 5 and 7 (66). In contrast, phosphate induces an upfield shift only on His-119, while 5′-CMP does so only on His-12.

The covalent modification of His-12 with 3′-(bromoacetamido)-3′-dT results in a complete inactivation of the enzyme, while a selective modification of His-

119 with 3'-(bromoacetamido)-3'-dU results in partial inactivation (67). The 1% residual activity is attributed to partial accessibility of the catalytic site.

His-119 assumes two discrete positions, A (major) and B (minor), in the crystal structure; it occupies the A site in the protein–purine derivative and the B site in the native enzyme (68). The centers of the His-12 and His-119 Im rings are separated by ~7 Å (46). In the crystal structure of RNase S complexed with 2'-fluoro-dUpA, the nominal distance between the 2'-F and N_δ of His-12 is ~3.0 Å (51). Replacement of the inhibitory sulfate anion (which is normally bound at the active site) by a water molecule leaves the geometry of the active site essentially unchanged except that the new water molecule is H-bonded to His-12 and to another water molecule (48).

Both X-ray diffraction and NMR data on the binding of mononucleotide inhibitors suggest that the phosphate group of 3'-CMP is particularly close to His-119, whereas His-12 is farther from the phosphate group and adjacent to the ribose ring (2'-OH) of the inhibitor (66). Purine nucleotides (2'-AMP and 3'-AMP) have little effect on either His-12 or His-119 (69).

(b) Basic residues. Arg-39 forms a salt bridge with the 5'-P of $d(pT)_4$ or $d(pA)_4$. Lys-41 is an essential residue which in crystal structures is positioned particularly close to the phosphate moiety of the bound nucleotide (47,48). Lys-41 is presumed to play the role of stabilizing the transient pentacoordinate phosphorane intermediate via charge interactions.

vi. *Physicochemical parameters.* Some of the physicochemical parameters are as follows (1): pI is 9.60, ε_{278} is 9700 M^{-1} cm^{-1}, and A_{278} (0.1%) is 0.71 at pH 7 and 25°C. The $s^\circ_{20,w}$ is 1.78S. Partial specific volume is 0.703 ml/g.

4. Genetics

RNase A is synthesized *in vivo* in a precursor form (pro-RNase A). The pro-RNase A contains a 25-residue signal peptide and is enzymatically as active as mature RNase A (70).

A complete gene (*bpr*) for RNase A [and RNase S as well (71)] has been synthesized and expressed in *E. coli* (72) and *B. subtilis* (73). The RNase A product from the synthetic gene is chromatographically, catalytically, and immunogenically identical to authentic RNase A.

As one of the extracellular enzymes, RNase A has long been regarded as playing a role in digestion, particularly in ruminant animals. Evidence suggests that the extracellular RNases, together with extracellular RNAs and RNase inhibitors, form a triad system necessary for cellular regulation (74,75).

5. Sources

A number of procedures have been developed for the purification of RNase A (1,2). Affinity chromatography using 5'-(4-aminophenylphosphoryl)-2'(3')-UMP-Sepharose has been shown to improve the yield and purity of RNase preparations (76).

Total chemical synthesis of functional RNase A (and RNase S as well) has been accomplished by applying a solid-phase synthesis method (77).

A synthetic *bpr* gene has been fused to the 3′ end of a sequence encoding a tetrapeptide bridge (Ile–Glu–Gly–Arg) to the *lacZ* gene (72). The expression of the fusion gene in *E. coli* (DH1), under the control of the *lacZ* promoter–operator, produces βGal-tetrapeptide–RNase A. The RNase A can be released from the fusion protein by a site-specific cleavage *in vitro* at the C-terminal Arg of the linker peptide using the blood-clotting factor Xa (78). [Note that an internal sequence, Cys^{154}–Asn–Gly–Arg^{157}, of RNase A was also found to be a site of slow cleavage, especially after extensive digestion (72).] The yield of pure enzyme from this procedure was 0.2 mg per liter of culture.

RNase A has also been expressed in *B. subtilis* as a free-standing secretory protein when the synthetic *bpr* gene was fused to the promoter and the signal sequence region of the *B. amyloliquefaciens* alkaline protease gene, *apr* (73).

References

1. Richards, F. M., and Wyckoff, H. W. (1971). "The Enzymes," 3rd ed., Vol. 4, pp. 647–806.
2. Blackburn, P., and Moore, S. (1982). "The Enzymes," 3rd ed., Vol. 15(B), pp. 317–433.
3. Davis, A. M., Regan, A. C., and Williams, A. (1988). *Biochem.* **27**, 9042–9047.
4. Kunitz, M. (1946). *JBC* **164**, 563–568.
5. Anfinsen, C. B., Redfield, R. R., Choate, W. L., Page, J., and Carroll, W. R. (1954). *JBC* **207**, 201–210.
6. Klee, W. A., and Richards, F. M. (1957). *JBC* **229**, 489–504.
7. Crook, E. M., Mathias, A. P., and Rabin, B. R. (1960). *BJ* **74**, 234–238.
8. Stark, G. R., and Stein, W. H. (1964). *JBC* **239**, 3755–3756.
9. Myers, R. M., Larin, Z., and Maniatis, T. (1985). *Science* **230**, 1242–1246.
10. Witzel, H., and Barnard, E. A. (1962). *BBRC* **7**, 295–299.
11. Usher, D. A., Richardson, D. I., Jr., and Eckstein, F. (1970). *Nature* **228**, 663–665.
12. Findlay, D., Mathias, A. P., and Rabin, B. R. (1962). *BJ* **85**, 134–139.
13. Puskas, R. S., Manley, N. R., Wallace, D. M., and Berger, S. L. (1982). *Biochem.* **21**, 4602–4608.
14. Berger, S. (1987). *Meth. Enzy.* **152**, 227–234.
15. Gonzalez, R. G., Haxo, R. S., and Schleich, T. (1980). *Biochem.* **19**, 4299–4303.
16. Skidmore, A. F., and Beebee, T. J. C. (1989). *BJ* **263**, 73–80.
17. Blackburn, P., Wilson, G., and Moore, S. (1977). *JBC* **254**, 5904–5910.
18. Blackburn, P., and Jailkhan, B. L. (1979). *JBC* **254**, 12488–12393.
19. Lee, F. S., Shapiro, R., and Vallee, B. L. (1989). *Biochem.* **28**, 225–230.
20. Shapiro, R., and Vallee, B. L. (1987). *PNAS* **84**, 2238–2241.
21. Shortman, K. (1961). *BBA* **51**, 37–49.
22. Lee, F. S., Fox, E. A., Zhou, H. M., Strydom, D. J., and Vallee, B. L. (1988). *Biochem.* **27**, 8545–8553.
23. Schneider, R., Schneider-Scherzer, E., Thurnher, M., Auer, B., and Schweiger, M. (1988). *EMBO J.* **7**, 4151–4156.
24. Cowgill, R. W. (1966). *BBA* **120**, 196–211.
25. Anfinsen, C. B., Harrington, W. F., Hvidt, A. A., Linderstrom-Lang, K., Ottesen, M., and Schellman, J. (1955). *BBA* **17**, 141–142.
26. Ukita, T., Waku, K., Irie, M., and Hoshino, O. (1961). *J. Biochem.* **50**, 405–415.
27. MacDonald, R., Swift, G. H., Przybyla, A. E., and Chirgwin, J. M. (1987). *Meth. Enzy.* **152**, 219–227.
28. Han, J. H., Stratowa, C., and Rutter, W. J. (1987). *Biochem.* **26**, 1617–1625.

29. Pace, C. N., Laurents, D. V., and Thomson, J. A. (1990). *Biochem.* **29**, 2546–2572.
30. Sambrook, J., Fritsch, E. F., and Maniatis, T. (1989). "Molecular Cloning: A Laboratory Manual," 2nd ed., B.17. Cold Spring Harbor Laboratory, Cold Spring Harbor, New York.
31. Stanley, W. M., Jr., and Bock, R. M. (1965). *Biochem.* **4**, 1302–1311.
32. Jensen, D. E., and von Hippel, P. H. (1976). *JBC* **251**, 7198–7214.
33. Birdsall, D. L., and McPherson, A. (1992). *JBC* **267**, 22230–22236.
34. Breslow, R., Huang, D.-L., and Anslyn, E. (1989). *PNAS* **86**, 1746–1750.
35. Findlay, D., Herries, D. G., Mathias, A. P., Rabin, B. R., and Ross, C. A. (1962). *BJ* **85**, 152–153.
36. Roberts, G. C. K., Dennis, E. A., Meadows, D. H., Cohen, J. S., and Jardetsky, O. (1969). *PNAS* **62**, 1151–1158.
37. Usher, D. A., Erenrich, E. S., and Eckstein, F. (1972). *PNAS* **69**, 115–118.
38. Winter, E., Yamamoto, F., Almoguera, C., and Perucho, M. (1985). *PNAS* **82**, 7575–7579.
39. Cotton, R. G. H. (1989). *BJ* **263**, 1–10.
40. Karpel, R. L., Yrttimaa, V. A., and Patel, G. L. (1981). *BBRC* **100**, 760–768.
41. von Hippel, P. H., Jensen, D. E., Kelly, R. C., and McGhee, J. D. (1977). In "Nucleic Acid-Protein Recognition" (H. J. Vogel, Ed.), pp. 65–89. Academic Press, New York.
42. Creighton, T. E. (1979). *JMB* **129**, 411–431.
43. Chiang, H. L., Terlecky, S. R., Plant, C. P., and Dice, J. F. (1989). *Science* **246**, 382–385.
44. Smyth, D. G., Stein, W. H., and Moore, S. (1963). *JBC* **238**, 227–234.
45. Kartha, G., Belle, J., and Harker, D. (1967). *Nature* **213**, 862–865.
46. Carlisle, C. H., Palmer, R. A., Mazumdar, S. K., Gorinsky, B. A., and Yeates, D. G. R. (1974). *JMB* **85**, 1–18.
47. Wlodawer, A., and Sjolin, L. (1983). *Biochem.* **22**, 2720–2728.
48. Campbell, R. L., and Petsko, G. A. (1987). *Biochem.* **26**, 8579–8584.
49. Wyckoff, H. W., Tsernoglou, D., Hanson, A. W., Knox, J. R., Lee, B., and Richards, F. M. (1970). *JBC* **245**, 305–328.
50. Wodak, S. Y., Liu, M. Y., and Wyckoff, H. W. (1977). *JMB* **116**, 855–875.
51. Pavlovsky, A. G., Borisova, S. N., Borisov, V. V., Antonov, I. V., and Karpeisky, M. Ya. (1978). *FEBS Lett.* **92**, 258–262.
52. Spackman, D. H., Stein, W. H., and Moore, S. (1960). *JBC* **235**, 648–659.
53. Neumann, H., Steinberg, I. Z., Brown, J. R., Goldberger, R. F., and Sela, M. (1967). *EJB* **3**, 171–182.
54. Sperling, R., Burstein, Y., and Steinberg, I. Z. (1969). *Biochem.* **8**, 3810–3820.
55. Harper, J. W., and Vallee, B. L. (1989). *Biochem.* **28**, 1875–1884.
56. Pigiet, V. P., and Schuster, B. J. (1986). *PNAS* **83**, 7643–7647.
57. Boniface, J. J., and Reichert, L. E., Jr. (1990). *Science* **247**, 61–64.
58. Richards, F. M., and Vithayathil, P. J. (1959). *JBC* **234**, 1459–1465.
59. Levit, S., and Berger, A. (1976). *JBC* **251**, 1333–1339.
60. Blackburn, P., and Gavilanes, J. G. (1980). *JBC* **255**, 10959–10965.
61. Wearner, S. J., and Creighton, T. E. (1989). *Proteins (S.F.G)* **5**, 8–12.
62. Svensson, L. A., Sjolin, L., Gilliland, G. L., Finzel, B. C., and Wlodawer, A. (1986). *Proteins (S.F.G)* **1**, 370–375.
63. Herries, D. G., Mathias, A. P., and Rabin, B. R. (1962). *BJ* **85**, 127–134.
64. Ramsden, E. N., and Laidler, K. J. (1966). *Can. J. Chem.* **44**, 2597–2610.
65. Markley, J. L. (1975). *Biochem.* **14**, 3546–3553.
66. Meadows, D. H., Roberts, G. C. K., and Jardetzky, O. (1969). *JMB* **45**, 491–511.
67. Nachman, J., Miller, M., Gilliland, G. L., Carty, R., Pincus, M., and Wlodawer, A. (1990). *Biochem.* **29**, 928–937.
68. Borkakoti, N. (1983). *EJB* **132**, 89–94.
69. Haar, W., Maurer, W., and Ruterjans, H. (1974). *EJB* **44**, 201–211.
70. Haugen, T. H., and Heath, E. C. (1979). *PNAS* **76**, 2689–2693.
71. Nambiar, K. P., Stackhouse, J., Stauffer, D. M., Kennedy, W. P., Eldredge, J. K., and Benner, S. A. (1984). *Science* **223**, 1299–1301.
72. Nambiar, K. P., Stackhouse, J., Presnell, S. R., and Benner, S. A. (1987). *EJB* **163**, 67–71.

73. Vasantha, N., and Filpula, D. (1989). *Gene* **76**, 53–60.
74. Deutscher, M. P. (1988). *TIBS* **13**, 136–139.
75. Strydom, D. J., Fett, J. W., Lobb, R. R., Alderman, E. M., Bethune, J. L., Riordan, J. F., and Vallee, B. L. (1985). *Biochem.* **24**, 5486–5494.
76. Wilcheck, M., and Gorecki, M. (1974). *Meth. Enzy.* **34**, 492–496.
77. Gutte, B., and Merrifield, R. B. (1971). *JBC* **246**, 1922–1941.
78. Nagai, K., and Thogersen, H. C. (1984). *Nature* **309**, 810–812.
79. Harper, J. W., Auld, D. S., Riordan, J. F., and Vallee, B. L. (1988). *Biochem.* **27**, 219–226.
80. Anderson, D. G., Hammes, G. G., and Walz, F. G., Jr. (1968). *Biochem.* **7**, 1637–1645.
81. Pollard, D. R., and Nagyvary, J. (1973). *Biochem.* **12**, 1063–1066.
82. Walz, F. G., Jr. (1971). *Biochem.* **10**, 2156–2162.
83. Sawada, F., Kanesaka, Y., and Irie, M. (1977). *BBA* **479**, 188–197.
84. Kim, E. E., Varadarajan, R., Wyckoff, H. W., and Richards, F. M. (1992). *Biochem.* **31**, 12304–12314.

B. Ribonuclease H
[EC 3.1.26.4]

Ribonuclease H (RNase H), also termed "hybridase" in some earlier literature, is an endoribonuclease that specifically degrades the RNA strand of DNA · RNA heteroduplexes (Scheme 3.3). The oligoribonucleotide products have 5'-P and 3'-OH termini.

$$\downarrow \quad\quad \downarrow$$

R–p–R–p–R–p–R–p–R–p–R–p–R–p–R–p–R–p–
D–p–D–p–D–p–D–p–D–p–D–p–D–p–D–p–D–p–

$\longrightarrow$ RpR + pRpRpRpR + pRpRpRp⁻ + ssDNA

SCHEME 3.3

Initially discovered in extracts of calf thymus as a contaminating enzyme during the purification of RNA polymerase (1,2), RNase H has since been found almost ubiquitously in lower and higher eukaryotes as well as in prokaryotes. A single cell type may have one or more species of RNases H residing in the cell nucleus and/or cytosol. Some RNases H have a single ribonucleolytic reaction specificity, whereas others are associated with DNA polymerase activities. Cellular RNases H, for example, from *E. coli* and calf thymus, are endonucleases; the retroviral reverse transcriptase-associated RNases H can function as both exo- and endoribonucleases depending on the type of substrates available.

RNase H is an important tool in cDNA synthesis, RNA mapping, and sequencing. Among all known RNases H, *E. coli* RNase H is probably the most widely studied and used enzyme in recombinant DNA technology.

This section focuses on *E. coli* RNase HI, which had been designated RNase H until a second RNase H (RNase HII) was isolated from *E. coli* (see Section on *Genetics* below). The RNase H activities associated with polymerases (DNA-directed DNA Pol and RNA-directed DNA Pol) are described as a subheading under the respective polymerases (see Chapter 6).

1. FUNCTIONS

a. Reaction conditions

i. Optimal reaction conditions. *E. coli* RNase HI has a broad pH optimum between pH 7.5 and 9.1. The half-maximal activity is observed at pH 6.9. A typical reaction mixture consists of 50 mM Tris–Cl (pH 8.0), 4 mM $MgCl_2$, 1 mM DTT (or 2-MSH), 50 μg/ml BSA, and hybrid substrates. The recommended incubation temperature is 30°C, although 37° or 16°C is often used.

ii. Kinetic parameters. For a nonanucleotide duplex prepared by mixing two complementary strands, i.e., nonaribonucleotide (rGGAGAUGAC) and nonanucleotide containing deoxyribonucleotides and 2′-O-methylribonucleotides (GmUmCm*ATCT*CmCm), the K_m and k_{cat} values at pH 8.0 and 37°C have been estimated to be 0.53 μM and 90 sec^{-1}, respectively (3).

iii. Metal ion requirements. RNase HI requires Mg^{2+} (2–4 mM) for optimal activity (4,5). Mn^{2+} (optimally 10–20 mM) can replace Mg^{2+} with equal efficiency. RNase HI is relatively insensitive to NaCl up to 0.1 M. At a concentration of 0.3 M, NaCl inhibits the enzyme activity by ~50% (6). Other monovalent cations such as NH_4^+ [NH_4Cl or $(NH_4)_2SO_4$] are inhibitory at concentrations above 0.2 M (4,7).

iv. Inhibitors and/or stabilizers. The chelation of Mg^{2+} by excess EDTA renders RNase HI inactive, but the enzyme can still bind to DNA·RNA hybrids (8). *E. coli* RNase HI is stabilized by SH reagents and inactivated by NEM (6). RNase HI is heat labile, showing a 50% loss of activity after 43 min at 50°C (4). When RNase HI is incubated without metal ions in 50 mM Tris–Cl (pH 7.5), 10 mM 2-MSH, 1 mM EDTA, and 10% glycerol, a 50% inactivation occurs after 15 min at 32°C, 10 min at 37°C, and 5 min at 42°C (9). [*Note:* A cloned, thermostable RNase H from *Thermus thermophilus* has at pH 5.5 a T_m (50% denaturation temperature) 34°C higher than that (52°C) of *E. coli* RNase HI (10).]

Dextran is selectively inhibitory on *E. coli* RNase HI: the degradation of RNA in φX DNA·RNA hybrid is unaltered, whereas poly(rA·dT) is rendered completely resistant (11). The inhibition is competitive with respect to nucleotide substrates, suggesting that, when there is inhibition, dextran interacts with the enzyme rather than with the substrate.

In contrast to the RNase H activities associated with reverse transcriptases, *E. coli* RNase HI is not inhibited by VRC. VRC thus provides a convenient way to protect the RNA components from undesirable degradations by other contaminating RNases while allowing digestion by RNase HI.

b. Activity assays and unit definition

RNase H activity is typically assayed under conditions similar to the optimal conditions described above except that labeled [^{3}H]poly(rA) · poly(dT) is used as substrates at 24 μM. The extent of RNase H digestion is commonly measured as the difference between the total radioactivity and that of the remaining substrates after digestion. The remaining substrate is either recovered as a TCA-insoluble fraction or selectively retained on filters (Whatman DE81, NC membrane) and counted.

Unit definition: One unit is defined as the amount of enzyme that produces 1 nmol of acid-soluble material in 20 min at 30°C (6).

An assay that selectively measures endonucleolytic (but not exonucleolytic) action uses *ccc* DNA (ColE1 plasmid) which contains ribonucleotides. The open-circular DNA product can be conveniently analyzed by agarose gel electrophoresis (12). RNase H activity can also be detected by *in situ* gel activity analysis (13).

c. Substrate specificities

i. Typical substrates. RNase H is an endonuclease that specifically degrades the RNA moiety in DNA · RNA hybrids. It is inactive on the DNA strand of the hybrid or on ssRNA or dsRNA. [*Note:* The RNase V$_1$ from cobra venom specifically cleaves dsRNA leaving 5'-P.]

ii. Product profile. The products of RNase HI digestion are 5'-P-oligonucleotides 2 to 9 nucleotides in length. Extensive digestion of (rA)$_n$ · (dT)$_n$ gives a series of 5'-oligonucleotides, (pA)$_{3-4}$ being the major products (5,6). The digestion produces very few ($\leq$4%) mononucleotides (5'-AMP). Dinucleotide products arise from the 5' and 3' ends of the substrate (14). The profile of product distribution remains invariant during all phases of the reaction and is not altered by changing temperature or ionic strength (8). The maximum extent of RNA strand digestion is ~95%.

iii. Substrate size and cleavage specificity. *E. coli* RNase HI shows no strong preference for the sequence of either synthetic or natural RNAs (6,15). Compared with restrictive cleavages performed by RTase-associated RNase H (see Section III,A, Chapter 6), *E. coli* RNase HI cuts the RNA strand at several prominent sites throughout the region of RNA in the heteroduplex.

The smallest oligodeoxynucleotide in a DNA · RNA hybrid that can serve as a substrate for RNase H is a tetradeoxynucleotide (16,17). The oligonucleotide-directed RNase H cleavage is site specific, although the cleavage pattern cannot be generalized. *E. coli* RNase HI is less discriminatory toward hybrid structure than the calf thymus enzyme. With a hybrid formed between RNA and a hexadeoxynucleotide (5'-dCGATGC-3') containing a single mismatch at the A position, the *E. coli* enzyme can cleave the RNA, whereas the calf thymus RNase H requires 50 times more enzyme to show any detectable activity.

With perfectly matching hybrids (RNA · oligodeoxynucleotide), the *E. coli* enzyme cuts the phosphodiester bond of RNA mainly between G and the non-H-bonded flanking nucleotide A, 5'-rGCAUCG|A-3'. In contrast, the calf thymus enzyme preferentially cleaves the internal phosphodiester bond between U and C, 5'-rGCAU|CGA-3'. With a 20-mer oligodeoxynucleotide hybridized to an RNA template, *E. coli* RNase HI introduces a site-specific cut into the RNA (5'-rCG|CUUUGAU-3') at a site two bases from the 3' end of the synthetic oligodeoxynucleotide (18).

The area of strong RNA secondary structures may block the hybridization with oligodeoxynucleotide and thus affect the cleavages by RNase H (see below). Unlike the calf thymus enzyme, *E. coli* RNase HI is active at low salt concentrations that tend to destabilize secondary structures. Therefore, the *E. coli* enzyme may be more useful with certain RNAs for making more specific cleavages.

iv. **Cleavage specificity in deoxynucleotide-containing oligoribonucleotide duplexes.** An oligoribonucleotide (e.g., GCGCGG*dAdT*CCGGCC) containing two consecutive deoxyribonucleotide residues allows *E. coli* RNase HI to make specific cleavages of the complementary RNA strand (14). Under partial digestion conditions (e.g., 10 min, pH 8.1, and 37°C), the cleavage occurs immediately 3' of the rU involved in the rU·dA base pair and after two consecutive ribonucleotides in the 3' direction: 5'-GGCCGGAU|C|C|GCGC-3'. A single deoxynucleotide substitution (dA, dT, or dU) does not allow cleavage of the RNA strand by RNase HI.

v. **Cleavage specificity with probes containing three or more deoxynucleotides.** By annealing chimeric probes which contain three or more deoxynucleotides linked to the 3' or 5' end or positioned in the middle of 2'-O-methylribonucleotides, *E. coli* RNase HI can cleave site specifically at a hairpin loop and even at a double-stranded stem region of the target RNA which contains secondary structures (19). The cleavage in the RNA strand is directed mainly at the nucleotide position either at or next to the 5' end of the complementary deoxynucleotide stretch. When the chimeric probe contains four deoxynucleotide residues, either in the middle or at the 5' end, the cleavage occurs at a single site (20). When the chimeric probe contains three or five deoxynucleotide residues, cleavages occur at more than one site.

vi. **Effects of modified phosphodiester bonds and sugar configuration**
(a) *Methyl phosphonate bond.* The cleavage of RNAs hybridized to certain oligodeoxynucleotides may be refractory when the oligodeoxynucleotide contains modified phosphodiester bonds. RNase HI can cleave the RNA strand when the strand is annealed to a 14-mer oligodeoxynucleotide or a derivative that contains alternating phosphorothioate bonds. When the same 14-mer oligodeoxynucleotide contains more than three methyl phosphonate internucleotide linkages, the hybrid RNA strand is progressively resistant to cleavage as a function of the number of phosphonate linkages (21).

(b) *α-Deoxyribonucleotides.* An α-oligodeoxynucleotide (hexamer), which differs from a natural oligodeoxynucleotide having a β-configuration at the anomeric carbon atom (C-1') of the sugar, can anneal with ssRNA (hexamer) in both parallel (80%) and antiparallel (20%) orientations (22). The parallel-stranded α-DNA·β-RNA duplex renders the RNA resistant to cleavage by RNase HI presumably because the heteroduplex adopts a B-type helix instead of the A-type helix normally adopted by a β-DNA·β-RNA duplex (23,24).

vii. **Cleavages at the RNA–DNA junction.** *E. coli* RNase HI was previously shown *in vitro* to cleave the 3' → 5' phosphodiester bond that links RNA and DNA, albeit inefficiently (6,25). On a model substrate, e.g., DNA-extended tRNA forming a duplex with a DNA strand, however, *E. coli* RNase HI was shown to cleave the phosphodiester bond not at the DNA–RNA junction but one nucleotide away from it (63). A substantial portion of the reaction products included DNA strands with two ribonucleotide 5'-phosphates at the 5' terminus.

d. Mechanism of RNase H action

Certain RNases, such as RNase A, T1, and U2, which cleave RNA via a 2',3'-cyclic phosphate intermediate, require the presence of the 2'-OH group of ribose. In contrast to this class of RNases, RNase H has no mechanistic requirement for the 2'-OH group because the hydrolysis product has a 5'-P. Therefore the catalytic mechanism of RNase H, which is optimally active at pH ~8, is rather unique and resembles those of some DNases. The catalytic groups of *E. coli* RNase HI have been identified as two Asp and one Glu residues that together form a unique "carboxyl triad."

i. **Role of DNA·RNA hybrid structure.** The characteristic conformation of DNA·RNA hybrids is the A-type helix as in DNA·DNA or RNA·RNA duplexes. However, the detailed conformation of one heteroduplex can be quite different from that of another: for example, poly(dA·rU), poly(dI·rC), and poly(rA·dT) are all heteromerous (26). In the A-type form, the major and minor grooves are similar in size and the bases are relatively inaccessible. The sugar–phosphate backbone, which is of particular importance for RNase H substrate specificity, is the most prominent part of the molecule. Unlike DNA·DNA and RNA·RNA duplexes, DNA·RNA hybrids are bipolar, with ribose in one strand and deoxyribose in the other. This polarity is presumed to play a critical role in the recognition and cleavage of the RNA moiety by the enzyme.

The 3-D structure of *E. coli* RNase HI shows an extensive hydrophobic cleft in the major domain, while the minor domain carries many basic amino acid residues (27,28). The hydrophobic interaction between the protein and a RNA·DNA substrate is mainly directed to the DNA moiety which can be as small as dinucleotides (14) or, more optimally, tetranucleotides (16,19). The RNA strand on the opposite side of the hybrid helix is correctly aligned with the catalytic residue(s), by virtue of the structural features of the minor groove and the electrostatic positioning effect of the polar, minor domain of the enzyme. In fact, a preliminary model complex between a crystalline enzyme and a hybrid decanucleotide substrate suggests that the putative scissile 3'O-P bond is positioned in the minor groove, near the two members (Asp-10 and Glu-48) of the carboxyl triad (28). Note that the 3' → 5'-exonuclease active site of the *E. coli* DNA Pol I also features a similar configuration with Asp, Glu, and a second Asp residue in the metal-binding site, although the catalytic function of the exonuclease involves two metal ions (29,30).

ii. **Role of active site residues.** The presumed active site residues (Asp-10, Glu-48 and Asp-70) are all positioned, together with the bound Mg^{2+}, in the center of the hydrophobic cleft of the major domain. How the catalysis actually occurs is not known, but two alternative mechanisms can be postulated. With the Mg^{2+} ion in close proximity to the scissile phosphate group, a carboxylate may directly attack the phosphate group as a nucleophile. Alternatively, the hydrolysis may be carried out by a nucleophilic OH^- ion that can be activated by coordination with the cation.

In support of the latter mechanism, the crystal structures of *E. coli* RNase HI are similar to those of DNase I in the dispositions of the β-strands and the cation-binding site. In addition to this structural similarity, the two enzymes share several catalytic features, including base nonspecificity, participation of a metal ion (Mg^{2+} for RNase H, Ca^{2+} for DNase I) in chemical catalysis, and generation of products with 5'-P.

2. APPLICATIONS

i. **Synthesis of second-strand cDNA.** *E. coli* RNase HI plays an important role in cDNA cloning strategies. Following the first-strand cDNA synthesis using a reverse transcriptase, a number of strategies are available for the synthesis of the second cDNA strand (refer to Section II, Chapter 6). Among various strategies, those incorporating an RNase H digestion step are probably the most efficient. The cDNA · mRNA hybrid can be converted to dsDNA by simultaneously digesting the RNA strand with the endonucleolytic *E. coli* RNase HI and synthesizing the second strand with *E. coli* DNA Pol I (31,32).

ii. **Removal of poly(A) in mRNA.** The poly(A) tail of eukaryotic mRNAs can be removed from the RNA by RNase H digestions following the hybridization of the mRNAs with oligo(dT) or poly(dT). Deadenylation of total cellular mRNAs reveals electrophoretically distinct mRNA size classes which would have otherwise been unobtainable due to the polydisperse nature of the native poly(A) tails (33). A selective removal of the poly(A) tail from viral RNA by poly(dT)-directed RNase H cleavage has been used to determine the biological importance of the poly(A) tail function in viral infectivity (34).

iii. **Quantitation of poly(A)-tailed mRNA.** Poly(A)-tailed mRNAs can be selectively quantitated from the total mRNAs or the mixtures with rRNAs. The method relies on two enzymatic steps: (a) poly(A) polymerase-catalyzed labeling of the 3' termini of all RNA species in the presence of an ATP analog chain terminator, cordycepin triphosphate; this reaction adds one or at most a few nucleotides from [α-^{32}P]ATP; and (b) RNase H digestion of the total sample in the presence of oligo(dT)$_{12-18}$, releasing ^{32}P-labeled poly(A) tails specifically from the poly(A)$^+$ RNA molecules. The difference between the amount of label that has incorporated into the RNA and that which remains TCA precipitable on glass fiber filters represents the fraction of polyadenylated samples and thus the concentration of poly(A)-tailed mRNA (35).

iv. **Specific fragmentation of RNA.** The observation that RNAs annealed to short synthetic deoxynucleotides can be site-specifically cleaved by RNase H serves as the basis for diverse applications of RNase H in RNA mapping and/or sequencing (16,17). The precise position(s) of ribonucleotides cleaved in a given DNA · RNA hybrid depends on the site of deoxynucleotide annealing, length of the deoxynucleotide probe, RNase H species (e.g., *E. coli* versus calf thymus), and, to some extent, on the RNA sequence. The RNase H mapping technique has been used to probe the mRNA heterogeneity arising from the presence of distinct 5′ and/or 3′ UTRs (36,37).

v. **Hybrid-arrested translation by antisense oligonucleotides.** Small antisense oligonucleotide-directed inhibition of gene expression or translation has emerged as a powerful technique for studying the control of gene expression *in vivo* and *in vitro*. The oligonucleotide-directed cleavage and destabilization of target mRNAs by RNases H has been found to be, at least in part, the mechanism of antisense oligonucleotide-directed inhibition of gene expression. The hybrid-arrested translation is a technique based on site-specific cleavages of target mRNA in the form of mRNA · oligodeoxynucleotide hybrids (38,39). This technique allows identification of specific cDNA clones for which no phenotypic assays exist but to which *in vitro* translation can be applied (40).

The antisense specificity or the selectivity of RNase H-mediated cleavages can be significantly improved according to certain oligonucleotide designs (64–66). For example, composite oligonucleotides in which the central phosphodiester (or phosphorothioate) part is flanked by two blocks of modified nucleotides (e.g., methylphosphonates) offer the following advantages: (i) they can be nuclease resistant and thus become longer-lived molecules compared with unmodified oligonucleotides, (ii) reduced affinity of the methylphosphonate analogs to complementary sequences rather destabilizes mismatching (or partial) duplexes, and (iii) RNase H-mediated cleavages are limited to the unmodified portion of the oligonucleotide, regardless of whether the flanking modified blocks are fully paired to the RNA.

The hybrid arrest of *in vitro* translation using some rabbit reticulocyte lysate preparations may not be reliable without added RNase H (41). Freshly prepared lysates are reported to contain 1–2% of the level of RNase H present in actively dividing cells, an amount sufficient to achieve near 100% cleavage of hybrid mRNAs (39). In contrast to some reticulocyte lysates, the wheat germ cell-free translation system apparently contains a sufficient amount of endogenous RNase H activity.

vi. **Detection of RNA-containing dsDNA structure.** The specificity of RNase H can be used to detect the presence of RNA in DNA structures. A covalently closed circular DNA that contains RNA can be converted to a relaxed form which, after denaturation, produces a single-strand circle and a single-strand linear molecule.

3. STRUCTURE

i. **General structure.** *E. coli* RNase HI is a monomeric enzyme composed of 155 amino acids (M_r 17,559) (42,43). The N- and C-terminal residues are Met and Val, respectively. The initiating N-terminal Met is presumed to be present. The RNase H may form a dimer under certain environments. The pI of the enzyme is 9.0 (9,44) [The calf thymus RNase H is an apparent dimer of 68-kDa subunits (45) and has a pI of 4.9 (12).]

E. coli RNase HI contains three Cys and four Met residues. Although none of the Cys residues are catalytically essential, the modification of either Cys-13 or Cys-133 results in inactivation, whereas the modification of Cys-63 does not. The sensitivity to NEM has been shown, by site-directed mutagenesis and chemical modification, to be due to steric hindrance by the incorporated modifying group (46). A mutant RNase H containing seleno-Met in place of all four Met residues is fully active (47).

The secondary structure consists of 28% α helix and 41% β sheet (27). The secondary structures and their topology have also been investigated by 3-D NMR (48).

According to structure–stability analyses with reference to *T. thermophilus* RNase H, a thermophilic enzyme (166 amino acid residues, 18.3 kDa) having 52% amino acid sequence identity, the thermostability of *E. coli* RNase HI can be increased, without affecting enzymatic activity, as much as 16.7°C (ΔG 3.66 kcal/mol) in a noncooperative manner by strategic replacement of amino acid residues in certain divergent regions (67). In particular, the substitution of Gly for Lys-95 results in an increase of stability by ~7°C (ΔG 1.9 kcal/mol) (68). The left-handed helical Lys-95 is located in a basic protrusion domain, R^{91}-TAE-G^{95} in *T. thermophilus* RNase H and K^{91}-TAD-K^{95} in *E. coli* RNase HI.

ii. **Physical parameters.** The A_{280} of a 0.1% solution is 2.02 (46). The thermal denaturation temperature of RNase HI is estimated to be 52°C at pH 5.5 in 1 M GuHCl (67). The 50% denaturation also occurs at 1.83 M GuHCl at pH 5.5 and 25°C. The thermal or GuHCl-induced denaturation proceeds in a single cooperative fashion.

iii. **Tertiary structure.** The 3-D structures have been determined by X-ray crystallography for free RNase HI (in the presence or absence of divalent cations but without substrates) (27) and for seleno-Met RNase HI (in the absence of both divalent cations and nucleotides) (28). They show that the enzyme is an α/β protein composed of 5 α helices and 5 β strands.

E. coli RNase HI has an irregular ellipsoidal shape with overall dimensions 50 Å × 45 Å × 40 Å. It consists of two domains, the "major" domain containing 4 α helices and all 5 β strands and the "minor" domain [residues from ~80 to ~100] containing 1 α helix with two turns followed by a 10-residue loop. An extensive hydrophobic cleft containing all 6 Trp residues runs through the 5 β chains and 3 α helices.

iv. Active site. Sequence alignments of E. coli RNase HI with corresponding C-terminal RNase H domains of retroviral and retrovirus-like reverse transcriptases have revealed that there are significant (18–28%) amino acid identities (3,28,43,46,47,49–51). Of the eight amino acid residues conserved in 26 sequences, four residues (Asp-10, Glu-48, Asp-70, and Asp-134) are invariant. In crystal structures, these four acidic residues form a cluster that serves as the binding site for the catalytically essential Mg^{2+} which lies 2.0–4.9 Å from the carboxyl oxygens. However, only three residues (the carboxyl triad, Asp^{10}/Glu^{48}/Asp^{70}, not Asp-134) are essential for RNase H activity. The triad is located at the center of a conserved depression with the carboxylate groups deployed in a triangle of 4–5 Å on an edge.

Site-directed replacements of these residues, e.g., D10N, E48Q, and D70N, result in almost a complete loss of activity (3). Replacements by Gly or Cys of the two conserved Asp residues of the MoLV RTase which correspond to Asp-10 and Asp-70 of the E. coli RNase HI also result in a 25- to 130-fold decrease in RNase H activity (52). In contrast, conservative substitutions in E. coli RNase HI, i.e., D10E, E48D, and D70E, result in mutant enzymes with 2–10% of the k_{cat} of the wild type, while the K_m remains invariant. Therefore, the carboxyl triad is involved in catalysis either directly or via bound Mg^{2+}. The minor domain contains many basic residues whose side chains are mostly exposed to solvent. This feature implies that the minor domain is most likely involved in the binding to the polar region of the DNA·RNA hybrids.

Single substitutions at His-124 (to Ala) and Asn-130 (to Asp) result in active enzymes but with reduced substrate binding affinity (4- and 7-fold higher K_m, respectively) and lower catalytic rates (45- and 10-fold lower k_{cat}, respectively). The changes at Ser-71 (to Ala) and Asp-134 (to Asn) hardly affect the RNase H activity (3).

4. Genetics

i. Gene structure and organization. The structural gene for RNase HI, *rnhA*, is located at the 5.1-min locus, 64 bp downstream of the *dnaQ* (or *mutD*) gene (for DNA polymerase III ε subunit) in the E. coli chromosome (43,53). The *rnhA* gene has been cloned (54), nucleotide sequenced (42,43), and expressed under the control of its own promoter or the λ p_L promoter. The two genes, *dnaQ-rnhA*, are organized face to face and are transcribed in the opposite direction from the overlapping promoters, a single promoter (P_{rnh}) for *rnhA* and two promoters for *dnaQ* (55). The 5' terminus of the *rnhA* RNA overlaps ~100 and 20 nt with the 5' termini of *dnaQ*-1 and *dnaQ*-2 RNAs, respectively. The transcription of *rnhA* starts over 10 nt upstream from the presumed −35 promoter sequences (TGCTCA). A translation initiation signal (RBS) is also present in the −1 to −35 sequence region.

A potential stem–loop transcription termination signal is found in the region [position 22–41] downstream from the 5′-UAA termination codon.

ii. Physiological roles of RNase H. Despite multiple roles proposed for RNase H, for example, the removal of RNA primers from Okazaki fragments during DNA replication, no *in vivo* roles have been clearly established in eukaryotes (56). In fact, the physical differences between eukaryotic and prokaryotic RNases H and among various RNases H subspecies from a single eukaryotic cell type suggest that their *in vivo* roles may be just as diverse.

Available evidence suggests that RNases HI in *E. coli* may not be indispensable for cell survival (57). A RNase H mutant, generated by ethylmethane sulfonate mutagenesis, exhibits less than 8% of the wild type activity but has no apparent phenotypic differences (53). A 15-fold overexpression of the cloned *rnh* gene has no detectable effect on cell growth (42). Although RNase HI-deficient mutants generally exhibit no grossly altered phenotype, they display some abnormalities such as (a) sustained DNA replication in the absence of protein synthesis, (b) lack of the requirement for *dnaA* protein and the origin of replication (*oriC*), and (c) growth sensitivity to rich media (58). The RNase H of *E. coli* is presumed to play a role in (i) the replication of chromosomal DNA, (ii) the replication of plasmid (ColE1) DNA at the precise replication origin (59,60) and functioning as a "renaturase" factor, and (iii) the transcript displacement by RNA polymerase (61). Perhaps some of these *in vivo* functions are substituted, when necessary, by other *E. coli* enzymes that are known to exhibit RNase H-type activity, albeit with less specificity and efficiency. The possibility of RNase H activity substitution has become more realistic in light of the discovery of a second RNase H gene in *E. coli*. In addition, DNA Pol I and exonuclease III can degrade either the DNA or RNA of the hybrids.

iii. A second RNase H of *E. coli*: RNase H isozymes. A second, distinct RNase H gene has been identified in *E. coli* K12 (62). The gene and its product have been named *rnhB* and RNase HII (M_r 23,225) in reference to the previously known *rnh* gene (now called *rnhA*) and RNase H (now *RNase HI*). The *rnhB* gene is located at the 4.5-min position between *lpxB* and *dnaE* (the α subunit of DNA Pol III) on the *E. coli* genetic map. RNase HII has a low specific activity [0.57 U/μg protein on poly(rA·dT)], which is less than 0.4% of the specific activity of RNase HI. RNase HII has only a 17% amino acid sequence identity with RNase HI in the N-terminal 155 amino acid region.

5. SOURCES

Although RNase H has been traditionally purified from *E. coli* cells (e.g., KS 351) (9), the enzyme is now more conveniently isolated from overexpression systems, e.g., from the *E. coli* cells harboring recombinant plasmid pPL801. In this plasmid, the *rnhA* gene is under the control of the λ p_L promoter. This expression system overproduces the enzyme 3000-fold (~8% of the total cytosolic protein or 7 mg/g wet cells) (43,44). A highly purified enzyme has a specific activity of 150,000 U/mg protein.

The *rnhA* gene cloned in another plasmid pDR600 has also been overexpressed under the control of the *tac* promoter (3).

E. coli RNase HI does not bind to DEAE-cellulose or DEAE-Sephacel, in contrast to the calf thymus enzyme. This is apparently due to the difference in isoelectric point. PMSF (0.1–1 mM) is often included in buffers to avoid possible proteolysis during purifications.

References

1. Stein, H., and Hausen, P. (1969). *Science* **166**, 393–395.
2. Hausen, P., and Stein, H. (1970). *EJB* **14**, 278–283.
3. Kanaya, S., Kohara, A., Miura, Y., Sekiguchi, A., Iwai, S., Inoue, H., Ohtsuka, E., and Ikehara, M. (1990). *JBC* **265**, 4615–4621.
4. Miller, H. I., Riggs, A., and Gill, G. N. (1973). *JBC* **248**, 2621–2624.
5. Keller, W. (1972). *PNAS* **69**, 1560–1564.
6. Berkower, I., Leis, J., and Hurwitz, J. (1973). *JBC* **248**, 5914–5921.
7. Henry, C. M., Ferdinand, F.-J., and Knippers, R. (1973). *BBRC* **50**, 603–611.
8. Crouch, R. J. (1981). *In* "Gene Amplification and Analysis" (J. G. Chirikjian and T. S. Papas, Eds.), Vol. 2, pp. 217–228. Elsevier, New York.
9. Arendes, J., Carl, P. L., and Sugino, A. (1982). *JBC* **257**, 4719–4722.
10. Kanaya, S. and Itaya, M. (1992). *JBC* **267**, 10184–10192.
11. Dirksen, M.-L., and Crouch, R. J. (1981). *JBC* **256**, 11569–11573.
12. Busen, W. (1980). *JBC* **255**, 9434–9443.
13. Huet, J., Sentenac, A., and Fromageot, P. (1978). *FEBS Lett.* **94**, 28–32.
14. Wyatt, J. R., and Walker, G. T. (1989). *NAR* **17**, 7833–7842.
15. Robertson, H. D., and Dunn, J. J. (1975). *JBC* **250**, 3050–3056.
16. Donis-Keller, H. (1979). *NAR* **7**, 179–192.
17. Donis-Keller, H. (1981). *Virol.* **110**, 43–54.
18. Oyama, F., Kikuchi, R., Crouch, R. J., and Uchida, T. (1989). *JBC* **264**, 18808–18817.
19. Shibahara, S., Mukai, S., Nishihara, T., Inoue, H., Ohtsuka, E., and Morisawa, H. (1987). *NAR* **15**, 4403–4415.
20. Inoue, H., Hayase, Y., Iwai, S., and Ohtsuka, E. (1987). *FEBS Lett.* **215**, 327–330.
21. Furdon, P. J., Dominski, Z., and Kole, R. (1989). *NAR* **17**, 9193–9204.
22. Gmeiner, W. H., Rao, E., Rayner, B., Vasseur, J.-J., Morvan, F., Imbach, J.-L., and Lown, J. W. (1990). *Biochem.* **29**, 10329–10341.
23. Gagnor, C., Rayner, B., Leonetti, J.-P., Imbach, J.-L., and Lebleu, B. (1989). *NAR* **17**, 5107–5114.
24. Bloch, E., Lavignon, M., Bertrand, J.-R., Pognan, F., Morvan, F., Malvy, C., Rayner, B., Imbach, J.-L., and Paoletti, C. (1988). *Gene* **72**, 349–360.
25. Darlix, J.-L. (1975). *EJB* **51**, 369–376.
26. Arnott, S., Chandrasekaran, R., Millane, R. P., and Park, H. S. (1986). *JMB* **188**, 631–640.
27. Katayanagi, K., Miyagawa, M., Matsushima, M., Ishikawa, M., Kanaya, S., Ikehara, M., Matsuzaki, T., and Morikawa, K. (1990). *Nature* **347**, 306–309.
28. Yang, W., Hendrickson, W. A., Crouch, R. J., and Satow, Y. (1990). *Science* **249**, 1398–1405.

29. Derbyshire, V., Freemont, P. S., Sanderson, M. R., Beese, L., Friedman, J. M., Joyce, C. M., and Steitz, T. A. (1988). *Science* **240**, 199–201.
30. Freemont, P. S., Friedman, J. M., Beese, L. S., Sanderson, M. R., and Steitz, T. A. (1988). *PNAS* **85**, 8924–8928.
31. Okayama, H., and Berg, P. (1982). *Mol. Cell. Biol.* **2**, 161–170.
32. Gubler, U., and Hoffman, B. J. (1983). *Gene* **25**, 263–269.
33. Vournakis, J. N., Efstratiadis, A., and Kafatos, F. C. (1975). *PNAS* **72**, 2959–2963.
34. Spector, D. H., and Baltimore, D. (1974). *PNAS* **71**, 2983–2987.
35. Krug, M. S., and Berger, S. L. (1987). *Meth. Enzy.* **152**, 262–266.
36. Goldstein, B. J., and Kahn, C. R. (1989). *BBRC* **159**, 664–669.
37. Irminger, J.-C., Rosen, K. M., Humbel, R. E., and Villa-Komaroff, L. (1987). *PNAS* **84**, 6330–6334.
38. Shuttleworth, J., and Colman, A. (1988). *EMB J.* **7**, 427–434.
39. Walder, R. Y., and Walder, J. A. (1988). *PNAS* **85**, 5011–5015.
40. Paterson, B. M., Robert, B. E., and Kuff, E. L. (1977). *PNAS* **74**, 4370–4374.
41. Minshull, J., and Hunt, T. (1986). *NAR* **14**, 6433–6451.
42. Kanaya, S., and Crouch, R. J. (1983). *JBC* **258**, 1276–1281.
43. Maki, H., Horiuchi, T., and Sekiguchi, M. (1983). *PNAS* **80**, 7137–7141.
44. Kanaya, S., Kohara, A., Miyagawa, M., Matsuzaki, T., Morikawa, K., and Ikehara, M. (1989). *JBC* **264**, 11546–11549.
45. Rong, Y. W., and Carl, P. L. (1990). *Biochem.* **29**, 383–389.
46. Kanaya, S., Kimura, S., Katsuda, C., and Ikehara, M. (1990). *BJ* **271**, 59–66.
47. Yang, W., Hendrickson, W. A., Kalman, E. T., and Crouch, R. J. (1990). *JBC* **265**, 13553–13559.
48. Yamazaki, T., Yoshida, M., Kanaya, S., Nakamura, H., and Nagayama, K. (1991). *Biochem.* **30**, 6036–6047.
49. Khudyakov, Yu. E., and Makhov, A. M. (1989). *FEBS Lett.* **243**, 115–118.
50. Ready, M. P., Katzin, B. J., and Robertus, J. D. (1988). *Proteins (S.F.G)* **3**, 53–59.
51. Johnson, M. S., McClure, M. A., Feng, D.-F., Gray, J., and Doolittle, R. F. (1986). *PNAS* **83**, 7648–7652.
52. Repaske, R., Hartley, J. W., Kavlick, M. F., O'Neill, R. R., and Austin, J. B. (1989). *J. Virol.* **63**, 1460–1464.
53. Carl, P. L., Bloom, L., and Crouch, R. J. (1980). *J. Bacteriol.* **144**, 28–35.
54. Horiuchi, T., Maki, H., Maruyama, M., and Sekiguchi, M. (1981). *PNAS* **78**, 3770–3774.
55. Nomura, T., Aiba, H., and Ishihama, A. (1985). *JBC* **260**, 7122–7125.
56. Crouch, R. J., and Dirksen, M.-L. (1982). In "Nucleases" (S. M. Linn and R. J. Roberts, Eds.), pp. 211–241. Cold Spring Harbor Laboratory, Cold Spring Harbor, New York.
57. Kogoma, T. (1986). *J. Bacteriol.* **166**, 361–363.
58. Ogawa, T., Pickett, G. G., Kogoma, T., and Kornberg, A. (1984). *PNAS* **81**, 1040–1044.
59. Itoh, T., and Tomizawa, J.-I. (1980). *PNAS* **77**, 2450–2454.
60. Polisky, B. (1989). *Cell* **55**, 929–932.
61. Kane, C. M. (1988). *Biochem.* **27**, 3187–3196.
62. Itaya, M. (1990). *PNAS* **87**, 8587–8591.
63. Furfine, E. S., and Reardon, J. E. (1991). *Biochem.* **30**, 7041–7046.
64. Agrawal, S., Mayrand, S. H., Zamecnik, P. C., and Pederson, T. (1990). *PNAS* **87**, 1401–1405.
65. Dagle, J. M., Walder, J. A., and Weeks, D. L. (1990). *NAR* **18**, 4751–4757.
66. Larrouy, B., Blonski, C., Boiziau, C., Stuer, M., Moreau, S., Shire, D., and Toulme, J.-J. (1992). *Gene* **121**, 189–194.
67. Kimura, S., Nakamura, H., Hashimoto, T., Oobatake, M., and Kanaya, S. (1992). *JBC* **267**, 21535–21542.
68. Kimura, S., Kanaya, S., and Nakamura, H. (1992). *JBC* **267**, 22014–22017.

C. Ribonuclease T1
[EC 3.1.27.3]

Ribonuclease T1 (RNase T1) is a guanine-specific endoribonuclease (also called guanyloribonuclease), which has been isolated from *Aspergillus oryzae*. It catalyzes the hydrolysis of phosphodiester bonds in ssRNA, producing 3'-P mononucleotides and oligonucleotides that end in Gp (Scheme 3.4).

$$\begin{array}{c} \quad\quad\quad\quad\downarrow\quad\quad\downarrow\quad\downarrow \\ 5'\,A-p-C-p-G-p-U-p-G-p-G-p-A-p-U-p-C\,3' \\ \quad\quad\quad\quad\quad\quad\quad\quad\quad\quad\quad\quad\longrightarrow\ ApCpGp + UpGp + Gp + ApUpC \end{array}$$

SCHEME 3.4

RNase T1 is similar to RNase A (see Section III,A, this chapter) in its catalytic mechanism in that a 2',3'-cyclic phosphate intermediate is formed and then hydrolyzed to yield the terminal 3'-phosphoguanosine. The molecular size of RNase T1 (11 kDa) is also similar to the 14-kDa RNase A. However, RNase T1 is highly specific for G bases as opposed to the pyrimidine-specific RNase A. RNase T1 has been studied extensively as an alternative model for RNases with a distinct base specificity and is an important tool in the analyses of RNA sequence, structure, and RNA–protein interactions.

1. FUNCTIONS

a. Reaction conditions

i. Optimal reaction conditions. For complete digestion of RNA, a typical reaction mixture contains 50 mM Tris–Cl (pH 7.5), 1 mM EDTA, and enzyme at 0.2 U/μg RNA. Incubation is carried out at 37°C for 30 min. The sample RNA is denatured by boiling for 2 min and chilled quickly before the addition of enzyme.

The activity of RNase T1 is subject to pH and temperature as follows.

(a) pH. The activity of RNase T1 on natural RNA substrates is optimal at pH 7.5 (1). The optimum pH for the hydrolysis of 2',3'-cyclic GMP is pH 7.0–7.2 (2). Note that the binding of NMPs to RNase T1 occurs optimally at pH 5.5 (3).

(b) Temperature. The activity at pH 7.5 increases linearly, according to Arrhenius equation [Eq. (1.11)], as a function of the incubation temperature up to 50°C. However, the incubation at or above 50°C results in a rapid drop in activity, most likely due to a conformational change in the enzyme (4). RNase T1 loses the ability to bind 3'-GMP at 60°C or in 8 M urea (3). The thermally induced unfolding of RNase T1 appears to be rapidly reversible. When assayed at 37°C after heating the enzyme to 100°C for 10 min in 10 mM citrate (pH 6), RNase T1 does not show any loss of activity (1). RNase T1 has maximal conformational stability at pH ~4.5 toward urea and GuHCl (5). The maximal energy of stabilization is ~9 kcal/mol, similar to that of RNase A.

ii. Kinetic parameters. The pH-independent values of k_{cat}/K_m are virtually identical for the following four dinucleotide substrates, while the individual values for k_{cat} and K_m range within an order of magnitude of each other and follow the sequence (6): GpC (350 sec^{-1}, 160 μM) > GpA (96 sec^{-1}, 55 μM) > GpG (62 sec^{-1}, 27 μM) > GpU (38 sec^{-1}, 22 μM). A 5'-P-dinucleotide, pGpC, is also a good substrate with a k_{cat} of 1260 sec^{-1} and a K_m of 536 μM (7). For homopolymers, the k_{cat} and K_m values measured at pH 6.2 and 25°C are: poly(I) (485 sec^{-1}, 0.83 mM), poly(A) (4.1 × 10^{-3} sec^{-1}, 2 mM), and poly(C) (not measurable) (57). Under the same conditions, the kinetics for 2',3'-cGMP give a k_{cat} of 0.35 sec^{-1} and a K_m of 0.44 mM.

When the transesterification step of dinucleotide mono-P substrates is compared at pH 5 and 25°C, the catalytic efficiency (k_{cat}/K_m) for GpN substrates is ~10^6-fold greater than the corresponding dinucleotide ApN and at least 10^8-fold greater than CpN and UpN (8). IpC and IpU are 10^2- to 10^3-fold less efficient substrates than the corresponding GpN. No quantifiable kinetics have been observed with CpN and UpN.

iii. Inhibitors and/or stabilizer

(a) Metal ions. RNase T1 does not require metal ions, which are rather inhibitory. MgCl$_2$ at 100 mM concentration is ~40% inhibitory, while CaCl$_2$ at 10 mM is ~30% inhibitory. Other metal ions such as Zn^{2+}, Fe^{2+}, and Cu^{2+} are strongly inhibitory, whereas EDTA is an activator probably because it chelates trace amounts of inhibitory heavy metal ions (9). Despite the strong inhibitory effect on the enzyme activity, Zn^{2+} (at 1 mM) does not affect the binding of 3'-GMP.

Binding of one Mg^{2+} or two Na$^+$ ions at the cation-binding site of RNase T1 significantly stabilizes the enzyme. NaCl up to 100 mM is noninhibitory to the enzyme.

(b) Mononucleotides. Many substrate analogs strongly bind and inhibit RNase T1. The binding strengths at pH 5.5 have been estimated by gel filtration methods as follows (3,10): 2'-GMP (K_d 6.5 μM) > 3'-GMP > 2'-IMP > 2'-XMP > 2'-AMP > 5'-GMP > guanosine (K_d 97 μM) > deoxyguanosine (K_d 700 μM) > 3'-CMP. (*Note:* X for xanthine.)

(c) Dinucleotides. Guanylyl-2',5'-guanosine (2',5'-GpG) is a specific inhibitor ($K_i \approx 0.165$ mM) and the strongest inhibitor among the 2',5'-GpPu (11). (*Note:* 2',5'-GpPy are not inhibitory.) The specific inhibition is attributed to the stacking interaction between the 3'-terminal guanine and the catalytically essential residue His-92 (12).

(d) Anions. Citrate or phosphate apparently binds to the active site of the enzyme and thereby inhibits activity. Binding of one HPO$_4^{2-}$ ion at the anion-binding subsite significantly stabilizes the enzyme (13).

RNase T1 is most stable at pH 4.5 (at 25°C) or at pH 7.0 (at -5°C) in the presence of 0.2 M Na$_2$HPO$_4$ (14).

(e) Organic solvents. Many organic solvents inhibit RNase T1 (4). At 20% (v/v), for example, dioxane, methanol, ethanol, and 1-propanol give 36, 41, 62, and 90% inhibition, respectively. However, the binding affinity for 3'-GMP is not significantly altered in the presence of the organic solvents.

b. Activity assays and unit definition

i. Acid-precipitation assay. This is the most common method used for RNase T1 activity assay. It measures the increase in acid-soluble products on RNase T1 digestion of yeast RNA (15). The assay mixture consists of 0.1 ml enzyme solution, 0.25 ml Tris–Cl (0.2 M, pH 7.5), 0.1 ml EDTA (20 mM), and 0.25 ml of RNA (12 mg/ml). After incubation for 15 min at 37°C, the reaction is stopped with 0.25 ml uranyl reagent [0.75% uranyl acetate in 25% (w/v) $HClO_4$]. The resulting precipitate is immediately removed by centrifugation. A 0.2-ml aliquot of the supernatant is diluted with 5.0 ml water, and the A_{260} is determined. The assay mixture without the nuclease is used as the blank.

Unit definition: One unit is the amount of enzyme that causes an absorbance increase of 1.0 at 260 nm under the assay conditions.

ii. Spectrophotometric assay. Another convenient assay for RNase T1 activity is to monitor the hydrolysis of GpN dinucleotide substrates by measuring the increase of A_{280} as the rapidly formed 2',3'-cyclic phosphate intermediate is converted to 3'-GMP.

iii. Titrimetric assay. This assay measures the hydrolase activity on 2',3'-cGMP by monitoring the consumption of alkali at 25°C in a pH-stat set at pH 7.0 (15). This assay is the most precise, but it is relatively slow and cumbersome.

c. Substrate specificities

i. Base and sugar specificities. RNase T1 cleaves ssRNA specifically at the G bases. However, the base specificity is not absolute: after a long incubation period at high concentrations of enzyme, cleavages at other bases occur to a measurable extent (9). The catalytic specificity for guanylyl groups is about 10^6-fold greater than for adenylyl residues. Uridylyl and cytidylyl nucleotides may well be considered to be no substrates for RNase T1 under the standard reaction conditions. RNA containing G bases modified at N7 is resistant to digestion.

RNase T1 specifically recognizes β-D-ribose. As a result, DNAs are not substrates.

ii. Cyclic mononucleotides. RNase T1 catalyzes the hydrolysis of 2',3'-cGMP to 3'-GMP (also see activity assay). In contrast, 6-thioguanosine 2',3'-cyclic phosphate is resistant to hydrolysis by RNase T1 (2).

iii. Dinucleotides. The bases in the position adjacent to G can be either Pu or Py, but they have a considerable effect on the rate of cleavage (6,16). The relative rates at pH 7.4 are as follows (17): GpCp [1100] > GpC [800] > GpA

[550] > GpG [450] > GpU [250] > IpC [150] > XpC [10] > glyoxal-GpC [5] > 2',3'-cGMP [2].

iv. Tri- and oligonucleotides. Although dinucleotides are substrates for RNase T1, catalytic efficiency is optimal with oligomers three or more nucleotides long (18). Transesterification reactions with trimeric substrates, ApGp|C and ApGp|U, are significantly different from reactions with dimeric substrates GpN (19): (a) a dramatic attenuation of the pH dependence of k_{cat}/K_m, (b) no discrimination for U and C nucleoside leaving groups, and (c) an opposite pH dependence for k_{cat}. The results indicate that the adenosine moiety of the trimeric substrates binds at an enzyme subsite and that it opens new reaction paths that parallel the catalytic pathway used for dimeric GpN.

d. Mechanism of RNase T1 action

RNase T1 cleaves P-O5' ester bonds in ssRNA, specifically at the 3'-P of the guanylic acid residues. As in RNase A (see Fig. 3.3), the catalysis occurs by a two-step mechanism, i.e., the formation of a terminal guanosine 2',3'-cyclic phosphate intermediate (transesterification step) and the hydrolysis of the cyclic ester to guanosine 3'-monophosphate (hydrolysis step). The transesterification step involves a general acid–base catalysis.

i. Specificity of interaction. Binding studies with various nucleoside mono-P and di-P substrate analogs suggest that the integrity of the guanine, ribose, and phosphate portions are important for the binding to the enzyme. The high specificity to the G base is partly attributed to a series of H-bonds between the G base and the enzyme, particularly the Asn-98 main-chain carbonyl and the Glu-46 carboxylate (20). The N1 and N7 as well as the 2-NH_2 and 6-oxo (or hydroxy) groups of the guanine base are important (3). The 2-NH_2 group contributes ~2 kcal/mol to the binding energy (10,21). The binding specificity is further enhanced by Tyr-42 and Tyr-45 residues which form a hydrophobic "sandwich" complex with the G base. RNase T1 discriminates guanine (in GpU) from nonionized xanthine (in IpU) by ~4.4 kcal/mol (21).

When 2'-GMP binds to RNase T1, it induces distinct conformational adjustments through a process termed the "snap-in" mechanism (22). In addition to the alteration of the H-bonding network, the conformational changes involve a closing of the active site with positional changes of the residues such as Asn[43]-Asn[44], Glu-46, Asn-98, and most prominently Tyr-45 (23). RNase T1 has some affinity for 2'-AMP, which binds the enzyme in a manner different from that of 2'-GMP; the adenosine moiety is not located in the guanosine recognition site and the phosphate group of 2'-AMP is H-bonded to active site residues different from those involved in 2'-GMP binding (55). The binding of 2'-AMP also introduces marginal but distinct conformational changes in RNase T1.

The phosphate group (of the GpG) is tightly H-bonded at the nonester oxygen atoms by the side chains of Tyr-38, Asn-36, Arg-77, and His-92, possibly activating the phosphorus for nucleophilic attack by O2' (24).

ii. Catalytic mechanism. The stereochemistry of transesterification follows a concerted in-line mechanism involving two distinct functional groups located on either side of the scissile bond (25). A general base withdraws the 2'-OH proton and enhances the nucleophilicity of the O2'. Nucleophilic attack of the O2' on phosphorus results in the pentacovalent phosphorane transition state. A general acid protonates the leaving O5', thereby completing the transesterification step. Various lines of evidence suggest that His-92 and most likely Glu-58 (or perhaps His-40) play catalytic roles (6,20,26,27): His-92 (pK_a 7.4) as a general acid and Glu-58 (or His-40) as a general base. The Glu-58 has a pK_a of ~4.1 in free RNase T1 (31) and 3.2–3.8 during transesterification reactions (29). The two catalytic residues exchange roles in each step, in a manner analogous to the catalytic roles played by His-12 and His-119 in RNase A.

In RNase T1, the identity of the residues acting as general acid and base has not been firmly established. For instance, a mutant RNase T1 in which Glu-58 is replaced with Ala (Glu58Ala RNase T1) maintains 5% residual activity whereas a His40Ala or His40Asp mutation results in a near complete loss of activity, suggesting that His-40 rather than Glu-58 might be the catalytic base (28). However, His40Lys RNase T1 retains a considerable (3.1% of the wild type) catalytic activity (29,30). Furthermore, His-40 in its protonated form (pK_a 7.7) is required for the optimal activity of the wild-type enzyme, suggesting that the charged His-40 participates in catalysis by providing electrostatic stabilization of the pentacovalent transition state (27,56). Yet in Glu58Ala RNase T1, His-40 may have acted as the base catalyst (pK_a 6.5), pairing with His-92 which acts as the acid catalyst (29).

2. APPLICATIONS

i. RNA sequencing. When a ^{32}P-labeled (either at 5' or 3' end) RNA is partially digested with RNase T1 and separated by PAGE, the ladder yields information on the ordering of the fragments terminated with the G residues (32). When this experiment is performed in conjunction with other base-specific RNases such as RNase U2 (A specific), RNase A (C + U specific) or *B. cereus* RNase (C + U specific) (33,34), and Phy M (U + A specific at 7 *M* urea, 50°C, pH 5.0) (35), RNA sequences can be read directly from the sequence ladder autoradiogram.

ii. Site-directed cleavage of RNA. The cleavage of ssRNA by RNase T1 can be directed to a specific site by the following strategy (36). The RNA strand is hybridized with a complementary DNA strand which has been site-specifically modified to contain mismatching bases at the target site of RNA. RNase T1 can then site specifically cleave the target G site of the RNA strand in the bulge region of the heteroduplex. The cleaved RNA strand can be purified, for example, by gel electrophoresis after denaturation of the heteroduplex, and used for other purposes such as the construction of recombinant RNA molecules.

iii. RNase T1 fingerprinting. Complete digestion of RNA with RNase T1 yields oligonucleotides ending in guanosine 3'-P. In high salt concentrations (e.g.,

0.3 M NaCl) which tend to promote a double-helical structure, RNase T1 selectively digests ssRNA while leaving dsRNA intact. Because the stem region of hairpin (or other secondary) structures is resistant to RNase T1, the digestion of ssRNA should be preceded by appropriate heat denaturation of the RNA (at 60°C for 10 min or 100°C for 1 min) and rapid cooling on ice. The digestion products of the radiolabeled RNAs can be separated in 2-D gel electrophoresis, giving an oligonucleotide pattern unique to the RNA, i.e., the "fingerprint." The RNase T1 fingerprinting is the most frequently used analytical technique for comparing and identifying the genomes of RNA viruses (37).

3. STRUCTURE

i. Primary structure. RNase T1 is a monomeric enzyme composed of 104 amino acids (M_r 11,085) (38,39). The enzyme occurs in two isozyme forms containing either Gln or Lys at position 25 (i.e., Glu^{25}-RNase T1 or Lys^{25}-RNase T1) (27). The most abundant form (Glu^{25}-RNase T1) is an extremely acidic protein with a pI of 3.8 (40), as expected from its high content of acidic residues (6 Asp, 6 Glu) versus only 1 Lys(-41) and 1 Arg(-77). The N- and C-terminal residues are Ala and Thr, respectively. RNase T1 contains three His, one Trp, and four Cys, but no Met. All Cys residues are involved in intramolecular disulfide bonds: C2/C10 and C6/C103. The C6/C103 brings the N and C terminus into close proximity. Reduction of the two disulfides causes the protein to unfold at low ionic strengths (41). The disulfides and folded conformation may be regenerated quantitatively under high salt conditions, by the elimination of reducing agents (e.g., DTT) or the addition of oxidized DTT (42).

The ε_{278} of RNase T1 is 21,200 M^{-1} cm^{-1} at pH 7.

ii. Tertiary structure. Crystallographic structures have been determined for RNase T1 complexed with 2'-GMP (20,27,43), 3'-GMP (44), 2',5'-GpG (12), or vanadate (22) and for mutant (His40Lys) RNase T1 complexed with 2'-GMP or phosphate (56).

RNase T1 has a compact globular structure in which the hydrophobic core is sandwiched between a 4.5-turn α helix and a 4-strand antiparallel β sheet.

(a) Active site. The active site of RNase T1 can be divided into two subsites: a base-binding site and a phosphate-binding site.

The guanine-binding site consists mostly of Asn-44, Glu-46, and the base-sandwiching Tyr-42 and Tyr-45. The substitution of Asn-44 by Asp or Ala reduces the activity to a few percent of the wild-type enzyme (45). The substitution of Tyr by Phe has no dramatic effect on the enzyme function.

The phosphate-binding site consists of several residues: His-40, His-92, Glu-58, Arg-77 (all located in the major β sheet), and Asn-98. These residues exhibit extensive H-bonding interactions with phosphate groups, whether from the backbone of RNA, the phosphate group(s) of nucleotides, inorganic phosphate (HPO_4^{-2}), or its analog tetrahedral vanadate anion.

The modification of RNase T1 with iodoacetate (but not with iodoacetamide) at pH 5.5 leads to inactivation of the enzyme concomitant with the specific esterification of Glu-58 (46). Under these conditions, no His or Lys residues are modified by iodoacetate. The site-specific substitution of Glu-58 with Gln (7) or Ala (28) results in a mutant enzyme with 1 or 5% residual activity, respectively. In contrast, the substitution of His-40 or His-92 by Ala results in a complete loss of activity.

The structure of a free enzyme formulated by molecular dynamics simulations has been shown to have an active site featuring a characteristic salt bridge triad, His^{40}–Glu^{58}–Arg^{77} (47).

(b) Metal-binding site. Although RNase T1 does not require a metal ion for catalytic activity, it has a cation-binding site for Mg^{2+} or Ca^{2+}. The bound Ca^{2+} is coordinated to both carboxylate oxygens of Asp-15 and to six water molecules (12, 22). The water molecules are in turn H-bonded to Ser-12, Cys-10, Ser-63, and Gly-94.

4. GENETICS

As the prototypic member of microbial RNases, RNase T1 shares a similar catalytic mechanism and varying degrees of primary and, when known, tertiary structural identities with other guanylo-RNases, such as RNase U1 from *Ustilago sphaerogena* (39), RNase N1 from *Neurospora crassa* (39), and barnase from *Bacillus amyloliquefaciens* (48).

The genes for RNase T1 and its mutant forms have been chemically synthesized and cloned into *E. coli* (45,49). Under the control of *E. coli trp* promoter, functional enzymes have been expressed as a fusion protein with the N-terminal part of human growth hormone (45).

5. SOURCES

Commercially available α-amylase (Sigma) or Taka-diastase (Sankyo) is an enzyme mixture from the fungus *A. oryzae*, and it has been a convenient source for the isolation of RNase T1. Note that Taka-diastase also contains nuclease S1 (Section IV,A this chapter) and another, less abundant RNase T2. The RNase T2 is a glycoprotein similar to RNase T1, but it has a higher MW (29,000), a p*I* of 5.0, and, most notably, no absolute base specificity. In nature, RNase T1 is an extracellular enzyme, similar to guanylo-RNases from various other bacterial and fungal organisms.

Since the first 370-fold purification of RNase T1 (1), a number of important improvements have been made to increase the yield, purity, and efficacy of purification (31). The improvements in separation techniques include various affinity chromatographies that rely on such adsorbents as 2′,3′-cGMP, aminophenylphosphoryl-GMP (50), 5′-GMP (40), and 2′,5′-GpG (51). In addition, the RNase T1 gene cloned into an expression–secretion vector efficiently produces the enzyme from *E. coli* hosts (yield of 20 mg enzyme per liter of culture) (49). In this system, the RNase T1 gene is expressed under the inducible control of the *lac* promoter–operator as a fusion protein with the signal peptide of OmpA, the major outer membrane protein of *E. coli*. The *in vivo* cleavage of the fusion protein liberates the enzyme (Lys^{25}-RNase T1) with an N-terminal 4 aa extension. The RNase T1 gene has also been expressed in *E. coli*, under the control of the *tac* promoter, as a fusion protein with the signal peptide of *E. coli* alkaline phosphatase (52). The RNase T1 secreted into the periplasmic space has the correct N-terminal sequence.

Functional RNase T1 and mutant enzymes have also been obtained by total solid-phase chemical synthesis using the method of peptide segment condensations (53,54).

References

1. Sato, K., and Egami, F. (1957). *J. Biochem.* **44**, 753–767.
2. Irie, S., Itoh, T., Ueda, T., and Egami, F. (1970). *J. Biochem.* **68**, 163–170.
3. Takahashi, K. (1972). *J. Biochem.* **72**, 1469–1481.
4. Takahashi, K. (1974). *J. Biochem.* **75**, 201–204.
5. Pace, C. N., Laurents, D. V., and Thomson, J. A. (1990). *Biochem.* **29**, 2564–2572.
6. Osterman, H. L., and Walz, F. G., Jr. (1978). *Biochem.* **17**, 4124–4130.
7. Nishikawa, S., Morioka, H., Fuchimura, K., Tanaka, T., Uesugi, S., Ohtsuka, E., and Ikehara, M. (1986). *BBRC* **138**, 789–794.
8. Walz, F. G., Jr., Osterman, H. L., and Libertin, C. (1979). *ABB* **195**, 95–102.
9. Uchida, T., and Egami, F. (1971). "The Enzymes," 3rd ed., Vol. 4, pp. 205–250.
10. Campbell, M. K., and Ts'O, P. O. P. (1971). *BBA* **232**, 427–435.
11. White, M. D., Rapoport, S., and Lapidot, Y. (1977). *BBRC* **77**, 1084–1087.
12. Koepke, J., Maslowska, M., Heinemann, U., and Saenger, W. (1989). *JMB* **206**, 475–488.
13. Pace, C. N., and Grimsley, G. R. (1988). *Biochem.* **27**, 3242–3246.
14. Pace, C. N. (1990). *TIBS* **15**, 14–17.
15. Uchida, T., and Egami, F. (1967). *Meth. Enzy.* **12**, 228–239.
16. Irie, M. (1968). *J. Biochem.* **63**, 649–653.
17. Whitfeld, P. R., and Witzel, H. (1963). *BBA* **72**, 338–341.
18. Watanabe, H., Ando, E., Ohgi, K., and Irie, M. (1985). *J. Biochem.* **98**, 1239–1245.
19. Osterman, H. L., and Walz, F. G., Jr. (1979). *Biochem.* **18**, 1984–1988.
20. Arni, R., Heinemann, U., Tokuoka, R., and Saenger, W. (1988). *JBC* **263**, 15358–15368.
21. Steyaert, J., Opsomer, C., Wyns, L., and Stanssens, P. (1991). *Biochem.* **30**, 494–499.
22. Kostrewa, D., Choe, H. W., Heinemann, U., and Saenger, W. (1989). *Biochem.* **28**, 7592–7600.
23. MacKerell, A. D., Jr., Nilsson, L., Rigler, R., Heinemann, U., and Saenger, W. (1989). *Proteins (S.F.G)* **6**, 20–31.
24. Lenz, A., Cordes, F., Heinemann, U., and Saenger, W. (1991). *JBC* **266**, 7661–7667.
25. Eckstein, F., Schulz, H. H., Ruterjans, H., Haar, W., and Maurer, W. (1972). *Biochem.* **11**, 3507–3512.
26. Ruterjans, H., and Pongs, O. (1971). *EJB* **18**, 313–318.
27. Heinemann, U., and Saenger, W. (1982). *Nature* **299**, 27–31.
28. Nishikawa, S., Morioka, H., Kim, H. J., Fuchimura, K., Tanaka, T., Uesugi, S., Hakoshima, T., Tomita, K., Ohtsuka, E., and Ikehara, M. (1987). *Biochem.* **26**, 8620–8624.
29. Steyaert, J., Hallenga, K., Wyns, L., and Stanssens, P. (1990). *Biochem.* **29**, 9064–9072.
30. Grunert, H.-P., Zouni, A., Beineke, M., Quaas, R., Georgalis, Y., Saenger, W., and Hahn, U. (1991). *EJB* **197**, 203–207.
31. Takahashi, K., and Moore, S. (1982). "The Enzymes," Vol. 15, pp. 435–468.
32. Brownlee, G. G. (1972). In "Laboratory Techniques in Biochemistry and Molecular Biology," Vol. 3 (PtI), pp. 1–265. North-Holland/American Elsevier, Amsterdam.
33. Simoncsits, A., Brownlee, G. G., Brown, R. S., Rubin, J. R., and Guilley, H. (1977). *Nature* **269**, 833–836.

34. Donis-Keller, H., Maxam, A. M., and Gilbert, W. (1977). *NAR* **4**, 2527–2538.
35. Donis-Keller, H. (1980). *NAR* **8**, 3133–3142.
36. Miele, E. A., Mills, D. R., and Kramer, F. R. (1983). *JMB* **171**, 281–295.
37. Kew, O. M., Nottay, B. K., and Obijesky, J. F. (1984). *Meth. Virology* **8**, 41–84.
38. Takahashi, K. (1965). *JBC* **240**, 4117–4119.
39. Takahashi, K. (1985). *J. Biochem.* **98**, 815–817.
40. Kanaya, S., and Uchida, T. (1981). *J. Biochem.* **89**, 591–597.
41. Oobatake, M., Takahashi, S., and Ooi, T. (1979). *J. Biochem.* **86**, 65–70.
42. Pace, C. N., and Creighton, T. E. (1986). *JMB* **188**, 477–486.
43. Sugio, S., Amisaki, T., Ohishi, H., and Tomita, K. (1988). *J. Biochem.* **103**, 354–366.
44. Sugio, S., Oka, K., Ohishi, H., Tomita, K., and Saenger, W. (1985). *FEBS Lett.* **183**, 115–118.
45. Ikehara, M., and 11 coauthors (1986). *PNAS* **83**, 4695–4699.
46. Takahashi, K., Stein, W. H., and Moore, S. (1967). *JBC* **242**, 4682–4690.
47. MacKerell, A. D., Jr., Nilsson, L., and Rigler, R. (1988). *Biochem.* **27**, 4547–4556.
48. Mossakowska, D. E., Nyberg, K., and Fersht, A. (1989). *Biochem.* **28**, 3843–3850.
49. Quaas, R., McKeown, Y., Stanssens, P., Frank, R., Blocker, H., and Hahn, U. (1988). *EJB* **173**, 617–622.
50. Jervis, L. (1974). *Phytochemistry* **13**, 723–727.
51. Ishiwata, K., and Yoshida, H. (1978). *J. Biochem.* **83**, 783–788.
52. Fujimura, T., Tanaka, T., Ohara, K., Morioka, H., Uesugi, S., Ikehara, M., and Nishikawa, S. (1990). *FEBS Lett.* **265**, 71–74.
53. Waki, M., Mitsuyasu, N., Terada, S., Matsuura, S., Kato, T., and Izumiya, N. (1974). *BBRC* **61**, 526–532.
54. Kaiser, E. T., Mihara, H., Laforet, G. A., Kelly, J. W., Walters, L., Findeis, M. A., and Sasaki, T. (1989). *Science* **243**, 187–192.
55. Ding, J., Koellner, G., Grunert, H.-P., and Saenger, W. (1991). *JBC* **266**, 15128–15134.
56. Zegers, I., Verhelst, P., Choe, H.-W., Steyaert, J., Heinemann, U., Saenger, W., and Wyns, L. (1992). *Biochem.* **31**, 11317–11325.
57. Both, V., Moiseyev, G. P., and Sevcik, J. (1991). *BBRC* **177**, 630–635.

IV. SINGLE-STRAND-SPECIFIC ENDONUCLEASES

Some endonucleases hydrolyze both DNA and RNA in a highly single-strand-specific manner. These nucleases degrade single-strand polynucleotides as well as single-strand regions in double-strand polynucleotides. Depending on the species of origin of an enzyme (fungal, bacterial, yeast, plant, or animal), a nuclease can be more or less specific for ssDNA or ssRNA.

The ability to discriminate between double-stranded and single-stranded regions of a polynucleotide makes the single-strand-specific endonucleases invaluable in fine structural analysis of nucleic acids. This property also makes the single-strand-specific nucleases an important tool in various manipulations of DNA and RNA in recombinant DNA work.

This section describes two representative single-strand-specific endonucleases: *nuclease S1* from a fungus *Aspergillus oryzae* and *mung bean nuclease* from mung bean sprouts. The two enzymes are very similar in many respects: both are thermostable, zinc-dependent glycoproteins of similar size and they share most of the substrate specificities. The most pronounced difference that exists between the two nucleases is optimum pH: pH 4 for nuclease S1 and pH ~5 for mung bean nuclease.

A. Nuclease S1
[*Aspergillus* nuclease S1, EC 3.1.30.1]

Nuclease S1, also called *S1 nuclease*, is an endonuclease highly specific for single-strand polynucleotides. Nuclease S1 is a glycoprotein (~35 kDa) from *A. oryzae*. Nuclease S1 displays (ss)DNase, RNase, and 3'-phosphomonoesterase activities. The nuclease activities give rise to 5'-mononucleotides and a small amount of 5'-oligonucleotides as final products (Scheme 3.5).

$$5' \text{ (d or r)N-p-N-p-N-p-N-p-N-p-N } 3' \longrightarrow$$

$$\text{(d or r)NpN + pNpNpN + pN}$$

SCHEME 3.5

Nuclease S1 also cleaves dsDNA at the single-stranded region(s) caused by a nick, gap, mismatch, or loop. Nuclease S1 is one of the most powerful probes for the analysis of nucleic acid structure and is also an important tool in the manipulation of nucleic acids.

1. FUNCTIONS

a. Reaction conditions

i. Optimal reaction conditions. A typical reaction mixture contains 50 mM NaAc (pH 4.5), 1.0 mM ZnCl$_2$, 0.15 M NaCl, ssDNA (or RNA), and enzyme (100–200 U/ml). The mixture is brought to a final reaction temperature (below 55°C) before adding the enzyme. The reaction can be stopped by chilling and adding excess EDTA (pH 8.0), preferably followed by phenol/chloroform extractions.

The activity of nuclease S1 is subject to pH, temperature, and ionic strength as follows.

(a) pH. Nuclease S1 is optimally active at pH 4.0–4.3 with Zn^{2+} ions in acetate buffer (1,2). It shows half-maximal rates at pH 3.3 and pH 4.9. Higher pH values (4.6–5.0) have been used to avoid possible nicking of DNA due to acid depurination. It is essentially inactive at pH >6.0.

At neutral pH values, e.g., pH 7.5 (50 mM Tris–Cl), nuclease S1 turns out to be active with Mg^{2+} as the metal cofactor (38).

(b) Ionic strength. Nuclease S1 is optimally active in 0.1 M NaCl (3). In 0.4 M NaCl, the enzyme is 55% as active (1), indicating that it is relatively insensitive

to ionic strengths. However, the extent of hydrolysis has been shown to decrease at higher ionic strengths.

(c) **Temperature.** Nuclease S1 is active toward ssDNA even at 0–5°C, and can be used at temperatures ranging from 0° to 65°C. The single-strand nuclease activity increases 10-fold between 37° and 65°C (5). Note that high incubation temperatures increase the probability of heat-induced single strandedness of dsDNA, rendering it susceptible to rapid degradation (the reason to avoid the attempt to heat inactivate nuclease S1). The nonspecific double-strand nuclease activity is further accentuated at high enzyme concentrations (e.g., 200 U/ml) (4). Certain RNA segments in looped DNA · RNA hybrids are more efficiently cleaved at 45°C than at 20°C.

ii. **Cofactor requirements.** Nuclease S1 requires Zn^{2+} (0.01–1.0 mM) for full activity (1,6,7). Other divalent cations, e.g., Co^{2+} and Hg^{2+}, can replace Zn^{2+}, but enzyme activity is reduced. Mg^{2+} is nonstimulatory at around pH 4. The metal ion requirement of nuclease S1 changes at neutral pH, and the nuclease can be activated with Mg^{2+} (optimum 20 mM) (38). The nuclease activated with Mg^{2+} at pH 7.5 is 10 times less efficient than the enzyme activated with Zn^{2+} at acid pH. In neutral pH conditions, however, Mg^{2+} activation represents a 100-fold stimulation over Zn^{2+} activation.

iii. **Inhibitors and inactivators**

(a) **Mononucleotides.** Mononucleotides, e.g., dAMP and dATP, competitively inhibit nuclease S1 activity, causing 50% inhibition at 85 and 1 μM, respectively. 5'-AMP is also a competitive inhibitor. However, cAMP has no effect even at 10 mM. Product inhibition of the nuclease becomes significant when enough DNA is digested.

(b) **Buffers.** Phosphate (pH 4.6) and PP_i inhibit nuclease S1, causing 50% inhibition at 2 mM and 20 μM, respectively (7). Citrate also inhibits the enzyme (1).

(c) **Metal chelators.** Nuclease S1 is fully active in 1 mM EDTA (pH 6.8). The enzyme is inactivated in excess EDTA or by extensive dialysis in EDTA (1 mM) solution. The addition of Zn^{2+} can restore at least 70% of its original activity (1).

(d) **Denaturants.** Nuclease S1 is remarkably stable toward denaturing agents. It is fully active in SDS up to 0.6% (8). In 9 M urea–0.1% SDS (a condition sometimes used for cell lysis) at 45°C, the enzyme retains ~30% of its maximal activity in 9 M urea alone or 7% of the activity in aqueous solution without urea (9). The presence of proteins such as albumin, at a ratio of 0.6 (albumin/SDS, w/w), prior to the addition of nuclease renders the enzyme up to 80% active even in 2.4% SDS.

Nuclease S1 remains active in aqueous solutions containing greater than 50% formamide, 50% Me_2SO, 30% DMF, or 2% formaldehyde. However, the activity as well as the extent of hydrolysis is reduced roughly in proportion to the content of the organic solvent (10).

(e) Temperature. Nuclease S1 is relatively thermostable. The enzyme does not lose activity at 65°C in the presence of substrate (6). It is resistant to brief heating up to 75°C in pH 4.6 acetate buffer (1). At neutral pH, the enzyme without substrate is inactivated at 60–65°C.

b. Activity assay and unit definition

Nuclease activity is assayed by measuring the release of acid-soluble products from heat-denatured DNA (e.g., calf thymus). A typical assay mixture (0.5 ml) contains 30 mM NaAc (pH 4.6), 50 mM NaCl, 1 mM zinc acetate, 5% (v/v) glycerol, 0.5 mg/ml heat-denatured DNA, and enzyme. The assay mixture is incubated for 10 min at 37°C.

Unit definitions: One ssDNase (or RNase) unit is usually defined as the amount of enzyme that liberates 1 μmol of acid-soluble nucleotides per minute under the assay conditions.

One unit of phosphomonoesterase activity is defined as the amount of enzyme that liberates 1 μmol of inorganic phosphate per minute from 3′-AMP at pH 4.6 and 37°C.

c. Substrate specificities

i. Substrate and product profiles. Nuclease S1 hydrolyzes both ssDNA and ssRNA. It is two- to three-fold more active on ssDNA than on ssRNA. The products of digestion are 5′-mononucleotides and a small amount of 5′-di- and 5′-oligonucleotides.

ii. Cleavage preferences. Nuclease S1 shows little or no sequence specificity, as long as the single-strand polynucleotides exist in a state of random coils (11). Apparent variation of nuclease S1 activity with different substrates probably reflects the sequence-dependent variation of secondary structures under the given conditions. Phage fd ssDNA is hydrolyzed twice as fast as heat-denatured calf thymus DNA, and heating RNA for 2 min at 100°C doubles the RNase activity compared with DNase activity.

iii. Single-strand specificity. Nuclease S1 can recognize and cleave single-stranded regions of duplex DNA or RNA. Non-H-bonded regions [e.g., anticodon loop in tRNA (12)], partially denatured or locally altered regions containing nicks, gaps, and mismatches are cleaved but with varying efficiency.

Single base pair mismatches, particularly dA·dG or dG·dG, are inefficiently cleaved by nuclease S1 (13,14). The nuclease sensitivity increases gradually as the mismatching length increases from 1 to 6 nt. Yet the cleavage at 4-bp mismatches occurs incompletely, and the efficiency varies depending on mismatching sequences (4).

The action on ssDNA is at least 75,000 times higher than that on native λ DNA (1,7). High salt concentrations (e.g., 0.3 M NaCl) that tend to help maintain the double helical structure of DNA are useful in ensuring a higher single-strand specificity.

iv. Circular DNAs. Nuclease S1 can introduce nicks into circular superhelical dsDNA, which is later converted to linear molecules. When the content of superhelical turns (e.g., in ColE1 and SV40 DNA) gets lower than 30–40% native DNA, a circular DNA molecule undergoes a transition to a negative supercoil, producing a localized unwinding of helical base pairs and rendering them hypersensitive to nuclease S1 (15). The nuclease S1-induced transitions of supercoiled plasmids to nicked circular molecules and to linear molecules can be readily monitored by agarose gel electrophoresis. Indeed the transition reaction can be used to find an optimized reaction time under a given reaction condition.

The cleavage of SV40 DNA occurs at one or two sites within unpaired or weakly H-bonded regions, depending on the ionic strength of the media (16). The cleavage generates unit length linear duplex molecules. Nicked, circular SV40 DNA can be cleaved at the opposite strand at or near the nick to yield linear molecules. Lower temperatures (0–20°C) can be effectively used to minimize the cleavage of dsDNA on the strand opposite a nick. (*Note: Neurospora* endonuclease does not cut the strand opposite the nick, whereas venom phosphodiesterase shows preferential activity toward it.)

v. Homopolymers. Under conditions of partial denaturation, e.g., in 40–50% DMF, a DNA insert cloned by dA · dT-homopolymeric tailing can be excised rather specifically by nuclease S1 (17). Regions of homopyrimidine · homopurine sequences, e.g., $d(TC)_n \cdot d(AG)_n$, under conditions of pH ≤6 and negative supercoiling are hypersensitive to nuclease S1 (and other single-strand-specific nucleases as well) (18,19). The S1 hypersensitivity is at least partially ascribed to the H-DNA conformation comprising a single-stranded region. In some other cases, S1 hypersensitivity shows a dependence on the homopyrimidine · homopurine length and its flanking sequences. The hypersensitivity thus correlates with certain aberrant base stacking and/or flexibility of the two strands (20). Probably for the same reason, nuclease S1 recognizes and cleaves, albeit inefficiently, the junction between contiguous right-handed B-DNA and left-handed Z-DNA (21).

vi. Phosphorothioate-containing DNAs. Oligodeoxynucleotides (from 15- to 28-mer) that contain two to five phosphorothioate groups at both 5′ and 3′ ends are digested by nuclease S1 2 to 45 times more slowly than normal phosphodiester oligodeoxynucleotides (22).

vii. Nuclease-associated 3′-phosphomonoesterase. Nuclease S1 has an associated phosphomonoesterase activity (mostly 3′-nucleotidase). It catalyzes the hydrolysis of 3′-AMP (and 3′-P or 3′,5′-ADP) and, to a lesser degree, 2′-AMP (23); 5′-AMP is not hydrolyzed. The 3′-nucleotidase activity is maximal at pH 6–7.

2. APPLICATIONS

For most of the applications requiring single-strand-specific endonucleases, both nuclease S1 and mung bean nuclease (nuclease MB) can be used interchangeably. Compared with mung bean nuclease, nuclease S1 may be generally regarded as more single strand specific. A distinct disadvantage of nuclease S1 has been that the enzyme is active only at low pH. The finding that, in the presence of Mg^{2+}, nuclease S1 is active at neutral pH significantly enlarges the utility of the nuclease not only in the manipulation of single-stranded termini but also in the studies of DNA and RNA structures and their interactions with proteins.

i. Cleavage of single-stranded termini. Duplex DNAs with either 3' or 5' protruding ends can be made blunt-ended by treatment with nuclease S1 or nuclease MB. The reaction is usually carried out at or below 25°C. Depending on reaction conditions, however, fully blunt-ended duplex molecules may coexist with equal or greater amounts of imperfect duplexes with +1 and +2 nucleotides. Nevertheless, the single-strand-specific nucleases are particularly useful in some DNA (sub)cloning and sequencing strategies. Nuclease S1 or MB has been used in combination with *E. coli* exonuclease III (see Section V,A, this chapter) to generate a nonrandom, nested set of deletion clones (24).

When the ability of nucleases S1 and MB to trim the single-stranded portion of the cohesive termini of λ DNA was compared (25), nuclease S1 (at pH 4.5) made a complete cleavage of the 12-nt ends at 10°C as well as at 30°C without any digestion of the dsDNA. Nuclease MB (at pH 5.0) cleaved off the single-stranded ends completely at 30°C but, at 5°C, ~4 nt remained undigested even at high enzyme concentrations. Under both reaction conditions, nuclease MB introduced some nicks into dsDNA.

ii. Opening of hairpin loops. During the synthesis of dsDNA complementary to RNA, the second-strand DNA can be synthesized using *E. coli* DNA Pol Ik by self-priming of the first cDNA strand from the 3'-terminal hairpin structure (26). Nuclease S1 has been instrumental in cleaving the hairpin loop to generate dsDNAs suitable for subsequent ligation into a cloning vector. This method results in low yields of cDNA and has the drawback of sometimes losing critical information at the 5' end region of starting RNA. Improved cloning methods that do not use single-strand-specific nucleases are now available such as that of Gubler-Hoffman (27) and those that employ vector primers (28) (see Section III,A,f in Chapter 6).

iii. RNA Secondary structure analysis

(a) Structural fingerprints. Following the nuclease S1 digestion of RNA, products are separated by electrophoresis. The fragments that are resistant to nuclease S1 due to certain secondary structures display unique electrophoretic patterns (29). The stability of nuclease S1 in high concentrations of organic solvents, e.g., 50% DMF, makes the enzyme particularly suitable for use in the study of secondary structures of nucleic acids (10).

(b) Assay of DNA·RNA hybrids. After the nuclease S1 digestion of a DNA·RNA hybrid, the percentage of hybridization can be estimated by monitoring the materials adsorbed to DEAE-cellulose filters (30). The method is particularly suitable for multisample analysis and is more efficient than the conventional method of hydroxyapatite gel adsorption and elution.

(c) DNA/RNA mapping. DNA homoduplexes or DNA·RNA heteroduplexes can be prepared by hybridization of nucleic acids with defined, labeled DNA probes. The selective removal of nonhybridized and hence ssDNA regions in the hybrid structure of RNA allows sizing and mapping of mRNAs, identification of intervening sequences, and mapping of 5′ and 3′ termini of the gene encoding mature mRNA (31–34). A more elaborate analysis of RNA splicing utilizes a combination of RNase H/nuclease S1 to circumvent artifacts inherent in conventional nuclease S1 analysis (35), i.e., uncleaved mismatches, hypersensitive rU·dA sites, and S1 resistance due to simultaneous hybridization of two different RNA molecules on a single DNA probe molecule (36).

(d) Mutation analysis. Hybridization between mutant and wild-type DNAs and subsequent digestion with nuclease S1 can be used to locate the areas of deletion and mismatch (37).

3. STRUCTURE

Nuclease S1 is a monomeric enzyme (29 kDa) consisting of 267 amino acid residues. It is glycosylated at Asn-92 and Asn-228, giving an overall molecular size of ~32 kDa. Purified enzymes may show two major forms: a pI 3.53 form (28%) and a pI 3.67 form (69%). Nuclease S1 is apparently synthesized as a precursor containing an N-terminal 20-amino acid signal peptide. The mature enzyme has Trp and Ser at the N- and C-terminus, respectively.

A single His residue (His-60 or His-125) is essential for the catalytic activity (39). A Lys is involved in the substrate binding. Four of the five Cys form disulfide bonds. Cys and Tyr have no apparent roles in the catalytic function.

4. GENETICS

The nuclease S1 from *A. oryzae* is encoded by a single copy chromosomal gene (*nucS*) which contains two short introns of 49 and 50 nt (11). The gene has an ORF of 963 bp and codes for a protein (287 amino acids) comprising the 20-residue signal peptide. The 5′ NC region of the gene contains a putative TATA box at −79 nt position, two CAAT-like sequences at −268 and −342 nt positions, and two response sites for carbon catabolite repression. A polyadenylation signal (AATAAA) is located 282 nt downstream of the UGA stop codon.

5. SOURCES

Commercially available crude extracts from *A. oryzae*, e.g., α-amylase (Sigma) or Taka-diastase (Sankyo), are a convenient source for the purification of nuclease S1 (1,6). A purification procedure including Con-A- and phenyl-Sepharose chromatographies was shown to give a 1600-fold purification at 32% yield (23).

DEAE-cellulose and/or SP-Sephadex chromatographies have proven useful in removing contaminating RNase T1 (pI ~2.9) and RNase T2 (pI ~5.0). Nuclease S1 is stable when stored in 50% glycerol at −20°C.

Expression in *A. oryzae* (strain *niaD* 300) of the structural gene of *nucS*, cloned in pNGS plasmid and fused to the glucoamylase gene (*glaA*) promoter, gives a ~100-fold overproduction (or ~25 mg/liter) of secreted nuclease S1 (11).

References

1. Vogt, V. M. (1973). *EJB* **33**, 192–200.
2. Vogt, V. M. (1980). *Meth. Enzy.* **65**, 248–255.
3. Sutton, W. D. (1971). *BBA* **240**, 522–531.
4. Brookes, A. J., and Solomon, E. (1989). *EJB* **183**, 291–296.
5. Rushizky, G. W. (1981). *In* "Gene Applification and Analysis" (J. G. Chirikjian and T. S. Papas, Eds.), Vol. 2, pp. 205–215. Elsevier, New York.
6. Ando, T. (1966). *BBA* **114**, 158–168.
7. Wiegand, R. C., Godson, G. N., and Radding, C. M. (1975). *JBC* **250**, 8848–8855.
8. Shishido, K., and Ando, T. (1982). *In* "Nucleases" (S. M. Linn and R. J. Roberts, Eds.), pp. 155–185. CSHL, CSH, New York.
9. Zechel, K., and Weber, K. (1977). *EJB* **77**, 133–139.
10. Hutton, J. R., and Wetmur, J. G. (1975). *BBRC* **66**, 942–948.
11. Lee, B. R., Kitamoto, K., Yamada, O., and Kumagai, C. (1995). *Appl. Microbiol. Technol.* **44**, 425–431.
12. Harada, F., and Dahlberg, J. E. (1975). *NAR* **2**, 865–871.
13. Silber, J. R., and Loeb, L. A. (1981). *BBA* **656**, 256–264.
14. Dodgson, J. B., and Wells, R. D. (1977). *Biochem.* **16**, 2374–2379.
15. Shishido, K. (1980). *FEBS Lett.* **111**, 333–336.
16. Beard, P., Morrow, J. F., and Berg, P. (1973). *J. Virol.* **12**, 1303–1313.
17. Hofstetter, H., Schambock, A., van den Berg, J., and Weissmann, C. (1976). *BBA* **454**, 587–591.
18. Johnston, B. H. (1988). *Science* **241**, 1800–1804.
19. Pulleyblank, D. E., Haniford, D. B., and Morgan, A. R. (1985). *Cell* **42**, 271–280.
20. Evans, T., and Efstratiadis, A. (1986). *JBC* **261**, 14771–14780.
21. Singleton, C. K., Klysik, J., and Wells, R. D. (1983). *PNAS* **80**, 2447–2451.
22. Stein, C. A., Subasinghe, C., Shinozuka, K., and Cohen, J. S. (1988). *NAR* **16**, 3209–3221.
23. Olesen, A. E., and Sasakuma, M. (1980). *ABB* **204**, 361–370.
24. Henikoff, S. (1984). *Gene* **28**, 351–359.
25. Ghangas, G. S., and Wu, R. (1975). *JBC* **250**, 4601–4606.
26. Efstratiadis, A., Kafatos, F. C., Maxam, A. M., and Maniatis, T. (1976). *Cell* **7**, 279–288.
27. Gubler, U., and Hoffman, B. J. (1983). *Gene* **25**, 263–269.
28. Deininger, P. (1987). *Meth. Enzy.* **152**, 371–389.
29. Flashner, M. S., and Vournakis, N. N. (1977). *NAR* **4**, 2307–2319.
30. Maxwell, I. H., van Ness, J., and Hahn, W. E. (1978). *NAR* **5**, 2033–2038.
31. Berk, A. J., and Sharp, P. A. (1977). *Cell* **12**, 721–732.
32. Favaloro, J., Treisman, R., and Kamen, R. (1980). *Meth. Enzy.* **65**, 718–749.
33. Weaver, R. F., and Weissmann, C. (1979). *NAR* **7**, 1175–1193.
34. Berk, A. J. (1989), *Meth. Enzy.* **180**, 334–347.
35. Sisodia, S. S., Cleveland, D. W., and Sollner-Webb, B. (1987). *NAR* **15**, 1995–2011.
36. Lopata, M. A., Sollner-Webb, B., and Cleveland, D. W. (1985). *Mol. Cell. Biol.* **5**, 2842–2846.
37. Shenk, T. E., Rhodes, C., Rigby, P. W. J., and Berg, P. (1975). *PNAS* **72**, 989–993.
38. Esteban, J. A., Salas, M., and Blanco, L. (1992). *NAR* **20**, 4932.
39. Gite, S., Reddy, G., and Shankar, V. (1992). *BJ* **288**, 571–575.

B. Mung Bean Nuclease
[EC 3.1.30.1]

Mung bean nuclease, abbreviated here as *nuclease MB*, is a single-strand-specific endonuclease similar to nuclease S1. Nuclease MB is a glycoprotein with a similar molecular size (39 kDa) and the functional properties of the two enzymes are also similar. For example, they produce 5'-mononucleotide and 5'-oligonucleotide products. For most practical purposes, nuclease MB can be used interchangeably with nuclease S1.

1. FUNCTIONS

a. Reaction conditions

i. Optimal reaction conditions. A typical reaction can be carried out at 37°C in a buffer composed of 30 mM NaAc (pH 5.0), 50 mM NaCl, 0.1 mM ZnCl$_2$, 1 mM L-Cys, and 5% (v/v) glycerol. Triton X-100 [0.001% (v/v)] can be included in the reaction mixture, especially when the enzyme concentration is below 6.7–13 μg/ml, to reduce the loss of enzyme activity due to surface denaturation and/or adsorption on containers (glass). The reaction can be stopped by raising the pH to 9.2 and/or adding SDS (to 0.01%) and cooling on ice.

The optimal pH for nuclease MB lies between pH 4.7 and 5.3, and is dependent on the NaAc concentration (1). Although the enzyme is less active at higher pH values, reaction conditions at neutral pH (7.0–7.5) may be more useful and even necessary in some applications. At neutral pH, the single-strand specificity of the enzyme, in terms of the relative cleavage rate of supercoiled versus relaxed phage PM2 DNA, increases substantially over that at low pH conditions.

ii. Cofactor requirement. Nuclease MB requires Zn^{2+} for its activity (2). Other divalent metal ions cannot replace Zn^{2+}.

iii. Inactivators and/or stabilizers

(a) Metal chelator. Excess EDTA results in an inactivation of the enzyme that can be restored by the addition of Zn^{2+} (but not by other divalent metal ions). Exhaustive dialysis in EDTA results in complete inactivation, which is irreversible regardless of whether Zn^{2+} or other metal ions are added.

(b) Thiol reagents. The activity of nuclease MB is stabilized or restored in the presence of thiol reagents such as glutathione, cysteine, DTT, and 2-MSH at the 1 mM level, alone or in combination with Zn^{2+} (0.1 mM). L-Ser can replace L-Cys and eliminates the problem of DNA nicking that can occur in the presence of SH compounds during long incubations at pH 5 (3).

(c) Other factors. Nuclease MB is instantaneously and completely inactivated in 0.01% SDS (pH 5.0). The enzyme is relatively stable to heat treatment (60–70°C) at pH 5.0 in the presence of both Zn^{2+} and SH compounds (4). In the absence of Zn^{2+}, nuclease MB is stable at pH 7.5, whereas it is rapidly inactivated at pH 5.0 (2).

b. Activity assay and unit definition

The assay mixture (0.5 ml) contains 10 mM NaAc (pH 5.0), 50 mM NaCl, 0.1 mM zinc acetate, 1 mM L-Cys, 5% (v/v) glycerol, 0.5 mg/ml denatured (calf thymus) DNA, and the nuclease. The mixture is incubated for 10 min at 37°C.

Unit definition: One unit is defined as the amount of enzyme that hydrolyzes 1 μg of denatured DNA to acid-soluble products in 1 min at 37°C. This unit definition is far more convenient than the earlier ones (4,5).

c. Substrate specificities

i. Substrate and product profiles. Nuclease MB catalyzes the hydrolysis of single-strand polynucleotides. The digestion products are 5'-mono- and 5'-oligonucleotides. Single-stranded RNAs (yeast) are hydrolyzed at approximately equal rates as heat-denatured DNA (calf thymus) (6).

ii. Cleavage preferences. Nuclease MB exhibits some cleavage preferences for AT-rich regions: A|pN (60%) and T(U)|pN (30%) in ssDNA (5,7). λ dsDNA is also cleaved most rapidly in the centrally located AT-rich region. Because of their thermodynamic instability, AT-rich regions are where sharp bends or unusual structures occur. A large distortion or non-B-DNA conformation makes such sites hypersensitive to single-strand-specific endonucleases, sometimes resulting in single-strand nicks and/or double-strand cleavages with a preciseness close to that of the restriction endonuclease (8).

Removal of short 5' extensions, e.g., at the restriction enzyme cut sites of dsDNA, is most efficient when the desired blunt end contains a G·C base pair at its terminus, regardless of the sequence of the overhang. When the desired blunt end is terminated with an A·T base pair, trimming of the 5' overhang is difficult. When the trimming occurs, it generates various degrees of overdigestions that result in deletions (9).

When compared with nuclease S1 under comparable conditions with a given substrate (e.g., 12-nt single-stranded termini of λ DNA), nuclease MB is less active and less single strand specific than nuclease S1 (10).

iii. Cleavages at mismatching sites. Like nuclease S1, nuclease MB cleaves the regions of single or double mismatches (dA·dG, dG·dG) very inefficiently, whereas three or more unpaired nucleotides are cleaved efficiently at high enzyme concentrations (11).

iv. Single-strand specificity. The nucleolytic activity on ssDNA versus dsDNA varies widely depending on the type of DNA and the reaction conditions: at 37°C under optimal conditions, 30,000-fold for T4 DNA, 65-fold for crab $d(AT)_n$, less than 2-fold for synthetic $d(AT)_n$, and no detectable hydrolysis with dG·dC homopolymer (7).

(a) Effect of pH. Single strand specificity increases as the pH becomes more basic (1). The supercoiled DNA of phage PM2, which contains a single AT-rich

region at 0.75 map units (12), is cleaved ≥10,000 times faster than the relaxed topoisomer (4). The initial products are almost exclusively DNA molecules that contain a single nick per molecule (13). Otherwise, nuclease MB may inflict more than one nick per molecule, and the number of nicks depends on ionic environments and the number of energetically "flexible" or unwinding sequence units present on the target DNA. The nicked circular form is predominant at neutral pH (7.0), while both nicked circular and unit-length linear forms are obtained at optimal pH 5. Extensive digestion of PM2 DNA results in small-sized DNA fragments containing approximately one double-chain cut for every three single-chain scissions.

(b) Effects of ionic strength and temperature. Under the conditions that tend to destabilize dsDNA, i.e., low ionic strength and high temperature, dsDNA can be completely degraded from their termini with the continuous accumulation of mono-, di-, and trinucleotides (1). The variation in cleavage specificity arises mostly from the environmental factors that affect the conformation of dsDNA and not the enzyme itself (14).

With T7 and PM2 DNA as substrates, nuclease MB initially introduces nicks (3). However, nicks do not accumulate, and they reach a maximum of three per T7 DNA molecule at 30 min. At this point, the number of double-strand scissions increases at a constant rate, reaching a limiting value of ~40 per molecule.

v. **Endonuclease-associated 3'-phosphatase.** Nuclease MB possesses an intrinsic 3'-monophosphatase activity which dephosphorylates 3'-mononucleotides (15). A dinucleotide dNpNp can be dephosphorylated first to dNpN, which can then be cleaved to dN + pdN by nuclease activity. The 3'-monophosphatase hydrolyzes rNp 50- to 100-fold faster than the corresponding dNp.

2. APPLICATIONS

Refer to Section IV,A,2 on nuclease S1.

3. STRUCTURE

Mung bean nuclease has an estimated MW of 39,000 and calculated 334 amino acid residues (1). It is a glycoprotein and has a 29% carbohydrate content. Removal of the carbohydrate chain by endo-β-N-acetylglucosaminidase H reduces the apparent MW to 31,000, but does not significantly alter the enzyme activity (16). The enzyme contains one SH group and three disulfide bonds. The single SH group is apparently essential for activity. Nuclease MB has a high content (12.6 mole %) of aromatic amino acid residues.

Approximately 70% of the enzyme molecules have a peptide bond cleaved at a single region in the protein. The nicked form of the enzyme is fully functional. Reduction by 2-MSH and subsequent SDS-PAGE result in the separation of the two polypeptides (25 and 15 kDa). After the endoglycosidase treatment, the two fragments separate as 18.7- and 12.5-kDa protein bands, respectively.

4. GENETICS

No genetic information is currently available for nuclease MB.

5. SOURCES

Mung bean nuclease is obtained from mung bean sprouts (5). Nearly homogeneous enzymes have been prepared by using various purification procedures (2,17).

References

1. Kroeker, W. D., Kowalski, D., and Laskowski, M., Sr. (1976). *Biochem.* **15**, 4463–4467.
2. Kowalski, D., Kroeker, W. D., and Laskowski, M., Sr. (1976). *Biochem.* **15**, 4457–4463.
3. Kroeker, W. D., and Kowalski, D., (1978). *Biochem.* **17**, 3236–3243.
4. Kowalski, D., and Sanford, J. P. (1982). *JBC* **257**, 7820–7825.
5. Sung, S. C., and Laskowski, M., Sr. (1962). *JBC* **237**, 506–511.
6. Johnson, P. H., and Laskowski, M., Sr. (1968). *JBC* **243**, 3421–3424.
7. Johnson, P. H., and Laskowski, M., Sr. (1970). *JBC* **245**, 891–898.
8. McCutchan, T. F., Hansen, J. L., Dame, J. B., and Mullins, J. A. (1984). *Science* **225**, 625–628.
9. Hammond, A. W., and D'Alessio, J. M. (1986). *FOCUS (BRL)* **8**(4), 4–6.
10. Ghangas, G. S., and Wu, R. (1975). *JBC* **250**, 4601–4606.
11. Wells, R. D., Blakesley, R. W., Hardies, S. C., Horn, G. T., Larson, J. E., Selsing, E., Burd, J. F., Chan, H. W., Dodgson, J. B., Jensen, K. F., Nes, I. F., and Wartell, R. M. (1977). *Crit. Rev. Biochem.* **4**, 305–340.
12. Sheflin, L. G., and Kowalski, D. (1984). *NAR* **12**, 7087–7104.
13. Wang, J. C. (1974). *JMB* **87**, 797–816.
14. Kowalski, D. (1984). *NAR* **12**, 7071–7086.
15. Mikulski, A. J., and Laskowski, M., Sr. (1970). *JBC* **245**, 5026–5031.
16. Trimble, R. B., and Maley, F. (1977). *BBRC* **78**, 935–944.
17. Laskowski, M., Sr. (1980). *Meth. Enzy.* **65**, 263–276.

V. DNA EXONUCLEASES

A. 3' → 5'-Exonuclease: *E. coli* Exonuclease III

[Exodeoxyribonuclease III (*E. coli*), EC 3.1.11.2]

Exonuclease III (Exo III) of *E. coli* is a monomeric multifunctional enzyme (31 kDa) that catalyzes the hydrolysis of at least four different types of phosphoester bonds in dsDNA (Fig. 3.4). The main enzymatic activity of Exo III is the 3' → 5'-exonuclease activity that carries out the successive release of 5'-P-mononucleotides from the 3' ends of dsDNA. The second activity is the DNA 3'-phosphatase activity that hydrolyzes 3'-terminal phosphomonoesters. In fact, Exo III was initially discovered as a DNA 3'-phosphatase in *E. coli* (1,2). Exo III has a third activity which degrades the RNA strand in a DNA·RNA heteroduplex, thus the RNase H activity. The fourth activity of Exo III is an AP endonuclease which cleaves phosphodiester bonds at apurinic or apyrimidinic sites.

Exo III is an invaluable tool in recombinant DNA technology because of its exonucleolytic activities that can be used for unidirectional or bidirectional synchronous digestions of dsDNA at controlled rates.

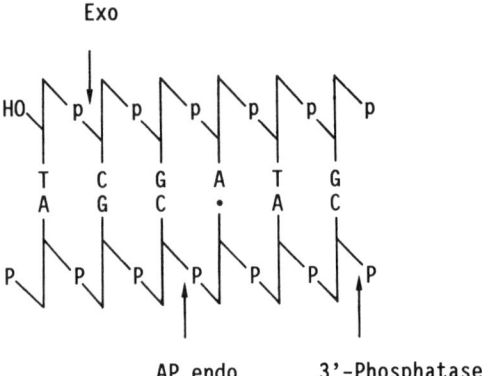

FIGURE 3.4 Reaction specificities of *E. coli* Exo III on dsDNA. Exo III exhibits three types of hydrolytic activities on dsDNA substrates: Exo (for 3' → 5'-exonuclease), 3'-phosphatase, and AP endo (for apurinic/apyrimidinic/urea endonuclease).

1. FUNCTIONS

a. Reaction conditions

i. Optimal reaction conditions. A typical reaction is carried out in 50 μl containing 50 mM Tris–Cl (pH 8.0), 10 mM $MgCl_2$, 1 mM DTT, 2 μg DNA, and 10 U of enzyme. The reaction mixture is incubated at 37°C for a desired length of time (1–30 min). The reaction is stopped by heating the mixture to 75°C for 10 min and/or by adding excess EDTA (20 mM final concentration).

Under the reaction conditions, one unit of Exo III removes ~200 nt from the 3' ends of 1 μg dsDNA (5 kbp) in 10 min. The rate of Exo III-catalyzed hydrolysis varies depending on reaction conditions, such as buffer component, ionic strength, and temperature, but not on substrate DNA (see below). The degradation of DNA by Exo III is more nonprocessive at 37°C than at 23°C. Nevertheless, a more synchronous digestion of both strands of DNA has been observed at 23°C, when the number of the nucleotides removed is within 750 nucleotides per end.

ii. pH Optimum. The pH optimum for *exonuclease* activity is 7.6–8.4 in Tris–Cl with 0.7 mM $MgCl_2$. With $MgCl_2$ close to its optimal concentration (10 mM), the pH optimum is 7.0–7.4 (2). The optimum pH for *phosphatase* activity is also ~7.0 in 50 mM K-P or Tris–maleate buffer (1). The optimum pH for *AP endonuclease* activity is 8.5 in 50 mM Tris–Cl and 10 mM $MgCl_2$ (3).

iii. Divalent cations. The *exonuclease* requires Mg^{2+} (2–10 mM) for optimal nucleolytic activity (4). Mn^{2+} is as effective as Mg^{2+} in the range from 0.2 to 0.7 mM (2). Mn^{2+} (10 mM) can replace Mg^{2+} with ~59% efficiency. The exonuclease activity shows a sharp optimum with 2 mM Ca^{2+}. With 10 mM Ca^{2+}, both exo- and endonuclease activities are inhibited more than 90%.

The *phosphatase* activity is maximal with 10 mM MgCl$_2$ (1). The AP *endonuclease* activity is optimal at 2–10 mM Mg^{2+} (4). Zn^{2+} (33 mM) inhibits Exo III (both exonuclease and phosphatase) by ~90%.

iv. Inhibitors and inactivators

(a) *NaCl.* Exo III activity decreases almost linearly with increasing NaCl concentration at 37°C. In unidirectional digestions, the rate decreases from ~400 nt/min in the absence of NaCl to 125 nt/min in 100 mM NaCl (5). In fact, NaCl can be used as a factor controlling the digestion rate of Exo III. At 5°C in Tris buffer (67 mM, pH 7.8) containing 70 mM or higher concentrations of NaCl, hydrolysis stops after the initial removal of 6 or 7 nt from a terminus and the enzyme apparently remains bound to DNA. The addition of more enzyme leads to the removal of another 6 or 7 nt by each newly bound enzyme molecule (6). Similar stepwise hydrolysis can be achieved by alternating temperature, for example, from 5° to 37°C and vice versa, and incubating for a specified length of time. At lower NaCl concentrations, an initial burst of hydrolysis is followed by a slow, constant rate of hydrolysis.

The inhibitory effect of NaCl suggests that, to use Exo III after a restriction enzyme digestion(s), DNA should be precipitated with ethanol and the pellet should be rinsed briefly with cold 70% (v/v) ethanol to remove as much NaCl as possible.

(b) *Other factors.* Exo III is inhibited by a Mg^{2+} chelator (e.g., EDTA), and is severely inhibited in K-P buffer. The enzyme is inactivated by PCMB (50–90% at 0.1 mM). Sulfhydryl compounds (2-MSH and DTT) stabilize the enzyme during long incubations. Exo III is rapidly inactivated by heat in the absence of substrate DNA with a half-life of 5–7 min at 37–40°C (2,3).

b. Activity assays and unit definitions

i. **Exonuclease assay.** The exonuclease assay measures the release of acid-soluble mononucleotides from uniformly radiolabeled DNA, e.g., phage T7 [^{3}H]DNA sonicated in the presence of unlabeled salmon sperm DNA. The assay mixture contains 50 mM Tris–Cl (pH 8.0), 1 mM DTT, 10 mM MgCl$_2$, and 0.1 mM (nucleotide equivalent) ^{3}H-labeled DNA (400–500 cpm/nmol).

Unit definition: One "exonuclease unit" is the amount of enzyme that releases 1 nmol of TCA-soluble nucleotides in 30 min at 37°C (2).

ii. **DNA 3′-Phosphatase assay.** The assay measures the release of Norit (activated charcoal)-nonadsorbable ^{32}P from the 3′-P-terminated DNA (1). The 3′-P DNA is prepared by partial micrococcal nuclease digestion of uniformly ^{32}P-labeled native DNA [e.g., *E. coli* or phage T7 (7)], followed by exhaustive dialysis to remove small oligonucleotides. The incubation mixture consists of 50 mM Tris–Cl (pH 7.0), 10 mM MgCl$_2$, 1 mM 2-MSH, 0.17 mM [3′-^{32}P]DNA, 60 μM yeast tRNA, and 0.04 to 0.4 U enzyme. After 30 min at 37°C, the reaction is stopped by adding ZnCl$_2$ (to 0.1 mM) and then chilling. The amount of [^{32}P]P$_i$

released is determined following TCA (10%) precipitation (1). The precipitates are removed by centrifugation (12,000 g for 5 min at 4°C), the supernatant is filtered through a Norit suspension, and the radioactivity of the supernatant is counted.

Unit definition: One "phosphatase unit" catalyzes the release of 1 nmol of acid-soluble, Norit-nonadsorbable [^{32}P]P$_i$ in 30 min.

iii. AP Endonuclease assay. This assay measures the nicking of partially depurinated supercoiled DNA (e.g., ϕX174 RFI) and consequent conversion to the open circular form (RFII) (8). The reaction mixture contains 33 μM freshly prepared DNA (18,000 cpm/nmol, 2–3 apurinic sites per molecule), 5 mM MgCl$_2$, 10 mM 2-MSH, 0.1 mg/ml BSA, 50 mM Tris–Cl (pH 8.0), and sufficient enzyme to cleave 5 to 20% of the substrate molecules. After 10 min at 37°C, an alkaline denaturation buffer (pH 12.0) is added to the reaction mixture. The mixture is then filtered through a NC membrane filter which allows passage of the rapidly renaturable RFI molecules while the denatured RFII DNA is adsorbed on the filter. The relative amount (in radioactivity counts) of retained material is used to calculate the number of nicks, assuming a Poisson distribution: $P(0) = e^{-r}$, where r is the average number of nicks per molecule and $P(0)$ is the relative amount of closed circular DNA remaining.

Unit definition: One "endonuclease unit" is the amount of enzyme that catalyzes the hydrolysis of 1 nmol of the phosphodiester bond in 30 min.

c. Substrate specificities: exonuclease

i. Double-strand specificity. Exo III is a dsDNA-specific $3' \rightarrow 5'$-exonuclease which sequentially releases 5'-mononucleotides as products. Double-stranded rAU copolymers are not effective substrates; the reaction proceeds at less than 0.1% of the rate observed with native DNA. RNA·DNA heteroduplexes or mixed polymers, such as a DNA with interspersed rC, are also cleaved by Exo III, although at reduced rates. Exo III can liberate a 5'-ribonucleotide when it terminates a DNA chain (2). Single-stranded DNA is not a substrate, although certain ssDNAs that form intra- or interstrand H-bonds are cleaved at 25–30% of the rate for dsDNA. The addition of ssDNA to "standard" reaction mixture produces no detectable inhibition. Single-stranded RNAs or rRNAs are not substrates.

ii. Base preference. Exo III hydrolyzes DNA with varying structures (e.g., *B. subtilis* DNA, *E. coli* DNA, and synthetic dAT copolymer) at similar rates (2). Although the rate of digestion is not greatly affected by base composition, there is some recognizable sequence dependence in the order of C > A ~ T > G (9). As a result, different termini may be degraded at different rates, especially on DNAs having biased nucleotide compositions. Nevertheless, the rate difference between C and G is a maximum of 3, and no composition effect extends beyond the one nucleotide to be cleaved.

iii. Extent of digestion. Partial digestion of dsDNA produces a duplex with protruding 5' tails. As the digestion proceeds, the length of the duplex region becomes progressively shorter. The remaining base-paired region eventually becomes too short to sustain a bihelical structure and it separates into single strands that are resistant to further digestion. The maximum extent of digestion of a natural dsDNA is 35–45%, and this decreases as the incubation temperature increases.

Duplexes composed of homopolymers $(dA)_n \cdot (dT)_n$ or of alternating copolymers $d(AT)_n$ can be digested almost completely (2), due to the unique property of such polymers in which the strands tend to slip or to creep over each other to maintain a maximum base-paired stability. At 25°C, Exo III can hydrolyze such strands up to the last 5'-terminal dinucleotide; at higher temperatures, longer 5'-terminal oligonucleotides are obtained as the limiting products (10).

iv. **Processivity and synchrony.** Exo III digests dsDNA in either a processive or a nonprocessive manner depending on temperature and divalent metal ions. At 23°C, Exo III removes at least 100 nt before it dissociates from the chain terminus (11). At 37°C, however, less than 50 nt are removed processively and thus Exo III behaves more like a nonprocessive enzyme (12).

When the DNA termini are saturated with excess enzyme for the digestion of ~300 nt at 23°C, the synchrony is as high as ±25 bases (11). A higher degree of synchrony is obtained at 23°C than at higher temperatures, such as 28° or 37°C, especially when the number of nucleotides removed is within 750 bases per end. In a deletion of over 1000 bases at 37°C, the termini dispersed over may be several hundred bases (11,13).

v. **Structure of 3' termini.** Exo III attacks DNA strands terminating in 3'-OH. The enzyme can also degrade the DNA strands terminating in 3'-P, apparently due to the prior action of 3'-phosphatase activity. Exo III can initiate hydrolysis internally at nicks on dsDNA at a rate comparable to that of the degradation from the termini of linear molecules (14). The hydrolysis at a nick is negligible at low temperatures and moderate salt concentrations, i.e., the conditions that favor base stacking of the residues holding the nick (6).

Exo III attacks the 3' terminus which is composed of either one or two paired or mismatched ribonucleotides (10). Exo III digests dsDNA from the 3' end when it contains up to three mispaired terminal nucleotides (15). However, Exo III fails to initiate digestion at the 3' ends of dsDNA terminating in 4 base protrusions (e.g., *Kpn*I and *Sst*I sites), although there may be some exceptions [e.g., *Pst*I site (16)]. This characteristic provides a useful means of digesting a dsDNA unidirectionally (13).

vi. **Altered or modified substrate structures.** Exo III can neither hydrolyze α-thiophosphate-containing phosphodiester bonds nor bypass them (17,18). The filling in of 5' cohesive ends using polymerase and dNTPαS and generation of one thiophosphate-containing 3' terminus thus provides another strategy to digest unidirectionally from the other 3' terminus.

DNAs containing glucosylated bases (usually at the 5-hydroxymethyl-C in T-even phages) are resistant to exonucleolytic activity, although they are substrates for AP endonuclease and phosphatase activities of the same enzyme (19).

Halogenated bases do not seem to affect the exonucleolytic activity of Exo III. This property has been used to increase the exposure of halogenated pyrimidines (e.g., Br-dU and I-dU) in ssDNA for immunochemical staining (20).

d. Substrate specificities: AP endonuclease

The endonuclease activity of Exo III produces single-strand breaks on the 5' side of the apurinic sites giving 3'-OH and 5'-P termini, just as the enzyme cleaves at the 5' side of the 3'-terminal nucleotide. The apurinic site is left as a base-free deoxyribose 5-P end group in the DNA molecule.

The AP endonuclease was previously called endonuclease II (21) or endonuclease VI (3). Apurinic sites introduced by acid or MMS treatment are substrates even after reaction with $NaBH_4$ (3), a treatment that reduces the free aldehyde groups on the base-free sugar residues and renders their 3'-phosphoesters stable to alkali. Apyrimidinic sites, which may be specifically generated by the enzymatic removal of U residues from U-containing DNA in a *dut* (dUTPase) mutant, are also cleaved by the enzyme (22). The endonuclease is also able to catalyze the hydrolysis of phosphodiester bonds 5' to the urea N-glycoside and other O-alkylhydroxylamine residues linked by a C-1'–N-1 (secondary amine) glycosidic bond in dsDNA (4).

e. Substrate specificities: DNA 3'-phosphatase

Double-stranded DNAs with 3'-P termini (e.g., the digestion products of micrococcal nuclease or calf spleen nuclease) are substrates for Exo III (1). The phosphatase activity releases inorganic P_i. In contrast, DNAs with 5'-P termini (e.g., the digestion products of *E. coli* endonuclease) are not substrates. Mononucleotides with 3'-P or 5'-P and small (or acid-soluble) 3'-P oligonucleotides such as d(TTT)p are not substrates. Exo III can also hydrolyze 3'-P glycolate ($-O-PO_2-OCH_2-COOH$) to 3'-OH in gamma-irradiated DNAs (23).

Ribosomal RNAs with 3'-P termini are not substrates. However, the enzyme attacks DNA with a ribonucleotide 3'-P terminus. After removal of the 3'-terminal phosphates, DNA or polynucleotides become natural substrates for the 3' → 5' exonuclease activity.

f. Substrate specificities: RNase H activity

Exo III degrades not only the DNA strand(s) in duplex DNA and DNA·RNA heteroduplexes, but it also degrades the RNA strand in heteroduplexes. Poly(rA·dT) is degraded at 20% of the rate for poly(dA·dT) (24). The enzyme has no measurable activity on either homopolymer strand in the absence of the other. The products of RNase H activity are a mixture of 5'-mono- and 5'-oligonucleotides, although 5'-rAMP is the major product (25). This suggests that the RNase H activity of Exo III is both exonuclease and, to some extent, endonuclease.

When the oligo(dT) in oligo(dT)·poly(dA) duplexes is labeled at the 3' ends with ribonucleotides, Exo III can cleave a single ribonucleotide or two consecutive ribonucleotides regardless of whether the ribonucleotides are base paired or mismatched, and at a rate similar to the hydrolysis of 3'-terminal deoxynucleotides (10).

g. Exonuclease reaction mechanism

i. Recognition of substrate. Exo III is thought to recognize substrate structure and catalyze the hydrolysis of phosphomonoester or phosphodiester bonds by a mechanism involving three-domain interactions in the active center (21,26). The first domain (or catalytic site) recognizes and cleaves the phosphoester bonds on the 3' side. A second domain recognizes the DNA duplex structure having recessed or blunt 3' ends or the 3' ends with up to three bases protruding. The third domain recognizes an inter-DNA space: the space beyond the end of a strand, created by unwinding of the 3'-terminal base in duplex, or previously occupied by a base at an AP site.

Based on the observed endonucleolytic activity on the phosphodiester bond 5' to the glycoside (e.g., urea and O-alkylhydroxylamine residues), a modified mechanism has been proposed (4). The modified mechanism relies on two factors regarding the substrate susceptibility: (i) the absence of base pairing and (ii) the presence of a secondary amine at the C-1'–N-1 glycosidic linkage. The secondary amine is presumed to facilitate the ring opening of the deoxyribose and to remove the steric hindrance via a configurational change, allowing the Exo III ready access to the scissile 5'-P linkage.

ii. Processivity. The exonucleolytic activity and processivity of Exo III are dependent on temperature and divalent metal ions. The temperature dependence (i.e., Arrhenius plot) of the activity in the presence of Mg^{2+} is discontinuous; the transition occurs at 25°C (4). In contrast, no discontinuity is observed in the presence of Ca^{2+}. (*Note:* The AP endonuclease activity shows no discontinuity in the presence of either Ca^{2+} or Mg^{2+}.) These data seem to provide a reasonable explanation on the previously observed temperature dependence of processivity; at temperatures below 25°C, Exo III has a high activation-energy conformation which binds tightly to the DNA substrate and displays high processivity. At high temperatures, Exo III assumes another conformation which affects the association/dissociation with the DNA so that it is no longer rate limiting and thus the enzyme becomes nonprocessive. With this conformer, the RDS is presumed to be the cleavage of the phosphodiester bond. In the presence of Ca^{2+}, Exo III apparently locks into the high activation-energy "processive" conformation.

2. Applications

The possibility of digesting the 3' termini of dsDNA either unidirectionally or bidirectionally at a controlled rate makes Exo III an extremely versatile and powerful tool in tailoring the size of DNA fragments. The following examples illustrate the basic principles and features of some of the wide-ranging applications of Exo III.

i. Applications in DNA sequencing

(a) Sequencing with Exo III-digested templates. Single-stranded DNA templates suitable for primed dideoxy DNA sequencing can be prepared by Exo III digestion of dsDNA (27). The repair synthesis in the presence of dideoxynucleotides with the DNAs that have been submitted to partial, controlled exonucleolytic digestion provides *sequence information* in an ordered manner (11,28).

An alternative, perhaps more advanced, form (called ExoMeth) of Exo III-based sequencing strategies is to incorporate 5-methyl-dCTP in place of dCTP during the repair synthesis. Digestion of the (5-Me-dCMP)-containing DNA with suitable "frequent cutter" restriction enzymes sensitive to the m^5C gives constant 5'-end points that allow one to obtain far more sequence information (up to 10 kb) from a nested set of Exo III-digested templates (29).

Another interesting application of Exo III for DNA sequencing utilizes the high resistance of thiophosphate bonds to degradation by the enzyme (30). In this application, the repair synthesis is carried out, for example, in the presence of dATPαS so that the thionucleotide is incorporated randomly at the A nucleotide positions. Similar reactions are performed using dCTPαS, dGTPαS, and dTTPαS. Subsequent Exo III digestions of the second strands result in a nested set of fragments with a pattern similar to that obtained from dideoxynucleotide chain-termination sequencing.

(b) Sequencing with Exo III-generated deletion mutants. Nonrandom sequencing of long stretches of DNA can be most efficiently performed using DNA templates from a nested set of deletion mutants. The mutant clones are generated by either bidirectional (31) or unidirectional Exo III digestion, followed by removal of the single-stranded portion of 5' ends using nuclease S1 (or mung bean nuclease or Bal31S) and blunt-end recircularization.

Unidirectional digestions can be performed using one of four available methods.

The *first method* is based on the observation that Exo III cannot digest 4-base protruding 3' termini (13). This method requires the digestion of recombinant plasmids with two restriction enzymes which differentially expose the 3' termini, one blunt or recessed terminus which is susceptible to Exo III digestion, and the other 4-base protruding terminus which is resistant to Exo III digestion. However, Exo III exhibits some sequence dependence and may thus digest some 3'-protruding 4-base termini such as those generated by *Apa*I (GGCC-3') and *Pst*I (TGCA-3'), but not by *Sph*I (CATG-3') (16).

The *second method* relies on the protection of one end by incorporating Exo III-resistant α-thio-substituted nucleotides into the 3' end of the strand complementary to the 5' overhang (17).

The *third method* utilizes the strong DNA:protein interaction that blocks the Exo III digestion. Introduction of a *lacO* sequence between the primer binding site and MCS of an M13 sequencing vector efficiently protects the end containing the *lacO* sequence from the Exo III digestion in the presence of the *lac* repressor while the other end is open to sequential deletions (32). Because this method depends on the strength of the *lacO:lac* repressor binding, it is useful to include ONPF (*o*-nitrophenyl-β-D-fucoside, an anti-inducer) which increases the half-life of repressor binding to 50 min. The mixture for Exo III digestion contains 50 mM Tris–Cl (pH 8.0), 1 mM $MgCl_2$, and 2.5 mM ONPF in 5% (v/v) Me_2SO.

The *fourth method* of generating deletion clones starts with a single-stranded phagemid template (33). The annealed primer is first extended using T4 DNA Pol, which leads to a fully double-stranded circular molecule with a nick or small gap just 5′ to the primer. The 3′ end of the primer extension is then progressively digested with Exo III as a function of time. The resulting single-strand region of the template strand is digested with a single-strand-specific endonuclease, generating a nested set of deletion molecules that can be circularized and used to transform competent cells.

ii. **Oligonucleotide-directed mutagenesis using phosphorothioate substitution.** Refer to Appendix A: Site-specific *in vitro* mutagenesis.

iii. **Polymerase repair synthesis.** Limited Exo III digestion of dsDNA leaves a 3′-OH terminus that can be used as the starting site for primed DNA synthesis by DNA polymerases. Such partially digested DNA molecules are excellent substrates for assaying DNA polymerase activities. The partially digested DNA molecules are also useful for the following applications.

(a) Localized mutagenesis. During the polymerase-catalyzed repair synthesis, mutagenic base analogs can be introduced to generate mutant sequences.

(b) DNA labeling. When radioactive nucleotides are incorporated during the DNA repair synthesis, the repair synthesis provides radiolabeled hybridization probes.

iv. **Applications in analytical methods**

(a) Detection of AP sites in DNA. The Exo III-associated AP endonuclease activity can be used to detect AP sites. The cleavage product has a base-free deoxyribose 5-P group at its new 5′ end.

(b) Mapping of mutation sites in DNA. Mutations can be induced into DNA by photochemical reactions with 4,5′,8-trimethylpsoralen. Extensive digestion of such mutated DNAs with Exo III leads to a final point where two single strands from different halves are joined by interstrand cross-linkage (34). Photoreversal of such cross-links is used to determine the hyperactive sites (e.g., B–Z-DNA junction) at the nucleotide level.

3. STRUCTURE

E. coli exonuclease III is a single polypeptide. The amino acid sequence deduced from a cloned gene contains 268 residues (M_r 30,921) (35); the N- and C-terminal residues are Met (the initiator) and Arg, respectively. The enzyme contains 3 Cys residues. The average hydropathy (−0.55) and the content of charged amino acid residues (Asp + Glu + Arg + Lys = 27.25 mol%) are typical of a soluble protein.

Exo III is a globular protein with $s^{\circ}_{20,w}$ of 2.92S.

4. GENETICS

i. In vivo role of Exo III gene and mutant phenotypes. *E. coli* contains, in addition to the exonucleases which are associated with DNA polymerases, a number of exo- and endo-DNases with varying but often overlapping substrate specificities.

The structural gene for Exo III is *xthA* (7,36), which is located at the 38-min position on the *E. coli* map. A number of *xth*$^-$ mutants (*ts* and deletion) have been isolated, all of which simultaneously affect the AP endonuclease, 3'-phosphatase, and exonuclease activities.

Phenotypic analysis of the mutants suggests that the gene product plays a role in multiple functions such as the repair of damaged or alkylated DNAs. However, none of the mutations at the single *xth* gene are conditionally lethal. The *xth* mutants are extremely sensitive to hydrogen peroxide and UV irradiations (37,38) and are unable to express a heat-shock response (39).

Wild-type *E. coli* K12 is estimated to contain ~3500 Exo III molecules, which are responsible for ~85% of the total cellular exonuclease activity, over 90% of the total 3'-phosphomonoesterase activity, and more than 80% of the total cellular AP endonuclease activity.

ii. Gene organization. The *xth* gene of *E. coli* K12 has been cloned as a 1.4-kbp *Hind*III–*Bam*HI fragment and the nucleotides sequenced (35). The gene consists of 269 codons including the initiator ATG and terminator TAA. The gene contains its own consensus promoter sequences, i.e., −35 (TTGACA), −10 (TCAAAT or TACCAT), and potential RBS (ATGG). Immediately downstream of the *xth* gene is a dyad symmetry sequence of 26 nt which can form a hairpin structure in RNA characteristic of a ρ-dependent terminator. These observations indicate that Exo III is expressed monocistronically. Nevertheless, there is some evidence that the expression of the *xth* gene may be regulated by a trans-acting element associated with a *katF* gene product (44 kDa) (37). The *katF* gene is a part of the hydroxyperoxidase gene operon that is positioned about 1 min distance from the *xth* gene.

5. SOURCES

The Exo III gene (*xthA*) of *E. coli* has been cloned in a pBR322-derived vector, pKC16, which contains a thermolabile repressor gene (cI857) from λ phage (40,41). Expression of the cloned gene (pSGR3) in *E. coli* (BE257) results in a 120-fold overproduction of the Exo III enzyme compared to the yield from a wild-type *E. coli*, or ~3% of the protein in the cell extract. From this source, Exo III has been purified to near homogeneity (160 U/μg protein) (24).

Exo III is relatively stable at 0–4°C, although the stability of diluted enzymes may vary depending on the extent of dilution and handling. Diluted enzymes can be stored at −20°C either frozen or in a 50% glycerol solution for several weeks without recognizable loss of activity.

References

1. Richardson, C. C., and Kornberg, A. (1964). *JBC* **239**, 242–250.
2. Richardson, C. C., Lehman, I. R., and Kornberg, A. (1964). *JBC* **239**, 251–258.
3. Gossard, F., and Verly, W. G. (1978). *EJB* **82**, 321–332.
4. Kow, Y. W. (1989). *Biochem.* **28**, 3280–3287.
5. Tomb, J. F., and Barcak, G. J. (1989). *BioTechniques* **7**, 932–933.
6. Donelson, J. E., and Wu, R. (1972). *JBC* **247**, 4661–4668.
7. Milcarek, C., and Weiss, B. (1972). *JMB* **68**, 303–318.
8. Clements, J. E., Rogers, S. G., and Weiss, B. (1978). *JBC* **253**, 2990–2999.
9. Linxweiler, W., and Horz, W. (1982). *NAR* **10**, 4845–4859.
10. Roychoudhury, R., and Wu, R. (1977). *JBC* **252**, 4786–4789.
11. Wu, R., Ruben, G., Siegel, B., Jay, E., Spielman, P., and Tu, C. P. D. (1976). *Biochem.* **15**, 734–740.
12. Thomas, K. R., and Olivera, B. M. (1978). *JBC* **253**, 424–429.
13. Henikoff, S. (1984). *Gene* **28**, 351–359.
14. Masamune, Y., Fleischman, R. A., and Richardson, C. C. (1971). *JBC* **246**, 2680–2691.
15. Brutlag, D., and Kornberg, A. (1972). *JBC* **247**, 241–248.
16. Promega Notes (1990). No. 24:5.
17. Putney, S. D., Benkovic, S. J., and Schimmel, P. R. (1981). *PNAS* **78**, 7350–7354.
18. Gupta, A. P., Benkovic, P. A., and Benkovic, S. J. (1984). *NAR* **12**, 55897–55911.
19. Richardson, C. C. (1966). *JBC* **241**, 2084–2092.
20. Dolbeare, F., and Gray, J. W. (1988). *Cytometry* **9**, 631–635.
21. Weiss, B. (1976). *JBC* **251**, 1896–1901.
22. Weiss, B. (1978). *In* "DNA Repair Mechanism: ICN-UCLA Symposia on Molecular and Cellular Biology" (P. C. Hanawalt, E. C. Friedberg, and C. F. Fox, Eds.), Vol. 9, pp. 191–194. Academic Press, New York.
23. Henner, W. D., Grunberg, S. M., and Haseltine, W. A. (1983). *JBC* **258**, 15198–15205.
24. Rogers, S. G., and Weiss, B. (1980). *Meth. Enzy.* **65**, 201–211.
25. Keller, W., and Crouch, R. (1972). *PNAS* **69**, 3360–3364.
26. Weiss, B. (1981). "The Enzymes," 3rd ed., Vol. 14, pp. 203–231.
27. Smith, A. J. H. (1979). *NAR* **6**, 831–840.
28. Guo, L. H., and Wu, R. (1982). *NAR* **10**, 2065–2084.
29. Sorge, J. A., Blinderman, L. A. Katayama, C., and Callan, W. (1990). STRATEGIES in Molecular Biology (Stratagene) **3(1)**, 3–9.
30. Labeit, S., Lehrach, H., and Goody, R. S. (1986). *DNA* **5**, 173–177.
31. Guo, L. H., Yang, R. C. A., and Wu, R. (1983). *NAR* **11**, 5521–5540.
32. Johnson, D. F., Nierlich, D. P., and Lusis, A. J. (1990). *Gene* **94**, 9–14.
33. Henikoff, S. (1990). *NAR* **18**, 2961–2966.
34. Kochel, T. J., and Sinden, R. R. (1989). *JMB* **205**, 91–102.
35. Saporito, S. M., Smith-White, B. J., and Cunningham, R. P. (1988). *J. Bacteriol.* **170**, 4542–4547.
36. White, B. J., Hochhauser, S. J., Cintron, N. M., and Weiss, B. (1976). *J. Bacteriol.* **126**, 1082–1088.
37. Sak, B. D., Eisenstark, A., and Touati, D. (1989). *PNAS* **86**, 3271–3275.
38. Saporito, S. M., Gedenk, M., and Cunningham, R. P. (1989). *J. Bacteriol.* **171**, 2542–2546.
39. Baek, K.-H., and Walker, G. C. (1986). *J. Bacteriol.* **165**, 763–770.
40. Rao, R. N., and Rogers, S. G. (1978). *Gene* **3**, 247–263.
41. Rogers, S. G., and Weiss, B. (1980). *Gene* **11**, 187–195.

B. General Exonuclease: Nuclease Bal31

The nuclease from a marine bacteria *Alteromonas espejiana* (Bal31) is an extracellular nuclease with wide-ranging substrate specificities. The *Alteromonas* nuclease is commonly known as Bal31 nuclease, nuclease Bal31, or more simply Bal31. The enzyme catalyzes the degradation of ssDNA both endo- and exonucleolytically and of linear dsDNA exonucleolytically from 3' termini generating mostly blunt ends. Products are predominantly 5'-mononucleotides.

Bal31, unless otherwise specified, is a mixture of two protein components with distinct kinetic properties, "fast" (F species or *Bal31F*) and "slow" (S species or *Bal31S*). Bal31F is a 3'-exonuclease with specificities for both ssDNA and dsDNA, whereas Bal31S is a single-strand-specific nuclease or a Bal31F that has lost most of the double-strand exonuclease activity. Bal31 has become an important enzyme with many applications in recombinant DNA technology for controlled deletion of DNA termini.

1. FUNCTIONS

a. Reaction conditions

i. Recommended reaction conditions. A typical reaction mixture consists of 20 mM Tris–Cl (pH 8.0), 0.60 M NaCl, 12 mM $MgSO_4$, 12 mM $CaCl_2$, 1 mM EDTA, DNA (~50 μg/ml), and 5–10 U/ml enzyme (1). Incubation is carried out at 30–37°C. The reaction can be stopped by adding excess EGTA or EDTA or by incubation at 75°C for 10 min. In the case when the DNA substrate is to be subsequently treated with other enzymes or precipitated with ethanol, a 0.2 M NaCl concentration rather than 0.6 M should be more convenient.

Enzyme amount and incubation time are critical parameters for successful digestions. For example, too much enzyme will only contribute, on gel analysis, to the smearing of DNA after only 5 to 10 min incubation. For best results, a given batch of Bal31 enzyme should be calibrated for its digestion rate with a standard substrate such as linearized plasmid DNA or, preferably, with the target DNA.

The activity of Bal31 is dependent on pH, ionic strength, and temperature.

(a) pH. The pH optimum for Bal31 in 10 mM Tris–Cl buffer is near pH 8.0 on dsDNA and pH 8.8 on ssDNA (1).

(b) Ionic strength. Bal31S is optimally active in 1 M NaCl (1). The enzyme is unique among all known nucleases in that it is active over a wide range of NaCl concentrations; in 0 and 2 M NaCl, the enzyme is ~90% active. The enzyme retains ~40% of its maximal activity in 4.5 M NaCl (2). In the presence of more soluble CsCl (7 M), the enzyme still exhibits 27% of the maximal activity.

(c) Temperature. The optimum temperature for Bal31S on ssDNA substrates is 60°C; the activity at the usual incubation temperature of 30° or 37°C is only 13 or 25% of the optimum. The enzyme rapidly loses activity at temperatures above 60°C.

TABLE 3.4 Steady-State Kinetic Parameters of Bal31 Nuclease

Substrate	Bal31F		Bal31S		Ref.
	k_{cat} (sec^{-1})	K_m	k_{cat} (sec^{-1})	K_m	
Linear dsDNA	72–103	60–96 nM	3.9	41 nM	3, 4
ssDNA	138	<5 µM	124	3 µM	3
Linear dsRNA	1.5	9.7 nM	—	—	4

ii. **Kinetic parameters.** The steady-state kinetic parameters of Bal31 for linear dsDNA, ssDNA, and linear dsRNA are listed in Table 3.4. Bal31F digests linear dsDNA about 19–27 times faster than Bal31S (3,4); Bal31F degrades dsDNA ~50 times more efficiently than dsRNA (4). Bal31S shows very little activity against dsRNA relative to that of the F enzyme. On ssDNA, Bal31F and Bal31S enzymes are 1.3 and 33 times more active than on dsDNA, respectively.

iii. **Requirements for divalent cations.** Bal31 requires both Ca^{2+} and Mg^{2+} for optimal activity (5). Nevertheless, both ssDNA and dsDNA (circular supercoiled) are cleaved in the absence of Mg^{2+} but at 1.5 and 13%, respectively, of the rates in 15 mM Mg^{2+}. Both F and S enzymes (at 10 U/ml concentration) display maximal activities on ssDNA and dsDNA substrates in 12.5 mM Mg^{2+} and 12.5 mM Ca^{2+} (6). At lower enzyme concentrations (e.g., 0.16 U/ml), however, 1 mM Ca^{2+} can exert full activity in the presence of 12.5 mM Mg^{2+}. Ca^{2+} is most likely to be a prosthetic group of the Bal31 enzyme, whereas the enzyme is most active on Mg-DNA substrates.

iv. **Inhibitors and inactivators.** Addition of EDTA (and EGTA as well) at a concentration in molar excess of Ca^{2+} irreversibly inactivates Bal31 (5). In contrast, the effect on Mg^{2+} is reversible.

With ssDNA, the presence of EtBr has no discernible effect on the Bal31 nuclease (7). The enzyme is almost fully active in 5% SDS in the presence of Ca^{2+} and Mg^{2+} (5). Bal31S retains over 40% of the maximal activity in the presence of 6 M urea. Bal31 is inhibited by DTT (1 mM) but not by PMSF (1 mM).

b. *Activity assay and unit definition*

The activity of Bal31 is most commonly assayed spectrophotometrically by measuring the release of nucleotides from ssDNA substrates (denatured calf thymus DNA). The assay conditions are essentially identical to the recommended reaction conditions described earlier. During the course of incubation, aliquots are taken, the reaction is stopped by adding excess EDTA and $HClO_4$, and large-size DNAs are filtered off on a NC filter. Clear filtrate is used to measure A_{260}. The measured value has to be corrected for blank values (any enzyme-associated DNAs), especially when crude enzyme preparations are assayed.

228   Chapter 3   Nucleases

Unit definition: One unit is the amount of enzyme that releases 1 μg of nucleotides from denatured calf thymus DNA in 1 min at 30°C. An A_{260} of 0.55 corresponds to 16.7 μg of nucleotides per ml in 6.7% (w/v) $HClO_4$.

c. Substrate specificities

i. Substrate and product profiles. Bal31 (F and S) is a 3' → 5'-exonuclease for linear dsDNA as well as ssDNA. Bal31 degrades ssDNA whether circular (e.g., φX174) or linear (after alkali denaturation). The products are predominantly 5'-mononucleotides (6,8). There is no evidence for a 5' → 3' mode of exonuclease action. Single-strand and double-strand exonuclease activities of Bal31F are kinetically distinct from those of Bal31S (see Table 3.4).

ii. Termini structure after Bal31 digestion. The terminally directed hydrolysis of linear dsDNA initially leaves 5'-protruding single strands, which are then cleaved by endonuclease activity and subsequently by the predominant (20-fold higher) 3'-exonuclease activity. Bal31 digestion results in an apparent gradual shortening of dsDNA from both the 3' and 5' ends, without introducing detectable scissions in the dsDNA away from the termini.

A significant fraction of the shortened ends may be flush at high enzyme concentrations (e.g., 30–40% at the enzyme concentration of 5–10 U/ml) but, at low enzyme concentrations (e.g., 0.5–0.8 U/ml), most of the termini carry 5' protruding single strands.

Regardless of the types of initial termini in linear duplex φX174 RF DNA, the single-strand protrusion after Bal31F (0.5 U/ml) digestion at 30°C carries about 66–80 nt per 100 nt removed, 120–140 nt per 300 nt removed, and 200–220 nt per 600 nt removed (6). Similarly, the single-strand protrusion after Bal31S (0.8 U/ml) digestion carries about 30–60 nt per 100 nt removed, 80–90 nt per 300 nt removed, and 140–170 nt per 600 nt removed.

The number of protruding nucleotides gets smaller in a nonlinear fashion as the nuclease concentration increases, apparently reaching a limiting value of 5–10 nt remaining per 100 nt removed.

iii. Processivity. The exonuclease activity of Bal31 is quasiprocessive. With dsDNA as a substrate, Bal31S and Bal31F remove 18 and 28 nt, respectively, per productive enzyme–substrate encounter (6).

iv. Sequence dependence and extent of reaction. The extent of the reaction is dependent on local sequences. Consecutive or a high content of G · C base pairs significantly retards the exonucleolytic activity (1), and most of the shortened sequences terminate in G · C base pairs. Subcloning of Bal31-digested DNA fragments has a tendency to underrepresent the DNA populations that terminate in AT-rich regions. Exonucleolytic degradation of $d(CG)_n$ is much slower, even after the correction for the salt dependency of the enzyme activity. (*Note:* Under high salt conditions, the polymer adopts a left-handed Z-conformation.)

v. DNA structure. Bal31 specifically cleaves dsDNA at the B–Z helix junction (2). The enzyme also cleaves the apparent dsDNA at the regions containing helical distortions following base modifications, e.g., with the carcinogen N-acetoxy-N-2-acetylaminofluorene which modifies dG or with nitrous acid which deaminates dC and dA and introduces interstrand cross-links into dsDNA (9). Digestion of dsDNA by Bal31 is arrested when the enzyme encounters interstrand cross-links mediated by photoreactive psoralens (10). A similar property has been observed with Exo III.

vi. Nuclease activity on apparent dsDNA. Bal31 cleaves dsDNA at single-strand breaks (the strand opposite the nick) and at gap or hairpin regions (5,11). It cleaves other nominal dsDNAs containing imperfect base pairing, apurinic sites, or other lesions generated by UV irradiation (9) and chemical methylation. Bal31F is ~2.5 times as efficient in cleaving apurinic lesions as Bal31S (12).

Nonaltered, nonstressed circular dsDNA is a very poor substrate for either the F or the S enzyme. Bal31 shows a virtual absence of activity against the positively supercoiled DNA which is converted from a negatively supercoiled form by EtBr intercalation. However, Bal31 (F and S) cleaves negatively supercoiled DNA to yield linear dsDNA (7), which can then be degraded exonucleolytically. The conversion of negatively supercoiled DNAs to linear duplex molecules occurs via a circular duplex intermediate that often contains a single interruption in one strand. This single interruption usually consists of a gap arising from the removal of ~6 nt (for F species) or 3 nt (for S species) after the introduction of the initial nick (13).

vii. RNase activity. Bal31F displays the unique property of catalyzing an exonucleolytic degradation of linear dsRNA (13). The exonucleolytic activity on linear dsDNA is ~50 times greater than that on linear dsRNA with a comparable G·C content. Bal31S shows very little activity against the dsRNA relative to that of the F enzyme.

d. Mode of Bal31 nuclease action

Based on the observations that a variety of altered duplex structures can elicit nucleolytic cleavages by Bal31, it has been postulated that Bal31 enzymes act on structures bearing relative thermodynamic instability (3). The canonical exonuclease activity against linear dsDNA may well be regarded as a manifestation of the activity at the thermodynamically unstable termini.

Bal31 attacks predominantly from the 3' termini. The nucleolytic action on protruding 5' single strands is presumed to involve two successive activities: initially an endonuclease activity and subsequently the exonucleolytic activity. During the course of Bal31 digestions of either ssDNA or dsDNA, however, possible intermediary oligonucleotide products have not been observed.

The nucleolytic activity is somewhere between the processive and distributive modes, and the degradation of DNA occurs quasi-uniformly from both 3' and 5' termini. Nevertheless, there are some sequence-dependent stops at the GC-rich regions and scatters at the end points of the termini. Indeed, a significant fraction of the Bal31-digested molecules carries protruding single-strand termini of variable lengths, especially when the digestion is performed at low enzyme concentrations.

2. APPLICATIONS

i. Progressive shortening of dsDNA. Nuclease Bal31 is widely used to produce controlled deletions from both 3' and 5' termini of DNA for various purposes including gene mapping, domain–function analysis, subcloning, and ordered nucleotide sequencing (8,14,15). Bal31 is especially useful in these applications because the digestion can be performed on any type of termini and results in progressively shortened molecules with blunt ends. However, a brief repair (or filling in) of the termini with DNA polymerases (T4 or *E. coli*) is generally required to increase the fraction of blunt ends.

For removal of a very limited number of nucleotides from the duplex termini, the use of Bal31S is preferred. For removal of several hundred nucleotides, Bal31F should be used. Mixed preparations of the two (F and S) species can be used for any of intermediary tasks, preferably preceded by calibrations of the digestive rate on the target DNA.

Deletion mutants are typically constructed as follows. (i) A vector (M13 or its derivatives) containing an insert is first linearized at a restriction site. Bal31 digestions are carried out on both "target arm" and "vector arm" in a controlled manner, and the termini are blunt ended by repair reactions. (ii) The other end of the insert is excised from the vector preferably at a restriction site which generates a cohesive end. (iii) The fragments with progressive deletions are isolated and ligated to a linearized subcloning vector by both cohesive-end and blunt-end ligations such that the Bal31-shortened target end is proximal to the sequencing primer site on the vector.

A strategy useful in generating small deletions relies on a marker gene inactivation (e.g., tetracycline resistance gene of pBR322) when an appropriately placed target DNA is digested by Bal31 beyond a certain distance. For example, a recombinant pBR322 plasmid which contains a DNA insert at the *Eco*RI site was constructed and linearized at a unique site 3' to the insert (16). When this DNA was digested with Bal31 and the subclones was screened by the Tet[r] phenotype, the clones containing sequential deletions of 20 to 100 bp in length could be readily obtained.

ii. Removal of nucleotides from dsRNA. The exonucleolytic activity of Bal31F has been shown to digest linear dsRNA. Progressively shortened dsRNA should prove useful in the construction and *in vitro* autocatalytic amplification of recombinant RNA molecules using an RNA vector, e.g., MDV-1(+)RNA, that carries a specific recognition sequence for Qβ replicase (17).

iii. Use as single-strand-specific endonuclease. Bal31 can be used to detect altered secondary structures in dsDNA (1), often in combination with or replacing nuclease S1 or mung bean nuclease, but with the advantage of a near-neutral optimum pH. The extent of unwinding of nonsupercoiled closed circular DNA can be accurately measured and correlated with the average lesions per DNA molecule by monitoring relative alterations in the electrophoretic patterns on agarose gels.

3. STRUCTURE

Nuclease Bal31 is a composite of two enzyme species, "fast" (Bal31F) and "slow" (Bal31S), which have distinct catalytic properties (3). Bal31F is a 109-kDa single polypeptide. Bal31S is a cleavage product (85 kDa) of the F enzyme. Both Bal31F and Bal31S have a pI of 4.2.

Proteolysis with pronase or subtilisin converts the Bal31F to an apparent Bal31S and drastically reduces the double-strand exonuclease activity while retaining the endonucleolytic single-strand nuclease activity in response to lesions in nominally dsDNA (18). Treatment of Bal31S with subtilisin also results in a drastic reduction in exonuclease activity on dsDNA but with retention of most of the single-strand nuclease activity. The internal cleavage of Bal31S becomes apparent only after the addition of urea (to 6 M) in the usual denaturation buffer.

4. GENETICS

Nuclease Bal31 (F and S) is produced as a precursor molecule (Pre-F, 120 kDa) from a single gene. This Pre-F nuclease is functionally active and can be isolated from the periplasmic space of *A. espejiana* only when cell growth is close to the stationary phase. This suggests that, in the early phase of cell growth, the Bal31 enzyme can be rapidly transported to culture medium with concomitant, sequential proteolysis to F species and then to S species (18).

5. SOURCES

Nuclease Bal31 is an extracellular enzyme excreted by a marine bacterium originally assigned to the genus *Pseudomonas* (19) but now reassigned to *A. espejiana* (ATCC 29659).

Homogeneous preparations of both Bal31F and Bal31S have been obtained after about 40-fold purifications from the large-scale culture supernatants of *A. espejiana* (3).

The enzyme is remarkably stable on storage, although repeated freezing and thawing can damage activity.

References

1. Gray, H. B., Jr., Winston, T. P., Hodnett, J. L., Legerski, R. J., Nees, D. W., Wei, C. F., and Robberson, D. L. (1981). *In* "Gene Amplification and Analysis" (J. G. Chirikjian and T. S. Papas, Eds.), Vol. 2, pp. 169–203. Elsevier, New York.
2. Kilpatrick, M. W., Wei, C. F., Gray, H. B., Jr., and Wells, R. D. (1983). *NAR* **11**, 3811–3822.
3. Wei, C. F., Alianell, G. A., Bencen, G. H., and Gray, H. B., Jr. (1983). *JBC* **258**, 13506–13512.
4. Bencen, G. H., Wei, C. F., Robberson, D. L., and Gray, H. B., Jr. (1984). *JBC* **259**, 13584–13589.

5. Gray, H. B., Jr., Ostrander, D. A., Hodnett, J. L., Legerski, R. J., and Robberson, D. L. (1975). *NAR* **2**, 1459–1492.
6. Zhou, X.-G., and Gray, H. B., Jr. (1990). *BBA* **1049**, 83–91.
7. Lau, P. P., and Gray, H. B., Jr. (1979). *NAR* **6**, 331–357.
8. Legerski, R. J., Hodnett, J. L., and Gray, H. B., Jr. (1978). *NAR* **5**, 1445–1464.
9. Legerski, R. J., Gray, H. B., Jr., Robberson, D. L. (1977). *JBC* **252**, 8740–8746.
10. Zhen, W. P., Buchardt, O., Nielsen, H., and Nielsen, P. E. (1986). *Biochem.* **25**, 6598–6603.
11. Barnes, W. M., and Bevan, M. (1983). *NAR* **11**, 349–368.
12. Wei, C. F., Legerski, R. J., Alianell, G. A., Robberson, D. L., and Gray, H. B., Jr. (1984). *BBA* **782**, 408–414.
13. Przykorska, A. K., Hauser, C. R., and Gray, H. B., Jr. (1988). *BBA* **949**, 16–26.
14. Talmadge, K., Stahl, S., and Gilbert, W. (1980). *PNAS* **77**, 3369–3373.
15. Poncz, M., Solowiejczyk, D., Ballantine, M., Schwarz, E., and Surrey, S. (1982). *PNAS* **79**, 4298–4302.
16. Henriquez, V., and Gennaro, M. L. (1990). *NAR* **18**, 6735–6736.
17. Miele, E. A., Mills, D. R., and Kramer, F. R. (1983). *JMB* **171**, 281–295.
18. Hauser, C. R., and Gray, H. B., Jr. (1990). *ABB* **276**, 451–459.
19. Espejo, R. T., and Canelo, E. S. (1968). *J. Bacteriol.* **95**, 1887–1891.

4
Restriction Endonucleases and Modification Methylases

I. GENERAL DESCRIPTION

Restriction endonucleases are a group of DNases that recognize specific nucleotide sequences and cut dsDNA in a site-specific or nonspecific manner. They were initially discovered in the early 1950s as a part of the restriction (R) and modification (M) systems that bacteria operate for their protection against invading bacteriophages and foreign genetic elements. Initially identified by phenotype, the R–M systems are now amenable to a wealth of genetic analysis. It is largely through this R–M system that bacteria destroy undesirable foreign DNAs that invade the cells through infection, conjugation, and transfection. The bacterial R–M systems are tantamount to "the prokaryotic equivalent of an immune system" (1).

R–M systems are found in microorganisms, mainly in bacteria, and are highly diverse even within a cell. The number of documented restriction enzymes has surpassed 2100, which include 17 type I, 179 type II, and 4 type III specificities (2), while over 190 DNA modification methyltransferases have been characterized (3).

R–M systems generally consist of two enzymatic activities: (a) a site-specific "restriction" endonuclease (R · ENase) that is responsible for digesting exogenous DNA and (b) a DNA "modification" methylase (or methyltransferase, M · MTase) with identical sequence specificity. The M· MTase is responsible for modifying and protecting endogenous DNAs from similar digestions by R · ENase. The R and M activities may reside in a single multisubunit enzyme or in physically separate enzymes. Despite the fact that restriction and cognate modification enzymes recognize identical DNA sequences, their amino acid sequences lack similarities, suggesting that the enzyme pairs recognize their target sequences by different mechanisms.

The major focus of this chapter is on type II R · ENases, which are commonly described as "restriction enzymes," and their cognate M · MTases because of their utmost importance in molecular biology and recombinant DNA technology. Other R–M systems such as type I and III R–M systems or restriction-independent MTases will be described only briefly for the sake of comparison.

1. TYPES OF RESTRICTION–MODIFICATION SYSTEMS

Restriction–modification enzymes fall into at least four types (type I, II, III, and IV) on the basis of enzyme composition, cofactor requirements, recognition sequence symmetry, and cleavage characteristics (Table 4.1). All MTase activities require AdoMet as the methyl group donor substrate.

Type I R–M enzymes are multienzyme complexes composed of three nonidentical subunits: R, M, and S. The S subunit determines specificity for both R and M. Type I R–M enzymes require Mg^{2+} and ATP as cofactors and cleave unmodified DNA at a relatively random location distant from the recognition site. Type I enzymes hydrolyze ATP as part of the restriction activity and site specifically methylate the target DNA in a manner competitive with the cognate ENase activity.

The *Type II* R–M system consists of physically separate R · ENase and M · MTase activities. The R · ENases are usually dimers having identical subunits, whereas M · MTases are monomeric enzymes. Type II R · ENases require only Mg^{2+} for their restriction activity. Type II R · ENase and M · MTase recognize the same specific nucleotide sequences. The majority of type II R · ENases cleave DNA at the recognition sequence, while a small number of enzymes, called *type IIS*, cleave the DNA at a defined distance from the recognition site.

Type III R–M enzymes are composed of two nonidentical subunits and require Mg^{2+} and ATP for restriction activity. Type III enzymes recognize short nonpalindromic sequences and cut DNA at fixed sites relative to the unmodified recognition sequence. Type III enzymes do not hydrolyze ATP. AdoMet is not required for restriction, but it stimulates ENase activity.

Another type of R–M system, classified as *Type IV*, represents an intermediate in the evolutionary pathway between the type II and type III enzymes (4). Type IV enzymes (e.g., *Eco*57I) consist of separate R · ENase and M · MTase, but the monomeric R · ENase possesses an additional methylase activity. The restriction-associated methylase activity is not strong enough to protect the host DNA *in*

I. General Description

TABLE 4.1 Classification of Restriction–Modification Systems

Features	Type I	Type II Prototype	Type IIS	Type III	Type IV
R–M active structure[a]	Single enzyme 3 s.u. (R,M,S) complex	Separate enzymes		Single enzyme 2 s.u. complex	Separate monomeric enzymes[b]
		R:dimer M:monomer	R:monomer M:monomer		
Cofactors	Mg^{2+} ATP[c]	Mg^{2+}		Mg^{2+} ATP[d]	Mg^{2+} (AdoMet)[e]
Recognition site	Asym (bipartite)	Palindromic	Asym	Asym	Asym
Cleavage[f]	Variable distance either side	Same site	Definite distance to 3' side	25–27 bp to 3' side	14 bp to 3' side
Methylation[g]	Two strands by M	Two strands	One MTase on each strand	One strand only	Two strands

[a] The structure representative for the majority of each type of enzymes. Asym, asymmetric.
[b] R-associated MTase activity is very weak.
[c] R activity requires the hydrolysis of ATP.
[d] R activity does not require the hydrolysis of ATP.
[e] R activity is stimulated by AdoMet.
[f] With respect to the recognition site.
[g] The methyl donor substrate is always AdoMet.

vivo. R · ENase and M · MTase recognize asymmetric sequences. R · ENase cleaves in the presence of Mg^{2+} target DNAs at a defined distance (with *Eco*57I for example, 16/14 nucleotides away) from the recognition site. The restriction activity is stimulated by AdoMet.

2. INDEPENDENT RESTRICTION OR MODIFICATION SYSTEMS

In addition to the four types of R–M systems just mentioned, some bacteria operate other systems such as M-independent R systems and R-independent M systems. A different category of R–M systems, which is of particular importance in recombinant DNA technology, is the *methylation-dependent restriction system* (MDRS) in which R · ENase preferentially cleaves DNA at methylated sites (5,6) (see Section III,A, this chapter). The MDRS consists of the gene products currently known to be linked to several independent loci in the *E. coli* K12 genome such as *mcrA*, *mcrBC* (the "mcr" standing for *m*odified *c*ytosine *r*estriction), and *mrr* (for *m*ethyl*a*denine *r*ecognition and *r*estriction). The nature of the restriction enzymes associated with the MDRS needs to be further elucidated.

Still another distinct M-independent R system is the *intron-encoded endonucleases*, the function of which is only beginning to be understood (7). This class of ENases is encoded in introns of diverse origin, is site specific on dsDNA, and probably functions in the site-specific intron transposition of mobile group I introns. In a manner similar to type I R–M enzymes, intronic ENases may be multifunctional enzymes possessing, in some cases, reverse transcriptase activity and/or *maturase* activity which is known to participate in the excision of the intron itself. Some of the better characterized intronic ENases, such as the ω (omega) and *coxI* gene (cytochrome oxidase subunit 1) intron 4 (aI4) products of yeast mitochondrial DNA (8) or *td* and *sunY* gene products of phase T4 (9), show defined cleavage specificities with respect to the site of the intron insertion called a "homing" site. The recognition sequences are relatively large, comprising about 18 or more bp, although only 6 to 9 noncontiguous bases may actually be necessary. The homing (or cleavage) site can be either at the recognition sequences or as many as 18 to 20 bp away. Intronic ENases cleave dsDNA, leaving 2-, 3-, or 4-base extensions at the 3'-OH termini.

Perhaps to complete the variety of R–M systems, bacteria also operate R-independent M systems which include *adenine-specific* (*Dam*) MTase and *cytosine-specific* (*Dcm*) MTase (see Section III,C, this chapter). The Dam and Dcm modification systems apparently play several utility functions *in vivo* through site-specific methylation and consequent modulation of DNA cleavage activities of certain type II endonucleases. Dam and Dcm MTases thus complement type II MTases which are often used strategically in molecular biology and recombinant DNA work.

3. Type II Restriction–Modification Systems

Restriction enzymes, especially type II, display unique nucleotide sequence specificities, reminiscent of the amino acid sequence specificities of some well-known proteases. The fidelity to a particular nucleotide sequence, the diversity of the sequences that the enzymes recognize, and the possibility of relatively easy restriction controls by the use of M · MTases make the type II enzymes indispensable tools in DNA manipulations. Indeed the burgeoning development of recombinant DNA technology is due largely to the discovery of site-specific restriction enzymes.

The genes for over 100 R and/or M systems have now been cloned and characterized (1). The genes for R–M systems are very heterogeneous in their arrangements, orientations, and size (Fig. 4.1). The cloning of the genes has made the purification of both R and/or M enzymes more convenient with higher yield and purity, thus promoting wider use of the enzymes at lower costs. This development has also enabled detailed studies to be carried out on reaction mechanisms and DNA–protein interactions at the molecular level. Type II (and type IIS) R · ENases, most of which are now commercially available, are extensively used as reagents to generate specific DNA fragments for gene cloning, DNA and chromosome mapping, DNA sequencing, and hybridization.

4. Nomenclature for Restriction–Modification Enzymes

The recommended nomenclature for R–M enzymes is largely based on the rules originally proposed by Smith and Nathans (10) and updated by Szybalski *et al.* (11).

I. General Description

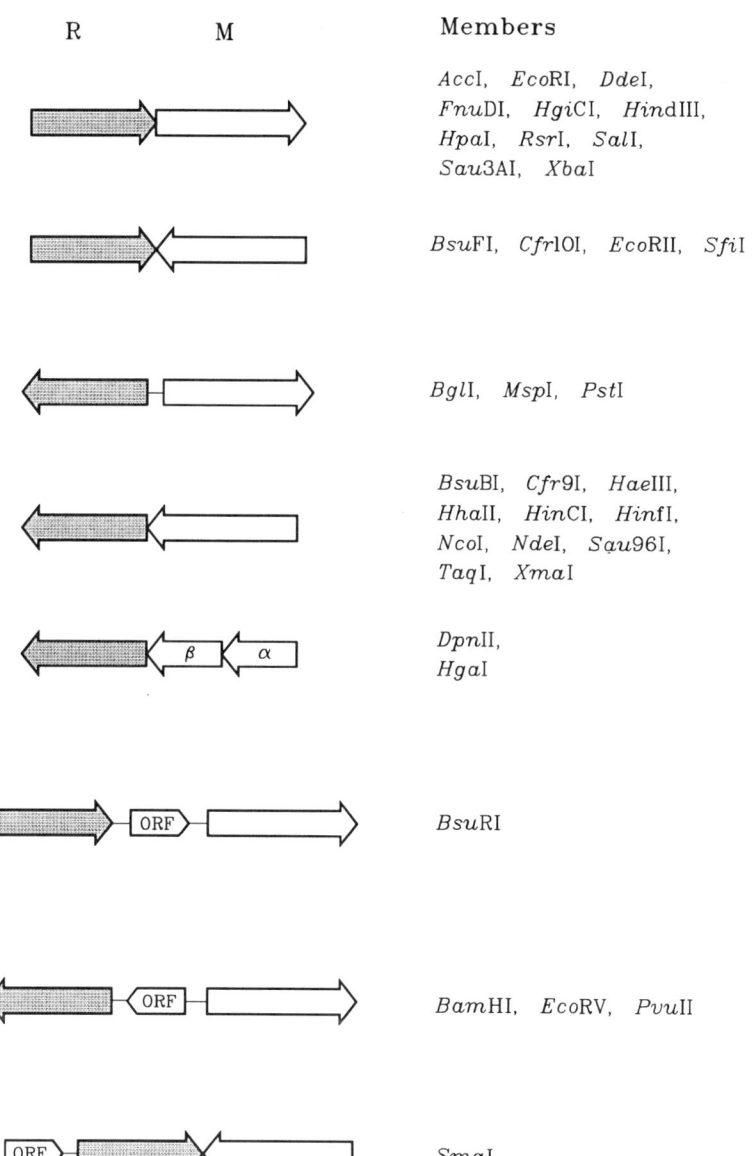

FIGURE 4.1 Gene organizations of typical type II restriction–modification systems: Schematic representation. The sizes of type II R and M enzymes vary widely with an average of 280 amino acids for R · ENases and 380 amino acids for M · MTases. The size of small ORF ranges from 84 to 102 amino acids. Exceptionally, the C terminus of the *taqIM* gene (421 codons) overlaps the N terminus of the *taqIR* gene (263 codons) by 13 codons.

(i) The first three letters (in italics) indicate the genus (first letter) and species (second and third letters) name of the source organism, e.g., *Eco* for *E. coli* and *Hin* for *Haemophilus influenzae*.

(ii) The first three letters are followed by strain or type identification in nonitalic symbols or arabic numerals; e.g., *Eco*K for *E. coli* K strain, *Hin*d for *H. influenzae* strain d, or *Sau*3A for *Staphylococcus aureus* 3A. Designation for the extrachromosomal elements such as virus or plasmid follows the same rule, e.g., *Eco*RI and *Eco*PI.

(iii) This is followed without space by roman numerals (I, II, III, etc.) to identify different enzymes produced by the same strain, e.g., *Hin*dII and *Hin*dIII.

(iv) To specify the restriction ENase or modification MTase, the capital letter R or M is placed in front of the enzyme symbol and separated by a centered dot, e.g., R · *Eco*RI and M · *Eco*RI. For the R · ENase, the symbol (e.g., *Eco*RI) with or without the letter R or endo R is generally accepted.

5. NOMENCLATURE FOR RESTRICTION–MODIFICATION GENES

The recommended nomenclature for the genes specifying type II R–M systems is based on the rules proposed by Szybalski *et al.* (11).

(i) The gene designation for type II R–M enzymes uses the same nomenclature used for the corresponding enzymes except that they are all written in italics.

(ii) The capital letter R (for ENase) or M (for MTase) follows the rest of the gene designation, e.g., *ecoRIR* and *ecoRIM*.

References

1. Wilson, G. G. (1991). *NAR* **19**, 2539–2566.
2. Roberts, R. J., and Macelis, D. (1992). *NAR* **20**(Suppl.), 2167–2180.
3. McClelland, M., and Nelson, M. (1992). *NAR* **20**(Suppl.), 2145–2157.
4. Janulaitis, A., Vaisvila, R., Timinskas, A., Klimasauskas, S., and Butkus, V. (1992). *NAR* **20**, 6051–6056.
5. Blumenthal, R. M. (1989). *FOCUS (BRL)* **11**(3), 41–46.
6. Raleigh, E. A., Benner, J., Bloom, F., Braymer, H. D., DeCruz, E., Dharmalingam, K., Heitman, J., Noyer-Weidner, M., Piekarowicz, A., Kretz, P. L., Short, J. M., and Woodcock, D. (1991). *J. Bacteriol.* **173**, 2707–2709.
7. Perlman, P. S., and Butow, R. A. (1989). *Science* **246**, 1106–1109.
8. Sargueil, B., Hatat, D., Delahodde, A., and Jacq, C. (1990). *NAR* **18**, 5659–5665.
9. Bell-Pedersen, D., Quirk, S., Clyman, J., and Belfort, M. (1990). *NAR* **18**, 3763–3770.
10. Smith, H. O., and Nathans, D. (1973). *JMB* **81**, 419–423.
11. Szybalski, W., Blumenthal, R. M., Brooks, J. E., Hattman, S., and Raleigh, E. A. (1988). *Gene* **74**, 279–280.

II. RESTRICTION ENDONUCLEASES

A. Type II Restriction Endonucleases: General Properties

[Type II Site-Specific Deoxyribonuclease, EC 3.1.21.4]

Type II restriction enzymes (R · ENases) are a large group of DNA endonucleases that are distinguished from other R–M systems by two characteristics: (i) the ENase activity for DNA restriction is physically and functionally separate from the cognate modification methyltransferases (M · MTases), and (ii) the nucleotide sequence specificities of the R and M enzymes are identical. Type II R · ENases recognize specific DNA sequences and cleave the DNA within or in the immediate vicinity of the sequences.

There are now over 1500 type II R · ENases with known recognition sequences isolated from various species of gram-positive and gram-negative bacteria (1,2). They represent at least 188 different site specificities. All R · ENases hydrolyze phosphodiester bonds in dsDNA according to specific base sequences and produce DNA fragments with 5'-P and 3'-OH termini.

This section covers the general features of type II R · ENases and then more specifically *Eco*RI, one of the typical and most thoroughly studied type II restriction enzymes. Because the major use of type II restriction enzymes is to cut dsDNA in a site-specific manner, their specific applications are described within each appropriate subsection.

1. FUNCTIONS

a. Substrate specifications

i. Site specificity. Type II restriction enzymes recognize specific nucleotide sequences along dsDNA chains and cut them at the recognition site (prototype type II enzymes) or at a defined number of nucleotides shifted from the recognition site (a subclass termed *type IIS* enzymes). The majority of type II R · ENases recognize tetra-, penta-, or hexanucleotides. There are barely over ten 8-base cutters known [the cleavage site is indicated in the upper sequence (5'-3') by a vertical bar]: e.g., *Asc*I (GG|CGCGCC), *Fse*I (GGCCGG|CC), *Not*I (GC|GGC-CGC), *Pac*I (TTAAT|TAA), *Sfi*I (GGCCN$_4$|NGGCC), *Srf*I (GCCC|GGGC), *Sse*8387I (CCTGCA|GG), and *Swa*I (ATTT|AAAT).

The specific set of nucleotide sequences most readily cleaved (the primary substrate) is designated as a *canonical* sequence. Those sequences that are similar to the canonical sequence and thus specifically recognized and cleaved by the enzyme but at a moderately reduced rate are generally called *noncanonical* sequences. Some RNases simply display multiple or "degenerate" recognition sequences.

Some enzymes, e.g., *Msp*I and *Hpa*II (both for 5'-C|CGG-3'), cleave ssDNA at appropriate cleavage sequences, albeit at a slower rate (3); however, this cleavage is believed to occur within a region of intramolecular base pairing. The formation of such bihelical structures may be transient and occurs either spontaneously or through the assistance of the enzyme (4).

ii. Termini structure of cleavage products. The cleavage involves the hydrolysis of two phosphodiester bonds (or one per strand) either sequentially or simultaneously. The 5'-P and 3'-OH termini can be either blunt (or flush) ends or 5'- or 3'-protruding (or cohesive or sticky) ends with various numbers (up to 6) of base extensions.

iii. Site structure. The recognition sequences of the majority of type II R · ENases consist of inverted complementary sequences called *palindromic* or *centrosymmetric* sequences. For these R · ENases, the cleavage occurs within the recognition sequence. In contrast, a subgroup of type II enzymes recognize *asymmetrical* 4- to 7-bp sequences and cleave DNA at 1–20 bp away from the recognition sequence. These "shifter" enzymes, now over 80 species representing over 35 sequence specificities, are called *type IIS* R · ENases (or *ENase-IIS*) (see Section II,C, this chapter). Some other type II R · ENases recognize *quasi-palindromic* sequences and cut within the recognition sequences: for example, *Aci*I (C|CGC), *Bsi*I (C|TCGTG), and *Bpu*10I (CC|TNAGC). These novel type II R · ENases have been classified as *type IIQ* restriction enzymes (5).

Some recognition sites are composed of an odd number of bases with the odd base in the center of the palindromic symmetry. Some other restriction sites contain up to nine nonspecific bases as a part of the specific recognition structure.

The base compositions of recognition sequences are in general highly biased toward GC-rich sequences. There are only a few enzymes known that require uniquely AT base pairs in their recognition sequences, e.g., *Aha*III (TTT|AAA), *Pac*I (TTAAT|TAA), and *Swa*I (ATTT|AAAT).

iv. DNA Conformation. Restriction enzymes are not only sequence specific but also structure sensitive and may exhibit either enhanced or inhibited cleavage activity. The restriction sites located in nuclease S1-sensitive, conformationally flexible junctions, for example, between contiguous B-DNA and Z-DNA or in non-B-DNA conformation, are resistant to cleavage by many restriction enzymes such as *Bam*HI, *Eco*RI, *Hin*dIII, and *Pst*I (6,7). The 5'-GCGC-3' site of *Hha*I is not cleaved when the sequence is within the stretch of alternating GC sequences that adopt a Z-DNA conformation (8). In contrast, *Mbo*I (GATC) has an enhanced cleavage reactivity at the B–Z junction (9). Modulation of DNA conformation by DNA intercalating drugs, oligonucleotides, or oligonucleotide analogs thus provides an efficient way for achieving controlled cleavages by restriction enzymes (see below).

v. DNA · RNA Heteroduplex. Several restriction enzymes are able to cleave DNA · RNA hybrids (for example, from the reverse transcription of RNA), albeit at slower rates or 20–50 times higher enzyme concentrations (20). These enzymes include *Eco*RI, *Hae*III, *Hha*I, *Hin*dIII, *Msp*I, *Sal*I, and *Taq*I. For *Hae*III, *Hha*I, and *Taq*I, the cleavage results in the fragmentation of the DNA · RNA hybrid to expected sizes. Although the cleavage of the DNA strand apparently occurs at the correct recognition sequences, the specific sites of cleavage in the RNA strand have not been established.

The reduced rate of cleavage by *Hin*dIII on the DNA · RNA heteroduplex compared with that on the DNA homoduplex has been utilized in the Okayama–Berg vector–primer cloning method (21).

vi. Isoschizomers. Often a given canonical DNA sequence is recognized and cleaved by more than one enzyme. Such groups of enzymes are generally referred to as *isoschizomers* (1). Isoschizomers recognize identical sequences, but the restriction patterns may not be the same. More specifically, those enzymes that cleave at a position different from their prototype are called *neoschizomers*. For example, the pair of enzymes *Sma*I and *Xma*I both recognize the sequence CCCGGG, but *Sma*I cleaves between the middle C and G (CCC|GGG) to give blunt-ended fragments, whereas *Xma*I cleaves at C|CCGGG to generate fragments with 4-base protruding 5′ ends. Some isoschizomers are differentially sensitive to methylation. For instance, *Mbo*I, *Sau*3AI, *Dpn*I, and *Dpn*II all recognize and cleave the basic sequence GATC. However, the methylation of the A base, (e.g., by Dam MTase) renders the site resistant to *Mbo*I and *Dpn*II, whereas it has no effect on *Sau*3AI. *Sau*3AI cannot cleave the sequence if the C base is methylated. *Dpn*I is in fact a distinct class II R · ENase that requires methylation (G^mA|TC) for cleavage. For a complete listing of R · ENases and their isoschizomers, readers should consult the annual update (2).

vii. Degenerate recognition sequences. Some R · ENases accommodate more than one base at a particular position in their recognition sequences and cleave them with equal efficiencies. This degeneracy can occur at any one or two base positions within palindromic sequences in such a way that the symmetry about the inversion (or helix-dyad) axis is usually maintained.

viii. Relaxation of site specificity. Type II restriction enzymes generally exercise rigorous sequence discrimination by a factor of $>10^6$ between the canonical site and other alternative cleavage sequences. However, under suboptimal conditions, a significant number of enzymes display moderate cleavage activities on slightly different recognition sequences as well as on recognition sequences containing a single base mismatch. We will consider here three variant forms of specificity relaxation.

(a) Tolerance of single base mismatch. Heteroduplex recognition sites that contain one mismatch of Pu:Py (or Py:Pu) type are cleaved, under optimal conditions, at moderately slower rates compared with the cleavage rate at the canonical sequence (10). The presence of original purine bases in mismatches appears to play an important role in the reaction rate (see *Eco*RI, Section II,B). The level of tolerance also depends on the exact position of single mismatches within the recognition sites. The strand containing the mismatching base is cleaved much more slowly than the normal strand, which is cleaved in turn more slowly than its counterpart in canonical sequences (11).

(b) Star activity. Under certain suboptimal reaction conditions, some enzymes show relaxed sequence specificities called "star" activity, which is commonly denoted by a superscript asterisk (e.g., *Eco*RI*). With the star activity, the recognition specificity is lowered, for instance, from a hexanucleotide to the core tetranucleotide or from a tetranucleotide to the core dinucleotide. Nevertheless, the cleavage by the star activity occurs at the same site as the canonical site.

Low ionic strengths and high pH (8.0–9.5) alone or in combination with organic solvents and/or Mn^{2+} generally favor the appearance of the star activity (see EcoRI, Section II,B). The star activity has been documented in a number of restriction enzymes, including BamHI, BstI, BsuI, EcoRI, EcoRV, HhaI, HindIII, PvuII, ScaI, and TaqI. HaeIII, the isoschizomer of BsuRI, does not show any star activity under the same conditions for BsuRI*.

(c) Secondary activity. Under suboptimal reaction conditions, many restriction enzymes display relaxed specificities and cleave, at moderate rates, the noncanonical sites that contain in the core tetranucleotide a single base change in a 5/6 homology context. This activity has been described as secondary activity in a manner loosely interchangeable with the star activity. We will adopt here the term "secondary activity" in a more restricted sense to make a distinction from the star activity that was defined earlier. Depending on reaction conditions, the cleavage rate at the next best noncanonical site can increase more than 500-fold, while the cleavage at the canonical site is reduced by about 10-fold. The efficiency of discrimination can thus be reduced by more than 10^3-fold (12). Characteristic manifestation of the secondary activities includes (i) appearance of a nicking activity and (ii) a lower affinity for Mg^{2+} (10 mM being far from saturation when compared with less than 1 mM to saturate the cleavage reaction at canonical sites).

ix. Increase of site specificity. Type II restriction enzymes maintain the capabilities of specific sequence recognition and cleavage, although the rigor of site specificity can be relaxed under certain reaction conditions. Indeed, site specificity can be increased under some other reaction conditions, by applying combination of strategies, and/or by modulating canonical (or flanking) sequences into sequences composed of higher number of nucleotides by means of methylation.

(a) Presence of spermidine. Spermidine is a trivalent organic cation that plays an essential role in an organism. Several type II R · ENases (BamHI, BsuRI, EcoRI, HindIII, PstI, and SalI) demonstrate a significant increase in cleavage accuracy in the presence of spermidine (3.3 mM) (13). The spermidine effect is observed regardless of whether the nonspecific cleavage is induced by high concentrations of enzyme under optimal buffer conditions or by high pH, low ionic strength, organic solvents, and Mn^{2+} ions.

The increased specificity is presumed to be due to an enhancement of the cleavage rate at the canonical site and a slowing down of the cleavage rate at other related sites. In some cases, a 40-fold activation can be observed (14). The spermidine concentration at which the activation occurs is dependent on salt concentrations, for example, 8 mM spermidine at a high salt (50 mM NaCl and 10 mM $MgCl_2$) vs 1 mM spermidine at a low salt concentration. Note that other polyamines (e.g., spermine) which are larger and/or branched are rather inhibitory to restriction enzymes due to their binding and stabilizing (or condensing) effect on DNA (15).

(b) Presence of activator DNA. Not all recognition sequences are cleaved equally; this is probably due to the differential influence of flanking sequences.

The sensitivity of two isoschizomers to these "resistant" or "slow" sites can also be different. For instance, certain CTCGAG sites are totally refractory to PaeR7, whereas they can be cut by XhoI (16).

For certain ENases, the cleavage at the "slow" or refractory sites can be significantly increased by the addition of cleavable DNA or spermidine (14,17). In the case of EcoRII, a short oligonucleotide (14-mer) duplex containing a single EcoRII site can exert an activator function (17). In addition to EcoRII, other enzymes that cut the target DNA much more efficiently in the presence of an excess (e.g., 100-fold in site concentration) of oligonucleotide duplexes containing enzyme-specific recognition sites include AtuBI, BspMI, Cfr9I, Eco57I, HpaII, Ksp632I, NaeI, NarI, SacII, and SauBMKI (14,18). These ENases are presumed to be among the group of proteins which interact with two DNA sites simultaneously and exhibit functional cooperativity. Activator DNA may increase V without altering K_m (e.g., for BspMI and NaeI) or decrease K_m without changing V (e.g., for HpaII, NarI, and SacII). DNAs devoid of the enzyme-specific recognition site do not interfere with activity.

(c) Increase of specificity by DNA methylation. The restriction activity of most type II enzymes is blocked by methylation on the recognition site by the cognate MTases. In addition, restriction enzymes are more or less sensitive to the recognition of overlapping sequences methylated by noncognate M · MTases. Therefore appropriate combinations of M · MTases and R · ENases provide a means to increase the site specificity (for details, see Section III,A, this chapter).

Site-specific DNA methylation can also be incorporated into other strategies in which recognition sites are reversibly modulated to offer highly selective cleavage capabilities to the restriction enzymes (see below).

x. **Reversible modulation of restriction activity.** The site specificity of restriction enzymes can be reversibly modulated by DNA-binding ligands (intercalators), DNA-binding proteins, or triplex-forming oligonucleotides.

(a) DNA intercalators. The action of ligands and intercalators depends on the nucleotide sequence of the recognition site and their local environment. DNA intercalators, such as G · C-specific actinomycin D, A · T-specific distamycin A or DAPI (6,4'-diamidino-2-phenylindole), and base-nonspecific daunomycin and propidium iodide, limit, but do not inhibit, the action of restriction enzymes (22,23). Different intercalators affect the restriction enzymes differently, blocking preferentially the site(s) the enzyme least prefers.

Intercalation of EtBr may inhibit the cleavage of DNA by restriction enzymes. Partial digestion in the presence of EtBr results in mixtures with one or both strands cleaved, among them a complete set of full-length permuted linear molecules. Such singly cleaved permuted linear molecules provide a convenient method for restriction mapping of DNA (24).

(b) DNA-binding proteins. Proteins such as *lac* repressor and phage λ repressor tightly bind DNA and protect a restriction site (HhaI or HaeII, respectively)

when it is embedded in the "operator" region, but leave all other sites available for cleavage. Such protein binding also inhibits modification by specific M · MTases, M · HhaI or M · HphI, while all other similar sites are methylated. Therefore the sequential use of DNA-binding proteins and MTases can direct the DNA restrictions uniquely to the protein-binding sites (19).

(c) Triplex-forming oligonucleotides. Oligonucleotides with defined sequences can bind a duplex and form a triplex either in the presence of recombinases or other appropriate conditions (see Section II,C, Chapter 1). When the site of triplex formation is within or flanking a restriction site, the cleavage of the dsDNA can be inhibited. Triplex-mediated inhibition has been observed with *Eco*RI (25) and *Ksp*632-I [a class IIS enzyme recognizing CTCTTC] (26). A target triplex site, for example, formed by a homologous oligonucleotide in the presence of RecA protein (60), is not amenable to methylation either. This permits all other regular recognition sites for a restriction enzyme (e.g., *Eco*RI) to be methylated and protected from *Eco*RI cleavages, while allowing a subsequent selective cleavage at the target site (refer to Fig. 4.5).

(d) Oligonucleotide analogs (peptide nucleic acids). Oligonucleotide analogs in which the deoxyribose–phosphate backbone is replaced by a peptide backbone consisting of (2-aminoethyl)glycine units, thus called *peptide nucleic acids* (PNA), are potentially useful agents for selectively protecting certain target sites. The binding of PNA to dsDNA in low salt buffers is stable ($t_{1/2} > 24$ hr), sequence specific, and involves two PNA molecules, $(PNA)_2$/dsDNA, resulting in strand displacement (61). Consequently, the binding of appropriately designed PNA to a target sequence immediately flanking restriction sites (e.g., *Bam*HI, *Sal*I, and *Pst*I) can cause complete inhibition of cleavages at these sites.

xi. Restriction sites containing base analogs. Irreversible site modifications such as base modifications and substitutions can seriously affect the action of certain restriction enzymes.

(a) Methylation. Site-specific DNA methylations by cognate M · MTases and noncognate or non-type II MTases that produce overlapping methylations exert varying effects on R · ENases (27,28). Canonical site-specific methylation always blocks the cleavage by R · ENases. Methylation at noncanonical sites inhibits the rate of duplex DNA cleavage at least 10-fold for some restriction enzymes whereas, for some other enzymes, noncanonical methylation poses no effect on restriction cleavage. For example, *Bam*HI cannot cut DNA methylated at GGATm^{4}CC or GGATm^{5}CC but it cuts DNA modified at GGATCm^{4}C or GGATCm^{5}C. With the same recognition sequences (e.g., CCGG), some isoschizomers (e.g., *Hpa*II and *Msp*I) differ in their sensitivity to methylation at particular sites: C^{m}CGG is cut by *Msp*I, but not by *Hpa*II, whereas ^{m}CCGG is refractory to *Msp*I or is cleaved by *Hpa*II very slowly and in a flanking sequence-dependent manner.

Restriction enzymes may also show distinct responses on hemimethylated recognition sites. For example, *Eco*RI and *Dpn*I cannot cleave hemimethylated sites.

In contrast, *Bsp*I and *Sau*3AI carry out single-strand cleavages within the unmethylated strand. *Dpn*I is unique in that it cleaves DNA (G^mA|TC) only when both strands are methylated on the A base.

DNAs containing m^5C can be conveniently synthesized by directly incorporating 5-methyl-dCTP using standard PCR procedures except that the denaturation temperature is elevated to 100°C (62).

(b) Base analog substitutions. Substitutions of pyrimidines (U for T) affect the restriction enzymes that recognize the A · T-containing site differently: *Hpa*I, *Hin*dII, and *Hin*dIII are slightly retarded, but *Eco*RI and *Bam*HI are not affected (29). 5-Hydroxymethyluridine in place of T retards the cleavage by enzymes whose recognition sequences contain A · T bp. The substitution has no effect on enzymes recognizing G · C-only sequences (30). Glucosylation on hm^5C renders the DNA (e.g., of T-even phages) resistant to cleavage by such R · ENases as *Bam*HI, *Eco*RI, *Hha*I, *Hin*dIII, and *Hpa*II (29). 5-Br-U affects both the rates and the sites of cleavages by *Eco*RI, *Bam*HI, *Hin*dIII, *Hpa*I, and *Sma*I. *Eco*RI and *Hin*dIII, both of which recognize nucleotide sequences that contain four T bases (or two T's per strand) GAATTC and AAGCTT, respectively, cleave fully substituted BrU-DNA at the same site but at a reduced rate. However, *Sma*I, whose recognition sequence (CCCGGG) does not contain T bases, cleaves some of the canonical sequence at approximately the same rate while some other sites are not cleaved, reflecting the influence of DNA structures outside of the canonical sequence (31).

The substitution of 7-deaza-dGTP renders many commonly used restriction sites refractory to cleavage by their restriction enzymes. Among the enzymes tested that recognize six-bases containing at least one G base, only two enzymes (*Hin*dIII and *Xba*I) are not inhibited by the presence of modified G residues (32). The enzymes which fail to cut 7-deaza-G-containing sequences include *Acc*I, *Bam*HI, *Eco*RI, *Pst*I, *Sal*I, and *Sma*I.

(c) Phosphorothioate substitutions. Thionucleotides (dNTPαS) having a S_p diastereomeric form can be efficiently incorporated into DNA by many DNA polymerases, resulting in a phosphorothioate internucleotide linkage of the R_p configuration in the new (−) DNA strand. When a phosphorothioate is appropriately placed at the cleavage position, some enzymes can cut the DNA, albeit at reduced rates, while other enzymes cannot. *Eco*RI, *Bam*HI, and *Hin*dIII cleave appropriate phosphorothioate-substituted DNA at a reduced rate and by an accentuated two-step process in which all of the DNA is converted to isolable nicked intermediates with the nick in the nonsubstituted (+) strand (33). In contrast, *Sal*I cleaves substituted DNA at a slightly lower rate than it cleaves unmodified DNA and without the formation of nicked intermediates. Some other enzymes, such as *Ava*I, *Eco*RV, *Hha*I, *Hin*dII, *Hpa*II, *Pst*I, *Pvu*II, and *Sac*I are only capable of cleaving the unmodified strand in the heteroduplex and not the phosphorothioate linkage (34). The differential sensitivity of phosphorothioate linkages toward certain restriction enzymes and exonucleases has been used as the basis for an

efficient site-directed mutagenesis method (see Appendix A) (35) and a DNA subcloning strategy for nonrandom DNA sequencing (see Section V, A, Chapter 3).

xii. Monitoring restriction activity. Sequence specificity of restriction enzymes can be differentiated from one another by gel analysis of their characteristic cleavage patterns on some standard DNAs, such as phage λ, SV40, ϕX174, or plasmid pBR322. Sequence specificity, as well as the activity of some restriction enzymes, is strongly dependent on reaction conditions such as the quality of DNA, ionic strength (i.e., salt concentration), pH, and temperature. Fortunately, some enzymes have similar optimal reaction conditions, which allows for grouping into several "restriction buffer systems." Indeed such a grouping is not only convenient for dealing with a wide variety of enzymes but, in a way, necessary to maintain consistency in data comparison among enzyme users.

b. Assay methods and unit definition

Restriction enzyme activities can be assayed by various methods including gel visualization assay, polynucleotide kinase exchange assay (36), and exonuclease-coupled assay (37). By far the most common of all assay methods is the visualization by EtBr staining of the digestion products separated by electrophoresis in an agarose gel (30,38). This method is straightforward and enables the detection of contaminating activities of endo- and exonucleases.

Unit definition: One restriction unit is usually defined as the amount of enzyme required to digest completely 1 μg of DNA (typically phage λ) in 1 hr. Because the definition relies on the activity at an end point of the reaction, it is distinguished from the usual kinetic definition of enzyme units based on the initial reaction rate.

c. Characterization of recognition and cleavage sites

The following two approaches are most commonly used to obtain information on the nature of recognition sequences and the position of cleavage sites.

i. Computer-assisted approach. For the majority of type II enzymes, especially type IIS enzymes, the recognition sequence can be identified most conveniently by means of a computer-assisted approach. Using DNAs of known sequences, the fragmentation pattern generated by a R·ENase of unknown specificity is compared and matched to a pattern predicted by a computer search of positions, frequencies, and distances of all possible palindromic sequences (39–41). One obvious limitation of the computer-assisted approach is when the enzyme recognizes a "new" sequence pattern not seen before or easily guessed. In this case, one is obliged to resort to direct nucleotide sequencing.

ii. Determination by direct nucleotide sequencing. The cleavage site characteristic of a restriction enzyme can be identified by direct nucleotide sequencing. One of the most useful strategies that can be applied to any type of fragment termini is to align the cut in each strand alongside an enzymatically or chemically produced ladder sequence (42).

d. Optimal reaction conditions

Despite the diversity of sources and specificities, type II restriction enzymes have remarkably similar, simple reaction conditions (4,43–45). The restriction enzymes can thus be subgrouped according to refined optimal conditions, leading to reproducible and efficient uses of restriction enzymes in specific DNA cleavages (46). [Most enzyme suppliers offer their conditioned buffers in 10× strength.]

A typical reaction mixture contains DNA substrates and 1–5 U/μg DNA of enzyme in a group-specific buffer (pH ~7.5) that contains Mg^{2+}, frequently Na^+, and other stabilizing agents, such as 2-MSH, BSA, and glycerol. The reaction mixture is incubated usually at 37°C for 1 hr.

To reduce potential nonspecific cleavages, spermidine (pH 7.0, 4 mM) can be added to the reaction mixture. The reaction can be stopped by adding excess EDTA (20 mM). Some thermolabile restriction enzymes can be simply inactivated by incubation for 10 min at 65° or 70°C. Inactivation of more stable enzymes requires phenol/chloroform extraction.

The reaction conditions can in fact be further optimized according to the specific properties of individual enzymes. Some of the readily adjustable factors are as follows.

(a) Temperature. Some enzymes are optimally active at temperatures higher or lower than the standard 37°C; for example, at 25°C for *Hin*dIII, 45°C for *Ava*I, 55°C for *Acc*I and *Hpa*II, 60°C for *Tha*I, and 65°C for *Hae*III and *Taq*I.

(b) Thermal stability. The stability of restriction enzymes during "normal" digestion or digestions longer than 1 hr varies widely depending on the enzyme and the presence or absence of DNA substrates (47). Some enzymes retain full activity after a 5-hr incubation while others are completely inactive after 1 or 2 hr. Among the six-base cutters, four enzymes (*Bam*HI, *Kpn*I, *Pst*I, and *Sma*I) have been shown to be the most stable under prolonged digestion conditions.

(c) Alternative buffer systems. A single buffer system for all R · ENases has also been devised that mimics the most abundant intracellular conditions of bacteria. This *universal buffer* system employs 50–100 mM potassium glutamate buffer (pH 7.6, KGB) (48) or its variant form KAc buffer (KAB) (49). It should be noted, however, that these "universal" buffers are not optimal for all R · ENases.

e. Strategies for double digestions

Double digestions of the MCS of M13mp and its derivative vectors (see Fig. 8.3) are a useful technique for the cloning of a DNA fragment into a predetermined orientation. However, because of the proximity of restriction sites in the MCS, the minimal substrate size becomes a critical factor. In fact, it is sometimes difficult to obtain complete digestion with the second enzyme. To make the situation more complicated, the small difference in the length of the singly digested vector compared to the double-digested vector does not allow easy monitoring of the progress of the second digestion nor purification of the double-digested vectors

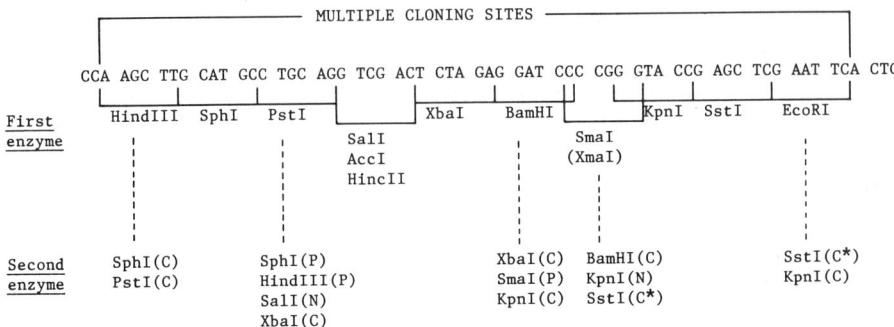

FIGURE 4.2 Efficiencies of double digestions at the multiple cloning sites of pUC19 plasmid. DNA (0.5 µg) was digested under the "standard" reaction conditions with 10 U of selected restriction enzymes for 1 hr. After the first digestion, DNA was precipitated with ethanol, resuspended, and subject to the second digestion. C, completion digestion; C*, evident star activity; P, partial digestion; N, no digestion. [After J. Crouse and D. Amorese (1986). *FOCUS (BRL)* 8(3), 9.]

by agarose gel electrophoresis. The difficulties can be circumvented, at least partially, by the following general approaches.

i. **Direct digestion approach.** The second digestion can be performed in general without much problem when the second cutting site lies beyond that immediately next to the primary cleavage site. When two selected sites are next to each other, the order of digestion becomes an important factor (Fig. 4.2). For example, to prepare a double-digested vector with *Sma*I and *Bam*HI termini, it is most efficient to digest first with *Sma*I and then with *Bam*HI instead of the other way around (50).

It is recommended that the double digestions be performed in succession rather than simultaneously, precipitating the DNA with ethanol after the first digestion. This is imperative when the two digestion buffers are incompatible. Even if the digestion buffers are compatible, the successive digestions avoid possible problems associated with competitive binding of the restriction enzymes at the two recognition sites.

ii. **Use of spacer-carrying vectors.** This approach entirely avoids the problems arising from the digestion of two proximal restriction sites. In this approach, a readily sizable DNA spacer (e.g., ~500 bp) is inserted in the middle of the MCS to make a utility vector (51). Two target sites are chosen such that one comes from each half of the MCS. Double digestion of such utility vectors is essentially identical to the situation of "standard" single digestions. Despite the initial efforts in preparing such a utility vector, the effort is amply rewarded by being able to separate the double-digested vector cleanly from the singly digested ones by agarose gel electrophoresis.

2. STRUCTURES

Type II restriction enzymes that have been isolated so far in homogeneous form are, possibly with a few exceptions, composed of two identical subunits. The sizes of type II enzymes vary widely with an average of ~280 amino acid residues per subunit (22–37 kDa). [The MTases have an average of 380 amino acid residues.] In a few exceptional cases, e.g., *Bgl*I (52) and *Bsp*I (53), the MW of the native form of the enzymes corresponds to that of the monomers. The majority of enzymes exist in solution as dimers and, at high concentrations, as oligomers and sometimes as higher aggregate states.

3. GENETICS

A number of restriction enzymes have been cloned either alone or together with M·MTases (54). The organization of the genes for type II R–M systems is extremely diverse (Fig. 4.1). The two genes bear very little, if any, homologies in their primary and secondary structures. Overall, the R and M genes are located more or less side by side, either in the same or opposite orientations and, in some cases, with a small ORF between or outside the R and M genes. The small ORF is presumed to play a regulatory function in the concerted expression of the R–M genes (55–57).

4. SOURCES

Restriction ENases have been found in both gram-positive and gram-negative bacteria. *Bacillus, E. coli, Haemophilus,* and *Streptomyces* are particularly fruitful genera as sources for restriction enzymes. Restriction enzymes are purified by various conventional (e.g., phosphocellulose, phenyl- or aminopentyl-Sepharose) and affinity (e.g., heparin, Cibacron blue) column chromatographic techniques as well as by HPLC. The combination of polyethyleneimine precipitation and chromatography on heparin–agarose is recommended as the first approach or the backbone of a general approach to purification schemes (58).

Matrix-bound enzymes: Some enzymes such as *Eco*RI and *Bam*HI have been immobilized on CNBr-activated Sepharose (59). Matrix-bound enzymes have enzymatic properties indistinguishable from those of soluble enzymes. However, immobilized enzymes have a higher thermal stability. Compared with soluble enzymes, immobilized enzymes offer advantages in their easy removal and repetitive use.

Storage: Purified restriction enzymes are generally stored at −20°C in buffers containing 50% glycerol. Activity loss can be further prevented by adding autoclaved gelatin or BSA to a final concentration of 50–100 µg/ml. Some enzymes (e.g., *Eco*RI and *Pst*I) are also stabilized by the presence of neutral detergents (e.g., 0.2% Triton X-100).

References

1. Roberts, R. J. (1990). *NAR* **18**(Suppl.), 2331–2365.
2. Roberts, R. J., and Macelis, D. (1993). *NAR* **21**, 3125–3137.
3. Yoo, O. J., and Agarwal, K. L. (1980). *JBC* **255**, 10559–10562.

4. Blakesley, R. W., Dodgson, J. B., Nes, I. F., and Wells, R. D. (1977). *JBC* **252**, 7300–7306.
5. Degtyarev, S. K., Rechkunova, N. I., Kolyhalov, A. A., Dedkov, V. S., and Zhilkin, P. A. (1990). *NAR* **18**, 5807–5810.
6. Singleton, C. K., Klysik, J., and Wells, R. D. (1983). *PNAS* **80**, 2447–2451.
7. Shouche, Y. S., Ramesh, N., and Brahmachari, S. K. (1990). *NAR* **18**, 267–275.
8. Vardimon, L., and Rich, A. (1984). *PNAS* **81**, 3268–3272.
9. Winkle, S. A. Aloyo, M. C., Morales, N., Zambrano, T. Y., and Sheardy, R. D. (1991). *Biochem.* **30**, 10601–10606.
10. Petranovic, M., Petranovic, D., Dohet, C., Brooks, P., and Radman, M. (1990). *NAR* **18**, 2159–2162.
11. Thielking, V., Alves, J., Fliess, A., Maass, G., and Pingoud, A. (1990). *Biochem.* **29**, 4682–4691.
12. Taylor, J. D., and Halford, S. E. (1989). *Biochem.* **28**, 6198–6207.
13. Pingoud, A. (1985). *EJB* **147**, 105–109.
14. Oller, A. R., Broek, W. V., Conrad, M., and Topal, M. D. (1991). *Biochem.* **30**, 2543–2549.
15. Kirino, H., Kuwahara, R., Hamasaki, N., and Oshima, T. (1990). *J. Biochem.* **107**, 661–665.
16. Gingeras, T. R., and Brooks, J. E. (1983). *PNAS* **80**, 402–406.
17. Pein, C.-D., Reuter, M., Cech, D., and Kruger, D. H. (1989). *FEBS Lett.* **245**, 141–144.
18. Reuter, M., Kupper, D., Pein, C.-D., Petrusyte, M., Siksnys, V., Frey, B., and Kruger, D. H. (1993). *Anal. Biochem.* **209**, 232–237.
19. Koob, M., Grimes, E., and Szybalski, W. (1988). *Science* **241**, 1084–1086.
20. Molloy, P. L., and Symons, R. H. (1980). *NAR* **8**, 2939–2946.
21. Okayama, H., and Berg, P. (1982). *Mol. Cell. Biol.* **2**, 161–170.
22. Kania, J., and Fanning, T. G. (1976). *EJB* **67**, 367–371.
23. Nosikov, V. V., Braga, E. A., Karlishev, A. V., Zhuze, A. L., and Polyanovsky, O. L. (1976). *NAR* **3**, 2293–2301.
24. Parker, R. C., Watson, R. M., and Vinograd, J. (1977). *PNAS* **74**, 851–855.
25. Hanvey, J. C., Shimizu, M., and Wells, R. D. (1990). *NAR* **18**, 157–161.
26. Francois, J.-C., Saison-Behmoaras, T., Thuong, N. T., and Helene, C. (1989). *Biochem.* **28**, 9617–9619.
27. Kessler, C., and Holtke, H. J. (1986). *Gene* **47**, 1–153.
28. Nelson, M., and McClelland, M. (1991). *NAR* **19**(Suppl.), 2045–2071.
29. Berkner, K. L., and Folk, W. R. (1979). *JBC* **254**, 2551–2560.
30. Sharp, P. A., Sugden, B., and Sambrook, J. (1973). *Biochem.* **12**, 3055–3063.
31. Marchionni, M. A., and Roufa, D. J. (1978). *JBC* **253**, 9075–9081.
32. Grime, S. K., Martin, R. L., and Holaway, B. L. (1991). *NAR* **19**, 2791.
33. Potter, B. V. L., and Eckstein, F. (1984). *JBC* **259**, 14243–14248.
34. Sayers, J. R., Olsen, D. B., and Eckstein, F. (1989). *NAR* **17**, 9495.
35. Sayers, J. R., and Eckstein, F. (1988). *Gen. Eng.* **10**, 109–122.
36. Berkner, K. L., and Folk, W. R. (1980). *Meth. Enzy.* **65**(I), 28–36.
37. Lackey, D., and Linn, S. (1980). *Meth. Enzy.* **65**(I), 26–28.
38. Aaij, C., and Borst, P. (1972). *BBA* **269**, 192–200.
39. Fuchs, C., Rosenvold, E. C., Honigsman, A., and Szybalski, W. (1978). *Gene* **4**, 1–23.
40. Gingeras, T. R., Milazzo, J. P., and Roberts, R. J. (1978). *NAR* **5**, 4105–4127.
41. Devereux, J., Haeberli, P., and Smithies, O. (1984). *NAR* **12**, 387–395.
42. Brown, N. L., and Smith, M. (1980). *Meth. Enzy.* **65**, 391–404.
43. Greene, P. J., Poonian, M. S., Nusbaum, A. L., Tobias, L., Garfin, D. E., Boyer, H. W., and Goodman, H. M. (1975). *JMB* **99**, 237–261.
44. Hinsch, B., and Kula, M. R. (1980). *NAR* **8**, 623–633.
45. Fuchs, R., and Blakesley, R. (1983). *Meth. Enzy.* **100**, 3–38.
46. Blakesley, R. W. (1987). *In* "Gene Amplification and Analysis" (J. G. Chirikjian, Ed.), Vol. 5, pp. 51–102. Elsevier, New York.
47. Crouse, J., and Amorese, D. (1986). *FOCUS (BRL)* **8**(3), 1–2.
48. McClelland, M., Hanish, J., Nelson, M., and Patel, Y. (1988). *NAR* **16**, 364.
49. Simcox, M. E., Davis, S. J., and Simcox, T. G. (1989). *Strategies* (Stratagene) **2**, 60–61.

50. Crouse, J., and Amorese, D. (1986). *FOCUS (BRL)* **8**, 9.
51. Eun, H. M., and Yoon, J. W. (1989). *BioTechniques* **7**, 992–997.
52. Lee, Y. H., and Chirikjian, J. G. (1979). *JBC* **254**, 6838–6841.
53. Koncz, C., Kiss, A., and Venetianer, P. (1978). *EJB* **89**, 523–529.
54. Wilson, G. G. (1991). *NAR* **19**, 2539–2566.
55. Tao, T., Bourne, J. C., and Blumenthal, R. M. (1991). *J. Bacteriol.* **173**, 1367–1375.
56. Heidmann, S., Seifert, W., Kessler, C., and Domdey, H. (1989). *NAR* **17**, 9783–9796.
57. Brooks, J. E., Nathan, P. D., Landry, D., Sznyter, L. A., Waite-Rees, P., Ives, C. L., Moran, L. S., Slatko, B. E., and Benner, J. S. (1991). *NAR* **19**, 841–850.
58. Pirrotta, V., and Bickle, T. A. (1980). *Meth. Enzy.* **65**(I), 89–95.
59. Lee, Y. H., Blakesley, R., Smith, L. A., and Chirikjian, J. G. (1980). *Meth. Enzy.* **65**(I), 173–182.
60. Ferrin, L. J., and Camerini-Otero, R. D. (1991). *Science* **254**, 1494–1497.
61. Nielsen, P. E., Egholm, M., Berg, R. H., and Buchardt, O. (1993). *NAR* **21**, 197–200.
62. Wong, K. K., and McClelland, M. (1991). *NAR* **19**, 1081–1085.

B. Type II Restriction Endonuclease: *Eco*RI
[Type II site-specific deoxyribonuclease, EC 3.1.21.4]

*Eco*RI restriction (R) and modification (M) enzymes are probably the best studied of all type II R–M systems in terms of structure and function. R · *Eco*RI specifically recognizes a palindromic hexanucleotide sequence, 5′-GAATTC-3′, in dsDNA and cuts the phosphodiester bond between G and A, producing a 5′ protruding four-base termini. *Eco*RI is the first restriction endonuclease (ENase) to be purified to homogeneity and in a large quantity. *Eco*RI ENase serves as the prototype of type II ENases in a detailed understanding of their endonucleolytic actions and protein–DNA interactions.

1. Functions

a. Reaction conditions

i. Recommended reaction conditions. A typical reaction mixture for *Eco*RI digestion consists of 50 mM Tris–Cl (pH 7.5), 50 mM NaCl, 10 mM MgCl$_2$, and 50 µg/ml BSA. The mixture is incubated for 1 hr at 37°C. The reaction can be stopped by adding EDTA (20 mM final concentration), by heating for 5 min at 70°C, or by extracting with phenol/chloroform.

The activity of *Eco*RI is dependent on pH and temperature. The pH optimum for *Eco*RI is 7.0–7.5 in 50–100 mM Tris–Cl (1). The temperature optimum for cleavage reactions is dependent on substrate structures: for a synthetic self-complementary octanucleotide (pTGAATTCA), the optimum temperature is 15°C, slightly below its T_m of 19°C (2). The optimum temperature is 37°C for a single *Eco*RI site-containing SV40 DNA or other large-size substrates. Incubation of *Eco*RI for 5 min at 60°C in the presence of λ DNA results in over 90% inactivation.

The activity of *Eco*RI is subject to inhibition by various DNA intercalators and enzyme modifiers (see "Modulation of site specificity" below).

ii. Kinetic parameters. ENase activity obeys Michaelis–Menten kinetics (3–5). The K_m is 2.5–30 nM (in *Eco*RI site concentration) for large-size DNAs

TABLE 4.2 Kinetic Parameters of EcoRI Cleavage Reactions

Parameters	Plasmid[a] (pBR322)	Oligonucleotides (13-mers)[b]
K_{d1} (k_{-1}/k_1, M)	—	$\leq 10^{-6}$–10^{-8}
k_2 (min^{-1})	50	22–78
k_3 (min^{-1})	27	2–4
k_4 (min^{-1})	—	16–240
K_{d5} (k_5/k_{-5}, M)	—	10^{-7}–10^{-8}
k_{cat} (min^{-1})	0.7–13[c]	6–36
K_m	2.5–30 nM[d]	7–16 μM[e]

[a] Temperature: 37°C (20). For definition of parameters, refer to Scheme 4.1.

[b] Ranges of values are for two different 13-mer substrates differing only in flanking sequences. Reaction at pH 7.2 (20 mM Tris–Cl), 50 mM NaCl, 10 mM MgCl$_2$, and 25°C (6,24).

[c] Ranges of values are dependent on $I = 0.059$ (25 mM NaCl) – 0.23 (200 mM NaCl). Reaction at pH 7.6 (20 mM Tris–Cl) and 5 mM MgCl$_2$ (21).

[d] For DNAs such as pMB9 (supercoiled form), SV40, ColE1, and phage λ (3–5,63).

[e] For octanucleotide substrates (2).

(pMB9, SV40, ColE1, and λ phage), and the k_{cat} is 1–4 dsDNA scission per min (Table 4.2). The K_m for synthetic octanucleotide substrates is 7–16 μM at optimal 15°C, ~200 times greater than that for natural DNAs. The k_{cat} for oligonucleotide substrates is 4–36 cleavages per min (2,6).

In reactions monitoring the cleavage of only the first phosphodiester bond, the following parameters were obtained: for pUC8 plasmid DNA, a K_m of 50 nM and k_{cat} of 19 min^{-1} (or k_{cat}/K_m of 6.3 μM^{-1} sec^{-1}) (7) and, for a tridecanucleotide, a K_m of 16 nM and k_{cat} of 18 min^{-1} (or k_{cat}/K_m of 19 μM^{-1} sec^{-1}) (8).

iii. **Metal ion requirements.** Like all type II ENases, the EcoRI enzyme requires Mg^{2+} (1–15 mM) for its activity. The k_{cat} increases linearly with increasing MgCl$_2$, reaching a plateau value at 20 mM (9). In the absence of Mg^{2+}, EcoRI is noncatalytic although it can bind and form stable sequence-specific complexes with substrate DNAs. For example, pBR322, which contains a single EcoRI recognition site, binds the enzyme with a K_d of 0.03–5.0 nM depending on the ionic strength of the medium (5,10). The K_d values for oligonucleotides range from 0.25 nM to 0.2 μM (8,11).

NaCl stimulates the enzyme activity; however, concentrations of NaCl greater than 0.1 M are inhibitory.

b. Substrate specificities

i. **Recognition sequence.** EcoRI recognizes the hexanucleotide palindromic sequence 5'-G|AATTC-3' in dsDNA and cleaves the phosphodiester bonds on both strands between the G and the A bases (12). The EcoRI cleavage produces four base-protruding 5' ends, pAATT. EcoRI has several isoschizomers, most notably *Rsr*I and *Sso*I which cut the recognition sequence at the same site.

ii. Site specificities

(a) At the canonical sequence. The site specificity for the canonical recognition sequence is estimated to be $\sim 10^7$ times higher than for alternative sequences on DNA (13) [or $\geq 10^5$ times higher at the DNA-binding level (8)].

(b) At noncanonical sequences containing a single base mismatch. When there is a single base mismatch in the recognition sequence contained in tetradecanucleotides, the double-strand cleavages occur in the following order, where each > sign denotes one order of magnitude difference (14): position 1, [G] >>> A > T > C; position 2, [A] >>> G ≈ C ≈ T; position 3, [A] >> C > G >> T; position 4, [T] >> G > C >> A; position 5, [T] >>>> C ≈ G ≈ A; and position 6, [C] >>> T > A > G. In the majority of cases, two strands are cleaved at considerably different rates. The strand containing the mismatching base is usually cleaved more slowly than the normal strand, which in turn is cleaved, with a few exceptions, more slowly (up to 100-fold) than the counterpart in double-stranded canonical sequences. The double-strand recognition sites containing a single base mismatch bind the *Eco*RI with 10- to 130-fold lower affinities than the canonical sequence.

In general, mismatching oligonucleotides are much better substrates than *Eco*RI* sites (see below).

(c) At noncanonical sequences in complementary double strands. Under other than normal reaction conditions, *Eco*RI shows a relaxed specificity, called *star activity* (*Eco*RI*), which recognizes the core tetranucleotide (5'-X|AATTY-3') from normal hexanucleotides (1). The recognition specificity is generally lowered under relaxed conditions, while the flanking nucleotides exert subtle effects on the cleavage by *Eco*RI*. When compared with the [G], the preference for the nucleotide (X) is A (1/2,000) > T (1/30,000) > C (1/400,000) (14). Noncanonical sequences that derive from a single base pair change in the core tetranucleotide are more or less subject to cleavage by *Eco*RI, which is termed *secondary activity*. The hierarchy of activity for the variants at the second position is, with respect to the [A], C (1/130,000) ≈ T (1/130,000) > G (<1/10^6), while at the third position [A], C (1/1) >>> G (1/7,000) >> T (<1/10^6). Hence under relaxed conditions, the canonical hexanucleotide sequence remains the most preferred restriction site.

The rates of cleavage in the two strands of *Eco*RI* or secondary sites can be widely different, i.e., from 1 to 700 times. The binding affinities of the *Eco*RI* sites range from 1/400 to 1/8000 of that of the canonical site.

Certain members (e.g., G|AATTN or T|AATTC) of the set of *Eco*RI* sites are subject to single-strand but not double-strand cleavage under the supposedly optimal conditions (15). This unusual activity has been termed *nicking activity*, *Eco*RI^n. Although the identity of nicking activity, which is present only in certain preparations of *Eco*RI enzyme (16), is not well established, nicking is considered to be an environment-sensitive intermediary step during the EcoRI catalysis and inherent to the enzyme (see below, catalytic mechanism).

The unmasking of numerous other secondary (or noncanonical) sites that would normally avoid cleavage under stringent conditions provides a useful means for DNA mapping. The star or secondary activity can also be exploited to cleave DNA molecules containing no canonical EcoRI site.

(d) Relaxed conditions. Star activities can be observed in any of the following conditions or combinations thereof: (i) alkaline conditions (pH 8.0–9.5) and low ionic strengths (≤ 25 mM Tris–Cl or NaCl, <2 mM Mg^{2+}) (1); (ii) presence of 2 mM Mn^{2+} (but not other divalent cations such as Ca^{2+}, Cu^{2+}, and Zn^{2+}) in place of or in addition to Mg^{2+} even under normal reaction conditions (17); (iii) presence of low MW organic solvents, e.g., 38–58% ethylene glycol (18), 40–50% glycerol (cf. no star activity below 20%), 4–6% Me_2SO or DMF; and (iv) high enzyme concentrations (19).

Under one or more of these "unusual" conditions, the sequence specificity of restriction enzymes in general is affected rather severely depending on the enzymes. The activities of some enzymes may also be severely inhibited due to altered reaction conditions. The cleavage by EcoRI* is usually limited even after long incubations (e.g., 16 hr at 37°C) under the conditions optimized for star activity.

Methylation of DNA at EcoRI* sites renders the sites largely resistant to EcoRI* cleavage. Actinomycin D (10 μM) which intercalates DNA preferentially at $G \cdot C$-containing sequences selectively suppresses the EcoRI* activity (20). The "normal" EcoRI activity may also be affected by the presence of actinomycin D depending on flanking $G \cdot C$ sequences.

iii. **Chain-length dependence.** EcoRI cuts its restriction site with a preference for either smaller or larger size polynucleotide depending on the ionic strength of the medium. At the low ionic strength of 0.059 (i.e., 25 mM NaCl, 5 mM $MgCl_2$ plus other buffer components), the cleavage rate on pBR322 (4.4 kbp) is 12 times higher than on a 34-bp fragment derived from pBR322 and containing the EcoRI site in a central location. At this ionic strength, the cleavage occurs at least partially in a processive mode by several hundred base pairs (21). No processivity is observed at high ionic strengths. At the high ionic strength of 0.23 (i.e., 0.20 M NaCl, 5 mM $MgCl_2$ plus other buffer components), the cleavage ratio of the pBR322/34-bp fragment is reversed to 0.94. The minimum size of cleavable oligonucleotide substrates is an octanucleotide ($T_m \approx 28°C$) (22).

iv. **Site preference and polarity.** All canonical sites in a given polynucleotide are not cut equally. With λ DNA, the site(s) nearest the right or left terminus is cleaved ~10 times as fast as the sites in the middle of the molecule (23,24). This apparent polarity is also shown with several other restriction enzymes and, in a varying degree, depends on substrate DNAs. The unequal cleavage frequencies, sometimes different by an order of magnitude, are attributed in part to nonspecific binding and to the effect of flanking DNA sequences.

In tridecanucleotides containing an EcoRI restriction site, the flanking A·T bp favors the cleavage of the EcoRI site by a factor of three compared with the flanking G·C bp (6).

v. Effect of methylation and base substitutions. When either one (hemimethylation) or both (full methylation) of the central A's are methylated by EcoRI MTase (see Section III,B, this chapter), EcoRI can no longer cleave the DNA. EcoRI is also sensitive to methylations at noncanonical sites and does not cleave at G^{m}AATTC or GAATT^{m}C. EcoRI shows a reduced rate of cleavage at hemimethylated GAATT^{m}C.

Glucosylation on C (actually on hm^{5}C in phage T2 and T4 DNAs) blocks site-specific binding and cleavage by EcoRI.

Selective replacement of dG by dI in the recognition sequence has only minor effects on EcoRI cleavage (25); the K_m increases about 4-fold while the k_{cat} decreases by ~30% (26). Deletion of the exocyclic amino group from the base A (position 2) reduces more than 10-fold the selectivity of the substrate largely via the increase in K_m (27). Substitution by dU of the T at position 4 (inner T) of the recognition sequence results in a 10-fold decrease of catalytic efficiency under star reaction conditions (46). Replacement of the T residue by 5-BrdU or by hm^{5}U also reduces the rate of EcoRI cleavage (28,29). Removal of 5-CH$_3$ of the T at position 5 (outer T) hardly changes the k_{cat} but slightly increases the K_m (46). Replacement of dC by 5-BrdC renders the substrates completely inactive toward EcoRI.

DNA substrates containing phosphorothioate substitution in the R_p configuration at the scissile bond can be cleaved by EcoRI, albeit at a rate that is 15 times slower.

vi. Modulation of site specificity. Like other type II restriction enzymes, EcoRI is not only sequence specific but also structure sensitive. Some DNA intercalators and ligands binding EcoRI reversibly modify the site preference of EcoRI and result in varying degrees of inhibition on EcoRI activity.

(a) DNA intercalators. The A·T-specific distamycin A (30,31) and DAPI (6,4'-diamidino-2-phenylindole) (32) block two out of five available EcoRI sites on λ DNA at different levels. Actinomycin D, which preferentially intercalates G·C-containing sequences, may interfere with EcoRI depending on flanking G·C sequences. Base-nonspecific intercalators, e.g., daunomycin, ethidium bromide (see below), or propidium iodide, have only partial or no effect on EcoRI sites. The reversible specificity modulation can be useful to probe the environment of DNA-binding proteins and also to generate overlapping DNA restriction fragments for mapping purposes.

(b) Intercalation of ethidium bromide. EtBr ($\varepsilon_{480} = 5450\ M^{-1}\ cm^{-1}$) is an intensely fluorogenic (emission at 595 nm, excitation at 335 nm) and highly mutagenic dye. When bound to dsDNA, the fluorescence quantum yield is ~5 times greater than that of the unbound EtBr (33). The fluorescence increases

linearly as a function of DNA concentration, providing a convenient method for DNA quantitation to a level of 1–5 ng (34). The fluorescence enhancement of EtBr on intercalating double-stranded nucleic acids is also the basis for sensitive fluorescence assays for DNA (or RNA) modifying enzymes and binding proteins (35). The rate of EtBr intercalation is far faster than the usual rates of DNA cleavage by restriction enzymes: the association and dissociation rate constants for linear DNA [d(AT)$_n$] are 8×10^6 M^{-1} sec^{-1} and 35 sec^{-1}, respectively (36). Open circular and linear forms of DNA bind one molecule of dye per 2.5–3 bp, ~2 times more than covalently closed DNA does. The binding of EtBr to the *ccc* DNA induces conformational changes in the plasmid from a negatively to a positively supercoiled form.

Intercalation of EtBr into DNA freezes the internal motion of proton and phosphorus localized to a 2-bp-long DNA region, and eventually alters the time-averaged conformation of DNA to a more rigid structure (37).

EtBr binding to a DNA does not affect *Eco*RI activity in transforming supercoiled plasmid (e.g., pMB9) to an open circular form, but the cleavage of the open circular form to linear products can be inhibited depending on the dye concentration (9). Binding of EtBr blocks the star activity differentially depending on the star site (17).

(c) Enzyme modifiers

(i) Heparin. Whether in its natural form of a highly negatively charged proteoglycan or as a commercial brand which mainly contains sulfated polysaccharide chains without the core protein, heparin binds to a variety of positively charged molecules including the *Eco*RI enzyme and inhibits the cleavage of DNA at certain *Eco*RI sites (38). Because heparin often contaminates DNA preparations, particularly from blood samples, care should be exercised in the interpretation of fragment length analyses following *Eco*RI restriction.

(ii) Polyphosphate. Polyphosphate is a ubiquitous metabolite in microorganisms including several fungal species of *Colletotrichum*. As a contaminant in DNA preparations from such organisms, polyphosphate (MW ~6000) interferes in the digestion of DNA with such restriction enzymes as *Eco*RI, *Bam*HI, *Bst*EII, *Hin*dIII, and *Pst*I (77). One unit of ENase activity can be completely inhibited by 4 μg polyphosphate. The polyphosphate apparently binds to the ENases because inhibition can be overcome by increasing the concentration of enzyme but not substrate. A recommended procedure for avoiding possible inhibition of ENases by polyphosphate is to purify the DNA further by precipitation at 25°C with 0.1 M NaCl and 2 vol of ethanol.

vii. DNA structure dependence

(a) Z-DNA. In contrast to the usual recognition sequence in B-DNA, the left-handed Z-DNA conformation within the recognition sequence or at a distance of four or less base pairs inhibits the cleavage by *Eco*RI (and *Hha*I and *Bam*HI as well) (39). Note that certain restriction enzymes, e.g., *Mbo*I (GATC), show enhanced cleavage activity at a B–Z junction.

(b) Triplex formation. Formation of a triple helix at pH 6.5 via Hoogsteen base pairing within or flanking the *Eco*RI recognition sequence inhibits the *Eco*RI activity at the particular site without affecting other sites (40). Such triplex-forming oligonucleotide-directed inhibition of DNA cleavages is site specific and reversible.

(c) Other structures. The endonucleolytic activity on dsDNA is the same regardless of its conformation, i.e., whether supercoiled *ccc* DNA (Form I), *oc* (open circular) DNA (Form II), or linear DNA molecules (Form III).

The degree of DNA curvature present in the recognition sequence may significantly affect the efficiency of *Eco*RI cleavage (26). Among the curvature-altering modifications to the purine functional groups (at C-2 of G and C-6 of A), deletion of the 2-NH_2 group from the G base (which corresponds to the change of dG to dI) most effectively increases the DNA curvature. Although the NH_2 group of the G base is positioned in the minor groove of the helix and is unlikely to contact the protein, its deletion results in an 83% reduction in the efficiency (k_{cat}/K_m) of *Eco*RI.

c. Mechanism of EcoRI action

i. DNA binding and recognition specificity. The recognition sequences of type II restriction enzymes usually contain twofold rotational (or palindromic) symmetry, although there are a few exceptions in the case of degenerate sites. Indeed crystal structures of DNA–*Eco*RI complex unambiguously show the twofold symmetry of the two enzyme subunits with the twofold axis of the bound DNA (41–43). The DNA is embedded in one side and the major groove is in intimate contact with the protein.

*Eco*RI binds nonspecific as well as specific DNAs with or without Mg^{2+}. In fact, the Gln-111 mutant enzyme lacks cleavage activity, yet it shows a higher than wild-type affinity for the *Eco*RI recognition sequence. The DNA binding involves an alternating array of complementary positive and negative charges (8 ion pairs) between basic amino acid (Arg) residues and six symmetry-related phosphates (three per strand) of the DNA backbone (44). The stringent discrimination for the recognition sequence from the high background of nonspecific sequences is mostly ascribed to 12 H-bonds between the bases of the canonical DNA sequence and the α-helical recognition modules of the protein. The incremental energetic contribution of each protein–base H-bond is 1–2 kcal/mol (45). In contrast, the introduction of any one incorrect natural base pair in the canonical sequence costs 6–13 kcal/mol in transition state interaction energy contributed by both protein–base interaction (direct readout) and protein–phosphate contacts and DNA conformation (indirect readout).

The position-dependent hierarchy of the *Eco*RI* activities is closely correlated with the maximum number of possible H-bonds between the enzyme and the particular *Eco*RI* sequence (41). Arg-200 forms two H-bonds with the G base, whereas Glu-144 and Arg-145 (from the other subunit) form four H-bonds to the adjacent AA bases at N-6s and N-7s, respectively. As reflected by the drastic loss of hexanucleotide recognition on methylation on N-6 of the central A, the negative charge on Glu-144 in addition to its two H-bonds provides an important electrostatic component that enhances the specific recognition in a nonadditive manner. On the other hand, Gln-115 makes hydrophobic interactions with the inner T residue of the recognition sequence and plays a crucial role for coupling specific DNA binding to catalysis (46).

The N-terminal "arm" has no role in sequence specificity but is essential for DNA cleavage because it stabilizes at least 50-fold the substrate DNA–enzyme complex via one electrostatic interaction between the DNA phosphate and either Lys-5 or Arg-9 (47). The "arm" of each subunit provides cooperative support to the scission of each DNA strand by the other subunit.

ii. Mechanism of specificity relaxation.

Both "star" and "secondary" activities belong to relaxed specificities. The exact mechanism by which the relaxed specificity arises remains to be further elucidated, although some possible explanations have been suggested. (i) The reduced ionic strength may contribute to the relaxation by stabilizing charge–charge interactions (salt bridges) between DNA phosphates and basic amino acid residues of protein (49); (ii) alkaline pH may stimulate the relaxed specificity by strengthening the sequence recognition through modified H-bond networks between DNA bases and some newly unprotonated group(s) of the protein; and (iii) Mn^{2+}, which in general binds more firmly to multidentate ligands than Mg^{2+}, may bind a suboptimal binding site for Mg^{2+} which is generated during complex formation between the enzyme and noncanonical DNA substrates, and may facilitate cleavages at "star" sites (46).

iii. Implication of structural modulations on EcoRI function.

Site-specific substitutions in R·EcoRI of the amino acid residue(s) implicated for substrate binding in the cocrystal structure generally result in mutant enzymes with comparable or reduced activity, depending on physicochemical similarities between the wild-type and the replacing amino acids. Surprisingly, none of the single site-specific mutants at such essential residues as E144, R145, and R200 exhibit alteration of the sequence specificity even under circumstances where the catalytic activity is drastically reduced (8,48,50). These observations suggest that the H-bond network identified from the crystallographic structure is not sufficient to explain EcoRI substrate specificity, and call for possible role of subtle factors that might originate from the sequences flanking the canonical one or might come into play at the moment of "allosteric activation" which converts the inactive zymogen-type conformation to a catalytically active one. Indeed, mutants have been identified that exhibit enhanced cleavage activities at EcoRI* sites under "normal" as well as "star" activity conditions. These mutants, obtained by genetic screening based on a phenotype that damages the host DNA despite the presence of M·EcoRI, bear two amino acid substitutions among the four possible amino acids (H114, A138, E192, and Y193) (51). The enhanced star activity mutations also suppress a certain group of EcoRI-binding site mutations (e.g., R200X, where X is T, M, A, and I). As an exception, single mutations (Q115A and Q115E) have been found that render the enzymes catalytically inactive under normal buffer conditions in part due to a diminished affinity toward DNA (46). Under star reaction conditions, however, Q115A turns out to be active, binds, and cleaves the canonical (-GAATTC-) and noncanonical (-GAAUTC-) sequences with equal affinity and catalytic efficiency. (*Note:* Wild-type *Eco*RI shows a 10-fold lower k_{cat}/K_m for cleavage of -GAAUTC- than for -GAATTC-.) These observations support the notion that the catalytic activity of *Eco*RI hinges on a conformational change(s) that couples DNA binding and cleavage.

iv. Role of DNA conformation

(a) Solution structure. The free solution structure determined by NMR spectroscopy of dodecanucleotides (<u>GAATTC</u>-<u>GAATTC</u>) has most of the features of right-handed B-DNA. However, significant structural differences in sugar pucker, sugar–phosphate geometry, and consequently larger intranucleotide distances between adjacent nucleotides have been observed at the scissile bond-containing G-A and the central TCG portions (52). A similar self-complementary DNA dodecamer (CGC-<u>GAATTC</u>-GCG)$_2$ has been shown by NMR to possess the elements of structural perturbation called "kink" at the C3-G4 and A6-T7 base steps (53). The term *kink* refers to an abrupt disruption of double-helical symmetry due to a sharp bending of DNA or a highly localized unwinding or overwinding or both. The intrinsic microscopic curvature originating from the dAATT tetramer in the *Eco*RI recognition sequence gives rise to an anomalous gel migration (26). This anomalous preexisting secondary structure may serve as the initial structural basis of sequence-dependent DNA recognition and it may further undergo interactive changes when bound to the enzyme.

(b) Crystal structure of free DNA and the DNA in DNA–enzyme complex. The crystal structures of the self-complementary dodecanucleotide d(CGT-<u>GAATTC</u>-ACG) and its parent dodecamer d(CGC-<u>GAATTC</u>-GCG) have a conserved B-form double-helix structure with an overall bend in the helix of 10° and an average helical twist of ~36.5° (54). Despite the globally similar conformations, the two dodecamers exhibit certain significant differences that are highly localized in a sequence-dependent manner.

The DNA (TCGC-<u>GAATTC</u>-GCG) bound to the enzyme in the absence of the required cofactor Mg^{2+} has been found to assume a unique non-B conformation in crystals at C4-G5 and A7-T8, a structural feature termed "neo-kink" (41). The kink or neo-kink at C4-G5, which is adjacent to the scissile bond G5-A6, induces unwinding (~25°) and widening (~3.5 Å in phosphate–phosphate distance) of the purine base-containing major groove. The base pairs do not significantly increase their interplanar separation although the base-stacking contacts are clearly changed. Breaking the screw symmetry at one point in a helix and twisting one part with respect to another would propagate its separative effect along the "threads" across the break, presumably leading to facilitated chemical cleavages by the ENase. In parallel with the specific "interactive" recognition at the DNA level (55), the binding of DNA also induces a conformational change in the enzyme, which can be monitored by CD measurements (22).

v. Parameters of DNA–protein interaction. EcoRI ENase binds DNA containing a single EcoRI site (e.g., pBR322, pMB9, and λ DNA derivatives) with a stoichiometry of 1:1 and a K_d of 10^{-9}–10^{-11} M (5,10,44,56). The K_d for nonspecific binding with large DNAs falls in the 10^{-6} M range. The substrate DNA–enzyme complexes are kinetically quite stable with a half-life longer than 15 min (57); this allows the capture of the complexes on a NC filter for binding assays. Analysis of the binding isotherm with pBR322 indicates that the specific DNA binding ($\Delta G = -15.9$ kcal/mol) is largely contributed by the entropy change ($\Delta S = 36.2$ cal/mol·deg) whereas ΔH contributes only -4.7 kcal/mol at 37°C (44). The binding affinity gets progressively smaller with smaller size DNAs; small oligonucleotides (hexamer and octamer) bind EcoRI with a K_d of 10^{-7} M. For small-size DNAs, the binding affinity of the canonical sequence is higher than those of nonspecific sequences or single-strand oligonucleotides only by a factor of 2 (22). Methylation of the single EcoRI site reduces the apparent affinity by a factor of 10^2–10^3 or, in terms of potential nonspecific binding sites, by ~10^6.

The ΔG of nonspecific interactions generally involves a large electrostatic component. Therefore, the increase in ionic strength reduces the binding of DNA to protein. The binding of DNA to EcoRI is indeed strongly dependent on the ionic strength of the media, with the maximum binding at 50–80 mM NaCl in the absence of Mg^{2+} (44); in the presence of Mg^{2+}, the binding of DNA to EcoRI is dependent on the ionic strength in a significantly different way. Therefore, Mg^{2+} is not used simply as a cofactor in the hydrolysis reaction, but it also affects the DNA–protein interaction. At high ionic strengths, the apparent affinity of the enzyme for the DNA is 10–100 times lower in the presence of Mg^{2+} than in its absence. High concentrations of Mg^{2+} actually increase the K_m for pMB9 DNA (9) and increase the rate of dissociation of the enzyme from DNA (33). The binding energy of the enzyme–DNA complex that is released by the Mg^{2+} ion is presumably channeled into the rate acceleration in the catalytic hydrolysis (56).

vi. Pathway to sequence-specific interactions. How EcoRI ENase locates its recognition sequence is best explained, at least *in vitro*, by a *facilitated diffusion* model. According to the model, a protein (e.g., *lac* repressor) initially binds to nonspecific DNA sequences and then transfers to its target sites (e.g., operator or promoter) by a linear diffusion process (58). Consistent with the model, both formation and dissociation rates of EcoRI–DNA complexes have been observed to increase up to eightfold as the DNA (single EcoRI site-containing pBR322 fragments) size increases from 34 bp [dissociation rate constant (k_d) of 5×10^{-3} min^{-1}, $t_{1/2} = 140$ min] to 6200 bp ($k_d = 4.6 \times 10^{-2}$ min^{-1}, $t_{1/2} \geq 15$ min) (57). The rate enhancements do not affect the intrinsic thermodynamic parameters of the site-specific interaction. The effective sliding distance (under the conditions of 75 mM NaCl and no Mg^{2+}, at pH 7.6) is calculated to be 1300 bp. Because the estimation of the sliding distance involves salt-dependent factors, the distance can decrease dramatically with increasing salt concentrations (Na^+ and/or Mg^{2+}). Indeed at or above 0.15 M NaCl (ionic strength of 0.17), there is no difference in k_d values (~1 min^{-1}) between 34- and 4361-bp DNA fragments. Perhaps these data explain why the *Eco*RI cleavage sometimes appears processive (at low ionic strengths) (57,59) and nonprocessive (at high salt concentrations) (59,60).

vii. Catalytic mechanism. With *Eco*RI (and some other R·ENases including *Hin*dIII), the endonucleolytic cleavage occurs via a two-step (or sequential) mechanism: an intermediate cleavage of a single strand followed by scission in the second strand. Although little is known about the chemical steps involved in the cleavage of the scissile phosphodiester bond, the reaction has been shown to proceed with an inversion of configuration at the reactive phosphorus (61). This observation implies that there is an odd number of chemical events during the hydrolysis—probably a direct enzyme-catalyzed nucleophilic attack of H_2O at phosphorus without involvement of a covalent enzyme intermediate.

The reaction mechanism can be described in the following kinetic scheme (62), which involves the enzyme (E), DNA (D), Michaelis complex (ED), nicked intermediate complex (E·D_I), and double-stranded scission product–enzyme complex (E·D_{II}) (Scheme 4.1).

$$E + D \underset{k_{-1}}{\overset{k_1}{\rightleftharpoons}} ED \xrightarrow{k_2} E \cdot D_I \xrightarrow{k_3} E \cdot D_{II} \xrightarrow{k_4} E + D_{II}$$

$$E \cdot D_I \underset{k_{-5}}{\overset{k_5}{\updownarrow}} E + D_I$$

SCHEME 4.1

The rate constants estimated for typical DNA substrates are summarized in Table 4.2. The essential features of the kinetic mechanism are that (a) the k_2 step is ~2 times faster than k_3 (6,21,63), (b) the k_{cat} increases with increasing ionic strengths in a manner consistent with a decrease in nonspecific affinity, and (c) depending on substrates and reaction conditions, the chemical catalysis step can be faster or slower than other steps that involve binding interactions (association and/or dissociation) between the enzyme and DNA. Based on the differences in the magnitude of the rate constants, the RDS is found primarily on association or dissociation, in particular the dissociation of ENase from nonspecific DNA sequences (21).

Dissociation of the enzyme from the nicked intermediate (k_5) depends on reaction temperature and substrate species. For example, a fivefold difference in K_{d5} (= k_5/k_{-5}) is observed with two different tridecanucleotide substrates (6). Smaller oligonucleotides favor the dissociation and so do low temperatures. At 37°C, dissociation of the nicked intermediate of ColE1 DNA is negligible. However, an extensive dissociation occurs at 0°C (3). With SV40 DNA, ~25% of the intermediate complexes dissociate at 37°C.

The sequential mechanism of dsDNA cleavage by R·*Eco*RI (and R·*Bam*HI as well) can be traced by gel assays where supercoiled, open circular, and linear forms of a plasmid (pMB9 or pBR322) containing a single *Eco*RI site can be readily separated and visualized (63). Simulation analysis of the time-resolved kinetics of DNA cleavage by *Bam*HI has shown that the kinetic scheme (Scheme 4.1) fits the data reasonably well (64) when the D_{II} release (k_4) is considered to be a rapid step.

Why the *Eco*RI enzyme, in which two identical subunits bind the recognition sequence with a matching twofold rotational symmetry, cleaves the dsDNA in a sequential manner instead of in a concerted manner is not clearly known. It should be noted that an apparent simultaneous cleavage of both strands by *Eco*RI can occur under certain limiting conditions, such as starting the reaction from the stage of enzyme–DNA complex by adding Mg^{2+} or with some different DNA substrates containing different sequences flanking the canonical site. On the other hand, other enzymes (e.g., *Sal*I, *Eco*RV) that cleave DNAs by an apparent concerted mechanism under optimal reaction conditions may simultaneously display a nicking activity under suboptimal or relaxed conditions (13,65). These observations indicate that one needs to consider a more elaborate kinetic mechanism that includes additional steps such as subunit conformational changes and nonspecific complex formations of the enzyme with reactant and product DNAs (21).

2. STRUCTURE

i. Primary and secondary structures. The *Eco*RI enzyme is active as a dimer having identical subunits. The complete 277 amino acid sequence (M_r 31,065/ monomer) has been deduced from the sequence of cloned DNAs (66,67). The N-terminal Met is apparently processed posttranslationally and the mature enzyme contains 276 amino acids. The new N-terminal residue is Ser, while the C-terminal residue is Lys. The polypeptide contains 1 Cys(−217), 8 Tyr, 11 Phe, and 2 Trp. Acidic amino acids outnumber basic amino acid residues.

Modification of the single SH group by NEM or PCMB has no effect on *Eco*RI activity (68). However, DTNB treatment gives a marginal reduction in activity. Specific modification of Lys by methyl acetimidate results in inactivation of the enzyme. Inactivation kinetics display two classes of modifiable Lys residues. The modification of the more sensitive class and consequent partial inactivation can be prevented by binding of either specific or nonspecific DNAs to the enzyme, in the presence or absence of Mg^{2+}.

The secondary structural contents calculated from a crystallographic structural model (43) are 35% α helix and 22% β sheet. These values are in excellent agreement with the previous estimates at low salt concentrations (11,22). Note that, at high salt concentrations, the secondary structures have been estimated to be ~40% α helix and ~16% β sheet (67).

ii. Physical parameters. The *Eco*RI enzyme is an acidic protein with a pI of 5.8. The A_{278} of a 1% solution is 8.30 (A_{286}/A_{260} = 1.86–1.90, ε_{278} = 5.2 × 10^4 M^{-1} cm^{-1}). The $s^{\circ}_{20,w}$ is ~1.7S for a monomer and 3.8S for a dimer. At high enzyme concentrations (0.2–0.5 mg/ml), the subunits associate as tetramers ($s^{\circ}_{20,w}$ = 5.4–6.8S).

For a dimer–tetramer transition the K_d is ~0.4 μM (69), indicating that the EcoRI enzyme is present mostly as a dimer under standard reaction conditions (~0.1 nM). Disruption of the subunit interactions by site-specific mutageneses at the subunit interface (e.g., Glu-144, Glu-152, and Gly-210) results in inactive EcoRI mutant enzymes (70).

iii. Tertiary structure.

The 3-D structure of the EcoRI enzyme was first determined in a cocrystalline form with an oligonucleotide substrate (TCGC-GAATTC-GCG) (41). This original model was subsequently revised by a refined structure of EcoRI complexed with a halogenated oligonucleotide, e.g., TCGC-GAAU^{I}TC-GCG (43). Note that these crystal structures have been obtained in the absence of Mg^{2+}, hence in a catalytically inactive form.

The EcoRI protein has a symmetrical globular structure with a diameter of ~50 Å. Each subunit is composed of a single principal domain organized according to α/β architecture. The domain consists of five-stranded β sheets which are associated with sequence recognition and the subunit interface, six α helices including two which play a key role in DNA sequence recognition, and a major extension called the "arm" which wraps around the bound DNA.

iv. Active site residues.

Based on the crystal structures of DNA–EcoRI complexes in the absence of Mg^{2+}, three amino acid residues (Glu-144, Arg-145, and Arg-200) on each subunit are presumed to make H-bonded contacts with the canonical sequence in double helix, in particular the Pu bases in the major groove. Glu-144 and Arg-145 (from opposite subunit) form four H-bonds with the two adjacent adenines of a half-site (GAA). Arg-200 forms two H-bonds with the terminal G. The part of the DNA segment containing the G-A scissile bond lies in a catalytic cleft.

The crystalline DNA–enzyme complex shows a solvent channel which runs between the DNA backbone and the protein, ending at the scissile bond. Presumably the cofactor Mg^{2+} ion enters the active site through this solvent channel and induces structural isomerization and eventually chemical catalysis. This conformational change would be a form of allostery in the sense that the recognition and cleavage sites are considered to be spatially separate. Gln-115, which makes a hydrophobic contact with the methyl group of inner T in the EcoRI recognition sequence, is presumed to play a critical role in activation of the catalytic center following specific DNA binding (46). Gln-115 serves to stabilize the structure of EcoRI via intramolecular contacts between various parts of the protein.

Glu-111 is another essential residue located at the C-terminal part of a β strand (β3) and near the scissile bond of the DNA (43). The negative charge on Glu-111 is required for DNA strand cleavage (71). The replacement of Glu-111 with Gln or Gly retains full DNA-binding specificity, but reduces the catalytic rate $3–6 \times 10^4$-fold. In contrast, the replacement of Glu-111 by Asp results in only a 100-fold decrease in cleavage rate constants (72).

Glu-144 and Arg-145 are also located at the subunit interface. The substitution of Glu-144 by Lys abolishes all apparent activity (70), whereas the Gln-144 mutant enzyme has a drastically reduced activity, mostly due to the decrease (about 100-fold) in k_{cat} (7). Other mutant enzymes with conservative substitutions that would preserve the charge and local structure (e.g., E144D, R145K, and R200K) retain the catalytic activity, albeit with altered rate and kinetic pathway. The activity of Lys-145 or Lys-200 mutant enzyme is ~2 orders of magnitude lower; this is due largely to the decrease in k_{cat} in the Lys-145 enzyme [cf. K_m is increased by about 2-fold] (7) or to an increased K_m in the Lys-200 enzyme [cf. the k_{cat} is not changed] (8). Accordingly, these mutant enzymes release the DNA and nicked intermediate at a higher rate than the wild-type enzyme (48). When Arg-200 is replaced by other amino acids (Cys, Pro, Val, Ser, and Trp) which disrupt the charge and H-bonding potential, the mutant enzymes show drastically reduced endonucleolytic activity. Other residues essential for enzyme–DNA interactions include Ala-139, Gly-140, Glu-144, and Arg-203 (70).

Among the essential residues that are not directly involved in the protein–DNA interface or protein–protein (subunit) interface are Gly-93 and Glu-96. These two residues are located close to the proposed Mg^{2+}-binding site, in (for Glu-96) or near (for Gly-93) the cluster of negatively charged surface residues, Asp-133, Asp-100, Asp-99, and Asp-26.

v. Structural relationship with isoschizomer, RsrI.
Determination of the complete nucleotide sequence and deduced amino acid sequence of *Rsr*I (from *Rhodobacter sphaeroides*) permits a close comparison between the two isoschizomeric enzymes. *Rsr*I has 276 amino acids (the initiating N-terminal Met is absent), exactly the same number of amino acids as *Eco*RI (73). The overall amino acid and nucleotide sequences show ~50% similarities. However, when the amino acid sequences are aligned so that the overlap between the two proteins is 266 amino acids, six regions 8 to 17 amino acids in length show from 75 to 100% identity. The bulk of the identity regions is concentrated around the regions that have been identified to be critical for sequence discrimination and catalysis.

3. GENETICS

The structural gene for *Eco*RI ENase resides naturally on the pMB1 plasmid and a series of its derivatives (4,74). The pMB1 plasmid originates from a clinical isolate of *E. coli* (strain 204) and is identical to the ColE1 plasmid except for a 1.95-kbp insert that contains the genes for both R·*Eco*RI and M·*Eco*RI. The *Eco*RI ENase gene (*ecoRIR*) is separated from the M·*Eco*RI gene (*ecoRIM*) by a 29-nt intercistronic region. The two genes are in the same orientation (see Fig. 4.1). The *ecoRIR* encodes a 277 residue polypeptide and apparently has its own promoter. The DNA sequence is 65% A + T, significantly higher than the originating plasmid or the host *E. coli*. The two genes are expressed from a single message. The two genes have been cloned in a pBR322-derived plasmid vector (pBH20) and expressed under the control of the *lac* promoter-operator (10).

Expression of the wild-type *ecoRIR* gene alone in cloned environments is lethal unless the cognate MTase is also present. In fact, the cell viability provides the basis for a positive selection method in screening *Eco*RI "null" mutants in a methylase-deficient background (70). The *viability factor,* defined by the ratio of the viable counts of *E. coli* cultures having or not having expressed the *ecoRIR* gene, gives a measure for the *in vivo* activity of mutant *Eco*RI enzymes. For example, wild-type *Eco*RI has a viability factor of 1/630,000, whereas mutants (Q115A and Q115E) have viability factors of 1/40 and 1/16, respectively (46).

4. Sources

The *Eco*RI enzyme has been purified from an *E. coli* strain (RY13) deficient in endonuclease I (69). The enzyme is now produced from an *E. coli* harboring a recombinant plasmid that contains both *ecoRIR* and *ecoRIM* genes. Expression of the genes, typically under the control of the *lac* promoter, is carried out in minimal media supplemented with glycerol rather than glucose as a carbon source to avoid catabolite repression. When induced by IPTG, the expression achieves an eightfold overproduction of both enzymes, compared with the typical level (1 mg protein/kg bacterial cells) of natural expression in *Eco*RI-producing bacteria. The *Eco*RI gene has also been cloned, overexpressed under the control of the λ p_L promoter and *ts* repressor, and the enzyme has been purified to homogeneity by a procedure consisting of successive chromatographies over phosphocellulose, DEAE-cellulose, and hydroxyapatite (75).

Affinity partitioning between two liquid (aqueous) phases has been used to isolate the *Eco*RI enzyme free from contaminating nuclease activities (76).

The *Eco*RI enzyme is irreversibly denatured on heating at 60°C. Compared to some other type II enzymes, *Eco*RI has a remarkably low solubility. At 0.5 μM concentration the enzyme has its maximal activity on and affinity for its recognition site.

References

1. Polisky, B., Greene, P., Garfin, D. E., McCarthy, B. J., Goodman, H. M., and Boyer, H. W. (1975). *PNAS* **72**, 3310–3314.
2. Greene, P. J., Poonian, M. S., Nussbaum, A. L., Tobias, L., Garfin, D. E., Boyer, H. W., and Goodman, H. M. (1975). *JMB* **99**, 237–261.
3. Modrich, P., and Zabel, D. (1976). *JBC* **251**, 5866–5874.
4. Goodman, H. M., Greene, P., Garfin, D. E., and Boyer, H. W. (1977). *In* "Nucleic Acid-Protein Recognition" (H. J. Vogel, Ed.), pp. 239–259. Academic Press, New York.
5. Jack, W. E., Rubin, R. A., Newman, A., and Modrich, P. (1981). *In* "Gene Amplification and Analysis" (J. G. Chirikjian, Ed.), Vol. 1, pp. 165–179. Elsevier, New York.
6. Alves, J., Urbanke, C., Fliess, A., Maass, G., and Pingoud, A. (1989). *Biochem.* **28**, 7879–7888.
7. Wolfes, H., Alves, J., Fliess, A., Geiger, R., and Pingoud, A. (1986). *NAR* **14**, 9063–9080.
8. Alves, J., Ruter, T., Geiger, R., Fliess, A., Maass, G., and Pingoud, A. (1989). *Biochem.* **28**, 2678–2684.
9. Halford, S. E., and Johnson, N. P. (1981). *BJ* **199**, 767–777.
10. Rosenberg, J. M., Boyer, H. W., and Greene, P. (1981). *In* "Gene Amplification and Analysis" (J. G. Chirikjian, Ed.), Vol. 1, pp. 131–164. Elsevier, New York.
11. Manavalan, P., Johnson, W. C., Jr., and Modrich, P. (1984). *JBC* **259**, 11666–11667.
12. Hedgpeth, J., Goodman, H. M., and Boyer, H. W. (1972). *PNAS* **69**, 3448–3452.
13. Maxwell, A., and Halford, S. E. (1982). *BJ* **203**, 93–98.
14. Thielking, V., Alves, J., Fliess, A., Maass, G., and Pingoud, A. (1990). *Biochem.* **29**, 4682–4691.

15. Bishop, J. O. (1979). *JMB* **128**, 545–559.
16. Bishop, J. O., and Davies, J. A. (1980). *MGG* **179**, 573–580.
17. Hsu, M. T., and Berg, P. (1978). *Biochem.* **17**, 131–138.
18. Mayer, H. (1978). *FEBS Lett.* **90**, 341–344.
19. Tikchonenko, T. I., Karamov, E. V., Zavizion, B. A., and Naroditsky, B. S. (1978). *Gene* **4**, 195–212.
20. Lane, F., Ankenbauer, W., Schmitz, G. G., and Kessler, C. (1990). *NAR* **18**, 3421.
21. Terry, B. J., Jack, W. E., and Modrich, P. (1987). In "Gene Amplification and Analysis" (J. G. Chirikjian, Ed.), Vol. 5, pp. 103–118. Elsevier, New York.
22. Goppelt, M., Pingoud, A., Maass, G., Mayer, H., Koster, H., and Frank, R. (1980). *EJB* **104**, 101–107.
23. Nath, K., and Azzolina, B. A. (1981). In "Gene Amplification and Analysis" (J. G. Chirikjian, Ed.), Vol. 1, pp. 113–130. Elsevier, New York.
24. Thomas, M., and Davis, R. W. (1975). *JMB* **91**, 315–328.
25. Modrich, P., and Rubin, R. A. (1977). *JBC* **252**, 7273–7278.
26. Diekmann, S., and McLaughlin, L. W. (1988). *JMB* **202**, 823–834.
27. McLaughlin, L. W., Benseler, F., Graeser, E., Piel, N., and Scholtissek, S. (1987). *Biochem.* **26**, 7238–7245.
28. Marchionni, M. A., and Roufa, D. J. (1978). *JBC* **253**, 9075–9081.
29. Berkner, K. L., and Folk, W. R. (1977). *JBC* **252**, 3185–3193.
30. Nosikov, V. V., Braga, E. A., Karlishev, A. V., Zhuze, A. L., and Polyanovsky, O. L. (1976). *NAR* **3**, 2293–2301.
31. Uchida, K., Pyle, A. M., Morii, T., and Barton, J. K. (1989). *NAR* **17**, 10259–10279.
32. Kania, J., and Fanning, T. G. (1976). *EJB* **67**, 367–371.
33. Paoletti, C., LePecq, J. B., and Lehman, I. R. (1971). *JMB* **55**, 75–100.
34. Sambrook, J., Fritsch, E. F., and Maniatis, T. (1989). "Molecular Cloning: A Laboratory Manual," 2nd ed., pp. E.5–E.7. CSHL Press, CSH, New York.
35. Morgan, A. R., Evans, D. H., Lee, J. S., and Pulleyblank, D. E. (1979). *NAR* **7**, 571–594.
36. Jovin, T. M. (1975). In "Biochemical Fluorescence: Concepts" (R. F. Chen and H. Edelhoch, Eds.), pp. 305–374. Marcel Dekker, New York.
37. Hogan, M. E., and Jardetzky, O. (1980). *Biochem.* **19**, 2079–2085.
38. Chen, J., Herzenberg, L. A., and Herzenberg, L. A. (1990). *NAR* **18**, 3255–3260.
39. Wohlrab, F., and Wells, R. D. (1987). In "Gene Amplification and Analysis" (J. G. Chirikjian, Ed.), Vol. 5, pp. 247–255. Elsevier, New York.
40. Hanvey, J. C., Shimizu, M., and Wells, R. D. (1990). *NAR* **18**, 157–161.
41. McClarin, J. A., Frederick, C. A., Wang, B. C., Greene, P., Boyer, H. W., Grable, J., and Rosenberg, J. M. (1986). *Science* **234**, 1526–1541.
42. Frederick, C. A., Grable, J., Melia, M., Samudzi, C., Jen-Jacobson, L., Wang, B. C., Greene, P., Boyer, H. W., and Rosenberg, J. M. (1984). *Nature* **309**, 327–331.
43. Kim, Y., Grable, J. C., Love, R., Greene, P. J., and Rosenberg, J. M. (1990). *Science* **249**, 1307–1309.
44. Terry, B. J., Jack W. E., Rubin, R. A., and Modrich, P. (1983). *JBC* **258**, 9820–9825.
45. Lesser, D. R., Kurpiewski, M. R., and Jen-Jacobsen, L. (1990). *Science* **250**, 776–786.
46. Jeltsch, A., Alves, J., Oelgeschlager, T., Wolfes, H., Maass, G., and Pingoud, A. (1993). *JMB* **229**, 221–234.
47. Jen-Jacobson, L., Lesser, D., and Kurpiewski, M. (1986). *Cell* **45**, 619–629.
48. Needels, M. C., Fried, S. R., Love, R., Rosenberg, J. M., Boyer, H. W., and Greene, P. J. (1989). *PNAS* **86**, 3579–3583.
49. Woodbury, C. P., Jr., and von Hippel, P. H. (1981). In "Gene Amplification and Analysis" (J. G. Chirikjian, Ed.), Vol. 1, pp. 181–207. Elsevier, New York.
50. Heitman, J., and Model, P. (1990). *Proteins (S.F.G)* **7**, 185–197.
51. Heitman, J., and Model, P. (1990). *EMBO J.* **9**, 3369–3378.
52. Chary, K. V. R., Hosur, R. V., Govil, G., Zu-Kun, T., and Miles, H. T. (1987). *Biochem.* **26**, 1315–1322.

53. Nerdal, W., Hare, D. R., and Reid, B. R. (1989). *Biochem.* **28**, 10008–10021.
54. Narayana, N., Ginell, S. L., Russu, I. M., and Berman, H. M. (1991). *Biochem.* **30**, 4449–4455.
55. Kim, R., Modrich, P., and Kim, S. H. (1984). *NAR* **12**, 7285–7292.
56. Halford, S. E., and Johnson, N. P. (1980). *BJ* **191**, 593–604.
57. Jack, W. E., Terry, B. J., and Modrich, P. (1982). *PNAS* **79**, 4010–4014.
58. Berg, O. G., Winter, R. B., and von Hippel, P. H. (1981). *Biochem.* **30**, 6929–6948.
59. Ehbrecht, H. J., Pingoud, A., Urbanke, C., Maass, G., and Gualerzi, C. (1985). *JBC* **260**, 6160–6166.
60. Langowski, J., Alves, J., Pingoud, A., and Maass, G. (1983). *NAR* **11**, 501–513.
61. Connolly, B. A., Eckstein, F., and Pingoud, A. (1984). *JBC* **259**, 10760–10763.
62. Rubin, R. A., and Modrich, P. (1978). *NAR* **5**, 2991–2997.
63. Halford, S. E., Johnson, N. P., and Grinsted, J. (1979). *BJ* **179**, 353–365.
64. Hensley, P., Nardone, G., Chirikjian, J. G., and Wastney, M. E. (1990). *JBC* **265**, 15300–15307.
65. Taylor, J. D., and Halford, S. E. (1989). *Biochem.* **28**, 6198–6207.
66. Newman, A. K., Rubin, R. A., Kim, S. H., and Modrich, P. (1981). *JBC* **256**, 2131–2139.
67. Greene, P. J., and Gupta, M., Boyer, H. W., Brown, W. E., and Rosenberg, J. M. (1981). *JBC* **256**, 2143–2153.
68. Woodhead, J. L., and Malcolm, A. D. B. (1980). *NAR* **8**, 389–395.
69. Rubin, R. A., and Modrich, P. (1980). *Meth. Enzy.* **65**(I), 96–104.
70. Yanofsky, S. D., Love, R., McClarin, J., Rosenberg, J. M., Boyer, H. W., and Greene, P. (1987). *Proteins (S.F.G)* **2**, 273–282.
71. King, K., Benkovic, S. J., and Modrich, P. (1989). *JBC* **264**, 11807–11815.
72. Wright, D. J., King, K., and Modrich, P. (1990). *JBC* **264**, 11816–11821.
73. Stephenson, F. H., Ballard, B. T., Boyer, H. W., Rosenberg, J. M., and Greene, P. J. (1989). *Gene* **85**, 1–13.
74. Betlach, M., Hershfield, V., Chow, L., Brown, W., Goodman, H. M., and Boyer, H. W. (1976). *Fed. Proc.* **35**, 2037–2043.
75. Geiger, R., Ruter, T., Alves, J., Fliess, A., Wolfes, H., Pingoud, V., Urbanke, C., Maass, G., Pingoud, A., Dusterhoft, A., and Kroger, M. (1989). *Biochem.* **28**, 2667–2677.
76. Vlatakis, G., and Bouriotis, V. (1991). *J. Chromatog.* **538**, 311–321.
77. Rodriguez, R. J. (1993). *Anal. Biochem.* **209**, 291–297.

C. Type IIS Restriction Endonucleases: *Fok*I Restriction–Modification System

[Type II site-specific deoxyribonuclease, EC 3.1.21.4]

As a subclass of type II enzymes, type IIS restriction enzymes (ENase-IIS) are characterized by (i) the asymmetry of 4 to 7 bp long recognition sequences and (ii) the cleavage at a defined distance (1 to 20 bp) away from their recognition sequence. As presented in Table 4.1, type IIS enzymes display a number of differences from type II enzymes. At present, ENase-IIS comprises more than 80 enzyme species representing at least 35 specificities, while 15 cognate MTase-IIS have been characterized (1).

Of the type IIS restriction endonucleases that have been reasonably well characterized, e.g., *Fok*I, *Hga*I, *Hph*I, and *Mbo*II, the *Fok*I R–M enzymes from *Flavobacterium okeanokoites* are probably the best studied, which is the reason for the focus of this section. *Fok*I R and M enzymes specifically recognize the sequence 5'-GGATG-3'/3'-CCTAC-5'. R·*Fok*I (66 kDa) cleaves the DNA at a site 9 and 13 nucleotides to the right of the top and bottom recognition sequences, respectively. The cognate type IIS modification methylase, M·FokI (76 kDa), methylates adenine residues in both strands of the recognition sequence.

1. FUNCTIONS

a. Reaction conditions

i. Recommended reaction conditions. FokI digestion is carried out in a reaction buffer composed of 10 mM Tris–Cl (pH 7.5), 10 mM $MgCl_2$, 20 mM KCl, 0.5 mM DTT, and 50 µg/ml BSA. The reaction mixture is incubated at 37°C. The pH optimum for FokI Enase is 7.5–8.5 (2).

ii. Metal ion requirements. FokI requires Mg^{2+} (10 mM) for optimal activity. However, in assays at a suboptimal pH of 5.7 (0.1 M KAc), other divalent cations such as Mn^{2+}, Co^{2+}, and Zn^{2+} can fully substitute for Mg^{2+} (2). Optimal activity is observed at 20–90 mM KCl or 60 mM NaCl and 37–42°C.

iii. Stability and stabilizer. BSA (0.01%) significantly stabilizes the enzyme: without BSA, a 60% loss of activity may occur on incubation at 37°C for 1 hr. The addition of 7 mM 2-MSH or 0.01% Triton X-100 does not increase the activity (3).

FokI ENase is fairly thermostable, i.e., for a few hours at 45°C. At higher temperatures, the enzyme rapidly loses activity.

b. Substrate specificities

i. Recognition sequence and cleavage site. FokI specifically recognizes a pentanucleotide sequence in unmodified dsDNA and cuts at a defined distance from the right-hand side of the recognition sequence (4). As in all type IIS enzymes, the recognition sequence is asymmetric and solely responsible for cleavage site specificity. The cleavage generates 5'-protruding four bases.

$$5'\text{-G G A T G (N)}_9\text{ -3'}$$
$$3'\text{-C C T A C (N)}_{13}\text{ -5'}$$

Replacement of G with 6-thioguanine in the recognition sequence completely abolishes the FokI activity, whereas a similar substitution at the scission site has no effect on the enzyme activity (5).

ii. Overlapping sequences. The FokI recognition sequence shares part of its nucleotides with two other type II restriction enzymes, i.e., AccIII (TCCGGA) and NsiI (ATGCAT). Hemimethylation of the FokI site can thus differentially block the cleavages by the two enzymes (6).

iii. Effect of methylation. Methylation of the FokI recognition sequence by M·FokI renders the DNA resistant to R·FokI. Hemimethylation of either strand of the FokI sequence protects the sequence from R·FokI digestion. Restriction by FokI is not inhibited by methylation at the cleavage site.

iv. Methylase specificity. Methylation by M·FokI occurs at the A bases in both strands of the recognition sequence (7). In buffers with low ionic strengths, M·FokI methylates only the top strand of the recognition sequence.

M·FokIA, a C-terminal truncated form of M·FokI, methylates the A base in only the top strand (hemimethylation) at all ionic strengths, while an N-terminal truncated form of M·FokI methylates only the A in the bottom strand (6). M·FokI does not methylate ssDNA.

v. Increase of recognition sequence. The recognition specificity of FokI can be increased from 5 to 7 bp by a double methylation (8). This strategy of increasing recognition specificity is based on the fact that noncognate methylation of specific nucleotides of the recognition sequence inhibits the cognate MTase but not the ENase of the same R–M system. MspI and FokI sites can overlap in the 7-bp sequence CCGGATC. Methylation of such sites by M·MspI (methylation specificity: ^{m}CCGG) produces a noncognate methylation of the overlapping FokI sites resulting in 5'-GGATG-3'/3'-C^{m}CTAC-5'. This sequence inhibits methylation by M·FokI, while the cleavage by R·FokI is unaffected. Therefore, a sequential methylation with M·MspI and then M·FokI directs the cleavage by R·FokI uniquely to the 7-bp sequences. The cleavage at such sites occurs to a maximum of 90%, reflecting that the noncognate methylation produces a minor inhibitory effect.

2. APPLICATIONS

Applications of FokI or other type IIS ENases as a site-specific DNA cutting tool are similar to those of type II restriction enzymes. However, the unique properties pertaining to ENase-IIS permit one to perform many novel operations: (i) the possibility of formulating a three-component system with a specially designed adapter or probe, (ii) repeated use of the recognition site, since it is not damaged during cleavage, and (iii) the possibility of studying enzyme–DNA binding without concurrent cleavage even in the presence of Mg^{2+} (1). Some of the typical applications of R·FokI (and M·FokI) are presented to illustrate the basic ideas and strategies that are relevant to other type IIS enzymes as well.

i. As "universal" restriction enzyme. The separation between the recognition site and the cleavage site has been exploited to make FokI a "universal" restriction enzyme for cutting DNA at any target site (9). The recognition specificity is provided by the self-complementary hairpin part of an oligonucleotide adapter, while the target specificity is afforded by the single-strand primer part that would be complementary to the target sequence (Fig. 4.3). When this bipartite oligonucleotide is hybridized to a ssDNA template, a site-directed cleavage by FokI is possible (10). One of many potential applications of this DNA cutting methodology would be for site-specific mutagenesis by oligonucleotide-directed double-strand break repair in plasmids (11).

ii. Gene synthesis. This method of gene synthesis is based on the unique property of FokI to generate four base-protruding termini which can be used for

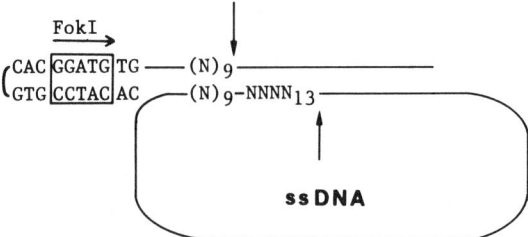

FIGURE 4.3 *Fok*I as a universal restriction enzyme. The universal restriction system consists of three components: an oligonucleotide adapter (which can also serve as a primer), a target ssDNA, and R · *Fok*I. The single-strand adapter(-primer) portion anneals with the complementary target sequence, providing a cleavage site for *Fok*I. The site-specifically cut DNA can be used as the templates for a DNA Pol in the synthesis of a second strand DNA. [After W. Szybalski (1985). *Gene* **40**, 169–173.]

ordered ligations of oligonucleotides. Briefly, a synthetic oligonucleotide (single-strand gene fragment) is first annealed between the two flanking *Fok*I sites in a linearized, denatured plasmid DNA (Fig. 4.4). Transformation of *E. coli* with the annealed DNA leads to an *in vivo* gap repair and amplification of the insert fragment (as long as 100 nt), a technique known as *bridge mutagenesis* (12). The recombinant plasmid is digested with *Fok*I, liberating a double-stranded fragment with 5′-protruding four base ends that can be used for cohesive-end ligation with other gene fragments similarly prepared. This *Fok*I method of gene synthesis offers flexibility characteristic of modular approaches in constructing a larger size gene from small fragments.

iii. As versatile tool for gene cloning. DNA cloning often involves insertion of a DNA fragment into a cloning vector via intermolecular ligations between compatible protruding ends. In some cases (e.g., *Bam*HI and *Bgl*II), type II restriction enzymes recognize different sequences yet produce identical protruding ends and facilitate the cloning tasks. More often, however, commonly used restriction enzymes (e.g., *Pst*I and *Eco*RI) generate protruding ends that are not compatible with those generated by other enzymes. When the DNA fragment to be cloned contains internal restriction sites, the situation requires either the consideration of alternative insertion strategies or inclusion of more complicated steps such as protection of internal sites by methylation.

Type IIS restriction enzymes may provide a simple, alternative strategy to such situations: a synthetic oligonucleotide linker (adapter/primer) can be designed by incorporating desired sequences at the cleavage termini. For example, *Eco*RI-compatible terminus (AATT) can be generated using one of the type IIS enzymes that produce four base 5′ overhang, e.g., *Fok*I (5′-GGATGN$_9$/CCTACN$_{13}$-5′) or *Bsa*I (5′-GGTCTCN/CCAGAGN$_5$-5′). [*Note:* By the same principle, four base 3′ overhangs compatible to any desired site can be generated within the *Bst*XI site (5′-CCAN$_5$|NTGG/GGTN|N$_5$ACC-5′).]

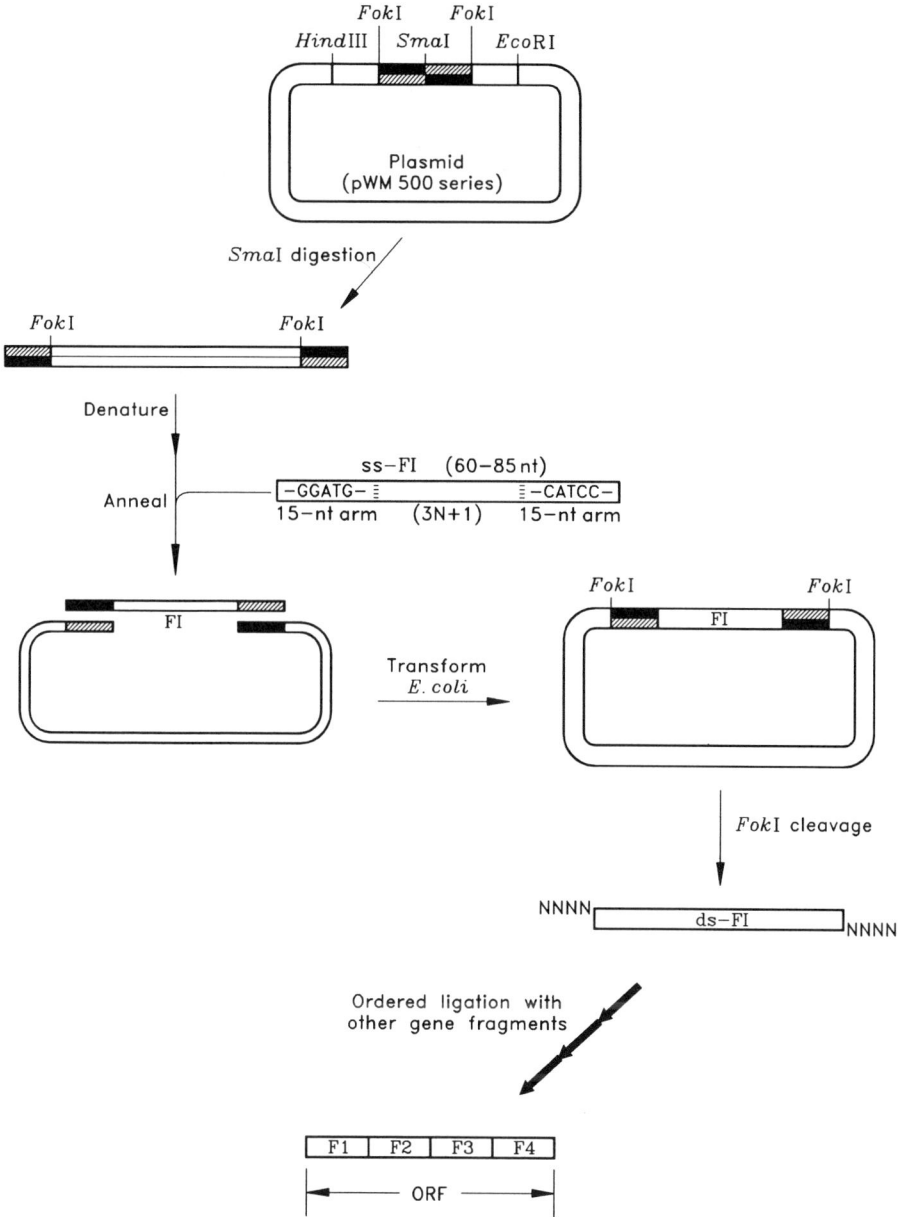

FIGURE 4.4 *Fok*I method of gene synthesis. This method is based on the generation of unique cohesive ends for each gene fragment as a result of *Fok*I digestion of each DNA fragment cloned in a specialized *Fok*I vector. The fragments can be linked together in an ordered manner by cohesive-end ligations. [After W. Mandecki and T. J. Bolling (1988). *Gene* **68,** 101–107.]

iv. **As versatile tool for gene trimming.** This technique employs a plasmid in which a unique ENase-IIS recognition site is present in or flanking the MCS. A DNA fragment cloned in the MCS can be progressively trimmed from one end. Larger deletions can be performed by multiple cycles of (a) cleavage, (b) mung bean nuclease digestion, and (c) religation. Employing the *Fok*I enzyme, 4 bp can be removed per cycle. The strategy is versatile and can be adapted to a cassette system of two type-IIS enzymes. The cassette system can be used to increase the number of nucleotides deleted per cycle, e.g., 12 bp for the *Mbo*II/*Fok*I recognition cassette sequence (13). In this context, the trimming method can be used in variable forms, allowing not only unidirectional but also bidirectional trimming when a pair of ENase-IIS sites are placed in opposite orientations in a vector.

An alternative use of such a cassette in MCS is as a fusion mediator between two genes that are cloned on each side of the cassette (14). The two genes are precisely trimmed as described earlier and then ligated after the cassette containing the unique type IIS restriction sites is excised and the termini blunt-ended.

v. **Oligonucleotide amplification reaction.** When the protruding 4-nt cohesive end generated by *Fok*I digestion is filled in using Pol Ik (+ 4 dNTPs), the termini can be recut by *Fok*I because the *Fok*I recognition site remains intact. In the presence of both Pol Ik and *Fok*I, the cycles continue, generating copious amounts of the tetramer which is complementary to the cohesive 4-nt end (1).

vi. **Locating methylated bases in DNA.** This method is based on the observation that type IIS restriction enzymes are not inhibited by methylation at the cleavage site (15). Briefly, a DNA fragment is chosen such that it contains the recognition site of the MTase under study and that the site appropriately overlaps with a type IIS enzyme cleavage site. The DNA is methylated by the MTase using [^{3}H]AdoMet and is then cut with the ENase-IIS. After separation of the asymmetrically cut fragments, the site of methylation is identified by chemical sequencing. Whether a methylation on cytosine is m^{4}C or m^{5}C can be further distinguished by different reactivities toward hydrazine: m^{4}C reacts with hydrazine at the same rate as cytosine, whereas m^{5}C reacts much more slowly.

vii. **Linker mutagenesis and DNA replacement.** A *Fok*I cassette, which contains a selectable marker and is amplifiable separately in a specialized vector, is placed in the linker area of a gene cloned in another specialized vector (16). By *Fok*I digestion, the cassette is released together with several nucleotides beyond the cassette/vector fusion sites. Cohesive-end ligation of such deleted double-stranded oligonucleotides via *Fok*I adapters allows the introduction of defined mutagenic sequences into the target gene.

3. STRUCTURE

*Fok*I R and M enzymes are larger (about twofold) than typical type II enzymes whose average molecular sizes are approximately 280 amino acids for R · ENase and 380 amino acids for M · MTase.

a. FokI endonuclease

R · FokI is a monomeric enzyme composed of 578 deduced amino acids (M_r 65,619) (6). The initiating N-terminal Met is apparently cleaved off by a posttranslational modification. The processed enzyme has Val and Phe as N- and C-terminal residues, respectively. R · FokI is a strongly basic protein with a pI of 8.9–9.4 (2,3).

When compared with many cloned restriction enzymes, R · FokI shares no apparent amino acid sequence similarities. It has no homology with the cognate M · FokI.

Tryptic digestion experiments suggest that FokI ENase consists of two distinct functional domains: the N-terminal "recognition" domain (41 kDa) apparently used for sequence-specific DNA binding and the C-terminal "cleavage" domain (25 kDa) used uniquely for DNA cleavage function (21). The two domains are connected by a linker region which is predicted to be a long stretch of α helix. If the two domains are functionally and structurally independent, extension of the linker region by one, two, or more helical turns would keep the two domains in proper orientation for sequence recognition and perhaps new cleavage specificities. Indeed, four and seven codon (amino acid) insertion mutants in the linker region have the same DNA sequence specificity as the wild-type enzyme but cleave at one nucleotide further away from the recognition site on both strands of the DNA substrates (22). Although infrequent, the mutant enzymes also cut the DNA at the same site as the wild-type enzyme.

b. FokI methyltransferase

M · FokI is composed of 647 deduced amino acids (M_r 75,622) (6,17). The N- and C-terminal residues are Met and Lys, respectively. M · FokI has no apparent amino acid sequence homology to most of the R–M enzymes. Among a few MTases that share some degrees of similarity, M · FokI displays a pronounced similarity with phage T4 Dam (recognition sequence GATC), especially in the C-terminal half. In contrast, the N-terminal half of M · FokI has a relatively strong homology with M · NlaIII (recognition sequence CATG).

In each half (positions 218–221 and 548–551) of the molecule M · FokI contains a DPPY (Asp–Pro–Pro–Tyr) sequence. This tetrapeptide sequence motif or its equivalent is regarded as the hallmark of N-methylating MTases because it is found by a single copy in all known m^{6}A-MTases and m^{4}C-MTases, but not in m^{5}C-MTases or ENases (18,19). Deletion mutant FokI MTases containing only a single copy of the DPPY motif are functionally competent; the N-terminal domain methylates only the top (GGATG) strand and the C-terminal domain only the bottom (CATCC) strand (6,20). Because the two halves of M · FokI bear distinct, but relatively sparse homology to each other, the *fokIM* gene is believed to have arisen from a tandem fusion of two ancestral M genes.

4. GENETICS

The genes for FokI ENase (*fokIR*) and FokI MTase (*fokIM*) have been cloned and sequenced. The FokI R–M system from *F. okeanokoites* is encoded by two adjacent chromosomal genes: *fokIR* and *fokIM* (6). The R gene (1740 bp) and

M gene (1941 bp) are large by type II system standards. The two genes are aligned in the same orientation: in the order of M, then R. They apparently form an operon, the expression of which is regulated by a promoter upstream from the *fokIM* gene. The compositions of M and R genes are 70 and 66% A + T, respectively. The two ORFs, each defined by their start and stop codons, are separated by a 79-nt interval. Preceding each gene is a pair of *Fok*I recognition sites. The sites nearest each gene form part of the presumptive RBS, AGG(A). The site nearest the *fokIM* forms the boundary of a perfect, 13-nt inverted repeat, allowing speculation that *Fok*I may participate in autoregulation of gene expression. Similarly, a stem–loop structure involving 50 bases is located in between the end of the M gene and the beginning of the R gene (3). When a half part of this putative stem–loop structure is deleted, the expression of the cloned *fokIR* gene is increased 30-fold.

5. Sources

R · *Fok*I can be purified either from *F. okeanokoites* (ATCC 33414) (2) or from the expression of cloned genes in *E. coli* (6). When the structural gene for R · *Fok*I is expressed in *E. coli* (UT481, a *lon*⁻ strain) under the control of the *tac* promoter in an expression vector pKK223-3, a 30-fold overproduction can be achieved (3). From this source, high purity (5600 U/mg) R · *Fok*I can be obtained at 58-fold purification. In another *E. coli* expression system (21), the *fokIR* gene was cloned into plasmid expression vectors in which alteration in both transcriptional signal (retroregulatory sequence) and translational signal (consensus *E. coli* RBS sequence) were introduced using a PCR procedure. Expression of the cloned gene, under the control of *lacUV5* or *tac* promoter, resulted in highly efficient productions of the enzyme at the level of 5–8% of total cellular protein (or 50 mg purified enzyme per liter of culture medium).

Because ENase-IIS can bind to the recognition sequence without cleaving the DNA (if the cutting site is absent or already cleaved), such a piece of DNA can be used as a component of affinity columns and for purification of type IIS enzymes in the presence of Mg^{2+} (1).

M · *Fok*I can be purified from the culture of *F. okeanokoites* (7) or from the expression of cloned genes in *E. coli* (6).

References

1. Szybalski, W., Kim, S. C., Hasan, N., and Podhajska, A. J. (1991). *Gene* **100**, 13–26.
2. Kaczorowski, T., Skowron, P., and Podhajska, A. J. (1989). *Gene* **80**, 209–216.
3. Kita, K., Kotani, H., Hiraoka, N., Nakamura, T., and Yonaha, K. (1989). *NAR* **17**, 8741–8753.
4. Sugisaki, H., and Kanazawa, S. (1981). *Gene* **16**, 73–78.
5. Iwaniec, L. M., Kroll, J. J., Roethel, W. M., and Maybaum, J. (1991). *Mol. Pharmacol.* **39**, 299–306.
6. Looney, M. C., Moran, L. S., Jack, W. E., Feehery, G. R., Benner, J. S., Slatko, B. E., and Wilson, G. G. (1989). *Gene* **80**, 193–208.
7. Landry, D., Looney, M. C., Feehery, G. R., Slatko, B. E., Jack, W. E., Schildkraut, I., and Wilson, G. G. (1989). *Gene* **77**, 1–10.

8. Posfai, G., and Szybalski, W. (1988). *NAR* **16**, 6245.
9. Szybalski, W. (1985). *Gene* **40**, 169–173.
10. Kim, S. C., Podhajska, A., and Szybalski, W. (1988). *Science* **240**, 504–506.
11. Mandecki, W. (1986). *PNAS* **83**, 7177–7181.
12. Mandecki, W., and Bolling, T. J. (1988). *Gene* **68**, 101–107.
13. Hasan, N., Kur, J., and Szybalski, W. (1989). *Gene* **82**, 305–311.
14. Kim, S. C., Posfai, G., and Szybalski, W. (1991). *Gene* **100**, 45–50.
15. Posfai, G., and Szybalski, W. (1988). *Gene* **74**, 179–181.
16. Vermersch, P. S., and Bennett, G. N. (1987). *Gene* **54**, 229–238.
17. Kita, K., Kotani, H., Sugisaki, H., and Takanami, M. (1989). *JBC* **264**, 5751–5756.
18. Narva, K. E., van Etten, J. L., Slatko, B. E., and Benner, J. S. (1988). *Gene* **74**, 253–259.
19. Chandrasegaran, S., and Smith, H. O. (1988). *In* "Structure and Expression" (M. H. Sarma and R. H. Sarma, Eds.), Vol. I, pp. 149–156. Adenine Press, New York.
20. Sugisaki, H., Kita, K., and Takanami, M. (1989). *JBC* **264**, 5757–5761.
21. Li, L., Wu, L. P., and Chandrasegaran, S. (1992). *PNAS* **89**, 4275–4279.
22. Li, L., and Chandrasegaran, S. (1993). *PNAS* **90**, 2764–2768.

D. Type I Restriction Enzymes

[Type I site-specific deoxyribonucleases, EC 3.1.21.3]

Type I enzymes are probably the most complex of the four R–M systems (Table 4.1): they are made up of three subunits, require two cofactors, and have at least three enzymatic activities (1,2). Several type I R–M systems have been identified in *E. coli* (e.g., *Eco*AI, BI, DI, and KI), *Salmonella* (*Sty*SBI, SJI, SPI, and SQI), and *Citrobacter* (*Cfr*AI). Type I enzymes fall into three discrete families depending on their gene organizations: the chromosomally encoded K- and A-family, and the plasmid-encoded family. The type I restriction enzymes are encoded by *hsd* genes (host specificity for DNA).

Among the type I enzymes, *Eco*KI (from *E. coli* K12) and *Eco*BI (from *E. coli* B) are most extensively characterized in terms of genetics, structure, and function. Based on the two K-family enzymes, which have proven themselves useful in recombinant DNA technology, this section describes the salient features of the type I R–M systems.

1. FUNCTIONS

a. Reaction specificities

A type I enzyme has an ENase activity on unmodified DNA. The same enzyme has two other activities: a site-specific DNA MTase and a DNA-dependent ATPase.

i. Restriction endonuclease The pH optimum for *Eco*BI ENase is 8 (3). It requires ATP, AdoMet, and Mg^{2+} for restriction activity (4). AdoMet is not used during the ENase reaction, but it is required as a cofactor, presumably as an allosteric effector for the ENase to bind DNA more favorably at the initial phase of reaction. The K_m for AdoMet is 0.3–0.4 μM for ENase activity (3,5). *S*-Adenosylethionine, an analog of AdoMet, is a potent inhibitor if present initially.

ENase activity is maximal at Mg^{2+} concentrations from 0.7 to 2 mM (3); Mn^{2+} and Ca^{2+} are somewhat effective. Although not absolutely required, DTT stabilizes the activity (and better than 2-MSH does).

ATP is also required for ENase activity. The ENase activity is half-maximal at ATP concentrations of 80–100 μM. ATP can be replaced by GTP and dATP with about 30% efficiency (3). ADP cannot replace ATP, nor is it inhibitory. The role of ATP in the restriction reaction remains to be clarified.

ii. ATPase. ATPase activity is associated with ENase and is thus AdoMet dependent (6). It hydrolyzes ATP to AMP and PP_i with a k_{cat} of 10^4 ATP molecules/min at 37°C (3). The ATP hydrolysis is not stoichiometric with ENase activity, but continues long after the DNA cleavage has gone to completion (7).

iii. Methyltransferase. *Eco*BI MTase has a pH optimum of 6. It requires only AdoMet as cofactor (8). The MTase is stimulated by ATP. The K_m for ATP is 250 times lower than that required for the ENase reaction. The k_{cat} is very low, with a turnover rate of 7 hr on unmodified substrate, while the rate is about 100-fold faster on hemimethylated substrates (9).

b. Activity assays

Because type I enzymes display three distinct activities, assay methods are also diverse. The following methods are among the most commonly used.

i. Endonuclease assay. A circular DNA substrate (e.g., phage fd RF DNA) preferably containing a unique restriction site is digested with enzyme. The activity is monitored by mobilities of the products in agarose gel electrophoresis or is measured by the decrease in infectivity on nonrestricting *E. coli* spheroplasts (10,11). Therefore, the assay measures a limiting value rather than a rate of reaction. However, this limit is proportional to enzyme concentration.

A typical assay mixture consists of 70 mM Tris–Cl (pH 8.0), 7 mM $MgCl_2$, 0.5 mM DTT, 1.4 mM ATP, 14 μM AdoMet, and 21 μM RF DNA. The reaction can be stopped by cooling on ice and adding EDTA to 20 mM final concentration (3).

Unit definition: One "restriction" unit is the amount of enzyme that reduces the infectivity of unmodified fd RF to 10% of its initial value. [*Note:* The RF of wild-type phage fd (6.6 kbp) has two sites for *Eco*BI restriction. Because the cleavages of any one or both sites contribute to the loss of infectivity, one enzyme *unit* is equivalent to 0.3 pmol of double-strand cleavages.]

ii. ATPase assay. The rate of [γ-^{32}P]ATP hydrolysis is determined in the ENase reaction by monitoring the quantity of ^{32}P in a charcoal-nonadsorbable form (3). Because type I ATPase is AdoMet dependent, it can be readily distinguished from other contaminating ATPase activities, especially during the early stages of purification, by carrying out a control reaction without AdoMet.

Unit definition: One "ATPase" unit degrades 100 nmol of ATP in 15 min at 37°C.

iii. Methyltransferase assay. Using [*methyl*-^{3}H]AdoMet as a cofactor in the reaction mixture, the incorporation of the [^{3}H]methyl group into TCA-precipitable DNA is monitored (12). A typical assay is carried out in 70 μl volume at pH 6.0 and 37°C in the presence of 7 mM MgCl$_2$, 0.5 mM DTT, 9 μM [*methyl*-^{3}H]Ado-Met, and 1.5 nmol of DNA.

c. Substrate specificities

i. Recognition sequences. DNA binding and cleavage by type I enzymes depend on the presence of "host specificity sites." Both *Eco*BI (13) and *Eco*KI (14), and in fact all type I R–M enzymes, recognize specific bipartite sequences [Scheme 4.2, asterisk (*) indicates methylation sites]. The general structure consists of a specific heptameric sequence divided into a trimer and a tetramer by nonspecific spacer sequences of a fixed but different length for the two enzymes. Neither sequence has any element of symmetry. The *Eco*BI and *Eco*KI recognition sequences can be aligned to match 4 bases out of 7 sequence-specific bp.

        *Eco*BI site   5'- T G A* -N$_8$- T G C T - 3'
                             3'- A C T -N$_8$- A$_*$ C G A - 5'

        *Eco*KI site   5'- A A* C -N$_6$- G T G C - 3'
                             3'- T T G -N$_6$- C A$_*$ C G - 5'

                                   SCHEME 4.2

ii. DNA cleavage. ENase action occurs only on unmodified dsDNA. The DNA cleavage takes place away from the recognition site: more than 400 bp away to as far as 7000 bp. The process of DNA cleavage at distant locations on either side of the specificity site most likely involves a loop-mediated translocation of DNA (15,16). The pattern of cleavage is not totally random but is localized in a way probably related to structural features of the DNA.

The cleavage occurs in a two-step process: the DNA is first nicked in one strand and then, several seconds later, in the second strand at a site very close to the first one on the opposite strand. Because each enzyme molecule apparently introduces no more than a single-strand scission in the presence of excess substrate, two enzyme molecules are required to produce a dsDNA cleavage.

iii. DNA methylation. For both *Eco*BI and *Eco*KI systems, the modification occurs on A bases (m^{6}A) on both strands (see Scheme 4.2). Two methyl groups are incorporated per site (17,18). For *Eco*BI, the top strand contains a single A and this must be the one that is methylated. The lower strand contains three adenines; the middle A is the one that is methylated. For *Eco*KI, the most likely sites of methylation are the single A in the lower strand and the second A in the top strand (19).

Type I MTases show a marked preference for half-methylated sites (9). Heparin or actinomycin D inhibits methylation, but has no effect on DNA cleavage.

d. Mechanism of DNA cleavage and methylation

As the multifunctional nature of the type I R–M enzymes suggests, DNA recognition and methylation or cleavage by type I enzymes occur through a complex series of interactions among the enzyme, cofactors, and DNA (19–21). Enzyme alone shows no detectable affinity for DNA. AdoMet induces a slow, allosteric conformational change to a form that can bind DNA (5). Once activated by AdoMet, the enzyme no longer requires the presence of free AdoMet for the rest of the reactions leading to restriction. The AdoMet-activated enzyme binds both nonspecific sequences and, more tightly, the recognition sequences whether the recognition sequences are modified ($t_{1/2}$ = 6 min) or nonmodified ($t_{1/2}$ = 22 min).

The addition of ATP stimulates methylation, releases the enzyme from the complex, or leads to DNA cleavage and further ATP hydrolysis; the choice of the route depends on the modification state of the recognition site. Methylation on unmethylated sites in the absence of ATP occurs at 6×10^{-5} sec^{-1} and the presence of ATP increases the rate twofold. The methylation on hemimethylated sites occurs faster than on unmethylated sites: at 10^{-4} sec^{-1} in the absence of ATP and at 3×10^{-3} sec^{-1} in the presence of ATP.

2. APPLICATIONS

Type I R–M enzymes have not found many applications in recombinant DNA technology for obvious reasons; namely, undefined cutting sites and difficult R–M controls *in vitro*. However, the sequence similarity of the recognition sites of the two type I R–M systems, *Eco*BI and *Eco*KI, has been exploited in an elegant "coupled or double-priming" method of oligonucleotide-directed site-specific mutagenesis (22) (refer to Appendix A).

The practical importance of the type I R–M system in recombinant DNA technology often lies in the choice of appropriate *E. coli* or other bacterial hosts whose genotypes are *hsdRMS*-minus (refer to Appendix B). This is to avoid inadvertant cleavages of the newly introduced vectors by the type I R–M system.

3. STRUCTURE

Type I R–M enzymes are complex molecules with a MW of ~400,000. Both *Eco*BI and *Eco*KI enzymes contain three nonidentical subunits: 135 kDa [R], 62 kDa [M], and 55 kDa [S]. For *Eco*KI, the structure is reported to be R_2M_2S (1). For *Eco*BI, however, at least three active oligomeric structures have been purified; the major form is $R_2M_4S_2$ (3). Purified subunits are not active.

i. R subunit. This subunit (1090 deduced amino acid residues in *Eco*KI, M_r 124,802) is required for all known ATP-dependent activities and is most likely to be the site for ATP binding. The primary structure of the R subunit contains two small domains commonly found in other ATP-binding proteins (23). The R subunit is also presumed to contain a domain responsible for ENase activity.

ii. M subunit. This subunit, with 529 deduced amino acid residues in *Eco*KI (M_r 59,289), has a binding site for AdoMet and assumes the role of adenine-MTase. At positions 266–269 the primary sequence contains a NPPF sequence motif typical of A-MTase domains (23).

The two subunits R and M of *Eco*KI and *Eco*BI are interchangeable. They are also immunologically highly cross-reactive (19).

iii. S subunit. This subunit confers sequence specificity to both R · ENase and M · MTase. The S subunit is composed of 464 deduced amino acid residues (M_r 51,336) in *Eco*KI and 474 amino acid residues in *Eco*BI (24). Comparison of the primary structures of *Eco*KI and *Eco*BI reveals two regions of high similarity: ~30 amino acids in the middle of the sequence and ~80 amino acids at the C-terminal. The two conserved regions are separated by two remarkably dissimilar regions, each approximately 200 amino acids in length. According to studies on *Sty*SBI and *Sty*SPI (25), the N-terminal variable domain is most likely responsible for specific recognition of the 5′-trinucleotide in the bipartite recognition sequences.

4. GENETICS

The R–M system of *E. coli* B is an allelic variant of *E. coli* K and shares extensive DNA sequence homology. Both systems involve three chromosomally located genes: *hsdR* for restriction, *hsdM* for modification, and *hsdS* for site recognition. These genes are mapped in the order *hsdR*, *hsdM*, and *hsdS*, followed by *serB* at the 98.5-min position on the *E. coli* map.

i. Gene organization. The genes are organized into two units that are transcribed in the same direction; *hsdR* (3270-bp coding sequence in *Eco*KI) is transcribed from one promoter, whereas *hsdM* (1587-bp coding sequence in *Eco*KI) and *hsdS* (1392-bp coding sequence in *Eco*KI and 1422-bp coding sequence in *Eco*BI) are transcribed from a second promoter (23,24,26). The *hsdR* and *hsdM* genes are separated by a 492-bp intergenic space, while the stop codon of *hsdM* overlaps the start codon of *hsdS* gene.

ii. Mutation and genetic analysis. A mutation in *hsdS* leads to a simultaneous loss of restriction and modification, (r^-m^-). Other mutants are either (r^-m^+) or (r^-m^-), but no (r^+m^-) mutants have been found presumably because they would be lethal. Some (r^-m^+) and (r^-m^-) mutants and their cell extracts can complement each other to give a (r^+m^+) wild type (10,27). This complementation is not due to subunit exchange but to the sequential catalysis of different steps in a complex reaction pathway. An *hsdM* mutation results in the loss of the AdoMet-binding capacity. An *hsdS* mutation allows normal AdoMet binding, but the transition to the activated enzyme is blocked by the loss of DNA recognition. An *hsdR* mutation does not interfere with DNA recognition or methylation, but it prevents ATP hydrolysis which is required for DNA cleavage.

*Eco*K restriction-defective *hsd* mutants of *E. coli* can be produced as part of the SOS response of the cell on UV irradiation and can be selectively enriched by transformation of the irradiated cells with plasmid that has not been modified by the *Eco*K system (28). Two-thirds of the r$^-$ mutants produced by this method are also modification defective (m$^-$).

5. SOURCES

*Eco*KI has been purified about 5000-fold to a form containing both ENase and MTase (15).

*Eco*BI has been purified to about 1000-fold by a somewhat different procedure which results in ENase deprived of MTase activity (3). The resolved MTase is purified separately. All steps are carried out in buffers containing both EDTA and 2-MSH (or DTT).

References

1. Meselson, M., Yuan, R., and Heywood, J. (1972). *Ann. Rev. Biochem.* **41**, 447–466.
2. Endlich, B., and Linn, S. (1981). "The Enzymes," 3rd ed., Vol. 14(A), pp. 137–156.
3. Eskin, B., and Linn, S. (1972). *JBC* **247**, 6183–91.
4. Meselson, M., and Yuan, R. (1968). *Nature* **217**, 1110–1114.
5. Hadi, S. M., Bickle, T. A., and Yuan, R. (1975). *JBC* **250**, 4159–4164.
6. Yuan, R., Heywood, J., and Meselson, M. (1972). *Nature NB* **240**, 42–43.
7. Linn, S., Lautenberger, J. A., Eskin, B., and Lackey, D. (1974). *Fed. Proc.* **33**, 1128–1134.
8. Haberman, A., Heywood, J., and Meselson, M. (1972). *PNAS* **69**, 3138–3141.
9. Vovis, G. F., Horiuchi, K., and Zinder, N. D. (1974). *PNAS* **71**, 3810–3813.
10. Lynn, S., and Arber, W. (1968). *PNAS* **59**, 1300–1306.
11. Kuhnlein, U., Linn, S., and Arber, W. (1969). *PNAS* **63**, 556–562.
12. Lautenberger, J. A., and Linn, S. (1972). *JBC* **247**, 6176–6182.
13. Lautenberger, J. A., Kan, N. C., Lackey, D., Linn, S., Edgell, M. H., and Hutchison, C. A., III (1978). *PNAS* **75**, 2271–2275.
14. Kan, N. C., Lautenberger, J. A., Edgell, M. H., and Hutchison, C. A., III (1979). *JMB* **130**, 191–209.
15. Yuan, R., Hamilton, D. L., and Burckhardt, J. (1980), *Cell* **20**, 237–244.
16. Rosamond, J., Endlich, B., and Linn, S. (1979). *JMB* **129**, 619–635.
17. Smith, J. D., Arber, W., and Kuhnlein, U. (1972). *JMB* **63**, 1–8.
18. Kuhnlein, U., and Arber, W. (1972). *JMB* **63**, 9–19.
19. Bickle, T. A. (1982). In "Nucleases" (S. M. Linn and R. J. Roberts, Eds.), pp. 85–108. CSHL Press, CSH, New York.
20. Yuan, R. (1981). In "Gene Amplification and Analysis" (J. G. Chirikjian, Ed.), Vol. 1, pp. 45–72. Elsevier, New York.
21. Yuan, R., Bickle, T. A., Ebbers, W., and Brack, C. (1975). *Nature* **256**, 556–560.
22. Carter, P., Bedouelle, H., and Winter, G. (1985). *NAR* **13**, 4431–4443.
23. Loenen, W. A. M., Daniel, A. S., Braymer, H. D., and Murray, N. E. (1987). *JMB* **198**, 159–170.
24. Gough, J. A., and Murray, N. E. (1983). *JMB* **166**, 1–19.
25. Fuller-Pace, F. V., and Murray, N. E. (1986). *PNAS* **83**, 9368–9372.
26. Sain, B., and Murray, N. E. (1980). *MGG* **180**, 35–46.
27. Hubacek, J., and Glover, S. W. (1970). *JMB* **50**, 111–127.
28. Hion, K., and Sedgwick, S. G. (1991). *NAR* **19**, 2502.

E. Type III Restriction Enzymes
[Type III site-specific deoxyribonuclease, EC 3.1.21.5]

Type III R–M system has so far four known members: one coded by *E. coli* prophage P1 (*Eco*PI), a second coded by a resident plasmid (p15B) of *E. coli* 15T⁻ (*Eco*P15I), and the third and fourth apparently coded by the chromosomes of *Haemophilus influenzae* Rf (*Hinf*III) and *S. typhimurium* LT (*Sty*LTI).

Type III enzymes recognize short, nonpalindromic sequences. Type III enzymes require Mg^{2+} and ATP as cofactors, but use them differently from the type I R–M enzymes (1). Unlike the type I enzymes, type III R–M enzymes are composed of two nonidentical subunits; they methylate only on a single strand of the recognition sequence (adenine present on only one strand) and cleave DNA at defined sites 3' to the unmodified recognition sequences (see Table 4.1).

1. FUNCTIONS

a. Reaction specificities

Type III enzymes require ATP and Mg^{2+} for ENase activity, but do not hydrolyze ATP. ATP behaves as an allosteric effector. The apparent K_m for ATP (of *Eco*P15) is 7 μM (2). The optimal concentration of Mg^{2+} is 5 mM. Nonhydrolyzable ATP analogs can substitute for ATP in the ENase reaction, albeit at slower rates (3). ENase activity does not require AdoMet, but is stimulated by it (4–6).

In the presence of ATP and AdoMet, type III enzymes act as both R·ENase and M·MTase. The two activities compete on the unmodified recognition site, although the methylation is a slower reaction. In the presence of AdoMet but without ATP, the enzymes function as pure MTases, which are stimulated by ATP. The apparent K_m values (of *Eco*P15I) for AdoMet depend on reaction type: 10 nM for restriction and 0.4 μM for methylation (2).

b. Endonuclease activities

i. Reaction conditions. Type III restriction enzymes cleave unmodified DNA in the presence of ATP and Mg^{2+} (2,4). This reaction is stimulated by the addition of AdoMet. The pH optimum is 6.7–7.0 in the absence of AdoMet. It is somewhat higher (pH 7.6) in the presence of AdoMet.

A typical restriction reaction with *Eco*P15I is carried out at 30°C in 100 mM Tris–Cl (pH 8.0), 0.2 mM EDTA, 14 mM 2-MSH, 5 mM $MgCl_2$, 0.5 mM ATP, 40 μM AdoMet, and DNA [e.g., superhelical SV40 (7)].

For mechanistic reasons, type III enzymes do not digest the substrate completely and some sites are always left uncut.

ii. Substrate specificities

(a) Recognition sequences. The recognition sequences for *Eco*PI (8,9), *Eco*P15I (10), *Hinf*III (11), and *Sty*LTI (12) are [cf. methylated adenine is marked with *] *Eco*PI (-AGA*CC-), *Eco*P15I (-CAGCA*G-), *Hinf*III (-CGAAT-), and *Sty*LTI (-CAGA*G-).

(b) DNA cleavage. Type III (*Eco*PI and *Eco*P15I) enzymes cleave DNA at about 25–27 bp (roughly 2.5 turns of the helix) 3' to the unmodified recognition sequence and generate fragments with two to four base 5'-protruding termini. The cleavage requires two unmodified recognition sites that can be separated by different distances, but which must be in inverse orientation (23). This explains why some sites are not restricted and why complete digestion is impossible even in the absence of AdoMet.

(c) Influence of flanking sequences. Type III R–M enzymes do not cleave or methylate each of their recognition sites with the same efficiency. *Eco*PI (3) and *Hin*fIII (13) show some dependence on the structure of the substrate DNA. Linear molecules of ColE1 DNA are poorer substrates for *Hin*fIII than the supercoiled molecules. λ DNA fragments with a MW $\leq 2 \times 10^6$ are not cleaved by the *Hin*fIII enzyme (but they can be methylated).

c. Methyltransferase activities

i. Reaction conditions. DNA methylation with *Eco*P15I is typically carried out at 30°C in 100 mM HEPES (pH 8.0), 0.25 mM EDTA, 6 mM MgCl$_2$, 12 mM 2-MSH, and 0.7 μM [*methyl*-^{3}H]AdoMet (73 Ci/mmol) (3).

ii. Substrate specificities. All type III R–M enzymes methylate on N-6 of adenine. Only one strand gets methylated both *in vivo* and *in vitro* because adenine is present on only one strand. Both M·*Eco*PI and M·*Eco*P15I methylate the second adenine in their recognition sequences, 5'-AGA*CC-3' and CAGCA*G, respectively. In contrast to the ENase activity, methylation is independent of the orientation of the recognition site (23). The methylation by M·*Eco*P15I of an overlapping *Alu*I restriction site, e.g., AGCTGCTG/CAGCA*GCT, renders the sequence insensitive to *Alu*I (14).

d. Mechanism of methylation

The methylation reaction apparently follows a Ping-Pong type of kinetic mechanism. The enzyme binds AdoMet, most likely forming a methylated enzyme intermediate (15). The methyl group is then transferred to the DNA which binds the enzyme independently.

2. Structure

Both *Eco*PI and *Eco*P15I are composed of two nonidentical subunits with MWs of 110,000 (large) and 73,000 (small) (16). The large subunit is responsible for restriction activity, whereas the small subunit is responsible for modification activity. For both enzymes, the modification activities can be isolated separately from the restriction activities. These modification enzymes are tetramers of a single subunit (73 kDa), the product of the *mod* gene (15,16). Antibodies prepared against *Eco*P15I cross-react with both subunits of *Eco*PI.

i. *Eco*PI. The small subunit (*mod* gene product) of *Eco*PI is composed of 646 amino acid residues (M_r 73,486). It contains ~40% α helix (15). In the form of a tetramer, the subunit is capable of modifying DNA *in vitro,* but not of restricting it. The small subunit apparently provides both specific sequence recognition and DNA methylation functions (17). The large subunit of *Eco*PI, the *res* gene product, is composed of 970 amino acid residues (M_r 111,459). The large subunit has the ENase activity (1,18).

ii. EcoP15I. The small subunit (*mod* gene product) of *Eco*P15I has 647 amino acid residues (M_r 74,223). Except for the central portion of ~250 amino acids, both N-terminal ~250 residues and C-terminal ~130 residues of *Eco*PI and *Eco*P15I *mod* gene products are highly conserved. The nonhomologous central region is thus believed to be the DNA recognition domain, whereas the conserved regions are likely to be involved in subunit interactions.

3. GENETICS

The genes coding for *Eco*PI and *Eco*P15I are allelic (19). In fact, the P1 prophage and the resident plasmid p15B (32.9 μm length DNA) of *E. coli* 15T⁻ have a high degree of homology (20). Both PI and P15I R–M systems involve only two genes, called *res* for restriction and *mod* for modification (1). The respective R–M systems are nonessential for phage development.

i. Gene organization. The two genes are contiguous in the order of *mod* (1938 bp) and *res* (2910 bp) (17). They are apparently transcribed in the same direction from a promoter located in front of the *mod* gene (21). The *res* gene is separated from the *mod* gene by 5 bp, including the *mod* gene stop codon, UAA.

ii. Mutation analysis. Both *res* (large subunit) and *mod* (small subunit) gene products are necessary for restriction, whereas modification requires only *mod*. Both (Res⁻Mod⁺) and (Res⁻Mod⁻) phenotypes have been obtained in transposition–mutation analyses. In addition, (Res⁺Mod⁻) mutants that present a clear plaque phenotype have been obtained (22). In these "clear" *Eco*PI mutants, DNA recognition and cleavage are normal, but methylation is defective, possibly due to a mutation in the AdoMet-binding site.

4. SOURCES

Both *Eco*PI and *Eco*P15I R–M enzymes have been purified to homogeneity from *E. coli* that contains the structural genes for the enzymes cloned on small, multicopy plasmids and overproduces the enzymes (16).

A cloned *Eco*P1 MTase gene has been overexpressed to about 20% of the total soluble protein in *E. coli* under the control of the phage λ p_L promoter and cI857 coding for thermolabile repressor (15). The M·*Eco*PI enzyme has been purified to homogeneity (about 10-fold purification or 4.5 mg protein per liter culture).

References

1. Bickle, T. A. (1982). In "Nucleases" (S. M. Linn and R. J. Roberts, Eds.), pp. 85–108. Cold Spring Harbor Laboratory, Cold Spring Harbor, New York.
2. Reiser, Y., and Yuan, R. (1977). *JBC* **252**, 451–456.
3. Yuan, R., Hamilton, D. L., Hadi, S. M., and Bickle, T. A. (1980). *JMB* **144**, 501–519.
4. Kauc, L., and Piekarowicz, A. (1978). *EJB* **92**, 417–426.
5. Haberman, A. (1974). *JMB* **89**, 545–563.
6. Yuan, R., and Reiser, J. (1978). *JMB* **122**, 433–445.
7. Risser, R., Hopkins, N., Davis, R. W., Delius, H., and Mulder, C. (1974). *JMB* **89**, 517–544.
8. Bachi, B., Reiser, J., and Pirrotta, V. (1979). *JMB* **128**, 143–163.
9. Hattman, S., Brooks, J. E., and Masurekar, M. (1978). *JMB* **126**, 367–380.
10. Hadi, S. M., Bachi, B., Shepherd, J. C. W., Yuan, R., Ineichen, K., and Bickle, T. A. (1979). *JMB* **134**, 655–666.
11. Piekarowicz, A., Bickle, T. A., Shepherd, J. C. W., and Ineichen, K. (1981). *JMB* **146**, 167–172.
12. De Backer, O., and Colson, C. (1989). *J. Cell. Biol. (Suppl.)* **13D**, 212.
13. Piekarowicz, A., and Brzezinski, R. (1980). *JMB* **144**, 415–429.
14. Meisel, A., Kruger, D. H., and Bickle, T. A. (1991). *NAR* **19**, 3997.
15. Hornby, D. P., Muller, M., and Bickle, T. A. (1987). *Gene* **54**, 239–245.
16. Hadi, S. M., Bachi, B., Iida, S., and Bickle, T. A. (1983). *JMB* **165**, 19–34.
17. Humberlin, M., Suri, B., Rao, D. N., Hornby, D. P., Eberle, H., Pripfl, T., Kenel, S., and Bickle, T. A. (1988). *JMB* **200**, 23–29.
18. Heilmann, H., Burkardt, H.-J., Puhler, A., and Reeve, J. N. (1980). *JMB* **144**, 387–396.
19. Arber, W., and Wauters-Willems, D. (1970). *MGG* **108**, 203–217.
20. Ikeda, H., Inuzuka, M., and Tomizawa, J.-I. (1970). *JMB* **50**, 457–470.
21. Iida, S., Meyer, J., Bachi, B., Stalhammar-Carlemalm, M., Schrickel, S., Bickle, T. A., and Arber, W. (1983). *JMB* **165**, 1–18.
22. Rosner, J. L. (1973). *Virol.* **52**, 213–222.
23. Meisel, A., Bickle, T. A., Kruger, D. H., and Schroeder, C. (1992). *Nature* **355**, 467–469.

III. METHYLTRANSFERASES (METHYLASES)
[Methyltransferases, EC 2.1.1]

A. General Description

Enzymes that catalyze the methylation of nucleotides in DNA are found in a wide variety of prokaryotes and eukaryotes. In prokaryotes, the DNA MTases are most often identified as elements of restriction–modification systems. In eukaryotes, DNA MTases have multiple functions, but their involvement in R–M is not clearly known. All DNA MTases use AdoMet as the methyl group donor (Scheme 4.3).

$$\text{DNA} + \text{AdoMet} \xrightarrow{\text{MTase}} \text{methylated DNA} + \text{AdoHcy}$$
SCHEME 4.3

Both cognate and noncognate DNA methylations provide a useful means of controlling restriction activities of many restriction enzymes. For this reason, modification MTases find important applications in gene cloning, chromosomal mapping, and other DNA restriction manipulations.

More than 100 methylation modification systems, the majority being type II MTases, have now been cloned in *E. coli* (1). Although methylation of RNA, which occurs on all four major base components of mRNA, tRNA, and rRNA, no doubt plays important physiological roles, relatively little is known about the enzymology and genetics of diverse RNA MTases. It may take a long while before these systems become useful tools in recombinant DNA technology.

This section focuses on the general characteristics of DNA MTases. In the following sections, two representative DNA MTase systems are examined in detail: *Eco*RI MTase which functions as part of the typical type II R–M system and *dam* and *dcm* MTases whose functions are apparently unrelated to any known R–M system.

1. Role of DNA Methylation

The DNA methylation occurring as a partial function of R–M systems is an important mechanism of self-defense in prokaryotes. In eukaryotic cells, however, DNA methylation apparently plays diverse biological functions. It has been shown to be tightly associated with DNA replication and to be implicated in mismatch repair, virus latency, and gene expression or differentiation (2–5). In mammalian DNA, approximately 90% of all m^5C occurs in the CpG dinucleotide. (*Note:* The CpG sequence is on average underrepresented in mammalian DNA.) Perhaps significantly, the sites of DNA methylation, in particular on cytosine, have been shown to be the "hot spots" of spontaneous base substitutions from C to T (6). Because some of the epigenetic defects caused by abnormal methylation can be inherited as a particular pattern of gene activities, DNA methylation has a role in oncogenesis, aging, and evolution (7).

DNA methylation is also involved in the activation of a restriction system called MDRS (methylation-dependent restriction system) which is site specific in restricting either adenine- or cytosine-methylated DNA (8,9). Expression of cloned MTase genes has been shown to induce SOS DNA repair responses in some *E. coli* strains (mcr^+ or mrr^+) (10).

2. Effects of DNA Methylation on DNA–Protein Interaction

The effect of DNA methylation on various aspects of cellular metabolism depends on the base (A or C) involved and its specific position in the genomic context. The stable alteration of local structure following DNA methylation is one certain way to alter the DNA–protein interaction. In fact, evidence suggests that methylation is not entirely a nondiscriminatory event throughout a given genome but is largely guided by certain, more specifically located signal sequences or a higher ordered local structure. The biological consequences of methylation may be mediated in some cases by direct steric and hydrophobic effects of the methyl group on particular protein bindings. Alternatively, methylation may facilitate the melting of DNA (11) and/or transition of DNA to a particular conformation such as the left-handed Z form (12,13) and the "alternating B" form (14,15), and thus serve as a partial determinant of specificity. In other cases, methylated

DNA is bound by a nuclear protein (e.g., methyl-CpG binding protein), which secondarily prevents other nuclear factors from interacting with the gene.

3. DIVERSITY OF METHYLTRANSFERASES

DNA MTases commonly use AdoMet as the methyl group donor. They differ in their acceptor specificity and may be highly specific to DNA sequences. They catalyze the transfer of a methyl group onto a specific base in a specific DNA sequence and can be categorized as either adenine-specific MTases or cytosine-specific MTases. In A-MTases such as M · EcoRI, M · HhaII, M · KpnI, and M · PstI, the methyl group is linked to an exocyclic amino group by C–N bonds, N^6-methyl-A (m^6A or ^mA). Cytosine-specific MTases can be further divided into the MTases generating 5-methyl-C (m^5C or ^mC) in which the methyl group is directly linked to the pyrimidine ring through a C–C bond or the MTases generating N^4-methyl-C (m^4C) in which the methyl group is linked by the C–N bond. Compared with the large number of m^5C-MTases, such as M · EcoRII, M · HaeIII, M · HhaI, M · HpaII, and M · MspI, there are only a small number of m^4C-MTases and they include M · BamHI, M · SmaI (and its isoschizomer M · Cfr9I), and M · PvuII.

Comparison of amino acid sequences of many cloned type II DNA MTases led to the identification of regionally conserved motifs characteristic for m^4C-, m^5C-, and m^6A-MTases. The base specificity for the majority of type II MTases is thus attributed to the presence of one of these motifs. In a subtype of type IIS MTases, which includes M · Alw26I, M · Eco31I, and M · Esp3I, the enzymes apparently possess sets of conserved domains characteristic for both m^6A- and m^5C-MTases. These subtype enzymes methylate both strands of asymmetric recognition sites, yielding m^5C in one strand and m^6A in the other strand (34).

4. TYPES OF METHYLTRANSFERASES

The MTase activities as part of type I and type III R–M systems have been briefly described earlier (see Section II,D and E, this chapter). In this section, type II modification enzymes and non-R–M MTases are described. These MTases are versatile reagents often used in molecular biology and recombinant DNA technology due to their relatively simple and easily controlled functional characteristics. Although it has not been determined whether all type II R · ENases are accompanied by cognate modification systems, in every known case, the two genes encoding R · ENase and M · MTase are contiguous and expressed in a concerted manner (see Fig. 4.1). The effect of site-specific DNA methylation on R · ENases differs widely depending on the R · ENase and the position of methylation in canonical or noncanonical sequences.

Type II DNA MTases (A- or C-specific) are a large group of enzymes that have site specificities identical to their cognate R · ENases. Type II MTases are generally composed of a single polypeptide having an average of 380 amino acid residues (~40 kDa). They exist in solution as a stable, functional monomeric form. Similar to type II R · ENases, type II modification enzymes are probably best characterized due to their simplicity in structure and function. The majority of type II M · MTases are *monospecific,* recognizing and methylating a single specificity site. In contrast, some MTases of phage origin are *multispecific* and they can recognize and methylate more than one independent sequence via independent target-recognizing domains (16).

Type IIS modification methylases are another group of interesting MTases due to the fact that their recognition sequences are asymmetric. In the case of M · *Fok*I, a single enzyme methylates adenine residues in both strands of the recognition site. Another type IIS methylase, M · *Hga*I, consists of two separate cytosine-MTases which are responsible for methylation of different DNA strands. A third type of type IIS MTase, M · *Mbo*II, modifies only one strand of the recognition site, like type III R–M enzymes. A fourth type of type IIS MTases, e.g., M · *Alw*26I and M · *Eco*31I, recognizes and methylates both strands but with different strand specificity, i.e., a cytosine residue on one strand and an adenine residue on the other.

5. METHODS OF LOCATING METHYLATED BASES IN DNA

Methylated bases and their positions in recognition sequences are usually determined by chemical or enzymatic hydrolysis of methylated DNA molecules, followed by chromatographic analysis. In a typical method (34), $5'$-^{32}P-labeled synthetic oligonucleotide duplexes containing suitable recognition sites are methylated with the corresponding methylases in the presence of [^{3}H]AdoMet. The double-labeled duplex substrates are then partially hydrolyzed with snake venom phosphodiesterase, a $3'$-exonuclease releasing $5'$-mononucleotides. The products are separated and identified by two-dimensional mapping: first, by electrophoresis (e.g., on cellulose acetate strip in pyridine acetate, pH 3.5) and second, by chromatography (e.g., on a DEAE-cellulose thin layer plate). The duplex substrates are designed so that the complementary oligonucleotides do not overlap with each other in the two-dimensional map. Alternatively, the ^{3}H-methylated substrates are enzymatically hydrolyzed to deoxynucleotides which are then separated and analyzed by HPLC using unlabeled methyldeoxynucleotides as standards.

Of particular concern is how to distinguish the m^4C from m^5C. The m^4C is reactive to hydrazine, but m^5C is less reactive to this reagent and to NaOH as well (17). The m^5C can thus be distinguished from m^4C using A > C and C-specific reactions of the chemical sequencing method.

In some cases, the methylation specificity can be deduced from the methylation-mediated blocking of cleavage at an overlapping site for a known R · ENase. With a DNA fragment containing the MTase recognition site, which appropriately overlaps with a class IIS restriction enzyme cleavage site, the methylated base can be identified by the asymmetric cleavage with the class IIS enzyme and then by chemical sequencing (18).

6. ASSAYS OF MTASE ACTIVITY AND UNIT DEFINITIONS

i. Methyl transfer assay. Activity is measured by the extent of [H^3]CH$_3$ group transfer from [^{3}H]AdoMet to DNA which can be quantitated either by acid (5% PCA or TCA) precipitation and collection on glass fiber disks or by binding to anion-exchange filters (e.g., Whatman DE81 paper) under conditions similar to the ones described next for the protection assay.

Unit definition: One "methyl transfer" unit catalyzes the transfer of 1 pmol of [^{3}H]methyl groups in 30 min at 37°C.

ii. **Protection assay.** The activity of MTase is often measured by a protection assay: MTase is incubated with 1 µg λ DNA for 1 hr at 37°C in 10 µl assay buffer composed of 50–100 mM Tris–Cl (pH 7.5–8.0), 80–100 µM AdoMet, 5–10 mM EDTA, 2 mM DTT (or 5 mM 2-MSH), and 100 µg/ml BSA. The extent of protection is determined by R·*Eco*RI (10 U) digestion for 30 min (or 1 hr) at 37°C with the assay buffer adjusted to 10 mM MgCl$_2$. The product is analyzed by agarose gel electrophoresis.

Unit definition: One "protection" unit protects 1 µg of λ DNA from the cleavage by *Eco*RI (or other R·ENases) under the incubation conditions.

7. VARIABLE EFFECTS OF METHYLATION ON DNA RESTRICTION

Although there are exceptions, type II restriction enzymes do not usually cleave the DNA methylated by the cognate MTase. Many, but not all, restriction enzymes may also be sensitive to noncanonical methylations produced by other DNA MTases (19). For example, *Sau*3AI (GATC) is unaffected by A methylation, but cannot cleave when C is methylated. *Bam*HI (GGATCC) is not inhibited by GG^{m}ATCC (by overlapping *dam* methylation) or GGATC^{m}C that derives from the overlapping methylation by M·*Hpa*II (C^{m}CGG), but is inhibited by GGAT^{m}CC which occurs on overlapping methylation by M·*Msp*I on the 5′-C of CCGG. In fact, almost all the enzymes, except *Acc*III, *Asu*II, *Cfr*9I, and *Xma*I, that can generate large fragments of mammalian DNA are inhibited by the ^{m}CpG modification at overlapping sites.

*Dpn*I cleaves DNA only when both strands are methylated at the sequence G^{m}ATC, whereas *Mbo*I cuts the same sequence with unmethylated adenines only. Neither *Dpn*I nor *Mbo*I cleaves hemimethylated DNA. (*Note:* As the noncognate methylation has differential effects on various R·ENases, so it does on cognate MTases. Some methylases are inhibited by a prior methylation at a secondary site within the recognition sequence.)

Restriction enzymes also display different sensitivities to noncanonically hemi-methylated DNAs which are created, for example, by incorporating m^{5}C during the second-strand DNA synthesis catalyzed by DNA polymerases (40). At high enzyme-to-substrate ratios, some enzymes (e.g., *Acc*I, *Bam*HI, *Pst*I, *Sal*I, *Sma*I, *Sst*I, and *Xho*I) are unable to cleave their recognition sites, whereas other enzymes (e.g., *Eco*RI, *Hin*dIII, *Dpn*I, *Pvu*II, and *Xba*I) are only partially sensitive to hemi-methylation.

8. APPLICATIONS

DNA MTases can be used singly for some applications that require methylation protection from certain restriction enzymes. In other applications, however, they can be used in various combinations to methylate specific subsets of overlapping restriction sites and thus alter and enhance apparent recognition specificities of R·ENases. The following examples illustrate the utilities of MTases in general as well as in specific applications.

i. **Cloning of DNA fragments.** When using *Eco*RI (or other restriction enzyme) linkers to insert cDNA or DNA fragments into the *Eco*RI site of a vector (plasmid, phage λ vectors), any potential internal *Eco*RI sites must first be protected by methylation. The linkers are then attached to the DNA by blunt-end ligation and are cleaved by *Eco*RI to open cohesive ends. The cohesive-ended DNAs are ligated to the vector and are transfected into a bacterial host (refer to [9] below).

Except for some special applications (see below), the conventional linker method of cloning is now largely superseded by a simpler and more efficient adapter method. In this improved method, two synthetic oligonucleotides are annealed to form a restriction site (e.g., *Eco*RI) in such a way that it has the necessary cohesive termini on one end, while the other end is blunt-ended and 5'-phosphorylated to allow ligation to the DNA (20,21).

A novel cDNA/PCR strategy for cloning rare mRNA species employs another kind of adapter–primer that has dual functions: one part as a PCR primer and the other as an adapter (22). The amplified cDNA products carry blunt ends as if linkers have been attached. Therefore the DNA molecules can be cloned in a way similar to the conventional linker-mediated cloning procedures mentioned earlier. For cloning unknown genes, the cDNA/PCR cloning procedure thus requires prior methylation to protect any internal restriction sites.

ii. **Enhancement of R·ENase cleavage site specificity.** Site-specific cleavage of DNA mediated by type II R·ENases forms the basis of modern recombinant DNA technology. Increasing the cleavage specificity of R·ENases thus has significant implications in amplifying the power of recombinant DNA technology. For this purpose, DNA methylation is one of the most important techniques that are used in various forms and strategies.

(a) Overlapping methylation. DNA MTases are known whose recognition sequences overlap only a subset of a given restriction site. Methylation of such an overlapping subsequence may be effective in blocking the cleavage action by some but not all R·ENases. For example, *Bam*HI cleavage at G|GATCC will not occur at a subset sequence ^{m}CCGGATCC when it has been methylated by M·*Msp*I (^{m}CCGG). A new addition to the useful panel of MTases is M·*Sss*I, a CpG methylase (-^{m}CG-) isolated originally from *Spiroplasma* sp. strain MQ1 (23).

M·MTases and R·ENases that recognize the same sequence can differ in their sensitivity to noncanonical methylation. For example, *Bam*HI can cut GGATC^{m}C efficiently, whereas M·*Bam*HI (normally GGATm^{4}CC) cannot methylate this modified sequence. Applying appropriate combinations of MTase and ENase pairs, e.g., sequential methylation with M·*Hpa*II (C^{m}CGG) and then M·*Bam*HI, *Bam*HI cleavage can be selectively directed to the octanucleotide sequence, GGAT-

CCGG (24). This is as if the cleavage sequence of BamHI has been increased from normal hexanucleotide to octanucleotide.

Similarly, the specificity of the FokI recognition sequence (GGATG) can be increased by prior methylation at a noncognate site with M · MspI. The resulting sequence, 5'-GGATG-3'/3'C^{m}CTAC-5', is inhibitory to M · FokI while it has no effect on R · FokI. Therefore a subsequent methylation by M · FokI directs the cleavage by R · FokI to the 7-bp recognition sites (CCGGATG) (25).

With restriction sites that contain degenerate sequences, site-specific methylation can be used to block a particular recognition sequence (e.g., TCG^{m}A by M · TaqI) while leaving other sequences cleavable by other R · ENases. Thus the cleavage specificity, for example, of HincII (GTPy|PuAC) can be increased by prior methylation with M · TaqI. In another example, methylations by M · TaqI and/or M · HpaII (C^{m}CGG) reduce the number of cleavage sites by AvaI (C| PyCGPuG).

(b) Shielding methylation. Sequence-specific cleavage of DNA can be performed by applying an efficient strategy based on overall methylation of DNA except at a predetermined cleavage site(s). In this so-called "Achilles cleavage" strategy (Fig. 4.5), all available recognition sites except for a target site for a given restriction enzyme are first methylated by a cognate MTase. The predetermined site is protected from the methylation by a sequence-specific DNA-binding protein or oligonucleotide. Subsequent removal of the blocking agent allows the DNA to be selectively cleaved at the target site by the R · ENase. Use of a pair of oligonucleotides enables one to generate a specific DNA fragment. Methylation-blocking agents that have proven to be useful include repressor (e.g., *lac*) proteins, eukaryotic transcription factors (e.g., GCN4), and synthetic oligonucleotides (or their analogs) capable of forming a triplex (or tight complex).

Among the available Achilles cleavage strategies, the one based on the use of the RecA protein–oligonucleotide complex as the methylation blocker is particularly useful for its versatility (26,35). Because the binding specificity is derived entirely from a synthetic oligonucleotide, virtually any restriction site for which the corresponding MTase is available can be converted to a unique cleavage site.

iii. **Use in partial cleavage of DNA.** Partial cleavage of DNA with R · ENase, e.g., NotI (GC|GGCCGC), is an important technique for genomic DNA mapping. The cleavage reaction can be controlled by the addition of competing MTase, e.g., M · BspRI (GG^{m}CC) (27). Partial DNA cleavages can also be controlled by using M · FnuDII (CG^{m}CG) or M · BepI (^{m}CGCG) since the methylation of the overlapping sequences, e.g., GCGGC^{m}CGC, blocks NotI cleavage (28).

iv. **Genome mapping of sites implicated in DNA–protein interaction.** Specific DNA–protein interactions play a key role in gene expression and DNA replication. Methylation provides an important approach to the identification and characterization of the DNA sequences involved in such interactions. The rationale is that endogenous or artificially introduced MTases methylate all genomic targets except

III. Methyltransferases (Methylases)   291

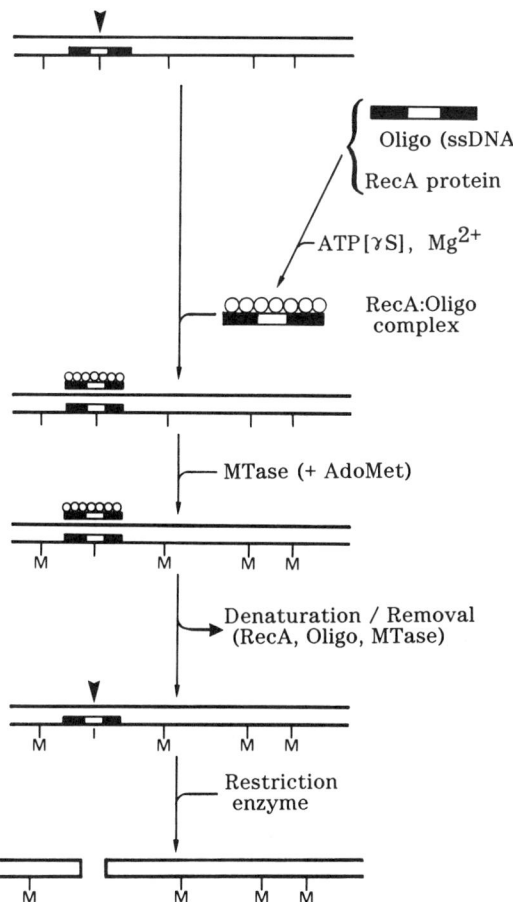

FIGURE 4.5 A schematic of RecA-mediated Achilles cleavage strategies. A target cleavage site (▼) for a chosen restriction enzyme (*Eco*RI) is selectively protected from methylation by using a methylation-blocking agent, e.g., a RecA:oligonucleotide filament. In the presence of ATP[γS], a nonhydrolyzable ATP analog, RecA proteins form a stable complex with ssDNA (30–70 nt long) at an optimal stoichiometry of 3–5 nt residues per RecA monomer. The nucleoprotein filament then binds the homologous dsDNA target, resulting in a triplex structure which interferes with methylation. All other *Eco*RI recognition sites (tic marks) are then subject to protective methylation (M) by the cognate MTase, M·*Eco*RI. RecA and MTase are subsequently inactivated and/or removed, and the DNA is selectively cleaved by *Eco*RI at the target site.

those protected *in vivo* by protein or nonprotein factors interfering with methylase action. The protected targets remain unmethylated in purified genomic DNA and can be identified using methylation-sensitive restriction enzymes singly or in combination with DNase I footprinting. For example, the role of sequences involving GATC has been examined using adenine-methylating Dam MTase which was endogenously expressed and the restriction enzymes (e.g., *Dpn*I, *Mbo*I, and

Sau3AI) which recognize the same GATC sequence but with different sensitivities to methylation (36).

9. METHYLATION-DEPENDENT RESTRICTION SYSTEM (MDRS) AND GENE CLONING

Introduction of methylated DNA into an *E. coli* host requires careful attention to the host genotype since the transformation efficiency can vary by 3 to 4 orders of magnitude depending on the strain. This widely varying efficiency can be the reason for success or failure in certain gene cloning, library construction, and plasmid rescue. The MDRS that is operational in some strains preferentially restricts the DNAs methylated at three target sequences: m^6A (e.g., in G^mAC and C^mAG modified by $M \cdot HhaII$ and $M \cdot PstI$, respectively) by Mrr, m^5CG (e.g., in ^mCG and C^mCGG modified by $M \cdot SssI$ and $M \cdot HpaII$, respectively) by McrA, and G^mC (including m^5C, m^4C, and hm^5C, for example, in Gm^5CGC modified by $M \cdot HhaI$) by McrB and McrC (8,9,29). *E. coli* K12 *mcrBC* genes are organized as an operon (37). The *mcrBC* operon specifies five proteins: three from *mcrB* and two from *mcrC*. The products of both *mcrB* and *mcrC* genes are required to restrict the methylated substrate DNA.

To remedy these situations or, more importantly, to avoid possible bias or skewing of cDNA libraries, methylated DNAs should be amplified initially in *E. coli* hosts that lack MDRS (30,31), preferably in Δ*mcrBCF* strains having a major deletion of the *mcr* region. The clones derived from methylated genomic DNA using a *mcr*$^+$ host (especially *recA*$^+$) may present significant frequencies of sequence artifacts (38). Some of the *mcr*$^-$ strains suitable as hosts include DH10B (BRL), KW251 (Promega), and SURE or SRB (Stratagene). An improved strain (e.g., PMC103) is recommended for the constructin of genomic libraries in λ and cosmid vectors using methylated DNAs that contain regions of potential secondary structures (39). Note that some commonly used *E. coli* strains (e.g., HB101 and RR1) are *mcrA*$^+$*mcrB*$^-$. Other strains such as Y1090 and JM107 are *mcrA*$^-$*mcrB*$^+$ (see Appendix B).

10. CLONING OF METHYLASE GENES

Based on the observation that the expression of site-specific DNA MTases induces the SOS response in *E. coli* strains that are *mcrA*$^+$, *mcrB*$^+$, or *mrr*$^+$ (10), a method has been developed that enables rapid identification of recombinant clones encoding MTases. The method employs *E. coli* strains that are *ts* for the Mcr and Mrr systems and contain a *lacZ* reporter gene fused to the damage-inducible *dinD* locus (32). The feature of this method is that the recombinant clones which express cloned MTase gene can grow and form colonies at a permissive temperature (42°C), but are unable to grow at a nonpermissive temperature (30°C) due to the damage to DNA caused by the M-dependent R system. This in turn induces the expression of the *dinD–lacZ* fusion protein producing X-Gal blue colonies. When the transformants initially grown at 42°C are incubated at 30°C, the clones carrying the MTase gene can be positively selected.

MTase genes, either alone or together with restriction enzyme genes, have been cloned using various other strategies as well, most notably by selecting *in vitro* for modified recombinants whose DNAs are resistant to cleavages by cognate ENases (33).

References

1. Wilson, G. G. (1988). *Gene* **74**, 281–289.
2. Doerfler, W. (1983). *Ann. Rev. Biochem.* **52**, 93–124.
3. Marinus, M. G. (1987). *Ann. Rev. Genet.* **21**, 113–131.
4. Modrich, P. (1987). *Ann. Rev. Biochem.* **56**, 435–466.
5. Adams, R. L. P. (1990). *BJ* **265**, 309–320.
6. Coulondre, C., Miller, J. H., Farabaugh, P. J., and Gilbert, W. (1978). *Nature* **274**, 775–780.
7. Holliday, R. (1987). *Science* **238**, 163–170.
8. Raleigh, E. A., and Wilson, G. (1986). *PNAS* **83**, 9070–9074.
9. Blumenthal, R. M. (1989). *FOCUS (BRL)* **11**(3), 41–46.
10. Heitman, J., and Model, P. (1987). *J. Bacteriol.* **169**, 3243–3250.
11. Engel, J. D., and von Hippel, P. H. (1978). *JBC* **253**, 927–934.
12. Behe, M., Zimmerman, S., and Felsenfeld, G. (1981). *Nature* **293**, 233–235.
13. Zacharias, W., Caserta, M., O'Connor, T. R., Larson, J. E., and Wells, R. D. (1988). *Gene* **74**, 211–224.
14. Klug, A., Jack, A., Viswamitra, M. A., Kennard, O., Shakked, Z., and Steitz, T. A. (1979). *JMB* **131**, 669–680.
15. Wu, H. Y., and Behe, M. J. (1985). *Biochem.* **24**, 5499–5502.
16. Lange, C., Jugel, A., Walter, J., Noyer-Weidner, M., and Trautner, T. A. (1991). *Nature* **352**, 645–648.
17. Ohmori, H., Tomizawa, J.-I., and Maxam, A. M. (1978). *NAR* **5**, 1479–1485.
18. Posfai, G., and Szybalski, W. (1988). *Gene* **74**, 179–181.
19. Nelson, M., and McClelland, M. (1991). *NAR* **19**(Suppl), 2045–2071.
20. Sartoris, S., Cohen, E. B., and Lee, J. S. (1987). *Gene* **56**, 301–307.
21. Haymerle, H., Herz, J., Bressan, G. M., Frank, R., and Stanley, K. K. (1986). *NAR* **14**, 8615–8624.
22. Akowitz, A., and Manuelidis, L. (1989). *Gene* **81**, 295–306.
23. Renbaum, P., Abrahamove, D., Fainsod, A., Wilson, G. G., Rottem, S., and Razin, A. (1990). *NAR* **18**, 1145–1152.
24. McClelland, M., and Nelson, M. (1988). *Gene* **74**, 169–176.
25. Posfai, G., and Szybalski, W. (1988). *NAR* **16**, 6245.
26. Koob, M., Burkiewicz, A., Kur, J., and Szybalski, W. (1993). *NAR* **20**, 5831–5836.
27. Hanish, J., and McClelland, M. (1990). *NAR* **18**, 3287–3291.
28. Qiang, B., McClelland, M., Poddar, S., Spokauskas, A., and Nelson, M. (1990). *Gene* **88**, 101–105.
29. Ross, T. K., Achberger, E. C., and Braymer, H. D. (1989). *MGG* **216**, 402–407.
30. Woodcock, D. M., Crowther, P. J., Doherty, J., Jefferson, S., DeCruz, E., Noyer-Weidner, M., Smith, S. S., Michael, M. Z., and Graham, M. W. (1989). *NAR* **17**, 3469–3478.
31. Raleigh, E. A., Murray, N. E., Revel, H., Blumenthal, R. M., Westaway, D., Reith, A. D., Rigby, P. W. J., Elhai, J., and Hanahan, D. (1988). *NAR* **16**, 1563–1575.
32. Piekarowicz, A., Yuan, R., and Stein, D. C. (1991). *NAR* **19**, 1831–1835.
33. Lunnen, K. D., and 12 coauthors (1988). *Gene* **74**, 25–32.
34. Bitinaite, J., Maneliene, Z., Menkevicius, S., Klimasauskas, S., Butkus, V., and Janulitis, A. (1992). *NAR* **20**, 4981–4985.
35. Ferrin, L. J., and Camerini-Otero, R. D. (1991). *Science* **254**, 1494–1497.
36. Wang, M. X., and Church, G. M. (1992). *Nature* **360**, 606–610.
37. Krueger, T., Grund, C., Wild, C., and Noyer-Weidner, M. (1992). *Gene* **114**, 1–12.

38. Williamson, M. R., Doherty, J. P., and Woodcock, D. M. (1993). *Gene* **124**, 37–44.
39. Doherty, J. P., Lindeman, R., Trent, R. J., Graham, M. W., and Woodcock, D. M. (1993). *Gene* **124**, 29–35.
40. Nelson, P. S., Papas, T. S., and Schweinfest, C. W. (1993). *NAR* **21**, 681–686.

B. Type II Methylase: *Eco*RI Methyltransferase
[Site-Specific DNA–Methyltransferase (Adenine-Specific), EC 2.1.1.72]

*Eco*RI MTase (M · *Eco*RI) is a member of the type II adenine-MTases whose structure and function have been relatively well characterized. It is a monomeric enzyme with a molecular mass of 38 kDa, an average size for type II MTases. M · *Eco*RI recognizes the same hexanucleotide sequence as R · *Eco*RI, and methylates the central adenine residues of both strands at the exocyclic N-6 amino group resulting in N^6-methyl-A (m^6A or ^mA): 5'-GA^{m}ATTC-3'/3'-CTT^{m}AAG-5' (1). The MTase requires as substrates only AdoMet and DNA that is unmodified in at least one strand.

1. FUNCTIONS

a. Reaction conditions

i. Optimal reaction conditions. The M · *Eco*RI reaction is typically performed in 50 mM Tris–Cl (pH 7.5), 80 μM AdoMet, 5 mM 2-MSH (or DTT), 1–10 mM EDTA, DNA (1–2 μg), and 3–10 U of MTase.

The optimum pH is 7.5–8.0 under the usual ionic strength of 50–100 mM NaCl. For a self-complementary synthetic octamer (T_m = 17–19°C) containing an *Eco*RI site, the optimal temperature of methylation is 12.5°C, slightly lower than the optimum temperature (15°C) for R · *Eco*RI activity (2). MTase activity is optimal at 37°C with SV40 DNA as substrate.

*Eco*RI MTase does not require divalent cations and, in many instances, especially with impure enzymes, the omission of Mg^{2+} helps to avoid the harmful effect of contaminating nucleases. Furthermore, EDTA increases MTase activity.

ii. Kinetic parameters. The methylation reaction obeys Michaelis–Menten kinetics with respect to both AdoMet and DNA substrates. The rate of the methyl transfer reaction is generally slow and is first order with respect to enzyme concentration. The kinetic parameters for an oligonucleotide (14-mer) and plasmid pBR322 are summarized in Table 4.3. The methyl transfer step is over 300-fold faster than k_{cat} (29). The rate of methylation on hemimethylated DNA (single turnover rate of 24 sec^{-1} at 20°C) is similar to that on unmethylated substrate (31 sec^{-1}). The MTase is over 45-fold more efficient than the cognate ENase. Considering the magnitude of k_{cat}/K_m (= 10^8 M^{-1} sec^{-1}), M · *Eco*RI is presumed to be one of the most efficient biocatalysts.

iii. Inhibitors and inactivators. *Salts:* At concentrations greater than 200 mM, NaCl inhibits MTase activity. M · *Eco*RI is also inhibited by MgCl$_2$: the enzyme shows only 50% activity in the presence of 4 mM MgCl$_2$.

TABLE 4.3 Kinetic Parameters of M·EcoRI[a]

Parameter	Oligonucleotide (14-mer)	Plasmid (pBR322)
k_{cat} (sec^{-1})	0.022	0.14
$K_{m, DNA}$ (nM)	1.3	0.35
$K_{m, AdoMet}$ (nM)	170	210
$k_{cat}/K_{m, DNA}$ (M^{-1} sec^{-1})	1.7×10^7	4.1×10^8
$K_{d, DNA}$ (μM)	4 (nM)[b]	1.0–4.0
$K_{d, AdoMet}$ (μM)	30	
$K_{d, AdoHcy}$ (μM)	220–250	

[a] The kinetic parameters are from assays at pH 8.0 (100 mM Tris–Cl) at 37°C (3,27). The oligonucleotide substrate used here is a 14-mer duplex containing the canonical recognition sequence hemimethylated at the second A residue in the bottom strand (27): 5'-GGCGGAATTCGCGG-3'. Hemimethylated and unmethylated oligonucleotides have similar kinetic parameters with the MTase.

[b] The K_d reported for oligonucleotide substrate is for ternary complex, MTase–DNA–sinefungin (27). The MTase has the same K_d for hemimethylated and unmethylated substrates.

Sulfhydryl reagents: The modification of SH groups by chemical reagents (e.g., NEM) or by spontaneous oxidation results in inactivation of the MTase.

Substrate analogs: S-Adenosylhomocysteine (AdoHcy), the reaction product of AdoMet (see Scheme 4.2), binds the enzyme to form a binary complex (K_d = 225 μM) and is a weak competitive inhibitor (K_i = 9.0 μM) (4). *Sinefungin,* a natural analog of AdoMet in which the methyl sulfonium (CH$_3$-S$^+$<) group is replaced with an amine methine (NH$_2$-CH<), is a potent inhibitor: the K_d of the binary complex is 4.3 μM and the K_i (competitive) is 10 nM (4). Inhibition with sinefungin occurs via formation of a stable, sequence-specific ternary complex (MTase–DNA–sinefungin). N-Methyl- and N-ethyl-AdoMet bind M·EcoRI weakly and are poor inhibitors.

Heat: M·EcoRI is readily heat denatured, losing more than 80% of its activity after a 10-min incubation at 50°C (5).

b. Substrate specificities

i. Recognition sequence. M·EcoRI recognizes the hexanucleotide sequence GAATTC in octanucleotides (e.g., TGGAATTCA) as well as in large DNA (2). It methylates the N-6 amino group of central adenine on both strands. M·EcoRI does not methylate single-strand or denatured dsDNA.

Note that M·RsrI recognizes the same sequence and methylates the same central adenine residue (6).

ii. Effect of base modification or substitution. Substitution of U or 5-hydroxymethyl-U for T (7) or I for G (8) within the EcoRI site inhibits methylation, suggesting the important role of the 5-methyl group of T and the 2-amino group of G in DNA site recognition. T4 phage DNA glucosylated on hydroxymethyl-C is as good a substrate for M·EcoRI as normal DNA (7), whereas the T4 DNA is resistant to R·EcoRI. Other single base substitutions within or flanking the canonical sequence give variable effects on catalytic efficiency of the MTase (refer to "Relaxation of site specificity").

iii. **Prior methylation.** M·EcoRI is sensitive to prior methylation at a noncanonical site, G^{m}AATTC. However, the DNA containing GAATT^{m}C is a substrate. DNAs hemimethylated at the canonical site are as good substrates for M·EcoRI as unmethylated DNAs.

iv. **Triplex site.** Formation of an oligonucleotide-directed triplex that partially overlaps the EcoRI recognition sequence inhibits the M·EcoRI activity site specifically (9).

v. **DNA structure.** The target sequence for M·EcoRI can be methylated when it is in a right-handed B conformation. When the target sequence is in or near a left-handed Z helix, however, the site is not methylated. Note that the site is not cleaved by R·EcoRI either. This observation has been used as the base of an assay for the *in vivo* existence of a non-B-DNA structure (10).

vi. **Relaxation of site specificity.** M·EcoRI exhibits a relaxed specificity under the same conditions as noted for EcoRI*, i.e., alkaline pH, low ionic strength, high glycerol, and/or high enzyme concentrations. Under these "star" conditions, the MTase displays 5- to 23,000-fold lower specificity and methylates noncanonical EcoRI sites (exclusively to adenine residues) containing single base pair substitutions (5,27). The methylation at noncanonical sites prevents them from EcoRI* cleavages. The minimal structural requirement for methylation is an AT/TA duplex. For instance, d(AT)n·d(AT)$_n$ accepts methyl groups slowly under the star conditions, whereas dA$_n$·dT$_n$ does not (11).

Although some noncanonical EcoRI sites are methylated *in vitro* with only fivefold lower specificities than the canonical site, the noncanonical sites (e.g., TAATTC, CAATTC, GTATTC, and GGATTC) remain unmethylated *in vivo* in the presence of M·EcoRI (28). The high *in vivo* specificity of M·EcoRI is partly attributed to the active removal of methylated sequences by DNA repair enzymes.

c. Mechanism of methyl transfer

i. **Kinetics of methylation.** M·EcoRI catalyzes the methyl group transfer from donor AdoMet to acceptor DNA by an ordered Bi–Bi kinetic mechanism in which AdoMet binds first, followed by DNA addition (3). In contrast, the binding of MTase with AdoMet and noncanonical DNA occurs randomly. The reaction product, AdoHcy, is an uncompetitive inhibitor with respect to DNA and a competitive inhibitor with respect to AdoMet. Therefore a ternary complex, MTase–DNA–AdoHcy, is a dead-end complex. More important in the reaction mechanism of M·EcoRI is the formation of ternary complexes at noncanonical sites.

The complex formation at noncanonical sites probably explains the observed lower K_m for DNA than for smaller oligonucleotide substrates. Assuming that the decreased K_m results from the increased rate of association, data imply that the MTase, analogously to ENase, locates canonical sites by processive movement along the DNA (3). This facilitated mode of transfer is presumed to contribute to the higher k_{cat} observed for plasmid DNAs than that for 14-mer oligonucleotides. Methyl transfer from the central complex (MTase–DNA–AdoMet) to products (MTase–^{m}DNA–AdoHcy) is over 300-fold faster than k_{cat}, suggesting that the steps after methyl transfer are rate limiting (29).

ii. **Course of methyl transfer.** As a monomeric enzyme, M·EcoRI transfers one methyl group at a time to the EcoRI site per binding event (12). The catalysis proceeds by a single, direct transfer from AdoMet to adenine (13). The group transfer accompanies an inversion of configuration (14). [cf. A similar inversion of configuration has also been observed in the methyl transfer catalyzed by M·HhaI, a m^{5}C-MTase (14).] These stereochemical results, together with data from methylation kinetics, rule out the possible involvement of a Ping-Pong mechanism in which the methyl group would be transiently transferred to a functional group on the enzyme. Dissociation of the enzyme from the DNA substrate is obligatory for the second methyl group transfer. Thus under the conditions of steady-state methyl transfer to unmodified DNA, hemimethylated molecules accumulate as a free intermediate species in solution, and the fully methylated product appears only after a lag period. The transfer of the second methyl group occurs independently or asymmetrically of the first event with similar kinetic parameters.

Contrary to R·ENases, the binding of MTase with DNA cannot be captured on a NC filter under noncatalytic conditions, i.e., in the absence of AdoMet or in the presence of a number of AdoMet analogs (15). The sequence-specific recognition of the DNA substrate by M·EcoRI is sensitive to the binding of AdoMet in the context of the ternary complex (MTase–DNA–AdoMet) (8). However, the decrease in k_{cat} for several substrates with base substitutions makes a disproportionately large contribution to the specificity difference, suggesting that discrimination is mediated mostly by the contact with catalytic residues rather than by binding interactions (27).

2. APPLICATIONS

Refer to Section III,A, this chapter.

3. STRUCTURE

i. **Primary structure.** EcoRI MTase is a monomeric enzyme. The amino acid sequence deduced from the cloned gene contains 326 residues (M_r 38,050) (16,17). The nascent N-terminal Met is cleaved off, and the mature M·EcoRI has N-terminal Ala–Arg- or more frequently Arg residue due to the cleavage of Ala as well (18). The C-terminal residue is Lys. The enzyme has seven Cys (no disulfide bridges) and two Trp residues.

M·EcoRI is a basic protein with a pI of 8.7. The $s^{\circ}_{20,w}$ is 3.0S. The A_{278} for a 1% (w/v) solution is 10.8 and A_{280}/A_{260} is ~2.0 (12).

A fourfold repeat of the Leu–Ile–Lys sequence is found within the primary structure, although the significance of this motif is unknown. Despite the ability to recognize the identical sequence that R·EcoRI does, M·EcoRI bears no apparent structural similarities with the ENase at the level of both primary and secondary structures. M·EcoRI bears no detectable amino acid identity with other DNA adenine-MTases and, more strikingly, bears little homology (15.6% when optimally aligned) with M·RsrI which recognizes and methylates the same nucleotide sequence (6). [*Note:* In contrast to A-MTases, m^5C-MTases display five highly conserved sequence motifs 10–20 amino acids long and five moderately conserved motifs. The variable region consisting of approximately 90 to 275 contiguous amino acid residues has been found to be wholly responsible for sequence-specific recognition and the choice of the base to be methylated (25,26).]

ii. Secondary and tertiary structures. In solution, M·EcoRI exhibits a very low content of secondary structure, although the sequence predicts ~18% α helix and 27–30% β sheets (16,17). M·EcoRI apparently has two domains connected by a highly flexible interdomain hinge (19). The putative interdomain encompasses the peptides implicated in AdoMet binding (20) and catalysis (21).

The MTase undergoes significant conformational changes on ligand binding. The AdoMet-bound enzyme is over 800 times more resistant to trypsinolysis than unbound enzyme, and the sequence-specific ternary complex (MTase–DNA–sinefungin) is 2000 times more resistant (19). Furthermore, the trypsinolysis of the ligand-bound form generates fragments different from those of unbound MTase.

iii. Active site. The 3-D structure of M·EcoRI is not yet available. Photoaffinity labeling of the enzyme with 8-azido-S-AdoMet, an analog of AdoMet and a competent substrate ($K_m \approx 0.71 \mu M$, $k_{cat} \approx 5$ min^{-1}), results in the labeling of a peptide region (residues 206–221) (20). However, the exact residue(s) that binds the cofactor remains to be determined.

M·EcoRI contains a tetrapeptide sequence motif N^{140}-P-P-F, which is commonly found in all known m^6A-MTases as DPPY or its equivalent forms (22).

Cys-223 is an active site residue that binds an arsenic(III) reagent with high affinity without affecting methylase activity (21). NEM modification of Cys-223 leads to inactivation of the MTase, most likely by a steric effect that reduces cofactor binding.

[*Note:* In m^5C-MTases, a single conserved Cys located in the Pro-Cys motif is presumed to play an essential role in methyl transfer by forming a covalent bond with the C-6 of target cytosine and activating the normally inert C-5 for electrophilic methyl transfer (23).]

4. Genetics

The MTase gene (*ecoRIM*) has been cloned together with the *ecoRIR* gene for R·*Eco*RI (16,17). It has been shown from selective cloning of the *ecoRIR* gene without *ecoRIM* or by *in vitro* mutagenesis (24) that the continued presence of an active M·*Eco*RI is necessary to maintain cell viability in the presence of R·*Eco*RI.

5. Sources

M·*Eco*RI has been purified to homogeneity (a specific activity of 76–91 U/μg protein) from *E. coli* RY13 (12). During the purification, the continued presence of thiol-reducing reagents (e.g., 20 mM 2-MSH) is essential.

References

1. Dugaiczyk, A., Hedgpeth, J., Boyer, H. W., and Goodman, H. M. (1974). *Biochem.* **13**, 503–512.
2. Greene, P. J., Poonian, M. S., Nussbaum, A. L., Tobias, L., Garfin, D. E., Boyer, H. W., and Goodman, H. M. (1975). *JMB* **99**, 237–261.
3. Reich, N. O., and Mashhoon, N. (1991). *Biochem.* **30**, 2933–2939.
4. Reich, N. O., and Mashhoon, N. (1990). *JBC* **265**, 8966–8970.
5. Berkner, K. L., and Folk, W. R. (1978). *NAR* **5**, 435–450.
6. Kaszubska, W., Aiken, C., O'Connor, C. D., and Gumport, R. I. (1989). *NAR* **17**, 10403–10425.
7. Berkner, K. I., and Folk, W. R. (1977). *JBC* **252**, 3185–3193.
8. Modrich, P., and Rubin, R. A. (1977). *JBC* **252**, 7273–7278.
9. Hanvey, J. C., Shimizu, M., and Wells, R. D. (1990). *NAR* **18**, 157–161.
10. Jaworski, A., Zacharias, W., Hsieh, W. T., Blaho, J. A., Larson, J. E., and Wells, R. D. (1988). *Gene* **74**, 215–220.
11. Woodbury, C. P., Jr., Downey, R. L., and von Hippel, P. H. (1980). *JBC* **255**, 11526–11533.
12. Rubin, R. A., and Modrich, P. (1977). *JBC* **252**, 7265–7272.
13. Pogolotti, A. L., Jr., Ono, A., Subramaniam, R., and Santi, D. V. (1988). *JBC* **263**, 7461–7464.
14. Ho, D. K., Wu, J. C., Santi, D. V., and Floss, H. G. (1991). *ABB* **284**, 264–269.
15. Jack, W. E., Rubin, R. A., Newman, A., and Modrich, P. (1981). In "Gene Amplification and Analysis" (J. G. Chirikjian, Ed.), Vol. 1, pp. 165–179. Elsevier, New York.
16. Newman, A. K., Rubin, R. A., Kim, S. H., and Modrich, P. (1981). *JBC* **256**, 2131–2139.
17. Greene, P. J., Gupta, M., Boyer, H. W., Brown, W. E., and Rosenberg, J. M. (1981). *JBC* **256**, 2143–2153.
18. Rubin, R. A., Modrich, P., and Vanaman, T. C. (1981). *JBC* **256**, 2140–2142.
19. Reich, N. O., Maegley, K. A., Shoemaker, D. D., and Everett, E. (1991). *Biochem.* **30**, 2940–2946.
20. Reich, N. O., and Everett, E. A. (1990). *JBC* **265**, 8929–8934.
21. Lam, W.-C., Tsao, D. H. H., Maki, A. H., Maegley, K. A., and Reich, N. O. (1992). *Biochem.* **31**, 10438–10442.
22. Chandrasegaran, S., and Smith, H. (1988). In "Structure and Expression" (M. H. Sarma and R. H. Sarma, Eds.), Vol. I, pp. 149–156. Adenine Press, New York.
23. Wyszynski, M. W., Gabbara, S., Kubareva, E. A., Romanova, E. A., Oretskaya, T. S., Gromova, E. S., Shabarova, Z. A., and Bhagwat, A. S. (1993). *NAR* **21**, 295–301.
24. Humphreys, G. O., Willshaw, G. A., Smith, H. R., and Anderson, E. S. (1976). *MGG* **145**, 101–108.
25. Klimasauskas, S., Nelson, J. L., and Roberts, R. J. (1991). *NAR* **19**, 6183–6190.
26. Mi, S., and Roberts, R. J. (1992). *NAR* **20**, 4811–4816.
27. Reich, N. O., Olsen, C., Osti, F., and Murphy, J. (1992). *JBC* **267**, 15802–15807.
28. Smith, D. W., Crowder, S. W., and Reich, N. O. (1992). *NAR* **20**, 6091–6096.
29. Reich, N. O., and Mashhoon, N. (1993). *JBC* **268**, 9191–9193.

C. Nontype II Methylases: Dam and Dcm Methyltransferases

DNA MTases are ubiquitous in nature. Although most of them function as part of the R–M systems as described in previous sections, there are still a good number of MTase systems that apparently function independently of any restriction system. Among the R-independent M systems are two MTases that are relatively well characterized in view of their potential physiological importance: A-MTase encoded by the *dam* (for DNA adenine methylase) gene and m^5C-MTase encoded by the *dcm* (for DNA cytosine methylase) gene.

The Dam of *E. coli* (and phages T2 and T4 as well) recognizes a tetranucleotide sequence GATC in dsDNA and methylates the adenosine, while the Dcm recognizes the sequence CC(A/T)GG and methylates the internal cytosine.

E. coli DNA is estimated to contain ~10 times more methyl groups than are required to prevent the action of endogenous R·ENases. Contents of methylated bases (m^6A and m^5C) are 0.5 mol% of the total bases (or 2% of the total adenines) and 0.25 mol% of the total bases (or 0.9% of the cytosines), respectively (1). The level of methylation is not affected by the presence or absence of the genes involved in type I R–M systems, *Eco*BI and *Eco*KI (2). Therefore the majority of the methylated bases in *E. coli* DNA may be attributed to the actions of nontype I and nontype II MTases which include most notably Dam and Dcm.

Although the *dam* and *dcm* genes are not required for viability of *E. coli*, DNA methylation has been implicated in various cellular functions such as DNA replication (especially in initiation timing), mismatch repair, and differential gene expression.

This section focuses on *E. coli* Dam (M·*Eco* Dam) and Dcm (M·*Eco* Dcm), although phage-encoded Dam and Dcm MTases are occasionally mentioned for the sake of comparison. M·*Eco* Dam and M·*Eco* Dcm provide a useful tool for site-specific *in vitro* methylation of DNA. They complement or extend the specificities of type II MTases.

1. FUNCTIONS

a. Reaction conditions for Eco Dam

i. Optimal reaction conditions. A typical reaction mixture (50 μl) contains 100 mM Tris–Cl (pH 8.0), 2 mM DTT, 100 μg/ml BSA, 10 mM EDTA, 100 μM AdoMet, and DNA substrates. The pH optimum is 7.6–8.0 in Tris–Cl buffer (20–100 mM) (3).

ii. Kinetic parameters. The methylation reaction by M·*Eco* Dam follows simple Michaelis–Menten kinetics. The initial rate of methyl transfer is first order in enzyme concentrations up to 6 nM. The K_m for AdoMet is 12.2 μM and the k_{cat} is 19 min^{-1} at 37°C (3). At 30°C, the K_m and k_{cat} for AdoMet are 6.5 mM and 3.6 min^{-1}, respectively (4). The apparent K_m for GATC sites in ColE1 DNA is 3.6 nM at the AdoMet concentration of 13.4 μM.

The kinetic parameters for methylation on hemimethylated DNA substrates are such that the methylation is, if not equal, slightly more favorable than on unmethylated DNA.

iii. **Inhibitors.** AdoMet analogs are competitive inhibitors of Dam. Adenosylornithine (sinefungin) and AdoHcy show K_i values of 2.0 and 7.0 μM, respectively, at 30°C when the K_m for AdoMet is 6.5 μM. At concentrations above 100 mM, NaCl significantly inhibits *Eco* Dam activity (3).

b. Methylase assays and unit definition

Methylase activity is usually determined by the methyl transfer assay (see Section III,A,6, this chapter) using calf thymus DNA as substrate.

The protection assay for Dam measures the susceptibility of λ DNA toward the cleavage by R·*Mbo*I following the methylation reaction.

Unit definition: One methylase unit is the amount of enzyme that catalyzes the conversion of 1 pmol of [^{3}H]methyl group into a form that binds to anion-exchange filters (e.g., Whatman DE81) in 30 min at 37°C (5).

c. Substrate specificities of M·Eco Dam

i. **Recognition sequence.** M·*Eco* Dam recognizes the tetranucleotide sequence 5′-GATC-3′ and methylates the A base (5,6). Both unmethylated and hemimethylated sites are substrates. Note that the same sequence is recognized by *Mbo*I and its isoschizomers *Dpn*II and *Sau*3AI, whereas G^{m}ATC (methylated on both strands) is recognized by *Dpn*I.

ii. **Site preference.** GATC sites are not methylated randomly, and the methylation is influenced by flanking sequences. For example, the gGATCa site is methylated in preference to the aGATCg (7). Among the 22 GATC sites available in pBR322 *dam*$^-$, preferential hemimethylation occurs both *in vivo* and *in vitro* at two particular sites which are flanked by three G·C base pairs on one side and two A·T base pairs on the other side (8).

iii. **DNA structure.** Double-stranded DNA is a better substrate than denatured DNA (3). The rate of methylation on denatured DNA is only 14% of that of the native DNA, although the extent of methylation can reach 60–70%.

iv. **Effect of Dam methylation on restriction enzymes.** The same GATC sequence is recognized by various type II restriction enzymes with different sensitivities to methylation. For example, *Mbo*I (|GATC) and *Dpn*II (|GATC) cleave only GATC sequences with unmethylated adenines, whereas *Dpn*I (G^{m}A|TC) cleaves the sequence only when both strands are methylated. *Sau*3AI (|GATC) and *Pvu*I (CGAT|CG) cleave the sequence regardless of its adenine-methylated status. Neither *Dpn*I nor *Mbo*I cleaves hemimethylated DNA. Wild-type *E. coli* DNA is readily degraded by methylation-requiring *Dpn*I and is virtually untouched by methylation-sensitive *Mbo*I and *Cpn*II, whereas the reverse is true for *dam*-3 (Dam⁻ mutant) DNA (9).

v. Comparison with phage Dam MTases. Phage T2 and T4 (but not T6) *dam* MTases recognize the same GATC and methylate the A base (10). In fact, their normal substrate is GAT^{hm}C which is found in place of GATC in viral genomes. The GAT^{hm}C is a poor substrate for M·*Eco* Dam. In contrast to M·*Eco* Dam which methylates exclusively on GATC, both T2 and T4 *dam* MTases also recognize noncanonical sequences, GAPy, at a lower frequency (11,12). T2 Dam is further distinguished from other *dam* MTases by its capacity to methylate certain GATA and GATT sequences.

d. Substrate specificities of M·Eco Dcm

M·*Eco* Dcm recognizes the sequence 5'-CCAGG-3'/3'-GGTCC-5' and methylates the internal C base (13). This sequence is also recognized and methylated at the same cytosine by M·*Eco*RII which is coded by the *E. coli* plasmid N3 (14,15). The Dcm recognition sequence also forms a subset of *Pfl*MI recognition sites (CCA-N$_5$-TGG) and, when modified by Dcm methylation, it becomes resistant to *Pfl*MI cleavage (16). On the other hand, *Apy*I cuts the sequence CC|AGG when the internal C (but not the first C) is methylated (17). Therefore the cleavage resistance of a DNA to *Eco*RII (|CCAGG) or *Pfl*MI and the sensitivity to *Apy*I provide a convenient method for screening the Dcm⁺ phenotype, as well as determining the extent of methylation at these sites.

e. Mechanism of methylation

i. M·Eco Dam. The functional form of M·*Eco* Dam is a monomer (3), although a dimer may be formed on interaction with substrates (18). Dam transfers one methyl group to the recognition site per binding event. The enzyme can bind and form complexes with dsDNA with or without the GATC recognition site. Dam recognizes the GATC site regardless of the presence of AdoMet. AdoMet strongly stimulates the binding of the enzyme to the GATC site, while the binding to nonspecific DNA is only slightly influenced. In the presence of 0.2 M NaCl, the enzyme is strongly inhibited and very little enzyme–substrate complex is formed. This is coincident with the loss of the allosteric effect of AdoMet (4). The minimal sequence requirement for MTase function is the element of symmetry at the GA sequence in both strands of the recognition site. Methylation of the recognition sequence is modulated by the asymmetry originating from the three base pairs that flank each side of the recognition site (8).

On one hand, AdoMet is an allosteric effector ($K_d < 100$ nM) which strengthens the affinity of the enzyme for the GATC site in the absence of catalysis (4). AdoMet analogs do not compete for binding at the allosteric site. On the other hand, AdoMet is a methyl donor substrate. AdoMet bound at the catalytic site can be exchanged by its competitive analogs. Simultaneous binding of AdoMet and the specific target sequence to the enzyme induces conformational changes in Dam MTase.

The process of the complex formation, MTase–AdoMet–DNA(GATC), is probably random, but the addition of each component tends to reinforce the stability and favor the formation of the active ternary complex. Whether the presumed allosteric site for AdoMet is separate from or the same as the catalytic site in the absence of DNA is not clearly known.

AdoMet can interact with Dam in the absence of DNA. The two successive steps of the activation of DNA binding by AdoMet and the conformational change induced by the interaction with the specific sequence suggest that the catalytic mechanism of Dam is, unlike M·EcoRI, closer to that of type I R–M systems such as EcoBI and EcoKI.

ii. **M·Eco Dcm.** The mechanism of Dcm methylation is not known. Presumably it resembles that of other C-MTases such as M·HhaI (G^{m}CGC) in which the reaction proceeds by an ordered Bi–Bi mechanism, i.e., DNA binds to the enzyme first, followed by AdoMet (19). After the methyl transfer, AdoHcy dissociates first from the enzyme, followed by the methylated DNA.

An enzyme nucleophilic group, most likely SH in the context of the Pro-Cys doublet conserved in all known m^{5}C-MTases, is believed to assist the chemical catalysis by attacking the C-6 of the substrate cytosine. Formation of a covalent intermediate apparently renders the otherwise inert C-5 electron rich and facilitates the electrophilic transfer of a methyl group from AdoMet. In the absence of AdoMet, M·HhaI catalyzes the exchange of the 5-H of the substrate cytosine for protons of water at a rate about sevenfold higher than the rate of methylation.

2. APPLICATIONS

The principles and range of applications involving Dam and Dcm are essentially identical to those of type II MTases (refer to Section III,A,8, this chapter).

3. STRUCTURE

a. **M·Eco Dam**

M·Eco Dam is a single polypeptide and exists in solution largely as a monomer ($s^0_{20,w}$ = 2.8S) (3). The primary structure contains 278 amino acids (M_r 32,000) (20). The N- and C-terminal residues are Met and Lys, respectively. M·Eco Dam has a pI of 5.1. The SH group modification by NEM inactivates the MTase.

The primary structure of M·Eco Dam contains the DPPY motif which is uniquely shared by m^{4}C-MTases and other A-MTases such as T4 Dam and the M·CpnII of *Streptococcus pneumoniae* (21–23). Despite the absence of any sequence homology with the phage T4 *dam* gene at the nucleotide level, M·Eco Dam shares four regions (11 to 33 residues long) of amino acid sequence homology containing 45 to 64% identity (21). In *Eco* Dam, the D^{181}-P-P-Y^{184} motif is located in the high homology region IV, possibly functioning as the AdoMet-binding site. Mutations introduced in the motif lead to loss of catalytic activity and greatly affect AdoMet cross-linking, which suggests that region IV is crucial for activity and stability of the MTase (24). Among the other three conserved regions, region

III is presumed to be the sequence recognition domain (25). Supporting evidence comes from the observation that a single amino acid change (Pro-126 to Ser) alters the sequence specificity of T4 and T2 *dam* MTases (26).

M·Eco Dam also shares a significant homology with both A- and C-MTases at the 23 amino acid region centering at positions 31–45, proximal to the N terminus (27).

b. M·Eco Dcm

M·Eco Dcm may be active as a dimer with a molecular mass of 46 kDa. Monomeric Dcm is 23 kDa (28). The protein contains 472 amino acid residues (29,30). In contrast to the m^6A-MTases and m^4C-MTases that possess a unique tetrapeptide motif (D/N)PPY, m^5C-MTases contain distinct blocks of highly conserved regions (31). A large variable-length segment of 90–270 amino acids is located in the C-terminal half of MTases and is probably responsible for specific sequence recognition.

4. GENETICS

a. M·Eco Dam

Eco Dam is the only enzyme in *E. coli* known to modify the A base in GATC sequences of dsDNA. In *E. coli* K12, the *dam* gene has been mapped to the 74-min position of the chromosome (32), with the *trpS* gene (tryptophanyl-tRNA synthetase or tryptophan-tRNA ligase) 3' proximal to it. The *dam* gene has been cloned in pBR322 (pDam118, pGG503), screened for a Dam$^+$ phenotype using growth resistance to 2-aminopurine and DNA resistance to *Mbo*I cleavage (20,22). The A+T content of the *dam* gene is 50%.

The promoter nearest the *dam* gene is located 1650–2100 bp upstream and within the *aroB* gene (32). The *dam* gene is apparently expressed as part of an operon with the 5'-proximal intervening ORF, named *urf*-74.3 (for unidentified reading frame at the 74.3 min locus), which is expressed in the same direction and predicted to encode a 46 kDa protein with 427 deduced amino acid residues. There is a 106-nt intergenic space between the UAA termination codon of *urf*-74.3 and the AUG start codon of the distal *dam* gene. The 3' end of the *dam* gene is flanked by a putative 23-kDa protein gene and then by a transcription termination site(s), followed by the *trpS* gene. *urf*-74.3 contains a cluster of 10 Dam methylation (GATC) sites which, together with a putative DnaA protein-binding site immediately 5' to the *dam* gene, may participate in the autoregulation of *dam* gene expression.

b. M·Eco Dcm

In *E. coli* K12, the *dcm* gene is at the 43-min position of the chromosome, in the order of *supD-dcm-flaA1* (33). The Eco Dcm gene, which was cloned in a pBR322 vector (34) and λ vector (λ 25D8) (35), has been nucleotide sequenced (29,30).

The *dcm* ORF contains 1416 nt. [*Note:* The *dcm* genes of *B. subtilis* temperate phages have also been cloned and sequenced (36).]

c. Mutations in dam and dcm genes

The deletion of the *dam* or *dcm* gene does not affect R–M systems (2). The Tn9 transposon–insertion mutations in the *dam* or *dcm* gene do not affect the viability of the *E. coli* cells either (37). However, mutations in the *dam* gene (e.g., *dam*-3 having 19% of the wild-type activity) result in undermethylation of *E. coli* DNA and render the cells more sensitive to UV irradiation (38).

Dam mutants generally show a pleiotropic phenotype (39): they are hypermutable, as are the strains that overproduce the MTase more than 10-fold (3,40); they are defective in their ability to restrict λ DNA (41) and grow poorly in the presence of certain base analogs like 2-aminopurine and 5-Br-U (42). However, the combination of *dam*-3 with *polA*, *recA*, *recB*, or *recC* is lethal (38). Available evidence suggests that Dam plays a role in mismatch repair and in the control of DNA replication in *E. coli* (43–45).

Dcm mutants with null MTase activity show no obvious phenotypic abnormalities (41). Besides, *E. coli* B lacks the Dcm MTase. The exact *in vivo* function of Dcm is not known.

5. Sources

a. M·Eco Dam

Eco Dam has been purified from a 10- to 20-fold overproducing strain (pGG503/*E. coli* GM31) which contains a cloned *dam* gene (3). At a specific activity of 10^6 U/mg protein, the enzyme is at least 95% pure. The wild-type intracellular concentration of Dam is very low (one to five molecules per cell) (5). Because high Dam methylation in *E. coli* provokes hypermutation as well as a SOS response, overproduction of Dam MTase from a cloned gene requires the gene to be under the control of a regulatable promoter. Expression of the *dam* gene cloned in a plasmid (pDDM6) containing the IPTG-inducible *tac* promoter in addition to the *lacI* gene has been shown to achieve a 20-fold induction of the MTase (46). A semisynthetic *dam* gene has also been introduced in a vector (pDOX1), which overproduces functional Dam at the level of milligram per liter under the heat-inducible control of the λ p_R-p_L double promoter (47).

b. M·Eco Dcm

Eco Dcm has been purified from *E. coli* K12 and also from cloned sources (34,35).

References

1. Vanyushin, B. F., Belozersky, A. N., Kokurina, N. A., and Kadirova, D. X. (1968). *Nature* **218**, 1066–1067.
2. Mamelak, L., and Boyer, H. W. (1970). *J. Bacteriol.* **104**, 57–62.
3. Herman, G. E., and Modrich, P. (1982). *JBC* **257**, 2605–2612.
4. Bergerat, A., and Guschlbauer, W. (1990). *NAR* **18**, 4369–4375.
5. Geier, G. E., and Modrich, P. (1979). *JBC* **254**, 1408–1413.
6. Hattman, S., Brooks, J. E., and Masurekar, M. (1978). *JMB* **126**, 367–380.

7. Doolittle, M. M., and Sirotkin, K. (1988). *BBA* **949**, 240–246.
8. Bergerat, A., Kriebardis, T., and Guschlbauer, W. (1989). *JBC* **264**, 4064–4070.
9. Lacks, S., and Greenberg, B. (1977). *JMB* **114**, 153–168.
10. Schlagman, S. L., and Hattman, S. (1983). *Gene* **22**, 139–156.
11. Schlagman, S. L., and Hattman, S. (1989). *NAR* **17**, 9101–9112.
12. Schlagman, S. L., Miner, Z., Feher, Z., and Hattman, S. (1988). *Gene* **73**, 517–530.
13. May, M. S., and Hattman, S. (1975). *J. Bacteriol.* **123**, 768–770.
14. Boyer, H. W., Chow, L. T., Dugaiczyk, A., Hedgpeth, J., and Goodman, H. M. (1973). *Nature NB* **244**, 40–43.
15. Schlagman, S., Hattman, S., May, M. S., and Berger, L. (1976). *J. Bacteriol.* **126**, 990–996.
16. Sturm, R. A., and Yaciuk, P. (1989). *NAR* **17**, 3615.
17. Razin, A., Urieli, S., Pollack, Y., Gruenbaum, Y., and Glaser, G. (1980). *NAR* **8**, 1783–1792.
18. Tuzikov, F. V., and 10 coauthors (1986). *Mol. Biol. (Russian)* **20**, 1002–1007.
19. Wu, J. C., and Santi, D. V. (1987). *JBC* **262**, 4778–4786.
20. Brooks, J. E., Blumenthal, R. M., and Gingeras, T. R. (1983). *NAR* **11**, 837–851.
21. Hattman, S., Wilkinson, J., Swinton, D., Schlagman, S., MacDonald, P. M., and Mosig, G. (1985). *J. Bacteriol.* **164**, 932–937.
22. Mannarelli, B. M., Balganesh, T. S., Greenberg, B., Springhorn, S. S., and Lacks, S. A. (1985). *PNAS* **82**, 4468–4472.
23. Lauster, R., Kriebardis, A., and Guschlbauer, W. (1987). *FEBS Lett.* **220**, 167–176.
24. Guyot, J.-B., Grassi, J., Hahn, U., and Guschlbauer, W. (1993). *NAR* **21**, 3183–3190.
25. Guschlbauer, W. (1988). *Gene* **74**, 211–214.
26. Miner, Z., Schlagman, S., and Hattman, S. (1988). *Gene* **74**, 275–276.
27. Lauster, R. (1989). *JMB* **206**, 313–321.
28. Mur'yanov, Y. I., Nesterenko, V. F., Kosykh, V. G., and Baev, A. A. (1981). *Dokl. Akad. Nauk. SSSR* **257**, 495–497.
29. Hanck, T., Gerwin, N., and Fritz, H. J. (1989). *NAR* **17**, 5844.
30. Roberts, R. J. (1989). Personal communication.
31. Posfai, J., Bhagwat, A. S., and Roberts, R. J. (1988). *Gene* **74**, 261–265.
32. Jonczyk, P., Hines, R., and Smith, D. W. (1989). *MGG* **217**, 85–96.
33. Marinus, M. G. (1973). *MGG* **127**, 47–55.
34. Bhagwat, A. S., Sohail, A., and Roberts, R. J. (1986). *J. Bacteriol.* **166**, 751–755.
35. Kohara, Y., Akiyama, K., and Isono, K. (1987). *Cell* **50**, 495–508.
36. Tran-Betcke, A., Behrens, B., Noyer-Weidner, M., and Trautner, T. A. (1986). *Gene* **42**, 89–96.
37. Parker, B., and Marinus, M. G. (1988). *Gene* **73**, 531–535.
38. Marinus, M. G., and Morris, N. R. (1974). *JMB* **85**, 309–322.
39. Bale, A., d'Alarcao, M., and Marinus, M. G. (1979). *Mut. Res.* **59**, 157–165.
40. Marinus, M. G., Poteete, A., and Arraj, J. A. (1984). *Gene* **28**, 123–125.
41. Marinus, M. G., and Morris, N. R. (1973). *J. Bacteriol.* **114**, 1143–1150.
42. Glickman, B., van den Elsen, P., and Radman, M. (1978). *MGG* **163**, 307–312.
43. Modrich, P. (1987). *Ann. Rev. Biochem.* **56**, 435–466.
44. Campbell, J. L., and Kleckner, N. (1990). *Cell* **62**, 967–979.
45. Boye, E., and Lobner-Olesen, A. (1990). *Cell* **62**, 981–989.
46. Guha, S., and Guschlbauer, W. (1992). *BBA* **1132**, 309–310.
47. Hulsmann, K.-H., Quaas, R., Georgalis, Y., Saenger, W., and Hahn, U. (1991). *Gene* **98**, 83–88.

5
Phosphatases and Polynucleotide Kinase

I. PHOSPHATASES (ALKALINE)
[Alkaline phosphatase, EC 3.1.3.1]

Alkaline phosphatases (APases) are nonspecific phosphomonoesterases with optimal activities at alkaline pH (≥ 9), as opposed to acid phosphatases which are optimally active at low pH (≤ 5). APases catalyze the hydrolysis of a wide variety of phosphomonoesters (ROP) to the corresponding alcohols (ROH) and inorganic phosphates (P_i) (Scheme 5.1).

APases can also act as phosphotransferases in the presence of a phosphate acceptor such as ethanolamine or Tris. (*Note:* Phosphatases cleave the P–O bond, whereas phosphorylases cleave the C–O bond in the transfer of a sugar to P_i.)

APases are widely distributed in prokaryotes as well as in eukaryotes (1)[*]. All purified enzymes are dimeric Zn(II)-metalloenzymes with an active-site Ser which forms a covalent phosphoserine intermediate during catalysis (2). The molecular sizes range from 80 to 220 kDa per dimer, and mammalian enzymes are generally

---

[*] For references cited, see Section I.A.

larger than bacterial APases. Plants are rich in nonspecific acid phosphatases, but not in alkaline phosphatases.

Although a number of physicochemical properties of mammalian APases (e.g., human and bovine APases) are similar to those of *E. coli* enzyme, their physiological functions are thought to be somewhat different.

In mammals, APases are glycoproteins essential in bone metabolism and phosphate transport. They are present as isozymes in different tissues including bone, intestine, kidney, and placenta. In *E. coli,* APase is a periplasmic enzyme which is not essential for phosphate metabolism. Note that *E. coli* also possesses a periplasmic acid phosphatase, a monomeric 45-kDa enzyme with active-site histidines. In contrast to the APase which is encoded by the *phoA* gene, the acid phosphatase is encoded by the *appA* gene.

APases are important tools in recombinant DNA work not only as the enzymes removing 3'- or 5'-terminal phosphates, but also as sensitive enzymatic reagents in the detection and screening of recombinant clones. Commercially available APases used for these purposes include the enzymes from *E. coli, B. licheniformis,* HK47 (an antarctic marine bacterium), shrimp, and calf intestine. When the structural genes of APases are fused to other genes in an expression vector, APases serve as a reporter in the expression of recombinant proteins (see Chapter 8). The signal sequence of the *E. coli* APase gene or a mammalian APase gene can also be attached to a foreign gene of interest, providing a means to express recombinant proteins in a secretable or membrane-bound form.

This section focuses on two most widely used APases, a bacterial alkaline phosphatase (BAP) from *E. coli* and a mammalian APase from calf intestinal mucosa (CIAP), to illustrate various strategies common to the application of APases. For the sake of general information on the diversity of mammalian APases, a brief description is given for human APases.

A. Bacterial Alkaline Phosphatase

The alkaline phosphatase from *E. coli* is a nonspecific phosphomonoesterase which catalyzes the hydrolysis of 3'- and/or 5'-terminal phosphates from DNA, RNA, and a wide variety of organic phosphomonoesters (Scheme 5.1). The phos-

$$
\begin{array}{l}
\downarrow \\
5'\ \text{p-ApCpGpTpApC}\ 3' \longrightarrow 5'\ \text{ApCpGpTpApC}\ 3' \\
\text{p-TpGpCpA-p} \text{TpGpCpA} + 3\ \text{P}_i \\
\uparrow \uparrow
\end{array}
$$

SCHEME 5.1

phatase reaction products are DNA (or RNA) with 3'-OH and/or 5'-OH termini and inorganic phosphates. *E. coli* APase, a prototype BAP, is a thermoresistant, unglycosylated, homodimeric 94-kDa enzyme requiring Zn^{2+} as a cofactor. The enzyme catalyzes the reaction via a covalent phosphoserine intermediate. In

E. coli, BAP is encoded by the *phoA* gene and is synthesized as a precursor carrying an N-terminal signal sequence which is processed upon secretion into the periplasmic space.

1. FUNCTIONS

a. Reaction conditions

i. Optimal reaction conditions. The dephosphorylation of dsDNA is typically carried out in a buffer containing 50 mM Tris–Cl (pH 8.0), 0.5 mM ZnCl$_2$, 0.5 mM MgCl$_2$, and ~50 U (ATP hydrolysis unit) of BAP per μg DNA (or ~70 U/pmol ends). The reaction mixture is incubated at 65°C for 1 hr.

The reaction can be terminated in various ways. If complete elimination of BAP is not necessary, the reaction can be stopped by adding either P$_i$ to 0.1 M final concentration or 0.1% (w/v) SDS plus 10 mM EGTA. To stop the reaction and remove BAP, usually phenol/chloroform extraction is performed twice, and the DNA is precipitated with ethanol.

The rate and efficiency of APase reactions are dependent on various factors, especially pH, ionic strength, and temperature.

(a) pH. The optimum pH for BAP activity is pH 8.0–9.5 with PNPP or *p*-phenylazophenyl phosphate as the substrate (3–5). This pH is generally optimal for BAP activity with nucleic acid substrates as well. Therefore, the BAP reaction can often be performed directly in the same reaction buffer employed for restriction enzyme digestions. BAP has no detectable activity below pH 6 and is rapidly denatured below pH 3.

(b) Ionic strength. The activity of BAP increases up to 10-fold as the concentration of salts, e.g., KCl (4) or NaCl (6), increases from ≤0.1 to 1 M.

(c) Temperature. The incubation at 65°C is especially recommended for dephosphorylating DNA (or RNA) at the nick, gap, or recessed ends. Dephosphorylation of RNA can be carried out in the same buffer at 45°C for 10–30 min with 1 U enzyme per pmol ends.

ii. Kinetic parameters. The overall rates of phosphatase reactions are remarkably similar for most substrates under a given assay condition (7). The K_m for PNPP is 5–22 μM at pH 8.0 and 25°C (3,8). The k_{cat} for the hydrolysis of phosphomonoesters ranges from 2 to 70 sec^{-1} per dimer. The K_m and k_{cat} values for transphosphorylations are ~12 μM and ~130 sec^{-1}, respectively, in the presence of 1 M Tris–Cl (pH 8.0) and 0.5 M NaCl (8). Both parameters vary as a function of pH. They may also vary considerably between enzyme preparations, reflecting the influence of Zn^{2+} and Mg^{2+} concentrations, enzyme purity, and P$_i$ concentrations. At acidic pH and/or with different cations, the k_{cat} is drastically reduced.

iii. Metal ion requirements. Zn^{2+} is required for catalytic activity as well as for structural stability of the enzyme. Under ordinary conditions, each monomer contains two tightly bound Zn^{2+} and one Mg^{2+} ions. Monomers containing one Zn^{2+} per molecule are inactive (63). Monomers are considerably less stable than native dimers. Further addition of Zn^{2+} results in dimerization and restores full activity. The substitution of Co^{2+} for Zn^{2+} results in a 10-fold less active enzyme (5). Cd^{2+} can replace Zn^{2+} but with ~100 times lower k_{cat} (6).

iv. Enzyme stability

(a) Thermostability. BAP is relatively thermostable, which is the reason that the phosphatase reactions can be performed at 65°C. In the presence of Zn^{2+} and Mg^{2+}, BAP retains 50% activity at 85°C for 15 min or at 95°C for 8 min. The apoenzyme is far more labile to heat denaturation (3,9). The binding of the first pair and second pair of Zn^{2+} increases the stability of the enzyme by ~70 and 30 kcal/mol, respectively (10). Addition of the third metal ion pair, e.g., one equivalent of Mg^{2+} per monomer, further stabilizes the metalloenzyme by 10 kcal/mol. The binding of P_i also produces a marked stabilization of the enzyme (11).

(b) Dilutions. The stability of BAP decreases rapidly on dilution, most likely due to an equilibrium shift between monomeric and dimeric enzyme structures. Diluted enzymes should not be stored more than 1 week.

v. Activators and inhibitors

(a) Guanidinium chloride. GuHCl at 1.5 M concentration stimulates the BAP activity three- to fourfold, apparently due to acceleration of the rate-limiting dissociation of the product P_i from the enzyme (12). The activity gradually decreases until the enzyme is totally inactivated in 6 M GuHCl. BAP is not denatured in 8 M urea at 0°C, although this concentration inhibits the enyzme activity when present in the assay solution.

(b) Phosphates and analogs. Orthophosphate is a competitive inhibitor ($K_i \approx 5$ μM) at pH 8.0 (3). Phosphate analogs such as *phosphonates* (P–C bond) are also competitive inhibitors ($K_i = 0.1-5.0$ mM) and are not hydrolyzed. Other competitive inhibitors include phosphorothioates ($K_i = 0.01-120$ μM).

Vanadate is a strong inhibitor of BAP with a K_i of 2–3 μM (13). Phenol enhances vanadate inhibition in the hydrolysis of PNPP, presumably by forming a phenyl vanadate ester ($K_i \approx 0.2$ μM) which has a greater affinity for the enzyme than the vanadate alone. Note that phenol by itself is a weak competitive inhibitor with a K_i of 180 mM. For other metal ions, inhibition decreases in the order of tungstate (WO_4^{2-}, $K_i \approx 6$ μM) > arsenate (AsO_4^{2-}, $K_i = 3-20$ μM) > molybdate (MoO_4^{2-}, $K_i \approx 0.3$ mM) > chromate (CrO_4^{2-}, $K_i \approx 5$ mM).

(c) Metal chelators. BAP is inhibited by the presence of metal chelating compounds such as EDTA, EGTA, and 8-hydroxyquinoline (3,9). The inhibition is instantaneously reversed by the addition of Zn^{2+} ions (14). At >1 mM concentrations and in the presence of 8 M urea, thiol compounds such as 2-MSH, Cys, and thioglycolic acid inactivate the enzyme. The inactivation is reversible.

b. Activity assays and unit definitions

APase activity can be assayed in a variety of ways ranging from conventional radiometric or spectrophotometric assays to chemiluminescent assays.

i. Phosphate release assay using [^{32}P]ATP. This is probably the most common assay used by commercial enzyme suppliers. APase activities are typically assayed in a reaction mixture containing 10 mM Tris–Cl (pH 8.0), 1.5 mM [γ-^{32}P]ATP and 2–8 U of enzyme.

Unit definition: One *ATP hydrolysis unit* is the amount of enzyme that hydrolyzes 1 nmol of ATP to ADP and P_i in 30 min at 37°C.

ii. Spectrophotometric assay using *p*-nitrophenyl phosphate. The spectrophotometric assay using PNPP as a substrate has been used from the early days of APase research (3,15). The release of PNP is determined at 400 nm using $\varepsilon = 1.68 \times 10^4\ M^{-1}\ cm^{-1}$ at pH 8.0 (or $\varepsilon_{401} = 1.84 \times 10^4\ M^{-1}\ cm^{-1}$ in 10 mM NaOH). The assay medium contains 1 mM PNPP and 0.5 M NaCl in either 1 M Tris–Cl (pH 8.0) or 0.1 M MOPS (pH 8.0). The activity measured in MOPS (or 10 mM Tris) is a hydrolase activity, whereas that in 1 M Tris buffer is the sum of the hydrolase and the phosphotransferase activities.

Unit definition: One "PNP" *unit* catalyzes the hydrolysis of 1 μmol of PNPP per minute at the assay temperature.

iii. Fluorometric assay using 4-methylumbelliferyl phosphate. This assay is essentially similar to the PNPP assay except that the reaction is carried out at pH 9.2 with a fluorescent substrate MUP (emission maximum at 380 nm). The MUP is dissolved in water and added to the assay mixture at a final concentration of 8 μM. The production rate of 4-methylumbelliferone (bluish white, emission maximum at 440 nm) is measured fluorometrically at excitation and emission wavelengths of 360 and 465 nm, respectively (16). The fluorescence intensity of the product is linear only between 0 and 8 μM. This assay is not as sensitive as the colorimetric assay (see below).

iv. Colorimetric dye precipitation assay. This assay uses BCIP as an APase substrate and NBT as a precipitating chromogenic reagent (17). The two reagents are relatively stable in aqueous solution. When the phosphate group of BCIP is cleaved, the indoxyl product tautomerizes with the ketone form which, under alkaline conditions, dimerizes to indigo white (Fig. 5.1). Two hydrogen ions released on the formation of indigo or dehydroindigo are consumed in reducing NBT to diformazan. The final product diformazan has an intense blue color (absorption maximum at 600 nm) and is insoluble in alcohols. This assay is very sensitive and widely used in enzyme-linked immunodetection in which the APase is conjugated to principal ligands such as avidin (or streptavidin), protein A, and antibodies (18,19). The stock solutions of both NBT and BCIP are prepared at 50 mg/ml in DMF and kept shielded from the light. Some commercial BCIP products may not be readily soluble, and they should not be used for this assay.

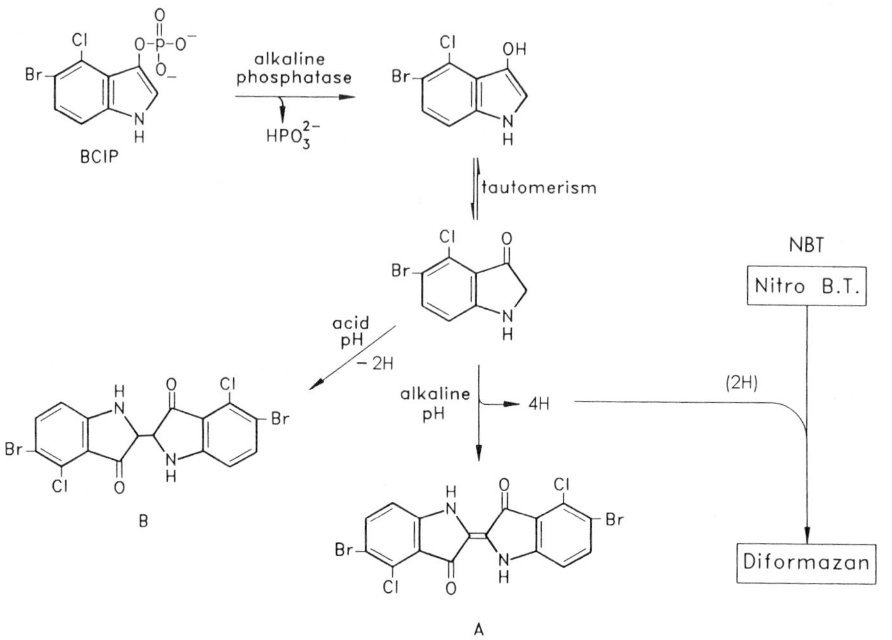

FIGURE 5.1 Reaction pathway of BCIP–NBT colorimetry. BCIP, 5-bromo-4-chloroindoxyl phosphate; NBT, nitroblue tetrazolium, (A) 5,5'-dibromo-4,4'-dichloro indigo white; (B) 5,5'-dibromo-4,4'-dichloro indigo. [After J. McGadey (1970). *Histochemie* **23**, 180–184.]

The reaction solution contains 5 ml APase buffer [0.1 M Tris–Cl (pH 9.0), 0.1 M NaCl, and 5 mM $MgCl_2$], 33 µl NBT, and 16.5 µl BCIP. Intense, stable blue color appears within minutes of the APase reaction at 25° or 37°C.

v. Bioluminescence-enhanced assay. This assay, originally developed for immunoblot detection (20), utilizes coupled enzyme reactions initiated by APase that releases D-luciferin from D-luciferin O-phosphate (Novabiochem AG). The D-luciferin subsequently reacts with ATP and oxygen in a reaction catalyzed by luciferase (from firefly *Photinus pyralis*), resulting in light emission that can be detected on a sensitive photographic film (see Section D, Chapter 8). The reaction is performed in a pH 8.0 buffer. In immunoblot detections, this assay achieves a sensitivity of 5–50 pg protein (or $2–20 \times 10^7$ IgG molecules).

vi. Chemiluminescent assay. This is a highly sensitive assay that employs a chemiluminescent substrate, adamantyl-1,2-dioxetane phosphate (AMPPD) or preferably its chlorinated derivative CSPD (21–23). The APase reaction, for example, in a coupled reaction system to detect the hybridization of DNA on a nylon membrane, is performed for 5 min in 0.1 M DEA (pH 10), 1 mM $MgCl_2$, 0.01% (w/v) NaN_3, and 0.25 mM AMPPD. Speed and sensitivity are significantly

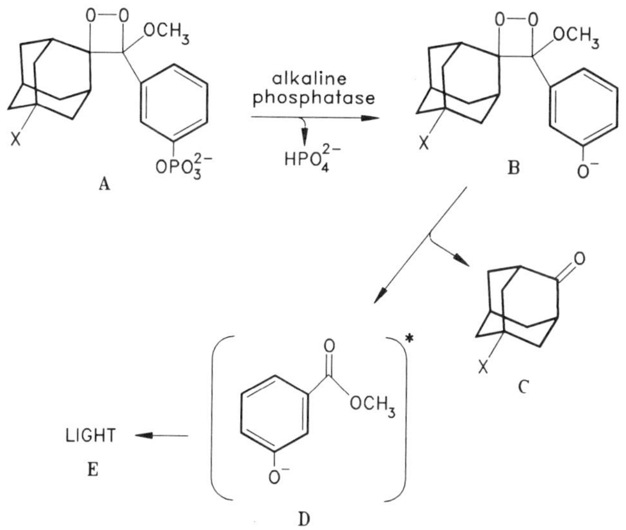

FIGURE 5.2 Dioxetane chemiluminescence reaction pathway. A, AMPPD (adamantyl 1,2-dioxetane phosphate) when X = H or CSPD when X = Cl; B, dioxetane anion; C, adamantanone; D, methyloxybenzoate anion; and E, light emission at 477 nm.

improved by performing the dephosphorylation reaction in DEA buffer at pH 10 rather than the usual carbonate buffer at pH 9.5 (24). DEA is an activating buffer for CIAP as well. CIAP catalyzes this reaction with the K_m and k_{cat} of 0.1 mM and 4100 sec^{-1}, respectively.

The product of dephosphorylation is a dioxetane anion (Fig. 5.2). The dioxetane anion is unstable ($t_{1/2}$ = 2–30 min) and decomposes to adamantanone and the excited state of the methyl m-oxybenzoate anion, the product emitting chemiluminescence at 477 nm. The chemiluminescence signal with 1,2-dioxetanes appears within 10 min and persists for several days on nylon membranes, permitting multiple film exposures. With CSPD as substrate, the maximum light emission is attained at ~5 hr, which is twice as fast as AMPPD. CSPD also produces two times higher amplitude of emission than AMPPD and thus higher sensitivity which is, in part, due to the reduced intermolecular aggregation and lower background signals of nonenzymatically produced breakdown products (27). The luminescent signal can be easily detected on Polaroid type 612 film or standard X-ray film (e.g., Kodak XAR-5). It can also be measured by a luminometer.

The chemiluminescence assay is fast and more sensitive than colorimetric (BCIP/NBT) or fluorimetric assays by 2–3 orders of magnitude. Compared with the detection techniques based on the ^{32}P radioisotope that usually require overnight exposure, the APase-based chemiluminescence detection systems coupled with biotin–avidin, digoxigenin–antidigoxigenin, oligonucleotide, or antibody recognition systems are stable, nonhazardous, and fast, reducing the detection time to a few hours. The sensitivity can also supercede that of ^{32}P-labeled probes, approaching near 20 fg of target DNA ($\sim 10^5$ molecules) (25), 60 fg of protein (26), or subattomoles of APase. This sensitivity allows detection of single copy genes or proteins from single cells in DNA hybridization or gene expression.

The chemiluminescence signal is severely quenched on NC or PVDF membranes unless the membranes are treated with hydrophobic enhancers and blocking buffers like PBS containing 0.5% SDS. In the presence of luminescence enhancers, e.g., fluoroescent micelles formed from CTAB (cetyltrimethylammonium bromide) and 5-(N-tetradecanoyl)aminofluorescein at pH 9.6, the chemiluminescence signal can be further increased 400-fold (21). The addition of fluorescein or rhodamine as a component of enhancer systems shifts the light emission maxima to 542 or 620 nm, respectively, as a result of energy transfer. The water-soluble quaternary ammonium polymers increase the emission efficiency of the reaction by partitioning the aqueous medium away from the excited state which resides in the hydrophobic environments, thus reducing quenching.

c. Substrate specificities

i. Phosphomonoesters. BAP is a nonspecific phosphomonoesterase. The 5'- and 3'-terminal phosphates of DNA, RNA, and oligonucleotides are substrates. PNPP and a host of other alkyl and aryl organic phosphates, such as 2,4-DNPP, are also substrates. The k_{cat} is similar regardless of the leaving group. In fact, the nonspecific phosphomonoesterase activity forms the basis for a number of sensitive detection methods relying on the conversion of phosphate derivatives to chromogenic [e.g., BCIP/NBT and FAD phosphate/D-amino-acid oxidase/horseradish peroxidase-coupled system (96)], fluorescent (e.g., MUP), bioluminescent (e.g., D-luciferin O-phosphate), or chemiluminescent (e.g., AMPPD and CSPD) organic compounds.

Phosphodiesters (e.g., diphenyl phosphate, NAD, and cAMP) and phosphotriesters (e.g., trimethyl phosphate and triphenyl phosphate) are not substrates.

ii. Substrate termini structures: Nicks and gaps. At temperatures below 37°C, BAP quantitatively hydrolyzes external phosphomonoesters (those located at either end of the duplex DNA molecules), but not the phosphomonoesters located at nicks (28). At elevated temperatures (e.g., 65°C), however, both types of phosphomonoesters are hydrolyzed.

When a gap is introduced into a nick, the phosphomonoesters are still resistant to hydrolysis; only 10% are hydrolyzed at 20°C. A single-strand region adjacent to the phosphomonoester apparently inhibits the action of APase. The 5'-phosphomonoesters on recessed ends of DNA molecules are also resistant to hydrolysis by APase at low temperatures (29).

iii. Phosphoric acid anhydrides. Simple acid anhydrides of orthophosphoric acid such as PP_i, PPP_i(-ATP), and short-chain metaphosphates are hydrolyzed in the same fashion and at the same rate as phosphate monoesters (30).

iv. Phosphate ester analogs

(a) Sulfur-substituted analogs. O-Phosphorothioates (P=S bond in RO-PO$_2$SNa$_2$) are hydrolyzed at a rate 100 times lower than PNPP (31). For example, the turnover rate of PNP-O-phosphorothioate (K_m = 0.13 mM at pH 8.0) or p-phenylazophenyl-O-phosphorothioate is ~100 times lower than its phosphate counterpart (5,31,32). O-Phosphorothioates are at best poor substrates and, more appropriately, are competitive inhibitors. The lower hydrolysis rate of phosphorothioates is attributed to the reduced susceptibility of the central phosphorus atom to nucleophilic attack.

In contrast to the O-substituted monoesters, S-substituted phosphates (S–P bond in RSPO$_3$Na$_2$), e.g., cysteamine S-phosphate, are hydrolyzed at a rate comparable to that of PNPP (32).

(b) N-Phosphoramidates. N-linked phosphates (R-NH-PO$_3$H$_2$) are substrates for BAP. The variations in structure of the side chain R from alkyl to aryl to acyl result in only minor differences in their enzymatic hydrolysis rates (4). The V is only 1.7- to 3.0-fold slower than that for PNPP at pH 8.0. However, the K_m values for the N-phosphoramidates are substantially higher than that for PNPP.

(c) Phosphates with C substitutions. Phosphonic (C–P) and phosphinic acids are not hydrolyzed by APases, but act as competitive inhibitors.

(d) Phosphates with F substitutions. The P–F bond in sodium monofluorophosphate is hydrolyzed by APase as readily as the P–O bond: FPO$_3^{2-}$ + H$_2$O → F$^-$ + HPO$_4^{2-}$ + H$^+$. The released fluoride ions provide a means of selectively detecting the APase activity by using a fluoride ion-selective electrode (33).

v. Phosphate acceptor substrates. In the presence of high concentrations of "acceptor" substrates such as Tris (1 M), ethanolamine, propanediol, and glycerol, BAP transfers the phosphate group from donor substrates to a hydroxyl group of the acceptors, resulting in transphosphorylation (2). The APase can synthesize PP$_i$ from P$_i$ (at 4 mM) via a similar phosphoryl transfer (36).

d. Reaction mechanism

i. Catalytic process. APase catalyzes the hydrolysis of phosphomonoesters and phosphoanhydrides. During the catalytic cycle, the enzyme forms a phosphoryl intermediate. The overall catalytic mechanism comprises four distinct, reversible steps (Scheme 5.2): (a) formation of a tight Michaelis–Menten complex betweeen APase (E) and the phosphomonoester substrate (ROP); (b) formation of a phosphoenzyme intermediate (E–P) by phosphorylation at a Ser residue and release of the first product, an alcohol (ROH or RO$^-$); (c) dephosphorylation of the phosphoserine forming a Michaelis–Menten complex of the enzyme and P$_i$ (or transphosphorylation of the P$_i$ group to an acceptor); and (d) dissociation of the noncovalent complex (E·P) to free enzyme and P$_i$.

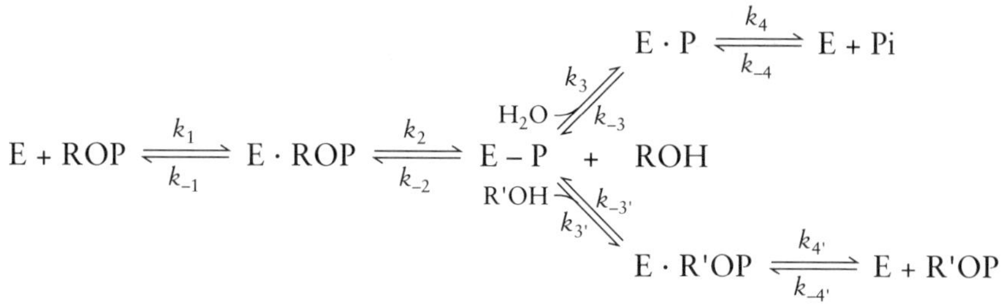

$$E + ROP \underset{k_{-1}}{\overset{k_1}{\rightleftharpoons}} E \cdot ROP \underset{k_{-2}}{\overset{k_2}{\rightleftharpoons}} \begin{matrix} H_2O \searrow k_{-3} \\ E - P \\ R'OH \searrow k_{-3'} \\ k_{3'} \end{matrix} + ROH$$

$$E \cdot P \underset{k_{-4}}{\overset{k_4}{\rightleftharpoons}} E + P_i$$

$$E \cdot R'OP \underset{k_{-4'}}{\overset{k_{4'}}{\rightleftharpoons}} E + R'OP$$

SCHEME 5.2

The overall reaction proceeds with the retention of configuration of the phosphorus atom which is consistent with the inversion of configuration at each step of in-line nucleophilic attacks (34,35).

When an acceptor nucleophile R′OH (e.g., Tris) is present at sufficiently high concentrations, APase transfers the product P_i to the R′OH resulting in transphosphorylation (2). Tris (1 M) is a more efficient acceptor than H_2O at pH 6–9 (6). Glycerol (3 M) is a poorer acceptor at pH ≤ 8 but a better acceptor at pH > 9.

Under acid conditions, the formation of phosphoenzyme is accompanied by a "burst" release of alcohol (ROH), that can be studied by transient kinetics when chromophoric or fluorescent substrates are used. The covalent E–P intermediate (phosphoserine) is stable at low pH and can be isolated (37,38). The bound P_i is not coordinated to any metal and is in a dianionic form throughout the pH range until unfolding occurs.

ii. Role of metal ions. *E. coli* APase is a homodimeric metalloenzyme requiring Zn^{2+} for its full activity. Each subunit contains three distinct metal-binding sites (two Zn^{2+} and one Mg^{2+} cations) which are sequentially occupied on progressive addition of the metal ions. The phosphate group in the E · ROP complex is coordinated to a Zn^{2+} ion (A site). The A-site metal also activates the active-site Ser-102 for nucleophilic attack at the phosphate group. The A-site metal is also believed to stabilize, perhaps assisted by Arg-166, the transition state of the phosphorus during bond formation and breakage. Once E–P (phosphoserine) is formed, the phosphate dianion coordinates with none of the metal ions. The catalytically important group in the subsequent dephosphorylation step is presumed to be a water molecule which is coordinated to the A-site metal ion (39). Hence the A-site metal ion has a major influence on the pK_a of the catalytic group.

The second (B site) Zn^{2+} ion has a much smaller effect on the dephosphorylation step. However, the B-site metal has a strong influence on the dissociation of P_i: the replacement of B-site Zn^{2+} by Cd^{2+} markedly slows the P_i dissociation.

Arsenate, a phosphate analog and enzyme inhibitor, binds between Ser-102 and Zn sites A and B. The guanidinium group of Arg-166 is within a H-bonding distance from the arsenate site. In the absence of any particular acid or base of the protein to facilitate the proton transfer, metals are apparently able to activate both nucleophiles, Ser-102 and H_2O, which participate in double in-line nucleophilic displacements of the phosphorus (40).

iii. Rate-determining step(s). Various kinetic studies suggest that the RDS changes depending on the metal ion, as well as the pH of the medium: for example, with Zn^{2+} ions and at alkaline pH (8.0), $k_3 \gg k_{-3}$ and E · P is the major equilibrium species. Under these conditions, the RDS is the dissociation of the tightly bound phosphate and $k_{cat} \approx k_4 \leq 30$ sec^{-1} (6,41,42). Consequently, the k_{cat} for the hydrolysis of a wide variety of phosphomonoesters remains relatively constant. Nevertheless, the k_{cat} can be enhanced three- to fourfold in the presence of 1.5 M GuHCl. GuHCl enhances the enzyme activity by abolishing negative cooperativity and by accelerating the rate-determining dissociation of the E · P complex (12). With Cd^{2+} ions or at a low pH, however, $k_{-3} \gg k_3$ and E–P is the major equilibrium species (43). Under these conditions, the RDS is the dephosphorylation of E–P.

2. APPLICATIONS

The application of BAP can be divided into four categories: (i) as simple phosphomonoesterase reagents for dephosphorylating terminal phosphates of nucleic acids, (ii) as detection tools in oligonucleotide- or protein-conjugated systems, (iii) as reporter enzymes in the expression of APase-fusion genes, and (iv) as the "leader" in extracellular secretion of fusion proteins.

i. As phosphomonoesterase. One major use of APases (BAP and CIAP) is for dephosphorylating 5'-P and 3'-P of DNA, RNA, and oligonucleotides. To dephosphorylate the terminal phosphates from dsDNA or dsRNA, the reaction mixture is incubated at 65°C for 1 hr with BAP or at 45–55°C with CIAP (~0.1 U/50 pmol ends). The dephosphorylation of single-stranded substrates is carried out at 45°C for 15–30 min with BAP.

In molecular cloning, prior dephosphorylation of the 5'-P groups at the vector termini is an important technique in preventing self-ligation of the vector thereby increasing the yield of intermolecular ligation products. The 5'-dephosphorylated DNA can be labeled with ^{32}P by the use of T4 PNK (see Section II, this chapter) and used for various purposes such as ligation, nucleotide sequencing, and DNA/RNA structural analyses. For the subsequent 5'-kination reactions, it is necessary to eliminate or inactivate the APase activity either by phenol/chloroform extraction or by heat denaturation. Another strategy is to employ K-P as a specific inhibitor of APase. The presence of 1–5 mM K-P inhibits BAP without significantly affecting T4 PNK and permits convenient ^{32}P labeling of the 5' ends by the T4 PNK in 10–15 min (45).

Because BAP is far more resistant to heat and denaturants than CIAP, which can be readily inactivated by heating at 70°C for 10 min, the use of CIAP is often recommended for applications that require serial enzymatic reactions.

The quest for heat-sensitive APases has led to the commercial development of two APases: one is *HK phosphatase* (from Epicentre) isolated from an antarctic marine bacterium HK47 (44), and the other is a *shrimp APase* (SAP, from USB) isolated from a cold-living sea shrimp. The HK phosphatase is completely and irreversibly inactivated by heating for 30 min at 65°C. The HK phosphatase reaction is performed at 30°C in a neutral to alkaline buffer containing 5 mM CaCl$_2$. The shrimp APase, whose reaction is performed at 37°C in Tris buffer (20 mM, pH 8) containing 10 mM MgCl$_2$, is completely inactivated by heating at 65°C for 15 min.

ii. As detection tools in nonisotopic hybridization, immunoscreening, and nucleotide sequencing. BAP can be easily conjugated to synthetic oligonucleotides or proteins via bifunctional reagents, e.g., disuccinimidyl suberate and glutaraldehyde. The oligonucleotides need to carry a modified base (e.g., thymidine analogs) with a "linker arm" terminating in a reactive primary amine. The linker arm approach is not only an efficient method for conjugating a reporter enzyme, but also provides easy control of the location and number of labeling sites (46). Cross-linking via the amine group results in oligomer–APase conjugates with full enzyme activity (47,48). The oligonucleotides also maintain the capacity to hybridize to the target sequence. The hybrid, on a nylon or NC membrane, can be detected by either colorimetric or chemiluminescent assays (see "Activity assays").

APases (e.g., BAP and human placental APase) can also be chemically conjugated to antibodies and other proteins such as avidin (or streptavidin) and protein A. As an attractive alternative, recombinant DNA technology can be applied to link an antigen-specific part of the Ig gene encoding F(ab)$_2$ to the BAP gene (between the *phoA* signal sequence and structural gene) and to produce a bifunctional F(ab)$_2$/APase conjugate in *E. coli* (98). The APase-conjugated system provides sensitive immunodetection capabilities to biotin- or digoxigenin-labeled probes (oligonucleotides or proteins) and secondary antibodies in ELISA, immunohistochemistry, and immunocytochemistry. Because of its sensitivity and color stability, the APase-based system is generally considered to be superior to an alternative system based on horseradish peroxidase in immunoscreening.

The APase chemiluminescent system affords similar detection advantages in dideoxynucleotide sequencing that employs biotinylated primers (27). A chemiluminescent detection strategy has also been applied to chemical nucleotide sequencing (49). In these sequencing applications, the sequence ladder generated on a gel is transferred to a nylon membrane and is detected either by the use of an APase-conjugated avidin system or by the procedures involving hybridization with DNA (oligonucleotide) probes labeled with APase or biotin.

iii. As reporter of fusion gene expressions. In these applications, the gene encoding a protein of interest is fused to either the N-terminal or the C-terminal region of the BAP gene (*phoA*) in an expression vector (50–53). A BAP gene engineered to facilitate the construction of translational fusions features several restriction sites, including a *Sma*I site at the N-terminal region and an *Eco*RV site at the C-terminal region (54). Recombinant constructs linking a promoter, a translation start site, and/or a complete heterologous or mutagenized signal sequence to the N-terminal part of the BAP gene are versatile and useful tools to study functional roles of the genetic elements involved in transcription and/or protein secretion (55,56). These experiments should employ an APase-minus *E. coli* strain, e.g., MZ-9 (57), CC118 or other *phoA* deletion mutants such as AD90 and SM547.

Several features make BAP an ideal reporter: (a) the expression of the BAP–fusion protein usually occurs at the periplasm, (b) BAP is active only when it is secreted into the periplasm, (c) BAP activity is readily detected in solution or on agar plates by various sensitive assays, (d) the N-terminal portion of the enzyme can be readily manipulated without affecting the enzyme activity, and (e) overproduction of BAP can be achieved when the BAP gene is carried on a multicopy plasmid.

iv. Use of BAP as exporter signal. In these applications, the leader sequence of BAP is fused to proteins of interest in order to direct the secretion to the periplasmic space. The BAP signal sequence (see "Structure") has been successfully used to achieve extracellular secretion of a variety of recombinant proteins in yeast, fungi, and eukaryotic cells, as well as in *E. coli* and other bacteria.

The signal sequence of BAP has also served as a prime model in the elucidation of the structure–function relationship of signal sequences in general in membrane transport of proteins. Mutant signal sequences which contain nine consecutive residues of Leu or Ile in the core region function as effectively as the wild-type sequence in the transport process (58). A largely polymeric, strongly hydrophobic α-helical sequence, $MKQST(L_{10})-(A_6)$, functions equivalently to the wild-type BAP signal peptide as well (59). Indeed the core region of the BAP signal sequence can be replaced with full competency by hydrophobic core regions of certain other signal peptides such as *E. coli* maltose-binding protein and OmpA (60).

Efficient transmembrane transport does not depend on the structural preference of the cleavage region (59). The presence of a Pro residue between the hydrophobic core and cleavage region is not essential for translocation either. In certain circumstances, however, the cleavage region as well as the mature part of the passenger protein plays a role in the efficiency of translocation. The heterologous export of normally cytoplasmic proteins may be particularly problematic, although small, compact globular proteins, like the soluble 12-kDa domain of mammalian cytochrome b_5 (99), can be effectively exported and processed. Other cytoplasmic proteins may not be secreted even when they are linked to the BAP signal sequence unless "stop transfer sequences" are removed or placed in other downstream sequence contexts (61). For the proteins that have naturally evolved to be secretion competent, the removal of "stop transfer sequences" can improve the efficiency of transport. For example, the removal of two positively charged amino acids (e.g., in Arg–Ile–Arg) at the N terminus of BAP results in a 50-fold increase in the export of the protein (62).

Translocation in *E. coli* of secretory proteins across the membrane is thought to be a streamlined process involving multiple interactions among Sec proteins, molecular chaperones, and signal peptidases as well as signal sequences. Overexpression in *E. coli* of recombinant proteins designed to be export competent often exceeds the transport capacity. As a consequence, it is not rare to see the majority of the recombinant proteins accumulating as immature proteins in the membranes and/or as aggregated inclusion bodies in the cytoplasm.

3. STRUCTURE

i. Primary structure. BAP is a homodimeric, unglycosylated metalloenzyme (M_r 94,058). The primary structure, initially determined by amino acid sequencing (64), consists of 449 amino acids/monomer (M_r 47,029). The N- and C-terminal residues are Thr (or Arg in another isozyme, see "Pre-BAP") and Lys, respectively. BAP has a high content (24.5%) of Gly plus Ala, and the acidic residues (26 Glu + 27 Asp) outnumber the basic residues (13 Arg + 28 Lys). All four Cys residues per monomer are involved in intrachain disulfide linkages, C168/C178 and C286/C336.

BAP shares some regional sequence identities (up to 50 and 70%) with mammalian APases, e.g., human placental APase and bovine intestinal APase, respectively (65).

Trp-109 is an inner core residue linked to the active-site pocket via a rigid α-helical rod. The phosphorescence of Trp-109 is unique among the three Trp residues in BAP and can serve as a sensitive internal probe for monitoring structural perturbations that occur in the active site during the binding of metal ion, substrate, or phosphates (66).

ii. Pre-BAP, isozymes, and truncated BAP

(a) Pre-BAP. Native BAP is synthesized as a precursor molecule of 471 amino acids. The N-terminal signal sequence consists of 21 highly hydrophobic amino acid residues (67–70): M^{-21}KEST IALAL LPLLF TPVTK $A^{-1}R^{1'}T^{+1}$. The signal sequence of BAP thus has the features common to most known signals, i.e., a basic N-terminal region, a central stretch of hydrophobic residues, and a basic, nonhelical C-terminal region encompassing the cleavage site. During the translocation across the membrane, the signal peptide is cleaved by the inner membrane enzyme signal peptidase (71,72). The monomers then dimerize and become catalytically active by acquiring the cofactor Zn^{2+} ions.

BAP is enzymatically active only when it is secreted into the periplasm (the space between the inner and outer membranes of gram-negative bacteria) since the formation of an active dimer involves disulfide bonding that apparently requires the oxidizing environment of the periplasm (51,73).

(b) BAP "isozymes." The cleavage of the signal sequence after Ala^{-1} results in a BAP (isozyme 1) with N-terminal Arg. The removal of the $Arg^{1'}$, probably by an exoprotease, gives rise to isozyme 3, whereas isozyme 2 is a heterodimer of the two types of chains. The three BAP isozymes can be resolved by nondenaturing PAGE. (*Note:* In a more restricted sense, isozymes designate the enzymes that have the same enzymatic activity but arise from different genes in a cell or organism.)

Over 98% of the initial dimers are inactive and are found exclusively as isozyme 1, which is converted during the stationary phase into isozymes 2 and 3 with concomitant activation.

(c) *Regional truncation and/or substitution of BAP.* The first N-terminal 34 amino acid residues (or 13 N-terminal residues beyond the signal sequence) of BAP are not absolutely required for activity (50). However, removal of the N-terminal decapeptide, for example, by trypsin cleavage at Arg^{10}–Ala^{11}, reduces the specific activity of the enzyme by ~20%, probably by affecting subunit interactions as well as the metal ion binding affinity (74). The deletion of 39 or more N-terminal residues of mature BAP results in a significant decrease in specific activity. The addition of ~150 foreign amino acids to the N terminus of BAP has little effect on the specific activity of the enzyme. Truncation or replacement of the C-terminal 22 amino acids of BAP may result in localization of the enzyme not in the periplasm but in the membrane (75).

iii. **Secondary and tertiary structures.** The secondary structure of BAP consists of 7% β sheets and 27% α helix (40). The X-ray crystallographic structure shows that BAP is a symmetric dimer of the α/β class (40,76). The core of the molecule consists of parallel and antiparallel β strands, surrounded by short α-helical segments and large regions of irregular secondary structures. Approximately 16% of the monomeric surface constitutes the subunit interface.

(a) Metal-binding sites. Three functional metal-binding sites (A, B, and C) have been located in BAP and all are very close to each other at the C-terminal part of a β sheet. The essential Ser-102 is located between metal ions A and B, slightly toward the A site. Two His residues (His-331, -412) and Asp-327 contribute to the A site which is 3.9 Å apart from the B site that is formed by His-370, Asp-369, and Asp-51. The C site, which is 5 and 7 Å apart from the B and A site, respectively, comprises Thr-155, Glu-322, and Asp-51. Asp-153 is an indirect ligand of the C-site Mg^{2+} via two water molecules. Mg^{2+} has little affinity for the A site and will not compete with Zn^{2+} for the site. However, Mg^{2+} binds to the B site competitively. Under normal conditions, two Zn ions occupy A and B sites.

(b) Active site. The size of the active pocket can barely accommodate inorganic phosphate. The two active sites in the dimer are 32 Å apart and there is no apparent physical overlap between the two subunits at the active sites. Nevertheless, the binding of P_i to the dimeric enzyme is apparently anticooperative, as shown by the P_i binding-associated structural stabilization that occurs at a P_i-to-enzyme dimer ratio of 1 (11).

BAP has 10 His/monomer, 3 of which are involved in Zn^{2+} coordination (two in the A site and one in the B site). His-372 contributes to catalysis via H-bonding interaction with the Zn-binding Asp-327 (77). The P_i of the E·P complex is anchored partly by the coordination with the A-site metal ion (76), and partly by the guanidinium group of Arg-166 which in turn interacts with the carboxyl group of Asp-101.

The active-site Ser-102, which is located within the Asp^{101}–Gly^{112} helix, is the essential residue that is phosphorylated in the covalent E–P intermediate (37,38). The phosphate of phospho-Ser is not coordinated to any metal. The substitution of Ser-102 by Cys (S102C) results in a mutant with 30–90% catalytic activity (78). The RDS of the S102C enzyme is presumed to be the formation of a phosphoryl enzyme intermediate.

Arg-166 lies within the H-bonding distance of the phosphate (or arsenate) site and forms a salt bridge with Asp-101 (40). Modification of the Arg by phenylglyoxal leads to the inactivation of the enzyme (79). However, the replacement of Arg-166 by Ala (80), Gln (81), or Lys (81) causes a 50-, 170-, or 4-fold decrease in catalytic efficiency, respectively. Although catalytically nonessential, these observations suggest that Arg-166 plays a role most likely in substrate–phosphate binding and release of the phosphate.

Residue Asp-101 is not necessary for catalytic activity, but is presumed to be involved in the positioning of Arg-166. The substitution of Asp-101 by Ser (D101S) renders the side chain of Arg-166 more flexible within very minor structural alterations in the active site (95). Compared with the wild-type enzyme, the D101S mutant performs a 10 times faster catalytic turnover (or 5-fold increase in k_{cat}/K_m), which is mostly attributed to a lower (30-fold) affinity toward the product phosphate. In remarkable contrast and for reasons not fully understood, the substitution of Asp-101 by Ala (D101A) results in a 2-fold decrease in phosphatase activity without changing k_{cat}/K_m (8). This mutation also induces a substantial decrease in thermal stability, but only a slight alteration in P_i binding.

iv. **Physical parameters.** The A_{278} of a 1-mg/ml solution equals 0.72 (82). The $s^o_{20,w}$ is 6.1 at pH 7.4 (9), and the pI is 4.5 (3).

4. GENETICS

i. ***phoA* gene structure.** The structural gene for BAP is *phoA*. The gene has been cloned and sequenced (69,83). The gene encodes a pre-BAP that has a total of 471 amino acid residues including a signal sequence of 21 amino acids. The codon for the first Met [−21] is not AUG but GUG. The S-D sequence GGAG is found 9 bases upstream from the GUG (67). At 45 bases upstream from the initiator codon is sequence TATAGTC, a homolog of the RNA polymerase-binding site in prokaryotic promoters (Pribnow box). The sequence between the Pribnow box and the initiator condon is highly A · T rich (80%), indicating that it is likely to be a highly efficient promoter. There are four dyad symmetrical sequences around the Pribnow box, all or any number of which may be involved in the regulation of BAP gene expression. The translation is terminated by an *ochre* codon (UAA), and this is immediately followed by a hairpin–loop structure, the putative transcriptional terminator. The *phoA* promoter can be used to efficiently express a gene cloned downstream of the promoter by strict phosphate induction (P_i at below 0.05 mM) (86).

The *phoA* genes of eight naturally occurring *E. coli* strains are polymorphic in 87 nucleotide sites, resulting in a 10 amino acid polymorphism (1 amino acid within the signal sequence and the remainder within the mature peptide) (84). This polymorphism is presumed to have arisen from genetic recombinations involving different segments of the *phoA* gene.

BAP is synthesized as a precursor molecule on polysomes bound to the inner membrane of the cell. The pre-BAP is then secreted into the periplasmic space with accompanying cleavage of the signal sequence (14,57).

A C-to-T transition (*phoA503* mutant) resulting in a substitution of Val for Ala-22 in the mature BAP gives a conditional phenotype: it dramatically reduces the synthesis of phosphatase activity during P_i starvation (85). The mutation does not alter the *phoA* gene expression, the dimer formation, or the translocation to the periplasm.

ii. **Phosphate regulon.** The *phoA* gene is located at the 9-min position on the *E. coli* linkage map, forming a part of the noncontiguous family of the genes in the *pho* regulon. The normal synthesis of BAP can be derepressed on P_i starvation or at a low level of P_i (10^{-7} M) (87), maximally reaching 6% of total cell protein (3). However, a high level of P_i available in rich growth medium represses the synthesis of BAP. The P_i regulation system of *E. coli* is now known to involve ~24 P_i-regulated promoters and ~15 proteins which function in P_i transport (the gene products of *pst* operon) and in intracellular regulation (88,89). In *E. coli*, passive diffusion of phosphate substrates from solution into the periplasm is the determining factor for the unconventional kinetic behavior of BAP *in vivo* and as much as a 1000-fold change in K_m (97). Because the diffusion occurs mainly through porins (e.g., OmpC, OmpF, and PhoE) in the outer membrane, differential expression of the porins under different growth conditions is also presumed to play a role in P_i regulation.

Expression of the *phoA* gene is regulated by bifunctional regulators, PhoR and PhoM, which act as both positive and negative regulators through modification of the PhoB protein. PhoM replaces the positive regulatory function of PhoR in PhoR mutants by cross-talking to PhoB, the principal response regulator of the *pho* regulon. PhoM is a histidine protein kinase and catalyzes the phosphorylation of PhoB, resulting in an activation of PhoB as a transcriptional activator of PhoA (90). PhoR, a putative multimeric protein induced by P_i starvation, is also a histidine protein kinase. PhoR is anchored to the cytoplasmic membrane (91).

In contrast to the positive regulation under conditions of phosphate starvation, the mechanism by which the *pho* regulon is repressed in a high phosphate environment remains relatively obscure. To repress the *pho* regulon, PhoR as well as intact Pst structures are required. In addition, another gene *phoU* located in the *pst* operon (83-min position) is also required to repress the *pho* regulon.

5. SOURCES

The gene for BAP, *phoA*, has been cloned into pBR322, and correctly processed, periplasmic APase has been overproduced ~10-fold on phosphate limitation (92). Some commercial BAP is prepared from *E. coli* K12 SW 1033 carrying a plasmid pKI-5.

BAP is located in the periplasm and can be conveniently and quantitatively released by osmotic shock or by conversion to spheroplasts in the presence of lysozyme and EDTA in a Tris buffer (pH 8.0) containing 20% sucrose. BAP has been purified efficiently on an ion-exchange HPLC column using carboxymethyl dextrans (93).

The fusion of a cellulose-binding domain of endoglucanase Cen A (of *Cellulomonas fimi*) to the *phoA* gene results in a hybrid protein which can be efficiently purified by affinity chromatography on cellulose (94).

References

1. McComb, R. B., Bowers, G. N., Jr., and Posen, S. (1979). In "Alkaline Phosphatase," pp. 1–986. Plenum Press, New York.
2. Coleman, J. E., and Gettins, P. (1983). *Adv. Enzymol.* **55**, 381–452.
3. Garen, A., and Levinthal, C. (1960). *BBA* **38**, 470–483.
4. Snyder, S. L., and Wilson, I. B. (1972). *Biochem.* **11**, 1616–1623.
5. Chlebowski, J. F., and Coleman, J. E. (1974). *JBC* **249**, 7192–7202.
6. Gettins, P., Metzler, M., and Coleman, J. E. (1985). *JBC* **260**, 2875–2883.
7. Reid, T. W., and Wilson, I. B. (1971). "The Enzymes," 3rd ed., Vol. 4, pp. 373–415.
8. Chaidaroglou, A., and Kantrowitz, E. R. (1989). *Prot. Eng.* **3**, 127–132.
9. Schlesinger, M. J., and Barrett, K. (1965). *JBC* **240**, 4284–4292.
10. Chlebowski, J. F., and Mabrey, S. (1977). *JBC* **252**, 7042–7052.
11. Chlebowski, J. F., Mabrey, S., and Falk, M. C. (1979). *JBC* **254**, 5745–5753.
12. Rao, N. M., and Nagaraj, R. (1991). *JBC* **266**, 5018–5024.
13. Stankiewicz, P. J., and Gresser, M. J. (1988). *Biochem.* **27**, 206–212.
14. Malamy, M. H., and Horecker, B. L. (1964). *Biochem.* **3**, 1889–1893.
15. Torriani, A. (1968). *Meth. Enzy.* **12(B)**, 212–218.
16. Cornish, C. J., Neale, F. C., and Posen, S. (1970). *Am. J. Clin. Path.* **53**, 68–76.
17. McGadey, J. (1970). *Histochemie* **23**, 180–184.
18. Blake, M. S., Johnston, K. H., Russel-Jones, G. J., and Gotschlich, E. C. (1984). *Anal. Biochem.* **136**, 175–179.
19. Knecht, D. A., and Dimond, R. L. (1984). *Anal. Biochem.* **136**, 180–184.
20. Hauber, R., and Geiger, R. (1987). *J. Clin. Chem. Clin. Biochem.* **25**, 511–514.
21. Schaap, A. P., Akhavan, H., and Romano, L. J. (1989). *Clin Chem.* **35**, 1863–1864.
22. Bronstein, I., Voyta, J. C., and Edwards, B. (1989). *Anal. Biochem.* **180**, 95–98.
23. Bronstein, I., Voyta, J. C., Lazzari, K. G., Murphy, O., Edwards, B., and Kricka, L. J. (1990). *BioTechniques* **8**, 310–314.
24. Bronstein, I., Voyta, J. C., Lazzari, K. G., Murphy, O., Edwards, B., and Kricka, L. J. (1990). *BioTechniques* **9**, 160–161.
25. Lanzillo, J. J. (1991). *Anal. Biochem.* **194**, 45–53.
26. Gillespie, P. G., and Hudspeth, A. J. (1991). *PNAS* **88**, 2563–2567.
27. Martin, C., Bresnick, L., Juo, R.-R., Voyta, J. C., and Bronstein, I. (1991). *BioTechniques* **11**, 110–113.
28. Weiss, B., Live, T. R., and Richardson, C. C. (1968). *JBC* **243**, 4530–4542.
29. Masamune, Y., Fleischman, R. A., and Richardson, C. C. (1971). *JBC* **246**, 2680–2691.
30. Heppel, L. A., Harkness, D. R., and Hilmoe, R. J. (1962). *JBC* **237**, 841–846.
31. Breslow, R., and Katz, I. (1968). *JACS* **90**, 7376–7377.
32. Neumann, H. (1968). *JBC* **243**, 4671–4676.
33. Venetz, W. P., Mangan, C., and Siddiqi, I. W. (1990). *Anal. Biochem.* **191**, 127–132.
34. Jones, S. R., Kindman, L. A., and Knowles, J. R. (1978). *Nature* **275**, 564–565.
35. Butler-Ransohoff, J. E., Kendall, D. A., Freeman, S., Knowles, J. R., and Kaiser, E. T. (1988). *Biochem.* **27**, 4777–4780.
36. Nayudu, R. V., and de Meis, L. (1989). *FEBS Lett.* **255**, 163–166.
37. Schwartz, J. H., Crestfield, A. M., and Lipmann, F. (1963). *PNAS* **49**, 722–729.
38. Milstein, C. (1963). *BBA* **67**, 171–172.

39. Gettins, P., and Coleman, J. E. (1984). *JBC* **259**, 4991–4997.
40. Sowadski, J. M., Handschumacher, M. D., Murthy, H. M. K., Foster, B. A., and Wyckoff, H. W. (1985). *JBM* **186**, 417–433.
41. Hull, W. E., Halford, S. E., Gutfreund, H., and Sykes, B. D. (1976). *Biochem.* **15**, 1547–1561.
42. Bloch, W., and Gorby, M. S. (1980). *Biochem.* **19**, 5008–5018.
43. Gettins, P., and Coleman, J. E. (1983). *JBC* **258**, 408–416.
44. Hoffman, L. M., and Jendrisak, J. (1990). *Gene* **88**, 97–99.
45. Chaconas, G., and van de Sande, J. H. (1980). *Meth. Enzy.* **65**, 75–85.
46. Ruth, J. L., Morgan, C., and Pasko. A. (1985). *DNA* **4**, 93.
47. Jablonski, E., Moomaw, E. W., Tullis, R. H., and Ruth, J. L. (1986). *NAR* **14**, 6115–6128.
48. Wahl, G. M. (1989). *Strategies* (Stratagene) **2**, 1–3.
49. Tizard, R., Cate, R. L., Ramachandran, K. L., Wysk, M., Voyta, J. C., Murphy, O. J., and Bronstein, I. (1990). *PNAS* **87**, 4514–4518.
50. Hoffman, C. S., and Wright, A. (1985). *PNAS* **82**, 5107–5111.
51. Matteucci, M., and Lipetsky, H. (1986). *Bio/Technology* **4**, 51–55.
52. Manoil, C., and Beckwith, J. (1986). *Science* **233**, 1403–1408.
53. San Millan, J. L., Boyd, D., Dalbey, R., Wickner, W., and Beckwith, J. (1989). *J. Bacteriol.* **171**, 5536–5541.
54. Kohl, J., Ruker, F., Himmler, G., Mattanovich, D., and Katinger, H. (1990). *NAR* **18**, 1069.
55. Manoil, C., and Beckwith, J. (1985). *PNAS* **82**, 8129–8133.
56. Farinha, M. A., and Kropinski, A. M. (1990). *J. Bacteriol.* **172**, 3496–3499.
57. Inouye, H., Pratt, C., Beckwith, J., and Torriani, A. (1977). *JMB* **110**, 75–87.
58. Kendall, D. A., and Kaiser, E. T. (1988). *JBC* **263**, 7261–7265.
59. Laforet, G. A., and Kendall, D. A. (1991). *JBC* **266**, 1326–1334.
60. Laforet, G. A., Kaiser, E. T., and Kendall, D. A. (1989). *JBC* **264**, 14478–14485.
61. Little, S., Campbell, C. J., Evans, I. J., Hayward, E. C., Lilley, R. J., and Robinson, M. K. (1989). *Gene* **83**, 321–329.
62. Li, P., Beckwith, J., and Inouye, H. (1988). *PNAS* **85**, 7685–7689.
63. Falk, M. C., Bethune, J. L., and Vallee, B. L. (1982). *Biochem.* **21**, 1471–1478.
64. Bradshaw, R. A., Cancedda, F., Ericsson, L. H., Neumann, P. A., Piccoli, S. P., Schlesinger, M. J., Shriefer, K., and Walsh, K. A. (1981). *PNAS* **78**, 3473–3477.
65. Coleman, J. E., and Besman, M. J. A. (1987). In "New Comprehensive Biochemistry" (L. Neuberger and K. Brocklehurst, Eds.), Vol. 16, pp. 377–406.
66. Cioni, P., Piras, L., and Strambini, G. B. (1989). *EJB* **185**, 573–579.
67. Inouye, H., Barnes, W., and Beckwith, J. (1982). *J. Bacteriol.* **149**, 434–439.
68. Kikuchi, Y., Yoda, K., Yamasaki, M., and Tamura, G. (1981). *NAR* **9**, 5671–5678.
69. Chang, C. N., Kuang, W. J., and Chen, E. Y. (1986). *Gene* **44**, 121–125.
70. Inouye, H., and Beckwith, J. (1977). *PNAS* **74**, 1440–1444.
71. Chang, C. N., Inouye, H., Model, P., and Beckwith, J. (1980). *J. Bacteriol.* **142**, 726–728.
72. Muller, M., and Blobel, G. (1984). *PNAS* **81**, 7421–7425.
73. Michaelis, S., Inouye, H., Oliver, D., and Beckwith, J. (1983). *J. Bacteriol.* **154**, 366–374.
74. Tyler-Cross, R., Roberts, C. H., and Chlebowski, J. F. (1989). *JBC* **264**, 4523–4528.
75. Gentschev, I., Hess, J., and Goebel, W. (1990). *MGG* **222**, 211–216.
76. Wyckoff, H. W., Handschumacher, M. D., Murthy, H. M. K., and Sowadski, J. M. (1983). *Adv. Enzymol.* **55**, 453–480.
77. Xu, X., Qin, X.-Q., and Kantrowitz, E. R. (1994). *Biochem.* **33**, 2279–2284.
78. Ghosh, S. S., Bock, S. C., Rokita, S. E., and Kaiser, E. T. (1986). *Science* **231**, 145–148.
79. Daemen, F. J. M., and Riordan, J. F. (1974). *Biochem.* **13**, 2865–2871.
80. Chaidaroglou, A., Brezinski, D. J., Middleton, S. A., and Kantrowitz, E. R. (1988). *Biochem.* **27**, 8338–8343.
81. Butler-Ransohoff, J. E., Kendall, D. A., Freeman, S., Knowles, J. R., and Kaiser, E. T. (1988). *PNAS* **85**, 7036–7040.
82. Simpson, R. T., Vallee, B. L., and Tait, G. H. (1968). *Biochem.* **7**, 4336–4342.
83. Shuttleworth, H., Taylor, J., and Minton, N. (1986). *NAR* **14**, 8689.

84. DuBose, R. F., Dykhuizen, D. E., and Hartl, D. L. (1988). *PNAS* **85**, 7036–7040.
85. Agrawal, D. K., and Wanner, B. L. (1990). *J. Bacteriol.* **172**, 3180–3190.
86. Lubke, C., Boidol, W., and Petri, T. (1995). *Enzyme and Microbial Technol.* **17**, 923–928.
87. Torriani, A. (1960). *BBA* **38**, 460–469.
88. Wanner, B. L. (1987). In "*Escherichia coli* and *Salmonella typhimurium*: Cellular and Molecular Biology" (F. C. Neidhart, J. Ingraham, K. B. Low, B. Magasanik, M. Schaechter, and H. E. Umbarger, Eds.), pp. 1326–33. American Society of Microbiology, Washington, D.C.
89. Torriani, A. (1990). *BioEssays* **12**, 371–376.
90. Amemura, M., Makino, K., Shinagawa, H., and Nakata, A. (1990). *J. Bacteriol.* **172**, 6300–6307.
91. Yamada, M., Makino, K., Shinagawa, H., and Nakata, A. (1990). *MGG* **220**, 366–372.
92. Inouye, H., Michaelis, S., Wright, A., and Beckwith, J. (1981). *J. Bacteriol.* **146**, 668–675.
93. Dunn, B. E., Edberg, S. C., and Torres, A. R. (1988). *Anal. Biochem.* **168**, 25–30.
94. Greenwood, J. M., Gilkes, N. R., Kilburn, D. G., Miller, R. C., Jr., and Warren, R. A. J. (1989). *FEBS Lett.* **244**, 127–131.
95. Chen, L., Neidhart, D., Kohlbrenner, W. M., Mandecki, W., Bell, S., Sowadski, J., and Abad-Zapatero, C. (1992). *Prot. Eng.* **5**, 605–610.
96. Harbron, S., Eggelte, H. J., Fisher, M., and Rabin, B. R. (1992). *Anal. Biochem.* **206**, 119–124.
97. Martinez, M. B., Schendel, F. J., Flickinger, M. C., and Nelsestuen, G. L. (1992). *Biochem.* **31**, 11500–11509.
98. Ducancel, F., Gillet, D., Carrier, A., Lajeunesse, E., Menez, A., and Boulain, J.-C. (1993). *Bio/Technology* **11**, 601–605.
99. Karim, A., Kaderbhai, N., Evans, A., Harding, V., and Kaderbhai, M. A. (1993). *Bio/Technology* **11**, 612–618.

B. Calf Intestinal Alkaline Phosphatase

The alkaline phosphatase from calf intestinal mucosa (CIAP) has many catalytic properties in common with the BAP (see Section I,A, this chapter). CIAP is also a dimeric, Zn-metalloenzyme, but it has a slightly larger molecular size (130 kDa). One notable distinction is that CIAP is heat labile compared with BAP. This property alone makes the CIAP a valuable alternative to BAP in many applications requiring successive treatments with other enzymes.

The gene coding for CIAP has now been cloned. As shown with BAP, the ability to express a cloned APase gene as a source of pure enzyme, as a reporter or as a secretion "leader" is an important quality that lends to a variety of applications in recombinant DNA technology. For the purpose of comparison, human APase isozymes are described briefly in the later part of the section.

1. FUNCTIONS

a. Reaction conditions

i. **Optimal reaction conditions.** For dephosphorylation of 5'-P, the reaction mixture typically contains 50 mM Tris–Cl (pH 8.0), 0.5 mM $ZnCl_2$, 0.5 mM $MgCl_2$, and 1 U of CIAP per 100 pmol end DNA or RNA. The mixture is incubated at 37°C for 30 min for external 5'-P or for 1 hr at 55°C for 5'-P at the nick or gap. The reaction is usually stopped by heating to 75°C for 10 min or by phenol/chloroform extractions.

The phosphatase reaction is significantly influenced by pH and buffer components.

(a) *pH.* The optimum pH of CIAP is 9.4–10.5 (1,2); this is slightly higher than for BAP (pH optimum of 8.0–9.5). CIAP is stable at pH 7.5–9.5, but is rapidly inactivated at acidic pH (3).

(b) *Buffer.* The activity of CIAP is significantly influenced by buffer substances and concentrations (4). DEA–HCl (1 M, pH 10.1) is a strongly activating buffer and is presumed to alter the environment of the active center Zn atoms. In carbonate (0.1 M, pH 10.1) and glycine–NaOH buffer (0.1 M, pH 10.1) CIAP activities are reduced to 57 and 9%, respectively, of that in DEA. The phosphatase activity decreases drastically with increasing buffer concentrations. N-Ethylaminoethanol–HCl buffer (1 M, pH 10.1) stimulates CIAP activity more than DEA. The stimulation reaches a maximum of 44% on 40 min preincubation. For APase assays (see below), however, 1 M DEA is recommended because it is more insensitive to slight variations in the reaction medium such as pH, buffer concentrations, and incubation times.

ii. Kinetic parameters. The K_m values for PNPP are 3.6 μM and 1.5 mM at pH 8.0 and 10.0, respectively (1). At pH 8.5, the K_m is 12 μM and V is 250 μmol/min/mg. The K_m for MUP is ~20 μM at pH 9.2 (2). At pH 10.5 and 25°C, the V of CIAP is 500 μmol/min/mg as compared to only 35 μmol/min/mg for BAP (1).

iii. Metal ion requirements. CIAP requires Zn^{2+} for its activity (3). Under normal conditions, it contains four Zn atoms per dimeric enzyme molecule. Mg^{2+} (10 mM) gives at least a 10-fold increase in phosphatase activity.

iv. Inhibitors and/or inactivators. *Chemicals:* CIAP is competitively inhibited by P_i ($K_i \approx$ 16.5 μM at pH 9.2) and by other phosphate analogs in a manner similar to BAP. *Levamisole* also inhibits CIAP but less effectively than P_i (5). *Inositol hexaphosphate* (phytic acid) or its Cu(II) complex inhibits CIAP competitively with a K_i of 0.26 mM at pH 8.0 and 25°C (6).
Heat: CIAP is inactivated by heating at 65°C for 45 min or 75°C for 10 min.
Metal chelators: The presence of sodium nitrilotriacetic acid (10 mM), EDTA, or EGTA inactivates the enzyme.

b. *Activity assay and unit definition*

Although APase activity can be assayed by various methods (see the "Activity assays" on BAP), the most common assay for CIAP is the hydrolysis of PNPP. A typical assay mixture consists of 1 M DEA–HCl (pH 9.8), 0.5 mM $MgCl_2$, 0.5 mM $ZnCl_2$, and 10 mM PNPP. The mixture is incubated with CIAP at 37°C.
Unit definition: One unit is the amount of enzyme that hydrolyzes 1 μmol of PNPP to PNP (ε_{405} = 18.2 mM^{-1} cm^{-1}) per minute. Two "DEA units" (DEA buffer, pH 9.8) are approximately equivalent to one "glycine unit" (glycine–NaOH buffer, pH 10.4).

c. Substrate specificities

i. Nucleotide substrates. CIAP catalyzes the hydrolysis of 5'-P and 3'-P of DNA, RNA, and oligonucleotides. Phosphodiesters are not substrates. CIAP also hydrolyzes nucleoside 5'-di- and 5'-triphosphates. The pyrophosphatase and triphosphatase activities are subject to the influence of $MgCl_2$ concentrations. At low concentrations of $MgCl_2$, there is a slight activation. At high concentrations (10 mM) of $MgCl_2$, a strong inhibition is observed, especially with pyrophosphate substrates (7,8). At 10 mM $MgCl_2$ and pH 9.2, however, the monoesterase activity is unaffected by 1 mM PP_i, suggesting that PP_i is neither an inhibitor nor a substrate under these conditions.

ii. Nonnucleotide substrates. CIAP hydrolyzes a wide variety of phosphomonoesters, such as the esters of primary and secondary alcohols, cyclic alcohols, phenols, and amines. The K_m values for MUP, fluorophosphate, PP_i, β-glycerophosphate, and ATP are all similar: ~20 μM at pH 9.2 (2). The V values for those monophosphates are almost the same.

iii. Phosphate analogs

(a) Phosphonates. N-substituted phosphates (P–N bond), e.g., p-chloroanilidophosphonate, are hydrolyzed only very slowly (1).

(b) Thiophosphates. CIAP hydrolyzes S-substituted phosphomonoesters (P–S bond, I) to P_i and the corresponding thioalcohols at a rate comparable to PNPP. However, it cannot hydrolyze O-substituted phosphorothioate (P=S bond, II) (9).

$$\underset{(I)}{R-S-\overset{\overset{O}{\|}}{\underset{\underset{O^-}{|}}{P}}-O^-} \qquad \underset{(II)}{R-O-\overset{\overset{S}{\|}}{\underset{\underset{O^-}{|}}{P}}-O^-}$$

2. APPLICATIONS

For most general applications involving dephosphorylation of DNA or RNA and as a reporter reagent in immunodetection (see Section I,A,2, this chapter), CIAP can be used interchangeably with BAP. However, the ease of inactivation by heat makes CIAP more useful than BAP in certain applications. Cloning of the full-length CIAP gene renders the enzyme more versatile to use.

Human APases (see further below) have also been cloned and molecularly well characterized. Like CIAP, thermolabile human intestinal APase (HIAP) may thus find general uses, particularly in the area of fusion gene expressions in eukaryotic cells. Human placental APase (10) and rat bone/liver/kidney APase (11) have been used as reporters in APase fusion expression systems. When combined with chemiluminescence technology, these enzymes offer a simple and convenient detection system which is at least as sensitive as the conventional CAT detection system.

3. STRUCTURE

i. General structure. The bovine IAP (BIAP) is a homodimeric glycoprotein (140 kDa) (3). A monomer (64.4 kDa) contains 514 deduced amino acid residues, and has Leu and Tyr as the N- and C-terminal residues, respectively (12,13). Acidic residues (Asp + Glu = 52) outnumber basic residues (Lys + Arg = 41).

BIAP is synthesized as a precursor protein with an N-terminal 19 amino acid signal peptide: MQGAC VLLLL GLHLQ LSLG. The mature BIAP contains four Zn(II) per dimer under normal conditions. The carbohydrate content is 8–12% of the total weight. Glycosylation has no apparent effect on enzyme activity nor on heat stability.

The A_{278} of a 1% CIAP solution is 7.6 (or $\varepsilon_{278} = 1.06 \times 10^5 \, M^{-1}cm^{-1}$) (3).

ii. Isozymes. CIAP, which is usually prepared from unweaned calf intestinal mucosa, is comprised of two developmentally distinct isozymes, C1 (higher subunit MW) and C2 (lower subunit MW) forms. The C1 form disappears during development. The CIAP (C1 form) has 69% sequence identity with the C2 form (BIAP) (15). Another isozyme of CIAP is a tissue-specific "intralumenal APase" that is obtained from the intralumenal space of calf intestine.

iii. Sequence homologies. BIAP shows numbering. BIAP shows on the BAP sequence three regions of particularly high amino acid sequence homology with BAP (15): 7 amino acids within 11 residues [residues 22–32], another 7 amino acids within 10 residues flanking the active site Ser-102, and still another 7 amino acids within 9 residues [residues 362–370]. BIAP displays a high degree of sequence identity with other mammalian IAPs such as human IAP (~80%) and mouse IAP (>60%). Among the conserved His residues that have counterparts in the metal-binding sites of BAP, one His is replaced in BAP by Asp-153, a ligand of the C-site Mg^{2+}. The substitution of Asp-153 by His results in a mutant BAP (D153H) in which the octahedral Mg^{2+} binding site is converted to a tetrahedral Zn^{2+} binding site (16). The mutant BAP exhibits many of the properties of mammalian IAPs including low activity in the absence of Mg^{2+}, enhanced activity induced in a time-dependent fashion by Mg^{2+}, and a shift in the pH of optimal activity. The altered activities are apparently due to slow replacement of Zn^{2+} by Mg^{2+} at the C site.

iv. Active site. The covalent E–P intermediate occurs via phosphoserine within the context of Asp–Ser–Ala (17), a common structural motif found in BAP and many serine proteases. At acidic pH values, the phosphorylation at one active site is very fast ($k > 1000 \, sec^{-1}$), while the other is slow ($k \approx 100 \, sec^{-1}$), showing a half of the sites reactivity (1).

v. Prepro-CIAP. BIAP contains a C-terminal hydrophobic domain [residues 488–514]. Like human placental APase that has a 17 amino acid C-terminal hydrophobic domain (18,19), CIAP is presumed to be processed *in vivo* at the putative C-terminal signal sequence (Asp-487) and anchored in the plasma membrane by a covalent linkage to glycosylphosphatidylinositol (14).

4. GENETICS

The gene for CIAP consists of 11 exons (2946 bp) and 10 small introns within a 5.4 kb genomic fragment (13). All exon-intron junctions conform to the GT–AG rule. The 5' NC region (1.5 kb) contains putative regulatory elements having homology to human and mouse IAP promoter sequences. The 3' NC region contains a putative poly(A) signal (AAATAAA) 781 bp downstream from the TAA stop of the 533-codon ORF.

5. SOURCES

CIAP is usually prepared from unweaned calf intestinal mucosa. The membrane-bound APase can be selectively released (up to 90%) by treatment with bacterial (*S. aureus*) phosphatidylinositol-specific phospholipase C (20). Affinity chromatography using a Sepharose-immobilized triazine dye analog that carries a phosphonate-terminal aminobenzene ring yields a 330-fold one-step purification of CIAP from a crude intestinal extract (21). The purity of CIAP, which can be specifically eluted from the column with a 5 mM P_i solution, is reported to be equivalent to that of a commercial "high purity" preparation.

When stored at 4°C in a buffer composed of 30 mM triethanolamine (pH 7.6), 3 M NaCl, 1 mM $MgCl_2$, and 0.1 mM $ZnCl_2$, CIAP is stable for at least 6 months.

Expression in CHO cells of a cDNA cloned in pcDNA1 (CMV-promoter) vector produces a recombinant CIAP having comparable properties and activities to natural enzymes (13).

6. HUMAN ALKALINE PHOSPHATASES

In humans, there are at least four distinct forms of APases: placental, placental-like (or germ cell), intestinal, and liver/bone/kidney (L/B/K) APases. The germ cell APase is normally found in trace amounts in the testis and thymus. The APase isozymes are clinically important markers for various diseases and cancers (22). Like CIAP, all human APases are dimeric Zn-metalloenzymes. They can be distinguished by immunological and various physicochemical properties. Those include thermal stability (see below), phosphatidate (phosphofatty acid) hydrolytic activity (23), and kinetics of denaturation in the presence of GuHCl (24) or urea (25).

Human APase isozymes are encoded by separate homologous genes which are believed to have been derived from a common ancestral APase gene. The genes for placental, germ cell, and intestinal APases are located in close proximity on the long arm of chromosome 2, and each gene contains 11 exons. In contrast, the L/B/K APase gene is located on chromosome 1 and contains 12 exons. Placental APase (HPAP) is closely related to germ cell APase (GCAP) and intestinal APase (HIAP), exhibiting 98 and 87% amino acid identity, respectively. L/B/K

APase is more evolutionarily distant from the other human APases, having 52 and 57% amino acid identity to HPAP and HIAP, respectively.

Under normal conditions, the expression of APase isozymes is strictly tissue specific. In exceptional cases such as human colon cancer (LoVo cells), however, both HIAP and HPAP can be simultaneously produced under apparently independent induction controls (26). Cell lines derived from choriocarcinoma of the placenta primarily express GCAP, suggesting that malignant transformation of the placenta led to switching of induction controls.

a. Human placental alkaline phosphatase (HPAP)

The HPAP gene is polymorphic in human populations. A comparison of the 5' flanking sequence (up to −540) of the HPAP gene with the analogous sequence of the HIAP gene shows several deletions and substitutions which might be implicated in tissue-specific expression of the APase gene (28).

HPAP is synthesized as a precursor protein containing N- and C-terminal signal peptides that are cleaved off during translocation. The prepro-HPAP consists of 530 amino acid residues per subunit, including a 17 amino acid N-terminal leader sequence and a 29 amino acid C-terminal signal sequence (35). The C-terminal signal peptide contains 17 consecutive hydrophobic residues (29,31). Mature HPAP is anchored *in vivo* to the plasma membrane via phosphatidylinositol glycan which is covalently linked to the newly exposed C-terminal Asp-484 following the processing of the C-terminal signal peptide (29,30).

Prepro-HPAP mutants with 13 or fewer stretches of C-terminal hydrophobic residues yield hydrophilic proteins that are no longer membrane bound but are efficiently secreted into the culture medium (32). Similarly, truncation of the C-terminal 24 amino acids results in a form of APase (64 kDa) that is efficiently secreted into the medium either from mammalian COS cells (10) or from insect cells that express abundant amounts of APase under the control of the baculovirus polyhedrin gene promoter (36). The truncated or "secreted" APase apparently requires glycosylation for secretion and enzymatic activity.

HPAP has a remarkable thermostability: no loss of activity occurs at 65°C for an hour or more. The activities of other isoenzymes are destroyed under these conditions.

Highly purified APase from human placenta has a specific activity of 620 U/mg at 30°C (37).

The placental-like (or GCAP) isozyme, which differs from HPAP by only 7 amino acids within their respective 484 residues (27), is more strongly inhibited by L-Leu, EDTA, and heat. Site-specific substitution of Gly-429 of GCAP by Glu (as the Glu-429 of HPAP) results in conversion of the phenotype of GCAP to that of HPAP (38).

b. Human intestinal alkaline phosphatase (HIAP)

Like HPAP, the HIAP gene has 11 exons (33). Multiple HIAP mRNAs (the major one being 2.6 kb in size) can be detected in the RNAs isolated from adult and fetal intestine and from the cells of that lineage, most likely due to the differential use of at least three of the four polyadenylation signals. HIAP is much less thermostable than HPAP, although it is much more stable than L/B/K APase at 56°C.

c. Liver/bone/kidney alkaline phosphatase

The gene for L/B/K APase apparently exists as a single copy on chromosome 1 in the haploid genome. However, its tissue-specific expression gives rise to liver-, bone-, and kidney-specific APase isozymes due to different posttranslational modifications. The L/B/K APase gene comprises 12 exons which are distributed over more than 50 kb (34). Mainly due to the size of introns, the L/B/K APase gene is at least five times larger than the HPAP and HIAP genes. Intron–exon junctions occur at analogous positions in all three genes, but there is an extra noncoding exon at the 5' end of the L/B/K APase gene. L/B/K APase is the most thermolabile of the four human APase isozymes.

References

1. Chappelet-Tordo, D., Fosset, M., Iwatsubo, M., Gache, C., and Lazdunski, M. (1974). *Biochem.* **13**, 1788–1795.
2. Fernley, H. N., and Walker, P. G. (1967). *BJ* **104**, 1011–1018.
3. Fosset, M., Chappelet-Tordo, D., and Lazdunski, M. (1974). *Biochem.* **13**, 1783–1788.
4. Moessner, E., Boll, M., and Pfleiderer, G. (1980). *Z. Physiol. Chem.* **361**, 543–549.
5. Prussak, C. E., and Tseng, B. Y. (1989). *BBRC* **159**, 1397–1403.
6. Martin, C. J., and Evans, W. J. (1989). *Res. Comm. Chem. Pathol. Pharmacol.* **65**, 289–296.
7. Morton, R. K. (1955). *BJ* **61**, 232–240.
8. Fernley, H. N., and Walker, P. G. (1967). *BJ* **104**, 1011–1018.
9. Neumann, H. (1968). *JBC* **243**, 4671–4676.
10. Berger, J., Hauber, J., Hauber, R., Geiger, R., and Cullen, B. R. (1988). *Gene* **66**, 1–10.
11. Yoon, K., Thiede, M. A., and Rodan, G. A. (1988). *Gene* **66**, 11–17.
12. Hua, J.-C., Berger, J., Pan, Y.-C. E., Hulmes, J. D., and Udenfriend, S. (1986). *PNAS* **83**, 2368–2372.
13. Millan, J. L. (1993). PCT Application: WO 93/18139.
14. Hoffmann-Blume, E., Garcia Marenco, M. B., Ehle, H., Bublitz, R., Schulze, M., and Horn, A. (1991). *EJB* **199**, 305–312.
15. Coleman, J. E., and Besman, M. J. A. (1987). In "New Comprehensive Biochemistry" (A. Neuberger and K. Brocklehurst, Eds.), Vol. 16, pp. 377–406. Elsevier, Amsterdam.
16. Murphy, J. E., Xu, X., and Kantrowitz, E. R. (1993). *JBC* **268**, 21497–21500.
17. Engstrom, L. (1964). *BBA* **92**, 79–84.
18. Caras, I. W., and Weddell, G. N. (1989). *Science* **243**, 1196–1198.
19. Low, M. G. (1987). *BJ* **244**, 1–13.
20. Yusufi, A. N. K., Low, M. G., Turner, S. T., and Dousa, T. P. (1983). *JBC* **258**, 5695–5701.
21. Lindner, N. M., Jeffcoat, R., and Lowe, C. R. (1989). *J. Chromatography* **473**, 227–240.
22. Fishman, W. H. (1990). *Clin. Biochem.* **23**, 99–104.
23. Sumikawa, K., Okochi, T., and Adachi, K. (1990). *BBA* **1046**, 27–31.
24. Lewis, W. H., Jr., and Rutan, S. C. (1991). *Anal. Chem.* **63**, 627–629.
25. Forsman, R. W., and O'Brien, J. F. (1991). *Clin. Chem.* **37**, 347–350.
26. Herz, F., and Halwer, M. (1989). *BBA* **1013**, 259–265.
27. Knoll, B. J., Rothblum, K. N., and Longley, M. (1987). *Gene* **60**, 267–276.
28. Knoll, B. J., Rothblum, K. N., and Longley, M. (1988). *JBC* **263**, 12020–12027.
29. Micanovic, R., Gerber, L. D., Berger, J., Kodukula, K., and Udenfriend, S. (1990). *PNAS* **87**, 157–161.

30. Ogata, S., Hayashi, Y., Takami, N., and Ikehara, Y. (1988). *JBC* **263**, 10489–10494.
31. Kam, W., Clauser, E., Kim, Y. S., Kan, Y. W., and Rutter, W. (1985). *PNAS* **82**, 8715–8719.
32. Berger, J., Howard, A. D., Brink, L., Gerber, L., Hauber, J., Cullen, B. R., and Udenfriend, S. (1988). *JBC* **263**, 10016–10021.
33. Henthorn, P. S., Raducha, M., Kadesch, T., Weiss, M. J., and Harris, H. (1988). *JBC* **263**, 12011–12019.
34. Weiss, M. J., Ray, K., Henthorn, P. S., Lamb, B., Kadesch, T., and Harris, H. (1988). *JBC* **263**, 12002–12010.
35. Millan, J. L. (1986). *JBC* **261**, 3112–3115.
36. Davis, T. R., Trotter, K. M., Granados, R. R., and Wood, H. A. (1992). *Bio/Technology* **10**, 1148–1150.
37. Chang, T.-C., Huang, S.-M., Huang, T.-M., and Chang, G.-G. (1992). *EJB* **209**, 241–247.
38. Watanabe, T., Wada, N., Kim, E. E., Wyckoff, H. W., and Chou, J. Y. (1991). *JBC* **266**, 21174–21178.

II. POLYNUCLEOTIDE KINASE: T4 POLYNUCLEOTIDE KINASE
[Polynucleotide 5′-hydroxyl-kinase, EC 2.7.1.78]

Kinases are an extremely diverse group of enzymes that catalyze the transfer of the γ-phosphoryl group from NTP to an acceptor molecule. Depending on the nature of the phosphate acceptor, kinases are largely divided into protein kinases, carbohydrate kinases, and polynucleotide kinases. Kinases are ubiquitously present in nature and in various tissues of an organism. They play essential roles in cellular metabolism, regulation, and replication.

This section focuses on polynucleotide kinase (PNK), especially from phage T4. The T4 PNK is perhaps the best characterized among all known PNKs from various sources such as T-even phages, rat liver, and calf thymus (1). No PNK with an explicit 5′-kinase activity has been found in bacteria.

T4 PNK, a tetrameric 140-kDa protein, is a multifunctional enzyme. It catalyzes as the principal activity the transfer of the terminal phosphate group of ATP to the 5′-OH groups of DNA, RNA, and oligonucleotides (Scheme 5.3). The PNK-catalyzed reaction is fully reversible and thus, under appropriate conditions, PNK catalyzes 5′-P dephosphorylation and exchange reactions as well. In addition, T4 PNK has a distinct 3′-phosphatase activity (see Scheme 5.3).

T4 PNK is widely used as a reagent to phosphorylate and/or label 5′-OH termini of nucleic acids and assumes an important role in recombinant DNA technology.

1. FUNCTIONS

a. Reaction conditions

i. Optimal reaction conditions. T4 PNK shows different pH optima depending on the substrates and also on the direction of the reaction. The pH optimum for the forward reaction (phosphorylation) is ~8.0 for DNA substrates and 9.5 for oligonucleotide substrates (2), while that for the reverse reaction is ~6.5 (3). The optimum temperature for kinase reaction is 30–35°C, yet the reaction is usually carried out at 37°C.

5'-kinase   5' HO–ApCpGpT 3' ⇌ p–ApCpGpT
            + ATP (or NTP)   + ADP (or NDP)

5'-P exchange   p–ApCpGpT + ADP ⇌ HO–ApCpGpT + ATP
                HO–ApCpGpT + [γ-^{32}P]ATP ⇌ p*–ApCpGpT + ADP
                ----
                p–ApCpGpT + ADP + ATP* ⇌ p–ApCpGpT + ATP + ADP

3'-phosphatase   pApCpGpT–p ⟶ pApCpGpT + P$_i$

SCHEME 5.3

The phosphorylation (or forward) reaction is typically carried out in a 30- to 50-μl volume containing 50 mM Tris–Cl buffer (pH 8.0), 10 mM MgCl$_2$, 5 mM DTT, 0.1 mM ATP, 50 μg/ml BSA, and 10–20 U PNK per μg DNA. The mixture is incubated at 37°C for 45 min. [The optimal reaction conditions for dephosphorylation and 5'-P exchange reactions are described in their respective sections of *substrate specificities* (see below).]

The T4 PNK reaction can be terminated in several ways, i.e., by adding EDTA (to 20 mM) or acid (0.4 M TCA or 20 mM PP$_i$) or by heating (78°C for 1 min or 65°C for 5 min).

ii. Kinetic parameters. The $K_{m,app.}$ for ATP at pH 8.0 ranges from 13 μM for dT(pT)$_9$ to 260 μM for phage T7 DNA (4,5). The true K_m for ATP with 3'-TMP as the acceptor is 40 μM. The V values relative to that of (dT)$_5$ are in the order of calf thymus DNA [16.7] > 3'-dTMP [10.5] > (dT)$_9$ [2.8] > (dT)$_{15}$ [2.5] > (dT)$_5$ [1.0] > *E. coli* tRNA [0.5]. The K_m and k_{cat} for calf thymus DNA at pH 8.0 are 18 μM and 25,000 min^{-1}, respectively. The K_m for ADP is 22 μM.

iii. Cofactor requirements and ionic strengths

(a) Divalent cations. T4 PNK requires Mg^{2+} for activity. In fact, the enzyme accepts only MgATP as the phosphate donor substrate, whereas free Mg^{2+} (up to 50 mM) or free ATP has no effect on the kinase reaction (6). The phosphorylation reaction is usually performed at 10 mM MgCl$_2$ (2). Mn^{2+} can replace Mg^{2+} but with ~50% efficiency (7).

(b) Monovalent cations. The kinase activity on single-stranded substrates increases with increasing concentrations of salts such as NaCl, KCl, and CsCl (8). LiCl and NH$_4$Cl also give similar levels of stimulation. The maximal stimulation (~5-fold) of phosphorylation activity is obtained at ~0.125 M, which corresponds to a total ionic strength of 0.19.

With substrates having 5'-OH at recessed ends or nicks, the addition of KCl results in a drastic decrease in the rate of phosphorylation, apparently due to the stabilization of the bihelical structure and the consequent reduction of fraying at the 5'-OH (5).

(c) Anions. Anions such as Cl^- and Br^- are highly stimulatory (~6-fold) while F^-, NO_3^-, HCO_3^-, and SO_4^{2-} are less so (3- to 4-fold) at the same concentration of 0.125 M and with Na^+ as a counterion in the presence of 9 mM $MgCl_2$ (8).

iv. Activators

(a) Polyamines. Spermine and spermidine increase the phosphorylation activity, with a maximum stimulation of threefold at a ~2 mM concentration (8). The stimulatory effect does not increase further with the addition of salt. Both salts and spermine promote the stabilization of the enzyme in the more active tetrameric form. Spermine has little influence on the kinase activity on 5'-OH at recessed ends or nicks.

(b) Polyethylene glycol. PEG (8000, 4–10%) has been shown to improve significantly (100- to 1000-fold) the efficiency of phosphorylation either in the forward reaction or in the exchange reaction (9). The stimulatory effect is general regardless of the type of termini. PEG is thus particularly useful for the exchange reaction with DNA substrates having unfavorable types of termini. The phosphorylation of DNA substrates less than 300 bp in length is less affected by the presence of PEG. The stimulatory effect of PEG is attributed partly to alteration of the DNA structure to a highly condensed state (10) and partly to macromolecular crowding. PEG also serves to stabilize the active form of the kinase.

v. Inhibitors and/or inactivators

(a) Phosphates. P_i and PP_i are competitive inhibitors for DNA substrates with K_i values of 29 and 2 mM, respectively (8). The $K_{d,app.}$ of P_i measured by fluorescence quenching is 1.4 mM (11).

(b) Sulfates. Sulfate-containing polymers, e.g., agar, dextran sulfate, and heparin, are powerful inhibitors (e.g., K_i for dextran sulfate is 0.05 μg/ml) (12). Nonsulfate polysaccharides have no effect on the reaction. The addition of cationic compounds such as polylysine and spermine counteracts the inhibitory effect. Compared with dextran sulfate, inorganic sulfate (Na_2SO_4) is much less inhibitory with a K_i (competitive against ATP) of ~4 mM, which is in the same range as that for P_i.

(c) Sulfhydryl reagents. T4 PNK is rapidly inactivated when SH protective reagents, e.g., DTT or 2-MSH, are not present. DTT (5 mM) is the most effective, whereas 2-MSH (10 mM) and glutathione (10 mM) have 80 and 70% of the efficacy of DTT, respectively (7).

b. Activity assay and unit definition

PNK activity is most commonly assayed using staphylococcal nuclease-treated DNA (5'-OH and 3'-P) as a phosphate acceptor and [γ-^{32}P]ATP as donor (7). Reactions are typically carried out at 37°C in 50 mM Tris–Cl (pH 7.6), 10 mM MgCl$_2$, 0.1 mM [γ-^{32}P]ATP, 5 mM DTT, and 0.2 mg/ml calf thymus DNA. After incubation, DNA is precipitated by TCA onto glass filters, exhaustively washed, and counted for radioactivity incorporation.

Unit definition: One kinase unit is usually defined as that amount of enzyme which catalyzes the transfer of 1 nmol of P$_i$ in 30 min at 37°C.

c. Substrate specificities: Phosphorylation

The kinase reaction catalyzed by T4 PNK is reversible. For the single-stranded dT(pT)$_9$ as the substrate at pH 8.0, the equilibrium has been estimated to be 50-fold in favor of the forward reaction. For practical purposes, therefore, each direction of the PNK-catalyzed reactions will be described hereafter as if they are irreversible.

i. Donor substrates. As phosphate donors, CTP, GTP, and UTP are equally effective as ATP with a $K_{m,app.}$ of ~20 μM (6). dATP also functions as an effective donor substrate.

ii. Acceptor substrates. All 5'-OH of DNA, RNA, and oligonucleotides are substrates for phosphorylation (Scheme 5.3). The minimal substrate is 3'-NMP which is converted to 3',5'-biphosphate products. Dinucleotides without 3'-P are also efficient substrates. They include the dinucleotides that have bulky mutagenic/carcinogenic chemicals (e.g., benzopyrene) covalently linked to the 5'-terminal bases (13). Certain structural alterations of the 5'-terminal base, e.g., cyclobutane–pyrimidine dimer formation by UV irradiation, in di- and trinucleotides render the 5'-OH insensitive to PNK (14).

Nucleosides (adenosine) and 2'-NMP (AMP) are not substrates (7).

iii. Termini structure of DNA. Preferred substrates are single-stranded oligonucleotides or double-stranded oligonucleotides with protruding 5'-OH termini. With these substrates, a rapid and essentially complete ($\geq$95%) phosphorylation is achieved at an ATP concentration of 0.2–1 μM. This ATP concentration is in 2- to 5-fold excess over the concentration of 5'-OH termini.

Blunt, recessed, or internal (nick or gap) 5'-OH termini of dsDNA or dsRNA are poor substrates, exhibiting 10 times slower rates than protruding 5'-OH termini. The level of phosphorylation is also reduced to 20–45% of that of the preferred substrates. In the presence of 20 μM ATP (~100-fold excess over the concentration of 5'-OH termini), however, phosphorylation becomes almost complete after ~45 min for all types of DNA substrates (5).

iv. Preferences for terminal nucleotides. T4 PNK displays some degree of preference toward 5'-terminal nucleotides (4). At a constant ATP concentration of 66 μM, the $K_{m,app.}$ for homodecanucleotides varies 15-fold from 1.8 μM for dT(pT)$_9$ to 29.6 μM for dG(pG)$_9$, while the V changes ~7-fold.

With synthetic double-stranded oligodeoxynucleotides containing two internal 5'-OH termini at G and T in the presence of 0.125 M KCl, the terminus ending with T is phosphorylated ~5 times faster than the one with G (3). At low salt concentrations or in the presence of spermine, however, this preference is not observed.

With 3'-NMP as substrates, the $K_{m,app.}$ and V are 2–5 times larger than those for homooligodeoxynucleotides. The K_m values for 3'-NMPs are in the order of rGp > rUp > rCp > rAp > dTp and the V values in the order of rGp > rCp > rAp > dTp > rUp (4).

The 5' termini of some E. coli tRNAs (pretreated with APase) are phosphorylated with $K_{m,app.}$ and V about 5% of the values for calf thymus DNA (15).

v. Modified terminal nucleotides

(a) Methylated G residue. When the ribose ring is opened by alkali treatment, 7-methyl-3'-dGMP is resistant to the 3'-phosphatase activity of nuclease P1 but is a good substrate for the 5'-kinase activity of T4 PNK (16). The ring-opened product is actually phosphorylated by T4 PNK at a rate ~10 times slower than the intact 7-methyl-dGMP (17). The nuclease P1-catalyzed dephosphorylation reaction of the ring-opened alkylated products and 3'-dNMPs (the normal products of staphylococcal nuclease digestion of DNA) has an enriching effect on the alkylated products which are effective substrates for T4 PNK in the subsequent ^{32}P postlabeling. These differential reactivities of nucleotide substrates toward nuclease P1 and T4 PNK can be used to analyze the methylated guanines in DNA (also see "Applications").

(b) Apurinic sites. T4 PNK shows distinct responses toward dinucleotides containing apurinic sites. It effectively phosphorylates the 5'-OH of dApS, where S represents an apurinic deoxyribose group, but not the isomer dSpA (18). This distinction serves as the basis for a ^{32}P postlabeling assay for apurinic sites in DNA following combined digestions of the DNA by snake venom phosphodiesterase (which cleaves dSpA but not dApS) and by an APase.

(c) Nucleotides with modified sugars. Thymidine cis-glycol-3'-P, an oxidative DNA damage product, is an acceptable substrate for T4 PNK in the presence of 1 mM BeCl$_2$ (a mutagenic metal ion in a salt form) (19).

vi. Nonnucleotide substrates.
Polycyclic aromatic hydrocarbon tetrols, e.g., tetra-OH derivatives of benzopyrene (BP) and chrysene, are substrates for T4 PNK, whereas diols, phenols, triols, and parent hydrocarbons are not (20). Compared with BP-tetrols, the BP-(diol epoxide)–DNA adduct is at least 2000 times more efficiently phosphorylated to approximately a 95% level of the theoretical maximum. This difference in sensitivity has been exploited for ^{32}P postlabeling analysis of the DNA covalently modified by carcinogenic aromatic substances.

d. Substrate specificities: Dephosphorylation

i. 5'-Phosphatase activity: reversal of 5'-kinase reaction. In the presence of excess nucleotide acceptor (e.g., ADP), T4 PNK dephosphorylates the 5'-P group of nucleic acids (DNA, RNA, oligonucleotides, and 3',5'-biphosphates) to yield 5'-OH and ATP in a reversal of the 5'-kinase reaction (Scheme 5.3). Dephosphorylation of internally situated 5'-P ends occurs in the order of T > A > G. The 5'-dephosphorylation reaction per se is not of much practical utility. Under the optimized conditions of a reversible reaction, however, the 5'-dephosphorylation turns into a valuable step toward the 5'-P exchange reaction (see below).

The pH optimum for the dephosphorylation reaction is 6.2–6.5 (3,21). At pH 7.4, the K_m and V for ADP are 0.2 mM and 4.3 pmol/min/μg, respectively. Note that the K_m and V for ATP in the phosphorylation reaction at this pH are 4 μM and 31.3 pmol/min/μg, respectively.

ii. 3'-Phosphatase and phosphohydrolase activities. T4 PNK has a 3'-phosphatase activity that selectively hydrolyzes the 3'-P of DNA (Scheme 5.3) (22–24). The 3'-P of RNA is also hydrolyzed but far less efficiently. The 3'-phosphatase activity requires Mg^{2+} (optimum 8 mM) and is optimally active at pH 6.0. The presence of 0.3 M NaCl has been found to promote a more complete removal of 3'-P while virtually eliminating the loss of 5'-P which supposedly occurs due to the presence of a contaminating nonspecific phosphatase (25). Other conditions such as 0.2 M NaCl or 0.3 M NaAc are less effective, whereas 0.4 M NaCl slightly inhibits 3'-phosphatase activity.

The 3'-phosphatase activity is functionally distinct from the 5'-kinase–phosphatase activity, although the two activities reside on the same molecule. The 3'-phosphatase activity can be specifically removed by a point mutation in the gene for T4 PNK. The mutant enzyme has an almost intact T4 PNK activity but is free of 3'-phosphatase activity (26).

T4 PNK is also known to act as a phosphohydrolase and catalyzes the hydrolysis of 2',3'-cyclic phosphodiesters to 3'-phosphomonoesters and eventually to 3'-OH.

e. Substrate specificities: 5'-P exchange reaction

i. Reaction profile. In a manner characteristic of a truly reversible reaction, T4 PNK catalyzes the 5'-P exchange reaction (e.g., 5'-P → 5'-[^{32}P]) by concomitant dephosphorylation and (re)phosphorylation in the presence of a phosphate acceptor (ADP) and a donor ([γ-^{32}P]ATP) (3,21,27,28) (see Scheme 5.3). In reactions involving prior restriction enzyme digestions, this feature allows the single-step labeling of DNA by T4 PNK in a single tube (29). The 5'-P exchange is proportional to the number of termini and not to the amount or size of the DNA substrates. With ATP as a phosphate acceptor instead of ADP, [δ-^{32}P]adenosine tetraphosphate can be produced.

ii. Optimal reaction conditions. 5'-P exchange reactions can be performed with all types of termini, although T4 PNK shows a preference for certain substrates (see below). The recommended reaction conditions are 12 μM ATP and 100–200 μM ADP in a buffer composed of 50 mM imidazole hydrochloride (pH 6.6), 20 mM $MgCl_2$, 5 mM DTT, 50 μg/ml BSA, and ~0.2 U/μl T4 PNK (28). The reaction is usually complete after incubation at 37°C for 15–30 min.

The rate of 5′-P exchange reaction is fairly sensitive to the following factors.

(a) pH. The pH optimum for the exchange reaction is ~6.5, similar to that for dephosphorylation. The initial rate of the exchange reaction drops rapidly at pH > 7: at pH 7.6 (50 mM Tris–Cl), the enzyme has 20% of its activity at pH 6.2 (50 mM imidazole–HCl).

(b) Salt and ionic strength. The exchange reaction for all types of substrate termini is highly sensitive to ionic strength. For example, the extent of exchange drops to 50% at a KCl concentration of 35–60 mM. Spermine has no significant effect on the exchange reaction.

(c) Macromolecules. PEG 8000 at ~6% concentration results in a remarkable increase of the extent of exchange phosphorylation (9). Note that the exact optimum concentration of PEG may vary depending on particular DNA substrates.

iii. Substrate preferences. The 5′-P of ssDNA or dsDNA with external termini are exchanged more rapidly and to a greater extent (50–70%) when compared with the exchange (15–20%) at the 5′-P of dsDNA with blunt or internal 5′-termini (30). T4 PNK shows a slight but measurable preference for DNA with termini ending in T to those ending in A or C (5). Termini with 5′-BrdU in place of T are labeled at equal rates when both are external termini but are labeled twice as fast as the T termini when the T termini are internal or blunt ended.

Some RNA species with 5′-P, e.g., in $tRNA^{Val}$ and $tRNA^{Phe}$, readily undergo ADP-mediated exchange reactions. However, the 5′-P of *E. coli* $tRNA^{His}$ has been shown to be resistant to 5′-P exchange. Prior dephosphorylation with APase renders it susceptible to 5′-kination by T4 PNK (31).

f. Mechanism of PNK action

i. Kinetic mechanism. The kinetics of the T4 PNK reaction involve two substrates (ATP, HO-DNA) and two products (ADP, P-DNA). ADP is a competitive inhibitor against ATP and an apparent noncompetitive inhibitor against HO-DNA (6). P-DNA is a competitive inhibitor against HO-DNA and a noncompetitive inhibitor against ATP. The enzyme can also bind one substrate in the absence of the other and can be protected against heat inactivation. These characteristics are consistent with the random Bi–Bi kinetic mechanism, most likely involving some dead-end complexes such as E(ATP · P-DNA) and E(ADP · HO-DNA) (refer to Section I,C, Chapter 1). Note that the kinetic data obtained from a slightly different set of reaction conditions have been interpreted as fitting to an ordered sequential Bi–Bi mechanism whereby the enzyme binds HO-DNA before binding ATP (1,4). Activators such as KCl and spermine which promote the formation of enzyme oligomers have been suggested to be responsible for altering the reaction mechanism from the ordered sequential type to the equilibrium random mechanism (1).

ii. Substrate preferences. Compared with ssDNA or dsDNA with protruding 5'-OH, dsDNAs with 5'-OH groups at recessed ends, nicks, or gaps are relatively poor substrates and require substantially higher concentrations of ATP to achieve a comparable rate and extent of phosphorylation. Presumably the DNAs with unfavorable termini structure bind the enzyme, partially blocking the ATP-binding site. The blocking appears to be relieved at higher ATP concentrations.

2. APPLICATIONS

The reversible nature of the T4 PNK-catalyzed reaction and the distinct 3'-phosphatase activity render the PNK extremely useful in a wide variety of applications. The reaction conditions can be readily adjusted for respective 5'-phosphorylation, dephosphorylation, and phosphate exchange reactions.

i. Phosphorylation or radiolabeling of 5'-OH of DNA and RNA. The 5'-labeling of nucleic acids by ^{32}P or ^{33}P using T4 PNK is often a key step in a number of important techniques in molecular biology, including chemical, enzymatic, and dideoxynucleotide sequencing of DNA and RNA (see Appendix A: "Nucleotide sequencing methods"), hybridization, and other general DNA manipulations involved in gene cloning and restriction mapping. The 5'-labeling is also an essential technique used in the analysis of nucleic acid structures by fingerprinting and neighboring nucleotide analysis.

The 5'-phosphorylation is also essential for DNA ligation catalyzed by DNA ligases (see Section I, Chapter 2). Therefore, synthetic oligonucleotide linkers, adapters, or more generally DNA fragments that lack 5'-P have to be phosphorylated by the use of T4 PNK. Insertion of ^{32}P or ^{33}P helps to monitor the efficiency of ligation using trace amounts of the ligated products.

In many 5'-end labeling experiments, the substrate DNA (or RNA) may already carry a 5'-P group. In these cases, the 5'-end can be labeled either by the phosphate exchange reaction with [γ-^{32}P]ATP or by the APase-catalyzed dephosphorylation of the 5'-P group followed by rephosphorylation using [^{32}P]ATP. The phosphate exchange reaction is convenient, but the labeling efficiency may be relatively low unless higher concentrations of the expensive radionucleotide are used. The alternative method circumvents the disadvantage of the exchange reaction but instead requires two separate enzymatic steps, i.e., dephosphorylation of 5'-P using APase and then 5'-kination using T4 PNK. One method streamlining the two successive enzymatic steps is to employ inorganic P_i (1–5 mM depending on the 5' termini type) to inhibit the APase (e.g., BAP) while leaving the PNK activity sufficiently intact to perform the kinase reaction with a labeling efficiency of 30–50% (32). More generally, however, dephosphorylation can be stopped by heat inactivation of the APase (e.g., CIAP), and the labeling reaction can be initiated in the same tube by adding fresh PNK and [^{32}P]ATP or [^{33}P]ATP.

Note that the 5′-end labeling of most eukaryotic mRNAs requires prior removal of the "cap" structure. T4 PNK is reported to have the capacity to "decap," generating the m⁷G(5′)pp and an mRNA with a 5′-P terminus (33). Alternatively, the cap can be removed by treatment with tobacco acid pyrophosphatase (34).

ii. Detection and/or assay of endonuclease activity. The T4 PNK-catalyzed 5′-P exchange with ^{32}P in DNA fragments is a sensitive and reproducible (within a given type of termini) method to assay endonuclease activity (21,28).

iii. ^{32}P Postlabeling assays. These assays permit detection at the femtomole level of base and sugar lesions of DNA caused by various exogenous factors, including radiations (X-ray, UV), alkylation, and other chemical additions. The assay is based on the ability of T4 PNK to phosphorylate the modified nucleotides or nucleosides with significantly different efficiencies (16,18,35,36). In combination with other enzymes, such as an endonuclease(s) and phosphatases, which display differential sensitivities to modified bases, the T4 PNK-catalyzed kination amplifies the concentration of target nucleotides and/or provides a detection signal for the modified nucleotides. The labeling by PNK is optimal at pH ~8.

iv. Use of 3′-phosphatase activity. T4 PNK can be used as a specific 3′-phosphatase (37). It has been employed in the studies of DNA damage and repair to determine the presence of 3′-P termini on 5′-^{32}P-end-labeled DNA fragments. Removal of the 3′-P is accompanied by a decrease in mobility of such fragments on sequencing gels, equivalent to the 0.5–1 nucleotide position.

v. Special applications of T4 PNK lacking 3′-phosphatase activity. This mutant enzyme is uniquely suitable for the preparation from 3′-NMP of 3′,5′-biphosphates which can be used for synthesis and labeling of DNA and RNA (38). The 5′-[^{32}P]pCp is a reagent commonly used for the 3′-end labeling of RNA with the T4 RNA ligase (39). The 3′-labeled RNA can then be used for RNA sequencing and hybridization, as well as for the analysis of RNA structure by fingerprinting.

3. STRUCTURES

i. General structure. T4 PNK is a tetrameric enzyme (140 kDa) composed of identical subunits (34 kDa) (11,40). Monomer and dimer are not enzymatically active. Several oligomeric species (monomers, dimers, tetramers, and perhaps higher-ordered oligomers) are in equilibrium under any given condition. A high ionic strength (e.g., 0.1 M KCl), addition of substrates (ATP, 3′-dTMP), or various ligands (1 mM spermine) shift the equilibrium toward the tetrameric form ($s^{\circ}_{20,w} = 6.2$ S) (11).

ii. Primary structure. T4 PNK consists of 301 deduced amino acid residues per 34-kDa monomer (41). The C-terminal residue is Phe. The N-terminal residue has been previously identified to be Phe which, in the deduced sequence, occurs at codon 22 (11). The putative "pro" sequence contains the presumed adenine-binding motif, Gly^{12}-Ser-Gly-Lys-Ser^{16}. A similar purine nucleotide-binding sequence motif has been found in other enzymes such as adenylate kinase (42), yeast T4 RNA ligase, and mammalian myelin 2',3'-cyclic nucleotide phosphohydrolase (43).

The deduced amino acid sequence of T4 PNK contains four Cys per monomer (41). T4 PNK absolutely requires reduced SH groups for activity. Note that only two SH groups could be titrated, one much faster than the other, by DTNB even in the presence of 6 M urea (11). The other two Cys residues are presumably located in the hydrophobic inner core of the molecule and are inaccessible to DTNB even in the presence of 6 M urea. The modification of SH groups is not affected by the presence of substrates such as 3'-TMP and ATP.

T4 PNK has six Trp residues per monomer, one or several of which are exposed to the solvent. On excitation at 280 nm, the enzyme fluoresces with an emission maximum at 340 nm (11). Substrates quench this fluorescence, which can be used to measure ligand-binding constants.

iii. **Secondary structure.** According to CD measurements (11), the secondary structure of T4 PNK contains 45–55% α helix and ~25% β sheet. The addition of polyamines significantly decreases the ellipticity at 222 nm.

iv. **Tertiary structure.** T4 PNK contains functionally separate domains. Genetic and biochemical evidence has shown that 5'-kinase and 3'-phosphatase reside on a single polypeptide and yet can function almost independently of each other. The 5'-kinase activity lies in the N-terminal domain, whereas the 3'-phosphatase activity is located in the C-terminal portion (44). Partial tryptic digestion produces a 29-kDa fragment with no kinase activity and nearly normal 3'-phosphatase activity.

4. GENETICS

i. **Primary structure and gene organization.** T4 PNK is encoded by the *pseT* (or *pnk*) gene, one of the early genes of phage T4 (17,45). The *pseT* gene is located close to gene 63, the structural gene for T4 RNA ligase (46), and between the thymidylate synthase and DNA ligase genes. It is contained within a 2.27-kb *Eco*RI fragment of the T4 genome.

The *pseT* gene has been cloned in a λ insertion vector (NM607) and its nucleotide sequence determined (41). The ORF comprises 906 nt (or 301 amino acid codons) including the UAA termination codon. Some possible promoter sequences corresponding to the typical -10 and -35 regions are present, but the -10 region immediately upstream of the initiator codon lacks a -35 region. Like many ORFs of phage λ, however, the initiator codon is preceded by a pronounced Pu-rich region.

ii. ***In vivo* function.** The *pseT* gene product is not essential because *pseT* deletion mutant phages grow well in most *E. coli* strains. Apparently a host gene product can substitute for the T4 *pseT* gene product *in vivo*. Most mutations in *pseT* inactivate both the 5'-kinase and 3'-phosphatase activities. However, a point mutation, *pseT1*, inactivates only the 3'-phosphatase activity (24,45), whereas a mutant *pseT47* lacks only the 5'-kinase activity (24). Both activities seem to be necessary for probable physiological functions such as mRNA processing and the repair of apurinic/apyrimidinic DNA (37) and of tRNA(s) cleaved by anticodon nuclease (47). Neither the *pseT1* nor the *pseT47* mutant is able to grow on *E. coli* CTr5X. Furthermore, a mixed infection of *pseT1* and *pseT47* mutants does not promote a phage growth in *E. coli* CTr5X, even though both kinase and phosphatase activities are present in the infected cells. These observations suggest that both activities must be present on the same molecule for proper biological functions.

5. Sources

T4 PNK has been traditionally purified from *E. coli* B (suppressor-minus) cells infected with phage T4 (40,48). A maximal level of PNK induction is achieved ~20 min after infection (2). Alternatively, amber mutants of the T4 phage such as *am*N82 which is defective in cell lysis, *am*XF1 which is defective in genes 41–45 (DNA polymerase and deoxycytidylate hydroxymethylase), and *am*H39X which is defective in gene 30 (DNA ligase) can be used to prepare the PNK. Mutant phages offer certain advantages in that either no premature cell lysis occurs or some of the contaminating enzymes are absent.

T4 PNK can now be produced by overexpressing the T4 PNK gene cloned in a λ (cI857) vector (41). Temperature induction and expression of the gene under the control of the λ late promoter P_R lead to >100-fold amplification of the enzyme, amounting to ~7% of the total cell protein. An even higher yield of T4 PNK activity has been obtained by expressing the *pseT* gene cloned in a multicopy plasmid (pT5) under the control of the *trp* promoter (49).

When stored in 50 mM Tris-Cl (pH 7.6), 0.1 M KCl, 2 mM DTT, 10 μM ATP, and 50% glycerol at −20°C, T4 PNK is stable for at least 2 years.

The 3'-*phosphatase-free T4 PNK* has also been prepared from mutant phage T4 *am*N81 *pseT1*-infected *E. coli* BB cells (23,26,50). The mutant enzyme has almost intact 5'-kinase activity but none of the 3'-phosphatase activity. The 3'-phosphatase-free T4 PNK is commercially available from BMB.

References

1. Kleppe, K., and Lillehaug, J. R. (1979). *Adv. Enzy.* **48**, 245–275.
2. Richardson, C. C. (1965). *PNAS* **54**, 158–165.
3. van de Sande, J. H., Kleppe, K., and Khorana, H. G. (1973). *Biochem.* **12**, 5050–5055.
4. Lillehaug, J. R., and Kleppe, K. (1975). *Biochem.* **14**, 1221–1225.
5. Lillehaug, J. R., Kleppe, R. K., and Kleppe, K. (1976). *Biochem.* **15**, 1858–1865.
6. Sano, H. (1976). *BBA* **422**, 109–119.
7. Richardson, C. C. (1971). *In* "Procedures in Nucleic Acid Research" (G. L. Cantoni, and D. R. Davies, Eds.), Vol. 2, pp. 815–828. Harper and Row, New York.
8. Lillehaug, J. R., and Kleppe, K. (1975). *Biochem.* **14**, 1225–1229.

9. Harrison, B., and Zimmerman, S. B. (1986). *Anal. Biochem.* **158,** 307–315.
10. Lerman, L. S. (1971). *PNAS* **68,** 1886–1890.
11. Lillehaug, J. R. (1977). *EJB* **73,** 499–506.
12. Wu, R. (1971). *BBRC* **43,** 927–934.
13. Randerath, K., Randerath, E., Danna, T. F., van Golen, K. L., and Putman, K. L. (1989). *Carcinogenesis* **10,** 1231–1239.
14. Weinfeld, M., Liuzzi, M., and Paterson, M. C. (1989). *JBC* **264,** 6364–6370.
15. Lillehaug, J. R., and Kleppe, K. (1977). *NAR* **4,** 373–380.
16. Hemminki, K., and Mustonen, R. (1990). *Teratogenesis Carcinog. Mutagen.* **10,** 223–230.
17. Hemminki, K., Peltonen, K., and Mustonen, R. (1990). *Chem. Biol. Interact.* **74,** 45–54.
18. Weinfeld, M., Liuzzi, M., and Paterson, M. C. (1990). *Biochem.* **29,** 1737–1743.
19. Hegi, M. E., Sagelsdorff, P., and Lutz, W. K. (1989). *Carcinogenesis* **10,** 43–47.
20. Masento, M. S., Hewer, A., Grover, P. L., and Phillips, D. H. (1989). *Carcinogenesis* **10,** 1557–1559.
21. Berkner, K. L., and Folk, W. R. (1977). *JBC* **252,** 3176–3184.
22. Becker, A., and Hurwitz, J. (1967). *JBC* **242,** 936–950.
23. Cameron, V., and Uhlenbeck, O. C. (1977). *Biochem.* **16,** 5120–5126.
24. Sirotkin, K., Cooley, W., Runnels, J., and Snyder, L. R. (1978). *JMB* **123,** 221–233.
25. Povirk, L. F., and Steighner, R. J. (1990). *BioTechniques* **9,** 562.
26. Cameron, V., Soltis, D., and Uhlenbeck, O. C. (1978). *NAR* **5,** 825–833.
27. Chaconas, G., van de Sande, J. H., and Church, R. B. (1975). *BBRC* **66,** 962–969.
28. Berkner, K. L., and Folk, W. R. (1979). *JBC* **254,** 2561–2564.
29. Oommen, A., Ferrandis, I., and Wang, M. J. (1990). *BioTechniques* **8,** 482, 484, 486.
30. Berkner, K. L., and Folk, W. R. (1980). *Meth. Enzy.* **65,** 28–36.
31. Allen, J. D., and Parsons, S. M. (1977). *BBRC* **78,** 28–35.
32. Chaconas, G., and van de Sande, J. H. (1980). *Meth. Enzy.* **65,** 75–85.
33. Abraham, K. A., and Lillehaug, J. R. (1976). *FEBS Lett.* **71,** 49–52.
34. Efstratiadis, A., Vornakis, J. N., Donis-Keller, H., Chaconas, G., Dougall, D. K., and Kafatos, F. C. (1977). *NAR* **4,** 4165–4174.
35. Randerath, K., Reddy, M. V., and Gupta, R. C. (1981). *PNAS* **78,** 6126–6129.
36. Weinfeld, M., and Soderlind, K.-J. M. (1991). *Biochem.* **30,** 1091–1097.
37. Bailly, V., Derydt, M., and Verly, W. G. (1989). *BJ* **261,** 707–713.
38. Ohtsuka, E., and 13 coauthors (1981). *PNAS* **78,** 5493–5497.
39. England, T. E., Bruce, A. G., and Uhlenbeck, O. C. (1980). *Meth. Enzy.* **65,** 65–74.
40. Panet, A., van de Sande, J. H., Loewen, P. C., Khorana, H. G., Raae, A. J., Lillehaug, J. R., and Kleppe, K. (1973). *Biochem.* **12,** 5045–5050.
41. Midgley, C. A., and Murray, N. E. (1985). *EMBO J.* **4,** 2695–2703.
42. Pai, E. F., Sachsenheimer, W., Schirmer, R. H., and Schulz, G. E. (1977). *JMB* **114,** 37–45.
43. Koonin, E. V., and Gorbalenya, A. E. (1990). *FEBS Lett.* **268,** 231–234.
44. Soltis, D. A., and Uhlenbeck, O. C. (1982). *JBC* **257,** 11340–11345.
45. Depew, R. E., and Cozzarelli, N. R. (1974). *J. Virol.* **13,** 888–897.
46. Snopek, T. J., Wood, W. B., Conley, M. P., Chen, P., and Cozzarelli, N. R. (1977). *PNAS* **74,** 3355–3359.
47. Levitz, R., Chapman, D., Amitsur, M., Green, R., Snyder, L., and Kaufmann, G. (1990). *EMBO J.* **9,** 1383–1389.
48. Richardson, C. C. (1981). "The Enzymes," 3rd ed., Vol. 14, pp. 299–314.
49. Campos, M., Ortega, M., Padron, G., Estrada, M. P., de la Fuente, J., and Herrera, L. (1991). *Gene* **101,** 127–131.
50. Soltis, D. A., and Uhlenbeck, O. C. (1982). *JBC* **257,** 11332–11339.

6

DNA Polymerases

I. GENERAL DESCRIPTION

i. **Primary functions of DNA polymerases.** DNA polymerases are a group of polymerases that catalyze the synthesis of polydeoxyribonucleotides from monodeoxyribonucleoside triphosphates (dNTPs), performing the most fundamental functions *in vivo* of DNA replication, repair, and, in some cases, cell differentiation. In fact, different types of DNA polymerases have been found in a single organism, for example, three (DNA Pol I, II, and III) in *E. coli* or five (DNA Pol α, β, γ, δ, and ε) in higher eukaryotes, which are believed to perform a specialized *in vivo* function(s). With somewhat different complexities from *in vivo* functions, DNA polymerases can, when given suitable conditions, also perform DNA synthesis *in vitro*. They require, in addition to dNTPs, an initiating oligonucleotide (or polynucleotide), called a *primer*, carrying a 3'-end hydroxyl group that can be used as the starting point of chain growth (1,2). DNA polymerases cannot initiate synthesis *de novo* from mononucleotides.

A primer can be a short or long piece of DNA or RNA which carries a free 3'-OH group. Primers provide a double-stranded structure to the DNA polymerase by annealing to a complementary region of the DNA or RNA strand called a *template*. The DNA polymerase moves along the DNA (or RNA) template, extending the primer in the 5' → 3' direction according to the Watson–Crick base pairing rule, i.e., A pairs with T (or U) and C pairs with G (see Section II, Chapter 1). The polarity of the newly synthesized chain is opposite (or *antiparallel*) to that of the template. Incorporation of a noncomplementary nucleotide is considered an "error." The *error frequency* (or *fidelity*) is an important characteristic of a polymerase (see below).

In addition to the major 5' → 3'-polymerase activity, a DNA polymerase may exhibit several other activities, such as 5'-nuclease, 3' → 5'-exonuclease, and/or RNase H activities, which are necessary for proper *in vivo* functions.

The initiation of cellular DNA replication takes place at a single site (e.g., *oriC* of *E. coli*) or multiple specific sites (in higher eukaryotes) of DNA called *origins of replication* (*ori*). The temporal site of dsDNA where the replication occurs is called a "replication fork." Because of DNA strand polarity, the bidirectional replication results in two distinct products, "leading" and "lagging" strands, according to the moving direction of the replication fork. The leading strand is synthesized as a single continuous chain, whereas the lagging strand is initially synthesized as small oligonucleotides, called *Okazaki fragments*, which are then ligated to form a continuous chain. Small RNAs play an important role as natural primers in the synthesis of both the leading strand and, in particular, the lagging strand.

Replication of duplex, RF DNA of single-stranded bacteriophages and of plasmids is initiated at a site of single-strand scission, called a *nick*, which is site specifically introduced by an endonuclease. The extension of a primer (or positive) strand on the template (minus) strand proceeds by a *rolling circle* mechanism in which a "daughter" strand longer than one full length of the template circle is produced and then processed.

ii. **Functional concepts of polymerase: Processivity and fidelity.** The functional roles of DNA polymerases in DNA replication, synthesis, and repair are based on a number of important concepts that define the enzymatic properties of the polymerases (1). Two such concepts pertinent to the polymerases catalyzing templated DNA synthesis are *processivity* and *fidelity*. Processivity and fidelity of a DNA polymerase are not only enhanced by the coordinate actions of polymerase and polymerase-associated activities but are also modulated by the interactions with other proteins, called *accessory proteins*, which form a *replication complex* with the DNA polymerase.

(a) ***Processivity.*** A DNA polymerase engaged in DNA synthesis may (or may not) be continuously bound and may progressively move along the template. In fact, DNA polymerases drop off the template after the addition of a certain number of nucleotides and become attached again to reinitiate polymerization. The rate of DNA chain elongation catalyzed by DNA polymerases is not uniform either. *Processivity* is the average number of nucleotides added by a polymerase molecule each time it binds a template. Processivity is an intrinsic property of the polymerase. A polymerase would be *nonprocessive* or *distributive* if it leaves the template after each nucleotide addition. A polymerase is called *processive* if it copies a long template maintaining uninterrupted contact with it. The number of nucleotides polymerized per binding–dissociation event gives a qualitative description of whether a polymerase has a "low" or "high" processivity. Kinetically, processivity is determined by whether a polymerase goes into a processive cycle (k_{cat}) by translocating to the next polymerization site in the DNA–polymerase complex or whether it dissociates into a free enzyme and DNA (k_{off}) to subsequently initiate polymerization on another template-primer. The relative k_{off} values for template–primers of random sequence may vary widely along the DNA, while k_{cat} remains relatively sequence independent. Thus the processivity of a polymerase varies along the DNA sequence in accordance with the relative ratio (k_{cat}/k_{off}) of the two rate constants. Processivity is also influenced by many factors that affect the secondary structure of the DNA and the conformation of the enzyme.

As a general rule, processivity decreases at high ionic strengths and low temperatures as a result of their influence on the structure of the template-primer and on the enzyme as well. Processivity is thus a feature kinetically related to the characteristics of DNA pause or termination sites. Processivity *in vivo* is subject to more complex interactions between the polymerase and various other protein and/or nonprotein factors, as typically exemplified by phage T4 accessory proteins (refer to Section B, II, this chapter).

The processivity of a polymerase can be estimated according to the following principles.

(i) Polymerase trap. In the presence of a polymerase-trapping agent, e.g., nonspecific DNA (or challenger templates) or more preferably heparin (3), polymerase molecules can be rapidly trapped after a single round of primer extension. This condition can be created by preincubating the test template-primers with a polymerase and initiating the reaction by adding dNTPs (and/or Mg^{2+}) and the trapping agent. Electrophoretic analysis of the size of the primary extension products gives the measure of processivity.

(ii) Limited reaction rate. On a given template-primer which is in large excess over the polymerase under a limiting condition of nucleotides, i.e., one, two, or three versus all four dNTPs, the polymerization rate of "processive" enzymes is sharply reduced, whereas the limiting condition has little effect on "nonprocessive" enzymes.

(b) Fidelity. Another important parameter that characterizes a polymerase is *fidelity*, i.e., the accuracy of nucleotide incorporation or the frequency of misincorporation. The implication of polymerase fidelity is that it can be correlated with the rate of mutation, as typically observed in retroviruses such as HIV.

Fidelity of a DNA polymerase is assured by the coordinate action of polymerase activity and $3' \rightarrow 5'$-exonuclease activity (4–6). The $3' \rightarrow 5'$-exonuclease activity, which is also called *proofreading* or *editing* exonuclease, may be either an integral part of the polymerase molecule as in *E. coli* DNA Pol I or it may be associated with the polymerase as a multisubunit complex as in *E. coli* DNA Pol III.

The fidelity-determining process consists of three components: (i) an improved base selection; (ii) when misinsertion occurs, a reduced rate of nucleotide sealing; and (iii) a higher probability for the removal of mismatching nucleotides by the $3' \rightarrow 5'$-exonuclease activity. The first two components are intrinsic properties of a polymerase, regardless of the presence or absence of the proofreading exonuclease activity.

Although the exact mechanism remains to be further elucidated, a polymerase is presumed to exercise discrimination between correct and incorrect nucleotide binding and chemical sealing primarily based on local DNA geometry and, to a lesser extent, on the energetics of competing reactions. Fidelities on the order of 10^4–10^6 are thus commonly observed with DNA polymerases which do not have or have no overtly apparent $3' \rightarrow 5'$-exonuclease activities, for example, reverse transcriptases, eukaryotic DNA polymerases (α, β, and γ), and mutant (exo$^-$) prokaryotic or phage DNA polymerases. The destabilization energy ($\Delta G°$) of nucleotide mispairing is merely 1–2 kcal/mol at 25°C, which predicts an error frequency of 10^{-2} or 1 in 100 nt. The relative binding affinities of a polymerase for matched and mismatched primer termini vary within around 10-fold. Therefore, the fidelity of a polymerase is largely due to a kinetic block which inhibits the extension of the termini following nucleotide misinsertion.

The frequency of misincorporation is not only characteristic of a polymerase but is also subject to a number of factors associated with reaction conditions. Most important is the nature of the nucleotide mismatches. The sequence context of the template–primer and other factors, such as symmetry of the mispair, rate of synthesis, and balance status of nucleotide (dNTP) pool, play a significant but secondary role. On the other hand, the mutagenic effect of Mn^{2+}, when it replaces Mg^{2+}, is primarily attributed to the perturbation in the mode of dNTP binding at the polymerase active site (7).

For a polymerase possessing $3' \rightarrow 5'$-exonuclease activity (e.g., *E. coli* DNA Pol I or T4 DNA Pol), the kinetic blocking at the mismatched termini is channeled into "proofreading" action by the exonuclease, resulting in an increase of fidelity by 2–3 orders of magnitude (8,9). The nucleotide excision probability depends on the following three factors: (i) the ratio of melted to annealed termini for both matched and mismatched primer termini in the exonuclease active site, (ii) the enzyme activity for removal of single-stranded primer termini, and (iii) the lag time required for polymerase to add the next correct nucleotide following a mismatch.

The lag time depends explicitly on the concentration of the next correct dNTP, giving rise to a "next nucleotide effect" (10,11). High concentrations of dNTPs (>200 μM) increase the error rate by driving the polymerase reaction (12); a decrease in lag time reduces the chance of error discrimination by the polymerase at each extension step and consequently the chance of mismatching nucleotide excision by the exonuclease. Polymerase fidelity tends to increase at lower pH (e.g., pH 6.5 versus pH 8.5) as a result of decreased efficiency of chain elongation. Note that a marked increase of error rate is also observed when the concentrations of dNTPs are too low (<25 μM) or unbalanced by having one or more dNTPs at lower concentrations (as is often encountered in radiolabeling reactions).

iii. **Structural features of polymerases: Domains, motifs, and multifunctions.** Template-dependent DNA polymerases generally possess multiple domains that have distinct activities but function coordinately *in vivo* to ensure a high rate of polymerization, maximum fidelity of replication, and regulation of polymerase activity. The most common combination of multifunctions is the polymerase and $3' \rightarrow 5'$-exonuclease activities as found in *E. coli*, T4, and T7 DNA polymerases. *E. coli* DNA Pol I has an important additional activity, i.e., $5'$-nuclease activity. In contrast, *Taq* DNA Pol has no apparent $3' \rightarrow 5'$-exonuclease activity, but has a strong $5'$-nuclease activity. Another common combination is the polymerase and RNase H activities that are found in reverse transcriptases. Despite their physically and functionally intimate associations, each function resides in a separate domain. For example, the $3' \rightarrow 5'$-exonuclease and/or RNase H activities can be selectively inhibited or inactivated without affecting the polymerase activity.

Alignments and comparisons of available polymerase sequences have allowed the polymerases to be allocated into related families and subfamilies based on amino acid sequence motifs (13,14). Certain primary structural motifs are conserved throughout polymerases, including RNA- and DNA-directed RNA and DNA polymerases. Note that the motif-based grouping of polymerases does not rule out the possibility that the subsets share common 3-D features. Taking the primary as well as the tertiary structures of *E. coli* DNA Pol I large (Klenow) fragment as prime references, DNA polymerases can be classified into four (A, B, C, and X) families.

Family A DNA polymerases (or *Pol I family*) include, in addition to the prototype *E. coli* DNA Pol I (*polA* gene product), DNA polymerases from *T. aquaticus* and from phages T7 and T5.

Family B (or *Pol-α type*) includes *E. coli* DNA Pol II (*polB* gene product), eukaryotic DNA polymerases (α and δ), archaebacterial DNA polymerases, phage T4 DNA Pol, and viral DNA polymerases from herpes, adeno- and vaccinia viruses.

Family C contains fewer members which include the DNA polymerases III from *E. coli*, *S. typhimurium*, and *B. subtilis*.

Family X (or *Pol-β type*) has two known members: DNA Pol β (rat) and terminal deoxynucleotidyltransferases.

iv. **Template-based division of DNA polymerases.** The majority of DNA polymerases use DNA as the preferred template in producing complementary (or second) strand DNA and thus are called *DNA-directed DNA polymerases* or, more commonly, "DNA polymerases." In contrast, some DNA polymerases show a distinct preference for RNA as the template in making a copy DNA (cDNA). This group of polymerases is classified as *RNA-directed DNA polymerase* or conventionally *reverse transcriptase*. It should be noted, however, that most DNA-directed DNA polymerases exhibit some intrinsic RNA-directed DNA polymerase activities and, under certain conditions (e.g., in the presence of Mn^{2+}), the reverse transcriptase activity can be substantially enhanced, thereby blurring the template-based classification of polymerases.

In contrast to the DNA polymerases that require templates, some DNA polymerases conventionally known as *template-independent DNA polymerases* (e.g., terminal deoxynucleotidyltransferases) catalyze polymerization reactions *in vitro* without a template in the presence of dNTPs and an initiator (3'-OH). [*Note*: In the enzyme classifications recommended by IUBMB, DNA polymerases that require a template and a primer belong to a subclass of *nucleotidyltransferases* (EC 2.7.7).]

Within each category of DNA polymerases, member enzymes can be grouped into phylogenetic families based on sequence similarity, and then subdivided according to biological functions and sensitivity to inhibitors (e.g., aphidicolin, phosphonoacetic acid, and cytosine arabinoside). Yet DNA polymerases can differ significantly from one another in various physicochemical properties such as molecular size, thermal stability, template and primer preference, fidelity, processivity, optimal reaction conditions, and catalytic efficiency. Therefore, each polymerase deserves due attention to its specific properties.

This chapter focuses on a few selected DNA polymerases that are most widely used in recombinant DNA technology. In the category of DNA-directed DNA polymerases (Section II), four DNA polymerases are described that originate from bacteria (*E. coli* and *T. aquaticus*) and bacteriophages (T4 and T7). In the category of RNA-directed DNA polymerases (Section III), two reverse transcriptases, one from AMV and the other from MoLV, are described. As the typical enzyme of template-independent DNA polymerases (Section IV), the terminal deoxynucleotidyltransferase (TDTase) from calf thymus is described.

References

1. Kornberg, A. (1980). "DNA Replication." Freeman, San Francisco.
2. Kornberg, A. (1982). 1982 Supplement to DNA replication. Freeman, San Francisco.
3. Reddy, M. K., Weitzel, S. E., and von Hippel, P. H. (1992). *JBC* **267**, 14157–14166.
4. Kunkel, T. A. (1988). *Cell* **53**, 837–840.
5. Kuchta, R. D., Benkovic, P., and Benkovic, S. J. (1988). *Biochem.* **27**, 6716–6725.
6. Patel, S. S., and Johnson, K. A. (1991). *Biochem.* **30**, 526–537.
7. Goodman, M. F., Keener, S., Guidotti, S., and Branscomb, E. W. (1983). *JBC* **258**, 3469–3475.
8. Kunkel, T. A., and Bebenek, K. (1988). *BBA* **951**, 1–15.
9. Preston, B. D., Zakour, R. A., Singer, B., and Loeb, L. A. (1988). *In* "DNA Replication and Mutagenesis" (R. E. Moses and W. C. Summers, Eds.), pp. 196–207. American Society for Microbiology, Washington, D.C.
10. Clayton, L. K., Goodman, M. F., Branscomb, E. W., and Galas, D. J. (1979). *JBC* **254**, 1902–1912.
11. Fersht, A. R. (1979). *PNAS* **76**, 4946–4950.
12. Eckert, K. A., and Kunkel, T. A. (1991). *In* "PCR: A Practical Approach" (M. J. McPherson, P. Quirke, and G. R. Taylor, Eds.), pp. 225–244. IRL Press, Oxford.
13. Blanco, L., Bernad, A., Blasco, M. A., and Salas, M. (1991). *Gene* **100**, 27–38.
14. Braithwaite, D. K., and Ito, J. (1993). *NAR* **21**, 787–802.

TABLE 6.1 Reaction Specificities of E. coli DNA Polymerase I[a]

Major reactions	
a. Polymerization (DNA-directed)	DNA$_n$-primer/DNA + dNTP → DNA$_{n+1}$ + PP$_i$
b. 3' → 5'-Exonuclease	dN$_1$N$_2$ ··· N$_n$/DNA $\xrightarrow{H_2O}$ DNA$_{n-1}$ + dN$_1$MP
c. 5'-Nuclease	dN$_1$N$_2$ ··· N$_n$/(DNA) $\xrightarrow{H_2O}$ DNA$_{n-1}$ + dN$_n$MP
Minor reactions	
d. Polymerization (RNA-directed)	DNA$_n$-primer/RNA + dNTP → DNA$_{n+1}$ + PP$_i$
e. Pyrophosphorylase	dN$_1$N$_2$ ··· N$_n$/DNA + PP$_i$ → DNA$_{n-1}$ + dN$_n$MP
f. Pyrophosphate exchange	dNTP/DNA + PP$_i^*$ → dNMP-PP* + PP$_i$
g. Nucleoside diphosphokinase	dNDP + ATP → dNTP + ADP
h. RNase H	RNA·DNA → RNA fragments + DNA

[a] All functions of DNA Pol I require the presence of Mg^{2+} ion.

II. DNA-DIRECTED DNA POLYMERASES

A. *E. coli* DNA Polymerase I
[EC 2.7.7.7]

E. coli DNA polymerase I (Pol I) is a monomeric (109 kDa), multifunctional enzyme that has three major activities: 5' → 3' DNA polymerase, 3' → 5'-exonuclease, and 5'-nuclease (1). Pol I, also known as the *Kornberg enzyme*, is probably the best characterized of all polymerases and serves as the prime model for DNA polymerases in the elucidation of polymerase structure and associated multifunction. Under the conditions of limited digestion with subtilisin, Pol I is cleaved into two fragments. The large (68 kDa) C-terminal fragment possesses the DNA polymerase and 3' → 5'-exonuclease activities. The small (34 kDa) N-terminal fragment contains 5'-nuclease activity. The large fragment, which is better known as the *Klenow fragment* (Pol Ik), can replace *in vitro* most of the functions of Pol I. In fact, the lack of 5'-nuclease activity renders Pol Ik more suitable than Pol I in many applications.

1. FUNCTIONS

a. Reaction specificities

E. coli Pol I is capable of carrying out various functions (Table 6.1). Its three major activities are (a) the 5' → 3' DNA polymerase which catalyzes the addition of dNTP to the 3'-OH terminus of a primer with a concomitant release of PP$_i$, (b) the 3' → 5'-exonuclease which hydrolyzes the 3' terminal nucleotides of ssDNA or dsDNA in the absence of dNTPs, and (c) the dsDNA-specific 5' → 3'-exonuclease–endonuclease which hydrolyzes the phosphodiester bonds from the 5' ends of DNA. Under special circumstances, Pol I exhibits various other "minor" activities, which are often expressions of the major activities in different forms. The minor activities include (see Table 6.1) (d) the RNA-directed DNA polymerase capable of using RNA templates, thus reverse transcriptase activity, (e) pyrophosphorylase which catalyzes the reverse reaction of polymerization, (f) pyrophosphate exchange with the β,γ-phosphates of dNTPs, which is equivalent to the removal by PP_i of the newly added nucleotide prior to its complete stabilization by the entry of the next dNMP (2), (g) nucleoside diphosphokinase activity which catalyzes the conversion of dNDP to dNTP using various ribo- and deoxyribonucleoside di- and triphosphates as substrates, and (h) RNase H activity which degrades the RNA of DNA · RNA hybrids.

b. Reaction conditions

i. Recommended reaction conditions. Polymerase reactions can be performed to suit a variety of applications in a controlled manner with respect to relative concentrations of nucleotide substrates, template-primer, and the polymerase. The rate and extent of polymerization are also profoundly influenced by such factors as pH, temperature, ionic strength, and buffer composition which affect not only the polymerase activity but also the conformation of template-primer. Furthermore, reaction conditions are often adjusted such that the polymerase reaction proceeds simultaneously with other enzymatic reactions such as RNase H (and/or *E. coli* DNA ligase) for RNA replacement synthesis and DNase I for nick translation.

For all practical purposes, the reaction conditions for Pol I and Pol Ik can be considered identical. The following is a typical reaction condition used to synthesize second-strand DNAs on ssDNA templates. The reaction mixture (20 μl) contains 50 mM Tris–Cl (pH 7.5), 5 mM $MgCl_2$, 50 mM NaCl, 50 μg/ml BSA, 0.2 mM each of dNTPs, 2 μg template-primers, and 5 U of Pol Ik (or Pol I). The mixture is incubated at 20–37°C for 1 hr. The reaction is terminated by adding EDTA (20 mM) or by heating to 75°C for 10 min.

The reaction conditions for repairing short 3' or 5' protruding ends to generate blunt ends are essentially identical to those described earlier except that the concentrations of dNTPs are reduced to 50 μM and the mixture is incubated for 10–15 min. The repair reactions can often be carried out directly in the buffer used for previous enzymatic reactions.

ii. pH and temperature dependence

(a) pH dependence

(i) Polymerase activity. Pol I is optimally active at pH 7.4 in potassium phosphate buffer (50 mM) with either native DNA or poly(dAT) as template–primers (3). At pH 7.0 and 7.8, the polymerase activity is 70% of the optimal value. At pH 7.0, the polymerase activity is approximately twofold higher than 5'-nuclease or 3' → 5'-exonuclease activity.

(ii) Exonuclease activities. The 3' → 5'-exonuclease activity is optimally active at pH 8.6 in Tris–Cl (67 mM) or pH 9.2 in sodium glycinate (67 mM). The 5'-nuclease activity is as strong as the 3' → 5'-exonuclease from pH 6.0 to 7.4 in potassium phosphate (67 mM). At higher pH values (e.g., pH 8.6), the 3' → 5'-exonuclease activity is about fourfold greater than 5'-nuclease activity which is essentially constant from pH 7.4 to 9.2 (4).

TABLE 6.2 Kinetic Parameters of *E. coli* DNA Pol I[a]

Substrate	K_m (μM)	K_d (μM)	k_{cat} (min^{-1})
dATP	38	33	156
dCTP	14	147	7
dGTP	11	12	204
dTTP	3	81	350[b]
d(AT)$_n$	4–10		
Mg^{2+}	830[c]	470	
ddTTP	89		(0.1)[b]

[a] Polymerization on homopolydeoxynucleotide template and homooligodeoxynucleotide primer at pH 7.4 (50 mM Tris–Cl) and 37°C (5). The K_m values for dNTPs are for Mg^{2+}–dNTP complex forms in the presence of 0.6 mM Mg^{2+} (for dATP and dCTP) and 2 mM Mg^{2+} (for dGTP and dTTP). The K_m values for free dNTPs are estimated to be at least 10 times lower.

[b] The initial rates (v_i) for dTTP and ddTTP incorporations on d(AT)$_n$ template-primer are ~600 and 0.1 min^{-1}, respectively (50).

[c] The K_m value is from the zero extrapolation of dTTP in reactions using poly(dA)·oligo(dT) as a template-primer (5).

(iii) Pyrophosphorolysis. The pyrophosphorylase activity is optimal at pH 6.5 (2). At pH 6.0, pyrophosphorolysis is 5 times more active than hydrolysis. However, at pH 9.2 (sodium glycinate), which is optimal for both 5'-nuclease and 3' → 5'-exonuclease activities, pyrophosphorolysis is ~150 times slower than the nuclease activities.

(b) Temperature. Polymerase reactions are usually carried out at 37°C, although higher temperatures (e.g., 40°C in DNA sequencing) and lower temperatures (e.g., 12°C in second-strand cDNA synthesis) may be employed for specific applications. The optimal temperature for polymerase activity can be significantly different from that of polymerase reaction due to the effect of temperature on other reaction components. For example, the optimal "priming" temperature is dependent on the stability of template-primer structure: for instance, near 0°C for hexamer (dAT)$_3$, ~10°C for (dAT)$_4$, ~20°C for (dAT)$_5$, and ~37°C for (dAT)$_6$ (1). The polymerase activity is stable at 37°C for many hours of incubation.

iii. Kinetic parameters. The kinetic parameters of polymerization with some typical substrates are shown in Table 6.2. The K_m and k_{cat} values of Pol Ik are virtually identical to those of Pol I. On homopolymeric templates, the K_m values of dNTPs range from 3 to 38 μM (5). The K_d values of dNTPs range from 12 to 150 μM. In PP_i exchange or pyrophosphorolysis on poly(dAT), PP_i is a substrate with a K_m of 0.6 mM (2).

The apparent K_m for poly(dAT) is 4–10 μM, but the K_m can be 3–4 orders of magnitude lower for longer DNA templates, e.g., K_m of 0.3 nM for T7 phage DNA (6) and K_m of 0.17 nM for E. coli DNA (7). The K_m is not markedly affected by base composition.

The V of Pol I on some simple templates is 350 min^{-1}, although the initial rates can be as high as 600 min^{-1}. The V is three times lower in 50 mM potassium phosphate buffer (pH 7.4) than in 50 mM Tris–Cl (pH 7.4) (5). The V of the $3' \rightarrow 5'$-exonuclease and $5'$-nuclease activities of Pol I are reported to be 16–19 and 27 min^{-1}, respectively (1).

iv. Metal ion requirements. Pol I requires as cofactors divalent cations such as Mg^{2+} or Mn^{2+} in addition to Zn^{2+}. In fact, Pol I (and Pol Ik as well) has multiple classes of divalent metal ion binding sites (8–10). Two atoms of Zn^{2+} are bound per molecule of Pol I (8). The addition of Zn^{2+} does not substitute for Mg^{2+} or Mn^{2+} in enzyme assays. The enzyme-bound Zn can be exchanged with exogenous Zn ions in buffer.

At optimum pH 7.4 in Tris–Cl (50 mM), the polymerase activity is sharply optimal at 1 mM Mg^{2+} with the poly(dA) · oligo(dT) template–primer (5). However, the optimal concentration of Mg^{2+} varies with the concentrations of template and dNTPs. The 7 mM concentration of Mg^{2+} previously observed to be optimal in potassium phosphate buffer (50 mM) apparently reflects the interaction of the phosphates with Mg^{2+} (3). In the potassium phosphate buffer, 0.33 and 33 mM Mg^{2+} produced 30 and 50% of the maximal activity, respectively. Note that Mg^{2+} is not required for the enzyme to bind DNA.

Mn^{2+} can replace Mg^{2+}. At its optimal concentration of 0.07 mM, Mn^{2+} is a better activator for the polymerase than 8 mM Mg^{2+} (9). At 1 mM concentration, however, Mn^{2+} gives a strong inhibition, which is likely to be overcome by the addition of isocitrate. Isocitrate (15 mM) has been shown to expand the effective range of Mn^{2+} concentrations significantly (11). The K_m of free Mn^{2+} at zero extrapolation of total dNTPs is 1.2 μM, while the average K_m for Mn-dNTP is ~4 μM. At 0.5–1.5 mM Mn^{2+}, Pol I not only incorporates higher levels of nucleotide analogs, such as ddNTP and rNTP, but also results in a higher frequency of dNTP misincorporation than with Mg^{2+} (12,13).

The $3' \rightarrow 5'$-exonuclease catalytic site has two binding sites for metal ions that are essential for enzymatic activity: a tight binding site (K_d for Mn^{2+} is 2.5 μM) and a weak binding site that is tightened 100-fold by TMP binding (10). A third metal-binding site is apparently associated with the polymerase activity. The binding of dGTP to the polymerase catalytic site creates one tight Mn^{2+}-binding site with a K_d of 3.6 μM. Mg^{2+} competes at this site with a K_d of 100 μM.

The 5'-nuclease activity also requires Mg^{2+} or Mn^{2+} ions.

TABLE 6.3 Noncomplementary Nucleotides as Competitive Inhibitors of Complementary Nucleotides in Polymerization Reactions[a]

Nucleotide	K_i (mM)	Complementary nucleotide
dCTP	>1	dTTP
dGTP	0.11	dTTP
dCTP	0.31	dATP
dGTP	0.38	dATP
dATP	0.28	dCTP
dTTP	>0.30	dCTP
dATP	>1	dGTP
dTTP	0.57	dGTP
NTP[b]	6–14	
ATP[c]	0.14	dATP
CTP[c]	0.53	dATP
ADP[c]	3.0	dATP
AMP[c]	7.8	dATP
PP_i^c	0.39	

[a] Polymerization reactions with homopolymeric templates and homooligomeric primers at pH 7.4 (50 mM Tris–Cl) and 37°C (5).

[b] Polymerization with Pol I on activated calf thymus DNA at pH 9.2 (66 mM glycine buffer) and 37°C (6).

[c] Polymerization with Pol Ik on poly(dT)·r(pA)$_{10}$ template-primer at pH 7.5 (50 mM Tris–Cl) and 37°C (104). The inhibition by PP$_i$ is noncompetitive.

v. Inhibitors and inactivators

(a) Metal chelators. E. coli Pol I is inhibited by EDTA (Mg^{2+} chelator) or o-phenanthroline (Zn^{2+} chelator). At saturating levels of DNA, however, Pol I is much less susceptible to the inhibition by o-phenanthroline, suggesting that DNA and the chelating agent compete for the enzyme-bound Zn^{2+} (8).

(b) Phosphates. High concentrations of P$_i$ (>100 mM) sharply reduce the DNA binding of Pol I. Pyrophosphate is a noncompetitive inhibitor (K_i = 0.7 mM) with respect to poly(dAT) or dNTPs.

(c) Noncomplementary nucleotides and nucleotide analogs. NTPs are competitive inhibitors of dNTPs in polymerization reactions (Table 6.3): the K_i values for NTPs are 6–14 mM (6). ADP, AMP, and deoxyadenosine have only weak inhibitory effects on Pol I.

The binding of dATP is competitively inhibited by other nucleotides: K_d(ATP) = 18 μM, K_d(3'-NTPs) = 38–280 μM, and K_d(5'-dNTP) = 400 μM.

Nucleotides such as dNDP, dNMP, and 3'-dNMP give only minor inhibition (<15%) to dATP binding.

An epoxide derivative, 2',3'-epoxy-ATP, inhibits the polymerase ($K_{m,app}$ = 16 μM) by forming an enzyme-catalyzed, tight binding epoxy-AMP-terminated inhibitor (14). The inactivated Pol Ik can still bind a second DNA molecule and carry out 3' → 5'-exonucleolytic hydrolysis (78).

Pol I is insensitive to aphidicolin (up to 0.4 mM), 2-(p-n-butylanilino)dATP (up to 0.2 mM), and N^2-(p-n-butylphenyl)dGTP (up to 0.4 mM). These nucleotide analogs are highly selective, potent inhibitors of the replication-specific mammalian DNA Pol α (15).

NMP is a selective inhibitor of 3' → 5'-exonuclease activity in both the absence and the presence of dNTP: AMP (5–10 mM) gives ~90 and ~50% inhibitions on the hydrolysis of oligo(dA) in poly(dT) · oligo(dA) in the absence and presence of dATP (1.5 μM), respectively (16). As a result of selective exonuclease inhibition, AMP (at 1 mM) stimulates the polymerase activity about twofold. dNMPs (5–20 mM) also inhibit the 3' → 5'-exonuclease activity.

In pyrophosphorolysis or hydrolysis, ATP or dATP is a competitive inhibitor with a K_i of 2 or 10–30 μM, respectively (6). The inhibition by ADP is a mixed type and is not reversed by increasing concentrations of DNA.

(d) Sulfhydryl reagents. Pol I is not inhibited by PHMB (0.5 mM) or PCMB. In fact, a stable modification of the unique sulfhydryl (Cys-907) of Pol Ik with a fluorophore produces no inhibitory effects on polymerase activity and has been used as a probe to measure the configuration of the polymerase active center (17).

(e) Other compounds. Captan (N-[(trichloromethyl)thio]-4-cyclohexene-1,2-dicarboximide) binds at the template-primer binding site and inhibits the polymerase and 3' → 5'-exonuclease activities, while the 5'-nuclease activity is enhanced (18).

NaF, at a concentration of 5–10 mM, is a selective inhibitor of 3' → 5'-exonuclease activity and consequently increases approximately 2.5-fold the polymerase activity of Pol Ik (16,19).

PALP at concentrations <0.5 mM inhibits the polymerase activity but does not affect the 3' → 5'-exonuclease activity (20).

c. Activity assay and unit definition

Polymerase activity is usually assayed using poly(dAT) as the template–primer. A typical assay mixture (100 μl) contains 50 mM K-P (pH 7.4), 6.6 mM $MgCl_2$, 1 mM DTT, 62.5 μg/ml poly(dAT), 33 μM each of dATP and dTTP, and 69 nM of [^{3}H]dTTP and polymerase. The mixture is incubated for 30 min at 37°C.

Unit definition: One polymerase unit is the amount of enzyme that incorporates 10 nmol of dNTPs into acid-precipitable material in 30 min.

d. Substrate specificities: Polymerase

i. Template structure. Pol I (and Pol Ik) catalyzes template-directed polymerization of dNTPs on a wide variety of natural and synthetic DNAs and RNAs. Depending on the structure of the template, polymerization reactions can be distinguished as follows (Fig. 6.1).

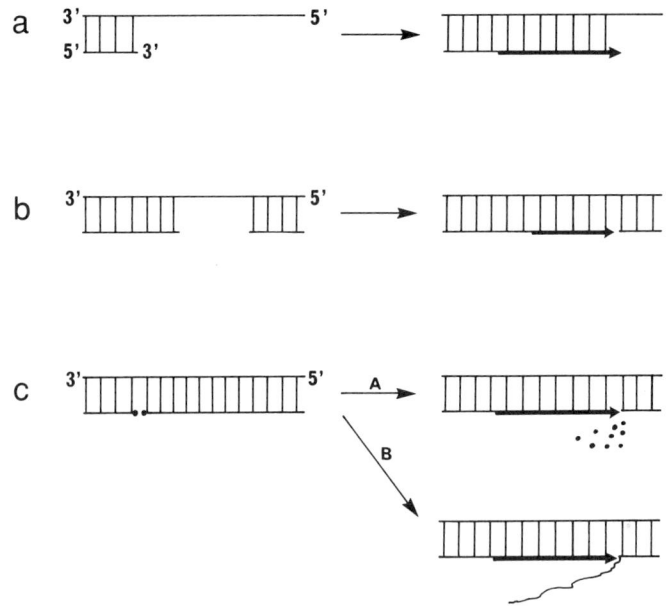

FIGURE 6.1 Template specificity of *E. coli* DNA Pol I and Pol Ik. (a) Typical second-strand synthesis on a single-stranded DNA template containing a primer-annealed double-stranded region, (b) filling-in reaction on a gapped duplex, and (c) polymerization reactions on nicked DNA which, depending on polymerases, may be a nick translation (Pol I, path A) or a strand displacement synthesis (Pol Ik, path B).

(a) Single-stranded DNA. On ssDNA or denatured dsDNA, Pol I synthesizes a complementary strand from the 3'-OH terminus of the primer. Either oligodeoxynucleotides or oligoribonucleotides can serve as a primer. For homopolymeric DNA templates, the maximal rate of synthesis varies by a factor of 31 depending on the nature of the base: $d(pC)_n > d(pT)_n > d(pA)_n \gg d(pG)_n$, indicating that Py templates are better substrates than Pu templates (5). The affinity (reflected by the K_m values) of the homopolymeric templates increases in the order of $d(pC)_n < d(pT)_n < d(pG)_n \leq d(pA)_n$.

(b) Gapped DNA. On gapped DNA, Pol I performs gap filling.

(c) Nicked DNA. Pol I carries out *nick translation* and/or *strand displacement synthesis* on nicked DNA. Nick translation is a covalent extension of the 3'-OH terminus with the concurrent 5'-nucleolytic degradation of the strand 3' to the nick (21) [For details, refer to Appendix A]. The strand displacement reaction is a polymerization uncoupled from the 5'-nuclease activity. Pol Ik catalyzes strand displacement synthesis more efficiently than Pol I. In fact, the strand displacement synthesis previously observed with Pol I (22) might have been due to the presence of contaminating Pol Ik (23,24).

(d) Discontinuous DNA template. A 3' → 5'-exonuclease-deficient Pol Ik can carry out a DNA synthesis *in vitro* on discontinuous templates (25). The addition of multiple nucleotides to the 3' end of a blunt-end duplex is templated by unlinked single-strand oligodeoxynucleotides, and the first nucleotide added to the duplex is complementary to the 3'-terminal nucleotide of the unlinked oligonucleotide. Both homopolymers and mixed sequence oligodeoxynucleotides can serve as discontinuous DNA templates.

(e) DNA·RNA hybrid. On the DNA·RNA heteroduplex which contains nicks in the RNA strand, Pol I carries out RNA strand displacement synthesis. This property has been incorporated into efficient procedures for the synthesis and cloning of ds-cDNA (26,27) (refer to "cDNA synthesis," Section III, this chapter). The overall feature of the reaction is similar to the "nick translation" performed on nicked dsDNA. Whereas the nick translation relies on the 5'-nuclease activity of Pol I, the RNA strand displacement synthesis relies on the strand displacement activity of Pol Ik.

(f) RNA templates. Pol I can also utilize polyribonucleotides (28) or natural RNAs (29) as templates to synthesize cDNAs. The reaction conditions are similar to the DNA-directed DNA synthesis. The rate of polymerization, for instance, with poly(rA)·poly(dT) as a template-primer, is higher than that of the comparable DNA-directed polymerization (5,28,30). Compared with AMV RTase, the rate of RNA-directed DNA synthesis by Pol I is more or less similar, although it can be higher with certain templates, for example, by a factor of 50 with poly(rA)·oligo(dT) (30).

The reverse transcriptase activity of Pol I is critically dependent on the amount of enzyme, which should exceed an enzyme/RNA molar ratio of four; above that, the reverse transcriptional activity rises exponentially. Labeled cDNAs of rabbit globin mRNA have been synthesized by Pol I and used successfully for molecular hybridizations (31).

ii. Primers

(a) Primer lengths. With homopolymeric DNA templates, mononucleotides (both dNMP and dNTP) as well as small oligodeoxynucleotides serve as primers of DNA synthesis by Pol Ik (32).

With poly(dA) as a template, when the length of the primer $d(pT)_n$ is increased by one unit in the interval from $n = 1$ to 10, the K_m changes from 40 to 0.2 μM, indicating that the affinity of the primers for Pol Ik increases by ~80% for each nucleotide (33). For $11 < n < 25$, the affinity decreases with the minimum for $d(pT)_{22-23}$. Note that the primers conventionally recommended for DNA or cDNA synthesis have been 12–18 nt long. Primers containing more than 50 nt exhibit an affinity for the enzyme which is comparable to that of $d(pT)_{10-11}$.

In a manner similar to the progressive changes of K_m values, the V also shows a dependence on primer chain length: the $\log(V)$ increases linearly up to $n = 10$ at a ~12% increment per nucleotide and then decreases, reaching a minimum at $n = 20-23$.

(b) **Mismatching primers.** The presence of noncomplementary nucleotides reduces the interaction of primers with Pol Ik. For the primers $d(pN)_n$ ($n \leq 10$), the deleterious effect is diminished not by the total length but by the number of bases from the 3'-end to the noncomplementary base. The K_m for $d(pT)_{10}pC$ ($K_m = 12$ μM) is ~120 times as great as for $d(pT)_{11}$ ($K_m = 0.1$ μM). The 3'-terminal nucleotide of the primer is thought to contribute a decisive amount of binding free energy to the polymerase both from the nucleoside and from the phosphate moiety. The 5'-end nucleotide is relatively unimportant for the binding, and it provides a convenient point for introducing various labels (intercalators). Nevertheless, some ligands may increase the affinity of the primer.

Selective inhibition of the 3' → 5'-exonuclease activity of Pol Ik by NaF leads to a ~30% increase in the V for primers containing noncomplementary bases and a 10–15% increase for fully matching primers without any change in their K_m values (33).

iii. Nucleotide substrates

(a) **Deoxynucleotides.** Pol I utilizes as the standard nucleotide substrates dNTPs with K_m values of 3–40 μM (Table 6.2).

For homopolymeric templates and homooligomeric primers, noncomplementary dNTPs are essentially competitive inhibitors against complementary dNTPs; K_i values are 1–2 orders of magnitude higher than the K_m values (Table 6.3).

(b) **Nucleotide analogs.** Pol I can incorporate various nucleotide analogs.

dNTPαS: α-Thionucleotides, specifically dATPαS, are incorporated into elongating DNA by Pol I at a 10–30% decreased rate compared with that for dATP (35,36). The K_m for dATPαS is higher than that for dATP by ~20%. The phosphorothioate-substituted DNA is insensitive to the 3' → 5'-exonuclease activity of Pol I (37).

ddNTPs: The incorporation of chain-terminating ddTTP on poly(dAT) is at least 1000 times slower than that of dTTP. The rate of 3' → 5' hydrolysis or PP_i exchange on ssDNA [poly(dT)], which is terminated with ddNMP, is also inhibited to a similar extent. Substituting Mn^{2+} (2 mM) for Mg^{2+} (5 mM) reduces the discrimination against ddNTP ~100-fold for Pol Ik, elevating the incorporation ratio (e.g., ddA/dA) to ~25% (11). A mutant Pol Ik (F762Y) actually prefers to utilize ddNTPs over dNTPs ~1.7-fold in the presence of Mg^{2+} (34).

Methylated nucleotides: N^6-Methyladenine (m^6A) is incorporated more slowly than the adenine residue. Its incorporation is random and proportional to the input ratio of m^6dATP and dATP. An eightfold decrease in rate is observed for pure m^6dATP (38).

α-Methylphosphonate analogs of dTTP, i.e., dTTP-α-CH_3 [-O-$P^α$(=O)(CH_3)O-ribose] and the chain-terminating 3'-fluoro-ddTTP-α-CH_3, are incorporated into DNA chains by Pol Ik, although at a very low rate (39). The internucleoside methylphosphonate groups are hydrolyzed by the 3' → 5'-exonuclease activity of Pol Ik but not by Exo III, a situation analogous to the responses to phosphorothioate groups.

Ribonucleotides: Pol I has the ability to incorporate certain ribonucleotides in place of the corresponding deoxynucleotides when Mg^{2+} is replaced by Mn^{2+}. Maximum rates and the extent of rNTP incorporation are observed with 0.5 to 1.5 mM Mn^{2+} (13). Of the four rNTPs, CMP is incorporated almost as fast as dCMP. GMP is incorporated to a lesser extent than dGMP at 10°C. At 37°C, however, GMP is incorporated at the same rate as dGMP. The incorporation of AMP occurs only slowly and to a very limited extent, while UMP is not incorporated at all.

Other nucleotides: The incorporation of a dATP analog, 2-aminopurine-dNTP (dAPTP), is decreased 3–10 times in the presence of an equal concentration of dATP (40). Although inefficient, Pol I can also incorporate various other nucleotide analogs such as dITP, 7-deaza-dGTP, biotin- or digoxigenin-labeled dUTP, and fluorescein-12-dUTP.

iv. Frequency of misincorporation. The frequency of mismatching nucleotide incorporation by Pol I has been estimated to be 1.5×10^{-6} (41). In the presence of Mn^{2+} (0.5–1.5 mM), in place of Mg^{2+}, Pol I (and Pol Ik) displays a substantial degree of specificity relaxation, permitting relatively efficient incorporations of rNTPs and mismatching dNTPs.

v. Polymerase processivity. The processivity of Pol I ranges between 18 and 50 nt at 37°C and low salt ($I = 0.085$) (42). With nicked DNA and gapped DNA templates, Pol I displays the processivities of 15–20 nt and 40–50 nt, respectively. The processivity can be as high as ~190 nt with a poly(dAT) template at 37°C ($I = 0.11$), but certain changes in reaction conditions can diminish it, for instance, to 3.3 nt at 5°C ($I = 0.31$). With poly(dA)·oligo(dT), the processivity of Pol I is 10–30 nt at 37°C (43).

vi. Rate of polymerization. The maximal rate of polymerization by Pol Ik on some homopolymeric templates is 350 min^{-1}, although the initial rate can be as high as 600 min^{-1}. The rate of elongation by Pol Ik is strongly dependent on the enzyme concentration, e.g., a ninefold increase in enzyme concentration results in an eightfold increase in the rate of elongation of each primer (44). The association–dissociation time on primer-template is very rapid with an average cycling time of less than 0.5 sec.

When the $3' \rightarrow 5'$-exonuclease activity of Pol Ik is inhibited by NaF (5–10 mM) or NMP (5–20 mM), the absolute rate of polymerization increases five- to ninefold relative to the decreased rate of hydrolysis (16).

e. Substrate specificities: $3' \rightarrow 5'$-Exonuclease

i. Substrate structure. The $3' \rightarrow 5'$-exonuclease activity of Pol I (and Pol Ik) requires free 3'-OH termini in ssDNA or dsDNA substrates. The 3'-P terminated or ddNMP-terminated DNA is not hydrolyzed by this activity.

DNAs containing α-phosphorothioate nucleotide analogs are not effectively hydrolyzed by the $3' \rightarrow 5'$-exonuclease activity. With Pol Ik, the thio linkage results in over a 100-fold reduction in the rate of exonuclease activity but not polymerase activity (36).

The $3' \rightarrow 5'$-exonuclease activity of Pol Ik is fourfold greater on ssDNA [e.g., poly(dT)$_{300}$] than on dsDNA [d(A)$_{4000}$·d(T)$_{300}$] (45). The $3' \rightarrow 5'$-hydrolytic activity is undetectable under the conditions of concurrent DNA synthesis.

ii. Product profile. The products of $3' \rightarrow 5'$-exonucleolytic action on ssDNA or dsDNA substrates are exclusively dNMPs.

iii. Rate of exonuclease activity. The turnover number of $3' \rightarrow 5'$-exonuclease activity on (dT)$_{300}$ is 16–19 min^{-1}, much less than that of T4 DNA Pol (~4000 min^{-1}) or T7 DNA Pol (16,000 min^{-1}). In the absence of dATP, the rate of the $3' \rightarrow 5'$-exonuclease activity of Pol Ik on poly(dT)·oligo(dA) template–primer is ~4% of the polymerization rate while, in the presence of dATP, it amounts to as much as 50–60% (16).

The rate of hydrolysis is markedly inhibited by high concentrations of NaCl or KCl (7): at 0.1 M NaCl, a 60–75% inhibition of exonuclease activity is observed and no exonuclease activity is detectable at 0.3 M NaCl.

f. Substrate specificities: 5′-Nuclease

i. Substrate structure. The 5′-nuclease activity, which is present in Pol I but not in its large fragment Pol Ik, requires dsDNA templates with a free 5′ end. The rate of hydrolysis from either end of native DNA is similar (4). Neither primer nor dNTPs are required for cleavage, although the presence of primer can influence the site and rate of cleavages. Denatured or ssDNA is a poor substrate with a k_{cat} of 3.6 min^{-1} on ssDNA [e.g., (dT)$_{170}$] as compared to 30 min^{-1} on dsDNA [(dT)$_{170}$·(dA)$_n$] (50). The 5′-nuclease activity has been shown to be stimulated ~10 times during concomitant DNA synthesis, i.e., in the presence of dTTP.

The 5′-nuclease activity is relatively nonspecific: DNAs with 5′-OH, monophosphate, diphosphate, or triphosphate termini are hydrolyzed at almost equal efficiencies. For certain substrates, the 5′-nuclease also functions as a substrate structure-specific endonuclease.

ii. Product profile. The products of 5′-nuclease-catalyzed exhaustive hydrolysis are predominantly 5′-mononucleotides and some oligonucleotides ($\leq$12 nt). 3′-P terminated DNAs give rise to 5′-dNMP (~80%) and smaller amounts of di-, tri-, and larger oligodeoxynucleotides carrying the 3′-P group. With a 5′-triphosphate-terminated substrate [pppT(dT)$_{300}$], the 5′ hydrolytic activity produces dinucleoside tetraphosphate (pppTpT) as the major product (46), whereas the principal product (~75%) of pT(dT)$_{300}$ is the mononucleotide, pT. The preference for cleavage at the penultimate rather than the terminal diester bond of pppT(pT)$_{300}$ is attributed to the binding of the 5′-triphosphate terminus to the enzyme site ordinarily occupied by dNTP (e.g., pppT) during polymerization.

iii. Substrates for endonuclease activity. The 5'-nuclease activity is presumed to excise the 5' single-stranded arm at the junction with the duplex, most likely between the first two base pairs of the duplex region (47). The 5' arm can be a frayed or displaced oligonucleotides. With DNA substrates containing thymidine dimers and mismatched sequences, the 5'-nuclease activity cleaves phosphodiester bonds as distant as 8 nt from the 5' terminus (48).

The 5'-nuclease activity is believed to be responsible for the degradation of the RNA in DNA · RNA hybrids, thus an RNase H activity (49). The RNase H activity has been demonstrated with *Taq* DNA Pol as well (47).

iv. Separate activity of 5'-nuclease domain. The 5'-nuclease activity is retained in the small N-terminal fragment (34 kDa) produced by limited proteolytic cleavages of Pol I. The small fragment resembles the 5'-nuclease of the intact Pol I in degrading DNA to mononucleotides and oligodeoxynucleotides and in its capacity to excise mismatched regions such as thymine dimers (51). The small fragment has the ability to hydrolyze dsDNA but not ssDNA. However, it differs from the intact Pol I in that dNTPs fail to stimulate the exonuclease or to increase the proportion of oligonucleotides among the products. When the small fragment is mixed with the large fragment (68 kDa) in the presence of nicked DNA and dNTPs, the 5'-nuclease activity can be stimulated in the same fashion as in Pol I, suggesting that the two fragments bind adjacent to each other on the nick and perform a coordinated polymerization and 5'-nuclease function. Individual fragments show no direct affinity to each other in the presence or absence of DNA.

g. Mechanism of action

i. Kinetic mechanism of polymerase action. Polymerization of nucleotides occurs in the 5' → 3' direction by nucleophilic displacement at the α-phosphorus of a dNTP by the 3'-hydroxyl of the DNA (or RNA) primer (52). The polymerization reaction proceeds by an ordered mechanism (Scheme 6.1) which consists of the following steps (53,54): a template DNA (D_n, containing n nucleotides) binds the polymerase (E) (step 1), a dNTP is selected from the pool of nucleotides to form a ternary complex (step 2), the polymerase undergoes a conformational change to a catalytically active form (E') (step 3), chemical catalysis takes place at a rate faster than the conformational change (step 4), a second conformational change occurs (step 5), the PP_i product is released from the ternary complex (step 6), and either the polymerization product (D_{n+1}) dissociates from the polymerase (step 7) or translocates to start new cycles (steps 2 to 5) of polymerization in the presence of dNTPs.

Individual rate constants for the Pol Ik-catalyzed reaction are listed in Table 6.4. In the kinetic mechanism (Scheme 6.1), the conformational change (step 3) of the ternary complex [E · DNA · dNTP] into a catalytically active form [E' · DNA · dNTP] is assigned to be a partial rate-determining step. If the first nonchemical step (k_3) limits the rate of single nucleotide incorporation, the second conformational change (k_5) after the chemical step (k_4) limits the rate of processive synthesis. The kinetic mechanism of exo$^-$ T7 DNA Pol is compatible with that

II. DNA-Directed DNA Polymerases 363

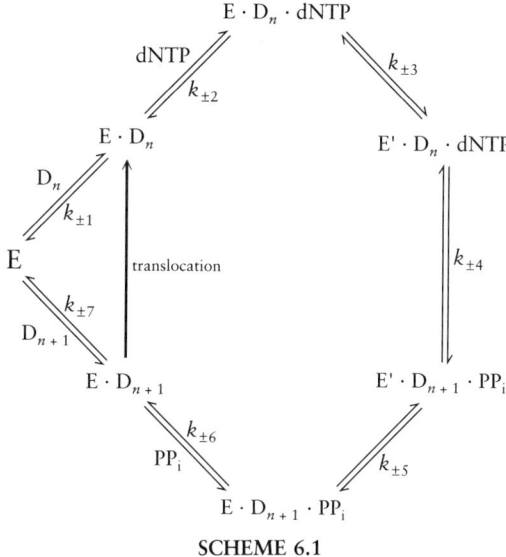

SCHEME 6.1

TABLE 6.4 Rate and Equilibrium Constants for Kinetic Mechanism of Pol Ik[a]

Parameter		Value
K_{d1}	k_{-1}/k_1	5 nM
K_{d2}	k_{-2}/k_2	5 μM
K_4	k_4/k_{-4}	2.5
K_{d6}	k_{-6}/k_6	100 μM
k_3		50 sec^{-1}
k_{-3}		0.5 sec^{-1}
k_4		1000 sec^{-1}
k_{-4}		400 sec^{-1}
k_5		15[b] sec^{-1}
k_{-5}		15[b] sec^{-1}
k_7		0.06 sec^{-1}
k_{-7}		$1.2 \times 10^7 \, M^{-1} \, sec^{-1}$

[a] Rate constants are from the best fit of single-turnover kinetics data to the reaction mechanism described by Scheme 6.1. The kinetic parameters are for the incorporation of dATP (correct nucleotide) on 13/20(T) primer/template at pH 7.4 (50 mM Tris–Cl) and 21°C (53,54).

[b] Rate constants from Dahlberg and Benkovic (55).

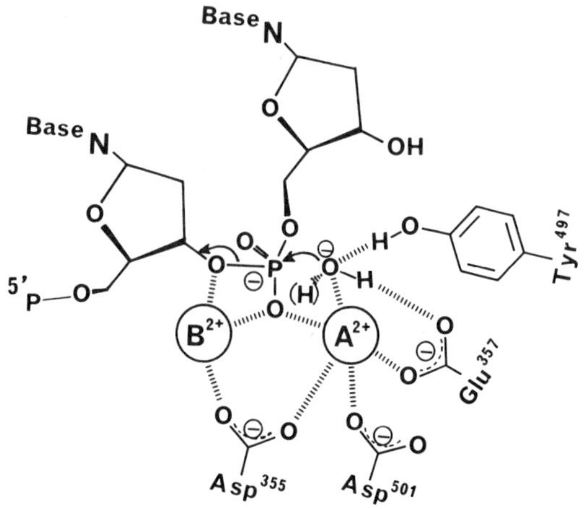

FIGURE 6.2 A two-metal catalytic mechanism proposed for the 3' → 5'-exonuclease reaction. Metal ion B (Mg^{2+}) stabilizes the pentacovalent transition state and the leaving 3' oxyanion. Metal ion A (Zn^{2+}) promotes the formation of hydroxide ion which is the attacking nucleophile. [After L. S. Beese and T. A. Steitz (1991). *EMBO J.* **10**, 25–33.]

of Pol Ik. In the T7 DNA Pol reaction, the RDS is also the conformational change of the enzyme in the ternary complex, although the rate is six times faster than that of Pol Ik (56). Under the conditions of multiple turnover, the synthesis of DNA is limited by the E · DNA dissociation rate (step 7). Translocation of the product to the next available polymerization site occurs at least five times faster in Pol Ik than in Pol I (57). The relative magnitude of the rate constants for each step of the polymerase reaction is dependent on the DNA sequence. The dissociation of E · D_{n+1} also exhibits a prominent DNA sequence dependence for up to 12 bp from the 3' terminus.

The polymerase and exonuclease reactions all proceed with a stereochemical inversion, suggesting that the chemical catalysis occurs via a single direct phosphoryl transfer between the donor and the acceptor substrates.

ii. Catalytic mechanism of 3' → 5'-exonuclease. The 3' → 5'-exonuclease activity is presumed to function by a two-metal ion enzymatic mechanism (58,59). The mechanism involves a catalytic water molecule in the hydrolysis of the phosphodiester bond of the 3'-terminal nucleotide positioned in the exonuclease site (Fig. 6.2). One of the metal ions, apparently Mg^{2+}, is coordinated with the phosphate oxygens, stabilizing the pentacovalent transition-state intermediate and assisting the leaving of 3' oxyanion from the penultimate nucleotide. In the absence of any protein side chain in the vicinity to assume a direct role in catalysis, the group that performs the nucleophilic attack is presumed to be the hydroxide ion from the water molecule which is activated by the metal ion A (presumably Zn^{2+}) and held in a catalytically appropriate position by two amino acid residues, Tyr-497 and Glu-357. The two metal ions are 3.9 Å apart in crystal structures.

The configuration and presumed roles of the two divalent metal ions are similar to those of alkaline phosphatase (see Section I,A, Chapter 5).

iii. Substrate shuttling between polymerase and exonuclease sites. A DNA (single stranded or double stranded) binds to the enzyme with the 3′-terminal 4 nt (spanning a distance of ~15 Å) in the exonuclease site and the remainder extending toward the polymerase site. The termini of duplex DNA apparently undergo melting by the protein which provides stabilizing hydrophobic interactions via Phe-473, Leu-361, and His-666 to the 3′-terminal 3 nt. Crystallographic data suggest that the DNA binding in the polymerase and exonuclease sites requires the DNA to meet a remarkable physical constraint of about 80° bending (105).

For cooperative functions of the polymerase and exonuclease, the 3′ terminus of DNA is presumed to shuttle rapidly without dissociation between the two active sites which are ~30 Å apart. The sliding path involves 4–5 bp of duplex DNA plus four to five bases of the single-stranded frayed end. Photolabeling studies with azido-DNA also suggest that the polymerase active site makes contacts with five to seven bases of duplex DNA (60). When a duplex contains unpaired bases in the primer strand, the 3′ primer terminus resides predominantly at the exonuclease site. The RDS for exonuclease activity of Pol Ik is thought to be the transfer of the primer terminus from the polymerase to the exonuclease site.

iv. Mechanism of polymerase fidelity. The high fidelity of Pol Ik is assured by a three-stage mechanism (53–55).

The first stage is the 10^4- to 10^6-fold discrimination primarily due to a dramatically reduced rate of phosphodiester bond formation for incorrect nucleotides. There is also a small contribution from selective dNTP binding. The identity of mismatching base pairs plays a major role in discrimination, with the surrounding sequence contexts playing a significant but secondary role. Under single-turnover conditions, for example, an exonuclease-deficient (exo$^-$) Pol Ik exhibits the following trends (63,106).

(1) Misincorporation of purines, especially opposite wobble base pairs (e.g., G·T and A·C), is relatively frequent, followed by some Pu·Pu and then by Py·Py. Overall, the k_{cat} ranges 2000-fold; the fastest one is dGMP incorporation opposite template T (G·T) which occurs at a rate 1/300 of the correct (dAMP) incorporation and the slowest one is dCMP incorporation opposite template C (C·C).

(2) Misincorporation efficiencies are "asymmetrical." For instance, dAMP is incorporated opposite template C (A·C) five times more efficiently than dCMP incorporation opposite template A (C·A).

(3) The K_m values vary within 50-fold, from 22 μM for dGTP (G·A) to ~1.2 mM for dCTP (C·T). The K_m for dCMP misincorporation represents a 240-fold increase over the K_m (= 5 μM) for correct incorporation. In fact, all cases of Py·Py misincorporation are, compared with Pu misincorporation, accompanied by highly increased K_m.

(4) The efficiency of dAPTP incorporation opposite template T is sensitive within a factor of 3 to the identity of the 5'-nearest neighbor base of the primer terminus and decreases in the order of G > C > A > T (107). In fact, the frequency of misincorporation is influenced by residues within two to three bases on either side of the mismatch.

Incorrect nucleotides that traverse the first-stage physical discrimination by a polymerase may get inserted but do not accumulate off the polymerase. Misinsertion results in a slower dissociation of the incorrect DNA product from the enzyme, providing a higher probability for the editing exonuclease to function and thus a fidelity increase of 4- to 61-fold. Finally, misinserted nucleotides are not locked in by addition of the next correct dNTP. The nucleotide sealing action occurs extremely slowly, resulting in a further fidelity increase of 6- to 340-fold. Ideal functioning of the triple-check process would lead to an error frequency of $\sim 10^{-10}$, close to the maximum fidelity estimated for *in vivo* DNA replication.

When viewed from the mismatched nucleotides inserted at the 3' primer termini, the fidelity ensuring process involves the following three factors: (i) the mismatch increases the probability of forming a frayed or single-stranded end in the exonuclease site, (ii) the mismatch renders the primer terminus a poorer substrate for additional DNA synthesis in the polymerase site, and (iii) an increase of the lag time favors the 3' → 5'-exonuclease to perform editing functions.

2. Applications

E. coli DNA Pol I and its large fragment Pol Ik are used as key reagents in a variety of applications which include molecular cloning, nucleotide sequencing, and preparation of DNA probes.

i. Second-strand DNA synthesis. Pol I is the first and one of the most important DNA polymerases used for the synthesis of second-strand DNA as part of the steps involved in molecular cloning of cDNA (refer to Section III, in this chapter). Depending on the strategy of cDNA synthesis, either Pol I or Pol Ik can be employed. Pol Ik is the enzyme of choice for many applications, including *in vitro* site-directed mutagenesis (see Appendix A).

ii. Nucleotide sequencing. Although much of the work involved in nucleotide sequencing is now shared with other DNA polymerases such as T7 DNA Pol (refer to Section II,C, this chapter) and *Taq* DNA Pol (see Section II,D, this chapter), Pol Ik (and not Pol I) is extensively used in the short-range nucleotide sequencing by Sanger's dideoxy chain termination method (64).

A unique variation of the nucleotide sequencing that exploits the relaxed specificity of Pol Ik in the presence of Mn^{2+} is the method of *partial ribonucleotide substitution* (65). A DNA synthesis is carried out in the presence of Mn^{2+}, all four dNTPs, and one of the four rNTPs under the conditions that result in $\sim 2\%$ ribonucleotide substitution. The cleavage of the products at the positions of partial ribosubstitution by alkali (0.3 M KOH + 10% piperidine) results in a ladder of fragments terminated by the respective ribonucleotides.

iii. DNA amplification reactions

(a) Polymerase chain reaction. Pol Ik was the key enzyme used during the early developmental stages of PCR technology, but it has now been replaced largely by the thermostable DNA polymerase, *Taq* DNA Pol. Nevertheless, Pol Ik remains to be a valuable enzyme for certain PCR applications that require polymerization reactions at lower temperatures.

(b) Strand displacement amplification reaction. The strand displacement activity of the Klenow fragment, especially exonuclease-deficient Klenow (exo$^-$ Klenow), has been exploited into an isothermal *in vitro* DNA amplification technique, called *strand displacement amplification* (SDA) (66). SDA is based on the ability of *Hinc*II (or other appropriate enzymes) to nick the unmodified strand of a hemiphosphorothioate form of its recognition site, which is provided as an integral part of the primers, and the ability of the exo$^-$ Klenow to extend the 3' end at the nick and to displace the downstream DNA strand. Briefly, the procedure for this isothermal amplification technique involves the following: (i) the target DNA sample is first heat denatured in the presence of two sets of primers (two outside and two inside primers), (ii) *Hinc*II and exo$^-$ Klenow are added, and (iii) the sample is incubated at 37°C in the presence of nucleotides dGTP, dCTP, TTP, and dATPαS. Repeated cycles of *Hinc*II nicking, primer extension/displacement, and automatic entrance of the displaced strand into SDA can result in an exponential amplification of the target DNA greater than 10^7-fold. The SDA technique with exo$^-$ Klenow is critically dependent on the length of the target sequence and is useful for the amplification of sequences only in a short range due to the limited processivity of the polymerase.

iv. Preparation of DNA hybridization probes

(a) Nick translation. This method of preparing labeled DNA probes employs Pol I (and not Pol Ik), exploiting the 5' → 3'-exonuclease activity together with the polymerase activity. The actual reaction includes a small amount of DNase I as the nicking agent (for details, see Appendix A).

(b) Random-hexamer method. This is an alternative to nick translation, which has proved to be simpler and more efficient than nick translation in the preparation of labeled probes (for details, see Appendix A).

v. Blunt-end formation of duplex DNA.
Duplex DNAs with cohesive termini can be blunted by using Pol Ik in one of the following ways.

When the DNA carries 3'-recessed ends, they can be filled in by the polymerase reaction in the presence of appropriate dNTPs. In the presence of limited, selectively complementary dNTPs, partial filling in can be obtained, allowing versatile use of the partially cohesive termini.

When the substrate DNA carries 3'-protruding ends, it can also be blunted by virtue of the 3' → 5'-exonuclease activity of Pol Ik. For a small number of nucleotides, e.g., 4 nt resulting from restriction enzyme digestions, Pol Ik is ideal.

To obtain blunt ends from the DNA with longer protruding 3' termini, the use of the more potent 3' → 5'-exonuclease activity of T4 DNA Pol is recommended.

vi. Labeling of duplex DNA. Pol Ik is one of the most frequently used enzymes in labeling the 3' ends of dsDNA fragments generated by restriction enzyme digestion. When the polymerase reaction is carried out on 3'-recessed ends in the presence of radiolabeled dNTPs, the filling-in reaction produces blunt, labeled dsDNAs suitable for DNA sizing or DNA sequencing by the Maxam–Gilbert chemical sequencing method (see Appendix A).

In some cases, the labeling reaction can be performed immediately after digestion with a restriction enzyme(s), preferably after heat denaturation of the restriction enzyme(s). However, removal of the restriction enzymes by phenol/chloroform extraction and ethanol precipitation of DNA is more preferable (67); this treatment produces uniform, superior quality labeling probably as a result of eliminating the restriction enzymes which would otherwise stick to the DNA even after the heating and consequently hinder the correct activity of Pol Ik.

Duplex DNA with blunt ends can be labeled by terminal nucleotide exchange with the radioactive nucleotides. In fact, for this end labeling, Pol Ik is much more reliable and easier to control than T4 DNA Pol (68).

Duplex DNAs with 3'-protruding ends can also be labeled using Pol Ik in a two-step procedure. First, the 3' ends are exonucleolytically digested in the absence of added dNTPs by the 3' → 5'-exonuclease activity and, second, dNTPs containing a labeled nucleotide are provided to start a replacement synthesis. Note that the replacement synthesis method of labeling can be more efficiently performed by the use of T4 DNA Pol whose exonuclease activity is substantially stronger.

vii. DNA polymerase-based *in vitro* mutagenesis. This method is based on the observation that, under the conditions of limiting nucleotides or polymerase, error-prone DNA polymerases can incorporate mismatching nucleotides. Furthermore, these polymerases have the tendency to pause at the sites, resulting in a single nucleotide addition. This misincorporation can then be trapped into the mutant strand by applying a second round of the polymerase reaction under favorable reaction conditions (41). This strategy of *in vitro* mutagenesis can be applied to DNA Pol I (and Pol Ik) as well as to other polymerases such as T7 DNA Pol (Sequenase) and reverse transcriptases (refer to Section III,A, this chapter), providing an alternative to the strategies used in site-directed mutagenesis (see Appendix A).

An ingenious variation of the strategy that increases the yield of mutagenesis up to 42% employs α-thiophosphate nucleotides to block the 3' → 5'-exonuclease activity. This method was initially applied to gap repair and thus became known as the *gap repair mutagenesis* (69). In this strategy, a short single-stranded gap(s) is created by limited exonuclease digestions at the nicks placed at either random or predetermined sites on plasmid DNA molecules. The Pol Ik reaction is then performed with single dNTPαS to force misincorporations, which is followed by gap filling in the presence of all four dNTPs.

viii. **Time-point restriction mapping.** The second-strand DNA synthesis catalyzed by Pol Ik provides a simple nonisotopic method for constructing restriction maps of an insert DNA cloned in ssDNA-producing vectors (70). The unidirectional DNA synthesis primed by commercially available sequencing primers allows the generation of time-dependent double-stranded structures suitable for restriction enzyme cleavages. When the digested products are separated on an agarose gel, the restriction bands appear in the order of their proximity to the priming site.

ix. **Other analytical applications.** In addition to the use of Pol I (and Pol Ik) in diverse applications oriented to recombinant DNA technology, the substrate specificities of Pol Ik have been exploited as valuable structural probes. In mapping hyperactive nucleotides in B–Z transition junctions (71), for example, thymine glycols which are derived from hyperactive T residues following a reaction with a Py-specific reagent OsO_4 serve as a stop signal in a primed DNA elongation by Pol Ik.

3. STRUCTURE

i. **Features of primary and secondary structures.** Pol I is a single polypeptide consisting of 928 amino acid residues (Met-1 to His-928, M_r 103,116) (72). It is folded into two domains in a roughly spherical shape with a 65-Å diameter. The polypeptide contains two Cys residues: one (Cys-262) in the "small" fragment and the other (Cys-907) in the "large" fragment. Pol I contains 43% α helix, 17% β sheets, 58 β turns, and several supersecondary structural elements (73).

E. coli DNA Pol I, especially the Klenow fragment, exhibits colinear regions of homology with the gene 5 subunit of T7 DNA Pol and small regions of similarity to the ε subunit of E. coli DNA Pol III. Pol I shares little homology with T4 DNA Pol. Nevertheless, sequence alignment with other DNA polymerases, including T4 and T7 DNA polymerases, suggests that the C-terminal region spanning ~340 amino acids contains six highly conserved motifs and critical residues associated with polymerization function (74).

The cleavage site in Pol I that gives rise to the large (68 kDa) and small (34 kDa) fragments on partial digestion with subtilisin is the peptide bond between Thr-323 and Val-324. The C-terminal large fragment (Pol Ik) contains both $5' \rightarrow 3'$-polymerase and $3' \rightarrow 5'$-exonuclease activities. The N-terminal small fragment contains a separate $5'$-nuclease activity.

ii. **Structural features of Klenow fragment.** Pol Ik (M_r 68,064) consists of 605 amino acid residues from 324 to C-terminal 928 (or 65%) of Pol I. The overall structure of Pol Ik can be divided into two subdomains (58,75,76,105). The N-terminal subdomain comprising approximately the first 200 amino acids (residues 324–517 on the Pol I sequence) corresponds to the $3' \rightarrow 5'$-exonuclease domain (89,108). This domain consists of four strands of β sheets and flanking α helices, and binds dTMP which is an exonuclease inhibitor. Site-directed mutagenesis, e.g., D424A single mutation or D355A/E357A double matuation, abolishes the $3' \rightarrow 5'$-exonuclease activity while fully maintaining the polymerase activity. The 3' → 5'-exonuclease domain also contains binding sites for two divalent metal ions, Zn^{2+} and Mg^{2+}.

The C-terminal 400 amino acid residues of the larger subdomain constitute the polymerase domain. It consists mainly of α helices which form a cleft of the configuration suitable for binding DNA and dNTPs. Part of the domain (residues 558–637) forms a thumb-like structure which, on contact with DNA, moves toward the 3' → 5'-exonuclease domain (105). The DNA-induced conformational change gives rise to a second cleft between the thumb and the exonuclease domain. The second cleft, into which the duplex DNA binds, runs at nearly right angles to the first cleft that contains the polymerase active site.

Binding of duplex DNA occurs exclusively through interactions of the protein with the DNA phosphate backbone, which is consistent with the requirement that the enzyme binds any DNA independent of its sequence. Duplex DNA containing the primer terminus is presumed to be positioned in the cleft proximal to the exonuclease domain, while the single-stranded template enters from the distal end of the cleft.

In polymerization mode, the primer strand is base paired to the template in the polymerase cleft and its 3' terminus is near the divalent metal ions. In editing mode, the frayed or mismatching 3' terminus of the primer is supposed to move from the polymerase site to the exonuclease site without dissociation.

iii. **Active sites.** For a multifunctional enzyme like *E. coli* DNA Pol I with functionally distinct structural domains, the active center can be subdivided into several subsites.

(a) Polymerase catalytic site. The polymerase catalytic site, located at the bottom of the large cleft, binds ssDNA or nicks and ends of dsDNA. It also binds dNTP (see below) and recognizes the 3'-OH primer terminus for dNMP addition. The contact with DNA occurs with ≤7 bp in the 3'-terminal duplex region (60,79) and in the order of $d(pC)_n < d(pT)_n < d(pG)_n \approx d(pA)_n$ (80). The single-stranded region of the template apparently binds the polymerase in a helical conformation which allows the strongest hydrophobic interactions at the primer 3' terminus (79).

The catalytic site contains a tight cluster of the most highly conserved residues that make contact with phosphates of DNA or dNTP, and with metal ions. Residues interacting with the primer strand include Asn-675, Asn-678, Lys-635, Arg-631, and Thr-609, while residues interacting with the template strand include Arg-682, His-734, and Ser-582 (105). Lys-635, a residue in the thumb subdomain, is particularly close to the primer strand and plays a role in processivity of the polymerase (90). Catalytically important residues, which are implicated in the binding of divalent metal ions, are Asp-705, Asp-882, and Gln-849 (81–83). The site-specific substitution of Asp-882, Arg-668, or Gln-849 results in a large decrease in k_{cat} without introducing significant changes in K_m for dNTPs.

(b) dNTP-binding site. As part of the polymerase active site, several residues such as Arg-754, Lys-758, Phe-762, Tyr-766, Arg-841, and His-881 are implicated in dNTP binding (83–88). Tyr-766 plays an important role in maintaining poly-

merase fidelity. Y766S mutation results in a fivefold reduction of catalytic efficiencies for matching nucleotide incorporation, whereas it increases the efficiencies for misincorporation (63). Phe-762 is another important residue for fidelity: F762Y mutation decreases discrimination against ddNTPs 250- to 2000-fold (34). Lys-758 not only participates in the dNTP binding but also is required for the translocation of the polymerase along the DNA template.

(c) $3' \rightarrow 5'$-Exonuclease active site. The exonuclease site recognizes the unpaired 3'-terminal nucleotide for exonucleolytic action. The exonuclease domain contains two binding sites for divalent metal ions: a low affinity site (B) for Mg^{2+} and a high affinity site (A) for Zn^{2+}. Site A consists of the carboxylate groups of Asp-355, Glu-357, and Asp-501 (59,75). The B site Mg^{2+} is coordinated with Asp-355. Both metal ions participate in exonucleolytic catalysis (Fig. 6.2). Mutagenesis of the carboxylic residues, e.g., Asp-355 (D355A), Asp-424 (D424A), and Asp-501, greatly reduces the exonuclease activity but not the polymerase activity (89). Mutations at Glu-357 decrease the exonuclease activity less severely, suggesting that Glu-357 plays a minor role in metal binding.

The binding of dTMP, an exonuclease inhibitor, involves its phosphate and ribose moieties but not the base; the 5'-P provides a fourth ligand to the Mg^{2+} ion while the 3'-OH of the ribose forms a H-bond with Thr-358.

(d) 5'-Nuclease active site. Amino acid sequence comparison of the 5'-nuclease domain of E. coli DNA Pol I with several bacterial polymerase-associated 5'-nucleases and independent $5' \rightarrow 3'$-exonucleases from T phages has revealed six conserved regions (motifs A to F) (108). Mutations in some of the conserved residues result in polymerases with defective 5'-nuclease activity: e.g., Y77C (*polA107*), G184D (*polA480ex*), and G192D (*polA214*). Gly-103, a residue not falling within a motif but highly conserved in bacterial polymerases and absent in T phage exonucleases, apparently plays a role in the interaction of the 5'-nuclease and polymerase domains. G103E mutation (*polA4113*) results in a DNA Pol I that is thermolabile for both 5'-nuclease and polymerase activities.

iv. $3' \rightarrow 5'$-Exonuclease-deficient Klenow. The larger subdomain (residues 515–928) of Pol Ik has been cloned and expressed in E. coli. The product retains the polymerase activity but no measurable $3' \rightarrow 5'$-exonuclease activity (77). This *Exo-minus Klenow,* available from USB, has a specific polymerase activity about 10-fold lower than parent Pol Ik.

v. Physical parameters. For Pol I, one A_{280} unit corresponds to 1.18 mg/ml, and the ratio of A_{280}/A_{260} is 1.81 in 10 mM NH_4HCO_3 (45). The $s^{\circ}_{20,w}$ is 5.6 S in 5 mM K-P (pH 6.8) (91). For Pol Ik, the A_{278} of a 1-mg/ml solution is 0.93 (or $\varepsilon_{278} = 6.32 \times 10^4\ M^{-1}\ cm^{-1}$) (23).

4. GENETICS

i. Structure and organization of E. coli DNA polymerase I gene. Pol I is the product of the *polA* gene positioned at the 87-min locus on the E. coli K12 linkage map (92). The *polA* gene is preceded by the ribosomal RNA operon, *rrnA*. The complete nucleotide sequence of the *pol*A gene was determined using plasmid subclones of *polA1* (72), which could be more stably propagated than the wild-type *polA* clone.

The ORF of the *polA* gene contains 928 codons and is terminated with UAA. The *polA* gene was initially cloned by ligating the *Hin*dIII-digested *E. coli* DNA to the phage λ vector and transducing the *polA1 E. coli* host strain (93). The upstream sequence of *polA* ORF contains many sites having some homology to the prototype bacterial promoter sequence. The most plausible promoter sequence is presumed to be one of the Pribnow-like boxes at −28 to −22 (CATAATC) or −150 to −144 (AATAATT). However, in neither case is there convincing homology at the −35 region, probably reflecting the reason for the low level expression of the *polA* gene. Potential base pairing with the 3'-end of 16S rRNA in the initiation complex is limited to the three base sequence GGA [−7 to −5], further suggesting the poor potential for initiation of translation.

Transcription of *polA* mRNA is thought to be terminated within the 100 nt downstream of the *polA* ORF. This region contains no obvious dyad symmetries analogous to the characteristic rho-independent terminators, suggesting that the termination may be rho dependent.

ii. **Mutant studies.** *E. coli* DNA Pol I is thought to function *in vivo* in excision repair and in the removal of RNA primers from Okazaki fragments during discontinuous replication. A number of *polA* mutants have been isolated, providing insight into the physiological roles as well as the structure of the enzyme. Some of the *polA* mutants with instructive phenotypes are as follows.

(a) ***polA1.*** This is an *amber* mutation but has a ~1% polymerase activity due to read-through of the mutation (94). The mutant enzyme (amber peptide) has a near normal level of 5'-nuclease activity. The *polA1* has been identified as the mutation at codon 342 from Trp (TGG) to amber (TAG) (72). The mutation greatly increases the sensitivity of the *E. coli* to UV irradiation. It is nonlethal and recessive to the wild-type gene in partial diploids. The *polA1* mutant grows at a normal rate, which led to the discovery of Pol II and Pol III.

(b) ***polA5.*** This is a nonthermolabile (at 43°C) mutation from Gly to Arg at position 850. The mutation impairs the interaction of the enzyme with templates during the template translocation step (95). The polymerase activity is reduced to less than 1% of the wild-type enzyme. The processivity is also reduced to one-fifth of the wild-type enzyme. The *polA5* polymerase has normal 3' → 5'-exonuclease activity. The 5' → 3'-exonuclease activity appears also normal in the absence of DNA synthesis, but it is sharply reduced under conditions of DNA synthesis such as nick translation.

(c) ***polA6.*** This is an Arg to His mutation at position 690. This mutation affects the DNA-binding capacity of the enzyme (96). Both 5'- and 3' → 5'-exonucleolytic activities are not significantly different from the *polA*$^+$ enzyme, although the *polA6* enzyme has a slightly higher specific activity. The *polA6* polymerase shows an altered pH optimum (at pH 8.5) for polymerization compared with the pH 7.5 optimum for the wild-type enzyme. The pH optimum for nucleolytic activities remains unchanged at about pH 8.6.

(d) polA12. This is a nonlethal *ts* mutant (43°C). The mutant enzyme shows a 10-fold decreased efficiency in nick translation at 30°C when its polymerase and 5' → 3'-exonuclease activities are not grossly defective (97). The *polA12* polymerase shows aberrant physical properties such as a slower electrophoretic mobility, a lower sedimentation coefficient, and instability at low ionic strengths, suggesting a significant alteration in the tertiary structure.

(e) polAex1. This is a temperature-sensitive, conditionally lethal mutation in the 5'-nuclease (98). Although its polymerase and 3' → 5'-exonuclease activities show no defects, the mutant cannot overcome the severe deficiency in the 5'-nuclease induced at 43°C.

iii. Relationship of Pol I to other *E. coli* DNA polymerases.

In addition to Pol I, two other DNA polymerases, namely Pol II and Pol III, numbered in the order of their discovery, have been purified from *E. coli*. The polymerases differ in their molecular size, ability to interact with accessory proteins, and biological functions.

Like mammalian Pol β, Pol I is primarily involved in the repair of damaged duplex DNA and in the processing of Okazaki fragments. Like mammalian Pol α, Pol II (120 kDa) and Pol III (165 kDa) are involved in DNA replication and in several key cellular repair processes as well.

Pol III designates the polymerase that has been isolated in a "monomeric" form and used to perform simple gap-filling reactions with a k_{cat} of 480 min^{-1} and a processivity of 10 nt (1,99). Pol III consists of three subunits: polymerase (α subunit, *dnaE* gene product), 3' → 5'-exonuclease [ε subunit, *dnaQ* (or *mutD*) gene product], and θ subunit of unknown function. *In vivo*, Pol III functions as a multisubunit complex ("holoenzyme") which contains, in addition to the aforementioned three subunits, a processivity factor (also called a "sliding clamp," β subunit), a polymerase dimerization factor (τ subunit), and accessory proteins (γ complex). The Pol III holoenzyme is highly processive (>5000 nt), accurate, and fast with a polymerization rate of 1200 sec^{-1}. Pol II is believed to be a subassembly product of the Pol III holoenzyme.

5. SOURCES

i. *E. coli* DNA Pol I (Kornberg enzyme).

In wild-type *E. coli*, which has been the traditional source of Pol I, there are ~400 Pol I molecules per bacterium. Pol I can now be more efficiently obtained from the cells infected with the λ *polA*-transducing phage that produce Pol I at an 80-fold higher yield (100) compared with the conventional yield of 10 mg per kg of cell paste from uninfected *E. coli* (101). The specific polymerase activity of purified Pol I is ~24,000 U/mg protein.

Commercial enzyme is usually supplied in 100 mM K-P (pH 7.0), 1 mM DTT (or 1 mM 2-MSH), and 50% glycerol. The recommended storage temperature is −20°C.

ii. Klenow fragment. The Klenow fragment was originally produced by limited proteolysis of Pol I using a bacterial protease, subtilisin, at pH 6.5 in K-P buffer (102). Some commercial Pol Ik are produced by the proteolytic digestion of the purified, cloned Pol I. The cloned gene for Pol I has also been modified to overproduce the Klenow fragment in *E. coli* directly (103). In this recombinant DNA work, the appropriate Klenow fragment "gene" was subcloned into an *E. coli* expression vector and expressed under the control of the λ p_L promoter and *E. coli lacZ* ribosome-binding sequence.

References

1. Kornberg, A., and Baker, T. A. (1991). "DNA Replication," 2nd ed. W. H. Freeman, New York.
2. Deutscher, M. P., and Kornberg, A. (1969). *JBC* **244**, 3019–3028.
3. Richardson, C. C., Schildkraut, C. L., Aposhian, V., and Kornberg, A. (1964). *JBC* **239**, 222–231.
4. Deutscher, M. P., and Kornberg, A. (1969). *JBC* **244**, 3029–3037.
5. Travaglini, E. C., Mildvan, A. S., and Loeb, L. A. (1975). *JBC* **250**, 8647–8656.
6. Beyersmann, D., and Schramm, G. (1968). *BBA* **159**, 64–74.
7. Lehman, I. R., and Richardson, C. C. (1964). *JBC* **239**, 233–241.
8. Slater, J. P., Mildvan, A. S., and Loeb, L. A. (1971). *BBRC* **44**, 37–43.
9. Slater, J. P., Tamir, I., Loeb, L. A., and Mildvan, A. S. (1972). *JBC* **247**, 6784–6794.
10. Mullen, G. P., Serpersu, E. H., Ferrin, L. J., Loeb, L. A., and Mildvan, A. S. (1990). *JBC* **265**, 14327–14334.
11. Tabor, S., and Richardson, C. C. (1989). *PNAS* **86**, 4076–4080.
12. Berg, P., Fancher, H., and Chamberlin, M. (1963). *In* "Symposium on Informational Macromolecules," p. 467. Academic Press, New York.
13. Van de Sande, J. H., Loewen, P. C., and Khorana, H. G. (1972). *JBC* **247**, 6140–6148.
14. Doronin, S. V., Nevinsky, G. A., Malygina, T. O., Podust, V. N., Khomov, V. V., and Lavrik, O. I. (1989). *FEBS Lett.* **259**, 83–85.
15. Khan, N. N., Wright, G. E., Dudycz, L. W., and Brown, N. C. (1985). *NAR* **13**, 6331–6342.
16. Potapova, I. A., Nevinsky, G. A., Khomov, V. V., and Lavrik, O. I. (1990). *FEBS Lett.* **277**, 109–111.
17. Allen, D. J., and Benkovic, S. J. (1989). *Biochem.* **28**, 9586–9593.
18. Freeman-Wittig, M.-J., Welch, W., Jr., and Lewis, R. A. (1989). *Biochem.* **28**, 2843–2849.
19. Mikhailov, V. C., Ataeva, D. O., Marlyev, K. A., and Atrazhev, A. M. (1989). *Mol. Biol. (USSR)* **23**, 306–313.
20. Modak, M. J. (1976). *BBRC* **71**, 180–187.
21. Kelly, R. B., Cozzarelli, N. R., Deutscher, M. P., Lehman, I. R., and Kornberg, A. (1970). *JBC* **245**, 39–45.
22. Masamune, Y., and Richardson, C. C. (1971). *JBC* **246**, 2692–2701.
23. Setlow, P., Brutlag, D., and Kornberg, A. (1972). *JBC* **247**, 224–231.
24. Lechner, R. L., and Richardson, C. C. (1983). *JBC* **258**, 11185–11196.
25. Clark, J. M. (1991). *Gene* **104**, 75–80.

26. Okayama, H., and Berg, P. (1982). *Mol. Cell. Biol.* **2**, 161–170.
27. Gubler, U., and Hoffman, B. J. (1983). *Gene* **25**, 263–269.
28. Karkas, J. D., Stavrianopoulos, J. G., and Chargaff, E. (1972). *PNAS* **69**, 398–402.
29. Loeb, L. A., Tartof, K. D., and Travaglini, E. C. (1973). *Nature NB.* **242**, 66–69.
30. Wells, R. D., Flugel, R. M., Larson, J. E., Schendel, P. F., and Sweet, R. W. (1972). *Biochem.* **11**, 621–629.
31. Gulati, S. C., Kacian, D. L., and Spiegelman, S. (1974). *PNAS* **71**, 1035–1039.
32. Nevinsky, G. A., Veniaminova, A. G., Levina, A. S., Podust, V. N., Lavrik, O. I., and Holler, E. (1990). *Biochem.* **29**, 1200–1207.
33. Nevinskii, G. A., Nemudraya, A. V., Levina, A. S., Lokhova, I. A., Gorn, V. V., and Khomov, V. V. (1990). *Mol. Biol. (USSR)* **24**, 96–103.
34. Tabor, S., and Richardson, C. C. (1995). *PNAS* **92**, 6339–6343.
35. Burgers, P. M. J., and Eckstein, F. (1979). *JBC* **254**, 6889–6893.
36. Gupta, A. P., Benkovic, P. A., and Benkovic, S. J. (1984). *NAR* **12**, 5897–5911.
37. Kunkel, T. A., Eckstein, F., Mildvan, A. S., Koplitz, R. M., and Loeb, L. A. (1981). *PNAS* **78**, 6734–6738.
38. Engel, J. D., and von Hippel, P. H. (1978). *JBC* **253**, 935–939.
39. Victorova, L. A., Dyatkina, N. B., Ju. Mozzherin, D., Atrazhev, A. M., Krayevsky, A. A., and Kukhanova, M. K. (1992). *NAR* **20**, 783–789.
40. Clayton, L. K., Goodman, M. F., Branscomb, E. W., and Galas, D. J. (1979). *JBC* **254**, 1902–1912.
41. Preston, B. D., Zakour, R. A., Singer, B., and Loeb, L. A. (1988). In "DNA Replication and Mutagenesis" (R. E. Moses and W. C. Summers, Eds.), pp. 196–207. American Society for Microbiology, Washington, D.C.
42. Bambara, R. A., Uyemura, D., and Choi, T. (1978). *JBC* **253**, 413–423.
43. Das, S. K., and Fujimura, R. K. (1979). *JBC* **254**, 1227–1232.
44. Tabor, S., Huber, H. E., and Richardson, C. C. (1987). *JBC* **262**, 16212–16223.
45. Brutlag, D., and Kornberg, A. (1972). *JBC* **247**, 241–248.
46. Cozzarelli, N. R., Kelly, R. B., and Kornberg, A. (1969). *JMB* **45**, 513–531.
47. Lyamichev, V., Brow, M. A. D., and Dahlberg, J. E. (1993). *Science* **260**, 778–783.
48. Kelly, R. B., Atkinson, M. R., Huberman, J. A., and Kornberg, A. (1969). *Nature* **224**, 495–501.
49. Baltimore, D., and Smoler, D. F. (1972). *JBC* **247**, 7282–7287.
50. Atkinson, M. R., Deutscher, M. P., Kornberg, A., Russel, A. F., and Moffatt, J. G. (1969). *Biochem.* **8**, 4897–4904.
51. Setlow, P., and Kornberg, A. (1972). *JBC* **247**, 232–240.
52. Mizrahi, V., and Benkovic, S. J. (1988). *Adv. Enzymol.* **61**, 437–457.
53. Kuchta, R. D., Benkovic, P., and Benkovic, S. J. (1988). *Biochem.* **27**, 6716–6725.
54. Kuchta, R. D., Mizrahi, V., Benkovic, P. A., Johnson, K. A., and Benkovic, S. J. (1987). *Biochem.* **26**, 8410–8417.
55. Dahlberg, M. E., and Benkovic, S. J. (1991). *Biochem.* **30**, 4835–4843.
56. Patel, S. S., Wong, I., and Johnson, K. A. (1991). *Biochem.* **30**, 511–525.
57. Mizrahi, V., Henrie, R. N., Marlier, J. F., Johnson, K. A., and Benkovic, S. J., (1985). *Biochem.* **24**, 4010–4018.
58. Freemont, P. S., Friedman, J. M., Beese, L. S., Sanderson, M. R., and Steitz, T. A. (1988). *PNAS* **85**, 8924–8928.
59. Beese, L. S., and Steitz, T. A. (1991). *EMBO J.* **10**, 25–33.
60. Catalano, C. E., Allen, D. J., and Benkovic, S. J. (1990). *Biochem.* **29**, 3612–3621.
61. Fersht, A. R., Knill-Jones, J. W., and Tsui, W.-C. (1982). *JMB* **156**, 37–51.
62. Joyce, C. M. (1989). *JBC* **264**, 10858–10866.
63. Carroll, S. S., Cowart, M., and Benkovic, S. J. (1991). *Biochem.* **30**, 804–813.
64. Sambrook, J., Fritsch, E. F., and Maniatis, T. (1989). "Molecular Cloning: A Laboratory Manual," 2nd ed., pp. 13.1–13.104. Cold Spring Harbor Laboratory Press, Cold Spring Harbor, New York.
65. Barnes, W. M. (1978). *JMB* **119**, 83–99.

66. Walker, G. T., Fraiser, M. S., Schram, J. L., Little, M. C., Nadeau, J. G., and Malinowski, D. P. (1992). *NAR* **20**, 1691–1696.
67. del Solar, G., and Espinosa, M. (1991). *NAR* **19**, 1956.
68. Zorbas, H., Fockler, R., and Winnacker, E.-L. (1990). *NAR* **18**, 5909–5910.
69. Shortle, D., Grisafi, P., Benkovic, S. J., and Botstein, D. (1982). *PNAS* **79**, 1588–1592.
70. Burton, F. H., Loeb, D. D., Hutchinson, C. A., III, and Edgell, M. H. (1986). *DNA* **5**, 239–245.
71. Rahmouni, A. R., and Wells, R. D. (1989). *Science* **246**, 358–363.
72. Joyce, C. M., Kelley, W. S., and Grindley, N. D. F. (1982). *JBC* **257**, 1958–1964.
73. Brown, W. E., Stump, K. H., and Kelley, W. S. (1982). *JBC* **257**, 1965–1972.
74. Blanco, L., Bernad, A., Blasco, M. A., and Salas, M. (1991). *Gene* **100**, 27–38.
75. Ollis, D. L., Brick, P., Hamlin, R., Xuong, N. G., and Steitz, T. A. (1985). *Nature* **313**, 762–766.
76. Yadav, P. N. S., Yadav, J. S., and Modak, M. J. (1992). *Biochem.* **31**, 2879–2886.
77. Freemont, P. S., Ollis, D. L., Steitz, T. A., and Joyce, C. M. (1986). *Proteins (S.F.G)* **1**, 66–73.
78. Catalano, C. E., and Benkovic, S. J. (1989). *Biochem.* **28**, 4374–4382.
79. Allen, D. J., Darke, P. L., and Benkovic, S. J. (1989). *Biochem.* **28**, 4601–4607.
80. Kolocheva, T. I., Nevinsky, G. A., Volchkova, V. A., Levina, A. S., Khomov, V. V., and Lavrik, O. I. (1989). *FEBS Lett.* **248**, 97–100.
81. Basu, A., Williams, K. R., and Modak, M. J. (1987). *JBC* **262**, 9601–9607.
82. Mohan, P. M., Basu, A., Basu, S., Abraham, K. I., and Modak, M. J. (1988). *Biochem.* **27**, 226–233.
83. Polesky, A. H., Steitz, T. A., Grindley, N. D. F., and Joyce, C. M. (1990). *JBC* **265**, 14579–14591.
84. Pandey, V. N., Kaushik, N. A., Pradhan, D. S., and Modak, M. J. (1990). *JBC* **265**, 3679–3684.
85. Pandey, V. N., Kaushik, N., and Modak, M. J. (1994). *JBC* **269**, 13259–13265.
86. Joyce, C. M., Ollis, D. L., Rush, J., Steitz, T. A., Konigsberg, W. H., and Grindley, N. D. F. (1986). *In* "Protein Structure, Folding and Design: UCLA Symposia on Molecular and Cellular Biology" (D. Oxender, Ed.), Vol. 32, pp. 197–205. A. R. Liss, New York.
87. Rush, J., and Konigsberg, W. H. (1990). *JBC* **265**, 4821–4827.
88. Pandey, V. N., Williams, K. R., Stone, K. L., and Modak, M. J. (1987). *Biochem.* **26**, 7744–7748.
89. Derbyshire, V., Freemont, P. S., Sanderson, M. R., Beese, L., Friedman, J. M., Joyce, C. M., and Steitz, T. A. (1988). *Science* **240**, 199–201.
90. Basu, S., Basu, A., and Modak, M. J. (1988). *Biochem.* **27**, 6710–6716.
91. Baldwin, R. L. (1964). *JBC* **239**, 231–232.
92. Bachman, B. J. (1987). *In* "*Escherichia coli* and *Salmonella typhimurium:* Cellular and Molecular Biology" (F. C. Neidhardt, J. L. Ingraham, K. B. Low, B. Magasanik, M. Schaechter, and H. E. Umbarger, Eds.), Vol. 2, pp. 807–876. American Society for Microbiology, Washington, D.C.
93. Kelley, W. S., Chalmers, K., and Murray, N. E. (1977). *PNAS* **74**, 5632–5636.
94. Lehman, I. R., and Chien, J. R. (1973). *JBC* **248**, 7717–7723.
95. Matson, S. W., Capaldo-Kimball, F. N., and Bambara, R. A. (1978). *JBC* **253**, 7851–7856.
96. Kelley, W. S., and Grindley, N. D. F. (1976). *NAR* **3**, 2971–2984.
97. Uyemura, D., and Lehman, I. R. (1976). *JBC* **251**, 4078–4084.
98. Konrad, E. B., and Lehman, I. R. (1974). *PNAS* **71**, 2048–2051.
99. Kornberg, T., and Gefter, M. (1972). *JBC* **247**, 5369–5375.
100. Kelley, W. S., and Stump, K. H. (1979). *JBC* **254**, 3206–3210.
101. Jovin, T. M., Englund, P. T., and Bertsch, L. L. (1969). *JBC* **244**, 2996–3008.
102. Klenow, H., and Henningsen, I. (1970). *PNAS* **65**, 168–175.
103. Joyce, C. M., and Grindley, N. D. F. (1983). *PNAS* **80**, 1830–1834.
104. Potapova, I. A., Nevinsky, G. A., Veniaminova, A. G., Khomov, V. V., and Lavrik, O. I. (1990). *FEBS Lett.* **277**, 194–196.
105. Beese, L. S., Derbyshire, V., and Steitz, T. A. (1993). *Science* **260**, 352–355.
106. Joyce, C. M., Sun, X. C., and Grindley, N. D. F. (1992). *JBC* **267**, 24485–24500.
107. Bloom, L. B., Otto, M. R., Beechem, J. M., and Goodman, M. F. (1993). *Biochem.* **32**, 11247–11258.
108. Gutman, P. D., and Minton, K. W. (1993). *NAR* **21**, 4406–4407.

B. T4 DNA Polymerase
[EC 2.7.7.7]

The DNA polymerase from bacteriophage T4 is a multifunctional, monomeric enzyme (104 kDa) which possesses both DNA-directed DNA polymerase and $3' \rightarrow 5'$-exonuclease activities, but not $5'$-nuclease activity. The $3' \rightarrow 5'$-exonuclease activity of T4 DNA Pol hydrolyzes both ssDNA and dsDNA in the absence of dNTPs. Compared with *E. coli* DNA Pol I (or Pol Ik), which is functionally a "repair" enzyme *in vivo*, the polymerase activity of T4 DNA Pol, which is a "replicative" polymerase, is approximately two times higher, and the $3' \rightarrow 5'$-exonuclease activity is two orders of magnitude stronger. These properties make T4 DNA Pol an extremely valuable tool in a variety of applications including DNA strand synthesis, manipulation, cloning, and nucleotide sequencing.

1. FUNCTIONS

a. Reaction conditions

i. Recommended reaction conditions. The optimal pH for the polymerase activity is the same as that for exonuclease activity: pH 7.5 in HEPES buffer, pH 8.6 in Tris–Cl (50 mM), or pH 9.0 in Na–glycinate (50 mM) (1,2). At pH 7.5 in Tris–Cl or pH 9.7 in K-P, ~50% of the optimal activity is observed.

A typical reaction mixture (50 μl) for second-strand DNA synthesis with T4 DNA Pol contains 50 mM Tris–Cl (pH 8.0), 50 mM KCl, 5 mM MgCl$_2$, 5 mM DTT, 40 μg/ml DNA, 0.2 mM each of dNTPs, 50 μg/ml BSA, and 10 U of T4 DNA Pol. The mixture is incubated at 37°C. The reaction can be stopped by adding EDTA to a 20 mM final concentration or by heating to 75°C for 10 min.

ii. Kinetic parameters. The kinetic parameters of T4 DNA Pol with some typical homopolymeric template/primer are listed in Table 6.5: the apparent K_m values for dATP and dTTP with poly(dA)·poly(dT) as template-primer are 17 and 6 μM, respectively, while the V is ~5 times higher than that with salmon sperm DNA (3). The K_m for poly(dA)·poly(dT) is also 6 μM. Under optimal conditions, the rate of polymerization is ~1200 min^{-1}, twice the rate of *E. coli* DNA Pol I under comparable conditions (4). The maximum rate of the $3' \rightarrow 5'$-exonuclease activity is ~4000 min^{-1}.

iii. Metal ion requirements and ionic strength. T4 DNA Pol has an absolute requirement for divalent cations, showing maximal activity at 6 mM Mg^{2+} (1). Mn^{2+} (3.3 mM) can substitute for Mg^{2+} (6.7 mM) but with about a 40-fold reduced catalytic efficiency (14). The substitution of Mn^{2+} for Mg^{2+} induces a 5- to 20-fold increase of misincorporation (5,6). At the same time, Mn^{2+} results in a 2-fold increase in the frequency of correct nucleotide removal (14).

T4 DNA Pol is optimally active at a total monovalent salt concentration of 100 mM (1). At a salt concentration near 0.3 M, the polymerase is 97% inhibited. The stimulatory and inhibitory effects are identical with NH$_4$Cl, NaCl, and KCl.

TABLE 6.5 Steady-State Kinetic Parameters of T4 DNA Polymerase

Parameter	$dATP^a$	$dTTP^a$	$dNTP^b$
$K_{m,app}$ (μM)	17	6	1.3–1.9
K_m(incorporation)	34	8	
K_m(hydrolysis)	12	2	
V_{app} (min^{-1})	83	116	23
V(incorporation)	29	93	
V(hydrolysis)	65	25	

[a] Reaction with poly(dA)·poly(dT) template-primer and single nucleotide substrates at pH 8.8 (5 mM NH$_4$HCO$_3$) and 30°C (3). The apparent kinetic constants were recalculated from the pmol/min/μg protein unit using 104 kDa as the MW of T4 DNA Pol. The polymerization reaction comprises two simultaneous reactions, i.e., stable incorporation of nucleotides and hydrolysis to dNMP. Although dATP and dTTP appear to be significantly different on poly(dA)·poly(dT), the two nucleotides are incorporated and turned over by the polymerase at equivalent rates with the alternating polymer, poly(dAT), as the template/primer.

[b] Reaction with denatured salmon sperm DNA (3).

iv. **Inhibitors and inactivators**

(a) *Nucleotide analogs.* The following nucleotide analogs are inhibitors: 2-(*p-n*-butylanilino)-dATP (K_i = 0.5 μM), N^2-(*p-n*-butylphenyl)-dGTP (K_i = 0.3 μM), and aphidicolin (K_i = 30 μM). Note that Pol Ik is not sensitive to these compounds within the range of concentrations that give appreciable inhibitions to T4 DNA Pol or mammalian Pol α.

T4 DNA Pol is, unlike HSV or vaccinia DNA polymerases which belong to the same B family polymerases, resistant to the PP$_i$ analogs phosphonoacetic acid (PAA) and phosphonoformic acid (PFA or forscarnet).

(b) *Sulfhydryl reagents.* As indicated by the high content of Cys residues in the T4 DNA Pol molecule, NEM or PHMB inactivates both the polymerase and the exonuclease activities. During the incubation, the polymerase requires the presence of 2-MSH (10 mM) or DTT to maintain its optimal activity.

(c) *Temperature.* When incubated at 42°C in 50 mM Tris–Cl (pH 7.5), 0.1 M (NH$_4$)$_2$SO$_4$, and 1 mg/ml BSA, T4 DNA Pol has a half-life of ~15 min (1).

b. **Activity assays and unit definition**

i. *Polymerase assay.* A typical assay mixture, which is essentially similar to the one originally described by Goulian *et al.* (1), consists of 67 mM Tris–Cl (pH 8.8), 16.6 mM (NH$_4$)$_2$SO$_4$, 6.7 mM MgCl$_2$, 6.7 μM EDTA, 10 mM 2-MSH (or 1 mM DTT), 1.6 mg/ml alkali-denatured salmon sperm DNA as the template/primer, and 0.33 mM each of dNTPs including 33 μM of [^{3}H]dTTP.

The mixture is incubated for 30 min at 37°C. The reaction is stopped by taking an aliquot onto a glass fiber filter which has been wetted with 10% TCA and dried. After exhaustive washing in ice-cold 10% TCA, H$_2$O, and ethanol, the radioactivity retained on the filter is counted.

Unit definition: One polymerase unit is the amount of enzyme that incorporates 10 nmol of dNTPs into acid-precipitable material in 30 min.

ii. **Exonuclease assay.** The assay conditions are identical to those used for the polymerase assay except that dNTPs are omitted and labeled polynucleotides (or oligonucleotides) are included as the substrate. The reaction is terminated by the addition of salmon sperm DNA (to 1 mg/ml) and cold perchloric acid (to 0.25 M). After centrifugation, the radioactivity in the supernatant is counted.

Unit definition: One exonuclease unit is the amount of enzyme that produces 10 nmol of acid-soluble dNMPs in 30 min.

c. *Substrate specificities: Polymerase*

i. **Template structure.** ssDNA with annealed primers and 3'-recessed dsDNA are both template-primer substrates for the polymerase activity. Homopolymeric templates/primers, e.g., poly(dA) · poly(dT), are four to five times more efficiently used as substrates than alkali-denatured salmon sperm DNA (3). Unlike *E. coli* DNA Pol I, T4 DNA Pol by itself is unable to use dsDNA containing a nick(s) as a template-primer (7).

ii. **Processivity and fidelity.** The processivity on the poly(dA) · oligo(dT) template-primer is 11–13 nt, which is in the same range as that of *E. coli* Pol I (7). The error rates (dNMP misincorporation) of T4 DNA Pol are estimated to be ~10^{-6} for dCMP·A, ~10^{-5} for dGMP·A, ~2×10^{-4} for dTMP·T, and $< 10^{-7}$ for dAMP·A (8).

iii. **Specificity alterations in presence of T4 ssDNA-binding protein.** In the presence of the helix-destabilizing (or ssDNA-binding) T4 gene 32 protein (gp32), T4 DNA Pol acquires the ability to catalyze the strand displacement synthesis on nicked dsDNA (9) and also displays an increased rate of synthesis, processivity, and fidelity. In the presence of gp32, the rate of synthesis with the wild-type enzyme is increased about 4-fold on poly(dAT) (10) or 5- to 10-fold on heat-denatured dsDNA (11). The strand displacement synthesis catalyzed by T4 DNA Pol (and also *E. coli* DNA Pol I) in the presence of gp32 results in products with multiple, double-stranded branch structures.

iv. **Nucleotide substrates and analogs.** T4 DNA Pol incorporates not only correct but also incorrect nucleotides and various nucleotide analogs. In the presence of Mn^{2+} (1 mM), T4 DNA Pol exhibits a 5- to 20-fold increased rate of nucleotide misincorporation (5,6). Remarkably, however, ribonucleotides are not incorporated into DNA even in the presence of Mn^{2+}.

N^6-*Methyl-dATP* (m^6dATP) is incorporated into DNA ~7 times less efficiently than dATP (13). The V is similar, but the apparent K_m is 39 μM for m^6dATP versus 6.1 μM for dATP.

2-Aminopurine deoxyribonucleotide triphosphate (2APTP), a base analog of dATP, is incorporated ~5 times less efficiently than dATP mostly due to differences in K_m, i.e., K_m (2APTP) is 24 μM when K_m (dATP) is 4.2 μM at pH 8.8 and 6.7 mM $MgCl_2$ (12,14,15). In contrast, the turnover rate of incorporated 2APMP is 3 times higher than that of dAMP. In the presence of Mn^{2+} (3.3 mM), when the V for dATP and 2APTP incorporations are equally reduced about 40-fold, the K_m for 2APTP becomes similar to the K_m (=3 μM) for dATP, indicating that the nucleotide analog is as efficiently incorporated as dATP (14).

d. Substrate specificities: 3' → 5'-Exonuclease

i. Preference based on DNA structure. In the absence of dNTPs, T4 DNA Pol exhibits a strong 3' → 5'-exonuclease activity on ssDNA and dsDNA. Single-stranded DNA is the preferred substrate and is hydrolyzed 3 times faster than dsDNA at 37°C and 40–100 times faster at 30°C. The V with ssDNA, $(dT)_{300}$, is 1100 min^{-1} at 37°C for 850 min^{-1} at 30°C (17). With another oligo(dT) substrate, a rate of 100 sec^{-1} has been estimated for the intrinsic exonuclease cleavage independent of the effects of duplex melting (12).

ii. Product profile. The products of hydrolysis are 5'-dNMPs and single dinu-cleotides derived from the 5' termini of polydeoxynucleotides. The exonucleolytic activity on dsDNA results in molecules with protruding 5' termini.

iii. Effect of chain length. The rate of hydrolysis is dependent on chain length even at saturating levels of polynucleotide termini. Polynucleotides of short chain lengths (e.g., 300 nt) are hydrolyzed 80-fold more rapidly than longer polymers (e.g., 3000 nt) of the same base composition and at the same termini concentration (2). The chain length dependence of the exonuclease activity is apparently due to the binding of the enzyme to internal nucleotides, which inhibits the hydrolytic activity that occurs only at the 3'-OH termini.

iv. Effect of thiophosphoryl linkages. Thiophosphoryl 3' → 5'-diester bonds in some DNAs have been shown to be resistant to T4 exonuclease activity (18,19) and sensitive in some other DNAs such as poly(dAT) (20). Note that Exo III or the 3' → 5'-exonuclease activity of Pol Ik is significantly inhibited by the presence of phosphorothioate linkages in poly(dAT).

v. Competition of exonuclease and polymerase activities. The exonuclease activity of T4 DNA Pol on the ssDNA substrate is not affected by the addition of dNTPs. However, it is inhibited by the presence of the ssDNA-binding protein gp32. When dsDNA is used as a substrate, the addition of complementary dNTPs causes a reduction in the rate of hydrolysis due to the emergence of competing polymerase activity.

Alternate functioning of the enzyme as a polymerase (nucleotide incorporation) and an exonuclease (nucleotide excision) in repeated cycles results in the conversion of dNTPs to dNMPs in a template-dependent manner, a phenomenon known as *idling turnover*. The K_m or V ratio of stable incorporation to turnover varies widely depending on templates and nucleotides. For example, dATP is turned over ~3 times as rapidly as it is stably incorporated, while 80% of the dTMP incorporated remains in polymer (3). Compared with the rate of dAMP turnover during polymerization, dTMP and dGMP are hydrolyzed 1.5–3 times more slowly, while dCMP is hydrolyzed 6–9 times more slowly (16).

Although less efficient, idling turnover is also observed with noncomplementary nucleotides as a result of transient incorporation. Compared with the complementary dNTPs, mismatching nucleotides are discriminated by the polymerase 10- to 100-fold (3).

Given a lengthy incubation time, especially at low concentrations of dNTPs, T4 DNA Pol can result in a complete degradation of dsDNA.

e. Mechanism of catalysis

i. Reaction mechanism. The kinetics of the T4 DNA Pol reaction features an ordered addition of a DNA substrate (D_n) and dNTP to the enzyme followed by a rapid conversion (400 sec^{-1}) of the ternary complex [E·D_n·dNTP] to [E·D_{n+1}·PPi] product species (Scheme 6.2). Except for the part concerning the polymerase-associated $3' \rightarrow 5'$-exonuclease activity, the kinetic mechanism, derived from rapid-quench experiments with a wild-type enzyme (12), fits equally well to a mutant T4 DNA Pol which is deficient in exonuclease activity (58).

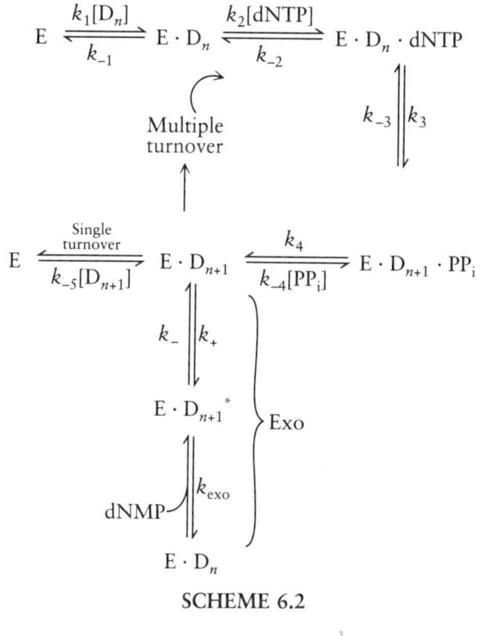

$$E \underset{k_{-1}}{\overset{k_1[D_n]}{\rightleftharpoons}} E \cdot D_n \underset{k_{-2}}{\overset{k_2[dNTP]}{\rightleftharpoons}} E \cdot D_n \cdot dNTP$$

SCHEME 6.2

Unlike Klenow and T7 DNA polymerases for which dissociation of the binary complex [E · D$_{n+1}$] is largely rate limiting, the dissociation rate constant (k_5) for T4 DNA Pol is threefold faster than the k_{cat} (Table 6.6). By analogy to both the Klenow and T7 polymerases, a conformational change step is thus likely to be involved following the chemical step (k_3), although this step is apparently bypassed in multiple incorporations which occur at rates >400 sec^{-1}. In the presence of a mere twofold elemental effect for the incorporation of dATP versus dATPαS, the assignment of RDS to either a putative conformational step or a chemical step remains ambiguous.

ii. Exonuclease activity and proofreading. Kinetic characterization of T4 DNA Pol suggests that the intrinsic property of polymerase plays a major role in fidelity mechanism. First, the polymerase discriminates incorrect nucleotides at the nucleotide-binding step by a factor of >10^3. Second, when an incorrect nucleotide is inserted, the polymerase becomes extremely inefficient in adding another nucleotide to the mismatched terminus. Indeed, extension at the mismatched termini occurs only after the removal of the mismatching base(s) (12,56). The efficiency of editing activity is independent of mismatch sequences. Third, there is a kinetic barrier which, although not absolute, contributes to the mismatch proofreading. The exonuclease activity of T4 DNA Pol functions via an activation step, E · DNA $\rightleftharpoons$ E · DNA*, whose rate constants reflect whether the 3′ terminus of the primer is matched or mismatched. For mismatched primer termini, the conversion to the exonuclease-active state (*) is favored by a factor of 5 (see Table 6.6). The exonuclease active site of T4 DNA Pol optimally accommodates only one or two unpaired nucleotides (56).

Replacement in T4 DNA Pol of divalent metal cofactor Mg^{2+} by Mn^{2+} introduces, as in T7 DNA Pol (59), significant perturbation in both polymerase and exonuclease functions. Kinetic analyses suggest that Mn^{2+} increases misincorporation frequencies (and hence mutation rates in T4 phages) by simultaneously affecting the two following processes (14). One is the differential increase in the binding of incorrect nucleotides to the enzyme–template complex. The other is that correctly paired nucleotides bind in distorted conformations, thereby increasing the frequency of exonucleolytic cleavages.

2. APPLICATIONS

T4 DNA Pol catalyzes polymerization reactions ~2 times faster than Pol Ik, while the processivity is similar. The real strength of T4 DNA Pol is its powerful 3′ → 5′-exonuclease activity as underscored by the following examples. For its high fidelity DNA synthesis, especially in the presence of T4 gp32, T4 DNA Pol should also prove to be valuable in high fidelity PCR at 37°C.

i. Blunt-end formation of duplex DNA. The construction of recombinant DNA molecules often involves blunt-end ligation between the DNA fragments generated by restriction enzyme digestions or synthesized by the use of DNA polymerases. Although Pol Ik can be adequately used to remove a small number

TABLE 6.6 Best-Fit Parameters for T4 DNA Pol Reaction Kinetics[a]

Polymerase		Exonuclease	
Parameter	Value	Parameter	Value
k_1	$8.5 \times 10^7 \ M^{-1} \ sec^{-1}$		
k_{-1}	$6 \ sec^{-1}$		
$K_{d1} \ (= k_{-1}/k_1)$	$70 \ nM$	k_+ (13-mer)	$1 \ sec^{-1}$
$K_{d2} \ (= k_{-2}/k_2)$	$20 \ \mu M$	(13T-mer)	$5 \ sec^{-1}$
k_3	$400 \ sec^{-1}$	k_- (13-mer)	$23 \ sec^{-1}$
k_{-3}	$0.5 \ sec^{-1}$	(13T-mer)	$20 \ sec^{-1}$
$K_{d4} \ (= k_{-4}/k_4)$	$20 \ mM$	k_{exo}	$100 \ sec^{-1}$
k_5			
13-mer	$6 \ sec^{-1}$		
13T-mer	$8 \ sec^{-1}$		
k_{cat}			
Single turnover	$2 \ sec^{-1}$		
Multiple turnover	$>400 \ sec^{-1}$		

[a] Parameters best fitting to the reaction mechanism (Scheme 6.2) were derived from the rapid quench experiments performed at pH 8.8 and 20°C with T4 DNA Pol and defined primer/template substrates (12). The substrate "13-mer" designates a 13/20-mer primer/template with the following sequences:

    5' TCGCAGCCGTCCA
    3' AGCGTCGGCAGGTTCCCAAA

Likewise, "13T-mer" designates a 14/20-mer substrate in which the first extension position of the 13-mer primer is occupied by a mismatching nucleotide T. (Note: Duplexes containing single-stranded termini more than seven bases long result in nonproductive binding of the polymerase.) Exo, exonucleolytic pathway.

of 3'-protruding nucleotides, T4 DNA Pol is often the enzyme of choice for this purpose because of its stronger $3' \rightarrow 5'$-exonuclease activity. DNA with a 5'-protruding end can also be blunt ended because, in the presence of dNTPs, T4 DNA Pol will readily fill in the recessed 3' ends.

The ability to generate blunt ends from both 3'- and 5'-protruding termini makes T4 DNA Pol a valuable tool for a wide range of applications. The "polishing reaction" can often be carried out by adding T4 DNA Pol directly into the previous reaction mixture (e.g., restriction enzyme digestions and cDNA synthesis), provided the reaction conditions are reasonably compatible.

In the presence of 0.1 mM of each dNTP (at 11°C, 20 min), the exonucleolytic action will stop when the polymerase reaches the duplex region in which the 3'-terminal base corresponds to the dNTP. In general, the exonuclease activity should be used as briefly as possible. Unnecessarily long incubations with T4 DNA Pol will simply consume dNTPs and result in degradations of dsDNA.

    ii. Directional cloning of PCR-amplified products. The powerful and precisely controllable $3' \rightarrow 5'$-exonuclease activity of T4 DNA Pol has given rise to

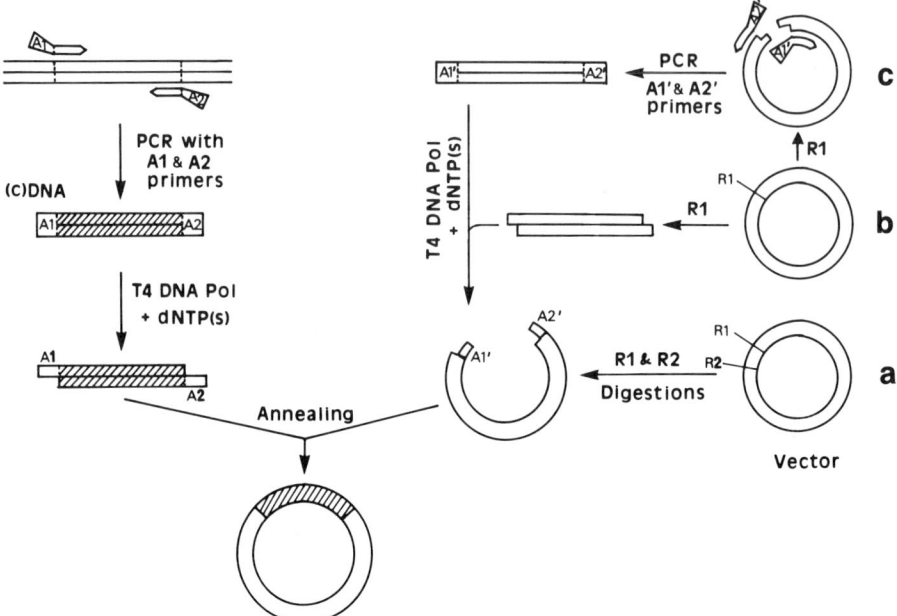

FIGURE 6.3 Outline of T4 DNA polymerase-based directional cloning strategies. DNA or cDNA is amplified by PCR using two adapter-primers, respectively, containing unique adapter sequences A1 and A2. The PCR products are digested with T4 DNA Pol in the presence of one (or two) dNTP, which is lacking in A1 and A2 sequences to expose desired 5'-protruding single-stranded termini A1 and A2, and annealed (and optionally ligated) to a vector carrying complementary single-stranded termini A1' and A2'. The cohesive-ended vector can be generated by a variety of ways: (a) when the A1' and A2' sequences are designed to be compatible directly with two unique restriction sites R1 and R2, (b) vector can be linearized at any suitable site and the A1' and A2' adapter sequences can be exposed by the similar treatment with T4 DNA Pol in the presence of a complementary set of dNTP(s), "complementary" with respect to the dNTP(s) used to expose A1 and A2 termini, and (c) the A1' and A2' adapter sequences can be simply created into the vector by means of PCR and exposed by T4 DNA Pol digestion in the presence of a complementary set of dNTP(s).

elegant and versatile methods for cloning PCR-amplified (c)DNAs (23–25). In the initial form of the "ligation-independent" cloning strategy employing the 3' → 5'-exonuclease activity of T4 DNA Pol, the termini of cDNAs are precisely trimmed back with T4 DNA Pol in the presence of two dNTPs to yield sticky ends fitting into a plasmid vector cleaved with two restriction enzymes (22). In the evolved forms of T4 DNA Pol-based cloning strategy, the procedure consists of the following steps (Fig. 6.3): (a) DNA amplification using two adapter-primers (with 3' site-specific and 5' adapter sequences), (b) digestion of the PCR products with T4 DNA Pol in the presence of one or two specific dNTPs which serve to block the exonucleolytic reaction and define the length and sequence of the 5'-protruding termini (e.g., 2–16 bases), (c) annealing (and optional ligation) with a vector carrying complementary single-stranded termini generated by similar T4

DNA Pol reaction, and (d) transformation of competent host cells with the annealed (or ligated) products.

The T4 DNA Pol-based cloning strategy offers a number of advantages over the conventional cloning methods (see Fig. 6.6): it does not require the use of a restriction enzyme(s), DNA ligase, alkaline phosphatase, or methylation of DNA, thus avoiding the pitfalls that often lead to low cloning efficiencies. By virtue of the unique, customized cohesive termini, the T4 DNA Pol-based cloning strategy ensures that the clone library consists of exclusively unidirectional recombinant clones. The versatility of the cloning strategy enables virtually any vectors with or without suitable preexisting cloning sites to be utilized for directional cloning (see Fig. 6.3).

In a streamlined approach using specially designed vectors, e.g., "prime" cloning system (ZymoGenetics, Seattle, WA) (25), the adapter part of the vector consists of unique 5' sticky ends 15 and 16 nt long ("prime" sequence) that are conveniently generated by the digestions of the vector with *Eco*RI and then with T4 DNA Pol in the presence of dTTP. Insert DNA, which carries the complementary "prime" sequence following a similar digestion with T4 DNA Pol in the presence of dATP, is cloned by hybridization (and optional ligation) to the vector DNA. Vector/primer systems employing slightly different but essentially similar ligation-independent cloning strategies are also commercially available from PharMingen (San Diego, CA).

iii. **In general method of gene trimming.** The single-strand-specific $3' \rightarrow 5'$-exonuclease activity of T4 DNA Pol has been used to develop a versatile gene-trimming method. Conventionally, in the absence of suitable restriction sites, a gene (or DNA) is trimmed to an exact size or position largely relying on a process known as *primer repair*. The basic procedure of primer repair comprises three steps: heat denaturing of dsDNA fragments, annealing with an oligonucleotide primer, and reacting with Pol Ik to synthesize the new second strand, while degrading the single strand extensions of the template upstream of the annealed primer. Although primer repair has been successfully used (26,27), it is an inefficient procedure for several reasons (28). Much of the gene-trimming work can now be performed by new or improved methods which typically include oligonucleotide-directed site-specific mutagenesis techniques (see Appendix A).

In an improved primer repair method (Fig. 6.4), ssDNA is used in place of the heat-denatured dsDNA as the template substrate (28). The single-stranded circular molecule is first linearized with a restriction enzyme in a duplex region which is provided by annealing an oligonucleotide (splinter), and a primer-directed second-strand synthesis is carried out with Pol Ik. Subsequently, T4 DNA Pol is used as the principal single-strand trimming tool. The ssDNA-initiated gene-trimming method has proven to be substantially more efficient than the conventional form of primer repair.

iv. **Generation of deletion subclones for nonrandom DNA sequencing.** DNA sequencing can be performed in a variety of ways based on either a dideoxy chain

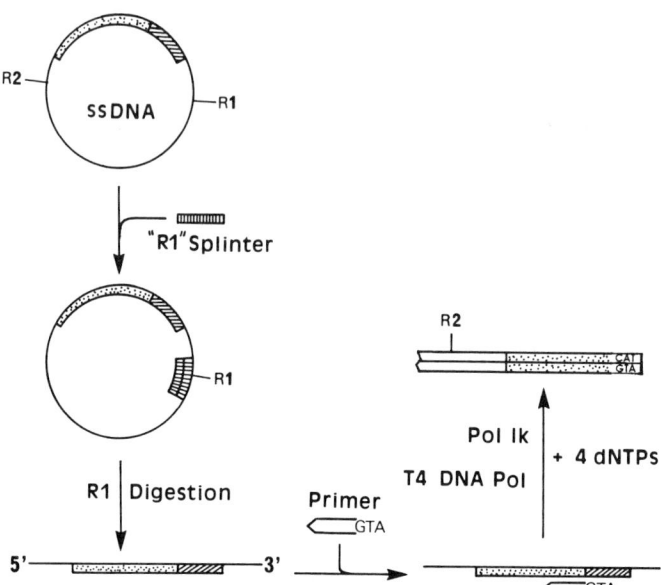

FIGURE 6.4 Single-stranded DNA-initiated gene trimming method. The gene (or DNA fragment) cloned in a phagemid vector is prepared in a single-strand form. R1 and R2 represent unique restriction sites present on the vector. The hatched region of the insert represents the region to be trimmed. The primer used here is designed to provide, as an example, an AUG initiator codon to the trimmed gene. Following the synthesis of the second strand with Pol Ik, T4 DNA Pol is added directly to ther reaction mixture. The trimmed dsDNA is subcloned in a suitable expression vector which has been linearized with restriction enzymes R2 and a blunt-end yielding enzyme (e.g., *Sma*I).

termination method or a chemical cleavage method (see Appendix A). Current separation techniques used in nucleotide sequencing are based largely on urea PAGE which can resolve maximally a sequence ladder of ~400 bases. Therefore, a short piece of DNA can be sequenced usually in a single set of experiments. Although a sequencing reaction per se may be considered routine, sequencing a long piece of DNA is not so straightforward and poses challenging problems as how to reach the region that lies beyond the resolution limit. Various techniques that have been developed to overcome this hurdle can be classified as random and nonrandom sequencing strategies. In contrast to the random sequencing strategies in which DNA fragments are randomly generated by enzymatic or mechanical means and subcloned in a sequencing vector, nonrandom sequencing strategies rely on directional deletion of the target DNA which can then be subcloned and sequenced using a "universal" primer (29). Nonrandom sequencing strategies also comprise the "primer walking" whereby specific sequencing primers are used in a sequential manner to generate sequence information from which the next primer can be synthesized.

Progressively shortened, nested sets of clones can be generated by various methods which are distinguished by the principal techniques and enzymes used,

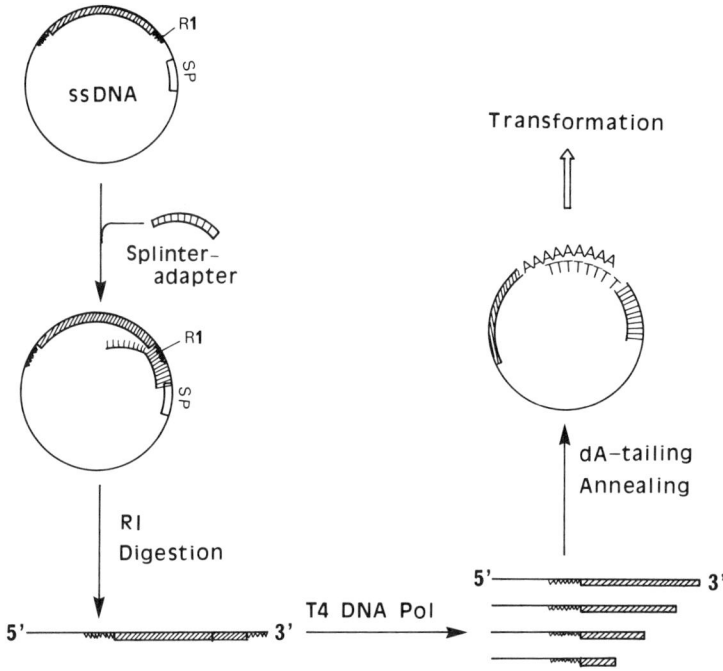

FIGURE 6.5 Outline of single-strand cloning procedure. The solid bar represents the insert and the sawtooth line represents the polylinker region of the M13 vector. R1 and SP denote a restriction enzyme (R1) site and the sequencing primer annealing site, respectively. The circular ssDNA molecule is linearized at a restriction site R1 assisted by a splinter–adapter. The DNA is then digested with T4 DNA Pol, taking aliquots at intervals to produce a series of overlapping deletions. The deletion products are tailed with dA (when R1 is *Hin*dIII) or dG (when R1 is *Eco*RI) by the use of terminal deoxynucleotidyltransferase, and the short homopolymer tail is annealed to the complementary part (adapter) of the splinter–adapter which serves to join the two ends of the molecule. The remaining nick is sealed with T4 DNA ligase and the product is used to transform competent E. coli cells to screen and prepare appropriate deletion clones. [Adapted from Dale *et al.* (1985). *Plasmid* 13, 31–40.]

for example, Exo III (see Section V,A, Chapter 3), Bal 31 nuclease (see Section V,B, Chapter 3), and DNase I ($+\text{Mn}^{2+}$) (see Section I, Chapter 3).

In an efficient method utilizing the $3' \rightarrow 5'$-exonuclease activity of T4 DNA Pol (30,31), programmed deletions are introduced into the ssDNA which has been linearized by a splinter/adapter-mediated cleavage with a restriction enzyme (Fig. 6.5). The nested sets of fragments are recircularized after homopolymeric tailing via the same splinter/adapter and are used for the transformation of competent E. coli cells.

Another efficient strategy for generating unidirectional deletions exploits the resistance of α-thiophosphate-containing phosphodiester bonds to the $3' \rightarrow 5'$-exonuclease activity of T4 DNA Pol (57). In this strategy, the 3' terminus of a linearized, duplex plasmid DNA is filled in with dNMPαS by using *E. coli* DNA Pol Ik. The other end of the target sequence is then cleaved at an appropriate restriction site and subjected to T4 exonuclease degradation. Digestion for different lengths of time followed by treatment with nuclease S1 (or mung bean nuclease) produces a nested set of deletion fragments suitable for intramolecular ligation.

v. 3′-End labeling of DNA. DNAs with 3′-recessed ends can be filled in with appropriate [α-^{32}P]dNTP (e.g., at 1–2 μM [α-^{32}P]dCTP at 11°C for 20 min). DNAs with blunt ends can also be labeled by terminal nucleotide exchange under similar reaction conditions. At higher temperatures, if only a subset of dNTPs is added, the strong 3′ → 5′-exonuclease activity may overcome the polymerase activity and end up degrading the template beyond the labeled nucleotide, resulting in a low specific activity. The ratio of polymerase to exonuclease activity can be maximized by using high concentrations (>100 μM each) of dNTPs.

The 3′ end of DNA can be labeled more extensively by means of replacement synthesis (32). In this method, duplex DNA is first digested in the absence of dNTPs by the 3′ → 5′-exonuclease activity, e.g., up to 30–40% of the length from each end. (Note that about 50% digestion will lead to dissociation of the two single-stranded halves and result in a loss of the template due to rapid degradation.) Subsequently, four dNTPs containing one [α-^{32}P]dNTP are added to the reaction mixture in order for the polymerase to extend the 3′ ends to the length of the template.

vi. Oligonucleotide-directed site-specific mutagenesis. T4 DNA Pol is often used for the primer-extension reaction on ssDNA templates or for gap filling in partially double-stranded DNA in site-specific mutagenesis (refer to Appendix A). An important advantage of using T4 DNA Pol as opposed to the use of Pol Ik is that T4 DNA Pol has no strand displacement activity, thus ensuring the presence of a mutagenic primer in the newly synthesized strand.

With certain templates, however, the DNA synthesis catalyzed by T4 DNA Pol (and some other DNA polymerases as well) may not proceed to full-length due to unpredictable secondary structures of the template DNA. For such templates, the addition of ssDNA-binding gp32 has proven to be highly effective in synthesizing full-length strands (33).

vii. Analytical application of T4 DNA polymerase: Detection of stable DNA lesions. The 3′ → 5′-exonuclease activity of T4 DNA Pol is unable to overcome bulky, stable DNA adducts such as thymine dimers generated by UV irradiation. This property allows detection and quantitation of stable DNA lesions on both ssDNA and dsDNA (34).

3. STRUCTURE

i. Primary and secondary structures. T4 DNA Pol is a single polypeptide of 898 amino acid residues (M_r 103,572) (35). The N- and C-terminal residues are Met (from the initiator codon) and Gly, respectively. The polymerase contains 15 Cys residues, compared with 2 Cys in *E. coli* DNA Pol I.

The secondary structure is predicted to contain ~24% α helix, 36% β sheets, and 40% random coil.

ii. Physical parameters. The A_{280} of a 1-mg/ml solution at pH 8 is 1.40 ($\varepsilon_{280} = 1.3 \times 10^5 \, M^{-1} \, cm^{-1}$), and A_{280}/A_{260} is 1.71 (1). The pI of T4 DNA Pol is 5.8.

iii. Activity domains. Analysis of a number of T4 DNA Pol mutants suggests that the $3' \to 5'$-exonuclease activity can function independently from the polymerase activity. The exonuclease activity is located at the N-terminal half of the molecule, whereas the dNTP-binding site and thus the polymerase active site are located in the C-terminal half (35,36). The two active sites are separated by a minimum distance corresponding to 2–3 nucleotides. The C-terminal region of T4 DNA Pol also interacts with the T4 polymerase accessory proteins (37).

T4 DNA Pol shares several regions of sequence identity with family B DNA polymerases such as herpes, adenovirus, and vaccinia virus polymerases and human DNA Pol α (35,36,39–41). In comparison, T4 DNA Pol shows little overall sequence similarity with family A DNA polymerases such as *E. coli* DNA Pol I and T7 DNA Pol. Within the N-terminal half of the protein, however, three small regions of significant similarity can be identified. These "exonuclease motifs" contain highly conserved acidic amino acid residues (e.g., Asp-112, Glu-114, Asp-219, and Asp-324) which are respectively located within sequences homologous to the sequences containing the residues (e.g., Asp-355, Glu-357, Asp-424, and Asp-501) essential for the exonuclease activity of the Klenow fragment. Indeed, a single D219A mutation or a double mutation (E191A and D324G) results in a T4 DNA Pol which has an intact polymerase activity but no or little (<1%) $3' \to 5'$-exonuclease activity (21,58).

Although the polymerase and $3' \to 5'$-exonuclease domains interact, the two domains are essentially separate in function. Indeed, a 27-kDa fragment (residues 96–331) generated by limited proteolysis or by gene cloning retains the $3' \to 5'$-exonuclease activity but not polymerase activity (38).

4. GENETICS

i. DNA polymerase gene structure and organization. T4 DNA Pol is the product of gene 43 of bacteriophage T4, a large-sized dsDNA (166 kbp) phage with typical morphology consisting of a polyhedral head, tail, and fiber legs reminiscent of the Apollo lunar landing module. Gene 43 belongs to the "middle" genes which encode other proteins involved in deoxyribonucleotide biosynthesis and phage DNA replication.

The polymerase gene has been cloned, and the complete nucleotide sequence (2694 base, 898 codons) has been determined (35). The 898 codon ORF is terminated by UGAUAG. Codon usage analysis shows that T4 gene 43 has a preference for T or A in the third codon position (72% T + A), presumably reflecting the relative abundance (67%) of T + A in T4 DNA.

ii. Regulation of gene 43 expression. In T4-infected cells, transcriptional activation of "middle" genes requires the phage-encoded *motA* protein, which is thought to modify the host transcriptional machinery for the recognition of T4 "middle" promoters (42). In the absence of *motA*, the gene products that are required for DNA synthesis are transcribed from other upstream promoters. Therefore the transcription of T4 gene 43 involves the synthesis of polycistronic and monocistronic mRNAs, and is subjected to RNA processing.

The major mRNA for T4 DNA Pol is apparently transcribed from the −38 position upstream from the first base of the AUG initiator codon (35). The upstream sequence of gene 43 contains the *E. coli* −10 consensus sequence (TATAAT) at the −45 to −50 positions and a *motA* promoter sequence (AAGGCTTCG) at positions −66 to −74 (43). Only 13 nt downstream from the UGA termination codon, there is a potential transcription termination site comprising a six-base (GGACCT) inverted repeat.

Subsequent to the transcription, the level of T4 DNA Pol is autonomously regulated at the translational level. T4 DNA Pol, as an RNA (and DNA)-binding protein, represses its own translation by selectively binding to its mRNA at the translational operator region (44). The binding site is a 36- to 40-nt segment 5' to the initiator AUG and contains the Shine–Dalgarno sequence and a putative RNA hairpin structure. The hairpin consists of a 5-bp stem and an 8-base loop. *In vitro* modulations of the hairpin structure either in the stem or the loop derepress the synthesis of the polymerase *in vivo*, suggesting an important role for the hairpin structure in the interaction with additional *in vivo* factors.

Expression in *E. coli* of a cloned gene 43 under its own promoter (~220-bp sequence upstream from the 5' end of the gene 43) does not provide a viable clone apparently due to the transcription of the gene by the host RNA polymerase and the production of T4 DNA Pol in an amount harmful to the cell.

iii. *In vivo* functions and accessory proteins of T4 DNA polymerase

(a) In vivo functions. T4 DNA Pol plays a central role in phage DNA replication. It is also the major determinant of spontaneous mutations in T4 (36,45). The polymerase–exonuclease activities of T4 DNA Pol are modulated *in vivo* via interactions with the products of several other T4 genes. In fact, T4 DNA Pol functions as a key component of the DNA replication complex in DNA replication.

(b) T4 replication complex. The complex, also known as a five-protein "holoenzyme," consists of T4 DNA Pol (gene 43), a ssDNA-binding protein (gene 32), and three other accessory proteins (the products of gene 44, 45, and 62) (12,37). The genes for the accessory proteins are clustered with the Pol gene in the T4 genome. The replication complex permits T4 DNA Pol to initiate new DNA strand synthesis from RNA primers *in vivo*, to synthesize DNA at or near the *in vivo* rate of ~600 nt/sec on ssDNA-oligoribonucleotide (6-8 nt) template-primers (12), and to initiate replication at a nick on dsDNA at a rate of 370 nt/sec and at physiological ionic strengths. The addition of T4 gene 41 protein (a DNA helicase) to the holoenzyme complex further increases the rate of polymerization and eliminates the pausing observed with the holoenzyme replication system alone (46). The accessory proteins greatly increase the processivity (>20,000 nt)

of the polymerase in an ATP-dependent manner. The ATP dependence is manifested at the initial assembly of T4 DNA Pol with 45 and 44/62 proteins on primer-template and not during the elongation process (47,48). The accessory proteins increase the DNA-bound lifetime of the holoenzyme complex relative to that of the polymerase alone.

The fact that the rate of nucleotide incorporation (>400 sec^{-1}) by T4 DNA Pol alone is comparable to the rate of DNA replication suggests that the stimulatory effect of accessory proteins is due solely to a change in processivity and not to an increase in the rate of polymerization per se. Accessory proteins apparently act to prevent dissociation of the enzyme from the mismatched 3' termini, thereby increasing the efficacy of proofreading by the exonuclease. Mutant T4 phages with a DNA polymerase defective only in 3' → 5'-exonuclease activity show 500- to 1800-fold higher frequencies of spontaneous mutation than the wild-type phages (21).

(c) T4 gene 32 protein. The gene 32 protein (gp32, 301 amino acids, M_r 33,487) is a prototype of ssDNA-binding proteins with an affinity (K_a) for ssDNA of 10^9 M^{-1}, several orders of magnitude greater than for dsDNA. Like the ssDNA-binding protein of *E. coli* (49), gp32 is essential for T4 DNA replication, recombination, and repair (50). gp32, a Zn-metalloenzyme, binds cooperatively to ssDNA with a cooperativity parameter of ~10^3. Each protein molecule has a binding region spanning 6–7 nt. In addition to being a DNA melting (or helix-destabilizing) protein, gp32 has an ability to catalyze DNA strand reassociation, presumably due to the destabilizing effect on hairpins that are known to impede DNA reassociation (51). The N-terminal region (residues 1–21) is responsible for cooperative gp32:gp32 interactions, whereas the C-terminal region (residues 254–301) is important for heterotypic interactions between gp32 and other T4 proteins involved in DNA replication. The core domain contains the intrinsic DNA-binding site.

gp32 is capable of transforming T4 DNA Pol *in vitro* to a much more active enzyme with wider substrate specificities. The stimulation of T4 DNA Pol by gp32 is greatest at low temperatures and high ionic strengths, suggesting that gp32 stimulates the polymerase by removing inhibitory secondary structures from the template DNA strand (11). gp32 also enhances the processivity of T4 DNA Pol and the fidelity of DNA synthesis, especially in the presence of Mn^{2+} (6). The accessory role of gp32 is not limited to T4 DNA Pol but extends to other DNA polymerases such as Pol Ik and T7 DNA Pol (Sequenase) as well, enabling those enzymes to overcome premature chain terminations (52). gp32 has also been used with *Taq* DNA Pol in PCR to increase the specificity and yield of the desired products.

iv. Phenotypes of T4 DNA polymerase mutants. Mutants of T4 DNA Pol have long been the subject of intensive studies, particularly in connection with their effects on polymerase fidelity. They have been classified into two major types: mutator and antimutator phenotypes (36,37,53). The mutations conferring *mutator* phenotype (i.e., an increased frequency of mutation) are located in the N-terminal half of the molecule. The mutator phenotype is generally associated with the reduced 3' → 5'-exonuclease activities and, consequently, has a much lower exonuclease to polymerase ratio than that of the wild type. As will be shown, a mutator phenotype also arises from a mutation that produced a DNA polymerase with a lower insertion specificity, which may be combined with a substantially reduced exonuclease activity. In contrast, the mutations conferring *antimutator* phenotype (i.e., decreased frequency of mutation) are located near the C terminus and generally are associated with reduced polymerase activities. Therefore, compared with the wild-type enzyme, the antimutator polymerase is characterized by a higher ratio of exonuclease to polymerase and a higher rate of nucleotide turnover. The antimutator phenotype is attributed in part to the slow rate of polymerase translocation along the DNA template (10).

Examples of the mutations described below present an overview of the structural and functional properties observed in many other T4 DNA Pol mutants.

(a) tsL141. This is a *ts* mutant with an antimutator phenotype. The mutation has been identified to be a point mutation at nucleotide position 2210 (codon 737) from GCA (Ala) to GTA (Val) (35). This C-terminal mutant exhibits a higher ratio of exonuclease to polymerase activity, typical of antimutator enzymes. The 10–100 times higher rate of nucleotide turnover (or hydrolysis) compared with that of the wild-type enzyme yields a 10–30% higher accuracy of nucleotide utilization in DNA synthesis, enough to give rise to the antimutator phenotype.

(b) amB22. This *amber* mutation arises from the change of the codon 731 from CAG (Gln) to UAG (*amber*). The polypeptide, which contains the N-terminal 80%, lacks polymerase activity but retains the exonuclease activity. In addition to the antimutator phenotype, the amB22 polypeptide is unable to interact with polymerase accessory proteins and is defective in DNA binding (54).

(c) ts meI74. This is a *ts* mutant with a high mutator activity as measured by the relative number of acriflavin-resistant mutants (36). Both *ts* and mutator phenotypes are due to the mutation at codon 317 which changes from TAC (Tyr) to TGC (Cys). The DNA polymerase of this mutant has a reduced exonuclease activity, but the reduction is not sufficient to account for the high mutator activity.

(d) Asp-219–Ala mutation. This is a mutator mutant constructed from site-specifically mutagenized T4 gene 43 (58). The T4 DNA Pol is deficient in 3' → 5'-exonuclease activity but has an intact polymerase activity. The mutant phage T4 displays, compared with the wild type, at least 760-fold higher mutation frequencies.

5. SOURCES

T4 DNA Pol has been prepared traditionally from the T4 phage-infected *E. coli* B which contains ~600 polymerase molecules per cell. T4 DNA Pol can now be more efficiently prepared from recombinant systems in which gene 43 has been cloned under the control of the λ p_L promoter on the pUC9-based plasmid (pTL43W) or of the *tac* promoter on the M13mp19-based plasmid (pTL43Q) (55).

Thermal induction in *E. coli* of the T4 DNA Pol from pTL43W at 42°C produces a large amount of polymerase as an insoluble aggregate. When the induction temperature is lowered to 40°C, however, the expressed polymerases are soluble. Induction of the cells harboring pTL43Q with IPTG at 40° and 42°C shows a similar dependence of polymerase solubility on culture temperatures. However, in contrast to pTL43W, the IPTG induction of pTL43Q produces a large quantity of soluble polymerase at temperatures as low as 30°C. The cloned systems provide about 1000-fold overexpression of the polymerase (~10% of the total cellular protein) compared with the amounts produced by the cells infected with the wild-type T4 phage.

It is to be noted that the cloned T4 DNA Pol has an Asn at position 214, whereas the polymerase from T4-infected cells contains Ser at this position (55). Apparently due to this mutation, the cloned enzyme has a twofold higher polymerase activity than the S214 polymerase. The two polymerases have the same level of $3' \rightarrow 5'$-exonuclease activity.

Commercial enzymes are usually supplied in 0.1 M K-P (pH 6.5), 10 mM 2-MSH, and 50% (v/v) glycerol. When stored at $-20°C$, the enzyme is stable for at least 10 months.

References

1. Goulian, M., Lucas, Z. J., and Kornberg, A. (1968). *JBC* **243**, 627–638.
2. Huang, W. M., and Lehman, I. R. (1972). *JBC* **247**, 3139–3146.
3. Gillin, F. D., and Nossal, N. G. (1975). *BBRC* **64**, 457–464.
4. Kornberg, A. (1980). "DNA Replication." Freeman, San Francisco.
5. Hall, Z. W., and Lehman, I. R. (1968). *JMB* **36**, 321–333.
6. Gillin, F. D., and Nossal, N. G. (1976). *JBC* **251**, 5225–5232.
7. Masamune, Y., and Richardson, C. C. (1971). *JBC* **241**, 2692–2701.
8. Creighton, S., and Goodman, M. F. (1995). *JBC* **270**, 4759–4774.
9. Nossal, N. G. (1974). *JBC* **249**, 5668–5676.
10. Gillin, F. D., and Nossal, N. G. (1976). *JBC* **252**, 5219–5224.
11. Huberman, J. A., Kornberg, A., and Alberts, B. M. (1971). *JMB* **62**, 39–52.
12. Capson, T. L., Peliska, J. A., Kaboord, B. F., Frey, M. W., Lively, C., Dahlberg, M., and Benkovic, S. J. (1992). *Biochem.* **31**, 10984–10994.
13. Mace, D. C. (1984). *JBC* **259**, 3616–3619.
14. Goodman, M. F., Keener, S., Guidotti, S., and Branscomb, E. W. (1983). *JBC* **258**, 3469–3475.
15. Clayton, L. K., Goodman, M. F., Branscomb, E. W., and Galas, D. J. (1979). *JBC* **254**, 1902–1912.
16. Hershfield, M. S., and Nossal, N. G. (1972). *JBC* **247**, 3393–3404.
17. Brutlag, D., and Kornberg, A. (1972). *JBC* **247**, 241–248.
18. Kunkel, T. A., Eckstein, F., Mildvan, A. S., Koplitz, R. M., and Loeb, L. A. (1981). *PNAS* **78**, 6734–6738.
19. Barcak, G. J., and Wolf, R. E., Jr. (1986). *Gene* **49**, 119–128.
20. Gupta, A. P., Benkovic, P. A., and Benkovic, S. J. (1984). *NAR* **12**, 5897–5911.
21. Reha-Krantz, L. J., Stocki, S., Nonay, R. L., Dimayuga, E., Goodrich, L. D., Konigsberg, W. H., and Spicer, E. K. (1991). *PNAS* **88**, 2417–2421.
22. Schmid, A., Cattaneo, R., and Billeter, M. A. (1987). *NAR* **15**, 3987.

23. Stoker, A. W. (1990). *NAR* **18**, 4290.
24. Aslanidis, C., and de Jong, P. J. (1990). *NAR* **18**, 6069–6074.
25. Kuijper, J. L., Wiren, K. M., Mathies, L. D., Gray, C. L., and Hagen, F. S. (1992). *Gene* **112**, 147–155.
26. Goeddel, D. V., Shepard, H. M., Yelverton, E., Leung, D., Crea, R., Sloma, A., and Pestka, S. (1980). *NAR* **8**, 4057–4074.
27. Lawn, R. M., Adelman, J., Bock, S. C., Franke, A. E., Houcke, K. M., Najarian, R. C., Seeburg, P. H., and Wion, K. L. (1981). *NAR* **9**, 6103–6114.
28. Eun, H. M., Kang, Y., Kang, S. M., Bae, Y. S., and Yoon, J. W. (1989). *BioTechniques* **7**, 506–510.
29. Sambrook, J., Fritsch, E. F., and Maniatis, T. (1989). "Molecular Cloning: A Laboratory Manual," 2nd ed., pp. 13.1–13.104. Cold Spring Harbor Laboratory Press, Cold Spring Harbor, New York.
30. Dale, R. M. K., McClure, B. A., and Houchins, J. P. (1985). *Plasmid* **13**, 31–40.
31. Dale, R. M. K., and Arrow, A. (1987). *Meth. Enzy.* **155**, 204–214.
32. Challberg, M. D., and Englund, P. T. (1980). *Meth. Enzy.* **65**, 39–43.
33. Craik, C. S., Largman, C., Fletcher, T., Roczniak, S., Barr, P. J., Fetterick, R., and Rutter, W. J. (1985). *Science* **228**, 291–297.
34. Doetsch, P. W., Chan, G. L., and Haseltine, W. A. (1985). *NAR* **13**, 3285–3304.
35. Spicer, E. K., Rush, J., Fung, C., Reha-Krantz, L. J., Karam, J. D., and Konigsberg, W. H. (1988). *JBC* **263**, 7478–7485.
36. Reha-Krantz, L. J. (1988). *JMB* **202**, 711–724.
37. Nossal, N. G. (1984). *Ann. Rev. Biochem.* **53**, 581–615.
38. Lin, T. C., Karam, G., and Konigsberg, W. H. (1994). *JBC* **269**, 19286–19294.
39. Blanco, L., Bernad, A., Blasco, M. A., and Salas, M. (1991). *Gene* **100**, 27–38.
40. Reha-Krantz, L. J. (1992). *Gene* **112**, 133–137.
41. Blanco, L., Bernad, A., and Salas, M. (1992). *Gene* **112**, 139–144.
42. Brody, G., Rabbusay, D., and Hall, D. H. (1983). *In* "Bacteriophage T4" (C. Mathews, E. M. Kutter, G. Mosig, and P. B. Berget, Eds.), pp. 174–183. American Society of Microbiology, Washington, D.C.
43. Guild, N., Gayle, M., Sweeney, R., Hollingsworth, T., Modeer, T., and Gold, L. (1988). *JMB* **199**, 241–258.
44. Andrake, M. D., and Karam, J. D. (1991). *Genetics* **128**, 203–213.
45. Reha-Krantz, L. J., Liesner, E. M., Parmaksizoglu, S., and Stocki, S. (1986). *JMB* **189**, 261–272.
46. Bedinger, P., Munn, M., and Alberts, B. M. (1989). *JBC* **264**, 16880–16886.
47. Capson, T. L., Benkovic, S. J., and Nossal, N. G. (1991). *Cell* **65**, 249–258.
48. Jarvis, T. C., Newport, J. W., and von Hippel, P. H. (1991). *JBC* **266**, 1830–1840.
49. Meyer, R. R., and Laine, P. S. (1990). *Microbiol. Rev.* **54**, 342–380.
50. Chase, J. C., and Williams, K. W. (1986). *Ann. Rev. Biochem.* **55**, 103–136.
51. Alberts, B. M., and Frey, L. (1970). *Nature* **227**, 1313–1318.
52. Kaspar, P., Zadrazil, S., and Fabry, M. (1989). *NAR* **17**, 3616.
53. Muzyczka, N., Poland, R. L., and Bessman, M. J. (1972). *JBC* **247**, 7116–7122.
54. Venkatesan, M., and Nossal, N. G. (1982). *JBC* **257**, 12435–12443.
55. Lin, T. C., Rush, J., Spicer, E. K., and Konigsberg, W. H. (1987). *PNAS* **84**, 7000–7004.
56. Reddy, M. K., Weitzel, S. E., and von Hippel, P. H. (1992). *JBC* **267**, 14157–14166.
57. Barcak, G. J., and Wolf, R. E., Jr. (1986). *Gene* **49**, 119–128.
58. Frey, M. W., Nossal, N. G., Capson, T. L., and Benkovic, S. J. (1993). *PNAS* **90**, 2579–2583.
59. Tabor, S., and Richardson, C. C. (1989). *PNAS* **86**, 4076–4080.

C. T7 DNA Polymerase
[EC 2.7.7.7]

T7 DNA polymerase is a heterodimeric 92-kDa enzyme composed of the phage gene 5 protein and the host-encoded thioredoxin in a 1:1 stoichiometry. In addition to the highly processive polymerase activity, the native T7 DNA Pol has a strong

$3' \to 5'$-exonuclease activity specific for both ssDNA and dsDNA. No $5'$-nuclease activity has been found with T7 DNA Pol. Modified forms of T7 DNA Pol from which the $3' \to 5'$-exonuclease activity has been removed are commercially available.

The gene 5 protein alone has the same high level of ssDNA exonuclease activity found in the complex, but only 0.5 to 5% of the DNA polymerase and dsDNA exonuclease activity. The addition of thioredoxin, which alone has neither DNA polymerase nor exonuclease activity, fully restores the DNA polymerase and double-strand exonuclease activities of the gene 5 protein. The high processivity, speed, and ability to incorporate nucleotide analogs make T7 DNA Pol particularly useful for long-range dideoxynucleotide sequencing.

1. FUNCTIONS

a. Polymerase reaction conditions

i. Standard reaction conditions. T7 DNA Pol is used in a variety of reactions involving polymerase and/or $3' \to 5'$-exonuclease activities, and reaction conditions can be optimized accordingly.

The following reaction conditions are typically used for second-strand DNA synthesis on ssDNA templates: 40 mM Tris–Cl (pH 7.5), 15 mM $MgCl_2$, 25 mM NaCl, 5 mM DTT, 0.25 mM each of four dNTPs, 1 μg template (plus primer) DNA, and 0.5–1 U of polymerase in a 20-μl reaction volume. The mixture is incubated at room temperature or 37°C for 1 hr. The reaction can be stopped by adding excess EDTA or by heating at 75°C for 10 min.

ii. Factors modulating polymerase activity. The polymerase activity of T7 DNA Pol is dependent on pH, temperature, and ionic strength. The optimal pH for polymerase activity is pH 7.6–7.8 in 80 mM K-P buffer. At temperatures below 37°C, the rates of polymerization and processivity are greatly reduced. High ionic strengths (e.g., 100 mM NaCl) drastically reduce the processivity but greatly increase the efficiency with which T7 DNA Pol binds the primer-template and initiates DNA synthesis (1). Low temperatures or high ionic strengths apparently affect the polymerase activity by increasing the stability of secondary structures in the DNA template.

iii. Kinetic parameters. The apparent K_m values for dNTPs are 80 μM with salmon sperm DNA as a template/primer (2). The k_{cat} is estimated to be $\sim$1800 min^{-1}. (For microscopic rate constants, refer to Table 6.7.)

iv. Metal ion requirements. T7 DNA Pol requires divalent cations, either Mg^{2+} or Mn^{2+}, for its activity. The optimal concentrations are 4 mM Mg^{2+} or 0.1 mM Mn^{2+} (3). Higher concentrations of Mn^{2+} are inhibitory to the polymerase activity, but the effect can be overcome by the addition of isocitrate. Isocitrate at a 15 mM concentration will maintain the free Mn^{2+} concentration at 30–300 μM over a 1–10 mM range of $MnCl_2$. The substitution of Mn^{2+} for Mg^{2+} results in important changes in the properties of T7 DNA Pol.

v. Inhibitors and inactivators

(a) *Metal chelator.* EDTA (Mg^{2+} chelator) inhibits the polymerase activity.

(b) *Temperature.* When incubated at 42°C in the absence of Mg^{2+}, both polymerase and exonuclease activities of T7 DNA Pol are rapidly inactivated with a half-time of ~3 min (4). In the presence of Mg^{2+}, T7 DNA Pol can be inactivated by heating to 75°C for 10 min.

(c) *Nucleotides and their analogs.* Nucleotides (dNTPs) are specific inhibitors of 3' → 5'-exonuclease activity. At 30 μM dNTPs (but not rNTPs), both the double-strand exonuclease activity of the holoenzyme and the single-strand exonuclease activity of the gene 5 protein are fully inhibited (4).

(d) *Modification reagents.* Modifications of the Tyr and Lys residues with acetic anhydride, the SH groups with NEM, and the His residues by a rose bengal-mediated photochemical reaction inactivate the 3' → 5'-exonuclease activity but not the polymerase activity (5). The polymerase activity is stimulated by 2-MSH (6).

b. *Activity assays and unit definitions*

i. **Polymerase assay.** Polymerase activity is typically assayed in 40 mM Tris–Cl (pH 7.5), 10 mM $MgCl_2$, 5 mM DTT, 50 mM NaCl, 0.2 mM each of dATP, dGTP, dCTP, and [^{3}H]dTTP (50 mCi/mmol), 1 μg M13mp19 DNA, and 0.5 pmol universal primer and polymerase. The mixture is incubated at 37°C. Some other assays employ phosphorus equivalents of denatured salmon sperm or calf thymus DNA instead of the M13 phage DNA and primer.

Unit definition: One polymerase unit is the amount of enzyme that incorporates 10 nmol of dNTPs into acid-insoluble products in 30 min.

ii. **Exonuclease assay.** A typical exonuclease assay mixture consists of 40 mM Tris–Cl (pH 7.5), 10 mM $MgCl_2$, 5 mM DTT, and 10 nmol of ^{3}H-labeled ssDNA (or dsDNA) and the enzyme. The mixture is incubated at 37°C.

Unit definition: One exonuclease unit catalyzes the release of 10 nmol of dNMPs into an acid-soluble form in 30 min.

c. *Substrate specificities: Polymerase*

The following properties are common to native and modified T7 DNA polymerases. However, the modified T7 DNA Pol (Sequenase) possesses certain properties that distinguish it from native polymerase.

i. **Template structure.** Single-stranded DNA, denatured dsDNA, or dsDNA that has been made partially single stranded by exonuclease action can serve as template-primers for the polymerase. The polymerase activity is sevenfold higher on primed M13 DNA than on denatured calf thymus DNA.

The polymerase activity shows some degree of sequence dependence, which is eliminated by using Mn^{2+} in place of Mg^{2+}.

ii. Processivity and fidelity. T7 DNA Pol is highly processive, allowing the polymerization of thousands of nucleotides without dissociating from the primer-template. The dissociation–reassociation rate is ~10 times slower than that of Pol Ik, undergoing at most two to three cycles in 10 sec in the absence of NaCl (1). T7 DNA Pol is a high fidelity enzyme with error rates $\leq 2.2 \times 10^{-6}$ for base substitutions, $\leq 3.7 \times 10^{-7}$ for $+1$ frameshifts, and $\leq 4.5 \times 10^{-7}$ for -1 frameshifts (7). T7 DNA Pol lacking $3' \rightarrow 5'$-exonuclease activity exhibits error rates >10-fold higher than the wild-type enzyme.

iii. Nicked DNA. Certain forms (Form II) of native T7 DNA Pol cannot initiate DNA synthesis at nicks on dsDNA, but other forms (e.g., Form I and modified T7 DNA Pol) can catalyze strand displacement synthesis. The strand displacement synthesis activity is greatly stimulated in the presence of the DNA-unwinding helicase (T7 gene 4 protein). The strand displacement synthesis on circular dsDNA is terminated by template switching, which results in duplex branch structures with less than 500 bp in length (8).

iv. Nucleotide substrates and analogs. T7 DNA Pol efficiently incorporates various nucleotide analogs such as ddNTPs, dNTPαS, dITP, and 7-deaza-dGTP. The incorporation efficiency is substantially enhanced when Mn^{2+} (2 mM) substitutes for Mg^{2+} (5 mM). The substitution reduces the discrimination against ddNTP approximately fourfold, enabling ddNTPs to be incorporated at almost the same rate as dNTPs (3). This specificity change is observed even when both Mg^{2+} and Mn^{2+} are present simultaneously.

d. Substrate specificities: $3' \rightarrow 5'$-Exonuclease

T7 DNA Pol has, in addition to the $5' \rightarrow 3'$ DNA polymerase activity, a relatively strong $3' \rightarrow 5'$-exonuclease activity. Unlike *E. coli* DNA Pol I or *Taq* DNA Pol, however, T7 DNA Pol has no polymerase-associated 5'-nuclease activity.

The $3' \rightarrow 5'$-exonuclease activity displays two distinct substrate specificities: one active on ssDNA and the other active on dsDNA (4,9). The phage gene 5 protein alone has the single-strand exonuclease. The gene 5 protein acquires the double-strand exonuclease activity in the presence of the thioredoxin subunit. The double-strand specific exonuclease activity is about 6-fold higher than the single-strand exonuclease activity (1) and can initiate hydrolysis at nicks and the 3'-OH termini of dsDNA. The double-strand exonuclease activity of native T7 DNA Pol is 5000–10,000 times stronger than that of Pol Ik. The single-strand exonuclease activity of the gene 5 protein is about 800-fold higher than that of Pol Ik. The products of hydrolysis of both ssDNA and dsDNA are >98% dNMPs; 5'-terminal dinucleotides are not produced.

e. Mechanism of catalysis

The polymerization reaction catalyzed by T7 DNA Pol is fully consistent with the mechanism (see Scheme 6.1) described for *E. coli* DNA Pol I (14). As in the mechanism of *E. coli* DNA Pol I (see Section II,A, in this chapter), the RDS has

TABLE 6.7 Kinetic Constants of T7 DNA Polymerase[a]

Parameter	Value	K_{eq}	K_d
k_1	11 μM^{-1} sec^{-1}	56 μM^{-1}	18 nM
k_{-1}	0.2 sec^{-1}		
k_2	>50 μM^{-1} sec^{-1}	0.054 μM^{-1}	18 μM
k_{-2}	>1000 sec^{-1}		
k_3	300 sec^{-1}	3	
k_{-3}	100 sec^{-1}		
k_4	>9000 sec^{-1}	0.5	
k_{-4}	>18,000 sec^{-1}		
k_5	1200 sec^{-1}	67	
k_{-5}	18 sec^{-1}		
k_6	>1000 sec^{-1}	2 mM	
k_{-6}	>0.5 μM^{-1} sec^{-1}		

[a] The rate constants represent the best fit of single turnover kinetics data to the reaction mechanism described by Scheme 6.1. The rate constants for exonuclease-minus T7 DNA Pol are virtually identical to those for wild-type T7 DNA Pol. Reaction conditions are 40 mM Tris–Cl (pH 7.5), 12 mM MgCl$_2$, 50 mM NaCl, 1 mM DTT, 0.1 mg/ml BSA, 25/36-mer primer/template (200 nM), and dTTP (0.5–100 μM) at 20°C (14).

been identified as the conformational change of the enzyme to a catalytically competent ternary complex, which occurs at a rate of 300 sec^{-1} on binding of the correct nucleotide (Table 6.7). The subsequent chemical step, measured with dTTPαS as the substrate, has an observed rate constant of 70 sec^{-1} and is presumed to be ≥9000 sec^{-1} with the dTTP substrate.

Incorporation by the exonuclease-deficient T7 DNA Pol of incorrect triphosphates dATP, dCTP, and dGTP occurs with k_{cat}/K_m values of 91, 23, and 4.3 M^{-1} sec^{-1}, respectively, indicating a discrimination in the polymerization step of 10^5–10^6 (15). The rates of misincorporation in all cases are linearly dependent on substrate concentrations up to 4 mM, beyond which severe inhibition occurs.

In contrast to the effect of thiophosphate linkages with a correct nucleotide(s), α-thiophosphate analogs of dATP, dCTP, and dGTP give moderate 19-, 17-, and 34-fold rate reductions, respectively, suggesting that the chemical step is only partially rate limiting. Incorporation of the next correct nucleotide, e.g., dCTP, on a mismatched DNA substrate is saturable with a K_m (for dCTP) of 87 μM and a k_{cat} of 0.025 sec^{-1}. A larger sulfur elemental effect of 60 is observed with dCTPαS, suggesting that, under the given condition, the chemical step is rate limiting. Besides, the pyrophosphorolysis on a mismatched 3′-end is undetectable, indicating that it does not play a proofreading role. The discrimination against incorrect nucleotides is thus attributed to an induced fit conformational change that does not favor the incorporation of the next nucleotide, particularly at the chemical catalysis step.

Both matched and mismatched DNAs bind tightly to the polymerase site with approximately equal affinities, i.e., K_d of 10 and 20 nM, respectively. Selective removal of the mismatch is governed by the rate of transfer of the DNA from the polymerase to the exonuclease site (16). The excision of the matched DNA is limited by a slow transfer rate (e.g., 0.2 sec^{-1}) from the polymerase to the exonuclease site relative to the high rate (300 sec^{-1}) of polymerization. The removal of mismatched DNA is facilitated by its faster transfer rate (e.g., 2.3 sec^{-1}) to the exonuclease site relative to the slow rate (0.012 sec^{-1}) of polymerization.

2. APPLICATIONS

i. **Dideoxynucleotide sequencing.** T7 DNA Pol, the 3' → 5'-exonuclease-deficient enzyme (Sequenase), is an extremely useful polymerase for long-range DNA sequencing by the dideoxy chain termination method (also see Appendix A). This is due to several excellent properties of the polymerase, i.e., the high processivity, speed, and efficient incorporation of dNTPαS and other nucleotide analogs such as 7-deaza-dGTP or dITP. Compared with *E. coli* DNA Pol Ik, Sequenase is particularly useful for sequencing heat- or alkali-denatured dsDNA templates.

The procedure recommended by the Sequenase supplier (USB) (using the reagents contained in the kit) is adequate for most sequencing reactions. However, the reaction conditions for heat-denatured templates can be further optimized with regard to the following set of parameters (17): a primer annealing temperature of −70°C, (b) a labeling reaction time between 15 and 45 sec, and (c) a primer/template ratio of 20.

The substitution of Mn^{2+} for Mg^{2+} offers important benefits: nucleotide analogs such as ddNTPs are incorporated as efficiently as dNTPs, and the incorporations are no longer sequence dependent, resulting in uniform termination of DNA extension products (3).

The use of pyrophosphatase (included in the sequencing kit jointly developed by Applied Biosystems and USB) provides further improvements in the sequencing techniques by reducing the level of PP$_i$ in the sequencing reaction mixture. A high accumulation of PP$_i$ can reverse the polymerization reaction due to pyrophosphorolysis and result in degradation products.

ii. **Site-specific mutagenesis.** The highly processive and fast polymerase activity of T7 DNA Pol provides important advantages over Pol Ik or T4 DNA Pol for the synthesis of mutagenic strands in oligonucleotide-directed mutagenesis (see Appendix A). For example, T7 DNA Pol can make complete copies of uracil-containing templates (of M13 vectors) within 30 min as compared to 90 min for T4 DNA Pol and, unlike T4 DNA Pol, does not require the ssDNA-binding gp32 to get through the hairpin loop region in the polylinker.

Although modified T7 DNA Pol (Sequenase), which is devoid of 3' → 5'-exonuclease activity, can be used similarly (18), the following pros and cons have to be considered. As a result of selectively removing 3' → 5'-exonuclease activity, Sequenase has a higher polymerase activity than the native enzyme and can easily overcome regions of strong secondary structures (5,19). However, the modified enzyme has strand displacement activity; it can displace the mutagenic oligonucleotide used for priming if the incubation time is slightly longer than necessary for the polymerization reaction. In contrast, the native T7 DNA Pol does not have strand displacement activity, but the polymerase is more susceptible to the secondary structures of the template in a manner analogous to T4 DNA Pol.

iii. Site-specific forced misincorporation mutagenesis.
Forced misincorporation mutagenesis is a technique for introducing a noncomplementary nucleotide(s) by using an error-prone polymerase either during the synthesis of DNA from a site-specifically annealed primer (20) or during the repair of short, site-specifically constructed gaps (21). The advantage of this "random" mutagenesis technique over oligonucleotide-directed mutagenesis is that any one of three possible mutagenic nucleotides can be incorporated per site during the extension of each primer.

Like retroviral reverse transcriptases (see Section III,A, in this chapter), modified T7 DNA Pol (Sequenase) is suited for such mutagenesis and is perhaps better than reverse transcriptases due to the high rate of polymerization. However, the mutagenesis efficiency with Sequenase is substantially lower than that with native T7 DNA Pol, apparently due to the known tendency of the modified T7 DNA Pol to perform strand displacement. Therefore a recommended procedure for misincorporation mutagenesis would consist of forced incorporation by modified T7 DNA Pol in the presence of one dNTP, brief extension of the terminally mismatched primer/template using the same enzyme in the presence of all four dNTPs, and finally chase/elongation using native T7 DNA Pol and T4 DNA ligase to produce heteroduplex DNA (22).

iv. Synthesis of DNA probes.
T7 DNA Pol can be used to produce labeled or unlabeled ssDNA probes from plasmid DNA templates (23). Rapid primer extension allows the production of long ssDNA probes, sparing the need to subclone restriction fragments into vectors capable of producing ssDNA templates. The ssDNA probe is suitable for nuclease S1 transcript mapping.

v. Blunt-end formation.
T7 DNA Pol can also be used to convert the ends of duplex DNA into blunt-ended structures. The 5'-protruding end is filled in by the polymerase activity, while the 3'-protruding end is digested, in the presence of dNTPs, by the 3' → 5'-exonuclease activity.

vi. Labeling of 3' termini.
T7 DNA Pol can be used in a manner similar to T4 DNA Pol for labeling 3' termini either by simple extension or by replacement synthesis in the presence of radiolabeled or nonisotopically labeled dNTP(s).

3. STRUCTURE

i. Subunit assembly.
T7 DNA Pol is a dimeric enzyme: a polymerase subunit (80 kDa) encoded by T7 phage gene 5 (6) and a thioredoxin subunit (12 kDa) encoded by the host *E. coli trxA* gene. Both subunits are essential for polymerase and double-stranded exonuclease activities. The phage-specified subunit contains the ssDNA-specific 3' → 5'-exonuclease activity present in the native enzyme,

but lacks polymerase and dsDNA-specific $3' \rightarrow 5'$-exonuclease activities. No polymerase-associated activities are detectable in the thioredoxin subunit.

Thioredoxin is considered to be a processive factor for the gene 5 protein. Thioredoxin binds tightly (K_d = 5 mM) and in a 1:1 stoichiometry to the gene 5 protein (25) and confers about 1000-fold higher processivity to the polymerase. Thioredoxin increases the stability of the gene 5 protein–DNA template complex 20- to 80-fold (24). The subunit interaction is salt insensitive but markedly temperature dependent, which is consistent with the involvement of a hydrophobic surface area in reduced thioredoxin (26). Native polymerase can be dissociated using 6 M GuHCl and resolved by gel filtration into two biologically active subunits (4).

The ε_{280} of native T7 DNA Pol and thioredoxin are 1.44×10^5 and $1.37 \times 10^4\ M^{-1}\ cm^{-1}$, respectively (27).

ii. **Phage gene 5 protein.** The gene 5 protein is composed of 704 amino acids (M_r 79,692) (10). The N- and C-terminal residues are Met and His, respectively. The gene 5 protein contains 10 Cys residues.

Comparison of the amino acid sequence of the gene 5 protein with that of Pol Ik shows significant (38–58%) homologies in eight of nine polypeptide segments lying in the large, C-terminal DNA-binding domain of Pol Ik (28,29). Glu-319 is presumed to be located at the interaction site with thioredoxin (44). The N-terminal half of the gene 5 protein shows little homology to Pol Ik, with the exception of a 10 residue peptide segment. Nevertheless, the N-terminal domain of 270 amino acid residues is presumed to contain the $3' \rightarrow 5'$-exonuclease active site. A double mutation D5A/E7A (14), a single mutation H123E (5), or a deletion mutation (Δ28 residues including His-123) (5) abolishes the $3' \rightarrow 5'$-exonuclease activity.

iii. **Thioredoxin.** Thioredoxin of *E. coli* is a multifunctional, small, redox protein composed of 108 amino acids (M_r 11,675) (30). It is a compact molecule with dimensions of 25 Å $\times$ 30 Å $\times$ 35 Å. The crystallographic structure of the oxidized form of thioredoxin has been determined to 1.7 Å resolution (31,32) and the structure of reduced thioredoxin has been derived from 2D NMR (33). Thioredoxin does not bind to DNA by itself and has neither polymerase nor exonuclease activity. Thioredoxin has a redox active S–S bridge (thioredoxin-S_2) in a conserved active-site motif of Cys^{32}-Gly-Pro-Cys^{35}-Lys. The active-site sequence (Cys–Gly–Pro–Cys) forms a protruding loop at the end of a β strand and is exposed to solvent. Thioredoxin does not require metal ions or cofactors and is inactivated by sulfhydryl reagents. Only reduced thioredoxin, thioredoxin-(SH)$_2$, but not oxidized thioredoxin (thioredoxin-S_2) can interact with the gene 5 protein and reconstitute the polymerase activity. Disulfide bond formation (or oxidation) apparently results in a significant conformational change that renders the thioredoxin incapable of interaction with the gene 5 protein (33). In contrast, mutant thioredoxins with alterations at either one or both active-site Cys residues can restore nearly full polymerase activity, albeit requiring higher concentrations than wild-type thioredoxin (25). A double mutant C32S/C35S thioredoxin (TrxA7) binds to the gene 5 protein 1/60th as tightly as the wild-type thioredoxin.

Substrates for the single-strand exonuclease enhance the reactivity of the SH groups with modifying reagents (e.g., NEM), while those for the polymerase or double-strand exonuclease function of holoenzyme provide protection (34). This observation suggests that the two SH groups may be located at or near the DNA-binding site of the polymerase. Nevertheless, the T7 DNA Pol complex shows thioredoxin activity in assays with ribonucleotide reductase or insulin (31), suggesting that the disulfide bond-containing active site of the thioredoxin unit may not be directly involved in the interaction with the gene 5 protein subunit. Alternatively, and perhaps more likely, the thioredoxin activity observed at 37°C is that of free thioredoxin dissociated from the polymerase complex.

G74D-thioredoxin (TrxA11 mutant) binds to the gene 5 protein at 1/200 strength of the wild-type thioredoxin (25). *E. coli* (SB2111) harboring the TrxA11 mutation does not support the growth of wild-type phage T7. Interestingly, however, it gives rise to T7 suppressor mutants at a frequency of 5×10^{-4}. One such suppressor mutant has a mutation at Glu-319 (E319K or E319V) of the gene 5 protein (44). G92D-thioredoxin (TrxA13 mutant) does not bind to the gene 5 protein, and the TrxA13 mutant *E. coli* (SB2113) does not support wild-type T7 phage growth. Substitution of the Trp-31 of thioredoxin by Tyr, Ala, or His results in a large decrease in T7 DNA Pol activity.

Thioredoxin is acidic (pI ~4.5), highly soluble, and unusually heat stable. It maintains about 80% of its activity after incubation for 5 min at 100°C (9). The high solubility of thioredoxin allows N- or C-terminal fusion proteins (e.g., eukaryotic proteins) to be expressed in *E. coli* as fully "soluble" proteins which would otherwise aggregate into "insoluble" inclusion bodies (45).

iv. **Modified T7 DNA polymerase.** T7 DNA Pol can be modified either chemically or genetically to give rise to a polymerase devoid of the $3' \rightarrow 5'$-exonuclease activity. The "modified" T7 DNA polymerases are available under the trade name of Sequenase (version 1.0 for chemically modified enzyme and version 2.0 for genetically modified enzyme from USB).

(a) *Chemical modification.* The chemical modification reagents consist of molecular oxygen, a reducing agent (DTT), and iron salts. The modification and inactivation of the exonuclease domain of the gene 5 protein occurs via localized generation of free radicals (35). The Fe^{2+} ion, which serves as the reaction center for oxidation and modification, apparently binds the N-terminal exonuclease domain, analogous to the $3' \rightarrow 5'$-exonuclease domain of Pol Ik which contains two metal ion-binding sites. The Fe^{2+}-catalyzed oxidative reaction also causes a slow decrease in polymerase activity as a function of reaction time, e.g., a 15% decrease in 4 days. The apparent 15% decrease of polymerase activity is presumably due to a composite of two opposing phenomena: (i) only 30% of the enzyme molecules remain active as DNA polymerase, but (ii) the specific activity of the modified polymerase is threefold higher than that of the unmodified one. When

99.5% inactivation of exonuclease activity occurs, the oxidation of the gene 5 protein results in ~25% degradation of and 25% cross-linking of the enzyme.

(b) Genetic modification. An alternative, more desirable approach to generate a T7 DNA Pol lacking the exonuclease activity would be to alter the polymerase subunit at the genetic level. This has been accomplished by deleting a 28 amino acid region (from Lys-118 to Arg-145) containing the His-123 residue and thus selectively removing the 3' → 5'-exonuclease activity (5).

The modification of the polymerase molecule entails notable changes in several enzymatic properties (5). (i) The modified polymerase lacks 3' → 5'-exonuclease activity, but it has a polymerase activity that is increased up to 9-fold (to 300 nt/sec) while retaining the same high processivity. (ii) Modified (but not native) T7 DNA Pol can incorporate nucleotide analogs, εdATP (1-N^6-etheno-2'-deoxyadenosine 5'-triphosphate). (iii) Modified T7 DNA Pol has the ability to polymerize nucleotides through the region containing strong secondary structures on the template. (iv) Modified polymerase can initiate strand displacement synthesis at nicks and can be stimulated by T7 helicase. Native T7 DNA Pol cannot catalyze the strand displacement synthesis. (v) Mutants of T7 DNA Pol deficient in exonuclease activity have a reduced burst size and about a 14-fold increased rate of spontaneous mutation.

4. GENETICS

T7 DNA Pol is a heterodimer composed of a polymerase subunit encoded by phage gene 5 and a thioredoxin subunit encoded by the *E. coli trx* gene. Together with gene 4 protein (a primase/helicase) and other phage-encoded proteins like gene 6 exonuclease, T7 DNA Pol is responsible for many events known to occur during the replication of T7 phage.

i. Structure and organization of phage T7 gene 5.
The genome of phage T7 is, like other similar medium-size T-odd phages, a linear, 40-kbp dsDNA with a coding capacity for ~50 genes (10). Phage gene 5 is a class II gene located at nucleotide positions 14,353–16,465 (or in T7 map units, 35.94–41.23). The transcription of gene 5 is carried out by phage T7 RNA Pol from the promoter ϕ4.7 at nucleotide position 13,915 (34.84 map unit) and is terminated at the Tϕ, a typical hairpin terminator located at a nucleotide position around 24,200 between gene 10B and 11. The major transcript after processing by RNase III is predicted to be 4648 nt long.

Gene 5 has a typical RBS (AGGAG) 10 nt upstream from the AUG codon and its ORF is terminated by the UGA codon.

ii. Cloning and expression of T7 gene 5.
Phage T7 gene 5 has been cloned and expressed in *E. coli* from a T7 RNA Pol-coupled plasmid system in which T7 gene 1 (RNA Pol) and gene 5 were placed under the control of the λp_L promoter and the T7 ϕ1.1 promoter, respectively (36). On heat induction at 42°C, the expression of the cloned gene 5 amounts to 30% of the cellular protein,

but only one-third is soluble in the absence of SDS. Expression of the gene 5 protein in some trx^+ (thioredoxin-positive) cells can be lethal, whereas $trxA^-$ cells are tolerant to the high level expression of the protein.

Cloning of T7 gene 5 in pBR322 and expression in a $trxA^-$ cell (e.g., E. coli BH215) result in an overproduction of the gene 5 protein (~5% of the total soluble protein) which is about 60-fold higher than the amount in T7-infected cells (27). Transcription of the gene 5 cloned in pBR322 apparently originates from a previously unknown E. coli RNA polymerase promoter located between nucleotides 14,290 and 14,320 of the T7 DNA, immediately upstream of the gene 5 ORF. The cloned gene 5 can also be stably expressed in E. coli trx^+ cells. However, its expression has been observed to be lower than in trx^- cells by a factor of 5–10. Because the two strains have the same plasmid copy number, the observed difference suggests an autoregulatory interaction of the T7 DNA Pol holoenzyme on the expression of T7 gene 5.

iii. **5'-Exonuclease encoded by phage T7 gene 6.** T7 DNA Pol has no polymerase-associated 5'-nuclease activity. Apparently, the *in vivo* role of polymerase-associated 5'-nuclease activity is fulfilled by the product of T7 gene 6, a gene adjacent to gene 5. T7 gene 6 encodes a 5'-exonuclease free of 3' → 5'-exonuclease and DNA polymerase activities. The gene 6 nuclease, a 347 amino acid single polypeptide (40 kDa), hydrolyzes dsDNA until ~50% of the DNA is acid soluble (10). The enzyme shows a marked preference for dsDNA. It also hydrolyzes ssDNA but very slowly.

The enzyme begins its exonucleolytic attack at the 5' terminus of DNA, liberating almost entirely 5'-dNMP (11). The 5'-exonuclease hydrolyzes 5'-phosphoryl- and 5'-OH-terminated DNA at equal rates (12). The enzyme releases a dinucleoside monophosphate from the 5'-OH terminus and NTP from the 5'-triphosphate terminus (13). In fact, the gene 6 nuclease has been shown, like the 5' nuclease of E. coli DNA Pol I, to function on model substrates as a structure-specific endonuclease (46). The gene 6 nuclease also degrades RNA in a variety of RNA·DNA hybrids and thus has an RNase H activity.

iv. **Thioredoxin gene**

(a) **In vivo functions.** Thioredoxin is present in E. coli in unusually high concentrations (~10^4 molecules/cell). As a ubiquitous protein, thioredoxin is one of the major hydrogen donors in many redox reactions (37). For instance, thioredoxin participates in the reduction of NDP to dNDP and of NTP to dNTP catalyzed by respective reductases. (*Note:* Oxidized thioredoxin is converted to the reduced form by the catalysis of thioredoxin reductase in the presence of a flavin cofactor and the reduced form of $NADP^+$.) Reduced thioredoxin also catalyzes the reduction of many exposed disulfides in proteins, thereby functioning as a general protein disulfide-reductase or protein disulfide-isomerase and as an active catalyst for protein folding (38).

Although thioredoxin is implicated in many biological functions, bacterial mutants with no detectable thioredoxin protein exhibit no phenotypic alterations except that they are unable to support replication of phage T7. This observation suggests that essential functions of thioredoxin can be substituted *in vivo* by other molecules. Mutations in the gene for thioredoxin reductase (*trxB*) and not for thioredoxin (*trxA*) have been shown to be necessary and sufficient to allow disulfide bond formation of, for example, signal-sequenceless alkaline phosphatase in the cytoplasm of *E. coli* (39).

(b) Organization of thioredoxin gene. The gene for thioredoxin (or TsnC protein) from *E. coli* (*trxA*, or formerly *tsnC*) has been cloned and sequenced (40–42). The *trxA* gene contains 127 amino acid codons followed by a UAA termination codon. Compared with the 108 amino acid sequence obtained by amino acid sequencing (30), the deduced sequence is 18 amino acids longer at the N terminus. This hydrophobic sequence is likely to be a signal peptide, and the codon AUG at coordinate +55 marks the beginning of the mature thioredoxin (41). The Met codons at coordinates +1 and +55 are both preceded by potential RBS. However, presently available data suggest that the RBS preceding the initiator codon at +55 is presumably more efficient and that the majority of thioredoxin is made without the leader peptide. Putative promoter sequences similar to CH-35 (TTGACA) and CH-10 (TATAAT) are located at −47 (TTTACG) and −23 (TAAAGT), respectively. Transcription is presumed to start from A at position −9. The *trxA* gene is mapped in the order of *ilv–trxA–uvrD*.

The cloned gene has also been overexpressed 150- to 200-fold (10^6 copies/cell or 40% of the total cell protein) in a $trxA^-$ *E. coli* host (40). The overproduction of thioredoxin is attributed to the high copy number of recombinant plasmid, suggesting a role for thioredoxin in plasmid DNA replication.

5. Sources

T7 DNA Pol is now purified from cloned, overexpression systems. When the T7 gene 5 protein–thioredoxin complex is purified from phage-infected cells ($\sim 10^4$ copies of T7 DNA Pol per cell), it can be isolated in one of two forms depending on the conditions of purification (43). When the complex is purified in the presence of EDTA (0.1 mM), the enzyme has high single-stranded and double-stranded exonuclease activities (Form II). The T7 helicase (gene 4 protein) does not stimulate this form of T7 DNA Pol on nicked DNA templates. The other form (Form I) is obtained by purification in the absence of EDTA or by irreversible conversion of the purified Form II when EDTA is removed by $FeSO_4$ or $CaCl_2$. The Form I enzyme has relatively low levels of exonuclease activity, less than 5% of that of Form II. Unlike Form II, Form I can initiate DNA synthesis at nicks, leading to strand displacement. The polymerase activity of Form I is stimulated by the T7 helicase on nicked DNA. The reasons for the physical differences between the two forms have not been fully determined.

The purified holoenzyme has a specific polymerase activity of 54,000 U/mg on M13 DNA, a double-strand exonuclease activity of 26,000 U/mg on T7 DNA, and a single-strand exonuclease of 4000 U/mg on T7 DNA (1). The purified gene 5 protein has a DNA polymerase (<100 U/mg) and a double-strand exonuclease

(120 U/mg), but the single-strand exonuclease activity remains the same as that of native T7 DNA Pol.

Commercial enzymes are usually supplied in 25 mM Tris–Cl or K-P (pH 7.5), 0.25 M NaCl, 1 mM DTT, and 50% (v/v) glycerol. The enzyme is stored at −20°C. When stored at 0°C, T7 DNA Pol is most stable between pH 9.0 and pH 10.0. Storage at room temperature, especially at high dilutions, can result in a 30% loss of activity in 1 week.

References

1. Tabor, S., Huber, H. E., and Richardson, C. C. (1987). *JBC* **262**, 16212–16223.
2. Nordstrom, B., Randahl, H., Slaby, I., and Holmgren, A. (1981). *JBC* **256**, 3112–3117.
3. Tabor, S., and Richardson, C. C. (1989). *PNAS* **86**, 4076–4080.
4. Adler, S., and Modrich, P. (1979). *JBC* **254**, 11605–11614.
5. Tabor, S., and Richardson, C. C. (1989). *JBC* **264**, 6447–6458.
6. Grippo, P., and Richardson, C. C. (1971). *JBC* **246**, 6867–6873.
7. Kunkel, T. A., Patel, S. S., and Johnson, K. A. (1994). *PNAS* **91**, 6830–6834.
8. Lechner, R. L., Engler, M. J., and Richardson, C. C. (1983). *JBC* **258**, 11174–11184.
9. Modrich, P., and Richardson, C. C. (1975). *JBC* **250**, 5508–5514.
10. Dunn, J. J., and Studier, F. W. (1983). *JMB* **166**, 477–535.
11. Kerr, C., and Sadwski, P. D. (1972). *JBC* **247**, 311–318.
12. Kerr, C., and Sadwski, P. D. (1972). *JBC* **247**, 305–310.
13. Shinozaki, K., and Okazaki, T. (1978). *NAR* **5**, 4245–4261.
14. Patel, S. S., Wong, I., and Johnson, K. A. (1991). *Biochem.* **30**, 511–525.
15. Patel, S. S., Wong, I., and Johnson, K. A. (1991). *Biochem.* **30**, 526–537.
16. Dolin, M. J., Patel, S. S., and Johnson, K. A. (1991). *Biochem.* **30**, 538–546.
17. Casanova, J.-L., Pannetier, C., Jaulin, C., and Kourilsky, P. (1990). *NAR* **18**, 4028.
18. Venkitaraman, A. R. (1989). *NAR* **17**, 3314.
19. Cariello, N. F., Thilly, W. G., Swenberg, J. A., and Skopek, T. R. (1991). *Gene* **99**, 105–108.
20. Zakour, R. A., and Loeb, L. A. (1982). *Nature* **295**, 708–710.
21. Shortle, D., Grisafi, P., Benkovic, S. J., and Botstein, D. (1982). *PNAS* **79**, 1588–1592.
22. Baldwin, J. E., Martin, S. L., and Sutherland, J. D. (1991). *Prot. Eng.* **4**, 579–584.
23. Sharrocks, A. D., and Hornby, D. P. (1991). *BioTechniques* **10**, 426–428.
24. Huber, H. E., Tabor, S., and Richardson, C. C. (1987). *JBC* **262**, 16224–16232.
25. Huber, H. E., Russel, M., Model, P., and Richardson, C. C. (1986). *JBC* **261**, 15006–15012.
26. Slaby, I., and Holmgren, A. (1989). *JBC* **264**, 16502–16506.
27. Reutimann, H., Sjoberg, B.-M., and Holmgren, A. (1985). *PNAS* **82**, 6783–6787.
28. Ollis, D. L., Kline, C., and Steitz, T. A. (1985). *Nature* **313**, 818–819.
29. Blanco, L., Bernad, A., Blasco, M. A., and Salas, M. (1991). *Gene* **100**, 27–30.
30. Holmgren, A. (1968). *EJB* **6**, 475–484.
31. Eklund, H., Cambillau, C., Sjoberg, B. M., Holmgren, A., Jornvall, H., Hoog, J. O., and Branden, C. I. (1984). *EMBO J.* **3**, 1443–1449.
32. Katti, S., LeMaster, D. M., and Eklund, H. (1990). *JMB* **212**, 167–184.
33. Dyson, H. J., Gippert, G. P., Case, D. A., Holmgren, A., and Wright, P. E. (1990). *Biochem.* **29**, 4129–4136.
34. Adler, S., and Modrich, P. (1983). *JBC* **258**, 6956–6962.
35. Tabor, S., and Richardson, C. C. (1987). *JBC* **262**, 15330–15333.
36. Tabor, S., and Richardson, C. C. (1985). *PNAS* **82**, 1074–1078.
37. Holmgren, A. (1985). *Ann. Rev. Biochem.* **54**, 237–271.
38. Pigier, V. P., and Schuster, B. J. (1986). *PNAS* **83**, 7643–7647.
39. Derman, A. I., Prinz, W. A., Belin, D., and Beckwith, J. (1993). *Science* **262**, 1744–1747.

40. Lunn, C. A., Kathju, S., Wallace, B. J., Kushner, S. R., and Pigiet, V. (1984). *JBC* **259**, 10469–10474.
41. Wallace, B. J., and Kushner, S. R. (1984). *Gene* **32**, 399–408.
42. Hoog, J. O., von Bahr-Lindstrom, H., Josephson, S., Wallace, B. J., Kushner, S. R., Jornvall, H., and Holmgren, A. (1984). *Biosci. Rep.* **4**, 917–923.
43. Engler, M. J., Lechner, R. L., and Richardson, C. C. (1980). *JBC* **258**, 11165–11173.
44. Himawan, J. S., and Richardson, C. C. (1992). *PNAS* **89**, 9774–9778.
45. LaVallie, E. R., DiBlasio, E. A., Kovacic, S., Grant, K. L., Schendel, P. F., and McCoy, J. M. (1993). *Bio/Technology* **11**, 187–193.
46. Lyamichev, V., Brow, M. A. D., and Dahlberg, J. E. (1993). *Science* **260**, 778–783.

D. *Taq* DNA Polymerase
[EC 2.7.7.7]

The DNA polymerase (*Taq* DNA Pol I) from *Thermus aquaticus* is a monomeric enzyme (94 kDa) with DNA polymerase and strong 5'-nuclease activities but no 3' → 5'-exonuclease activity (1,2). *Taq* Pol is highly thermostable and processive and has a high rate of polymerization. These properties, together with the availability of the polymerase from cloned sources, make it a key enzyme in many applications involving high-temperature, automated polymerase chain reactions.

1. FUNCTIONS

a. Polymerase reaction conditions

i. Standard reaction conditions. *Taq* DNA Pol has a broad temperature optimum centered at 75°C. A commonly used reaction buffer is 20 mM Tris–Cl, pH 8.3, at 25°C (or pH ~7 at 72°C). Although there are no conditions that are singly optimal for all PCR, the following "standard" conditions provide a useful start (3,4).

A typical reaction mixture (100 μl) contains 20 mM Tris–Cl (pH 8.3), 50 mM KCl, 1.5 mM $MgCl_2$, 0.2 mM each of four dNTPs, 2 ng DNA, and 0.1 μM of each primer. The mixture may optionally include 5% (v/v) Me_2SO, 0.01% (v/v) gelatin, and/or 0.05% (v/v) NP-40/Tween 20. The polymerization reaction is initiated by adding 2.5 U *Taq* Pol. [*Note:* In "hot-start" PCR (see further below), the polymerase is made to come into contact with the rest of the mixture at or above the primer annealing temperature (5). To prevent evaporation during PCR, the reaction mixture is overlaid with two or three drops of mineral oil (e.g., Nujol) or one pellet of paraffin wax (e.g., AmpliWax of Perkin-Elmer)].

The amplification scheme on a thermal cycler typically consists of a 2- to 10-min predenaturation at 95°C and then cycles of 1 min denaturation at 95°C, 1–2 min annealing at 50–60°C, and 1–5 min extension at 72°C for ~30 cycles. Products can be readily separated from mineral oil and recovered as a droplet by adding 100–200 μl of water-saturated chloroform.

ii. Kinetic parameters. The apparent K_m values for each of four dNTPs are 10–20 μM (2). The rate of polymerization is ≥3600 nt/min on M13mp18 ssDNA at 70°C, which is at least six times faster than *E. coli* DNA Pol I. At the optimum temperature of 75–80°C, the rate is estimated to be near 9000 nt/min. At lower temperatures, i.e., 55°, 37°, and 22°C, the polymerization rate drops to about 1400, 90, and 15 min^{-1}, respectively.

iii. Requirements for metal ion and ionic strength.
The polymerase activity of *Taq* DNA Pol is maximally stimulated at 2.0 mM $MgCl_2$ in assays using activated salmon sperm DNA as a template and 0.7–0.8 mM total dNTPs (6). Higher concentrations of Mg^{2+} are inhibitory, giving a 40–50% inhibition at 10 mM $MgCl_2$. Note that the actual concentration of Mg^{2+} required to maximally activate the enzyme is dependent on dNTP concentrations.

Modest concentrations of KCl stimulate the polymerase activity by 50–60% with an apparent optimum at 50 mM (6). Higher concentrations of KCl are inhibitory, and no significant activity is observed at ≥75 mM KCl in DNA-sequencing reactions or at ≥200 mM KCl in a 10-min incorporation assay. NaCl at a 50 mM concentration stimulates the polymerase activity by 25–30%.

iv. Activators, inhibitors, and stabilizers

(a) Detergents. Nonionic detergents such as NP-40, Tween 20, and Triton X-100 do not inhibit the reaction at concentrations as high as 5% (7). In fact, the 0.05% Tween 20/NP-40 combination stimulates polymerase activity (2). At 0.5% each of Tween 20/NP-40, the inhibitory effect of 0.01% SDS can be instantaneously reversed. A nonionic detergent, *n*-octylglucoside, shows a moderate level of inhibition at concentrations above 0.4% (w/v).

Ionic detergents such as sodium deoxycholate, sarkosyl, and SDS are inhibitory to the polymerase reaction at concentrations higher than 0.06, 0.02, and 0.01% (w/v), respectively (7).

(b) Organic solvents and chemicals. Low concentrations of DMF (≤5%), formamide (≤10%), and Me_2SO have no undesirable effects on *Taq* Pol (6). Me_2SO is a particularly interesting cosolvent whose primary effect is believed to be the reduction of denaturation temperature (T_m) in a linear concentration-dependent manner [0.6°C/1% (v/v)]. Me_2SO as high as 10% has thus been used in *Taq* PCR with excellent amplification results (8,9). The optimal conditions for using formamide should be determined experimentally because formamide produces multiple effects not only on polymerase but also on other organic materials, not to mention DNA denaturation and primer annealing. Whole bacteria or blood samples can be efficiently amplified by *Taq* PCR in the presence of 18% (v/v) formamide and at a reduced incubation temperature (10). Use of cosolvents requires the effective cycling temperature to be adjusted as a function of the cosolvent concentration. The product yield is usually highest at or within several degrees of the midpoint T_m of a given primer pair (11).

(c) Polysaccharides. Most of the polysaccharides, including dextran and starch, are not inhibitory to PCR with *Taq* Pol (8). However, two of the acidic polysaccharides (dextran sulfate and gum ghatti) are inhibitory.

The addition of 0.5% Tween 20 reverses the inhibitory effect of gum ghatti (polysaccharide:DNA ratio of 500:1). The inhibitory effect of dextran sulfate (50:1) can be reversed by the addition of Tween 20 (0.25% or 0.5%), Me$_2$SO (5%), or PEG 400 (5%), but none of the three additives are effective at a 100:1 ratio of dextran sulfate to DNA.

Heparin, as an anticoagulant, displays a potent inhibitory effect on PCR, affecting both *Taq* Pol and DNA (13). Note that heparin is used in enzymology as an efficient trap for "free" DNA polymerases. DNA samples from the blood long stored in heparinized tubes have often been found to be unsuitable for PCR amplification, even after removal of the heparin. EDTA or citrate can be used as the substitute anticoagulant to store the whole blood with no deleterious effects on DNA.

(d) Stabilizers. The presence of gelatin, glycerol, and BSA stabilizes *Taq* Pol. Gelatin, as the most commonly used stabilizer, does not affect the polymerization reaction per se, but it produces distortion during electrophoresis. Nonionic detergents such as Tween 20 and/or NP-40 (at 0.05–0.1%) may also be considered as stabilizers.

(e) Reverse transcriptases. Both AMV and MoLV RTases, which are frequently used for the reverse transcription (RT) of mRNA in coupled RT–PCR, inhibit *Taq* Pol activity (14). Heat inactivation of RTase prior to the *Taq* Pol addition eliminates the inhibition. Increasing the unit ratio of *Taq* Pol to RTase to over 1:1 improves the efficiency of *Taq* PCR.

(f) Temperature. *Taq* Pol is thermostable, with a half-life ($t_{1/2}$) of 40 min at 95°C in the absence of stabilizing agents. The half-life of *Taq* Pol is >2 hr and 10 min at 92.5° and 97.5°C, respectively. Under standard reaction conditions, *Taq* Pol retains about 70% of activity after a 30 cycle PCR.

v. **Cycling parameters for PCR.** The standard PCR procedure consists of three-step cycling: template (dsDNA) denaturation, primer annealing, and primer extension. The number of cycles depends mainly on the starting concentration of target DNA and the desired final concentration of the amplified DNA. The number of cycles is thus intimately related with the amplification efficiency. For a given number of cycles, each step is carried out at different temperatures and for different lengths of time (3,4,15). In fact, two-step temperature cycling, which consists of (i) denaturation and (ii) annealing/extension, has proved to give as good, if not better, results as the standard three-step cycling PCR.

(a) Denaturation. Denaturation of dsDNA by brief (≤1 min) heating to 90–95°C is generally sufficient unless the target sequence is unusually GC rich. Denaturation of GC-rich DNA may better be performed in the presence of denaturants or cosolvents such as 5% (v/v) formamide or Me$_2$SO. Elongated thermal denaturation may result in accelerated inactivation of the polymerase and also template damages via depurination.

(b) Primer annealing. The time (20 sec to 2 min) and temperature (40–70°C) for primer annealing are interdependent as well as primer dependent. The two parameters critically affect the specificity and yield of amplification, and need to be optimized for each PCR application.

As long as the concentrations of primers remain substantially higher than that of the template, an annealing time of 20 sec is adequate. As the primers become exhausted, a longer time should be allowed so that the primer annealing is maximized with respect to the template self-annealing.

(c) Primer extension. The reaction with *Taq* Pol is usually performed at 72°C, close to the optimum temperature of the polymerase. The time allowed for polymerization varies according to the size of the DNA fragment to be amplified. Under standard PCR conditions, the rate of polymerization is estimated to be 35–100 nt/sec, indicating that a time of 30–60 sec would be sufficient for a 1-kbp DNA fragment. A separate extension time is not necessary for target sequences smaller than 150 bp, but it may have to be over 20 min for optimal amplification of >20-kbp fragments.

(d) Ramping time. Another factor that may be of special importance in some cases is the *ramping* (or *transition*) time, the time taken to change from one temperature to another. Ramping time is dependent on several parameters: (i) heating/cooling mechanics of a thermal cycler, (ii) dynamics of heat transfer to the sample involving parameters like thickness and material type of the reaction vessels, and (iii) mass and fluidity of the reaction mixture. The ramping time is generally short enough to be of no serious concern in maintaining a reaction at a specific temperature for the required amount of time. However, it may be a critical parameter when short primers (e.g., ~15 bases), which require low annealing temperatures, are used and especially when the polymerase is added last at or above the annealing temperature. In these cases, ramping time may better be considered as a part of the extension time and the reaction temperature should be adjusted in a stepwise manner.

b. Activity assay and unit definition

A typical assay, which is commonly adopted by commercial enzyme suppliers, is based on the incorporation of labeled dNTPs by nick translation. The mixture (50 μl) contains 50 mM Tris–Cl (pH 9.0 at 25°C), 50 mM NaCl, 10 mM MgCl$_2$, 0.2 mM each of dATP, dCTP, and dGTP, 50 μM [^{3}H]dTTP, and 12.5 μg activated (DNase I-treated) calf thymus DNA. The mixture is incubated at 74°C.

Unit definition: One unit is the amount of enzyme that incorporates 10 nmol of dNTPs into acid-insoluble products in 30 min at 74°C.

Note: The typical assay just described is not reliable in measuring polymerase activity in crude cell extracts. More critically, the assay cannot be applied to polymerases lacking 5'-(exo)nuclease activity. A polymerase may better be assayed uniquely by its ability to catalyze a 5' → 3' strand synthesis reaction. One such assay employs a synthetic oligonucleotide as a highly sensitive, nuclease-resistant, self-priming substrate (16). The oligonucleotide comprises a 34 (or longer)-base inverted repeat designed to have a stable hairpin with a GC clamp at the 3′ terminus and a 5′ oligo-T single-stranded tail to which is linked biotin. Reaction products can be readily captured by means of streptavidin-labeled magnetic beads, which provide all the advantages of a solid-phase assay.

c. Substrate specificities: Polymerase

i. Template structure

(a) DNA template. The standard template for *Taq* Pol in primed synthesis reactions is ssDNA, which can be a synthetic polymer, heat-denatured dsDNA, or DNAs from single-strand phages.

A duplex DNA with nicks can also serve as an efficient template in a manner resembling nick translation. This property is due to the presence of a strong 5′-nuclease activity associated with *Taq* DNA Pol.

(b) RNA template. *Taq* Pol is able to use RNA as a template substrate and synthesize cDNA, thus exhibiting a reverse transcriptase activity. The reverse transcription by *Taq* Pol has an optimal Mg^{2+} requirement of 2–3 mM. As an alternative to AMV RTase or MoLV RTase, *Taq* Pol has been used directly in coupled RT–PCR to make and then amplify cDNAs from cellular mRNAs, often from subnanogram quantities (17–19). [For actual RT–PCR applications, *Tth* DNA Pol from *T. thermophilus* is recommended because its reverse transcriptase activity is over 100-fold more efficient than *Taq* Pol (20). *Tth* DNA Pol is also better suited for *in situ* transcription using fluorescein-dUTP.]

ii. Primers. The efficiency of polymerization reactions catalyzed by DNA polymerases is dependent on specific, efficient priming afforded by the primer. Primer design should thus take into account a number of variables such as length, GC content, the degree of complementarity with the target and nontarget sites, and the degree of self-complementarity. A number of computer programs are commercially available that help alleviate the complexities involved in primer or probe design with optimum selectivity/sensitivity.

(a) Matching primers. The length of primers suitable for the polymerization reaction at 72°C should be a minimum of 16 bases and preferably around 25 bases with at least a 50% GC content. The primer configuration should avoid repetitive sequences and stretches of polypurines, polypyrimidines, or other unusual sequences, particularly at the 3′ end. It is also important to avoid any complementarity between the two bracketing primers because 3′ overlaps will create "primer dimer" which not only reduces the actual concentration of the primers available for DNA amplification but also serves as the major artifactual template-primer substrate.

Whereas good annealing is of prime importance at the 3′ end of the primer, the 5′ end need not be complementary. This property allows exogenous sequences (e.g., linker/adapter and promoter) to be added and integrated into the double-stranded PCR products, providing versatile termini with restriction/adapter sites, promoters, and other regulatory sequences.

(b) Degenerate primers. In many instances of molecular cloning, one is often obliged to start with limited sequence information which is obtained from a partial amino acid sequence(s) of the gene product, the DNA sequence of the gene, or a regulatory region. The sequence information may also be derived from the region(s) of reasonably high sequence homology to the proteins or genes of phylogenetically related species. In such cases, *mixed oligonucleotide* (or *degenerate*) priming is probably most appropriate (21,22). The need for degenerate priming arises from the fact that the reverse translation generates multiple sets of nucleotide possibilities, particularly at the third base of amino acid codons. Among the combinations of degenerate sequences, one sequence can be expected to match the target gene correctly. In fact, *Taq* Pol is relatively tolerant to primer mismatches: as the minimum requirement for successful priming, primers (17–20 nt) need at least three matching nucleotides at their 3' end (23). Furthermore, an annealing temperature lower than that which would normally be used for fully matching primers is adopted to increase the chance of priming at the candidate gene by the degenerate sequences. By virtue of the primer design, however, degenerate priming increases the probability of nonspecific amplifications.

To create primer degeneracy, (i) a complete set of nucleotides can be used at the degenerate positions, avoiding if possible the sequences that involve multicodon amino acids; (ii) degenerate positions can be filled in by deoxyinosine which base pairs relatively promiscuously, although there is a preferential order for pairing, i.e., $I \cdot C > I \cdot A > I \cdot G \sim I \cdot T$ (24,25); (iii) a combination of the two aforementioned methods at different positions in a single degenerate pool, for example, deoxyinosine at positions where codon degeneracy is greater than two and pairs of nucleotides at twofold degenerate positions (26); or (iv) a "universal" base analog, e.g., 1-(2'-deoxy-β-D-ribofuranosyl)-3-nitropyrrole, which possesses a high base-stacking capacity in addition to a low H-bonding capacity (thus less base discriminating than dI), can be incorporated into one or more positions with little variation in T_m (lowering within a 3°C range) (27).

(c) Mutagenic primers. The primers used for substitution or insertion mutagenesis typically contain one or more noncomplementary nucleotides flanked on both sides by 10–15 complementary bases. In deletion mutagenesis, the mutagenic primer consists of bipartite sequences which flank the segment of the DNA to be deleted.

In some particular instances of PCR with *Taq* DNA Pol, correctly matching primers may behave as mutagenic primers due to the addition of an untemplated nucleotide at the 3' end of the products (see "Product profile" below). The 3' additional nucleotide is almost invariably an A residue. A method to overcome the potential mutagenic effect of the untemplated addition is to use a primer that is designed in such a way that the first 5' nucleotide of the primer follows a T residue in the same strand of the template sequence (28) (also see "Applications" below).

(d) Primers with 3'-terminal base mismatch. Because of the lack of 3' → 5'-exonuclease activity, *Taq* Pol can extend the primers with 3'-terminal base mismatches. The efficiency of such extensions depends on the identity of mismatching bases as well as the sequence contexts. When carried out under permissive conditions, the PCR with 3'-terminal mismatching primers provides an alternative means of introducing site-specific mutations.

Perhaps the most important use of the 3'-terminal base mismatching primers, however, is their ability to be discriminated from fully complementary primers under stringent, competitive priming conditions. This provides a sensitive means for detecting point mutations and/or polymorphism of genes (29,30). In principle, mismatching bases can be embedded in the primer sequence and yet be successfully discriminated from the fully matching primers when the primer length is small (e.g., 12 bases for PCR with Pol Ik or 16 bases for PCR with *Taq* Pol). However, the discrimination of embedded mismatches becomes less efficient as the primer length increases, whereas the 3'-terminal mismatch, when used with polymerases lacking 3' → 5'-exonuclease activity, consistently gives the strongest discrimination regardless of the primer length. The discrimination against the 3'-terminal mismatch extension can be further enhanced by using low concentrations of dNTP (<20 μM).

iii. Nucleotide substrates

(a) dNTPs. The preferred nucleotide substrates are dNTPs. At high concentrations (4–6 mM total) of dNTPs, however, the polymerase rate decreases by as much as 20–30%. Low, balanced concentrations (e.g., <0.2 mM each) of dNTPs give optimal results in polymerization. The concentrations of dNTPs also affect the fidelity of polymerization (see below).

(b) Nucleotide analogs

(i) dNTPαS. This nucleotide analog (*Sp*-diastereomer) is efficiently incorporated into DNA under standard PCR conditions (31). The polymerization rate is lower than with dNTP, but the efficiency of incorporation can be significantly increased by lowering the temperature to 55°C and increasing the polymerization time (e.g., 7 min for a 957-bp fragment length) (32).

(ii) ddNTPs. *Taq* Pol incorporates ddNTPs ~3000 times less efficiently than the corresponding dNTPs. Remarkably, a mutant *Taq* Pol (F667Y) can utilize ddNTPs ~ twofold more efficiently than dNTPs (33).

(iii) dGTP analogs. 7-Deaza-dGTP is incorporated with essentially the same kinetics as dGTP (2). dITP is also incorporated, but fourfold less efficiently than dGTP. The incorporation of dITP results in a high frequency of chain termination; its use is not recommended in DNA sequencing reactions.

(iv) dCTP analogs. *Taq* Pol can incorporate dm^{5}CMP into DNA correctly at the positions of cytosine nucleotide (34). The PCR incorporation of dm^{5}CMP is efficient only when the temperature of the denaturation step is increased to 100°C or dITP is used to destabilize the dm^{5}C·dG base pairs.

(v) dTTP and nonisotopic nucleotide analogs. *Taq* Pol can efficiently utilize dUTP as a substrate in place of dTTP (35), thus providing efficient means of

generating dU-containing DNA templates for site-specific mutagenesis or as substrates for uracil-DNA glycosylase in carryover contamination controls during PCR amplifications. *Taq* Pol can also efficiently utilize various nucleotide analogs such as biotin-11-dUTP, biotin-21-dUTP, digoxigenin-11-dUTP (36,37), bromo-dUTP, and fluorescently labeled dNTPs (in particular rhodamine-dUTP as compared with fluorescein-dNTPs) as substrates, thereby facilitating nonisotopic detection and quantitation of the products or templates. Compared with *Taq* DNA Pol, other thermostable DNA polymerases, e.g., Vent DNA Pol and *Pfu* DNA Pol, are less efficient in incorporating dUMP (35).

iv. Processivity. *Taq* Pol is highly processive at 70°C with a 10-fold molar excess of template/primer to polymerase, extending most of the initiated primers to a full 7-kb length prior to reinitiation on new template/primer (2). Under different conditions which still produce full-length DNAs as the major product, however, *Taq* Pol has been shown to generate a wide array of prematurely terminated extension fragments, suggesting that the polymerase can exhibit a low processivity (32). Under standard PCR conditions, the processivity is taken to be 50–60 nt and the extension rate to be ~3000 base/min.

v. Product profile. *Taq* Pol has a pronounced terminal transferase activity which adds an extra nontemplated nucleotide (usually A) to ~95% of the products under single as well as multiple cycle reaction conditions (38–41). *Taq* Pol can use any one of the four dNTPs to generate +1 products when the nucleotides are supplied individually. In the presence of equimolar concentrations of all four dNTPs, however, *Taq* Pol uses dATP much more efficiently than the other dNTPs. Similar nontemplated nucleotide additions have been observed with *E. coli* Pol Ik (42), T7 DNA Pol (Sequenase), and other polymerases such as mammalian DNA polymerases and AMV RTase (38).

vi. Efficiency of PCR

(a) Amplification factor. Under ideal conditions with short templates, one PCR cycle produces a twofold amplification of the template sequence, resulting in an exponential accumulation of the product by a factor of 2^n for n number of cycles. Typically, however, the "gain" or amplification efficiency per cycle during the exponential phase of accumulation is 1.8–1.95, which is close to the theoretical maximum of 2.

At the outset of standard PCR conditions, primers and dNTPs are 10^7 and 10^{10} times, respectively, in excess over the template, while *Taq* Pol is in the lowest molar excess of 10^5. After a 10^6-fold amplification of target, enzymes become totally occupied and the primer to template ratio decreases, promoting self-annealing of the strands. Eventually the reaction begins to saturate; the exponential rate of accumulation is attenuated to a linear rate at a product concentration near 10 nM (1 pmol in the standard reaction) and finally reaches a "plateau" at ~100 nM (43). Whether the efficiency decreases rapidly or slowly depends on several variables: (a) initial concentrations and respective balance of dNTPs and/or primers, (b) stability of dNTPs and polymerase, (c) degree of template denaturation and template reannealing, (d) degree of specific and nonspecific primer annealing, and (e) product inhibition by pyrophosphates and nonspecific products.

Under limited or suboptimal reaction conditions, the onset of linear and plateau phases may occur prematurely, rendering the yield of PCR arbitrary with respect to the number of cycles. In fact, the gain can be controlled by carefully adjusting the reaction conditions, giving rise to such useful techniques as "asymmetric" PCR, whereby daughter strands are produced by linear amplification (2,44).

(b) Amplification of "long" DNA. The "standard PCR" conditions, with an optimized set of primers, can usually be counted on to amplify DNA fragments efficiently with sizes smaller than 2 kbp, preferably ≤1 kbp. Amplification of longer DNA by *Taq* PCR encounters increasing inefficiencies, although there are "episodes" of successful long extensions. Some of the factors critically contributing to prematurely terminated strand synthesis are presumed to be: (i) mismatched 3' ends arising from spontaneous nucleotide misinsertion by *Taq* Pol, (ii) chemically impaired templates due to depurination and/or nick introduced during the course of thermal denaturation and/or template preparation, and (iii) inhibition of polymerase activity due to denaturation or other inhibitory conditions.

Reduction or elimination of the presumed limiting factor(s) has been shown to lead to successful amplification of DNA fragments as long as 40 kbp (8,45). The following are some of the "long PCR" conditions:

(i) Addition of a small amount (~1/160th polymerase unit) of 3'-exo$^+$ polymerase (e.g., *Pfu* or Vent DNA Pol) in the *Taq* Pol (or 5'-exo$^-$ *Taq* Pol) PCR. The PCR with a 3'-exo$^+$ polymerase may give longer extension products than with *Taq* Pol alone, but it is inferior to the reaction using a combination of the two polymerases.

(ii) Short denaturation time (e.g., 2–20 sec at 94°C) and higher pH (optimally at pH 9.1 for Tris buffer at 25°C). Denaturation for 1 or 2 min at 95° or 98°C does not produce long extensions. Use of an alternative buffer system with a lower thermal coefficient can also be envisaged, e.g., Bicine (pK_a = 8.35 with ΔpK_a/°C = −0.018) as compared to Tris (pK_a = 8.1 with ΔpK_a/°C = −0.031).

(iii) Addition of cosolvents, e.g., glycerol (10–12%, v/v) and Me$_2$SO (5–10%, v/v), which help to lower melting and strand separation temperatures. Combinatory use of the two cosolvents, for instance, 5% (v/v) Me$_2$SO and 7% (v/v) glycerol, is deemed to be more efficient because the stabilizing effect (as much as twofold) of glycerol on *Taq* Pol can counteract the untoward effect of Me$_2$SO which reduces the thermal stability of *Taq* Pol. Formamide is not helpful in the combinatory system.

(iv) Use of a more thermostable, modified *Taq* Pol such as Stoffel fragment or Klentaq. Note that glycerol is apparently not tolerated by the Klentaq and *Pfu* Pol combination.

(v) Adjustment of polymerase extension time, e.g., ≥5 min for >10-kb fragment, ≥10 min for >15-kb fragment, and ≥20 min for ~40-kb fragment.

vii. Specificity and temperature. Because of the immense power of PCR in amplifying DNA, typically over a millionfold in a 30-cycle experiment, the specificity of polymerization critically affects the yield and purity of the products. For high specificity PCR, good primer design is primordial. Also factors such as pH, temperature, time, and concentration of each reaction component should be carefully optimized. There is no single rule that defines the highest specificity; the "standard" reaction conditions should be taken only as a guide in the search for "optimal" reaction conditions. Under "standard" or compromised reaction conditions, nonspecificity falls into two main categories: (a) presence of heterogeneous nucleic acids in the starting material or contamination during PCR, and (b) amplification of nontarget sequences present on a homogeneous starting material.

Assuming that contamination by exogenous DNA is a factor that can be controlled and that primers are appropriately designed to be able to distinguish them, the specificity of PCR is largely influenced by the nonspecific amplification of homogeneous samples. Since a DNA polymerase will extend all available 3'-OH termini suitably H-bonded to a template, the problem of nonspecificity is essentially a problem of hybridization stringency, i.e., the conditions that discriminate a primer annealed to a perfectly matching (target) site from those annealed to nonperfectly matching (nontarget) sites. For well-designed primers with unique sequences, the main determinant of hybridization stringency in the standard, automated PCR is the annealing temperature. Indeed, the specificity of PCR can be significantly enhanced when an optimized annealing temperature (T_a) is applied. Alternatively, the stringency of hybridization and thus the priming efficiency can be further improved by employing formamide ($\leq 5\%$, v/v) (46) or tetramethylammonium chloride (50–100 μM) (47). For short (15-mer) and/or AT-rich primers, inclusion of Me$_2$SO at low concentrations (0.9–2.0%, v/v) can greatly improve priming efficiency without affecting polymerase activity (9).

In the following, the specificity enhancement by T_a optimization and by two other important PCR techniques is briefly described.

(a) Optimization of annealing temperature. Following the heat denaturation of a dsDNA template at $\sim 95°C$, the optimal annealing temperature (T_a^{OPT}) is usually determined by running preliminary PCR experiments at varying temperatures ($T_a \approx 40–70°C$). The importance of T_a optimization is underscored by the observations that if the T_a is too low or too high, either nonspecific or no DNA amplification occurs. The more precise the information about the T_m of a primer on its target sequence is, the more specific the primer annealing and extension can be. The optimal T_a, "optimal" in terms of product yield and purity, can be calculated with reasonable accuracies, i.e., $\pm 0.7°C$, sparing the effort of experimental determinations (48). The T_a^{OPT} is a function of the T_m of the less stable template–primer pair and also of the PCR product according to the empirical formula (Eq. 6.1):

$$T_a^{OPT} = 0.3\ T_m^{primer} + 0.7\ T_m^{product} - 14.9. \qquad (6.1)$$

The T_m^{primer} can be calculated from Eq. (1.14) (see Chapter 1) using the thermodynamic parameters in Table 1.8 and taking C (the total molar concentration of the annealing oligonucleotides) as 250 pM. Because the product is a long DNA molecule to which Eq. (1.14) is not applicable, the $T_\text{m}^\text{product}$ is calculated from the base composition according to Eq. (1.12).

(b) Application strategies of annealing temperature. A PCR can be performed using either a fixed annealing temperature at T_a^OPT or variable annealing temperatures. When amplifying a DNA fragment less than 300 bp, a *variable annealing temperature method* is useful. As the products get larger, the difference between the variable and constant T_a methods becomes less obvious. In the variable T_a method, the T_a used for the first cycle is 8°C lower than the T_m of the less stable primer. The T_a is then increased by 1° every other cycle. In a typical PCR amplifying 135 bp DNA, a 33% better yield can be obtained by this method.

(c) Techniques of specificity extension in Taq PCR
(i) "Hot-start" PCR. This technique is based on the principle of initiating the polymerase reaction at or above the primer annealing temperature, thereby preempting possible extensions of the primers nonspecifically annealed at ambient temperature (20°–55°C) at which *Taq* Pol retains partial activity (5). For some applications, hot-start PCR may be the only way to amplify successfully the desired product(s) with high yield and specificity.

In practice, a hot-start PCR condition can be attained by adding the polymerase (or any other essential reagent like dNTPs or Mg^{2+}) only when the mixture has reached a temperature higher than T_a and then starting cycles of primer annealing and extension. The use of a wax pellet (e.g., AmpliWax of Perkin-Elmer), which can provide at room temperature a solid-phase barrier between the polymerase and the rest of the reaction components, has made it substantially easier and safer to perform the technique in an uninterrupted PCR operation. Introduction of thermolabile antibodies which reversibly block the polymerase provides an alternative efficient technique for hot-start PCR: as the thermal cycling starts and temperature of the complete PCR mixture is raised, the inhibitory antibody is selectively denatured (within 30 sec at >85°C), thereby releasing fully active *Taq* Pol that initiates the polymerization reaction under high stringency priming conditions (49). (*Note:* One such antibody, with a trade name of *TaqStart,* is commercially available from Clontech.)

(ii) Nested priming PCR. This is a two-tiered approach in which the correct first product is given the greatest chance to be amplified in the second round of PCR, thereby enhancing specificity. After an initial amplification, the interim products are appropriately diluted (e.g., 10^5-fold) and the second round of PCR is performed using another primer pair defining a target that lies within the sequence specified by the original primer pair (50–52). The nested priming strategy can be doubly useful: when the second primer pair is designed such that they contain specific ligand-binding sequences or modified (e.g., biotinylated) nucleotides, they can serve as tools for detection and/or purification of the PCR products (53).

viii. Fidelity of *Taq* DNA polymerase

(a) Frequency of misincorporation. *Taq* DNA Pol lacks the proofreading $3' \rightarrow 5'$-exonuclease activity and has a relatively high error rate. Although the estimates vary depending on reaction conditions, experimental systems, and methods of calculation, single-base substitution errors are estimated to occur at a rate of $1–7 \times 10^{-4}$ per base pair per cycle (54–58). Note that an error rate of 1 in 10^4 bp per cycle corresponds to one error in 400 bp in the final product after 25 cycles of PCR. The most common error observed with *Taq* Pol is a C-to-T transition (50%), followed by a T-to-C transition (36%) (59). Transversions are rare, but do occur: A to C (5%), A to T (5%), and C to G (2%). Frameshift errors occur 10 times less frequently than substitution errors (60).

In a set of PCR sequencing reaction conditions, *Taq* Pol was estimated to exhibit a total error rate of about 1% per position over the first 350 nt after 48 cycles (61). This error rate is in fact similar to that of Sequenase ($3'$-exo⁻ T7 DNA Pol) in standard sequencing reaction conditions. In other estimates, *Taq* Pol showed an error rate of 8.9×10^{-5} per bp, while Sequenase had 4.4×10^{-5} per bp and a $3'$-exo⁺ polymerase (Vent DNA Pol) had 2.4×10^{-5} per bp (62).

Fidelity of a polymerase is an outcome of multiple interactions between the polymerase and various factors, including dNTP concentrations, pH of the medium, and the concentration of $MgCl_2$ relative to the total concentration of dNTPs. At pH 7.2, for example, *Taq* Pol is six times more error prone in the buffer containing 1 mM Mg^{2+} and 0.25 mM each of four dNTPs than in the buffer containing 4 mM Mg^{2+} and 1 mM each of four dNTPs. At the same pH and 10 mM Mg^{2+}, however, the error frequency varies only twofold between 1 mM and 1 μM dNTPs. *Taq* Pol is generally presumed to exhibit a higher fidelity at low pH conditions (pH 5–6 at 70°C) or when $MgCl_2$ and dNTPs are present at equimolar concentrations (60). The low pH conditions reduce the polymerase activity such that the amplification of large fragments is impossible. Interestingly, however, some alkaline conditions are known, e.g., 25 mM TAPS (pH 9.3), 2 mM $MgCl_2$, 50 mM KCl, 1 mM 2-MSH, and 0.2 mM dNTPs (63), which makes *Taq* Pol a high fidelity enzyme (2×10^{-6} errors/bp) comparable to or better than Vent DNA Pol or *Pfu* DNA Pol.

*(b) **Higher fidelity with a $5'$-nuclease-deficient Taq DNA Pol.*** A modified *Taq* Pol in which the region responsible for $5'$-nuclease activity is truncated (see "Structure" below) turns out to be approximately twofold less error prone than the wild-type enzyme (54). The increased fidelity of the $5'$-nuclease-deficient ($5'$-exo⁻) polymerase, called *Klentaq*, is attributed to the reduced processivity of the modified *Taq* Pol. Although Klentaq can catalyze the amplification of short DNA fragments as efficiently as the full-length *Taq* Pol, the amplification of longer fragments requires more units of Klentaq. Excess Klentaq has been shown to result in more efficient amplifications of DNA.

d. Substrate specificities: $5'$-Nuclease

i. **Optimal conditions.** The $5'$-nuclease activity of *Taq* Pol requires a divalent cation (Mg^{2+} or Mn^{2+}) (64). The activity is optimal at 4–8 mM $MgCl_2$ or 3 mM $MnCl_2$; Mn^{2+} is the preferred cation. Neither Zn^{2+} nor Ca^{2+} ions support the nuclease activity. The presence of KCl influences nuclease activity, giving maximum stimulation at 50 or 20 mM in the presence or absence of primer, respectively (65).

The 5'-nuclease activity is optimal at pH 9.0 in Tris–Cl (50 mM) (64). At pH 8.5 and pH 9.5, the activities are reduced to 89 and 96%, respectively, of the maximal activity. *Taq* Pol shows a near maximal activity in HEPES–KOH buffer (50 mM, pH 8.5).

ii. **Template structure.** The 5'-nuclease activity associated with *Taq* DNA Pol is, like the 5'-nuclease of *E. coli* DNA Pol I, a structure-specific endonuclease that requires a duplex structure as a part of the substrate. The preferred structure for cleavage is the displaced DNA or a bifurcated structure with a free 5' end (65,66). The single-stranded DNA or RNA is cleaved at the bifurcated end of a base-paired duplex, preferentially at the phosphodiester bond between the first two base pairs from the junction. The length of a 5' single-stranded arm does not interfere with the cleavage, as long as the 5' substrate arm is held single stranded either by the absence of a complementary sequence or of adventitious secondary structures or by the presence of a primer complementary to the template strand.

The presence of a primer (or "pilot" DNA) influences the site and rate of cleavage. Indeed, 5'-nuclease activity can be coaxed to perform a targeted (or sequence-specific) cleavage into any single-stranded DNA or RNA through hybridization with a pilot DNA that converts the desired cleavage site into a substrate. Cleavage can occur with a long (at least up to 2 kb) or short (a single mismatching nucleotide) 3' arm of the template strand. A certain DNA substrate containing a template strand with no 3' arm may be resistant to 5'-nuclease activity. In contrast, an RNA substrate with the same sequence can be cleaved in the presence of a pilot oligonucleotide without a 3' arm, indicating that the 5'-nuclease of *Taq* DNA Pol can function as a structure-specific RNase H that cleaves at a single site opposite the 3'-end of the pilot DNA.

Single-stranded DNAs (e.g., heat-denatured dsDNA or free oligonucleotides) or duplex DNAs that lack a free 5' end (e.g., circular DNAs) are not efficient substrates.

iii. **Termini structure.** The 5'-P oligonucleotide annealed to a DNA strand is hydrolyzed twice as efficiently as the same oligonucleotide with a 5'-OH group (64). Hydrolysis of a 5'-terminal residue from a nick is preferred three times over the same 5'-end of duplex DNA.

iv. **Product profile.** The 5'-nuclease activity of *Taq* Pol cleaves 5' terminal nucleotides of duplex DNA and, in a manner similar to the 5'-nuclease activity of *E. coli* DNA Pol I, releases mono- and oligonucleotides terminated with 3'-OH.

v. **Sequence dependence.** Although the 5'-nuclease of *Taq* Pol is in general regarded as a sequence-independent, structure-specific endonuclease, size distribution of the cleavage products reveals that the nuclease is not entirely independent on base composition and sequence contexts of the 5'-end sequences. For instance, GC-rich sequences carrying three 5'-terminal noncomplementary bases give rise to primarily three or four base cleavage products (66). In contrast, AT-rich sequences similarly carrying three 5'-terminal noncomplementary bases give rise to primarily a six base product.

vi. Relationship with polymerase activity. The 5'-nuclease and polymerase activities function concurrently. In reactions resembling a nick translation, polymerase activity has been shown to be unaffected by the simultaneous 5'-nuclease activity. In fact, precise orientation of the 5'-nuclease on a substrate is dominated by the interaction of the polymerization domain of the polymerase with the primer (65).

2. APPLICATIONS

The highly thermostable *Taq* DNA Pol has become a key ingredient of PCR technology, serving as the workhorse for an ever-widening range of applications (3,4,43,67,68). The thermostability of *Taq* Pol not only contributes to higher specificities and yields of PCR but has made the automation of PCR possible, leading to greater efficiency, productivity, and general acceptance of the technology. The amount of literature testifies to the importance and capability of PCR, which owes much to its marvelous ability to amplify a DNA sequence over a millionfold (for example, from nanogram to milligram DNA) in a matter of hours with minimal efforts. The synthesis of cDNA has also been subjected to a similar revolution through a coupled process known as reverse transcription–PCR (RT–PCR). Apart from the applications oriented toward the detection and/or analysis of new, rare, or mutant genes (and proteins) and DNA sequences, *Taq* PCR has been applied to recombinant DNA technology in a number of important ways. They include: (a) mass production of DNA material for the construction of special, as well as routine, recombinant DNAs; (b) quantitation of a minute amount of target DNA or RNA with a sensitivity down to a single copy, an increasingly important technique referred to as "quantitative PCR"; (c) molecular cloning; (d) nucleotide sequencing; and (e) site-specific oligonucleotide-directed mutagenesis.

The following is a brief summary of the techniques and problems particularly associated with nucleotide sequencing and molecular cloning. Readers are advised to consult other references for more exhaustive treatment of these and other subjects regarding principles and methods (3,4,67) and step-by-step procedures (15). For principles and methods of PCR-based site-specific mutagenesis, refer to Appendix A.

i. Nucleotide sequencing. DNA fragments, whether PCR amplified or not, can be used as templates in dideoxynucleotide sequencing reactions using *Taq* Pol as the sequencing enzyme. Alternatively, PCR-amplified DNA fragments can be first subcloned and then subjected to standard sequencing reactions with such polymerases as *E. coli* DNA Pol I (Klenow) or T7 DNA Pol (Sequenase). In fact, direct sequencing of PCR products is recommended because it provides a "consensus" sequence, reducing the error frequency to 1 in 4000–5000. In contrast, the subcloning routine requires sequencing several subclones to minimize the probability of sequencing a fortuitous error-laden subclone.

Taq DNA Pol (5'-exo$^-$) is ideal for direct sequence analysis of PCR products either in a classical single round sequencing reaction or, more preferably, by a new strategy called *cycle sequencing* in which multiple rounds of sequencing reaction are performed under PCR thermal cycling conditions. The ability of *Taq* Pol to function at high temperatures and low salts permits heat destabilization of the secondary structures in the template and reduces sequencing artifacts thanks to more stringent primer annealing. A modified *Taq* DNA Pol (5'-exo$^-$, F667Y), called "Thermo Sequenase" (Amersham), should be particularly useful for nucleotide sequencing due to its enhanced ability ($\geq$3000-fold over the wild type) to utilize ddNTPs as opposed to dNTPs in the presence of Mg^{2+}.

Taq Pol can also be employed as an auxilliary enzyme which just fulfills the need to prepare ample amounts of template DNA (either ssDNA or dsDNA). Single-stranded DNA templates can be readily generated by a nonstandard PCR strategy known as "asymmetric PCR." Compared with "symmetric PCR," which in essence corresponds to the standard PCR, the asymmetric PCR employs one of the two primers about 100-fold more than the other primer such that, after a certain number of exponential amplification which exhausts the limiting primers, linear amplification of ssDNA is attained.

Single-stranded DNA templates can also be readily prepared using the standard protocols of symmetric PCR, when provided with some ingenuity. If one of the two primers carries the 5'-terminal phosphate group, the phosphorylated strand of PCR products can be selectively digested from the phosphorylated 5' ends by the action of λ exonuclease (70). (Note that the PCR products generated with nonphosphorylated primers can be digested with a suitable restriction enzyme to expose the 5'-terminal phosphate group on one strand.) This treatment leaves ssDNA that can be used as templates in standard dideoxy sequencing reactions using any suitable DNA polymerases. When one of the two primers in the symmetric PCR is labeled at the 5' termini with biotin, the biotinylated strand of the PCR products can be captured by means of streptavidin-coated magnetic beads. The dsDNA products are denatured by heat or alkali, enabling a separate purification(s) of ssDNAs which can be used in uni- or bidirectional dideoxy sequencing reactions (71).

ii. **Molecular cloning.** PCR-amplified DNA or cDNA fragments can be used as suitable material for molecular cloning via either cohesive-end ligation or blunt-end ligation strategies.

(a) Blunt-end ligation strategies. Blunt-end ligation/cloning is intrinsically less efficient than cohesive-end ligation/cloning, but it is often more convenient. A particular problem associated with blunt-end ligation/cloning is that about 95% of the DNA molecules amplified with *Taq* Pol have "ragged" ends rather than perfect blunt ends. To increase the cloning efficiency of the "blunt-ended"

molecules, the following alternative strategies should be considered. (i) The termini of PCR products should be repaired before blunt-end ligation by a brief treatment with Pol Ik or T4 DNA Pol. (ii) Counting on the portion of +1 products carrying a 3′-overhanging A residue, a partial cohesive-end ligation can be attempted on a ddT-tailed vector (41). (iii) Other DNA polymerases which possess a 3′ → 5′-exonuclease activity, e.g., T7 DNA Pol or more preferably Vent DNA Pol (72), can be used to amplify the blunt-ended products suitable for direct cloning. Perhaps an important additional advantage of the last alternative is that the chances of selecting mutant clones can be substantially reduced.

(b) Cohesive-end ligation strategies. PCR products can be most efficiently cloned by cohesive-end ligation largely due to the ease of introducing a restriction site or a suitable sequence as a part of PCR primers. If the cloning methods based on the use of a restriction-site primer are taken to be "standards," other nonstandard, efficient cloning methods exist. One of the nonstandard "ligase-free cloning" methods is T4 DNA Pol-mediated directional cloning whereby single stranded adapter arms of the insert (as well as vectors) are generated using the 3′ → 5′-exonuclease activity of T4 DNA Pol (see Fig. 6.3, Section II,B, this chapter). Annealing of 5′-protruding termini to the vector DNA containing complementary 5′ ends results in chimeric molecules which can be transformed, with high efficiency, without *in vitro* ligation. Another elegant ligase-free cloning method is uracil-DNA glycosylase (UDG)-mediated cloning (73). This method is based on the use of dU-containing primers during PCR and selective degradation of the dU residues in the PCR products with UDG resulting in disrupted base pairing. The UDG treatment generates 3′ overhangs which can then anneal to similarly treated, complementary ends of cloning vectors. The annealed chimeric molecules can be directly transformed into suitable competent cells.

Cohesive-end ligation of the restricted fragments to a linearized vector often suffers low efficiency due to the persistent association of polymerase molecules with the product DNA throughout the routine procedure of phenol–chloroform extraction with vortexing, ethanol precipitation, and drying. Because dNTPs can partially survive these same treatments, the DNA polymerase, which is present and active during the subsequent restriction enzyme digestion of the PCR product, can fill in the 3′-recessive termini.

Two alternative strategies are available to increase the efficiency of cloning PCR products. One is the proteinase K treatment of the DNA products to remove the DNA-bound polymerase. This can increase the clonability by 20- to 30-fold (74). The other, which is claimed to be more efficient and reliable, is to precipitate the DNA with PEG and high salt concentrations (by adding 1/2 volume of 30% (w/v) PEG/1.5 *M* NaCl) (54).

iii. 3′-End labeling of DNA probes. *Taq* DNA Pol provides an efficient method to generate phosphatase-insensitive 3′-radiolabeled DNA probes (76). The method is based on the template-independent terminal transferase activity that adds a single radiolabeled nucleotide to the 3′ ends of blunt end DNA molecules. Where appropriate, DNA fragments with 5′ overhangs can be labeled simultaneously at the 5' ends by a fill-in reaction. Note that DNA fragments can also be conveniently labeled at the 5' ends using T4 PNK. However, 5'-phosphorylated DNAs are highly sensitive to endogenous phosphatases present in the crude or partially purified protein extracts that are typically used in DNA/protein-binding reactions. DNA fragments (e.g., 50 ng in 20 μl) are labeled in standard PCR reaction buffers in the presence of 10 μCi dNTP (3000 Ci/mmol, preferably [α-^{32}P]dATP or dGTP) by incubating for 2 hr at 70°C.

iv. Applications of 5'-nuclease activity. The 5'-nuclease activity associated with *Taq* DNA Pol and also with other DNA polymerases provides some unique applications, for example, as a tool for detecting/quantifying specific PCR products and also for site-specifically cleaving DNA or RNA.

(a) As tool for product (or template) DNA quantitation. In quantitative PCR, the detection probe is designed as an oligonucleotide with a nonextendable 3' end (3'-phosphorylated) and a labeled 5' end (66). When the probe hybridizes to a specific target sequence within the amplified region, it turns into a suitable substrate for 5'-nuclease. The amount of released 5'-end labels serves as a quantitative indicator of target sequence amplification.

(b) As primer-directed site-specific cutter. The 5'-nuclease activity can be configured as a versatile, site-specific cleavage reagent for single-stranded DNA or RNA by virtue of its sequence-nonspecific but structure-specific endonucleolytic activity (65). Essentially any sequence can be targeted for selective cleavages by annealing it with appropriate oligonucleotides (or pilot DNA). The primer-directed cleavage is carried out at 75°C, a stringent hybridization condition that minimizes the possible interference of long 5' arms.

3. STRUCTURE

i. Overall features. *Taq* DNA Pol I consists of 832 amino acids (M_r 94,000) (1). The N- and C-terminal residues are predicted to be Met and Glu, respectively. The enzyme contains no Cys.

Taq DNA Pol shows ~50% sequence homology with *E. coli* DNA Pol I within the C-terminal ~400 residues that constitute the polymerase domain. *Taq* Pol has an independent 5'-nuclease activity in the N-terminal domain (residues 1–290) but has no 3' → 5'-exonuclease activity.

The crystal structures of *Taq* Pol (69) and Klentaq1 (75) reveal that the C-terminal polymerase domain is identical in fold to that of Klenow Pol. Regarding the 3' → 5'-exonuclease domain (residues 324–515) of Klenow Pol, *Taq* Pol is markedly different by having deletions of four loops of lengths 8 to 27 residues. All four of the acidic residues (D424, D501, D355, E357) known to be essential for divalent metal binding and exonucleolytic catalysis in Klenow Pol are replaced by residues incapable of binding metal ions (L356, R405, G308, V310) in the vestigial 3' → 5'-exonuclease domain of *Taq* Pol.

The N-terminal 5'-nuclease domain of *Taq* Pol has a deep cleft that contains at its bottom a cluster of acidic residues highly conserved in 5'-nuclease domains of the Pol I family. Seven of the carboxylate groups are presumed to form three divalent metal-binding sites (a triangle of ~5 Å × ~10 Å × ~10 Å). Two residues (R25 and R74) are essential for 5′-nuclease activity.

Among other notable differences from Klenow Pol, Klentaq1 has 19 opposite-charge substitutions, suggesting a global charge redistribution which is likely to play a role in the thermostability.

ii. Active site. The amino acid residues identified to be in the active center of *E. coli* Pol I, i.e., Arg-682, Lys-758, Tyr-766, Arg-841, and His-881, are all conserved in *Taq* Pol I. Phe-667 is an important residue for dNTP binding; F667Y mutation decreases discrimination against ddNTPs ~6000-fold (33).

Taq Pol I contains all of the four amino acid residues which, if replaced in *E. coli* Pol I, result in polymerase mutants defective in 5′-nuclease activity, i.e., Y77C, G103E, G184D, and G192D.

iii. Modified *Taq* DNA polymerases. Modified forms of *Taq* Pol have been generated by genetic engineering that lack the N-terminal domain responsible for 5′-nuclease activity of the wild-type *Taq* Pol.

Klentaq5: This 5′-nuclease-deficient polymerase starts from Met-236 as a result of the N-terminal truncation of 235 amino acid residues and is called *Klentaq* in analogy to the Klenow fragment of *E. coli* DNA Pol I (54). (Note: Klentaq5 is available from USB/Amersham under the trade name of Δ*Taq*.) As notable features, the modified enzyme displays an enhanced thermostability ($t_{1/2}$ of 60 min at 95°C) and a higher polymerase fidelity (~2-fold) than the wild-type *Taq* DNA Pol. Another version of the modified polymerase, *Klentaq1* (available from Ab Peptides, St. Louis, MO) has residues 281–832 with a 7-amino-acid N-terminal extension (MGKRKST). Klentaq1 has a fidelity 1.5 times that of wild-type *Taq* Pol (45).

Stoffel fragment: This is another genetically modified *Taq* Pol (61 kDa) that has been generated by deletion of the N-terminal 289 amino acid residues. (cf. The Ampli*Taq* DNA Pol Stoffel fragment is available from Perkin-Elmer.) Compared with the full-length *Taq* Pol, the truncated polymerase (Stoffel fragment) lacks the 5′-nuclease activity, has an enhanced thermostability (by approximately twofold, i.e., $t_{1/2}$ of 80 min at 95°C or 20 min at 97.5°C), and exhibits optimal activity over a broader range of Mg^{2+} concentrations (2–10 mM) and at lower ionic strength. The Stoffel fragment has a polymerase activity (~3 kb/min at 70°C) similar to that of the full-length enzyme, but it displays a lower processivity (5–10 nucleotides).

4. GENETICS

The structural gene for *Taq* DNA Pol I contains 2496 bp (or 832 codons). The gene, which was initially cloned in the λgt11 vector, was subsequently recloned, sequenced, and expressed in *E. coli* under the control of the *lac* promoter (1). The gene contains 67.9% GC compared to a 52.0% GC content for the *E. coli* Pol I gene. A heavy bias toward G and C in the third codon position (91.8% C and G) translates into bias for Arg over Lys and Leu over Ile. The expression of cloned *Taq* Pol gene in *E. coli* can be over 100-fold increased by changing the first 6 codons to simple AT-type codons favored in *E. coli*.

Although the overall nucleotide sequence homology may be low, the *Taq* DNA Pol I gene shows four regions of significant homology to *polA* (*E. coli* DNA Pol I), with 19 bases being the longest stretch.

5. Sources

Taq DNA Pol has been conventionally purified at a level of 0.01–0.02% of the total cellular protein from *T. aquaticus* (strain YT1 originally isolated in 1965 by Thomas Brock from a Yellowstone hot spring). The polymerase is now more conveniently purified from *E. coli* recombinant overexpression systems, for instance, using pTaq plasmids under the control of the IPTG-inducible *tac* promoter (77,78). Purification procedures consisting of a heat treatment (75°C, 1 hr) of the cell lysate, polyethyleneimine (0.05–0.8%) or ammonium sulfate (30 g per 100 ml lysate) precipitation, and ion-exchange (Bio-Rex 70) chromatography are reported to yield 40–50 mg purified protein per liter of culture. Further improvements in the yield and purity can be achieved by optimization of such factors as host bacterial strain, e.g., *E. coli* INV 1 αF' (a strain from Invitrogen Corp.) as opposed to *E. coli* DH1, and induction time (12 hr at 37°C and 125 μg/ml IPTG).

The specific activity of purified *Taq* Pol is ~250,000 U/mg. Commercial stocks of *Taq* Pol are usually supplied in a buffer containing 20 mM Tris–Cl (pH 8.0), 100 mM KCl, 0.1 mM EDTA, 1 mM DTT, autoclaved gelatin (200 μg/ml), 0.5% (v/v) NP-40, 0.5% (v/v) Tween 20, and 50% (v/v) glycerol. *Taq* DNA Pol, at ~10 mg/ml and in buffers containing no detergents, gelatin, or other proteins, can be safely stored at −20°C for over 12 months or at −70° for over 18 months.

References

1. Lawyer, F. C., Stoffel, S., Saiki, R. K., Myambo, K., Drummond, R., and Gelfand, D. H. (1989). *JBC* **264**, 6427–6437.
2. Innis, M. A., Myambo, K. B., Gelfand, D. H., and Brow, M. A. D. (1988). *PNAS* **85**, 9436–9440.
3. Innis, M. A., Gelfand, D. H., Sninsky, J. J., and White, T. J. (Eds.) (1990). "PCR Protocols: A Guide to Methods and Applications." Academic Press, New York.
4. Erlich, H. A. (Ed.) (1989). "PCR Technology: Principles and Applications of DNA Amplification." Stockton Press, New York.
5. Chou, Q., Rusel, M., Birch, D., Raymond, J., and Bloch, W. (1992). *NAR* **20**, 1717–1723.
6. Gelfand, D. H. (1989). *In* "PCR Technology: Principles and Applications of DNA Amplification" (H. A. Erlich, Ed.), pp. 17–22. Stockton Press, New York.
7. Weyant, R. S., Edmonds, P., and Swaminathan, B. (1990). *BioTechniques* **9**, 308–309.
8. Cheng, S., Fockler, C., Barnes, W. M., and Higuchi, R. (1994). *PNAS* **91**, 5695–5699.
9. Filichkin, S. A., and Gelvin, S. B. (1992). *BioTechniques* **12**, 828–830.
10. Panaccio, M., Georgesz, M., and Lew, A. M. (1993). *BioTechniques* **14**, 238–243.
11. Chester, N., and Marshak, D. R. (1993). *Anal. Biochem.* **209**, 284–290.
12. Demeke, T., and Adams, R. P. (1992). *BioTechniques* **12**, 332–334.
13. Satsangi, J., Jewell, D. P., Welsh, K., Bunce, M., and Bell, J. I. (1994). *Lancet* **343**, 1509–1510.
14. Sellner, L. N., Coelen, R. J., and Mackenzie, J. S. (1992). *NAR* **20**, 1487–1490.

15. Sambrook, J., Fritsch, E. F., and Maniatis, T. (1989). "Molecular Cloning: A Laboratory Manual," 2nd ed., pp. 14.1–14.35. Cold Spring Harbor Laboratory Press, Cold Spring Harbor, New York.
16. Day, D. J., Saul, D. J., Reeves, R. A., and Bergquist, P. L. (1993). *Anal. Biochem.* **211**, 174–176.
17. Jones, M. D., and Foulkes, N. S. (1989). *NAR* **17**, 8387–8388.
18. Shaffer, A. L., Wojnar, W., and Nelson, W. (1990). *Anal. Biochem.* **190**, 292–296.
19. Tse, W. T., and Forget, B. G. (1990). *Gene* **88**, 293–296.
20. Myers, T. W., and Gelfand, D. H. (1991). *Biochem.* **30**, 7661–7666.
21. Lee, C. C., Wu, X. W., Gibbs, R. A., Cook, R. G., Muzny, D. M., and Caskey, C. T. (1988). *Science* **239**, 1288–1291.
22. Girgis, S. I., Alevizaki, M., Denny, P., Ferrier, G. J. M., and Legon, S. (1988). *NAR* **16**, 10371.
23. Sommer, R., and Tautz, D. (1989). *NAR* **17**, 6749.
24. Knoth, K., Roberds, S., Poteet, C., and Tamkun, M. (1988). *NAR* **16**, 10932.
25. Case-Green, S. C., and Southern, E. M. (1994). *NAR* **22**, 131–136.
26. Moremen, K. W. (1989). *PNAS* **86**, 5276–5280.
27. Nichols, R., Andrews, P. C., Zhang, P., and Bergstrom, D. E. (1994). *Nature* **369**, 492–493.
28. Kuipers, O. P., Boot, H. J., and de Vos, W. M. (1991). *NAR* **19**, 4558.
29. Newton, C. R., Graham, A., Heptinstall, L. E., Powell, S. J., Summers, C., Kalsheker, N., Smith, J. C., and Markham, A. F. (1989). *NAR* **17**, 2503–2516.
30. Gibbs, R. A., Nguyen, P.-N., and Caskey, C. T. (1989). *NAR* **17**, 2437–2448.
31. Nakamaye, K. L., Gish, G., Eckstein, F., and Vosberg, H.-P. (1988). *NAR* **16**, 9947–9959.
32. Olsen, D. B., and Eckstein, F. (1989). *NAR* **17**, 9613–9620.
33. Tabor, S., and Richardson, C. C. (1995). *PNAS* **92**, 6339–6343.
34. Wong, K. K., and McClelland, M. (1991). *NAR* **19**, 1081–1085.
35. Sluppaug, G., Alseth, I., Eftedal, I., Volden, G., and Krokan, H. E. (1993). *Anal. Biochem.* **211**, 164–169.
36. Finckh, U., Lingenfelter, P. A., and Myerson, D. (1991). *BioTechniques* **10**, 35–39.
37. Duplaa, C., Couffinhal, T., Labat, L., Moreau, C., Lamazière, J.-M. D., and Bonnet, J. (1993). *Anal. Biochem.* **212**, 229–236.
38. Clark, J. M. (1988). *NAR* **16**, 9677–9686.
39. Hemsley, A., Arnheim, N., Toney, M. D., Cortopassi, G., and Galas, D. J. (1989). *NAR* **17**, 6545–6551.
40. Mole, S. E., Iggo, R. D., and Lane, D. P. (1989). *NAR* **17**, 3319.
41. Holton, T. A., and Graham, M. W. (1991). *NAR* **19**, 1156.
42. Clark, J. M., Joyce, C. M., and Beardsley, G. P. (1987). *JMB* **198**, 123–127.
43. Bloch, W. (1991). *Biochem.* **30**, 2735–2747.
44. Gyllensten, U. B., and Erlich, H. A. (1988). *PNAS* **85**, 7652–7656.
45. Barnes, W. M. (1994). *PNAS* **91**, 2216–2220.
46. Sarkar, G., Kapelner, S., and Sommer, S. S. (1990). *NAR* **18**, 7465.
47. Hung, T., Mak, K., and Fong, K. (1990). *NAR* **18**, 4953.
48. Rychlik, W., Spencer, W. J., and Rhoads, R. E. (1990). *NAR* **18**, 6409–6412.
49. Sharkey, D. J., Scalice, E. R., Christy, K. G., Jr., Atwood, S. M., and Daiss, J. L. (1994). *Bio/Technology* **12**, 506–509.
50. Haqqi, T. M., Sarkar, G., David, C. S., and Sommer, S. S. (1988). *NAR* **16**, 11844.
51. Schowalter, D. B., and Sommer, S. S. (1989). *Anal. Biochem.* **177**, 90–94.
52. Kaneko, S., Feinstone, S. M., and Miller, R. H. (1989). *J. Clin. Microbiol.* **27**, 1930–1933.
53. Kemp, D. J., Smith, D. B., Foote, S. J., Samaras, N., and Peterson, M. G. (1989). *PNAS* **86**, 2423–2427.
54. Barnes, W. M. (1992). *Gene* **112**, 29–35.
55. Saiki, R. K., Gelfand, D. H., Stoffel, S., Scharf, S. J., Higuchi, R., Horn, G. T., Mullis, K. B., and Erlich, H. A. (1988). *Science* **239**, 487–491.
56. Tindall, K. R., and Kunkel, T. A. (1988). *Biochem.* **27**, 6008–6013.
57. Ennis, P. D., Zemmour, J., Salter, R. D., and Parham, P. (1990). *PNAS* **87**, 2833–2837.
58. Keohavong, P., and Thilly, W. G. (1989). *PNAS* **86**, 9253–9257.

59. Chen, J., Sahota, A., Stambrook, P. J., and Tischfield, J. A. (1991). *Mut. Res.* **249**, 169–176.
60. Eckert, K. A., and Kunkel, T. A. (1990). *NAR* **18**, 3739–3744.
61. Koop, B. F., Rowan, L., Chen, W.-Q., Deshpande, P., Lee, H., and Hood, L. (1993). *BioTechniques* **14**, 442–447.
62. Cariello, N. F., Swenberg, J. A., and Skopek, T. R. (1991). *NAR* **19**, 4193–4198.
63. Brail, L., Fan, E., Levin, D. B., and Logan, D. M. (1993). *Mut. Res.* **303**, 171–175.
64. Longley, M. J., Bennet, S. E., and Mosbaugh, D. W. (1990). *NAR* **18**, 7317–7322.
65. Lyamichev, V., Brow, M. A. D., and Dahlberg, J. E. (1993). *Science* **260**, 778–783.
66. Holland, P. M., Abramson, R. D., Watson, R., and Gelfand, D. H. (1991). *PNAS* **88**, 7276–7280.
67. McPerson, M. J., Quirke, P., and Taylor, G. R. (Eds.) (1991). "PCR: A Practical Approach." IRL Press, Oxford.
68. Erlich, H. A., Gelfand, D., and Sninsky, J. J. (1991). *Science* **252**, 1643–1651.
69. Kim, Y., Eom, S. H., Wang, J., Lee, D.-S., Suh, S. W., and Steitz, T. A. (1995). *Nature* **376**, 612–616.
70. Takagi, S., Kimura, M., and Katsuki, M. (1993). *BioTechniques* **14**, 218–221.
71. Hultman, T., Bergh, S., Moks, T., and Uhlen, M. (1991). *BioTechniques* **10**, 84–93.
72. Garrity, P. A., and Wold, B. J. (1992). *PNAS* **89**, 1021–1025.
73. Rashtchian, A., Buchman, G. W., Schuster, D. M., and Berninger, M. S. (1992). *Anal. Biochem.* **206**, 91–97.
74. Crowe, J. S., Cooper, H. J., Smith, M. A., Sims, M. J., Parker, D., and Gewert, D. (1991). *NAR* **19**, 184.
75. Korolev, S., Nayal, M., Barnes, W. M., Di Cera, E., and Waksman, G. (1995). *PNAS* **92**, 9264–9268.
76. Niedenthal, R. K., and Hegemann, J. H. (1993). *NAR* **21**, 4413.
77. Engelke, D. R., Krikos, A., Bruck, M. E., and Ginsburg, D. (1990). *Anal. Biochem.* **191**, 396–400.
78. Pluthero, F. G. (1993). *NAR* **21**, 4850–4851.

III. RNA-DIRECTED DNA POLYMERASES (REVERSE TRANSCRIPTASES)
[RNA-Directed DNA Polymerase, EC 2.7.7.49]

A. General Description

Reverse transcriptase (RTase) is a group of DNA polymerases that catalyze the synthesis of complementary DNAs from RNA templates efficiently, thus performing reverse transcription. Reverse transcriptase was first found in purified virions of murine leukemia virus (MLV) by Baltimore (1) and in Rouse sarcoma virus (RSV) by Temin and Mizutani (2). This enzyme is proteolytically derived from the precursor encoded by the viral *pol* gene. RTase uses viral RNA as the natural template to synthesize double-stranded proviral DNA *in vivo*, thereby playing a central role during the life cycle of retroviruses (3–6). Reverse transcriptase and the phenomenon of reverse transcription have since been discovered in many other organisms, including DNA viruses (hepatitis B virus, caulimovirus), bacteria (*Myxococcus xanthus,* some clinical strains of *E. coli* B), yeast (Ty retrotransposon), fungi, invertebrates (*copia*-like element of *Drosophila*), and plants (7,8).

Although the newly discovered RTases of nonretroviral origin are exciting sources of new features and diversities for this class of polymerases, this section focuses on two typical retroviral RTases, i.e., *avian myeloblastosis virus* (*AMV*) *RTase* and *Moloney murine leukemia virus* (*MoLV*) *RTase*, largely due to their thorough characterization as well as their widespread use in recombinant DNA technology. Some DNA-dependent DNA polymerases, including *E. coli* DNA Pol I (9), also possess RNA-dependent DNA polymerase activities. Nevertheless, the RTase of retroviral origin is still the enzyme of choice for the synthesis of cDNAs from RNA templates and thus is an essential tool in recombinant DNA technology. There is also a surge of interest in retroviral RTases as a potential chemotherapeutic target in various retrovirus-associated diseases such as leukemia/lymphoma, sarcoma, other malignant tumors, and acquired immune deficiency syndrome (AIDS).

a. Reaction specificities

Retroviral RTases are multifunctional enzymes that exhibit at least three distinct enzymatic activities: (i) RNA-directed DNA polymerase, (ii) DNA-directed DNA polymerase in the conversion of ssDNA to dsDNA, and (iii) RNase H that selectively removes the RNA moiety from a RNA · DNA heteroduplex. Reverse transcriptases lack both $3' \rightarrow 5'$- and $5' \rightarrow 3'$-exonuclease activities, and exhibit relatively high frequencies of nucleotide misincorporation. The RNA- and DNA-dependent polymerase activities are physically inseparable. Reverse transcriptases from avian retroviruses, but not from mammalian retroviruses, exhibit a fourth activity, DNA endonuclease, that is essential for the integration of the dsDNA intermediate into the host chromosome and for the productive infection of the retrovirus.

In a manner similar to DNA polymerases, RTase requires Mg^{2+} (or Mn^{2+}), dNTPs, an RNA template, and a primer with a 3'-OH terminus for polymerization reactions. The nucleophilic attack on the α phosphate of the dNTP by the oxygen atom of the ribose 3'-OH of the primer strand is thought to be metal-mediated. As a reversal of polymerization, RTases carry out pyrophosphorolysis and pyrophosphate exchange reactions (10,11). AMV RTase has been shown to carry out strand displacement synthesis in the RNA · DNA heteroduplex and in the DNA · DNA homoduplex as well (12,13).

The RNase H activity associated with purified RTases specifically degrades the RNA strand of an RNA · DNA hybrid independently of concurrent DNA synthesis (14). The RNase H activity generates as major products fragments 4–30 bases long with 5'-P and 3'-OH ends (16,17). Like the *E. coli* RNase H that acts as an endonuclease (15), the RTase-associated RNases are essentially endonucleases which are active on RNA · DNA duplex (RNase H) (18,19), as well as on the RNA · RNA duplex (RNase D).

b. Template specificities

Reverse transcriptase is active on a wide variety of synthetic and natural RNA and DNA templates, provided a complementary oligodeoxyribonucleotide primer is present (3). The level of activity is usually 10- to 1000-fold greater when a homopolymer, rather than viral RNA, is used as a template. Homopolyribonucleotides are generally better templates than polydeoxyribonucleotides (20). The identity of primers also influences the efficiency of the templates. In fact, the optimal reaction conditions (e.g., pH, ionic strength, and Mg^{2+} concentration) for a given RTase differ significantly depending on the template-primer pair. Furthermore, the pattern of preferred or most active templates is characteristic for specific polymerases regardless of viral or cellular origin.

The unique ability of RTases to use poly(2'-O-methyl-C)·oligo(dG) as an effective template-primer distinguishes them from DNA-dependent DNA polymerases and has been used as a specific assay for RTases (21).

c. Template switching

RTases (AMV, MoLV, and HIV) have unique abilities to switch templates during the course of reverse transcription *in vitro* as well as *in vivo*. The *template switching*, also called *strand transfer*, is an intrinsic property of RTases and plays an essential role in replication and genetic recombination of retroviruses (refer to Fig. 6.7).

Template switching designates the process in which a RTase initiates primer extension on one template and then switches to another template for additional elongation, often resulting in a transcript longer than the individual templates on which the transcript was made (70,116,117). The strand-transfer synthesis requires a sequence overlap of at least 10 bases between the donor and acceptor templates, and the efficiency of transfer increases almost linearly up to a 100-nt overlap, reaching as high as 83% (117). Template switching is thus a homology-dependent intermolecular recombination. Nonhomologous recombination is estimated to occur at a rate of about 0.1 to 1% rate of homologous recombination (118). Template switching is favored by polymerase pausing that occurs under various limiting conditions, for instance, at low (0.25 μM) concentrations of dNTP (119). Higher incubation temperatures (e.g., 50°C versus 37°C) and a longer incubation time significantly increase template-switching efficiencies (120). During the strand transfer, the polymerase may incorporate additional bases beyond the 5' "termini" of the donor RNA template, resulting in a point mutation at the recombinant site (121). Template switching also results in deletions in recombinant molecules. *In vivo*, the dimeric association of retroviral RNAs via "dimer linkage sequences" is presumed to contribute to efficient template switching. RTase-catalyzed template switching is observed between RNA templates, between DNA donor and RNA acceptor templates, and, although at lower efficiencies, between DNA templates as well.

The efficiency of strand-transfer synthesis by RNase H-minus MoLV RTase (with RNase H domain deletion) is substantially reduced on heteropolymeric RNA or DNA templates. Under conditions that simulate a higher processivity to the RNase H$^-$ RTase, however, the level of strand-transfer synthesis is restored, suggesting that processivity plays a determinant role in template switching. On homopolymeric RNA templates, HIV-1 RTase with deficient RNase H activity can also catalyze strand-transfer reactions efficiently (119).

d. Primer requirements

With poly(A) as the template, the oligo(dT) primer is physically incorporated via a 3' → 5'-phosphodiester linkage into the 5' end of the poly(dT) products. Although both ribo- and deoxyribooligomers can serve as primers, oligodeoxyribonucleotides are considerably more efficient as primers on homopolymeric as well as heteropolymeric templates *in vitro* (22). This is in contrast to the fact that the natural primer is a tRNA: $tRNA^{Trp}$ for AMV RTase (23), $tRNA^{Pro}$ for MoLV RTase (24), and $tRNA^{Lys,3}$ for HIV-1 RTase.

e. cDNA products

In vitro DNA transcripts of viral RNA using a specific tRNA primer are generally short and vary in length from 100 to 180 bases with very few full-length sizes (3). The length of the short DNAs, called "strong-stop (−) DNAs," is characteristic of the 5' termini of retroviral RNAs (25).

The cDNAs synthesized from poly(A)-tailed mRNA in the presence of the oligo(dT) primer range widely in size from 200 bases to several kilobases, all depending on reaction conditions. Reverse transcriptases lacking RNase H activities produce longer cDNAs and at substantially higher yields.

Duplex DNA can also be synthesized from ssRNA *in vitro*. The dsDNA synthesized by the use of RTases, whether short or full-length, have been found to carry hairpin structures (26). Hairpin-primed dsDNA or the first-strand cDNAs carrying the hairpin generally represent 5 to 10% of the total products (27). Duplex DNAs obtained from self-primed synthesis can be rendered suitable for molecular cloning after treatment with nuclease S1 (28). Because the conformation, dynamics, and thus the priming capacity of hairpin loops are sensitive to the nucleotide composition of the loop and stem sequences, the yield of double-stranded cDNA generated by hairpin priming varies depending on RNA templates. Regardless of the situation, self-primed synthesis is much less efficient than the synthesis with added oligodeoxyribonucleotide primers. Actinomycin D prevents the synthesis of hairpin loop and dsDNA.

f. Factors affecting efficiency of cDNA synthesis and cloning strategies

Slight variations in reaction conditions such as the amount of RNA template, total reaction volume, reaction temperature, enzyme concentration, and other factors may profoundly affect the yield as well as the length of the cDNA products. Optimal reaction conditions also vary significantly from one RTase to another. The following factors need to be considered to "optimize" the reaction conditions for obtaining full-length transcripts.

i. RNA templates. One of the most critical factors for full-size cDNA synthesis is the integrity and purity of RNA templates. Because the quality of cDNAs cannot be better than that of the RNA, it is essential not only to start with full-length RNA but also to prevent any possible degradation of RNA during the reaction. Therefore, the cDNA synthesis is usually carried out in the presence of RNase inhibitors and at a high concentration (>50 μM) of each dNTP. To prevent potential damages to RNA templates by RNases that sometimes accompany "purified" RTases, RNase inhibitors are added to the mixture prior to the addition of RTase: VRC (to 10 mM) or more preferably RNasin (to 0.5–1.0 U/μl) (see Section III,A,1,d in Chapter 3). In cDNA synthesis using virion-purified AMV RTase, VRC has been shown to contribute to a higher yield of full-length cDNA products. Nevertheless, VRC at 2 mM is known to inhibit RTase in cDNA synthesis unless the concentrations of all four dNTPs exceed 1 mM. RNasin is probably the preferred choice for most cDNA synthesis reactions regardless of RTase species.

To avoid potential RNase contaminations from other sources, especially from hands, clean gloves should be worn for all RNA manipulations. Water can be made RNase free by treatment with DEPC (0.1%), followed by autoclaving.

The presence of contaminating rRNA and/or tRNA in mRNA samples may render the cDNA synthesis inefficient with certain RTases (e.g., AMV RTase) but not with other RTases (e.g., MoLV RTase) (29).

Certain secondary structures in an RNA template may force the RTase to halt and/or terminate reverse transcription, which is often the cause of inefficient cDNA synthesis. To disrupt potential secondary structures, RNA templates are customarily denatured either by brief heating (30–60 sec at 100°C) or by treatment with methylmercuric hydroxide (CH_3HgOH, 2.5 mM) followed by neutralization of the reagent with at least three times excess 2-MSH immediately prior to the addition of RTases (30). Despite the initial disruption of secondary structures by either of the two methods mentioned earlier, RNA templates tend to regain, at least partially, their original secondary structures during the period of incubation. The elevation of the reaction temperature, sometimes up to 55°C with AMV RTase, helps retard the process of refolding, while the elevated temperature increases the rate of polymerization by the RTase. However, the optimal temperature has to be compromised with the stability of the individual RTase.

Applications of thermostable DNA polymerases (e.g., *Taq* DNA Pol and *Tth* DNA Pol) further demonstrate that the reverse transcription (and subsequent DNA amplification) performed at 72°C efficiently produces longer cDNAs at higher yields under the "standard" PCR conditions (see Section II,D, this chapter).

cDNA synthesis on retroviral RNAs needs special attention because retroviral RNAs tend to form dimers in the presence of salts via a short stretch of sequences, called "dimer linkage sequence" (DLS), located at the 5' ends. The dimerization, which is mediated by a "guanine quartet" in an overall parallel orientation of the two RNA strands, is favored by monovalent cations such as Na^+, K^+, and in particular NH_4^+ (122,123). Efficient full-length cDNA synthesis of retroviral RNA is possible only with monomerized RNA templates which can be prepared by deionization of the templates, e.g., by passage through a Sephadex G-50 spin column. Furthermore, cDNA synthesis on dimerized RNA is presumed to result in recombinant molecules via a phenomenon known as *template switching* (or *strand transfer*). Also, for the reasons of possible template switching, cautious analysis of cDNA products is necessary for RNAs containing repetitive sequences.

ii. Primers. To synthesize cDNAs for the purpose of cloning and/or library construction, synthetic oligodeoxyribonucleotide primers of various size and configuration can be used depending on subsequent cloning strategies (Fig. 6.6). Regardless of whether the oligodeoxyribonucleotide is designed to be complementary to the poly(A) tail of mRNAs or to a specific region of an RNA sequence, oligodeoxyribonucleotide primers used for cDNA synthesis fall into three major categories: (a) free primer, (b) linker/adapter-primer, and (c) vector-primer.

(a) Free primer. One classical example of free primers is the oligo(dT)$_{12-18}$ which is most commonly used for the synthesis of cDNAs from the poly(A) tail of the mRNA(s) (31–33). In this strategy, the cDNA synthesis starts from the 3′ end of the mRNAs and, consequently, the fraction of cDNAs representing the 5′-end RNA sequences is progressively smaller. The bias against the 5′-end RNA sequences can be critical, especially with long or polycistronic mRNAs. To overcome this problem, *random primers* (oligodeoxyribonucleotide hexamer or octamer) have been used to obtain an equal representation of 5′- and 3′-end and inbetween sequences (34,35). In fact, the combined (sequential) use of oligo(dT) primers in the initial phase of cDNA synthesis, followed by additional random hexamer priming, is probably the best formula for synthesizing full-length cDNAs. Random primer-primed cDNA synthesis also allows a cDNA library to be constructed from poly(A)⁻ RNAs. Nevertheless, there is a tradeoff in using random oligodeoxyribonucleotide priming strategies: the risk of contamination by the cDNAs from often more abundant rRNAs and the burden of screening.

When cloning a gene where the sequence information is at least partially available, site-specific primers (~20 nt long) containing either single or degenerate codons can be used to pick up specific mRNA species for reverse transcription from a pool of mRNAs containing other nonspecific RNAs.

The use of free primers requires that the double-stranded form of the cDNA synthesized be subsequently ligated to a linearized vector by one of the following three methods [see Fig. 6.6(I)]. The first is a direct, low efficiency blunt-end ligation between double-stranded cDNA and a blunt-ended vector. This is the most straightforward method to clone cDNA from abundant mRNA species and is not suitable for cDNA library construction.

Alternatively, double-stranded cDNA can be ligated to a vector by cohesive-end ligation via restriction linkers or adapters [Fig. 6.6(Ib)]. Conventionally, the linker ligation method requires a series of cumbersome procedures consisting of linker phosphorylation, low efficiency blunt-end ligation of linkers to cDNA, a restriction enzyme digestion(s) to open the cohesive ends (with possible prior methylation of inner restriction sites), purification, and final intermolecular ligation with a vector linearized at the same restriction site(s). A streamlined procedure that does not require a restriction enzyme digestion (nor methylation) and thus

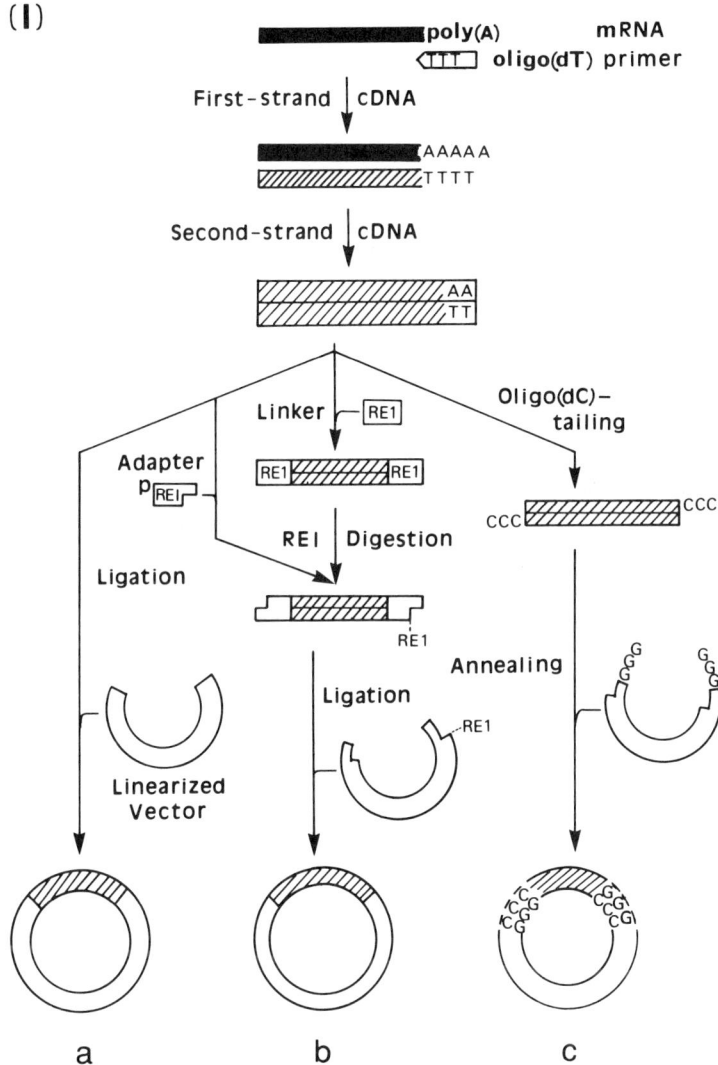

FIGURE 6.6 cDNA cloning strategies. Major strategies for cloning cDNAs can be grouped into three types based on the format of primers used for the initiation of first strand cDNA synthesis: (I) free primers, (II) linker (or adapter)-primers, and (III) vector-primers. Following the synthesis of first-strand cDNA, cloning strategies are further divided into subgroups according to the methods of second-strand DNA synthesis, ligation with, and/or circularization of recombinant molecules. The cloning vector may be a plasmid, phagemid, or phage vector. The strategies presented here illustrate the basic designs of some widely used cDNA cloning strategies. There are a number of other "variant or hybrid" strategies with special features and distinct advantages (see references cited in the text). (Ia) Blunt-end ligation strategy, (Ib) linker or adapter ligation strategy, and (Ic) homopolymeric tail strategy; (IIa) directional cloning strategy with double adapters, (IIb) RT-PCR cloning strategy, and (IIc) adapter-homopolymeric tail cloning strategy; and (IIIa) simplified vector–primer cloning strategy and (IIIb) Okayama–Berg cloning strategy.

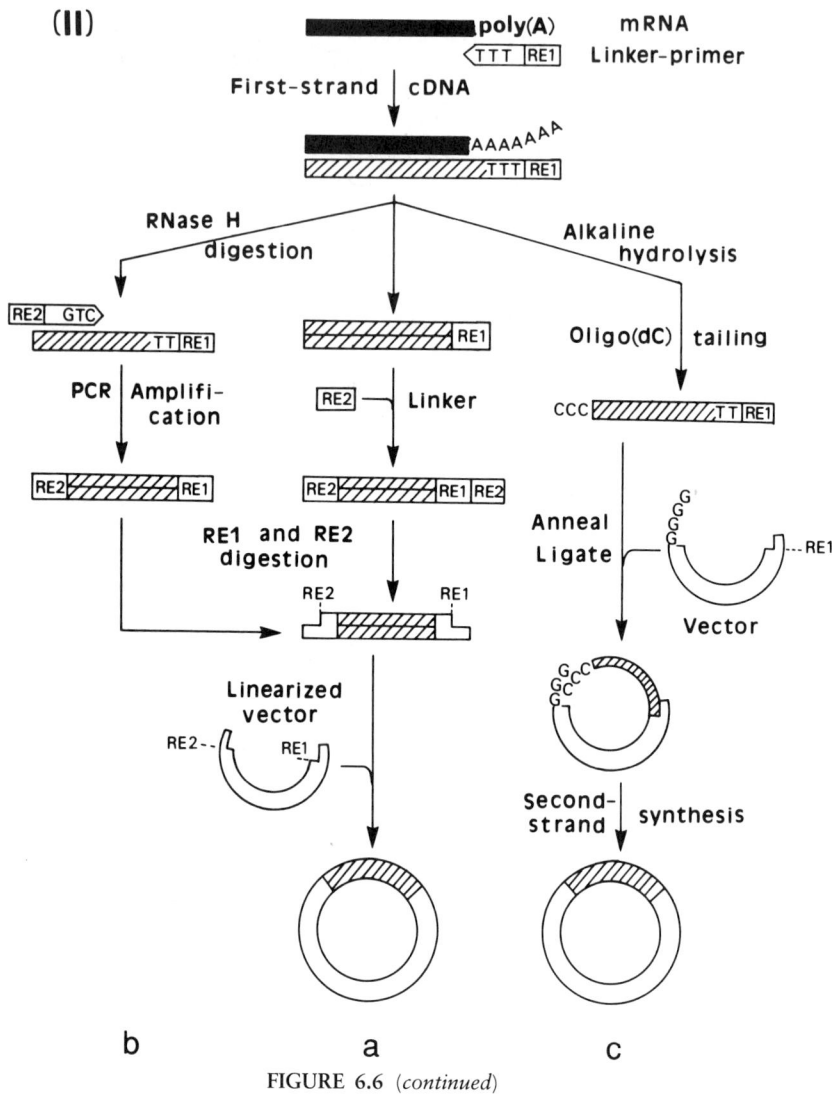

FIGURE 6.6 (continued)

substantially improves the efficiency of cloning employs a cohesive-ended adapter that is either nonphosphorylated (34) or phosphorylated only on the 5' end of the blunt-ended terminus (36) and thus allows a single round of ligation. The ligation of such a nonphosphorylated or asymmetrically phosphorylated adapter to cDNA leaves the cohesive termini available for direct ligation with a vector following brief heat denaturation, removal of the annealed adapters, and 5'-end kination.

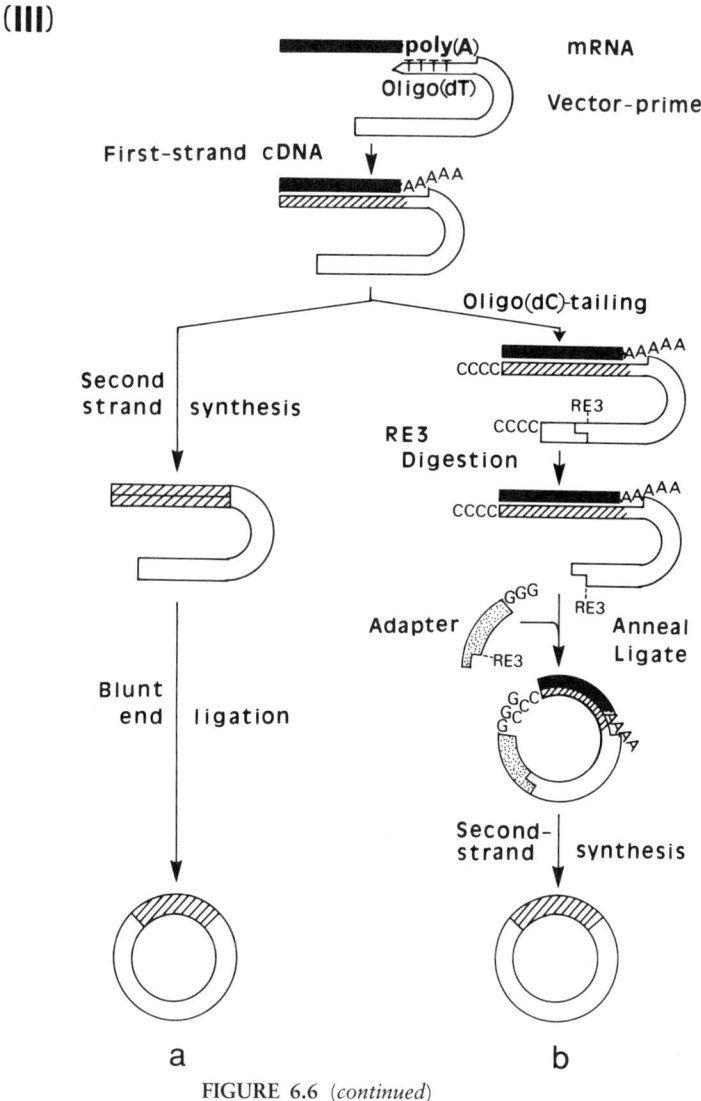

FIGURE 6.6 (continued)

The third alternative is the homopolymeric tailing of cDNA using terminal deoxynucleotidyltransferase and annealing with the vector carrying complementary homopolymeric tails (see Section IV, this chapter).

Free (universal or random) primers are readily available, inexpensive, and easy to use at the stage of cDNA synthesis. However, subsequent cloning steps may require additional care and efforts. Cloning via a free-primer strategy usually results in inserts occurring in both orientations in equal amounts.

(b) Linker/adapter-primer. This is a bipartite oligodeoxyribonucleotide of around 25 bases usually consisting of a restriction site and a primer sequence [see Fig. 6.6(II)]. Because of its versatile configuration, the linker/adapter-primer provides a convenient way to insert double-stranded cDNA fragments into a vector via cohesive-end ligation (37,38). Attachment of a different restriction linker on the other end of the dsDNA permits the cDNA to be cloned directionally. The linker-primer has also proved to be an extremely practical format for molecular cloning of the cDNA products amplified by PCR technology.

A restriction sequence should be selected based on the efficiency of opening the cohesive termini not only of the cDNA but also of the cloning vector. Restriction enzymes have different cleavage characteristics at terminal sequences, and methods, such as the ligate-and-cleave strategy (39), have been devised to increase the efficiency of cleavages. The linker-primer strategy requires that the potential inner restriction site(s) be protected by methylation prior to restriction enzyme digestion and that appropriate host cells be chosen, e.g., MDRS (methylation-dependent restriction system)-minus *E. coli* cells (refer to Section III,A, in Chapter 4).

A new generation of a directional, ligation-independent cloning strategy based on the $3' \rightarrow 5'$-exonuclease activity of T4 DNA Pol now provides the most efficient cloning methods which circumvent the use of restriction enzymes, polynucleotide kinase, phosphatase, or DNA ligase (see Fig. 6.3).

For the synthesis of cDNAs using either free primer or linker-primer, the optimum ratio of primer to RNA template ranges from 5:1 to 10:1. The use of primer at higher ratios would only contribute to nonspecific priming and consequently to a lower yield of desired cDNA products.

(c) Vector-primer. Vector-primer is an efficient alternative that incorporates two important steps in cDNA cloning: priming function and subsequent circularization of the duplex cDNA by intramolecular ligation [see Fig. 6.6(III)]. Since its initial conception featuring oligo(dT)-tailed vectors (40,41), vector-primer cloning strategy has engendered a number of variations with increased efficiency and versatility (42–48). Apart from the improvements brought into the cloning process per se, one of the limiting steps in the vector-primer cloning strategy has been the preparation of the vector-primer itself. One of the latest variations providing versatility to the vector-primer cloning strategy, especially in the preparation of vector-primer, employs an improved direct adapter-primer ligation method (49). In the direct ligation method, a linearized, dephosphorylated vector with a cohesive end is annealed and ligated with the 5'-end of a phosphorylated adapter-primer which has a temporary double-stranded form with a 6- to 10-mer supporter oligonucleotide complementary to the adapter-primer. This method is generally applicable to most commonly used vectors with MCS and allows either oligo(dT)-tailed or custom-tailored primer sequences to be attached. The cloning efficiencies of the vector-primer strategy are high enough to allow an mRNA/primer (-vector) ratio of 3 to 0.5 to be used, providing the maximum efficiency when the availability of the mRNA is limited.

iii. **Reverse transcriptase.** Only the highest purity RTase without contaminating nucleases can ensure the highest yield of full-length cDNA products. In addition to this extrinsic factor, the interplay of polymerase and the polymerase-associated RNase H is another important factor that affects the yield of first-strand cDNA (50,51). First, there is competition between the initiation of DNA synthesis and the RTase/RNase H-mediated deadenylation or, more generally, cleavages in mRNA when using oligo(dT) or other site-specific primers. Second, when a RTase encounters secondary structural barriers (or pause sites) during reverse transcription, the RNase H activity introduces nicks to the RNA at pause sites. Indeed, RNase H$^-$ RTase produces cDNA products with a yield up to 60–80%, a level far beyond the 20–40% yield usually observed with wild-type RNase H$^+$ RTases.

Ideally, the molar ratio of enzyme to template at about one gives a higher rate of synthesis than at other ratios. In reality, the size and yield of cDNA products tend to increase at high concentrations of RTase, partly due to compensating effects on low polymerase processivity and thermal denaturation of the RTase. A higher enzyme to template ratio also disfavors the strand transfer synthesis which results in abnormal products. Nevertheless, an excess amount of enzyme should be avoided as it could bind along the template and hinder chain elongation. The use of a high amount of RTase also increases the probability of synthesizing self-primed dsDNA unless such a process is blocked by using actinomycin D or Na-PP$_i$.

Reverse transcription followed by PCR amplification of cDNAs, a process known as RT–PCR, can be efficiently performed with AMV RTase (or MoLV RTase) and *Taq* DNA Pol in sequential reactions. Use of *Tth* DNA Pol in both reverse transcription (in the presence of Mn^{2+}) and amplification (in the presence of Mg^{2+}) of cDNAs has further streamlined the RT–PCR techniques, with all the advantages of performing cDNA synthesis at high temperatures (e.g., 72°C).

iv. Other factors affecting cDNA synthesis

(a) Actinomycin D. Actinomycin D, an antitumor drug and carcinogen at the same time, binds dsDNA (more strongly than dsRNA) and thus inhibits primarily cellular transcription. Actinomycin D (at 50 μg/ml) can be effectively used to prevent the genesis of a hairpin structure at the end of the first-strand cDNA synthesis and to block the RTase-catalyzed dsDNA synthesis (52–54).

The intercalation of actinomycin D occurs preferentially at the 3' side of the guanine residue, particularly in the dinucleotide site GpC. Nevertheless, flanking sequences (e.g., X and Y in -XGCY-) can have a profound effect on the binding characteristics of actinomycin D. Oligomeric duplexes that lack GpC sites do not exhibit high affinities for the drug. However, a high affinity, sequence-dependent binding of the drug may occur with certain non-GpC duplexes, e.g., [d(CGTCGACG)$_2$], apparently via cooperative binding of two actinomycin D molecules (55). Actinomycin D (M_r 1255) has an ε_{440} of 21,900 M^{-1} cm^{-1}. The A_{440} of a 1-mg/ml solution is 0.182.

(b) Detergents. Reverse transcriptases are stable and fully active in the presence of nonionic detergents such as NP-40 and Triton X-100 (0.2%, v/v). In the presence of either NP-40 (0.02%) or more preferably melittin (25 μg/ml), a 26-

amino-acid cationic peptide and the major component of bee venom, the AMV virion envelope can be made permeable and the RTase can be activated to perform endogenous cDNA synthesis reactions (56).

(c) Spermidine and spermine. The presence of spermidine (0.5 mM) has been shown to result in as much as a two-fold increase in full-length cDNAs synthesized by AMV RTase in the presence or absence of PP_i or actinomycin D (57). In contrast, spermidine inhibits MoLV RTase. Spermine has also been shown to enhance the polymerase activity of AMV RTase on certain templates in the presence of suboptimal Mg^{2+} concentrations (58).

(d) Glycerol. High concentrations of glycerol, which is commonly used at a 50% (v/v) concentration as an enzyme stabilizer, affect the quality of cDNA. Dilutions of AMV RTase in buffered 10% glycerol are reported to give higher RTase activities than undiluted enzymes (57). However, some enzymes (e.g., MoLV RTase) may be highly unstable on dilution and therefore should be used in the originally supplied form.

g. Rate and fidelity of polymerization

The initial rate of polymerization by AMV RTase *in vitro* is ~300 nt/min, a rate nearly equal to that of *E. coli* DNA Pol I on some templates (9,50). The rates of DNA synthesis by RSV RTase *in vivo* and in the endogenous reactions have been reported to be ~30 nt/min (59), while the rate of DNA synthesis on the poly(A) template by HIV-1 RTase is 60–90 nt/min (60).

Reverse transcriptases generally lack $3' \rightarrow 5'$ or $5' \rightarrow 3'$-exonuclease activity. At least partly due to this, RTases exhibit relatively high frequencies of misincorporation. Reverse transcriptases do not catalyze dNTP turnover (10). They utilize mispaired primer termini complexed to either RNA or DNA templates as functional sites for polyermization (61). Substitution of Mn^{2+} for Mg^{2+} increases the error frequency of RTases (62,63). The error rates for AMV and MoLV RTases are similar. On a poly(A) or poly(C) template, the two RTases exhibit an error rate of 1/3500–1/7400 at a dNTP concentration of 250 μM, the condition often employed for cDNA synthesis reactions *in vitro* (64). On DNA template-primers, the frequencies of misincorporation at 50 μM dNTPs are 10^{-4}–10^{-5} (65–67). On certain RNA templates, however, AMV RTase shows higher error rates than MoLV RTase (68).

h. Proccessivity of polymerase activity

Using an RNA template (272 base) and a 20-nt DNA primer in a template challenge assay, the processivities of AMV and MoLV RTases have been estimated to be approximately 97 and 69 nt, respectively (69). (Under the same reaction conditions, the processivity of HIV-1 RTase has been measured to be ~85 nt.) Note that processivity can vary widely with the template employed. On the $poly(A)_{1000} \cdot oligo(dT)_{20}$ template-primer, the processivity has been shown to be over 1000 nt for all three RTases (70). For HIV RTase, the processivity increases from a few to >300 nt in the order of poly(dA) < dsDNA < ssDNA < ssRNA

< poly(rA) (60). In general, RTases display higher processivities at lower salt, higher dNTP concentrations, and higher temperatures.

i. RNase activities of reverse transcriptase

Since the initial discovery of RTases, RNase H activity has been found as a constituent of the virion of all retroviruses (14,71). Although RNase H and DNA polymerase activities are found on the same polypeptide (16,72), they reside on separate domains and their functions are not strictly coupled. The two activities show differential sensitivities to thermal inactivation (73). The RNase H activity of AMV RTase is more sensitive than the polymerase activity to inhibition by NaF (74), whereas the polymerase activity is substantially more sensitive than the RNase H activity toward other reagents such as PALP (75), DTNB, and PHMB (76). Proteolytic cleavage of MoLV RTase yields a small fragment possessing only RNase H activity and so does the proteolysis of AMV RTase. Chelation of the polymerase-bound zinc inhibits DNA synthesis but not RNase H activity (77).

The general features of RTase-associated RNase activities are as follows.

i. Metal ion requirements. RTase/RNase H activity requires a divalent cation. AMV RNase H prefers Mg^{2+} (14,16), whereas MoLV RNase H prefers Mn^{2+} for activity (17). A recombinant RNase H of HIV-1 RTase is more active with Mn^{2+} than with Mg^{2+} (124).

ii. Reaction specificities. RTase/RNase is a double-strand-specific endonuclease active on RNA·DNA (RNase H) and RNA·RNA (RNase D). It cleaves an RNA chain at the 3' side of 3' → 5'-phosphodiester bonds to yield products containing 5'-P and 3'-OH ends, and degrades substrates in either the 5' → 3' or 3' → 5' direction. It cleaves substrates with 5'-triphosphorylated termini, 3'-OH, 3'-P, and 3'-end hairpin structures (19). It can cleave relaxed, circular, or covalently closed duplex plasmids in which a large part of the strand is RNA (18) or a covalently closed ssRNA hybridized to a synthetic oligodeoxyribonucleotide (18,19). Single-stranded RNA is resistant to degradation. The RNase D activity is apparently as strong as the RNase H activity. The RNase D activity resides in the RNase H domain, although the two activities are functionally and genetically distinct (125,126).

iii. Cleavage specificities. RTase/RNase H cleaves a template RNA at least once during DNA synthesis. AMV RTase generates mostly large segments (100–200 nt) of the RNA·DNA hybrid and virtually no small RNA cleavage products (69). MoLV RTase generates more small degradation products than AMV RTase, but still leaves much of the potentially degradable hybrid undigested. The RNase H function is much less active than the polymerase function during processive DNA synthesis.

iv. Temporal relationship between RNase H and polymerase activities. RNase H activity appears both in the presence and in the absence of dNTPs. It cleaves the RNA template under the annealed primer and during the elongation, generating a unique set of RNA products. Although the proportion of cleaved templates may be approximately equal in the presence or absence of dNTPs, the products can be distinguished according to two modes of cleavages: polymerase independent and polymerase dependent (78,127).

In the *polymerase-independent* mode (i.e., in the absence of dNTPs), HIV-1 RTase/RNase H can catalyze the hydrolysis of an RNA strand at many sites from 14–15 nt up to 5–6 nt behind the 3'-OH of the DNA primer. The cleavage site may show some sequence preference. Under single turnover conditions (e.g., using heparin trap), RNase H cleavage occurs at (and not below) 18 or 19 nt from the 3'-OH of the primer.

In the *polymerase-dependent* mode (i.e., in the presence of dNTPs), the specificity of HIV-1 RTase/RNase H is defined by the binding of the polymerase active site to the primer 3' terminus. The RNase H active site is then positioned 18–19 nt from the primer 3' terminus. In this coupled mode, RTase/RNase H needs to be positioned only transiently at a site along the elongating primer to effect hydrolysis of the RNA template, generating oligoribonucleotide products 2–30 bases long.

v. Appearance of initial RNase H activity. After forming a stable complex with the DNA-primed RNA template (with a half-life of 30 and 15 sec for AMV RTase and MoLV RTase, respectively), RTase/RNase H activities cleave the hybrid portion of the RNA before dissociating. In the presence of the oligo(dT) primer, RNase H-mediated deadenylation takes place in the mRNA and competes with the RTase-mediated polymerization, resulting in lower yields of cDNA (50).

vi. Processivity of RNase H activity. AMV and MoLV RNase H activities are partially processive; RNase H remains bound to a template through a finite number of hydrolytic events, but dissociates before the RNA is completely degraded (79).

j. Kinetic mechanism of reverese transcriptase action

The polymerization reaction catalyzed by RTase involves two reactants (primer/template and dNTPs) and two products (elongated primer/template and PP_i). For HIV RTase, steady-state kinetic analysis indicates an ordered Bi–Bi mechanism in which the primer/template (P/T) binds first, followed by the addition of dNTPs (80). Analyses of pre-steady-state and binding kinetics using defined primer/template substrates suggest that the reaction mechanism of HIV RTase is generally similar to that of *E. coli* DNA Pol Ik (refer to Scheme 6.1, Section II, this chapter). Although the kinetic relevance of certain intermediates remains controversial (128–133), essential features of the mechanism can be described as follows. (i) For a single turnover reaction (e.g., incorporation of dTMP opposite a template dA), the RDS is the dissociation of DNA from the enzyme. (ii) Binding of P/T to the enzyme apparently occurs in a two-step process. The second, slow step is presumed to be a conformational change of the enzyme. The DNA/RNA

(P/T) substrate is more stably bound to RTase than the DNA/DNA (P/T) substrate, largely due to differences in the rate of dissociation (k_{off}). (iii) The enzyme–substrate complex is further stabilized, by a factor of at least 10, when a cognate dNTP or a nucleoside inhibitor is bound. (iv) Mg^{2+}, although not required for the binding of primer/template, strongly enhances the productive interaction of RTase with the DNA substrate. Presumably Mg^{2+} stabilizes the form of RTase that can bind the duplex substrate. The presence of Mg^{2+} results in a decrease (about 10-fold) in K_d for the E · DNA complex. (v) Kinetic data are most consistent with a model involving a conformational change of the enzyme prior to the chemical step (E · D_n · dNTP $\rightleftharpoons$ E' · D_n · dNTP). No significant phosphorothioate elemental effect has been observed with dTTPαS or dATPαS. The conformational change is apparently rate limiting in multiple turnover kinetics where the rate of polymerization (k_{pol}) ranges from 30 to 80 sec^{-1}. The forward conformational change is strongly favored over the reverse change, suggesting that the nucleotide addition is kinetically irreversible. (vi) The equilibrium constant for the chemical step (E' · D_n · dNTP $\rightleftharpoons$ E' · D_{n+1} · PP_i) in dTMP incorporation is estimated to be 3.8, slightly favoring the forward reaction. (vii) The kinetics with DNA/RNA (P/T) are essentially identical to the corresponding DNA/DNA (P/T) except for some minor differences in the rate constants. For instance, the rate of polymerization is 2- to 20-fold greater on RNA than on DNA templates. (viii) The polymerase activity is independent of RNase H activity. For example, the rates of polymerization vary from 4 to 70 sec^{-1} depending on dATP concentrations (2 to 200 μM), whereas the RNA cleavage occurs at a constant rate of 10 sec^{-1}. (ix) The processivity, as kinetically defined by the ratio (k_{pol}/k_{off}), is 10–20 for both RNA and DNA templates.

k. Structure and genetics

i. Precursor forms of reverse transcriptases. Retroviral RTases constitute a part of the polyprotein product encoded by the retroviral *pol* gene (refer to Fig. 6.8, MoLV RTase). In avian sarcoma–leukosis viruses (ASLV, e.g., myeloblastosis, leukosis, Rous sarcoma), the polyprotein precursors, Gag-Pol, are produced as a result of a translational (-1) frameshift at the *gag–pol* junction. In mammalian retroviruses (e.g., MoLV and HIV), however, the Gag-Pol precursor is produced via an in-frame read through of a termination codon at the end of the *gag* gene.

The Gag-Pol polyprotein contains, in the order of occurrence within the precursor: *gag* gene products (major structural proteins), protease (PRO), RTase/RNase H, and endonuclease (or integrase, IN). All known retroviral *pol* genes display a similar order of functional domains. Mature RTases are derived from the Gag-Pol precursor by proteolytic cleavages performed by the virion-encoded protease. In ASLV, the *pol* gene product consists of RTase/RNase H-IN.

The viral proteases are produced in two precursor forms: as a C-terminal domain of the Gag precursor and as an embedded domain within the Gag-Pol precursor. The functional PRO is autocatalytically derived from the Gag precursor when the Gag and Gal-Pol precursors are packaged into an immature virion. The embedded form of PRO is unable to carry out any detectable processing of the precursor polyproteins (81,82). In the precursor form, ASLV RTases have no polymerase activity and the active polymerase is generated only by polyprotein processing.

In MoLV and HIV, the PRO-coding sequence is located within the *pol* gene so that PRO is synthesized only in the form of a Gag-Pol fusion protein. In these cases, the *pol* gene product consists of PRO-RTase/RNase H-IN, which subsequently gives rise to mature RTase (RTase/RNase H). Native Gag-Pol precursors and those packaged in the immature particles exhibit polymerase activity.

ii. **Mature forms of reverse transcriptases.** Mature RTases from different families of retroviruses show different subunit structures. For example, HIV-1 RTase is a heterodimer (p51/p66); the smaller subunit is derived from the larger subunit by a proteolytic cleavage of the RNase H domain (p15). In contrast, MoLV RTase is a monomer (~80 kDa) consisting of two domains: the N-terminal polymerase domain and the C-terminal RNase H domain. Avian RTases are unique in the sense that they have DNA endonuclease activity in addition to the polymerase and RNase H activities. For example, AMV RTase is an $\alpha\beta$ heterodimer (~160 kDa); the larger β subunit (~90 kDa) consists of three domains: RNase H (p24), polymerase, and integrase (IN protein, pp32). The smaller α subunit (~60 kDa) is a truncated form of the β subunit and lacks the integrase domain.

Crystallographic data from HIV-1 RTase suggest that the overall folding pattern of the two subunits (p66 and p51) is similar, but the spatial arrangements of the subdomains are dramatically different (134,135). In contrast to p66, p51 has no apparent DNA-binding cleft but is presumed to form part of the tRNA primer binding site.

(a) *Polymerase domain.* In HIV-1 RTase, the polymerase domain of p66 has an overall shape featuring, in analogy to a hand, four subdomains named fingers, palm, thumb, and connections (134). The fingers, palm, and thumb subdomains form a large cleft to which the template-primer binds. The template-primer, the 3′ primer terminus of which is bound to the polymerase active site, is presumed to have a bend (40–45°) separating the substrate into A-DNA (between the primer terminus and the RNase H active site) and B-DNA regions.

RTases have four conserved motifs centering on or around such amino acids as Asp, Gly, (Tyr-X-)–Asp–Asp, and Lys (83,84). The palm subdomain of p66 contains the "catalytic triad" (Asp-110, Asp-185, and Asp-186) which structurally superimposes on the catalytically important Asp-705, Asp-882, and Glu-883 of the Klenow polymerase and also on the three essential Asp residues [150, 224, and 225] of the MoLV RTase. The Asp-224 and -225 of the MoLV RTase are coordinated to a metal ion (Mn^{2+}) which is presumed to play a key role in polymerase catalysis (85). Site-specific mutations of each Asp residue in the Asp^{185}-Asp^{186} motif to Asn or Gln render the RTase catalytically inactive, although the mutant enzymes can still bind the template-primer complex (86). The mutation of Asp-110 also abolishes the RTase activity. The "ddNTP-resistant" mutations that cause lower fidelity and reduced processivity to RTases are often located in the thumb and/or finger subdomains, mostly affecting template-primer binding.

(b) RNase H domain. In all retroviral RTases with known sequences (e.g., MoLV, RSV, and HIV), the RNase H domain is located at the C-terminal region of ~150 amino acid residues. Retroviral RNases H have in general a low sequence homology (~20%) with *E. coli* RNase H. Among the conserved amino acid residues are two Asp(-443 and -498) and one Glu(-478) residues, called "carboxyl triad," which are catalytically essential (87,88) and serve as the Mg^{2+}-binding site in *E. coli* RNase H (89,91). In HIV-1 RTase, the polymerase and RNase H active sites are estimated to be 18–19 nt length (in A-form duplex) apart.

Despite the variations in the primary structure, the RNase H domains of HIV-1 RTase (92) and MoLV RTase have tertiary structures strikingly similar to that of *E. coli* RNase H. The structural similarity includes the positions of the conserved acidic active site residues, two metal ion-binding sites, and other residues implicated in hydrophobic and H-bonding interactions (93). RNase H activity is presumed to cleave RNA by the same two-metal ion mechanism that has been postulated for the $3' \rightarrow 5'$-exonuclease of *E. coli* DNA Pol I.

RNase H domains, whether proteolytically isolated from RTase or genetically expressed from a cloned gene, have independent enzymatic activities of varying strengths depending on the RTase species. Separate AMV and MoLV RTase/RNase H domains are fully active, whereas the HIV-1 RNase H domain is practically inactive. The HIV-1 RNase H domain regains its full activity when supplemented with the polymerase domain (p51) or when provided with some additional amino acid residues (e.g., six histidines) at the natural N-terminal cleavage site.

(c) DNA endonuclease domain. Genetic and enzymatic evidence suggests that the C-terminal region of avian retroviral *pol* gene products forms another distinct domain exhibiting DNA endonuclease activity. The endonuclease domain (pp32), also called integrase (IN or INT protein), has been found in virus-infected cell extracts as well as in infectious virions. The IN protein is responsible for proviral integration into the host chromosomal DNA, thus playing an essential role in the retroviral life cycle. The endonuclease domain may be present as an integral part of the RTase molecule as in AMV RTase or may be proteolytically processed from the polyprotein as a free-standing molecule. For example, RTases from MoLV and HIV-1 do not possess DNA endonuclease activity.

The viral integrase recognizes and cleaves specific sequences of the terminal-inverted repeat within LTR, producing a linear molecule with 2-nt recessed 3' termini (94–97). A staggered cut is then made in the target DNA and the 5'-P overhangs are covalently joined to the recessed 3'-OH ends of the viral DNA, resulting in a gapped intermediate. Integration is completed by a gap repair, which results in a 4-bp duplication of a target DNA sequence flanking the integrated DNA. The cleavage–ligation reaction does not require an exogenous energy source and, given a suitable substrate, the integrase can promote the reversal of the integration reaction (called "disintegration").

l. In vivo role of reverse transcriptase

Retroviruses contain two copies of the (+)-sense ssRNA genome. Soon after a retrovirus enters a permissive cell and releases its encapsidated RNAs, the RTases, which are carried in with the virion core, direct the synthesis of a dsDNA copy (98,99). Reverse transcription involves a complex but well-orchestrated series of steps including two distinct strand transfer reactions (Fig. 6.7). The dsDNA is then integrated into the host chromosome as a *provirus* and undergoes either a dormant or an active replication phase. Viral DNA intermediates (linear and/or circular forms) detectable in virus-infected cells are longer at both ends than the viral RNA because of unique reiterative copying of selective regions of the RNA during the reverse transcription. The unique repeating structure (300–1200 nt) found in provirus is known as the *long terminal repeat* (LTR).

m. In vivo role of reverse transcriptase-associated RNase H

The RNase H activity of viral RTase plays an essential role in retroviral replication. Site-specific mutations that severely depress the RNase H activity but retain the polymerase function produce the replication-defective phenotype with concurrent accumulation of the (−) strand strong-stop DNA (88,100).

As illustrated in the model of reverse transcription (Fig. 6.7), the RTase-associated RNase H activity is implicated in releasing the tRNA primer covalently bound to the (−) strand DNA (101), in nicking and degrading viral RNA first to promote the transfer of the (−) strand strong-stop DNA to the 3' end of the genome (100) and then to generate a specific purine-rich oligoribonucleotide that primes the synthesis of (+) strand DNA (102,103), and in removing the RNA primer from the DNA extension to which it is linked (104).

n. Applications

i. Synthesis of first-strand cDNA.
Reverse transcriptase plays a key role in the synthesis of cDNA from RNA templates, the primordial step in cDNA cloning. When the cDNAs are labeled during synthesis, they can be used as hybridization probes as well.

(a) Conventional first-strand cDNA synthesis. Synthesis of the first-strand cDNA can be achieved in a number of ways, depending on cloning strategies (see Fig. 6.6). The conventional or "standard" method of cDNA synthesis is described under the respective sections on AMV and MoLV RTases.

During the course of cDNA synthesis, products can be labeled with either radioactive nucleotides or nonradioactive labeling nucleotides.

(b) Nucleic acid amplification. In an isothermal amplification strategy called "nucleic acid sequence-based amplification" (NASBA), alternate steps of DNA synthesis from an RNA template and RNA synthesis from a DNA template are performed using AMV RTase and T7 RNA Pol, respectively (90).

RTases (both AMV and MoLV) also perform efficient synthesis of cDNAs under buffer conditions similar to those of "standard" PCR: 20 mM Tris–Cl (pH

III. RNA-Directed DNA Polymerases   445

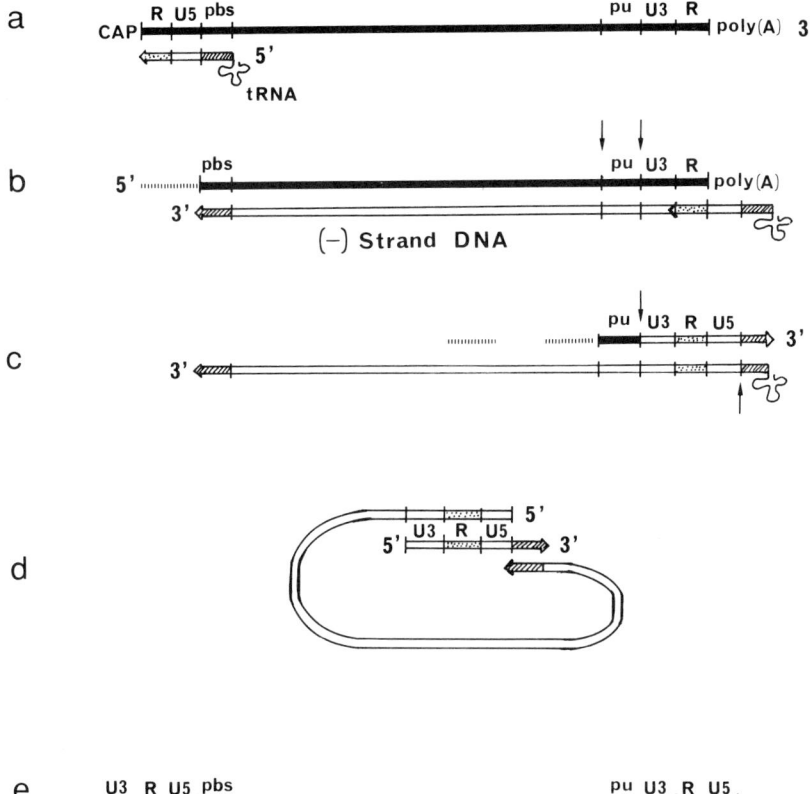

FIGURE 6.7 A model of reverse transcription on the retroviral RNA genome. (a) A specific tRNA (e.g., tRNA^{Lys,3} for HIV-1, tRNA^{Trp} for AMV, or tRNA^{Pro} for MoLV) anneals to the primer-binding site (pbs) and primes the synthesis of (−) strand "strong-stop" DNA. An RNase H activity of RTase then degrades the U5 and R of the template RNA, exposing the short, nascent (−) DNA. (b) The (−) strand strong-stop DNA is transferred, either intra- or intermolecularly, to the 3′ end of the RNA genome mediated by the complementarity between the R regions. The (−) strand DNA synthesis continues. (c) Specific nicking of the RNA strand by the RNase H activity provides a Pu-rich segment (<20 nt) which is used as the primer for the synthesis of (+) strand "strong-stop" DNA. The RTase/RNase H activity continues to introduce nicks and degrade the RNA strand, producing largely ssDNA. (d) Intramolecular transfer of the strong-stop (+) strand DNA presumably by the complementarity to the pbs region. Elongation of both (+) and (−) strand DNAs continues using each other as templates. The (+) strand DNA synthesis is sometimes discontinuous, with some RNase H-nicked viral RNAs functioning as internal primers; this discontinuous synthesis leaves gaps in the (+) strand DNA, which are filled after the transfer of the (+) strand strong-stop DNA either by replacement of the previously synthesized fragments or by ligation of the existing products. (e) Complete dsDNA copy of the retroviral RNA with a LTR on each end. Solid lines, RNA; double lines, DNA; vertical arrows, specific cleavages by RNase H; pu, purine-rich sequence; R, U5, and U3, repeated sequence and unique sequences at the 5′ and 3′ ends, respectively, at the LTR of the RNA genome. The tick marks delimiting each region are not to scale.

8.3), 50 mM KCl, 2.5 mM MgCl$_2$, 1 mM each of dNTPs, 100 µg/ml BSA, 0.1–1 µg RNA, 25 µg/ml adapter-primer, and RTase (200 U of MoLV or 10 U of AMV), with incubation at 42°C for 1 hr. The combination of reverse transcription followed by DNA amplification by PCR, known as *RT–PCR*, has thus emerged as a powerful technique in obtaining cDNA products.

For efficient RT-PCR with RNase H$^-$ RTase (e.g., SuperScript RTase, BRL), the RNA template needs to be digested with *E. coli* RNase H after the cDNA synthesis and prior to the DNA amplification using *Taq* Pol (105). To further increase the PCR efficiency, either the RTases have to be heat inactivated prior to the addition of *Taq* Pol or *Taq* Pol has to be added to a concentration higher than a 1 : 1 ratio to RTase (106). Intact RTases (AMV and MoLV) interfere with *Taq* Pol activity.

Although AMV and MoLV RTases remain as important reagents in RT–PCR, *Tth* DNA Pol has emerged as an efficient alternative for the RTases (107).

ii. Nucleotide sequencing. Reverse transcriptases (both AMV and MoLV) complement DNA-dependent DNA polymerases (e.g., Pol Ik) in the dideoxynucleotide sequencing of DNA, especially at the regions of high GC content and/or secondary structures (108). Reverse transcriptase has been shown to be particularly useful in sequencing with fluorescence tag-modified ddNTPs that are not substrates for Pol Ik (109).

Reverse transcriptases are also valuable reagents in the direct dideoxynucleotide sequencing of RNA (110). In fact, RTases have been instrumental in obtaining nucleotide sequence information directly from viral RNA genomes, 16S rRNA (111), and mRNAs (112). The higher thermostability of AMV RTase compared with that of MoLV RTase makes the AMV RTase more useful for nucleotide sequencing at elevated temperatures.

iii. Site-specific forced misincorporation mutagenesis. Two site-specific mutagenesis methods have been developed that use RTases as error-prone polymerases: one is the *forced misincorporation mutagenesis* (also see Pol Ik) and the other is a method of simply extending the primers with mispaired termini. The latter method is based on the ability of RTase to use mispaired primer termini as functional sites for polymerization (62).

The forced misincorporation mutagenesis method exploits the relatively high frequency of nucleotide misincorporation by RTases under limiting conditions that further amplify the rate of misincorporation. In the approach developed by Zakour and Loeb (113), a site-specifically annealed primer is extended to a designated position on the ssDNA template by sequentially adding complementary nucleotides using a high-fidelity DNA polymerase (e.g., Pol Ik). Then polymerization reaction is switched to incorporate a noncomplementary nucleotide(s) using an error-prone DNA polymerase (e.g., AMV RTase), which is switched again to a high-fidelity synthesis in the presence of all four dNTPs. Another approach, initially developed by Shortle *et al.* (114) and called *gap repair mutagenesis*, employs an error-prone DNA polymerase (e.g., RTase or Sequenase) to fill in short single-stranded gaps in the presence of either unbalanced or noncomplementary dNTPs. The gaps can be created by site-specific nicking of dsDNA with restriction enzymes followed by brief treatment with a polymerase-associated $3' \rightarrow 5'$-exonuclease activity (e.g., Pol Ik or *M. luteus* DNA Pol).

The sites of misincorporation may also be controlled by applying the following forced conditions: (a) using one dNTP at a limiting concentration or (b) providing a high ratio of one nucleotide to the other three dNTPs (115).

References

1. Baltimore, D. (1970). *Nature* **226,** 1209–1211.
2. Temin, H. M., and Mizutani, S. (1970). *Nature* **226,** 1211–1213.
3. Temin, H. M., and Baltimore, D. (1972). *Adv. Virus Res.* **17,** 129–186.
4. Verma, I. M. (1981). "The Enzymes,' 3rd ed., vol. 14(A), pp. 87–103.
5. Gerard, G. F. (1983). *In* "Enzymes of Nucleic Acid Synthesis and Modification" (S. T. Jacob, Ed.), Vol. 1, pp. 1–38. CRC Press, Florida.
6. Coffin, J. M. (1990). *In* "Virology" (B. N. Fields, D. M. Knipe, R. M. Chanock, M. S. Hirsch, J. L. Melnick, T. P. Monath, and B. Roizman, Eds.), 2nd ed., pp. 1437–1500. Raven Press, New York.
7. Skalka, A. M., and Goff, S. P. (Eds.) (1993). "Reverse transcriptase." Cold Spring Harbor Laboratory Press. Cold Spring Harbor, New York.
8. Varmus, H. E. (1989). *Cell* **56,** 721–724.
9. Travaglini, E. C., Dube, D. K., Surrey, S., and Loeb, L. A. (1976). *JMB* **106,** 605–621.
10. Seal, G., and Loeb, L. A. (1976). *JBC* **251,** 975–981.
11. Srivastava, A., and Modak, M. J. (1980). *JBC* **255,** 2000–2004.
12. Boone, L. R., and Skalka, A. M. (1981). *J. Viriol.* **37,** 117–126.
13. Matson, S. W., Fay, P. J., and Bambara, R. A. (1980). *Biochem.* **19,** 2089–2096.
14. Moelling, K., Bolognesi, D. P., Bauer, H., Buesen, W., Plassmann, H. W., and Hausen, P. (1971). *Nature NB* **234,** 240–243.
15. Keller, W., and Crouch, R. (1972). *PNAS* **69,** 3360–3364.
16. Baltimore, D., and Smoler, D. (1972). *JBC* **247,** 7282–7287.
17. Verma, I. M. (1975). *J. Viriol.* **15,** 843–854.
18. Krug, M. S., and Berger, S. L. (1989). *PNAS* **86,** 3539–3543.
19. Oyama, F., Kikuchi, R., Crouch, R. J., and Uchida, T. (1989). *JBC* **264,** 18808–18817.
20. Baltimore, D., and Smoler, D. (1971). *PNAS* **68,** 1507–1511.
21. Gerard, G. F., Rottman, F., and Green, M. (1976). *Biochem* **13,** 1632–1641.
22. Verma, I. M. (1977). *BBA* **473,** 1–38.
23. Panet, A., Haseltine, W. A., Baltimore, D., Peters, G., Harada, F., and Dahlberg, J. E. (1975). *PNAS* **72,** 2535–2539.
24. Peters, G., Harada, F., Dahlberg, J. E., Panet, A., Haseltine, W. A., and Baltimore, D. (1977). *J. Viriol.* **21,** 1031–1041.
25. Haseltine, W. A., and Kleid, D. G. (1978). *Nature* **273,** 358–364.
26. Swanstrom, R., Varmus, H. E., and Bishop, J. M. (1981). *JBC* **256,** 1115–1121.
27. D'Alessio, J. M., and Gerard, G. F. (1988). *NAR* **16,** 1999–2014.
28. Efstratiadis, A., Kafatos, F. C., Maxam, A. M., and Maniatis, T. (1976). *Cell* **7,** 279–288.
29. Gerard, G. F. (1987). *FOCUS (BRL)* **9,** 5–6.
30. Payvar, F., and Schimke, R. T. (1979). *JBC* **254,** 7636–7642.
31. Verma, I. M., Temple, G. F., Fan, F., and Baltimore, D. (1972). *Nature NB* **235,** 163–167.
32. Kacian, D. L., Spiegelman, S., Bank, A., Terada, M., Metafora, S., Dow, L., and Marks, P. A. (1972). *Nature NB* **235,** 167–169.
33. Gubler, U., and Hoffman, B. J. *Gene* **25,** 263–269.

34. Haymerle, H., Herz, J., Bressan, G. M., Frank, R., and Stanley, K. K. (1986). *NAR* **14**, 8615–8624.
35. Gruber, C. E., Cain, C., and D'Alessio, J. M. (1991). *FOCUS (BRL)* **13**, 88–91.
36. Sartoris, S., Cohen, E. B., and Lee, J. S. (1987). *Gene* **56**, 301–307.
37. Coleclough, C., and Erlitz, F. L. (1985). *Gene* **34**, 305–314.
38. Han, J. H., Stratowa, C., and Rutter, W. J. (1987). *Biochem.* **26**, 1617–1625.
39. Kaufman, D. L., and Evans, G. A. (1990). *BioTechniques* **9**, 304–306.
40. Rabbitts, T. H. (1976). *Nature* **260**, 221–225.
41. Okayama, H., and Berg, P. (1982). *Mol. Cell. Biol.* **2**, 161–170.
42. Heidecker, G., and Messing, J. (1983). *NAR* **11**, 4891–4906.
43. Alexander, D. C., McKnight, T. D., and Williams, B. G. (1984). *Gene* **31**, 79–89.
44. Falkenthal, S., Parker, V. P., Mattox, W. W., and Davidson, N. (1984). *Mol. Cell. Biol.* **4**, 956–965.
45. Krawetz, S. A., Connor, W., Cannon, P. D., and Dixon, G. H. (1986). *DNA* **5**, 427–435.
46. Palazzolo, M. J., and Meyerowitz, E. M. (1987). *Gene* **52**, 197–206.
47. Deininger, P. L. (1987). *Meth. Enzy.* **152**, 371–389.
48. Perbal, B. (1988). "A Practical Guide to Molecular Cloning," 2nd ed., pp. 553–569. Wiley, New York.
49. Eun, H. M., and Yoon, J. W. (1989). *BioTechniques* **7**, 992–997.
50. Berger, S.L., Wallace, D. M., Puskas, R. S., and Eschenfeldt, W. H. (1983). *Biochem.* **22**, 2365–2372.
51. Gerard, G. F., D'Alessio, J. M., and Kotewicz, M. L. (1989). *FOCUS (BRL)* **11**, 66–69.
52. McDonnell, J. P., Garapin, A.-C., Levinson, W. E., Quintrell, N., Fanshier, L., and Bishop, J. M. (1970). *Nature* **228**, 443–435.
53. Manly, K. F., Smoler, D. F., Bromfeld, E., and Baltimore, D. (1971). *J. Virol.* **7**, 106–111.
54. Gerard, G. F. (1985). *FOCUS (BRL)* **7**, 1–3.
55. Snyder, J. G., Hartman, N. G., D'Estantoit, B. L., Kennard, O., Remeta, D. P., and Breslauer, K. J. (1989). *PNAS* **86**, 3968–3672.
56. Boone, L. R., and Skalka, A. (1980). *PNAS* **77**, 847–851.
57. Krug, M. S., and Berger, S. L. (1987). *Meth. Enzy.* **152**, 316–325.
58. Aoyama, H. (1989). *Biochem. Int.* **19**, 67–76.
59. Varmus, H. E., Heasley, S., Kung, H.-J., Opperman, H., Smith, V. C., Bishop, J. M., and Shank, P. R. (1978). *JMB* **120**, 55–82.
60. Huber, H. E., McCoy, J. M., Seehra, J. S., and Richardson, C. C. (1989). *JBC* **264**, 4669–4678.
61. Battula, N., and Loeb, L. A. (1976). *JBC* **251**, 982–986.
62. Sirover, M. A., and Loeb, L. A. (1977). *JBC* **252**, 3605–3610.
63. Mizutani, S., and Temin, H. M. (1976). *Biochem.* **15**, 1510–1516.
64. Gerard, G. F. (1986). *FOCUS (BRL)* **8**, 12.
65. Preston, B. D., Poiesz, B. J., and Loeb, L. A. (1988). *Science* **242**, 1168–1171.
66. Roberts, J. D., Bebenek, K., and Kunkel, T. A. (1988). *Science* **242**, 1171–1173.
67. Ricchetti, M., and Buc, H. (1990). *EMBO J.* **9**, 1583–1593.
68. Huang, P., Farquhar, D., and Plunkett, W. (1990). *JBC* **265**, 11914–11918.
69. DeStefano, J. J., Buiser, R. G., Mallaber, L. M., Myers, T. W., Bambara, R. A., and Fay, P. J. (1991). *JBC* **266**, 7423–7431.
70. Buiser, R. G., DeStefano, J. J., Mallaber, L. M., Fay, P. J., and Bambara, R. A. (1991). *JBC* **266**, 13103–13109.
71. Grandgenett, D. P., Gerard, G. F., and Green, M. (1972). *J. Viriol.* **10**, 1136–1142.
72. Grandgenett, D. P., Gerard, G. F., and Green, M. (1973). *PNAS* **70**, 230–234.
73. Verma, I. M., Mason, W. S., Drost, S. D., and Baltimore, D. (1974). *Nature* **251**, 27–31.
74. Brewer, L. C., and Wells, R. D., (1974). *J. Viriol.* **14**, 1494–1502.
75. Modak, M. J. (1976). *BBRC* **71**, 180–187.
76. Gorecki, M., and Panet, A. (1978). *Biochem.* **17**, 2438–2442.
77. Modak, M. J., and Srivastava, A. (1979). *JBC* **254**, 4756–4759.
78. Furfine, E. S., and Reardon, J. E. (1991). *JBC* **266**, 406–412.

79. Gerard, G. F. (1981). *Biochem.* **20**, 256–265.
80. Majumdar, C., Abbotts, J., Broder, S., and Wilson, S. H. (1988). *JBC* **263**, 15657–15665.
81. Craven, R. C., Bennett, R. P., and Wills, J. W. (1991). *J. Virol.* **65**, 6205–6217.
82. Stewart, L., and Vogt, V. M. (1991). *J. Virol.* **65**, 6218–6231.
83. Poch, O., Sauvaget, I., Delarue, M., and Tordo, N. (1899). *EMBO J.* **8**, 3867–3874.
84. Doolittle, R. F., Feng, D.-F., Johnson, M. S., and McClure, M. A. (1989). *Q. Rev. Biol.* **64**, 1–30.
85. Georgiadis, M. M., Jessen, S. M., Ogata, C. M., Telesnitsky, A., Goff, S. P., and Hendrickson, W. A. (1995). *Structure* **3**, 879–892.
86. Lowe, D. M., Parmar, V., Kemp, S. D., and Larder, B. A. (1991). *FEBS Lett.* **282**, 231–234.
87. Johnson, M. S., McClure, M. A., Feng, D.-F., Grey, J., and Doolittle, R. F. (1986). *PNAS* **83**, 7648–7652.
88. Repaske, R., Hartley, J. W., Kavlick, M. F., O'Neill, R. R., and Austin, J. B. (1989). *J. Virol.* **63**, 1460–1464.
89. Kanaya, S., Kohara, A., Miura, Y., Sekiguchi, A., Iwai, S., Inoue, H., Ohtsuka, E., and Ikehara, M. (1990). *JBC* **265**, 4615–4621.
90. Sooknanan, R., Howes, M., Read, L., and Malek, L. T. (1994). *BioTechniques* **17**, 1077–1085.
91. Katayanagi, K., Miyagawa, M., Matsushima, M., Ishikawa, M., Kanaya, S., Nakamura, H., Ikehara, M., Matsuzaki, T., and Morikawa, K. (1992). *JMB* **223**, 1029–1052.
92. Davies, J. F., II, Hostomska, Z., Hostomsky, Z., Jordan, S. R., and Matthews, D. A. (1991). *Science* **252**, 88–95.
93. Nakamura, H., Katayanagi, K., Morikawa, K., and Ikehara, M. (1991). *NAR* **19**, 1817–1823.
94. Brown, P. O., Bowerman, B., Varmus, H. E., and Bishop, J. M. (1989). *PNAS* **86**, 2525–2529.
95. Fujiwara, T., and Carigie, R. (1989). *PNAS* **86**, 3065–3069.
96. Ishimoto, L. K., Halperin, M., and Champoux, J. J. (1991). *Virol.* **180**, 527–534.
97. Katzman, M., Katz, R. A., Skalka, A. M., and Leis, J. (1989). *J. Virol.* **63**, 5319–5327.
98. Gilboa, E., Mitra, S. W., Goff, S., and Baltimore, D. (1979). *Cell* **18**, 93–100.
99. Hu, W.-S., and Temin, H. M. (1990). *Science* **250**, 1227–1233.
100. Blain, S. W., and Goff, S. P. (1995). *J. Virol.* **69**, 4440–4452.
101. Omer, C. A., and Faras, A. J. (1982). *Cell* **30**, 797–805.
102. Luo, G. X., Sharmeen, L., and Taylor, J. (1990). *J. Virol.* **64**, 592–597.
103. Rattray, A. J., and Champoux, J. J. (1989). *JMB* **208**, 445–456.
104. Champoux, J. J., Gilboa, E., and Baltimore, D. (1984). *J. Virol.* **49**, 686–691.
105. Hu, A.-L. W., D'Alessio, J. M., Gerard, G. F., and Kullman, J. (1991). *FOCUS (BRL)* **13**, 26–29.
106. Sellner, L. N., Coelen, R. J., and Mackenzie, J. S. (1992). *NAR* **20**, 1487–1490.
107. Myers, T. W., and Gelfand, D. H. (1991). *Biochem.* **30**, 7661–7666.
108. Graham, A., Steven, J., McKechnie, D., and Harris, W. J. (1986). *FOCUS (BRL)* **8**, 4–5.
109. Prober, J. M., Trainor, G. L., Dam, R. J., Hobbs, F. W., Robertson, C. W., Zagursky, R. J., Cocuzza, A. J., Jensen, M. A., and Baumeister, K. (1987). *Science* **238**, 336–341.
110. Hahn, C. S., Strauss, E. G., and Strauss, J. H. (1989). *Meth. Enzy.* **180**, 121–130.
111. Lane, D. J., Field, K. G., Olsen, G. J., and Pace, N. R. (1988). *Meth. Enzy.* **167**, 138–144.
112. Geliebter, J. (1987). *FOCUS (BRL)* **9**, 5–8.
113. Zakour, R. A., and Loeb, L. A. (1982). *Nature* **295**, 708–710.
114. Shortle, D., Grisafi, P., Benkovic, S. J., and Botstein, D. (1982). *PNAS* **79**, 1588–1592.
115. Liao, X., and Wise, J. A. (1990). *Gene* **88**, 107–111.
116. Goodrich, D. W., and Duesberg, P. H. (1990). *PNAS* **87**, 2052–2056.
117. Luo, G. X., and Taylor, J. (1990). *J. Virol.* **64**, 4321–4328.
118. Zhang, J., and Temin, H. M. (1993). *Science* **259**, 234–238.
119. Buiser, R. G., Bambara, R. A., and Fay, P. J. (1993). *BBA* **1216**, 20–30.
120. Ouhammouch, M., and Brody, E. N. (1992). *NAR* **20**, 5443–5450.
121. Peliska, J. A., and Benkovic, S. J. (1992). *Science* **258**, 1112–1118.
122. Awang, G., and Sen, D. (1993). *Biochem.* **32**, 11453–11457.
123. Weiss, S., Hausl, G., Famulok, M., and Konig, B. (1993). *NAR* **21**, 4879–4885.
124. Smith, J. S., and Roth, M. J. (1993). *J. Virol.* **67**, 4037–4049.
125. Ben-Artzi, H., Zeelon, E., Gorecki, M., and Panet, A. (1992). *PNAS* **89**, 927–931.

126. Blain, S. W., and Goff, S. P. (1993). *JBC* **268**, 23585–23592.
127. Gopalakrishnan. V., Peliska, J. A., and Benkovic, S. J. (1992). *PNAS* **89**, 10763–10767.
128. Hsieh, J.-C., Zinnen, S., and Modrich, P. (1993). *JBC* **268**, 24607–24613.
129. Kati, W. M., Johnson, K. A., Jerva, L. F., and Anderson, K. S. (1992). *JBC* **267**, 25988–25997.
130. Reardon, J. E. (1993). *JBC* **268**, 8743–8751.
131. DeStefano, J. J., Bambara, R. A., and Fay, P. J. (1993). *Biochem.* **32**, 6908–6915.
132. Divita, G., Mueller, B., Immendorfer, U., Gautel, M., Rittinger, K., Restle, T., and Goody, R. S. (1993). *Biochem.* **32**, 7966–7971.
133. Kruhoffer, M., Urbanke, C., and Grosse, F. (1993). *NAR* **21**, 3943–3949.
134. Kohlstaedt, L. A., Wang, J., Friedman, J. M., Rice, P. A., and Steitz, T. A. (1992). *Science* **256**, 1783–1790.
135. Jacobo-Molina, A., and 12 coauthors (1993). *PNAS* **90**, 6320–6324.

B. AMV Reverse Transcriptase
[EC 2.7.7.49]

The reverse transcriptase from the avian myeloblastosis virus (AMV RTase) is a prototype of avian retroviral RTases. The enzyme is a dimer (155 kDa) composed of nonidentical subunits α and β. AMV RTase has multiple enzymatic activities including RNA-directed DNA polymerase (reverse transcriptase), DNA-directed DNA polymerase, RNase H, and DNA endonuclease. AMV RTase does not have $3' \rightarrow 5'$- or $5' \rightarrow 3'$-exonuclease activity. AMV RTase is a key reagent in cDNA synthesis and nucleotide sequencing, although its role is now partially shared with cloned MoLV RTase.

1. FUNCTIONS

a. Polymerase reaction conditions

i. Recommended reaction conditions. Current protocols for the synthesis of cDNA from mRNA usually involve two distinct reactions, i.e., first-strand and second-strand syntheses, although the succession from the first to the second reaction may be more or less separate depending on the strategies used for cDNA synthesis and cloning (1,2). During the polymerization reactions, cDNA can be labeled by incorporating an appropriately labeled nucleotide(s) in the first and/or second reaction(s).

(a) First-strand cDNA synthesis. The polymerase activity at 41°C is optimal at pH 8.3 (or pH 8.1 at 25°C) in Tris–Cl buffer (3). No significant differences exist in the synthesis of full-length cDNA at 37° and 41°C. The following reaction conditions have been shown to be optimal for the synthesis of long cDNA transcripts and not necessarily for dNTP incorporation into acid-insoluble products. Slightly different but essentially similar reaction conditions have been recommended by other investigators (4).

A typical first-strand cDNA synthesis reaction (50 μl) is carried out in 50 mM Tris–Cl (pH 8.3 at 42°C), 150 mM NaCl (or 100 mM KCl), 10 mM MgCl$_2$, 1 mM DTT, 50 μg/ml BSA, 50 U RNasin, 0.5 mM spermidine hydrochloride, 4 mM Na-PP$_i$, 0.2 mM each of dNTPs, 1–5 μg poly(A)$^+$ mRNA, 0.5–2.5 μg (dT)$_{12-18}$ (or random hexamers), and 10 U of RTase. [*10× buffer*: 0.5 M Tris–Cl (pH 8.3), 1.5 M NaCl, 0.1 M MgCl$_2$, 10 mM DTT, and 0.5 mg/ml BSA.] The mixture is incubated at 42°C for 1 hr.

AMV RTase is stable under the reaction conditions and continuous polymerization occurs up to 1 hr. The yield of cDNA products is generally between 20 and 40%, depending on mRNA species.

(b) Second-strand DNA synthesis. The cDNA products of the first-strand synthesis reaction are in the form of a DNA · RNA hybrid. A typical method for second-strand DNA synthesis, known as RNA strand replacement synthesis (5,6), consists of nicking the RNA in the hybrid with *E. coli* RNase H. This treatment leaves small RNA fragments with free 3'-OH annealed to the cDNA, which are subsequently used as primers for DNA synthesis by *E. coli* DNA Pol I.

The first-strand cDNA products, which are purified twice by "standard" methods of ethanol precipitation, are resuspended in the second-strand synthesis buffer [2× *buffer*: 40 mM Tris–Cl (pH 7.5), 10 mM MgCl$_2$, 20 mM (NH$_4$)$_2$SO$_4$, 200 mM KCl, and 100 μg/ml BSA]. The following reagents are added to the reaction mixture (100 μl containing ≤1 μg hybrid): 0.4 mM each of dNTPs, 2 U of *E. coli* RNase H, and 25 U Pol I. The mixture is incubated at 12°C for 1 hr and then at 22°C for 1 hr. Although a large fraction of the dsDNA products may be blunt ended at this point, the products are customarily treated with T4 DNA Pol (5 U) for an additional 5–10 min at 37°C. The reaction is terminated by adding EDTA (to 20 mM), and the products are precipitated with ethanol.

(c) Labeling cDNAs. The cDNA products may be labeled for various reasons. To monitor the progress of the cDNA synthesis reaction or to check the yield and size of the products, either the first- or the second-strand DNA can be labeled. To use the cDNAs as a hybridization probe, the first-strand DNA is labeled with either a radioactive nucleotide(s) or a nonradioactive labeling nucleotide(s).

(i) Radiolabeling. The nucleotide of choice for the synthesis of labeled cDNA is [α-^{32}P]dATP or [^{35}S]dATPαS. Alternatively, [α-^{32}P]dCTP can be used for labeling but a slightly higher concentration is needed (7). Of particular concern in the synthesis of labeled cDNA is the relationship between the concentration of radionucleotides and the yield and quality (length) of the cDNA products. In reactions (20 μl) with AMV RTase (or MoLV RTase) and 2.3-kb RNA templates, cDNAs can be labeled to a sufficiently high specific activity of $>1 \times 10^7$ cpm/μg when 2.5 μCi [α-^{32}P]dATP is present together with the other three dNTPs at 0.5 mM. The minimum concentration of [α-^{32}P]dATP or [^{35}S]dATPαS to achieve greater than 25% yield of full-length cDNA is 25 μM for AMV RTase and 10 μM for MoLV RTase. When [α-^{32}P]dCTP is used in place of [^{32}P]dATP, the minimal concentration of [^{32}P]dCTP has to be increased at least twice. Higher concentrations of dATP at the same total 2.5 μCi give a higher yield of full-length cDNA, although the specific radioactivity of cDNAs may drop due to a dilution effect. Thus to obtain full-length cDNAs at a higher yield and specific activity, higher concentrations of the isotope(s), e.g., 10 μCi at 100 μM, should be employed. Note that [α-^{33}P]dNTP is now a preferred substitute for [α-^{32}P]dNTP.

TABLE 6.8 Kinetic Parameters of AMV Reverse Transcriptase

Template-primer	dNTP	K_m (μM)	k_{cat} (min^{-1})	Condition
$(A)_n \cdot (dT)_{12-18}$			1200	a
Globin mRNA $\cdot (dT)_{12-18}$			400	a
Ribohomopolymer	dTTP	36		b
	dGTP	2		b
RNA	dATP	15		c
Deoxyribohomopolymer $(dC)_n \cdot (dG)_{12-18}$	dGTP	0.25		
Activated DNA	dNTPs	4		

[a] Reaction at pH 8.3, 37°C (9).
[b] Reaction at pH 8.0, 37°C (8).
[c] Reaction at pH 8.3, 37°C (7).

(ii) Nonisotopic labeling. Advances in the nonisotopic labeling technology permit cDNAs to be efficiently labeled for the preparation of hybridization probes. AMV RTase can incorporate biotin- or digoxigenin-labeled dUTP in place of dTTP. The nonisotopic labels can be detected typically by using alkaline phosphatase (or peroxidase)-conjugated avidin (or streptavidin) or the antidigoxigenin Ab detection system. The cDNA synthesis with digoxigenin-11–dUTP is optimal at a molar ratio (labeled dUTP/dTTP) of 1/2. This labeling gives a cDNA probe with a detection sensitivity reaching ~80 ng.

ii. **Kinetic parameters.** The kinetic parameters for some typical substrates are listed in Table 6.8; the K_m values for dNTPs are 2–40 μM (7,8), while the k_{cat} ranges from 300 to 1200 min^{-1} (9).

iii. **Metal ion requirements.** Divalent metal ions, Mg^{2+} and Zn^{2+}, are required for RTase activity. Mg^{2+} but not Zn^{2+} is required for RNase H activity. The optimum concentration of Mg^{2+} is 6 mM for polymerase activity (3,10) and ~20 mM for RNase H activity (11). Excess Mg^{2+} is not inhibitory for RNase H activity. With Mn^{2+}, the polymerase activity is maximal at 0.6 mM Mn^{2+}, although the maximal activity amounts to ~65% of that with Mg^{2+} (12).

iv. **Ionic strength.** Monovalent cations strongly influence the polymerase activity. The yield of long cDNA transcripts is maximal at ~150 mM NaCl (or 100 mM KCl) which is near physiological concentrations (3). Increasing the NaCl concentrations to 214 mM significantly inhibits the synthesis of long transcripts. At 150 mM KCl, RNase H activity is inhibited greater than 80% (13).

v. **Inhibitors and inactivators**
(a) Metal chelators. AMV RTase is, like other DNA polymerases, a Zn-metalloenzyme and is thus inhibited by the Zn^{2+} chelator, 1,10-phenanthroline (14,15). The inhibition is initially reversible ($K_i \approx 70$ μM) but later becomes

irreversible. Both polymerase and RNase H activities can be inhibited by EDTA, a Mg^{2+} chelator.

(b) Sulfhydryl reagents. The AMV polymerase activity is very sensitive to the modification of SH groups by hydrophobic reagents such as DTNB and PHMB but is resistant to hydrophilic sulfhydryl reagents (16).

AMV RNase H is resistant to both hydrophobic and hydrophilic sulfhydryl reagents. NEM, a reagent specific to amino and SH groups, inactivates both polymerase and RNase H, but the RNase H activity is fourfold more resistant. The polymerase activity lost by DTNB or NEM modification can be partially (40–90%) recovered by incubation with 20 mM DTT (16).

(c) Flavonoid and quinoid compounds. Many flavonoid and quinoid compounds have been found to inhibit various retroviral RTases more or less specifically (17,18). Ficetin, a naturally occurring flavonoid compound, is one of the inhibitors more potent to AMV RTase than to other RTases. At 10 µg/ml, ficetin inhibits 94% of the polymerase activity of AMV RTase on poly(A) · oligo(dT), whereas HIV RTase and mammalian DNA Pol α are inhibited 10 and 43%, respectively (19).

(d) Phosphate, pyrophosphate, and analogs. Phosphate at or near physiological concentrations is inhibitory to AMV RTase in the synthesis of long transcripts (3).

Pyrophosphate (Na-PP_i) at a 0.5 mM concentration reduces the RTase activity to 50%. Among PP_i analogs, phosphonoformate (or foscarnet) is the strongest inhibitor to AMV RTase (8,20). The AMV RNase H activity is unaffected by the phosphonoformate. Na-PP_i has also been reported to have no inhibitory effect on RNase H activity (21). The mode of RTase inhibition by phosphonoformate is noncompetitive ($K_i \approx 5$ to 100 µM) with respect to and depending on nucleotide substrates. The structurally related phosphonoacetate is not a RTase inhibitor.

In some instances, however, PP_i has been shown to block the degradation of RNA in RNA · DNA hybrids and to impede the synthesis of anticomplementary DNA (22). Na-PP_i (≤4 mM) has also been shown to increase the yield of full-length cDNAs significantly (9,23), making it a common practice to include PP_i in the cDNA synthesis reaction with AMV RTase (but not with MoLV RTase). The simultaneous use of PP_i and actinomycin D results in an inhibition of the synthesis of both large and small cDNAs (4).

Phosphorothioate oligodeoxynucleotides are inhibitors of RTases (AMV and HIV-1) competing against the primer-template for binding with the enzyme. In particular, phosphorodithioate-$(dC)_{14}$ is a potent inhibitor with an ID_{50} of 0.25 µM for AMV RTase (or a K_i of 9.6 nM for HIV-1 RTase) (92).

(e) Non-mRNAs. AMV RTase is inhibited by the presence of rRNA and tRNA. At 1, 2, and 5 µg of rRNA (calf thymus or yeast), the yield of cDNA synthesis from a RNA template (7.8 kb, 1 µg) has been shown to be reduced by 8, 24, and 43%, respectively (24). The presence of tRNA (yeast) is even more

inhibitory than rRNA. In contrast to the AMV RTase, MoLV RTase is not inhibited by any of the ribosomal or transfer RNAs tested.

(f) Other compounds inhibitory to reverse transcriptase. Adriamycin inhibits AMV RTase noncompetitively against dNTPs but competitively against the 3' termini of the template (25).

Aurochloric acid (AuCl$_4$H) is an inhibitor of AMV RTase with an ID$_{50}$ of 18–100 μM depending on the template-primer (26). It is a competitive inhibitor to dTTP and an uncompetitive inhibitor to template-primer (A)$_n$ · (dT)$_{12-18}$, but is uncompetitive to dGTP and noncompetitive to (C)$_n$ · (dG)$_{12-18}$.

Diphosphates of N-(2-phosphonylmethoxyethyl) derivatives of heterocyclic bases have varying inhibitory effects on AMV RTase (27). The 2-amino-adenine derivative is the most potent inhibitor with an ID$_{50}$ of $\sim$1 μM. (*Note*: This is a more potent inhibitor than either AZT triphosphate or ddTTP.)

VRC ($\geq$5 mM), but not RNasin, inhibits the RNase H activity of AMV RTase (28). Note that the vanadyl complexes do not inhibit *E. coli* RNase H.

NaF (20 mM) inhibits the RNase H activity with no or only a slight effect on cDNA synthesis (13,29). Preincubation with NaF at 27–30 mM inhibits the RNase H activity by 80–100%.

(g) Inhibition by antisense oligodeoxynucleotides. The cDNA synthesis using AMV RTase can be inhibited by an oligodeoxyribonucleotide (either unmodified or linked at the 5' end to an intercalating agent) which is complementary to a region downstream from the primer. The inhibition is due to the degradation of the RNA template in the antisense oligodeoxyribonucleotide · RNA hybrid by the RNase H activity of RTase (30).

(h) Thermal inactivation. AMV RTase is inactivated by preincubation at and above 41°C, but not in the presence of template-primers. AMV 70S RNA protects the enzyme better than poly(C) · oligo(dG) (31). The half-time of inactivation at 57°C is $\sim$1 min for the polymerase activity, 2 min for the RNase H activity, and 4.4 min for the DNA endonuclease activity (32). In contrast to the dimeric form ($\alpha\beta$) of the RTase, the α subunit is not protected at all from heat inactivation by the template-primer.

b. Activity assays and unit definition

i. Polymerase assay. A typical assay mixture (0.1 ml) contains 50 mM Tris–Cl (pH 8.3), 6 mM MgCl$_2$, 80 mM KCl, 10 mM DTT, 0.1 mg/ml BSA, 0.2 mM (A)$_n$ · (dT)$_{12-18}$, 0.5 mM [^{3}H]dTTP (10–20 cpm/pmol), and polymerase. The mixture is incubated at 37°C. Reaction is stopped by adding 10 μl EDTA (0.2 M, pH 8.0).

Unit definition: One unit is the amount of enzyme that incorporates 1 nmol of dTMP into acid-insoluble products in 10 min at 37°C.

ii. RNase H assay. The standard RNase H assay is similar to the polymerase assay except that (a) MgCl$_2$ is replaced with 2 mM MnCl$_2$ and (b) the template-primer is replaced with 3 μM [^{3}H]poly(A) (10^4 cpm) and 1.5 μM poly(dT). After 10 min at 37°C, calf thymus DNA (30 μg) is added as a carrier and the reaction is terminated by adding 10% TCA (90 μl). The radioactivity of the acid-soluble material in the supernatant is counted after centrifugation at 12,000 g for 15 min.

c. Substrate specificities: polymerase

i. Templates

(a) Homopolymeric templates. The initial rate of DNA polymerization decreases in the order: poly(A) $\approx$ poly(C) > poly(I) > poly(U). Homopolyribonucleotides are better templates than deoxyribo counterparts, e.g., by a factor of ~50 for poly(A) over poly(dA) and 1-2 for poly(C) over poly(dC) (10,33). With the exception of poly(dC), other homopolydeoxyribonucleotides are extremely poor templates. The particular property of AMV RTase in using a poly(dC) template as efficiently as poly(C) provides an important alternative to Pol Ik in nucleotide sequencing of the regions of DNA template containing a high GC content and/or secondary structures (also see below).

Under the conditions optimal for poly(A) · d(pT)$_{10}$, relative values of the initial rate for different template-primers are as follows: poly(A) · d(pT)$_{10}$ [100], poly(dA) · (pU)$_{10}$ [40], poly(A) · (pU)$_{10}$ [33], and poly(dA) · d(pT)$_{10}$ [4.7] (34).

(b) Native RNA templates. In copying natural RNAs with oligodeoxyribonucleotide primers, the rates of polymerization catalyzed by AMV RTase and *E. coli* DNA Pol I vary depending on priming sites (33). The rates of dCMP and dTMP incorporation are higher with oligo(dT) priming than with oligo(dG) priming. However, the ratio of dCMP to dTMP incorporation is disproportionate between the two enzymes. AMV RTase prefers copying the poly(C) regions of the RNA template to which the oligo(dG) primer is annealed. This preference of AMV RTase over that of Pol I is dramatic (>7 × 10^5-fold) with the homopolymeric poly(C) · oligo(dG) template-primer.

(c) Template switching. AMV RTase carries out strand-transfer synthesis *in vitro* as well as *in vivo*. (Refer to Section III,A, in this chapter.) Template switching occurs largely in a template homology-dependent manner. The switching efficiency is dependent on incubation temperature (93). For example, chimeric cDNA molecules are generated within 30 min at high incubation temperatures, with an increasing efficiencies from 42° to 50°C (~5% of total cDNA). At lower temperatures (e.g., 37°C), such products are detectable only after much longer incubation times (e.g., 90 min).

(d) Methylated templates. Poly(2'-O-methyl-C) · oligo(dG) serves as an efficient template-primer for many RTases including both avian and mammalian enzymes (35). The methylated homopolymers do not serve as templates for DNA-dependent DNA polymerases such as *E. coli* DNA Pol I and *M. luteus* DNA Pol. AMV RTase cannot transcribe through N^6-dimethyladenine (36).

(e) DNA templates. Although homopolymeric ribonucleotides are preferred templates for RTase, homopolydeoxyribonucleotides have also been shown, with

a few notable exceptions, to be utilized efficiently as templates. Note that DNA is a part of the natural templates for RTases during *in vivo* synthesis of duplex cDNA from viral RNA genome.

On primed ssDNA templates, AMV RTase has an initial polymerization rate comparable to that of Pol I. AMV RTase can use nicked or "activated" dsDNA as templates for DNA synthesis, most likely by a strand displacement mechanism (25). Compared with the primed ssDNA templates, nicked or activated DNA is used at ~60% efficiency. Compared with natural RNAs, certain DNAs (e.g., activated calf thymus DNA) are over 100-fold more efficiently used as templates for AMV RTase (33). With DNA templates that require strand displacement or exonucleolytic activity, *E. coli* DNA Pol I is ~100 times more active than AMV RTase.

The comparison of RNA and DNA template efficiencies is informative, but it should be recognized that the reaction condition optimal for one substrate may be suboptimal for another. In fact, the optimal conditions for a polymerization reaction with AMV RTase on RNA templates have been shown to be different from those on DNA templates (34,37).

The synthesis of poly(dT) on the poly(dA) template is sharply optimal at pH 7.6 (50 mM Tris–Cl) in contrast to the broad pH optimum around pH 8.1–8.4 on poly(A). Varying the KCl concentrations from 0 to 60 mM has no significant effect on the rate of polymerization on poly(dA) in contrast to the reaction with poly(A) which is stimulated by 50–100 mM KCl. The efficiency of synthesis on poly(A) at the optimum MnCl$_2$ concentration (0.7 mM) is ~60% of that with MgCl$_2$ (4 mM). In contrast, the synthesis on poly(dA) is maximal at 5 mM MnCl$_2$ or 18 mM MgCl$_2$; the preferred cation is Mn^{2+} (100%) and not Mg^{2+} (30–40%). The RNA or DNA template-dependent differences are also manifested on the utilization of primers with different lengths (see below). Compared with the DNA · DNA which adopts the B helix, the RNA · DNA heteroduplex which adopts the A-form apparently allows more effective interactions with RTase.

On templates containing abasic lesions, AMV RTase dose not catalyze significant insertion and extension of the primer strand (94). (In contrast, HIV-1 RTase readily performs such reactions by preferentially utilizing the A base opposite the abasic site.)

ii. Primers

(a) Native primer. AMV RTase utilizes tRNA^{Trp} as the natural primer for DNA synthesis *in vivo* (38). The tRNA primer anneals to the viral RNA (70S or heat-denatured 35S RNA) over a 5'-end region of ~18 nt, called *pbs* (primer-binding site). AMV RTase has a binding site for tRNA^{Trp}, and the tRNA binds with the highest affinity of all chicken cell tRNAs. AMV RTase can bind the tRNA in the absence of template RNA.

(b) Synthetic primers. AMV RTase can utilize as primers homo- and hetero-oligoribo- and deoxyribonucleotides of varying lengths. With homopolymeric templates, e.g., poly(A) and poly(dA), dTTP can serve both as the minimal primer

and as a nucleotide substrate (37). In contrast to DNA-dependent DNA polymerases which use dNMP as well as dNTP as primers (39), AMV RTase does not use dNMP as a primer.

The K_m values for $d(pT)_n$ series primers on the poly(A) template decrease as the length of the primers increases. For $n = 2-10$, the K_m for the primer decreases by a factor of approximately 3 for each additional nucleotide, e.g., from 110 μM for $d(pT)_2$ to 0.008 μM for $d(pT)_{10}$ (34,37). On poly(dA) templates, under a set of optimal conditions which differ from those on poly(A), the K_m for primers decreases by a factor of approximately 2 per each additional nucleotide, e.g., 540 μM for $d(pT)_2$ to 22 μM for $d(pT)_{10}$. Further increases in the length of the primer ($n = 10-25$) bring no practical change in the value of K_m. The V values do not depend on the length of the primers; however, the values of V for poly(A) template are 3.5–4 times higher than those for poly(dA).

When the temperature is lowered from 37° to 25°C, the K_m values for $d(pT)_n$ increase by a factor of 1.5 on the poly(A) template. The temperature dependence of the initial rates shows a reasonable fit to the Arrhenius plot [see Eq. (1.11)], which gives a ΔE_a of 12 kcal/mol. The temperature dependence of the initial rates with poly(dA) templates deviates from the Arrhenius relationship, although an apparent ΔE_a of 50 kcal/mol can be estimated from the linear portion of the plot.

(c) Mismatching primers. AMV RTase utilizes mispaired primer termini as functional sites for polymerization when they are complexed to ribo- and deoxyribohomopolymeric templates (40).

With primers containing 3'-terminal mismatching nucleotides, AMV RTase extends the G(primer) · T mispair almost 20% as efficiently as it extends A · T (41). The G · G mispair is extended with the lowest efficiency (10^{-6}) compared with G · C. Other types of mispairs are extended with intermediate efficiencies; e.g., 2.1×10^{-3} and 1.3×10^{-3} for C · T and T · T, respectively.

Using a series of primers which contain noncomplementary bases in different positions from the 3' end, the K_m and V values of the primers have been shown to increase. Compared with $d(pT)_{10}$ on the poly(A) template, the K_m values for $d(pT)_9(pC)$, $d(pT)_8pCpT$, $d(pT)_7pC(pT)_2$, $d(pT)_4pC(dT)_5$, and $d(pT)_3(pT)_7$ are higher by a factor of 93, 338, 425, 95, and 15, respectively (42). The V values for these primers were 1.2–1.5 times higher than for $d(pT)_{10}$.

The observed changes in the affinity (or K_m values) of AMV RTase toward the primers containing single noncomplementary bases indicate that the discrimination between correct and incorrect primers is most sensitive at the second or third position from the 3' end. Note that Pol Ik discriminates most effectively the 3'-terminal mismatching nucleotides (43).

(d) Priming of second-strand DNA synthesis in vitro. AMV RTase can synthesize dsDNAs from synthetic or natural RNA templates following the first-strand cDNA synthesis with an RNA primer. This is reminiscent of its role *in vivo* (see Fig. 6.7). The anticomplementary or (+) strand synthesis by RTase is presumed to occur by a strand displacement mechanism, as evidenced by the presence of multiple ssDNA branches (44). In reconstituted cDNA synthesis using

oligodeoxyribonucleotides as primers, the second-strand DNA has been shown to be synthesized apparently by a hairpin structure-mediated priming of the first strand, although such a loop-back mechanism is inefficient (45,46). With RSV 70S RNA as a template for AMV RTase, the synthesis of a hairpin DNA could be attributed to a 5-nt-long inverted repeat located in the middle of the RNA (47). The formation of hairpin DNA is blocked in the presence of actinomycin D, and so is the second-strand synthesis.

iii. **Nucleotide substrates and analogs.** Like all DNA polymerases, dNTPs are the standard nucleotide substrates for RTases. AMV RTase can also efficiently incorporate a variety of nucleotide anlogs such as dNTPαS, ddNTP, dITP, 7-deaza-dGTP, biotin-11-dUTP, digoxigenin-11-dUTP, and fluorescein-12-dUTP.

Many other nucleotide analogs, notably those of dTTP, have been developed as potential RTase inhibitors having pharmacological value. For example, 3′-azido-3′-deoxythymidine 5′-triphosphate (N_3dTTP) is incorporated by AMV RTase (and HIV-1 RTase as well) into the T sites of DNA and terminates chain elongation in a dose-dependent manner (48). The N_3dTTP, which is converted from 3′-azido-3′-deoxythymidine (AZT) by cellular kinases, is believed to be the active metabolite producing the potent inhibitory effect of AZT on HIV-1 replication.

Other dTTP analogs such as 3′-fluoro-2′,3′-dideoxythymidine 5′-(α-methylphosphonyl)-β,γ-diphosphate and 2′-deoxythymidine 5′-(α-methylphosphonyl)-β,γ-diphosphate are incorporated into DNA chains by AMV RTase (and also by HIV-1 RTase) (49). The 3′-fluoro analog is a chain terminator whereas the second analog is not. AMV RTase incorporates the second compound about seven times, forming a DNA that contains approximately seven internucleoside methylphosphonate groups bearing no negative charges. Both dTTP analogs are effective inhibitors of DNA synthesis.

iv. **Processivity.** The polymerase processivity of AMV RTase on the RNA template with oligodeoxyribonucleotide primers is ~97 nt (50). The processivity is ~26 nt on the nicked dsDNA template and is over 1000 nt on the poly(A) template.

v. **Frequency of misincorporation.** AMV RTase displays a high error rate, under certain conditions, exceeding more than 10 times that of *E. coli* DNA Pol I when trascribing homopolymeric RNA, DNA, or single-strand phage DNA templates (51,52). In the presence of 250 μM dNTPs, a reaction condition often employed in cDNA synthesis, the error rates of AMV RTase (and MoLV RTase as well) on $(A)_n \cdot (dT)_{12-18}$ and $(C)_n \cdot (dG)_{12-18}$ template-primers have been determined to be $\sim 2 \times 10^{-4}$ (1/5000) (53). Although the fidelity of AMV RTase can vary depending on templates and the nucleotide sequence(s), the observed error rates reflect the *in vivo* mutation rate of AMV, i.e., 3×10^{-4} per site per virus passage (54).

With poly(dAT) as the template-primer, the error rate is ~1/2500; the fidelity increases up to fivefold when the *E. coli* ssDNA-binding protein is added to the reaction (55).

On natural DNA template-primers, AMV RTase shows a misincorporation frequency of 1/10,000–1/26,000 for dC opposite a template dA (56,57). The misincorporation of dAMP or dGMP also occurs at a rate of ~1/30,000 for both AMV and MoLV RTases (58).

Extension of single base mismatches by AMV RTase occurs with varying efficiencies, ranging from 10^{-2} for G·T (primer/template) to 10^{-6} for G·G compared with G·C. The discrimination in mismatch extension is ascribed primarily to a kinetic effect rather than to the difference in polymerase binding affinity to the 3' termini of the mispaired primer (95).

vi. Pyrophosphate exchange and pyrophosphorolysis. AMV RTase has no 3'→5'- or 5'→3'-exonuclease activity. With high concentrations of PP_i, however, AMV RTase catalyzes pyrophosphorolysis and PP_i exchange (59,60). The PP_i exchange is dependent on template-primers (both DNA and RNA) and is base specific. In contrast to the polymerase activity, the PP_i exchange activity is more efficient with the DNA template. The PP_i exchange does not convert dNTP to dNMP.

The PP_i exchange activity is a variant manifestation of the polymerase activity, whereas pyrophosphorolysis activity is a part of the RNase H activity. The functional grouping is based on the responses of catalytic functions to various site-specific inhibitors such as the Zn^{2+} chelator and the modification reagents for SH and other groups (60).

d. Substrate specificities: RNase H

i. Substrate and product profiles. AMV RNase H degrades the RNA strand of RNA·DNA hybrids, e.g., poly(A)·poly(dT), in both the 3' → 5' and 5' → 3' directions at the same rate (61,62). The digestion products are oligoribonucleotides which range in size from 2 to 30 nt with an averge of 10 to 12 nt (62,63). In fact, the size of the RNase H digestion products varies widely depending on template sequences and assay conditions. The products have 5'-P and 3'-OH termini.

ii. RNA endonuclease activity. AMV RNase H is essentially an endonuclease. It can cleave relaxed, circular, and covalently closed plasmids in which 770 consecutive residues of one strand are ribonucleotides (28). It also cleaves single-stranded circular RNA when a synthetic oligodeoxyribonucleotide is hybridized to the tRNA (64). It deadenylates the 5'-capped globin mRNA in the presence of oligo(dT)$_{12-18}$ even when the 3' end of the mRNA is blocked by covalent modifications with pCp.

The RNA endonuclease activity is operational with or without concomitant polymerase activity. As the cDNA synthesis progresses, the length of RNA decreases. Either during or on completion of cDNA synthesis, heterogeneous oligoribonucleotides having 7–16 bases remain hybridized to the 3' terminus of the cDNA strand (64). Gel analysis of the RNA samples aliquoted during cDNA synthesis reactions shows only a few discrete bands of 100–200 nt (50), suggesting that the RTase-associated RNase H has some cleavage site preference.

iii. Cleavage at RNA–DNA junction. RNase H can cleave *in vitro* the junction between RNA and DNA, an action suited for its *in vivo* role in removing RNA primers (65). Examination of tRNA-primer excisions *in vitro* using model substrates (tRNA with DNA extensions) shows that AMV RTase/RNase H hydrolyzes the phosphodiester bond at the DNA–RNA junction, whereas MoLV (and also HIV-1) RTase/RNase H hydrolyzes the substrate to leave a single ribonucleotide 5'-P at the 5' terminus of the model DNA genome (66).

iv. Processivity of RTase/RNase H. The RNase H activity of AMV RTase ($\alpha\beta$ complex) is partially processive *in vitro*, whereas the mode of RNase H action of the α subunit enzyme is distributive (61,62,67).

e. Substrate specificities: DNA endonuclease

i. Divalent cation requirements. Virion-purified AMV RTase ($\alpha\beta$ complex) possesses a physically associated DNA endonuclease activity. Under certain assay conditions (e.g., with 5–10 mM Mg^{2+}), the endonuclease activity is hardly detectable. The activity appears when activated (10-fold) in the presence of Mn^{2+} (2 mM).

ii. Effect of ATP. ATP ($\geq$2.5 mM) stimulates the Mn^{2+}-activated DNA-nicking activity of DNA endonuclease threefold (68). The endonuclease, which is undetectable in the presence of 10 mM Mg^{2+}, is also stimulated by ATP to a level comparable to Mn^{2+} activation. Both ADP and dATP are stimulatory, albeit at lower levels than ATP. GTP has no apparent effect and AMP is rather inhibitory.

iii. Preferred substrate. Supercoiled dsDNA (e.g., ColE1 plasmid) is the preferred substrate (69). AMV RTase also introduces nicks on linear dsDNA, ssDNA, and highly structured mammalian 5.5S RNA but with much less efficiency. The products have 5'-P and 3'-OH termini.

iv. Sequence preference. In the presence of Mg^{2+}, the DNA endonuclease, in the form of a proteolytic fragment (pp32), introduces a single nick on supercoiled DNA (ColE1) regardless of the ratio of the enzyme to DNA (70). With a supercoiled DNA containing LTR circle junction sequences, the pp32 enzyme nicks the DNA at one or the other of two sites (indicated by *) each mapped 2 nt away from the circle junction, CA*TT|AA TG/GT AA|TT*AC (71). The endonuclease activity has also been shown to be selective on the nearly perfect 15-base terminal inverted repeat of AMV DNA LTR, creating 2-nt recessed 3' termini which serve as the immediate precursor to provirus integration (72).

The majority of nicking activities on other LTR DNA sites are adjacent to or within the dinucleotide CA. The apparent site specificity shown with Mg^{2+} is less restricted with Mn^{2+}.

v. **DNA-binding activity.** AMV RTase demonstrates a strong dsDNA-binding activity, strong enough to retain the enzyme on a NC filter with bound ColE1 DNA (73). This dsDNA-binding activity is presumed to be a prerequisite to the RTase-associated endonuclease and the unwinding activities. The endonuclease activity is responsible for viral integration into the host chromosomal DNA (74).

2. APPLICATIONS

The reader is referred to Section III,A in this chapter.

3. STRUCTURE

i. **General features of AMV RTase molecule.** AMV RTase is a dimeric ($\alpha\beta$) enzyme (155 kDa) with three functional domains. It is composed of two nonidentical subunits of 92 kDa (β subunit) and 63 kDa (α subunit) (75, 76). The larger β subunit is cleaved in the virion, resulting in the smaller α subunit (572 amino acids) and pp32, a phosphoprotein (286 amino acids) having a DNA endonuclease (IN) activity (77, 78). The N termini of both α- and β-polypeptides are identical, Thr–Val–Ala–Leu–His– (79). The C termini of α and β subunits are Tyr and Ala, respectively (80).

According to CD measurements, the secondary structure of AMV RTase contains 16% α helix, 25% β sheet, 24% β turn, and 36% undefined structures (81).

ii. **Distinct domains and multifunctions.** The α and β subunits can be dissociated in the presence of NP-40 and Me_2SO (30%) or 1,4-dioxane and separated by phosphocellulose chromatography (82). The α subunit displays both polymerase and RNase H activities as does the $\alpha\beta$ complex (76). The α subunit lacks, however, any detectable DNA endonuclease activity. The α subunit contains RNase H and polymerase activities in distinct N- and C-terminal domains, respectively. Compared with the α subunit, the β subunit is either enzymatically inactive or possesses limited activities. The β subunit is believed to mainly augment the stability of enzyme–template interactions, thus increasing the processivity of the α subunit activities (31,61).

Mild digestion of AMV RTase by chymotrypsin generates a 24-kDa polypeptide that exhibits only RNase H activity (77). Only the RNase H activity of the α subunit, but not that of $\alpha\beta$ complex, can be inhibited by captan, suggesting that there is some conformational change around the active center on $\alpha\beta$-complex formation (83). If the subunit structure of AMV RTase is considered similar to that of HIV-1 RTase (p66/p51 heterodimer), there is only one functional polymerase site within the asymmetric complex (96).

iii. **DNA endonuclease domain.** Proteolytic cleavage of the β subunit by chymotrypsin generates two polypeptides: a 63-kDa fragment still possessing the DNA polymerase and RNase H activities and a C-terminal 32-kDa fragment (pp32 or IN protein) with Mg^{2+}- and Mn^{2+}-activated DNA endonuclease activity (78). Both $\alpha\beta$-heterodimeric RTase and the separately purified IN protein show DNA endonuclease activities *in vitro*. The C terminus of pp32 has the sequence

—Ser—Pro—Leu—Phe—Ala (84). The IN protein has a secondary structure composed of 17% α helix, 32% β sheet, 18% β turns, and 33% disordered structure (85).

iv. **Physical parameters.** The pI of AMV RTase is 6.5 (86). The $s^{\circ}_{20,w}$ of the enzyme is 7.1S in 0.3 M K-P (pH 7.8), representing an $\alpha\beta$ heterodimer (81). In Tris–Cl (pH 7.9) buffer containing 0.46 M NaCl and 4% glycerol, the enzyme sediments with an apparent s°_{20} of 10.1S (or 11.9S when the solvent effect of glycerol is taken into account). Therefore, the native enzyme dimerizes to a $(\alpha\beta)_2$ form in the presence of 4% glycerol at 20°C. At 4°C, the $\alpha\beta$ form predominates.

4. Genetics

i. **Phenotypes of AMV.** AMV is a replication-defective, type C retrovirus (70S RNA) that causes acute myeloblastic leukemia in chickens and transforms myelomonocytic hematopoietic cells in culture. The oncogenic potential of AMV is most likely attributed to a genetic locus, v-*myb*, near the 3' end of the viral genome. The genome of AMV fails to produce functional RTase as a result of the v-*myb* insertion (87). The manner of insertion of the v-*myb* (1.2 kb), a truncated from of c-*myb*, has damaged two structural genes of the virus (88). The bulk of *env* has been deleted and the C-terminal domain (36 amino acid residues) of the *pol* gene has been altered such that the polyprotein precursor to RTase (Pr180$^{gag\text{-}pol}$) is not processed into active enzyme protomers. As a consequence, AMV can replicate only in the presence of nondefective helper retroviruses, MAV (myeloblastosis-associated viruses), which are isogenic with the oncogenic AMV except for the part corresponding to the *myb*. To be exact, the AMV RTase should thus be called *MAV RTase*.

ii. **Gene structure and organization.** The full 7226-nt sequence of AMV proviral DNA has been determined, including the *pol* gene (nt positions 2803–5376) (80). For other general features, refer to Section III,A, in this chapter.

5. Sources

AMV RTase molecules naturally reside in the core of the virion. Virions (AMV BAI strain A) are purified from the AMV-infected myeloblasts suspended in tissue culture or from blood plasma of AMV-infected chicks at the stage of full-blown leukemia. When inoculated with 10^{10} virus particles, the multiplication factor per bird of the AMV is ~300 (86). Reverse transcriptase comprises ~3% of the MAV virion proteins or ~70 polymerase molecules per virion (89).

The virions are lysed with nonionic detergents (0.2% Triton X-100 or NP-40) in K-P buffer (pH 7.3) containing 2 mM DTT and 10% glycerol. The detergent-solubilized, active RTase is fractionated by one or a combination of column chromatographic techniques (86,90,91). The yield of >95% pure enzyme varies from 20,000 to 35,000 U/g of virus. The specific activity of ≥95% pure enzyme is 35,000–60,000 U/mg protein (86).

AMV RTase loses less than 10% of activity in 1 year when stored at −20°C in phosphate buffer containing glycerol. The enzyme can be stored more stably in small aliquots at −70°C.

References

1. Sambrook, J., Fritsch, E. F., and Maniatis, T. (1989). "Molecular Cloning: A Laboratory Manual," 2nd ed., pp. 8.1–8.86. Cold Spring Harbor Laboratory Press, Cold Spring Harbor, New York.
2. Perbal, B. (1988). "A Practical Guide to Molecular Cloning," 2nd ed., pp. 550–577. Wiley, New York.
3. Retzel, E. F., Collett, M. S., and Faras, A. J. (1980). *Biochem.* **19**, 513–518.
4. Krug, M. S., and Berger, S. L. (1987). *Meth. Enzy.* **152**, 316–325.
5. Okayama, H., and Berg, P. (1982). *Mol. Cell. Biol.* **2**, 161–170.
6. Gubler, U. (1987). *Meth. Enzy.* **152**, 330–335.
7. Gerard, G. F. (1988). *FOCUS (BRL)* **10**, 12–13.
8. Eriksson, B., Stening, G., and Oberg, B. (1982). *Antiviral Res.* **2**, 81–95.
9. Gerard, G. F. (1985). *FOCUS (BRL)* **7**, 1–3.
10. Baltimore, D., and Smoler, D. (1971). *PNAS* **68**, 1507–1511.
11. Baltimore, D., and Smoler, D. (1972). *JBC* **247**, 7282–7287.
12. Sirover, M. A., and Loeb, L. A. (1977). *JBC* **252**, 3605–3610.
13. Brewer, L. C., and Wells, R. D. (1974). *J. Virol.* **14**, 1494–1502.
14. Auld, D. S., Kawaguchi, H., Livingston, D. M., and Vallee, B. L. (1974). *BBRC* **57**, 967–972.
15. Poiesz, B. J., Battula, N., and Loeb, L. A. (1974). *BBRC* **56**, 959–964.
16. Gorecki, M., and Panet, A. (1978). *Biochem.* **17**, 2438–2442.
17. Ono, K., Nakane, H., Fukushima, M., Chermann, J.-C., and Barre-Sinoussi, F. (1990). *EJB* **190**, 469–476.
18. Spedding, G., Ratty, A., and Middleton, E., Jr. (1989). *Antiviral Res.* **12**, 99–110.
19. Take, Y., Inouye, Y., Nakamura, S., Allaudeen, H. S., and Kubo, A. (1989). *J. Antibiotics* **42**, 107–115.
20. Margalith, M., Falk, H., and Panet, A. (1982). *Mol. Cell. Biochem.* **43**, 97–103.
21. Srivastava, A., and Modak, M. J. (1979). *BBRC* **91**, 892–899.
22. Myers, J. C., and Spiegelman, S. (1978). *PNAS* **75**, 5329–5333.
23. Kacian, D. L., and Myers, J. C. (1976). *PNAS* **73**, 2191–2195.
24. Gerard, G. F. (1987). *FOCUS (BRL)* **9**, 5–6.
25. Matson, S. W., Fay, P. J., and Bambara, R. A. (1980). *Biochem.* **19**, 2089–2096.
26. Semba, M., Tanaka, N., Yamada, M.-A., Takeuchi, F., Matsuta, K., Miyamoto, T., Hanaoka, F., and Ui, M. (1990). *JPN. J. Cancer Res.* **81**, 1259–1264.
27. Votruba, I., Travnicek, M., Rosenberg, I., Otmar, M., Merta, A., Hrebabecky, H., and Holly, A. (1990). *Antiviral Res.* **13**, 287–293.
28. Krug, M. S., and Berger, S. L. (1989). *PNAS* **86**, 3539–3543.
29. Watson, K. F., Schendel, P. L., Rosok, M. J., and Ramsey, L. R. (1979). *Biochem.* **18**, 3210–3218.
30. Loreau, N., Boiziau, C., Verspieren, P., Shire, D., and Toulme, J.-J. (1990). *FEBS Lett.* **274**, 53–56.
31. Panet, A., Verma, I. M., and Baltimore, D. (1975). *CSHSQB* **39**, 919–923.
32. Golomb, M., Grandgenett, D. P., and Mason, W. (1981). *J. Virol.* **38**, 548–555.
33. Travaglini, E. C., Dube, D. K., Surrey, S., and Loeb, L. A. (1976). *JMB* **106**, 605–621.
34. Lokhova, I. A., Nevinsky, G. A., Gorn, V. V., Veniaminova, A. G., Repkova, M. V., Kavsan, V. M., Rudenko, N. K., and Lavrik, O. I. (1990). *FEBS Lett.* **274**, 156–158.
35. Gerard, G. F., Rottman, F., and Green, M. (1974). *Biochem.* **13**, 1632–1641.
36. Hagenbuchle, O., Santer, M., Steitz, J. A., and Mans, R. J. (1978). *Cell* **13**, 551–563.
37. Lokhova, I. A., Nevinskii, G. A., Bulychev, N. A., Gorn, V. V., Levina, A. S., Rudenko, N. K., Kavsan, O. I., and Lavrik, O. I. (1990). *Mol. Biol. (USSR)* **24**, 396–407.
38. Panet, A., Haseltine, W. A., Baltimore, D., Peters, G., Harada, F., and Dahlberg, J. E. (1975). *PNAS* **72**, 2535–2539.

39. Nevinsky, G. A., Veniaminova, A. G., Levina, A. S., Podust, V. N., Lavrik, O. I., and Holler, E. (1990). *Biochem.* **29**, 1200–1207.
40. Battula, N., and Loeb, L. A. (1976). *JBC* **251**, 982–986.
41. Mendelman, L. V., Petruska, J., and Goodman, M. F. (1990). *JBC* **265**, 2338–2346.
42. Lokhova, I. A., Nevinskii, G. A., Levina, A. S., and Kavsan, V. M. (1990). *Mol. Biol. (USSR)* **24**, 984–992.
43. Nevinskii, G. A., Nemudraya, A. V., Levina, A. S., Lokhova, I. A., Gorn, V. V., and Khomov, V. V. (1990). *Mol. Biol. (USSR)* **24**, 96–103.
44. Boone, L. R., and Skalka, A. M. (1981). *J. Virol.* **37**, 117–126.
45. Swanstrom, R., Varmus, H. E., and Bishop, J. M. (1981). *JBC* **256**, 115–1121.
46. Efstratiadis, A., Kafatos, F. C., Maxam, A. M., and Maniatis, T. (1976). *Cell* **7**, 279–288.
47. Swanstrom, R., Varmus, H. E., and Bishop, J. M. (1981). *JBC* **256**, 1115–1121.
48. Huang, P., Farquar, D., and Plunkett, W. (1990). *JBC* **265**, 11914–11918.
49. Victorova, L. S., Dyatkina, N. B., Ju. Mozzherin, D., Atrazhev, A. M., Krayevsky, A. A., and Kukhanova, M. K. (1992). *NAR* **20**, 783–789.
50. DeStefano, J. J., Buiser, R. G., Mallaber, L. M., Myers, T. W., Bambara, R. A., and Fay, P. J. (1991). *JBC* **266**, 7423–7431.
51. Battula, N., and Loeb, L. A. (1975). *JBC* **250**, 4405–4409.
52. Gopinathan, K. P., Weymouth, L. A., Kunkel, T. A., and Loeb, L. A. (1979). *Nature* **278**, 857–859.
53. Gerard, G. F. (1986). *FOCUS (BRL)* **8**, 12.
54. Coffin, J. M., Tsichlis, P. V., Barker, C. S., and Voynow, S. (1980). *Annals N.Y.A.S.* **354**, 410–425.
55. Kunkel, T. A., Meyer, R. R., and Loeb, L. A. (1979). *PNAS* **76**, 6331–6335.
56. Preston, B. D., Poiesz, B. J., and Loeb, L. A. (1988). *Science* **242**, 1168–1171.
57. Roberts, J. D., Bebenek, K., and Kunkel, T. A. (1988). *Science* **242**, 1171–1173.
58. Roberts, J. D., Preston, B. D., Johnson, L. A., Soni, A., Loeb, L. A., and Kunkel, T. A. (1989). *Mol. Cell. Biol.* **9**, 469–476.
59. Seal, G., and Loeb, L. A. (1976). *JBC* **251**, 975–981.
60. Srivastava, A., and Modak, M. J. (1980). *JBC* **255**, 2000–2004.
61. Grandgenett, D. P., and Green, M. (1974). *JBC* **249**, 5148–5152.
62. Leis, J. P., Berkower, I., and Hurwitz, J. (1973). *PNAS* **70**, 466–470.
63. Verma, I. M. (1975). *J. Virol.* **15**, 843–854.
64. Oyama, F., Kikuchi, R., Crouch, R. J., and Uchida, T. (1989). *JBC* **264**, 18808–18817.
65. Champoux, J. J., Gilboa, E., and Baltimore, D. (1984). *J. Virol.* **49**, 686–691.
66. Furfine, E. S., and Reardon, J. E. (1991). *Biochem.* **30**, 7041–7046.
67. Keller, W., and Crouch, R. (1972). *PNAS* **69**, 3360–3364.
68. Nissen-Meyer, J., Raae, A. J., and Nes, I. F. (1981). *JBC* **256**, 7985–7989.
69. Golomb, M., and Grandgenett, D. P. (1979). *JBC* **254**, 1606–1613.
70. Grandgenett, D. P., Vora, A. C., and Schiff, R. D. (1978). *Virol.* **89**, 119–132.
71. Grandgenett, D. P., and Vora, A. C. (1985). *NAR* **13**, 6205–6221.
72. Katzman, M., Katz, R. A., Skalka, A. M., and Leis, J. (1989). *J. Virol.* **63**, 5319–5327.
73. Matson, S. W., and Bambara, R. A. (1981). *BBA* **652**, 29–38.
74. Quinn, T. P., and Grandgenett, D. P. (1988). *J. Virol.* **62**, 2307–2312.
75. Kacian, D. L., Watson, K. F., Burny, A., and Spiegelman, S. (1971). *BBA* **246**, 365–383.
76. Grandgenett, D. P., Gerard, G. F., and Green, M. (1973). *PNAS* **70**, 230–234.
77. Lai, M. H. T., and Verma, I. M. (1978). *J. Virol.* **25**, 652–663.
78. Grandgenett, D. P., Golomb, M., and Vora, A. C. (1980). *J. Virol.* **33**, 264–271.
79. Copeland, T., Grandgenett, D. P., and Oroszlan, S. (1980). *J. Virol.* **36**, 115–119.
80. Baluda, M. A., and Reddy, E. P. (1994). *Oncogene* **9**, 2761–2774.
81. Lin, T.-H., Quinn, T., Walsh, M., Grandgenett, D., and Lee, J. C. (1991). *JBC* **266**, 1635–1640.
82. Grandgenett, D. P. (1976). *J. Virol.* **17**, 950–961.
83. Freeman-Wittig, M.-J., and Lewis, R. A. (1990). *Mol. Cell. Biochem.* **94**, 9–17.
84. Grandgenett, D. P., Quinn, T., Hippenmeyer, P. J., and Oroszlan, S. (1985). *JBC* **260**, 8243–8249.
85. Lin, T.-H., Quinn, T. P., Grandgenett, D., and Walsh, M. T. (1989). *Proteins (S.F.G.)* **5**, 156–165.
86. Houts, G. E., Miyagi, M., Ellis, C., Beard, D., and Beard, J. W. (1979). *J. Virol.* **29**, 517–522.

87. Duesberg, P. H., Bister, K., and Moscovici, C. (1980). *PNAS* **77**, 5120–5154.
88. Klempnauer, K.-H., Gonda, T. J., and Bishop, J. M. (1982). *Cell* **31**, 453–463.
89. Panet, A., Baltimore, D., and Hanafusa, T. (1975). *J. Virol.* **16**, 146–152.
90. Verma, I. M. (1981). "The Enzymes," 3rd ed., Vol. 14(A), pp. 87–103.
91. Golomb, M., Vora, A. C., and Grandgenett, D. P. (1980). *J. Virol. Meth.* **1**, 157–165.
92. Marshall, W. S., and Caruthers, M. H. (1993). *Science* **259**, 1564–1570.
93. Ouhammouch, M., and Brody, E. N. (1992). *NAR* **20**, 5443–5450.
94. Cai, H., Bloom, L. B., Eritja, R., and Goodman, M. F. (1993). *JBC* **268**, 23567–23572.
95. Creighton, S., Huang, M.-M., Cai, H., Arnheim, N., and Goodman, M. F. (1992). *JBC* **267**, 2633–2639.
96. Kohlstaedt, L. A., Wang, J., Friedman, J. M., Rice, P. A., and Steitz, T. A. (1992). *Science* **256**, 1783–1790.

C. MoLV Reverse Transcriptase
[EC 2.7.7.49]

The reverse transcriptase (RTase) from the Moloney murine leukemia virus (MoLV) possesses both DNA polymerase and RNase (H and D) activities. It is monomeric (80 kDa) in solution but is presumed to function as a dimer. The ratio of the polymerase to RNase H activities and many of MoLV RTase functions are similar to those of AMV RTase. In contrast to AMV RTase, however, MoLV RTase does not possess DNA endonuclease activity. The gene for MoLV RTase has been cloned and expressed in *E. coli*, and both wild-type RTase (RNase H$^+$) and modified RTase (RNase H$^-$) are commercially available. MoLV RTase is thus a reliable reagent which shares with AMV RTase much of the role of cDNA synthesis in molecular cloning. Modified MoLV RTase has proven to be particularly useful in the synthesis of full-length cDNAs.

1. FUNCTIONS

Unless otherwise specified, the reaction conditions and substrate specificities described below apply equally to wild-type MoLV RTase (RNase H$^+$) and modified RTase (RNase H$^-$).

a. Reaction conditions

i. Recommended reaction conditions. MoLV RTase is optimally active at pH 7.6 and 37–42°C. The polymerase activity is maximal in the presence of 60–80 mM NaCl (1). At 20 and 120 mM NaCl, the polymerase activity decreases to 64 and 56% of the maximum, respectively.

Although optimal conditions may vary depending on particular applications, the following is a typical set of conditions suitable for cDNA synthesis.

(a) First-strand cDNA synthesis. A typical 20-μl reaction mixture (with RNase H$^-$ RTase) contains 50 mM Tris–Cl (pH 8.1 at 25°C), 3 mM MgCl$_2$, 75 mM KCl (or NaCl), 10 mM DTT, 100 μg/ml BSA, 0.5 mM each of dNTPs including [α-^{32}P]dCTP (1100 cpm/pmol), 20 U RNasin, 50 μg/ml actinomycin D, 0.5 μg oligo(dT)$_{12–18}$, 1 μg poly(A)$^+$ mRNA, and 200 U MoLV RTase. The mixture is incubated at 37°C for 1 hr.

When second-strand DNA is synthesized by the "one-tube" method (see below), first-strand cDNA synthesis is performed without actinomycin D and the first reaction mixture is directly adjusted with the reaction buffer for second-strand synthesis. When second-strand DNA is synthesized by the conventional procedures (see *AMV RTase*), the first reaction is terminated by adding 2 μl of SDS (10%, w/v) and 2 μl EDTA (0.5 M, pH 8.0), and the cDNA products are precipitated with ethanol.

The 20-μl reaction volume is suitable for amounts of RNA up to 7.5 μg. For mRNA below 1 μg, 200 U of enzyme is still recommended. For copying 5–10 μg of RNA, which is often necessary for the construction of a cDNA library, the reaction conditions are proportionally upgraded to a 50-μl volume. For 1–7.5 μg mRNA, the usual amount of primer would be 0.5–3.8 μg oligo(dT)$_{12-18}$ or linker-primer, which roughly corresponds to a primer-to-RNA molar ratio of 10. Vector-primer can be efficiently used at a molar ratio of near 1:1. For random hexamer-primed cDNA synthesis, 50–150 ng hexamers per 2 μg RNA have been shown to be optimal (2). The use of large amounts of MoLV RTase (e.g., 1000 U/50-μl reaction) tends to increase the yield of full-length cDNAs.

(b) Second-strand DNA synthesis. The methodology used for the synthesis of second-strand cDNA is closely tied to the cloning strategies as a whole (see Fig. 6.6). Conventionally, the most efficient method for second-strand DNA synthesis is the RNA strand replacement synthesis using both *E. coli* RNase H and DNA Pol I on the RNA·cDNA heteroduplex (3,4).

According to the "one-tube" second-strand DNA synthesis protocol recommended by BRL, the first-strand cDNA reaction mixture (20 μl) is put on ice and the following reagents are added directly in the following order: (i) H$_2$O to a final volume of 160 μl, (ii) 32 μl 5× *second-strand buffer* [100 mM Tris–Cl (pH 6.9), 450 mM KCl, 25 mM MgCl$_2$, 0.75 mM β-NAD$^+$, and 50 mM (NH$_4$)$_2$SO$_4$], 3 μl of 10 mM dNTPs, 6 μl of 100 mM DTT, 15 U of *E. coli* DNA ligase, 40 U of Pol I, and 1.4 U of *E. coli* RNase H. The mixture is incubated at 16°C for 2 hr. Then T4 DNA Pol (10 U per μg first-strand cDNA) is added and the mixture is incubated at 16°C for 5 min.

The blunt-ended, double-stranded cDNA products are ready for ligations with linkers/adapters or directly with cloning vectors. As a result of RNA priming, the dsDNAs generated by the strand replacement procedure described earlier tend to miss some 5'-end RNA sequences (8–21 bases) (5). One way to retain the full 5'-terminal RNA sequence information is to use a cDNA cloning strategy involving homopolymeric tailing (6). In such a strategy, the first-strand cDNA is tailed with oligo(dC) using terminal deoxynucleotidyltransferase, and then second-strand synthesis is performed using (dG)$_{12-18}$ or adapter-primers containing oligo(dG). The DNA can be synthesized either by the RNA strand replacement synthesis protocol or by using Pol Ik on the ssDNA template following alkaline hydrolysis of the RNA template.

The second-strand DNA can also be synthesized by *Taq* DNA Pol and the cDNAs can be amplified by PCR (RT–PCR). Note that the high concentrations of MoLV RTase used in the first reaction may significantly inhibit *Taq* Pol unless the RTases are heat denatured prior to the addition of *Taq* Pol. The amplified products can then be efficiently cloned by the T4 DNA Pol-based cloning strategy (see T4 DNA Pol, Section II,B, in this chapter).

Either the first- or the second-strand DNA products can be labeled as described previously (see Section III,B on AMV RTase).

ii. **Kinetic parameters.** With $(A)_n \cdot (dT)_{12-18}$ and globin mRNA $\cdot (dT)_{12-18}$ as template-primers, the rates of polymerization catalyzed by MoLV RTase are 2400 and 700 min^{-1}, respectively (7). The K_m values for dATP and dCTP on 2.3-kb RNA are 24 and 31 μM, respectively (8).

iii. **Metal ion requirements.** MoLV RTase requires divalent metal ions for the polymerase activity. At optimum Mn^{2+} (1 mM), the polymerization on poly(A) $\cdot$ oligo(dT) is three times more efficient than at 2 mM Mg^{2+} (9). However, the polymerization on poly(C) $\cdot$ oligo(dG) or poly(dC) $\cdot$ oligo(dG) is four to eight times more efficient at the optimum Mg^{2+} (10 mM) than at the optimum Mn^{2+}. Mn^{2+} at a concentration higher than 2 mM significantly inhibits the polymerase activity (1).

The MoLV RNase H activity is more efficient with Mn^{2+} than with Mg^{2+} both at the same optimum concentration of 1–2 mM.

iv. **Inhibitors and inactivators**

(a) Phosphates and analogs. *Inorganic phosphate* is inhibitory, resulting in 27 and 54% inhibitions at 5 and 20 mM concentrations, respectively (1).

Pyrophosphate: Unlike AMV RTase, MoLV RTase is very sensitive to Na-PP$_i$. At 0.1 and 0.5 mM Na-PP$_i$, the RTase activity is inhibited by 32 and 50%, respectively, while at 1.0 mM Na-PP$_i$, only 6% of the maximal activity is observed (1).

Phosphonoformate, a PP$_i$ analog, is a strong inhibitor of MoLV RTase (IC$_{50}$ = 8–250 μM) (10). The inhibition is noncompetitive with respect to nucleotide substrates ($K_i \approx 4.7$ μM). Phosphonoformate inhibits neither the RNase H activity of AMV RTase nor that of MoLV RTase.

(b) Non-mRNAs. In remarkable contrast to AMV RTase, MoLV RTase is not inhibited by non-mRNAs such as rRNAs and tRNAs (11). MoLV RTase would thus be particularly useful for synthesizing cDNAs with low purity or crude mRNAs.

(c) Primer analogs. *Phosphorothioate oligodeoxynucleotide,* e.g., S(dC)$_{28}$, strongly binds RTase and competes against the template-primer, poly(A) $\cdot$ oligo(dT), with a K_i of 2.8 nM (12).

The α-*anomeric oligonucleotide* hybridizes to the complementary region of an RNA template in a parallel orientation and blocks the polymerase activity of MoLV RTase. In the presence of 0.01 μM of the normal primer (12-mer β-oligodeoxyribonucleotide) for the rabbit globin mRNA template, an 88% inhibition of DNA synthesis is observed with 0.5 μM α-oligodeoxyribonucleotide (12-mer) (13). The α-DNA · β-RNA hybrids are not substrates for RNase H, although they can displace the enzyme from its natural substrate (14).

Peptide nucleic acid (PNA), an oligodeoxynucleotide analog in which the phosphodiester backbone is replaced with a polyamide consisting of N-(2-aminoethyl)-glycine units, can strand invade duplex DNA, causing a D-loop. Hybridization of an antisense PNA to an AT-rich target region of RNA results in a PNA dose-dependent termination of cDNA synthesis by MoLV RTase (64). Unlike oligodeoxynucleotides, PNA induces inhibition of reverse transcription in an RNase H-independent manner because of the highly stable nature of the RNA · PNA heteroduplex.

*(d) **Other inhibitory compounds.*** *Digallic acid* (gallic acid 5,6-dihydroxyl-3-carboxyphenyl ester), but not gallic acid, is a potent inhibitor of MoLV RTase, resulting in a ~90% inhibition at the 0.5-μg/ml concentration (15). Digallic acid is partially competitive with respect to the template-primer $(A)_n \cdot (dT)_{12-18}$ and noncompetitive with respect to dTTP.

Podoscyphic acid, a fungal metabolite with the structure $CH_3(CH_2)_{10}$-$C(C=O)$-$CH=CH$-CO_2H, is an inhibitor having an IC_{50} of 10–20 μg/ml (16).

Sulfhydryl reagent and protease: MoLV RTase is rapidly inactivated in the presence of NEM or by treatment with proteinase K.

*(e) **Thermal sensitivity.*** MoLV RTase is relatively thermolabile and is rapidly and completely inactivated by heating at 70°C for 15 min (1). The polymerase activity is three times more susceptible than the RNase H activity to heat inactivation at 45°C. The addition of template-primers during preincubation does not protect the polymerase from thermal inactivation. Note that Rauscher MLV RTase, a monomeric enzyme similar to MoLV RTase, is protected from heat inactivation (43°C) by substrates, e.g., poly(C) · oligo(dG) and AMV 70S RNA (17).

The half-life of RNase H⁺ RTase at 37°C is ~15 min, which means that 10–20 times more MoLV RTase than AMV RTase is required to synthesize an equivalent amount of cDNA products (7). RNase H⁻ RTase is reported to be slightly more thermostable at 45°C, producing a better yield of full-length products than at 37°C (18).

b. Activity assays and unit definition

i. Polymerase assay. A typical assay mixture (50 μl) contains 50 mM Tris–Cl (pH 8.3), 10 mM DTT, 6 mM Mg^{2+}, 40 mM KCl, 100 μg/ml BSA, 0.5 mM [³H]dTTP (50 cpm/pmol), 0.1 mM poly(A), and 0.1 mM $(dT)_{12-18}$. The mixture is incubated at 37°C.

Unit definition: One unit is the amount of enzyme that incorporates 100 pmol of dTMP into acid-insoluble products in 15 min.

ii. RNase H assay. The reaction conditions for the RNase H activity assay are identical to those for the polymerase assay described earlier, except that the template-primer is replaced by [³H]poly(A) · poly(dT).

c. Substrate specificities: Polymerase

i. Template structure. Like AMV RTase, MoLV RTase utilizes a variety of templates (RNA, DNA, ribo- and deoxyribohomopolymers) to synthesize DNA. The relative efficiency of homopolyribonucleotide templates with cognate homo-oligodeoxyribonucleotide primers decreases in the order: poly(A) > poly(C) > poly(U) > poly(I) (19). Poly(dC) serves as an exceptionally efficient template for MoLV RTase; the rate of synthesis is approximately three times higher than that of poly(C). In contrast, other homopolydeoxynucleotides, e.g., poly(dA), poly(dI), and poly(dT), are very poor substrates.

The polymerase activity of MoLV RTase, but not of AMV RTase, is almost completely blocked on sequences beyond an oligo(dA)$_{16}$ insert in an M13 DNA template, whereas other homopolymers such as (dG)$_{16}$, (dT)$_{16}$, and (dC)$_{16}$ can be traversed without much difficulty (20). (This property of MoLV RTase is uniquely shared with HIV-1 RTase.)

ii. Templates for second-strand DNA synthesis. In the cDNA synthesis with globin mRNA and oligo(dT) primers, over 90% of the DNA products are found to be resistant to nuclease S1, suggesting that the products are double stranded and that the second-strand DNA was synthesized from the first-strand cDNA most likely by a loop-back mechanism (1). The second-strand DNA synthesis is inhibited in the presence of actinomycin D, which does not affect the RTase in the synthesis of the first-strand cDNA.

iii. Template switching. MoLV RTase can catalyze strand transfer synthesis during reverse transcription *in vitro* (21–23). (Refer to Section II,A, in this chapter.)

iv. Primers. MoLV RTase utilizes synthetic oligodeoxyribonucleotides as efficient primers. Oligodeoxyribonucleotides as short as four to eight bases can serve as primers for DNA synthesis. The natural primer for DNA synthesis by MoLV RTase is tRNA^{Pro}, although other tRNAs can be used in place of tRNA^{Pro} (24).

v. Nucleotide substrates. Although the preferred nucleotide substrates for MoLV RTase are dNTPs, the absence of 3' → 5'-exonuclease activity allows the enzyme to incorporate a variety of nucleotide analogs such as dNTPαS, dITP, 7-deaza-dGTP, biotin- or digoxigenin-labeled dUTP, and fluorescein-12-dUTP relatively efficiently.

A chain-terminating dTTP analog, 3'-azido-3'-deoxythymidine 5'-triphosphate (N$_3$dTTP), is also efficiently incorporated by MoLV RTase into the T sites of elongating DNA in a dose-dependent manner (25).

vi. Processivity. The processivity of RNase H$^+$ RTase on an RNA template with an oligodeoxyribonucleotide primer is ~69 nt (26). The processivity of

RNase H⁻ RTase (in which the RNase H domain is deleted) is substantially lower than that of RNase H⁺ RTase (65).

vii. Frequency of misincorporation. In the presence of 250 μM dNTPs, a concentration close to that used in cDNA synthesis, the error rates of MoLV RTase on $(A)_n \cdot (dT)_{12-18}$ and $(C)_n \cdot (dG)_{12-18}$ have been estimated to be ~1/3500 and ~1/7400, respectively (27). These error rates are similar to those of AMV RTase. On certain RNA templates, however, MoLV RTase has been shown to be less error-prone than AMV RTase (25). On DNA template-primers, the error frequency of MoLV RTase is $10^{-4}-10^{-6}$, which is also similar to that of AMV RTase (28–30). A·C error occurs at a rate of 4×10^{-6}, while T·G occurs at a rate of 7×10^{-5}.

d. Substrate specificities: RNases H and D

i. Preferred substrates. MoLV RTase/RNase H selectively cleaves the 3' → 5'-phosphodiester bonds of the RNA moiety in the RNA·DNA hybrid from both the 5' → 3' and 3' → 5' directions. Single-stranded RNA, ssDNA, and dsDNA are not substrates (9). When a model genomic substrate (DNA-extended tRNA) is presented in a duplex form with a complementary DNA strand, MoLV RNase H hydrolyzes the substrate and leaves a single ribonucleotide 5'-P at the 5' terminus (31). With the same model substrate, AMV RTase/RNase H hydrolyzes the phosphodiester bond at the DNA–RNA junction. MoLV RTase/RNase is also capable of degrading RNA in RNA·RNA duplexes (RNase D), as assayed by *in situ* gel techniques (67). The apparent specific activity of MoLV RNase D on RNA·RNA duplexes is similar to that of RNase H on RNA·DNA hybrids.

ii. Influence of RNase H activity on cDNA synthesis. The RNase H activity cleaves the poly(A) tail when a primer [e.g., oligo(dT)] anneals to it, reducing the chance for the polymerase to initiate polymerization. The RNase H activity also degrades the RNA template, leading to a possible termination of transcription, especially at the regions at which RTase pauses. Removal of the RNase H domain renders the RTase significantly more efficient in cDNA synthesis, producing longer cDNAs at higher yields (see below).

iii. Processivity of RNase H. MoLV RNase H activity is partially processive in a way similar to that of AMV RTase/RNase H (32).

iv. Product profile. With $(A)_n \cdot (dT)_n$ as a substrate, MoLV RNase H initially generates products ranging in length from 80–90 nt to less than 10 nt. An extended incubation results in digestion products 3 to 15 nt long (33). The digestion products have 5'-P and 3'-OH ends. The degraded products are readily separable by sucrose gradient centrifugation, suggesting that they are not H-bonded to the DNA strand.

e. RNase H⁻ reverse transcriptase

The RNase H domain of MoLV RTase can be genetically removed without affecting the polymerase activity. Nevertheless, RNase H⁻ RTase (e.g., *SuperScript* RTase, BRL) has some similar and also dissimilar properties compared with RNase H⁺ RTase. Similar properties include thermostability (in fact, RNase H⁻ RTase is slightly more thermostable at 45°C), mono- and divalent ion optima, fidelity of dNTP incorporation with homopolymeric templates, and insensitivity to stimulation by polyanions. The polymerase activity of RNase H⁻ RTase is almost equal to that of the wild-type enzyme on certain substrates, e.g., $(C)_n \cdot (dG)_{12-18}$. On other substrates, e.g., $(A)_n \cdot (dT)_{12-18}$, however, the polymerase activity is only ~23% (34) and the polymerase has a tendency to pause at certain regions of the RNA template, probably reflecting a reduced processivity. In fact, on the $(A)_{1000} \cdot \text{oligo}(dT)_{20}$ template-primer, the processivity of RNase H⁻ RTase is 80–90 nt at maximum, in contrast to over 1000 nt for RNase H⁺ RTase (23). Other mutant RTases lacking an RNase H structural domain or with impaired RNase H functions also display reduced processivities on heteropolymeric DNA templates (65). Despite the reduced processivity, RNase H⁻ RTase, at high concentrations, can synthesize longer cDNA products at higher yields than RNase H⁺ RTase. For instance, the RNase H⁻ RTase synthesizes first-strand cDNA at a ~2 times higher yield than AMV RTase (35).

2. APPLICATIONS

The reader is referred to Section III,A, in this chapter.

3. STRUCTURE

i. Primary and secondary structures. MoLV RTase is an 80-kDa monomeric enzyme (36). It has an apparent sedimentation coefficient of 4.6S in glycerol gradient media. The primary structure deduced from the nucleotide sequence of the *pol* gene contains 671 amino acid residues with N-terminal Thr and C-terminal Leu (37,38). The overall sequence identity for MoLV and HIV-1 RTases is ~25%.

MoLV RTase contains a high degree of ordered structure: 44% α helix and 37% β sheet (39).

ii. Polymerase and nuclease domains. On the basis of amino acid sequence identity (40), mutant analysis (41,42), and *in vitro* mutagenesis (43), the polymerase domain has been assigned to the N-terminal two-thirds region of the single polypeptide and the nuclease domain to the C-terminal one-third region of ~160 amino acids. In fact, respective single domains expressed separately from cloned genes exhibit the expected single activities (41). A polymerase fragment (residues 10–278) obtained by limited proteolysis with trypsin is catalytically active, although its activity is reduced ~6000-fold relative to the whole molecule (66). This catalytic fragment consists of two subdomains, fingers and palm, resembling those of HIV-1 RTase. On the other hand, an ~30-kDa fragment, apparently arising from a proteolytic cleavage of the MoLV RTase in cell lysates, has an RNase H activity but not polymerase activity (33).

The RTase domain contains four consensus sequence motifs that have a highly conserved amino acid(s) such as Asp^{15}–Asp^{16}, Lys-28, Gly-32, and Asp-270 (44).

The MoLV RNase H domain shares a mere 26% amino acid sequence homology with *E. coli* RNase H but the two molecules are thought to have strikingly similar 3-D structures (45); the whole backbone structures (five α helices and five β strands) are the same and the presumed active site residues as well as the residues which form the hydrophobic core and H-bonds are all located in the same positions.

Compared with the structural model of the RNase H domain of MoLV RTase, the HIV RTase/RNase H domain lacks the third α helix. Deletion of the α helix, i.e., 11 amino acids from 593 (Ile) through 603 (Leu), from MoLV RNase H has thus been predicted to make it structurally similar to that of the HIV RNase H domain. Such a deletion mutant of MoLV RTase was apparently fully active in polymerase and RNase H activity assays *in vitro*, but showed a defect in the polymerase function *in vivo* (46). Thus the third α helix of the RNase H domain is apparently crucial for cooperative interactions with the polymerase domain, presumably affecting the processivity of the polymerase.

The RNase H domain also exhibits a functionally distinct nuclease activity on RNA · RNA duplexes. Although the double-stranded RNase activity (RNase D) is genetically separable from the RNase H activity, neither the DNA polymerase domain nor the RNase H domain expressed individually exhibit the RNase D activity (67).

iii. Active sites

(a) Polymerase active site. The polymerase active site is located at the junction of the fingers and palm domains, in a 3-D configuration almost superimposable with that of HIV-1 RTase (66). Residues that interact with the template strand include Lys-103, Arg-110, and Tyr-222, while residues that interact with the primer strand include Asp-114, Art-116, Asn-119, Gly-270, and Tyr-271. Lys-329 is also involved in template-primer binding (47). The catalytic site residues include three highly conserved Asp (-150, -224, and -225). Asp-224 and Asp-225 are coordinated to a metal ion (Mn^{2+}) which is presumed to play a key role in polymerase catalysis. Lys-103 but not Lys-421 is the essential residue involved in substrate dNTP binding of RTase (48,49).

(b) Nuclease active site. By analogy to the presumed active site residues of *E. coli* RNase H (i.e., Asp-10, Glu-48, and Asp-70) (50) and also of HIV-1 RTase/RNase H (i.e., Asp-443, Glu-478, and Asp-498) (51), the homologous conserved triplet of MoLV RNase has been identified as Asp-524, Glu-562, and Asp-583. In tertiary stuctural models, these and several other conserved residues are superposable with those of *E. coli* RNase H (45). Asp-524 and Glu-562 are thus thought to be catalytically essential residues while the Asp-583 is a structurally essential residue.

Site-specific mutations in the conserved Asp residues (e.g., D524G and D583C) in the C-terminal region of MoLV RTase depress the RNase H activity 130- and

25-fold, respectively, with little or no effect on polymerase activity (43). The D524N mutant exhibits about 10–25% of the wild-type RNase H activity.

Mutational analysis suggests that RNase H activity is genetically separable from RNase D activity (67). For example, R657S mutant RTase has a wild-type RNase H activity but <1% RNase D activity, whereas the Y586F mutant has the opposite phenotype with about 5% RNase H activity but full RNase D activity.

4. GENETICS

i. Phenotype of MoLV. MoLV is a prototype, replication-competent mammalian type C retrovirus originally isolated from a sarcoma and passaged in BALB/c mice (52). In newborn mice, the virus results in a generalized lymphocytic (mostly T cell origin) leukemia within 3–6 months after infection. In newborn rats, MoLV induces T-lymphoblastic lymphomas. Each tumor contains multiple (5 to 10) integrated MoLV copies. The integration sites are grouped into three sites: *Mlvi*-1, -2, and -3. All mice (e.g., Mov-3 substrain) undergoing the Mendelian mode of segregation with the endogenous MoLV genome activate the virus and develop viremia and leukemia through a mechanism involving a differential degree of DNA methylation (53).

The virion contains 70S genomic RNA, which is a heat-dissociable complex containing two molecules of 35S RNA and at least 15 low MW (4 to 7S) RNAs, including 4S tRNAs. Unintegrated double-stranded viral DNA and the integrated proviral DNA have been molecularly cloned and are biologically infectious. Mutations of MoLV during replication, as measured by the frequency of inactivation of the HSV thymidine kinase gene cloned in a MoLV-based vector, occur at a high rate (~8% per cycle), largely due to gene rearrangements and deletions (68).

ii. Genome organization. The MoLV genome is a (+)-sense RNA composed of 8332 nt (counting from the 5′-capped nucleotide) (37). The proviral DNA is organized as 5′-LTR-*gag-pol-env*-LTR-3′ (Fig. 6.8). The LTR (594 bp) is produced during the course of reverse transcription (54) and serves as the site of integration into host DNA (55). The LTR houses a number of upstream regulatory sequences that are responsible for the disease specificity and latency of the virus (56). The retroviral genome contains additional 5′-cap (m^7Gppp) typical of eukaryotic mRNAs and 3′ poly(A) tail (100–150 bases long).

iii. Expression of *pol* gene: polyprotein. In the MoLV genome, the *gag* and *pol* genes are in the same reading frame and are separated by an amber codon (UAG) at nucleotide position 2235. A read-through translation of the genome-sized mRNA yields, in addition to the Pr65^{gag}, a larger precursor designated Pr180$^{gag-pol}$ in an amount 4–10% of that for Pr65^{gag} (Fig. 6.8). MoLV RTase (80 kDa) is one of several mature products derived by proteolytic processing from the precursor Pr180$^{gag-pol}$. Note that the complete processing of the precursor is not required for the functional activation of the RTase (57). The N-terminal Thr of the RTase is coded by ACC which starts at position 2598, while the C-terminal

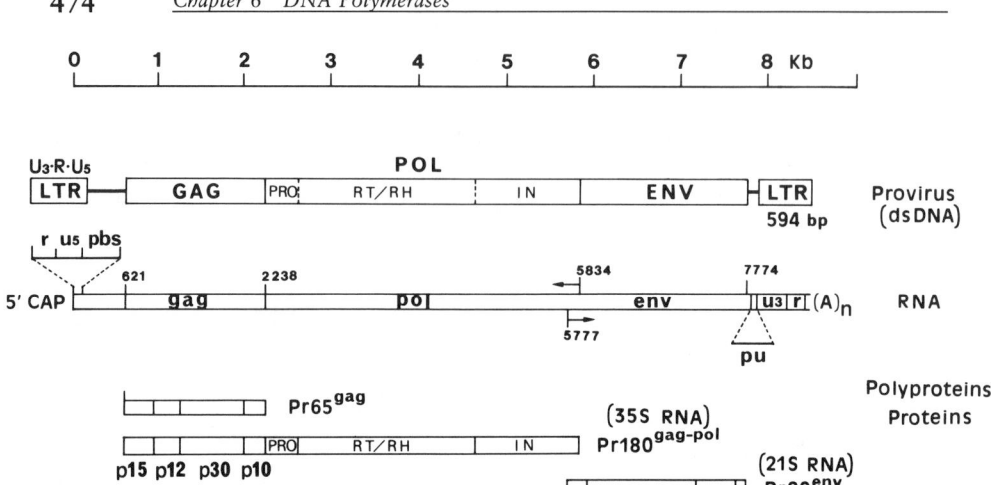

FIGURE 6.8 Structure of MoLV provirus, genomic RNA, polyproteins, and proteins. The 5' end of the genome (nt 1) is the methylated G to which 5'-cap (m^{7}Gppp) is added (37). The 3' end is presumed to be a dinucleotide CA to which the poly(A)$_{100-150}$ tail is attached. Precursor proteins Pr65^{gag} and Pr180$^{gag-pol}$ are translated from a 35S mRNA, and Pr80^{env} from a spliced 21S mRNA. The *env* gene overlaps the 3' end of the *pol* gene. Pr65^{gag}, which alone can produce noninfectious virus particles, is cleaved by the viral protease (PRO) into four mature Gag proteins: matrix (MA, p15), p12, capsid (CA, p30), and nucleocapsid (NC, p10). Pr80^{env} gives rise to the envelope protein complex consisting of receptor binding or surface protein (gp70) and transmembrane protein (p15E). A cellular protease removes the putative N-terminal leader sequence and the C-terminal 16 amino acids (R peptide). The mature products of the *pol* gene include protease (PRO), reverse transcriptase (RT) with associated RNases (RH/RD), and integrase (IN). GAG, *gag* gene and/or products; POL, *pol* gene and/or products; ENV, *env* gene and/or products; pbs, primer (tRNA^{Pro}) binding site; pu, polypurine tract; r, direct repeat sequence; u$_5$ and u$_3$, unique 5' and 3' sequences, respectively.

Leu is coded by CTC that terminates at position 4610 (38). The *pol* reading frame encodes a 1199 amino acid protein (~132 kDa). The processing products corresponding to the extra coding capacity of the *pol* gene are now known to be a viral protease [PRO] with 125 amino acid residues (M_r 13,315) and an integrase [IN], in the order of [PRO-RTase/RNase H-IN]. The protease is located at the 5' end of the *pol* gene (from nucleotides 2233 to 2597), and the first four amino acid codons overlap with the 3' end of the *gag* gene (58). The fifth residue is Gln, which is inserted by an in-frame read through of the UAG codon at the *gag–pol* junction. (*Note*: This mechanism is in contrast to the ribosomal frameshift found in the translation of AMV RTase.) The integrase is the only protein required to accomplish the integration of double-stranded proviral DNA into the host chromosome (59).

iv. RNase H$^-$ reverse transcriptase

(a) *Effect on in vivo function.* Mutant genomes encoding RTases which retain polymerase function but lack RNase H activity are replication defective (43,60). Analysis of the reverse transcription reactions carried out *in vitro* by mutant virions shows that (−) strand strong-stop DNA does not efficiently translocate to the 3' end of the genome and thus accumulate; the DNA is retained in RNA · DNA hybrid form. Plus-strand strong-stop DNA is not detected, suggesting that RNase H normally promotes strong-stop translocations, perhaps by exposing ssDNA sequences for base pairing.

(b) Effect on in vitro structure and function. The C terminus of RTase strongly influences the stability of cloned RTase. For example, when the universal terminator sequence GC(TTAA)$_3$GC is introduced at a number of positions inside and outside the RTase-coding region, the level of RTase expression varies in direct relation to the stability (from 2 to 35 min) of the protein in *E. coli* K12 (39). However, the deletion of all of the 3' end of the gene downstream from the *Eco*RV site (amino acid position 503) results in an active polymerase without RNase H activity (34). This RNase H$^-$ RTase, which has a functional profile somewhat different from that of the wild-type RTase (RNase H$^+$), has been developed into a commercial enzyme named *SuperScript* RTase (BRL).

5. Sources

The MoLV virions grown in Swiss high-passage mouse embryo cells or NIH 3T3 cells contain RTase molecules associated with the genomic RNA. The RTase is active in the detergent (0.05% NP-40)-disrupted virus particles and has been purified using various chromatographic procedures under nondenaturing conditions (24). Freezing and thawing produce spontaneous degradations of RTases inside the virion, similar to that of the RTase/RNase H complex (84 kDa, single polypeptide) from Friend MLV (61).

The RTase gene contained in the *pol* gene segment (nucleotide position 2574–4588) has been cloned into pBR322-derivative plasmids, and a high level expression of stable, soluble RTase has been attained in *E. coli* as a TrpE–RTase fusion protein (71 kDa) (1,62). A homogenous preparation of the fusion protein exhibiting both RTase and RNase H activities requires only 22-fold purification in a procedure involving multiple column chromatographic steps such as phosphocellulose and poly(C)–agarose.

One major difference between the cloned RTase and the viral RTase is the low activity of the cloned RTase on poly(A) · oligo(dT) in the presence of Mg^{2+}. The ratio of dTMP incorporation in the presence of Mn^{2+} versus Mg^{2+} for the authentic viral RTase is reported to be 3.5:1 (9), whereas the ratio for the cloned RTase is 49:1.

Modified MoLV RTases devoid of either the RNase H activity or the polymerase activity have also been produced (34,41,63).

Cloned MoLV RTase can be stored at −20°C at concentrations of 200 U/μl for over 2 years without loss of activity (11). Exposure of the enzyme to 4°C for 7 days has no apparent adverse effects on the polymerase activity of MoLV RTase.

References

1. Roth, M., Tanese, N., and Goff, S. P. (1985). *JBC* **260**, 9326–9335.
2. Gruber, C. E., Cain, C,. and D'Alessio, J. M. (1991). *FOCUS (BRL)* **13**, 88–91.
3. Okayama, H., and Berg, P. (1982). *Mol. Cell. Biol.* **2**, 161–170.

4. Gubler, U., and Hoffman, B. J. (1983). *Gene* **25**, 263–269.
5. D'Alessio, J. M., and Gerard, G. F. (1988). *NAR* **16**, 1999–2014.
6. Land, H., Grez, M., Hauser, H., Lindenmaier, W., and Schuetz, G. (1981). *NAR* **9**, 2251–2266.
7. Gerard, G. F. (1985). *FOCUS (BRL)* **7**, 1–3.
8. Gerard, G. F. (1988). *FOCUS (BRL)* **10**, 12–13.
9. Verma, I. R. (1975). *J. Virol.* **15**, 843–854.
10. Margalith, M., Falk, H., and Panet, A. (1982). *Mol. Cell. Biochem.* **43**, 97–103.
11. Gerard, G. F. (1987). *FOCUS (BRL)* **9**, 5–6.
12. Majumdar, C., Stein, C. A., Cohen, J. S., Broder, S., and Wilson, S. H. (1989). *Biochem.* **28**, 1340–1346.
13. Lavignon, M., Bertrand, J.-R., Rayner, B., Imbach, J.-L., Malvy, C., and Paoletti, C. (1989). *BBRC* **161**, 1184–1190.
14. Bloch, E., Lavignon, M., Bertrand, J.-R., Pognan, F., Morvan, F., Malvy, C., Rayner, B., Imbach, J.-L., and Paoletti, C. (1988). *Gene* **72**, 349–360.
15. Nakane, H., Fukushima, M., and Ono, K. (1990). *J. Natural Prod.* **53**, 1234–1240.
16. Erkel, G., Anke, T., Velten, R., and Steglich, W. (1991). *Z. Naturforsch. C.* **46**, 442–450.
17. Panet, A., Verman, I. M., and Baltimore, D. (1974). *CSHSQB* **39**, 919–923.
18. Gerard, G. F., D'Alessio, J. M., and Kotewicz, M. L. (1989). *FOCUS (BRL)* **11**, 66–69.
19. Baltimore, D., and Smoler, D. (1971). *PNAS* **68**, 1507–1511.
20. Williams, K. J., Loeb, L. A., and Fry, M. (1990). *JBC* **265**, 18682–18689.
21. Goodrich, D. W., and Duesberg, P. H. (1990). *PNAS* **87**, 2052–2056.
22. Luo, G. X., and Taylor, J. (1990). *J. Virol.* **64**, 4321–4328.
23. Buiser, R. G., DeStefano, J. J., Mallaber, L. M., Fay, P. J., and Bambara, R. A. (1991). *JBC* **266**, 13103–13109.
24. Peters, G., Harada, F., Dahlberg, J. E., Panet, A., Haseltine, W. A., and Baltimore, D. (1977). *J. Virol* **21**, 1031–1041.
25. Huang, P., Farquar, D., and Plunkett, W. (1990). *JBC* **265**, 11914–11918.
26. DeStefano, J. J., Buiser, R. G., Mallaber, L. M., Myers, T. W., Bambara, R. A., and Fay, P. J. (1991). *JBC* **266**, 7423–7431.
27. Gerard, G. F. (1986). *FOCUS (BRL)* **8**, 12.
28. Preston, B. D., Poiesz, B. J., and Loeb, L. A. (1988). *Science* **242**, 1168–1171.
29. Roberts, J. D., Preston, B. D., Johnston, L. A., Soni, A., Loeb, L. A., and Kunkel, T. A. (1989). *Mol. Cell. Biol.* **9**, 469–476.
30. Ricchetti, M., and Buc, H. (1990). *EMBO J.* **9**, 1583–1593.
31. Furfine, E. S., and Reardon, J. E. (1991). *Biochem.* **30**, 7041–7046.
32. Gerard, G. F. (1981). *Biochem.* **20**, 256–265.
33. Gerard, G. F. (1981). *J. Virol.* **37**, 748–754.
34. Kotewicz, M. L., Sampson, C. M., D'Alessio, J. M., and Gerard, G. F. (1988). *NAR* **16**, 265–277.
35. D'Alessio, J. M., Gruber, C. E., Cain, C., and Noon, M. C. (1990). *FOCUS (BRL)* **12**, 47–50.
36. Ross, J., Scolnick, E. M., Todaro, G. J., and Aaronson, S. A. (1971). *Nature NB* **231**, 163–167.
37. Shinnick, T. M., Lerner, R. A., and Sutcliffe, J. G. (1981). *Nature* **293**, 543–548.
38. Copeland, T. D., Gerard, G. F., Hixson, C. W., and Oroszlan, S. (1985). *Virol.* **143**, 676–679.
39. Gerard, G. F., D'Alessio, J. M., Kotewicz, M. L., and Noon, M. C. (1986). *DNA* **5**, 271–279.
40. Johnson, M. S., McClure, M. A., Feng, D.-F., Gray, J., and Doolittle, R. F. (1986). *PNAS* **83**, 7648–7652.
41. Tanese, N., and Goff, S. P. (1988). *PNAS* **85**, 1777–1781.
42. Schwartzberg, P., Colicelli, J., and Goff, S. P. (1984). *Cell* **37**, 1043–1052.
43. Repaske, R., Hartley, J. W., Kavlick, M. F., O'Neill, R. R., and Austin, J. B. (1989). *J. Virol.* **63**, 1460–1464.
44. Poch, O., Sauvaget, I., Delarue, M., and Tordo, N. (1989). *EMBO J.* **8**, 3867–3874.
45. Nakamura, H., Katayanagi, K., Morikawa, K., and Ikehara, M. (1991). *NAR* **19**, 1817–1823.
46. Telesnitsky, A., Blain, S. W., and Goff, S. P. (1992). *J. Virol.* **66**, 615–622.
47. Nanduri, V. B., and Modak, M. J. (1990). *Biochem.* **29**, 5258–5264.
48. Basu, A., Nanduri, V. B., Gerard, G. F., and Modak, M. J. (1988). *JBC* **263**, 1648–1653.

49. Basu, A., Basu, S., and Modak, M. J. (1990). *JBC* **265**, 17162–17166.
50. Kanaya, S., Kohara, A., Miura, Y., Sekiguchi, A., Iwai, S., Inoue, H., Ohtsuka, E., and Ikehara, M. (1990). *JBC* **265**, 4615–4621.
51. Mizrahi, V., Usdin, M. T., Harington, A., and Dudding, L. R. (1990). *NAR* **18**, 5359–5363.
52. Moloney, J. B. (1960). *J. Nat. Cancer Inst.* **24**, 933–951.
53. Harbers, K., Schnieke, A., Stuhlmann, H., Jaehner, D., and Jaenisch, R. (1981). *PNAS* **78**, 7609–7613.
54. Van Beveren, C., Goddard, J. G., Berns, A., and Verma, I. M. (1980). *PNAS* **77**, 3307–3311.
55. Brown, P. O., Bowerman, B., Varmus, H. E., and Bishop, J. M. (1989). *PNAS* **86**, 2525–2529.
56. Hanecak, R., Pattengale, P. K., and Fan, H. (1991). *J. Virol.* **65**, 5357–5363.
57. Crawford, S., and Goff, S. P. (1985). *J. Virol.* **53**, 899–907.
58. Yoshinaka, Y., Katoh, I., Copeland, T. D., and Oroszlan, S. (1985). *PNAS* **82**, 1618–1622.
59. Craigie, R., Fujiwara, T., and Bushman, F. (1990). *Cell* **62**, 5645–5648.
60. Blain, S. W., and Goff, S. P. (1995). *J. Virol.* **69**, 4440–4452.
61. Moelling, K. (1976). *J. Virol.* **18**, 418–425.
62. Tanese, N., Roth, M., and Goff, S. P. (1985). *PNAS* **82**, 4944–4948.
63. Mann, R., Mulligan, R. C., and Baltimore, D. (1983). *Cell* **33**, 153–159.
64. Hanvey, J. C., and 14 coauthors (1992). *Science* **258**, 1481–1485.
65. Telesnitsky, A., and Goff, S. P. (1993). *PNAS* **90**, 1276–1280.
66. Georgiadis, M. M., Jessen, S. M., Ogata, C. M., Telesnitsky, A., Goff, S. P., and Hendrickson, W. A. (1995). *Structure* **3**, 879–892.
67. Blain, S. W., and Goff, S. P. (1993). *JBC* **268**, 23585–23592.
68. Varela-Echavarria, A., Prorock, C. M., Ron, Y., and Dougherty, J. P. (1993). *J. Virol.* **67**, 6357–6364.

IV. TEMPLATE-INDEPENDENT DNA POLYMERASE: TERMINAL DEOXYNUCLEOTIDYLTRANSFERASE

[DNA nucleotidylexotransferase, EC 2.7.7.31]

Terminal deoxynucleotidyltransferase is a group of DNA polymerases which, unlike DNA- or RNA-directed polymerases, require neither DNA or RNA templates but catalyze linear condensation polymerization reactions. Terminal deoxynucleotidyltransferase is traditionally abbreviated as *TdT*. However, this book will break with the tradition and use *TDTase* to avoid any possible confusion with the dinucleotide TdT. TDTase, which was first isolated from the calf thymus gland, is found almost exclusively in the thymuses of a wide variety of mammals and birds and in certain lymphoid cells. TDTase from calf thymus is a heterodimer (44 kDa) and both subunits are derived from a single polypeptide precursor (~60 kDa).

TDTase catalyzes *in vitro* the polymerization of dNTPs onto the 3′-OH group of initiator oligo- or polynucleotides (Scheme 6.2).

$$d(pX)_m\text{-OH} + n\, dNTP \xrightarrow{Mg^{2+}} d(pX)_m(pN)_n + n\, PP_i$$
SCHEME 6.2

The apparent equilibrium constant of the reaction is 99 (at pH 7.0 and 35°C) when both initiator and nucleoside are adenosine (1). The reaction is thus practically irreversible and no PP_i exchange has been demonstrated. When carried out in the presence of single dNTPs and Co^{2+}, the TDTase reaction produces homopolymeric tails on the 3' ends of all types of dsDNA. TDTase from calf thymus, also known as the "Bollum enzyme," is a valuable tool for a variety of recombinant DNA work, including synthesis of polynucleotides, nucleotide sequencing, and gene cloning.

1. FUNCTIONS

a. Reaction conditions

i. Recommended reaction conditions. The optimum pH for TDTase is ~7.2 in potassium cacodylate, Tris–Cl, or HEPES buffer. Although reaction conditions can be further optimized for tailing with specific nucleotide substrates (2), the following reaction conditions seem to be adequate for most purposes.

A typical reaction mixture (20 μl) for the addition of a small number of nucleotides to the 3'-protruding termini of restriction DNA fragments contains 0.2 M potassium cacodylate (pH 7.2), 25 mM Tris, 2 mM $CoCl_2$ (or $MnCl_2$ for dG tailing), 0.2 mM DTT, 0.1–1.0 mM dNTP, 2 μg plasmid DNA, and 500 U/ml TDTase (3–5). For homopolymeric tailing, the mixture is usually incubated at 22° or 30°C for 1 to 60 min.

For DNA with 3'-protruding ends, a dNTP/DNA ratio of ~20 is optimal. For DNA with blunt or 3'-recessed ends, a dNTP/DNA ratio of ~100 is recommended. The reaction can be stopped by either adding EDTA (20 mM) or heating at 75°C for 5 min.

ii. Kinetic parameters. The K_m values for dNTPs with $d(pA)_4$ as the initiator are 0.1–0.5 mM and the k_{cat} is 1–14 min^{-1} (1). The K_m values for nucleotides vary significantly depending on initiator and other reaction conditions (Table 6.9). Under the "standard" assay conditions (see below), k_{cat} values range between 70 and 290 min^{-1}. The K_m values for initiator oligonucleotides range between 0.1 and 7.0 mM for the incorporation of dATP into the polymer (1). The K_m values tend to decrease with increasing length of oligomers in a homologous series, approaching a limiting value characteristic of the base at a chain length of 5–7 nt. The K_m values for initiators with fixed chain lengths are dependent on base composition (see Table 6.9).

iii. Metal ion requirements. Divalent cations are essential for TDTase activity. Although the maximal activity with a given cation varies with the nucleotides and buffer used in the reaction, Mg^{2+} at ~1 mM is the optimum. Co^{2+}, Mn^{2+}, and Zn^{2+} can replace Mg^{2+}, but the activity is substantially reduced (1,6). For dsDNAs with blunt or recessed 3' ends, TDTase is more active with $CoCl_2$ than with $MgCl_2$ (7,8).

iv. Inhibitors and inactivators

(a) Nucleotides. Nucleotides such as ATP, GTP, and 3'-deoxy-ATP (cordycepin triphosphate) act as competitive inhibitors against dNTPs. When dATP is the nucleotide substrate for TDTase, ATP inhibits the polymerization with an apparent K_i of 43 μM (1).

TABLE 6.9 Kinetic Parameters of Calf Thymus TDTase

Substrate	K_m^a (mM)	V^a (min^{-1})	K_m^b (μM)	V^d (min^{-1})
dNTPs				
dATP	0.11	14	4	73
dCTP	0.52	4.1	71	220
dGTP	0.11	11^c	114	290
dTTP	0.39	1.5	248	220
Initiator d(pN)$_5$; for N =				
dI	0.14			
dA	0.23			
dT	0.95			
dC	2.8			

^{a} Reaction measuring radioactive nucleotide incorporations into polymer in 40 mM potassium cacodylate (pH 6.8), 8 mM MgCl$_2$, and 6.6 μM d(pA)$_4$ initiator at 35°C (1). The V is converted from the original μmol/min/mg unit using 44 kDa as the MW of TDTase.

^{b} Reaction with d(pA)$_{12-18}$ initiator at 37°C under the following respective conditions (14): dATP (100 mM HEPES, pH 8.0, 0.1 mM Mn^{2+}, and 10 mM KCl), dCTP (100 mM HEPES, pH 8.0, 0.5 mM Mn^{2+}, and 30 mM KCl), dGTP (100 mM K-P, pH 6.5, 5 mM Mn^{2+}, and 50 mM KCl), and dTTP (100 mM K-P, pH 6.0, 0.5 mM Co^{2+}).

^{c} The value decreases with time.

^{d} Under the conditions for "standard" activity assay.

(b) Metal chelators. Metal chelators such as EDTA, heparin, o-phenanthroline, and cysteine inhibit TDTase by competitively interfering with the initiator binding to the enzyme.

(c) Anions. High concentrations of anions such as phosphate, sulfate, and chloride are inhibitory (9). The inhibitory effect of phosphate is a specific ion effect; the phosphate apparently interferes with the initiator binding on the enzyme (1). Pyrophosphate, which is also a product of the polymerization reaction, gives a 31% inhibition at the 1 mM concentration (1). It acts as a noncompetitive inhibitor against the initiator and a competitive inhibitor against dATP.

(d) Polymers. Synthetic polymers which resemble the initiator and other polymers such as X-irradiated ssDNA, heparin, and pyrans act as inhibitors of TDTase.

(e) Sulfhydryl reagents. PCMB and IAA inactivate TDTase. Although the enzyme does not require 2-MSH as a protective agent, the PCMB-inactivated enzyme can be reactivated by adding excess 2-MSH.

(f) Other factors. TDTase is inactivated at temperatures above 40°C and when diluted below 50 μg/ml (9). Protein denaturants such as urea, SDS, and organic solvents rapidly destroy the enzyme activity.

b. Activity assay and unit definition

A typical assay mixture consists of 0.2 M K-cacodylate (pH 7.2), 4 mM MgCl$_2$, 1 mM 2-MSH, 50 μg/ml BSA, 10 μM d(pA)$_{50}$, 1 mM [^{3}H]- or [^{14}C]dATP (10 Ci/mol), and enzyme. The assay is carried out at 37°C.

Unit definition: One unit is the amount of enzyme that catalyzes the addition of 1 nmol of dATP into acid-insoluble products in 1 hr.

c. Substrate specificities

i. Initiators. The 3'-OH termini of oligo- or polydeoxyribonucleotides serve as the initiator for polymerization. The minimum length of initiator substrates is three phosphate residues. Trinucleotides with 5'-P are suitable initiators, but trinucleoside diphosphates are not.

RNA is a poor initiator. An oligoribonucleotide, $r(A)_6$, can be made to act as an efficient initiator if one or two deoxynucleotides are added at the 3' end, $r(A)_6$-dT or $r(A)_6$-dT-dT, by the use of polynucleotide phosphorylase (10). Otherwise, an extended incubation, especially in the presence of the Co^{2+} ion, results in polymerization of dNTPs onto the 3'-OH terminus of the RNA molecule. The rate-determining step in this reaction is the addition of the first dNMP.

ii. 3'-OH termini structure

(a) Structural preference. TDTase requires single-stranded 3'-OH termini under usual reaction conditions with the Mg^{2+} ion. 3'-Phosphoryl or 3'-O-acetyl oligonucleotides are inactive as initiators. The 3'-OH of an internal nucleotide (or nick) does not function as an acceptor. Polynucleotides that have a tendency to form strong secondary structures are not good initiator substrates. Duplex DNAs with 3' overhangs are good substrates (>90% efficiency in the percentage of tailed molecules), whereas blunt or recessed 3' ends are poor substrates (~30% efficiency) (11).

(b) Effect of Co^{2+} ion. The priming efficiency for some oligonucleotide initiators if five- to sixfold higher with Co^{2+} than with the Mg^{2+} ion. The initiator specificity of TDTase is also altered so that dsDNAs with 3' blunt or recessed termini are labeled reasonably well, although they are still poor substrates compared with the 3'-protruding termini (7).

(c) Base sequence preference. The tailing efficiency at the 3' blunt ends is dependent on the base composition of the sequences adjacent (up to 6–8 bp) to the terminus. The efficiency is higher with AT-rich sequences than with GC-rich sequences (8,12).

iii. Unusual initiator substrates. In the presence of various alcohols under conditions resembling the "standard" polymerization reaction, the enzyme TDTase catalyzes a novel reaction in which phosphodiesters are synthesized and inorganic PP_i is released. With glycerol and dATP as substrates, the product has been identified as the 2,3-dihydroxypropyl ether of 2'-deoxyadenosine 5'-monophosphate (13).

iv. Nucleotide substrates

(a) dNTPs. dNTPs are the preferred nucleotide substrates in producing either homo- or heteropolymeric tails depending on whether single dNTP or all four dNTPs are supplied. The efficiency of polymerization is remarkably higher with

Co^{2+} than with Mg^{2+}: a 10- to 15-fold increase of initial rates for dAMP addition or a 50- to 60-fold increase for dTMP addition (7). Nevertheless, TDTase has been observed to have stronger affinity for the adenine base of all of the ribo-, deoxyribo-, and dideoxyribonucleoside triphosphates than for the other bases of these nucleotides (5).

(b) Ribonucleotides. Ribonucleoside triphosphates are polymerized at substantially reduced rates: 2500 times less than dATP in the presence of the Mg^{2+} ion (1) and in the order: GTP > CTP > ATP > UTP (7). The extent of the addition is also limited to dinucleotides, and the formation of mono- or dinucleotide addition products is strongly dependent on the ratio of initiator to nucleotide substrate (3,7).

When Mg^{2+} is replaced by Co^{2+}, the efficiency of initiator utilization increases significantly and multiple additions of all four ribonucleotides occur on a given initiator (7). Under such conditions, Pu nucleotides (ATP, GTP) are more efficiently incorporated than Py nucleotides (CTP, UTP) (3).

(c) Dideoxynucleotides. 2′,3′-Dideoxy-NTPs or cordycepin triphosphate produce chain-terminating single nucleotide additions. In the presence of dNTPs, ddNTPs are inhibitory due to stronger competitive binding to the enzyme: the K_i values determined with $(dA)_{12-18}$ as the initiator are 1.3 μM (ddATP, ddGTP), 2.8 μM (ddCTP), and 17.5 μM (ddTTP) (14). DNAs labeled with 3′-deoxy-AMP are completely resistant to degradation by either exonucleases or phosphatases (3).

(d) Modified nucleotides. The analogs of dTTP, such as biotin-11-dUTP, digoxigenin-11-dUTP, fluorescein-12-dUTP, and digoxigenin-11-ddUTP, can be incorporated at the 3′ termini. On oligonucleotide substrates, biotin-7-dUTP and -dCTP show consistently better incorporation than their dATP counterpart (56). Biotin-7-dUTP can be added to give primarily a single residue addition product, whereas biotin-7 (or 14)-dCTP is added in multiple numbers.

In contrast to the deoxynucleotide counterpart, TDTase incorporates preferentially and almost exclusively a single fluorescein- or biotin-riboUTP at the 3′ terminus of oligonucleotides (57). The nonisotopically labeled oligonucleotides (or primers) are suitable for use in nucleic acid hybridization, DNA sequencing, and PCR priming.

2′-Deoxy-6-thioguanosine 5′-triphosphate (S6dGTP), a metabolite of the antileukemia agent 6-thioguanine, is a relatively good substrate for TDTase. The tailing efficiency with the thioanalog is lower than with dGTP (15). In the presence of dNTPs, S6dGTP acts as a competitive inhibitor with a K_i value 6–20 times lower than the K_m values for dNTPs.

8-Azido-dATP, a photoaffinity analog of dATP, is an efficient nucleotide substrate for TDTase with a K_m of 53 μM (when the K_m for dATP is 140 μM) (16).

A novel nucleotide, dTTP-α-CH_3 [-O-P^α(=O)(CH_3)O-ribose], containing an α-phosphorus as the methyl phosphonate is a substrate (17). TDTase incorporates the nucleotide only one or two times and also incorporates the chain-terminating 3′-fluoro-ddTTP-α-CH_3 (18). The methyl phosphonate-linked oligomers are resistant to hydrolysis by Exo III and snake venom phosphodiesterase I, but are not resistant to the 3' → 5'-exonuclease activity of Pol Ik.

The following modified nucleotides are not substrates: nucleotides with a substitution on the 3'-OH of deoxyribose, halogenated dUTP or dCTP, and arabinonucleotides.

v. **Product profiles: Tail lengths.** Synthesis of poly(dA) and poly(dT) with chain lengths up to 20,000 nt proceeds reasonably well, as does the synthesis of poly(dC) with chain lengths up to 3000 nt. In contrast, the addition of dG is self-limiting to about 20–30 nt (19–21). The dG tailing is also peculiar in that the reaction is more efficient in Mn^{2+}-containing buffer than in Co^{2+} buffer (8,22). The rate of dGMP polymerization is linear for only a brief period, after which the reaction ceases (19,20). The dGMPs added to the initiator are resistant to exonuclease degradation. Poly(dG) can be obtained by an alternative route: first, prepare poly(N-Ac-dG) by TDTase-catalyzed polymerization of N-Ac-dGTP, and then deacylate the preliminary product with 10 N NH_4OH (19). In contrast to the poly(N-Ac-dG) which stays as a single chain, poly(dG) tends to exist in aggregated forms (refer to G4 DNA, Section II, in Chapter 1).

d. Mechanism of TDTase action

i. **Distribution of product chain lengths.** The polymerization reaction catalyzed by TDTase is a linear condensation between 3'-OH of the growing polymer chain and activated nucleotidyl residues. The mode of addition is "distributive" (11). Under ideal conditions where all molecules in the population have an equal and complete reactivity, the chain length in the products follows a Poisson distribution [Eq. (6.2)]

$$N_x = e^{-m}m^x/x!, \tag{6.2}$$

where N_x is the mole fraction of the degree of polymerization x, m is the average number of monomers per initiator ($m = M/I$, which corresponds to the chain length of the products at the ideally complete reaction), and x is the class of molecules with 0, 1, 2, etc., additions (9). The distribution has a standard deviation (σ) of $m^{1/2}$. Many practical cases of TDTase-catalyzed synthesis obey this rule, particularly when the initiator chain length is greater than six (4).

However, the average length tends to be at the lower end of the distribution range. The range of distribution, for example, in dC tailing substantially increases at higher temperatures: 10–90 nt (with an average of 30 nt) at 37°C compared with 10–50 nt (with an average of 20 nt) at 22°C (23). At 22°C, the distribution is further narrowed using higher cacodylate concentrations (e.g., 0.2 M compared with 0.1 M).

ii. **Determination of average chain length.** Several factors, including substrate concentrations and ratio (dNTP/DNA), incubation temperature, and buffer conditions, influence not only the reaction rate but also the tail length and its distribution (23). The progress of a tailing reaction can be most convincingly monitored by gel electrophoresis of the product DNA from timed aliquots. The resolution of the gel would be maximized if the polymerized tail parts are cleaved to give appropriate sizes (e.g., <500 bp) at the nearest possible convenient restriction site.

Alternatively, an average tail length (L) can be determined from the amount of radionucleotide incorporation according to the formula [Eq. (6.3)]

$$L = \frac{\text{TCA (cpm)} \times 325n}{c(2.2 \times 10^3)SaM}, \qquad (6.3)$$

where TCA (cpm) is the total radioactivity of the TCA-precipitated sample; 325 Da is the average mass of a dNMP (Na-salt); n is the number of nucleotides in the probe; c is the counting efficiency of the radionuclide (~100% for ^{32}P and ^{32}S and ~35% for ^{3}H); 2.2×10^3 is the conversion factor (dpm/nCi); Sa is the specific radioactivity in nCi/pmol (= Ci/mmol); and M is the mass in picograms of the oligonucleotide probe (24).

iii. **Reaction kinetics.** At saturating yet noninhibitory concentrations of Mg^{2+}, the reaction of TDTase is reduced to simple bisubstrate kinetics of the oligonucleotide initiator and nucleotide monomer. Available data with human TDTase suggest a random Bi–Bi kinetic mechanism, and the binding of one substrate does not influence the binding of the other substrate. The K_d values are 110 μM for dATP and 0.4 μM for p(dA)$_{50}$ (6). Pyrophosphate, which is a product of the reaction, acts as a noncompetitive inhibitor against the initiator and a competitive inhibitor against dATP.

2. APPLICATIONS

i. **Homopolymeric tailing.** Homopolymeric tailing by means of TDTase was the first and one of the most important techniques used in molecular cloning (see Fig. 6.6). Regarding the cloning strategies which involve homopolymeric tailing, many variations exist either in the synthesis of cDNA or in the annealing and ligation of cDNA with a cloning vector or both.

(a) *Tailing strategies.* The most critical consideration here is to determine what homopolymeric tails are to be used and when they should be employed during the cloning steps. Most commonly, the DNA fragment to be inserted is tailed in a reaction containing a single dNTP (e.g., dCTP), while a linearized cloning vector is tailed with complementary nucleotides (e.g., dGTP) (25,26). Alternatively, oligo(dT)-tailed vector-primers can be used for both cDNA synthesis and circularization either by blunt-end ligation or through mediation of an adapter (27–31). Another strategy is to synthesize cDNAs starting from an oligo(dT)-tailed restriction adapter and subsequently join the double-stranded cDNA with a tailed vector (32). The homopolymeric tail can also be attached selectively on the 3' end of a DNA chain in an RNA · DNA heteroduplex, taking advantage of the discriminating properties of TDTase against the RNA initiator (27).

The tailing reaction is relatively simple, although several factors have to be considered for optimal results. Regarding the criteria for "optimum," the transformation efficiency of bacteria gives a good estimate of the effect of tail lengths. The highest transformation efficiency or the "optimum" tail length is obtained when the complementary homopolymeric tails are approximately of equal length: ~100 nt for dA/dT tails or ~20 nt for dC/dG tails (21). Tailing reaction conditions should be optimized to attain these lengths by adjusting the incubation time, temperature, and/or enzyme concentration. The length of the tails is also strongly influenced by the ratio of nucleotide monomers to initiator DNA and follows a Poisson distribution under ideal conditions. In the presence of the Co^{2+} ion, TDTase can add dNMPs to any 3' terminus structure (blunt, recessed, or protruding), albeit at different efficiencies (7).

(b) Choice of homopolymeric nucleotides. The nucleotides used for homopolymeric tailing are selected in principle to maximize the utility of the original restriction site in the cloning vector such that it can be regenerated after the tailing. This provides a convenient means to excise the insert whenever desired. For example, the dG tailing at the 3'-protruding end of the *Pst*I site is most commonly used for many cloning vectors since it restores the *Pst*I cleavage site (CTGCA|G). Following the same line of strategy, other 3'-protruding termini can be extended by C tailings, for example, on the *Kpn*I site (GGTAC|C), *Hha*I site (GCG|C), and *Sst*I site (GAGCT|C) or G-tailing on the *Sst*II site (CCGC|GG) (3). In the case of 3' recessed termini, they are first converted to blunt ends using a DNA polymerase (e.g., Pol Ik) and dNTPs, and then extended with a suitable nucleotide, e.g., T-tailing on a *Hin*dIII site (A|AGCTT) or C-tailing on an *Eco*RI site (G|AATTC).

The multiple cloning site (MCS) usually found on cloning vectors (see Fig. 8.3, in Chapter 8) provide flexibility in inserting DNA fragments by means of a homopolymeric tailing strategy. The insert of such recombinant clones can be directionally excised using any two flanking restriction sites.

(c) Formation of chimeric DNAs and transformations. The annealing between homopolymer-tailed inserts and vectors with complementary tails produces open circular DNA molecules capable of transforming competent *E. coli* cells. The conditions for optimal annealing are ~57°C and 30 min for dA-dT tails or up to 2 hr for dC-dG tails (21). Transformation efficiency depends on the genetic background of the host strain and, for a given strain (e.g., RR1), it is markedly influenced by the length of the tail and the quality (double strandedness) of the annealing. The efficiency of transformation, by the conventional $CaCl_2$ method (33) or an improved PEG transformation method (34), is optimal (10^7–10^8 transformants per µg DNA) when the homopolymeric tails are approximately equal in length, having ~100 nt for dA-dT tails or ~20 nt for dC-dG tails.

(d) Homopolymeric tailing versus other cloning strategies. In addition to ease and convenience, homopolymeric tailing offers a few particular advantages: (i) essentially any DNA fragment can be cloned by means of homopolymeric tailing regardless of the availability of suitable restriction sites and (ii) all recombinant DNAs contain a single copy of the insert since the DNAs carrying the same

type of tail cannot hybridize to one another. On the other hand, homopolymeric tailing presents the following potential disadvantages: (i) DNA inserts carry heterogeneous homopolymeric tails at either one or both termini. Unless the tails are relatively short (e.g., 20 bp for dG·dC or <50 bp for dA·dT), they may turn out to be a structural barrier which DNA polymerases (e.g., Pol Ik) cannot or can barely overcome in dideoxy nucleotide sequencing. (ii) Under circumstances where the tail lengths are widely distributed, the long homopolymeric additional sequences may inhibit the transcription of the cloned gene *in vivo* and *in vitro*. Taken together, the cloning strategies employing homopolymeric tailing have to be weighed against alternative cloning strategies relying on linkers/adapters or vector-primers.

ii. **3′-End labeling.** TDTase is extremely useful for labeling DNAs at the 3′ ends either with radionucleotides (^{32}P, ^{35}S) or nonradioactive nucleotides such as rNTP (7), ddNTP (35), and 3′-deoxy-ATP (36). The enzyme efficiently incorporates nonisotopic labeling nucleotides as well. The 3′ end-labeled DNAs are suitable substrates for chemical nucleotide sequencing (37), nearest-neighbor analysis (20), and molecular hybridization for various screening and/or detection purposes.

(a) *Molecular hybridization probes.* The 3′-end labeling is particularly useful for hybridization probes because the addition of one or two specific ribonucleotides complementary to a given template increases the hybridization efficiency of short oligonucleotides, thereby rendering the oligonucleotide probes significantly more sensitive. The 3′-end tailing usually has no undesirable effect on the specificity of oligonucleotide probes. Hybridization with oligonucleotide probes labeled at their 3′ ends with [^{32}P]dAMP residues can, in some cases, be as much as 30 times more sensitive than that using 5′ end-labeled probes (24).

(b) *Special applications for nucleotide sequencing.* For nucleotide sequencing by chemical methods, single nucleotide labeling of DNA with ddNTP or 3′-deoxy-ATP is usually preferred to the labeling with rNTPs. Ribonucleotide labeling often gives rise to heterogeneously labeled DNAs, although the DNA products labeled with multiple ribonucleotides can be converted to homogeneous monoaddition products by treatment with alkali (0.3 N KOH) followed by alkaline phosphatase. In fact, rNTP labeling provides some unique advantages in the determination of short terminal sequences by nearest-neighbor analysis. The insertion of rNMP at the 3′ end of an oligonucleotide can be extended by *E. coli* DNA Pol I when the ribonucleotide is base paired to a DNA template. The ribonucleotide provides an alkali-sensitive linkage from which the initiator and product sequences can be separated and analyzed (3,38).

(c) *Artifact resolution in nucleotide sequencing.* TDTase provides a unique method for resolving sequence ambiguities which are often encountered in dideoxynucleotide sequencing of RNA (with RTases) or DNA: the appearance of radioactive bands of cDNA that extend over all four lanes on a sequencing gel.

Such ambiguity arises from the premature termination of polymerization usually due to certain secondary structures. Elegant methods that can resolve such an ambiguity rely on TDTase which, when introduced in the dNTP chase step, would continuously add dNMPs to the spurious termini caused by reasons other than the insertion of a ddNMP (39,58).

iii. **Misincorporation mutagenesis.** Because TDTase catalyzes a linear polymerization of dNTP without template direction, a mismatching nucleotide(s) can be added to the 3′ termini of DNA, especially in the presence of the Co^{2+} ion. The misincorporation can be made fairly site specific, even to a single nucleotide level, when 3′-OH termini are exposed by controlled digestion with either Exo III or a restriction enzyme (22). The "mutagenized" DNA is then repaired *in vitro* using DNA polymerases (e.g., AMV RTase) lacking 3′-exonuclease activity and is used for transformation and amplification. Alternatively, a thionucleotide (dNTPαS) can be used as the mutagenic nucleotide, which not only renders the new 3′ termini exonuclease resistant but also enables one to purify and enrich the sulfur-labeled DNA from unlabeled DNA using a mercury glass bead affinity column.

3. STRUCTURE

i. **General features.** Calf thymus TDTase is a small (44 kDa) heterodimeric ($\alpha\beta$) enzyme (33-kDa β subunit and 11-kDa α subunit). It is a basic protein with a p*I* of 8.6 (40). The two subunits are derived by specific proteolytic cleavages from a 60-kDa polypeptide precursor (41). The precursor is enzymatically active. Note that the TDTase was originally purified from calf thymus as a catalytically active complex of 24- and 8-kDa subunits (40), most likely the products of proteolytic degradation during purification. [Human TDTase (42) and mouse TDTase (43) are single polypeptides (~60 kDa) with deduced amino acid sequences of 508 and 529 residues, respectively.]

ii. **Primary structure and homology.** The primary structure deduced from the ORF of a cloned TDTase gene contains 520 amino acids (M_r 59,678) (43). The β subunit has 289 amino acid residues (M_r 32,857) with N- and C-terminal residues of Ser and Asn (44). The α subunit contains 89 amino acid residues (M_r 10,721) with Gln-290 and Ala-378 as N- and C-terminal residues, respectively. In the precursor form, the β subunit is preceded by a presequence of 142 amino acids which contains a nuclear localization signal, i.e., Pro^{22}–Arg–Lys–Lys–Arg–Pro–Arg^{28} (43,45). The N-terminal regions of human and mouse TDTases also carry the signals, consistent with the fact that TDTase is localized in the nucleus.

Calf thymus TDTase is immunologically cross-reactive with various mammalian (e.g., human, rat, mouse, and chicken) TDTases, suggesting that the polypeptide structure of TDTase is highly conserved among species (42,46). Indeed the amino acid sequence of the calf thymus enzyme shares over 90% identities with those of human TDTase (45) and mouse TDTase (43).

iii. **Active site.** Cross-linking studies with photoaffinity-labeled (5-azido-dUMP) DNA (47) or with [α-^{32}P]dTTP (48) suggest that the substrate-binding domain contains two tryptic peptides with the following sequences: Asp^{221}–Lys^{231} and Cys^{234}–Lys^{249}.

4. GENETICS

i. **Gene structure and organization.** A full-length cDNA coding for the calf TDTase has been isolated and sequenced (43). It has an ORF of 1560 bp. The sequence upstream from the AUG initiator codon of the cDNA clone has only 22 nt, whereas the sequence downstream from the UAG termination codon comprises 332 nt which include the poly(A) signal (AAUAAA). A single species of TDTase mRNA has been shown to be 2100–2200 nt long.

Compared with the ORF for calf TDTase, the ORF for mouse TDTase contains an additional 60 bp in the 3' end region. Indeed similar insertion sequences occur in calf TDTase as well, suggesting that TDTase isoforms apparently arising from alternative splicing of mRNA may exist in bovine thymus (59). Overall, the gene for calf TDTase shares ~80% sequence identity with the genes for human and mouse TDTases.

ii. *In vivo* **function of TDTase.** TDTase activity is uniquely found in the cortical lymphocytes of thymus and not in any other organs or nonlymphoid tissues (42,51). TDTase is expressed in human bone marrow T cell precursors and possible lymphoid stem cells, serving as the earliest available marker for lymphoid differentiation (49). An unusually high level expression of TDTase accompanies the proliferation of certain lymphoblastic leukemia, providing a useful marker for neoplastic cells in certain human diseases (50). In mice, TDTase has been shown to be essential for adding template-independent (or N region) nucleotides during the rearrangement of variable (V), diversity (D), and joining (J) segments of Ig and T cell receptor variable region genes (52,53). TDTase thus plays, as a tissue-specific component of V(D)J recombinase, a role of expanding the diversity of variable regions beyond the germ-line-encoded repertoire. TDTase is not required for lymphocyte differentiation nor for cell viability.

5. SOURCES

The usual source of TDTase is the calf thymus gland. It contains 5–10 mg of TDTase per kg of fresh tissue. The enzyme is also found at high levels (100–200 U/10^8 cells) in certain leukemic lymphocytes (e.g., human leukemia Molt-4 and mouse thymoma P1798) and at a detectable level in bone marrow. TDTase activity is not found in prokaryotes nor in unicellular eukaryotes.

The TDTase activity in calf thymus is localized in both cytoplasm and nucleus (1). In immature thymocytes and neoplastic cells, TDTase is observed almost exclusively in the nucleus (54). The structural gene for calf TDTase has been cloned and efficiently expressed from a plasmid expression vector in COS7 (monkey fibroblast) cells (43).

TDTase can be purified by means of various conventional chromatographies (40), oligo(dT)$_7$-cellulose affinity chromatography (6), and an anti-TDTase IgG-Sepharose immunoadsorbent column chromatography (46,55). TDTase is quite stable at pH 4.5, thus providing one of the most useful steps in the purification.

Purified TDTase has a specific activity of 100–400 μmol dNTP polymerized per hour per mg protein.

References

1. Kato, K., Goncalves, J. M., Houts, G. E., and Bollum, F. J. (1967). *JBC* **242**, 2780–2789.
2. Nelson, T., and Brutlag, D. (1979). *Meth. Enzy.* **68**, 41–50.
3. Roychoudhury, R. (1981). In "Gene Amplification and Analysis" (J. G. Chirikjian and T. S. Papas, Eds.), Vol. 2, pp. 41–83. Elsevier, New York.
4. Bollum, F. J. (1978). *Adv. Enzy.* **47**, 347–374.
5. Eschenfeldt, W. H., Puskas, R. S., and Berger, S. L. (1987). *Meth. Enzy.* **152**, 337–342.
6. Coleman, M. S., and Deibel, M. R., Jr. (1983). In "Enzymes of Nucleic Acid Synthesis and Modification" (S. T. Jacob, Ed.), Vol. I, pp. 93–118. CRC Press, Florida.
7. Roychoudhury, R., Jay, E., and Wu, R. (1976). *NAR* **3**, 863–877.
8. Deng, G., and Wu, R. (1981). *NAR* **9**, 4173–4188.
9. Bollum, F. J. (1974). *Enzymes* **10**, 145–171.
10. Feix, G. (1972). *BBRC* **46**, 2141–2147.
11. Michelson, A. M., and Orkin, S. H. (1982). *JBC* **257**, 14773–14782.
12. Roychoudhury, R., Tu, C.-P. D., and Wu, R. (1979). *NAR* **6**, 1323–1333.
13. Rybalkin, I. N., Chidzhavadze, Z. G., Florent'ev, V. L., and Bibilashvili, R. Sh. (1991). *Mol. Biol. (USSR)* **25**, 223–230.
14. Ono, K. (1990). *BBA* **1049**, 15–20.
15. Ling, Y. H., Nelson, J. A., Cheng, Y. C., Anderson, R. S., and Beattie, K. L. (1991). *Mol. Pharmacol.* **40**, 508–514.
16. Evans, R. K., and Coleman, M. S. (1989). *Biochem.* 707–712.
17. Higuchi, H., Endo, T., and Kaji, A. (1990). *Biochem.* **29**, 8747–8753.
18. Victorova, L. S., Dyatkina, N. B., Mozzherin, D., Atrazhev, A. M., Krayevsky, A. A., and Kukhanova, M. K. (1992). *NAR* **20**, 783–789.
19. Lefler, C. F., and Bollum, F. J. (1969). *JBC* **244**, 594–601.
20. Roychoudhury, R., and Wu, R. (1980). *Meth. Enzy.* **64**, 43–62.
21. Peacock, S. L, McIver, C. M., and Monahan, J. J. (1981). *BBA* **655**, 243–250.
22. Deng, G., and Wu, R. (1983). *Meth. Enzy.* **100**, 96–116.
23. Whitted, B. E., and Chan, S. Y. (1985). *BioTechniques* **3**, 118–122.
24. Collins, M. L., and Hunsaker, W. R. (1985). *Anal. Biochem.* **151**, 211–224.
25. Gubler, U., and Hoffman, B. J. (1983). *Gene* **25**, 263–269.
26. Land, H., Grez, M., Hauser, H., Lindenmaier, W., and Schutz, G. (1983). *Meth. Enzy.* **100**, 285–292.
27. Okayama, H., and Berg, P. (1982). *Mol. Cell. Biol.* **2**, 161–170.
28. Heidecker, G., and Messing, J. (1987). *Meth. Enzy.* **154**, 28–41.
29. Falkenthal, S., Parker, V. P., Mattox, W. W., and Davidson, N. (1984). *Mol. Cell. Biol.* **4**, 956–965.
30. Krawetz, S. A., Connor, W., Cannon, P. D., and Dixon, G. H. (1986). *DNA* **5**, 427–435.
31. Deininger, P. L. (1987). *Meth. Enzy.* **152**, 371–389.
32. Coleclough, C. (1987). *Meth. Enzy.* **154**, 64–83.
33. Hanahan, D. (1983). *JMB* **166**, 557–580.
34. Chung, C. T., and Miller, R. H. (1988). *NAR* **16**, 3580.
35. Olson, K., and Harvey, C. (1975). *NAR* **2**, 319–325.
36. Tu, C.-P. D., and Cohen, S. N. (1980). *Gene* **10**, 177–183.

37. Maxam, A. M., and Gilbert, W. (1977). *PNAS* **74**, 560–564.
38. Sekiya, T., van Ormondt, H., and Khorana, H. G. (1975). *JBC* **250**, 1087–1098.
39. De Borde, D. C., Naeve, C. W., Herlocher, M. L., and Maassab, H. F. (1986). *Anal. Biochem.* **157**, 275–282.
40. Chang, L. M. S., and Bollum, F. J. (1971). *JBC* **246**, 909–916.
41. Chang, L. M. S., Plevani, P., and Bollum, F. J. (1982). *JBC* **257**, 5700–5706.
42. Chang, L. M. S., and Bollum, F. J. (1986). *CRC Crit. Rev. Biochem.* **21**, 27–52.
43. Koiwai, O., Yokota, T., Kageyama, T., Hirose, T., Yoshida, S., and Arai, K. (1986). *NAR* **14**, 5777–5792.
44. Evans, R. K., Beach, C. M., and Coleman, M. S. (1989). *Biochem.* **28**, 713–720.
45. Peterson, R. C., Cheung, L. C., Mattaliano, R. J., White, S. T., Chang, L. M. S., and Bollum, F. J. (1985). *JBC* **260**, 10495–10502.
46. Bollum, F. J., and Chang, L. M. S. (1981). *JBC* **256**, 8767–8770.
47. Farrar, Y. J. K., Evans, R. K., Beach, C. M., and Coleman, M. S. (1991). *Biochem.* **30**, 3075–3082.
48. Pandey, V., and Modak, M. J. (1988). *JBC* **263**, 3744–3751.
49. Gore, S. D., Kastan, M. B., and Civin, C. I. (1991). *Blood* **77**, 1681–1690.
50. Almasri, N. M., Iturraspe, J. A., Benson, N. A., Chen, M. G., and Braylan, R. C. (1991). *Am. J. Clin. Pathol.* **95**, 376–380.
51. Alt, F. W., Blackwell, T. K., DePinho, R. A., Reth, M. G., and Yancopoulos, G. D. (1986). *Immunol. Rev.* **89**, 5–30.
52. Komori, T., Okada, A., Stewart, V., and Alt, F. W. (1993). *Science* **261**, 1171–1175.
53. Gilfillan, S., Dierich, A., Lemeur, M., Benoist, C., and Mathis, D. (1993). *Science* **261**, 1175–1178.
54. Janossy, G., Thomas, J. A., Bollum, F. J., Grangers, S., Pizzolo, G., Bradstock, K. F., Wong, L., McMichael, A., Ganeshaguru, K., and Hoffbrand, A. V. (1980). *J. Immunol.* **125**, 202–212.
55. Nakamura, H., Tanabe, K., Yoshida, S., and Morita, T. (1981). *JBC* **256**, 8745–8751.
56. Flickinger, J. L., Gebeyehu, G., Buchman, G., Haces, A., and Rashtchian, A. (1992). *NAR* **20**, 2382.
57. Igloi, G. L., and Schiefermayr, E. (1993). *BioTechniques* **15**, 486–494.
58. Fawcett, T. W., and Bartlett, S. G. (1990). *BioTechniques* **9**, 46–48.
59. Takahara, K., Hayashi, N., Fujita-Sagawa, K., Morishita, T., Hashimoto, Y., and Noda, A. (1994). *Biosci. Biotech. Biochem.* **58**, 786–787.

7

RNA Polymerases

I. DNA-DIRECTED RNA POLYMERASES

A. General Description

1. Transcriptional Machinery

DNA-directed RNA polymerases are the key enzymes in the transfer of genetic information from DNA to RNA. During this process of information transfer, called *transcription*, an RNA chain (messenger, transfer, or ribosomal) is copied by RNA polymerase from a dsDNA template. RNA polymerases elongate RNA chains by forming 3'-5' phosphodiester linkages using ribonucleoside 5'-triphosphates (rNTPs or NTPs) as substrates and releasing pyrophosphate as a reaction product. The mRNA transcripts are subsequently translated into proteins in the ribosome, a process called *translation* (see Section I, Chapter 1). As the primary step in the process of gene expression, transcription is subject to various regulatory mechanisms. The correct balance of regulatory mechanisms leads to normal cellular functions, differentiation, and development, whereas an imbalance is responsible for disarrayed cellular functions, abnormal development, and/or oncogenesis.

Expression of molecularly cloned genes in bacterial or eukaryotic cells is governed by the same regulatory mechanisms. Understanding the mechanism of transcription and its regulation is therefore a major issue of molecular biology and has tremendous impact on all areas relating to natural or artificial gene expressions. Indeed, the controlled expression of a gene is no more a distant dream but is a reality, thanks to remarkable advances in our understanding of the complex processes involved in transcription initiation, elongation, and termination. In this section, a brief account of these processes will be presented with particular emphasis on the fundamental aspects of DNA elements and protein factors involved in the initiation of transcription.

a. Basic transcriptional machinery

The basic transcriptional machinery that operates in certain bacteriophages consists of only two components: the monomeric RNA polymerase (Pol) and the DNA template. The polymerase carries out *basal* transcription, "basal" implying that no accessory proteins (or factors) are necessary. In prokaryotes, the transcriptional machinery requires auxiliary molecules generally known as *transcription factors* or *activators* in addition to the RNA polymerase which is composed of six subunits. The transcriptional machinery of eukaryotes is even more complex than that of prokaryotes in both the number of transcription factors involved and the number of subunits (10–13 s.u.) composing the RNA polymerase.

In the most fundamental scheme of transcription, which is believed to be common to all transcriptional machineries, RNA Pol binds to DNA at the start of a gene (*initiation*), moves along in a single direction transcribing RNA from DNA (*elongation*), and stops the transcription at or beyond the end of the gene, dissociating from the DNA template (*termination*). A DNA sequence signaling a transcription start is called a *promoter*. In phages and bacteria, a promoter consists of a stretch of approximately 20 and 60 nt, respectively. A promoter is the basal element to which RNA polymerase binds and, either alone or in combination with transcription factors, initiates transcription.

In physiological settings, transcription is under the complex control of *regulatory proteins*. Of all the phases of transcription, however, the initiation of transcription is the prime target of regualtion, and the regulatory factors involved in the initiation are called *transcription factors*. Some transcription factors (*activators*) exert positive effects (increased transcription), some factors (*repressors*) exert negative effects (repressed or no transcription), and some transcription factors exert either positive or negative effects depending on the physiological context.

A promoter may consist of a single continuous block of sequences (as in phage DNAs) or several *regulatory elements* (as in bacterial and eukaryotic DNAs). Regulatory elements are the sites at which regulatory proteins interact. Depending on the nature of their interactions, transcription may be activated or repressed.

Analyses of promoter sequences suggest that the basis of interaction with RNA polymerases does not reside solely in the primary structure but may also involve certain secondary and/or higher ordered structures. A number of promoters are

AT- or GC-rich sequences, which are known to have propensities for assuming unusual conformations including kinks, Z-DNA, cruciforms, triple helices, or supercoils (see Section II, in Chapter 1). The inherent curvature, as well as protein-induced bending of DNA, is believe to be part of the driving force in the initiation of transcription.

Our knowledge on promoters, transcription factors, and gene regulation in general has advanced at a remarkable pace largely because of the application of recombinant DNA technology. One notable experimental approach has been the *in vitro* fusion of DNA fragments containing a specific transcriptional regulatory element(s) to a marker/reporter gene allowing rapid assay or selection (see Chapter 8). Gene cloning and deletional or mutational analyses have further provided the opportunities to dissect the structure and function of regulatory elements as well as transcription factors at the nucleotide level in controlled environments.

b. Terms used to describe functional aspects of transcription

When a regulatory element controls a gene(s) with which it is contiguous, it is called *cis-acting*. Cis-acting elements usually show *polarity,* in the sense that the regulatory effect is stronger on the gene(s) located closer to the regulatory element. Promoters are good examples of cis-acting elements. In contrast, *trans-acting* elements exert their regulatory effects on a gene(s) regardless of its location. Repressor proteins are typical examples of trans-acting regulators.

The transcription initiation site (or the site corresponding to the first nucleotide of an RNA transcript) is called the *start site* or *start point*. It is conventionally designated as +1. The sequence of DNA that signals the RNA polymerase to end transcription is called a *terminator*. The length of DNA that extends from the promoter to the terminator is defined as a *transcription unit*. A transcription unit gives rise to a single RNA (or primary transcript) which may contain only one gene (*monocistronic*) or several genes (*polycistronic*). The primary transcript is almost always unstable and undergoes rapid modification, degradation, and/or processing to smaller, mature units. When cryptic termination sites are utilized under certain physiological conditions, multiple transcripts of different lengths may result from a single promoter.

The single-strand DNA which serves as the natural *template* for transcription is complementary to the transcribed RNA. When the transcript (RNA) is coding for or translatable into a protein or proteins, the transcript is called an mRNA which is also qualified as *plus* (+) or *positive sense* RNA. The DNA strand originally used as the template for RNA polymerase is called *noncoding, antisense,* or *minus* (−) strand, in analogy to an RNA strand whose sequence cannot be directly converted to amino acid codons.

Sequences prior to the start point are described as *upstream* of it, and the base positions are marked by negative numbering. Sequences after the start point are called *downstream* of it and the base positions are given positive numbers. Regions close to the promoter are described as *proximal;* those remote from the promoter are described as *distal*.

2. RNA Polymerases: General Properties
a. Prokaryotic RNA polymerases

In bacteria (*E. coli*), one RNA polymerase transcribes almost all bacterial genes, albeit at different levels and different occasions. Part of these controls (environment-responsive, developmental) are ascribed to different σ *factors* that recognize distinct promoter sequences and direct the *core enzyme* (RNA Pol without σ factor) to initiate more favorably at those particular promoters. RNA Pol makes contacts with the promoter DNA at the -35 and -16 regions (on the same side at two full helix turns) as well as the areas extending through the -10 region (1).

E. coli RNA Pol is probably the best characterized among all known prokaryotic RNA polymerases. An *E. coli* cell contains ~3000 RNA polymerase molecules. The holoenzyme (~480 kDa) consists of five types of subunits (β, β', α, ω, and σ) which can be separated into two components: the core enzyme ($\alpha_2\beta\beta'\omega$) and the σ factor (σ protein). The core enzyme is catalytically competent, but binds nonspecifically to DNA. The σ factor, which alone does not bind DNA, confers the promoter-binding specificity to the core enzyme. The elongation of the chain is about 40 nucleotides/sec at 37°C or 13–20 nucleotides/sec at 20°C (2).

The frequency of misincorporation is estimated to be $\sim 10^{-5}$. Unlike DNA Pol I, however, RNA Pol lacks a $3' \to 5'$-proofreading exonuclease activity. The accuracy is assured in part by the intrinsic property of the RNA Pol which, on incorporation of a mismatched nucleotide, slows down the addition of new NMP and enters a kinetically "unactivated" state (3). The transcription "cleavage" factor GreA also plays a role in the fidelity of transcription by preferentially cleaving the transcripts which contain misincorporated residues.

Although the exact functions of the subunits need to be further elucidated, the structural and functional roles assignable to individual subunits may be summarized as follows.

The β *subunit*, a 150-kDa protein encoded by *rpoB*, has binding sites for NTPs and the Zn^{2+} ion and carries out phosphodiester bond formation. It is the target of two antibiotics: rifampicin which blocks transcription initiation just after the formation of the initial phosphodiester bonds, and streptolydigin which inhibits phosphodiester bond formation noncompetitively during any stage of initiation or elongation.

The β' *subunit*, a 155-kDa protein encoded by *rpoC*, makes contacts with σ and β subunits and is responsible for binding DNA as well. Heparin, a polyanion that competes with DNA for binding to the polymerase, binds to the β' subunit and inhibits transcription *in vitro*.

The α *subunit*, the gene product (329 amino acids, M_r 36,512) of *rpoA*, is required for the assembly of the core enzyme and plays a role in promoter recognition. When phage T4 infects *E. coli*, an Arg residue of the α subunit is modified by ADP-ribosylation which results in a reduction of the affinity of the promoter to the holoenzyme. The carboxyl-terminal domain (CTD, 99 amino acids) is regarded as the contact site for transcription activators, e.g.,

CRP (cAMP receptor protein), and for the "upstream" promoter elements (4,17). The N-terminal region (up to residue 235) is involved in the assembly of the active enzyme molecule.

The function of the ω subunit ($M_r \approx 10,000$) and its stoichiometry in the holoenzyme complex have not been fully determined.

The σ factor confers on the core RNA Pol a binding specificity to σ-dependent promoters. It destabilizes nonspecific binding of the holoenzyme and increases specific binding to the promoter DNA by a factor of $\sim 10^4$. The σ factor loses contact with the 3' end of the nascent RNA after synthesis of a tetranucleotide (5) and is released from the core enzyme when RNA synthesis enters an elongation phase after the incorporation of 8–12 bases. Four different σ subunits are encoded by the *E. coli* genome, each acting in combination with the core enzyme to recognize a specific subset of promoter sequences (6). The σ^{70} (613 amino acids, the *rpoD* gene product) is the predominant σ factor found in *E. coli*. It is directly involved in the recognition of two promoter regions: the −35 element by one (region 4.2) of its C-terminal conserved regions and the −10 element by another (region 2.4) of its conserved regions (7). The other less abundant σ subunits promote transcriptions of different subsets of genes that are coordinately regulated for growth in unusual conditions: σ^{32} for the heat-shock response genes, σ^{54} for the genes involved in nitrogen metabolism, and σ^{28} most likely for the genes involved in motility and chemotaxis (8). In contrast to the σ^{70} factor, the σ^{54} factor recognizes a different set of basal elements located at −12 and −24, and interacts with distant activators in a manner reminiscent of the transcriptional regulatory systems of eukaryotic RNA polymerase II promoters (9,10).

b. Eukaryotic RNA polymerase II

In contrast to prokaryotes, all eukaryotic cells contain three forms of RNA Pol which perform specialized tasks: RNA Pol I (or RNA Pol A) for rRNA synthesis, RNA Pol II (or RNA Pol B) for mRNA synthesis, and RNA Pol III (or RNA Pol C) for synthesis of tRNA, 5S RNA, and small nuclear RNAs. Despite similarities in size (500–600 kDa) and multimeric structure (9–14 polypeptides), the three polymerases have characteristic differences at the subunit level and in the set of associated transcription factors (11). The RNA Pol II transcription system, which is the main subject of this section, is the best characterized of the three polymerase transcription systems.

Like mammalian RNA Pol II, the RNA Pol II of yeast is a multimeric complex composed of two large subunits, 220 kDa (B220) and 150 kDa (B150), and 8–11 small subunits with sizes ranging from 10 to 50 kDa (12). Cloning and sequencing of the structural genes for 10 subunits of yeast RNA Pol have revealed some intriguing similarities and differences between the eukaryotic RNA polymerase and its simpler prokaryotic counterpart (13,14). Most viral RNA polymerases of DNA viruses (e.g., vaccinia virus) also have complex subunit structures (15).

Although the exact functions of each component subunit are as yet poorly defined, the second largest subunit (B150, the *rpb2* gene product) of yeast RNA Pol II contains sequences implicated in binding purine nucleotides and a Zn ion.

In addition to this functional similarity to the *E. coli* RNA Pol β subunit, B150 exhibits highly significant sequence similarities as well (16). The largest subunit (B220, the *rpb*I gene product of yeast), which is homologous to the β' of *E. coli* RNA Pol, has at the C terminus tandem copies of a conserved heptapeptide sequence (Ser–Pro–Thr–Ser–Pro–Ser–Tyr); the number of copies varies from 26 in yeast to 52 in mice. At least 13 of the heptad repeats are essential for full polymerase activity, for the response to transcriptional enhancer signals, and also for the viability of the yeast cell (18). Evidence suggests that the heptad repeat can directly bind DNA by intercalation (19). The C-terminal domain, which contains an acidic subdomain as well as (un)phosphorylated heptad repeats, binds basal transcription factors TFIID (TBP) and TFIIB.

In higher eukaryotes, the basal transcriptional machinery requires at least six distinct "general" transcription factors (TFII A, B, D, E, F, and H) which form a *preinitiation complex* in concert with RNA Pol II. The basal transcription is regulated by interaction with various "specific" transcription factors.

c. Phage RNA polymerases

Although *E. coli* RNA Pol has been extensively studied, its multimeric structure and complex interaction with a variety of host factors make the enzyme less practical as a tool in molecular biology. Compared with the large, multimeric *E. coli* RNA Pol and the even larger eukaryotic RNA Pol II, phage RNA polymerases (e.g., T3, T7, and SP6) are much smaller ($\leq$100 kDa) single polypeptides. The phage RNA polymerases are able to carry out transcription in the absence of any additional protein factors, thus representing the minimum apparatus necessary for transcription. Despite their similarities in size and amino acid sequence, phage RNA polymerases demonstrate remarkable differences in promoter specificity (21). Furthermore, the promoter-specific transcription is highly efficient, synthesizing RNA transcripts at a rate of ~200 nt/sec at 37°C, which is nearly five times faster than that of *E. coli* RNA Pol. The initiation reaction differs very little among the phage RNA polymerases. Indeed the simplicity, specificity, and efficiency of the phage RNA polymerases provide the basis for the use of *portable* (or *artificial*) *promoters* in a wide variety of hybrid gene expression and *in vitro* transcriptions. In Sections B, C, and D of this chapter, phage (T7, SP6, and T3) RNA polymerases and their functions are described in greater detail.

3. TRANSCRIPTION INITIATION

The promoter regions of both prokaryotic and eukaryotic genes often contain widely dispersed regulatory sequences (or elements) that interact, either directly or indirectly, with transcription factors and RNA polymerase and consequently regulate (activate or repress) the initiation of transcription. Although the eukaryotic transcription systems are substantially more complex, and therefore less well defined, than the prokaryotic transcription systems, the basic mechanism of transcription initiation is believed to be similar between the two systems.

The initiation of transcription can be described as a four-step process: (i) initial binding of RNA Pol to the promoter, forming a "closed" complex; (ii) polymerase isomerization and DNA melting ("open" complex); (iii) intermediate or abortive RNA production (initiating complex); and (iv) clearance of the promoter (elongation complex).

What are the features of the promoters and regulatory sequences? What are the structural and functional characteristics of transcription factors? How do the protein factors interact with DNA elements?

a. Features of prokaryotic regulatory elements

i. Prokaryotic promoters. A prokaryotic promoter consists of several short sequences around the transcription start site (Fig. 7.1a). Comparison of a large number of available promoter sequences has revealed a few *consensus sequences* in proximal promoter regions (nucleotide positions −65 to +20). The consensus promoter sequence may vary depending on organisms or subsets of genes within an organism. In *E. coli*, the promoters for the RNA Pol containing the predominant σ^{70} subunit have the following features (8,10,22).

(a) The start point is usually a Pu, which is most commonly (>90%) the A positioned as the central base in the (C/T)A(T/C) sequence. RNA Pol covers ~44 bases upstream from the start site along the antisense strand. The start site selection by RNA Pol is strongly influenced by any change of the sequence in a rather broad range around the initiation site.

(b) Just ~10 bp (varies from −18 to −9) upstream of the start point is a 6-bp consensus sequence *TATAAT*. This consensus hexamer (CH) is also known as the *−10 sequence, CH-10,* or *Pribnow box*. The −10 sequence provides the initial unwinding (or local melting) of the dsDNA, converting the initially inactive, "closed" RNA Pol–DNA complex to the active, "open" initiation complex. In the presence of Mg^{2+}, the unwound or single-stranded region in the open complex spans at least 14 bases, extending from the middle of the Pribnow box to the start point (+1,+2).

(c) A second element of *E. coli* promoters consists of a highly conserved consensus hexamer, *TTGACA*, also called the *−35 sequence* or *CH-35* because it is located ~35 bp upstream from the start site. The −35 sequence is recognized by a helix–turn–helix motif located in the carboxy-terminal region 4 of σ^{70}. The sequences of CH-35 may often lack significant resemblance to the consensus sequence, implying that CH-35 has a weak intrinsic binding activity and may even be dispensable. In the context of an "extended −10 promoter" featuring TGNTATAAT, CH-35 is nonessential for initiation of transcription.

(d) No base in any consensus hexamer is strictly conserved, but wild-type promoters usually carry at least half of the consensus bases in the appropriate positions. Most of the mutations in promoters result in decreased transcriptions, thus called *down mutations*. Some mutations result in increased transcriptions and are called *up mutations*. The "up" mutations usually occur at positions which increase the homology to the consensus hexamer.

(e) The distance between CH-10 and CH-35 is important and in fact serves as part of the signal; 17 ± 1 bp (spacer) is optimal. The spacer is AT or GC rich,

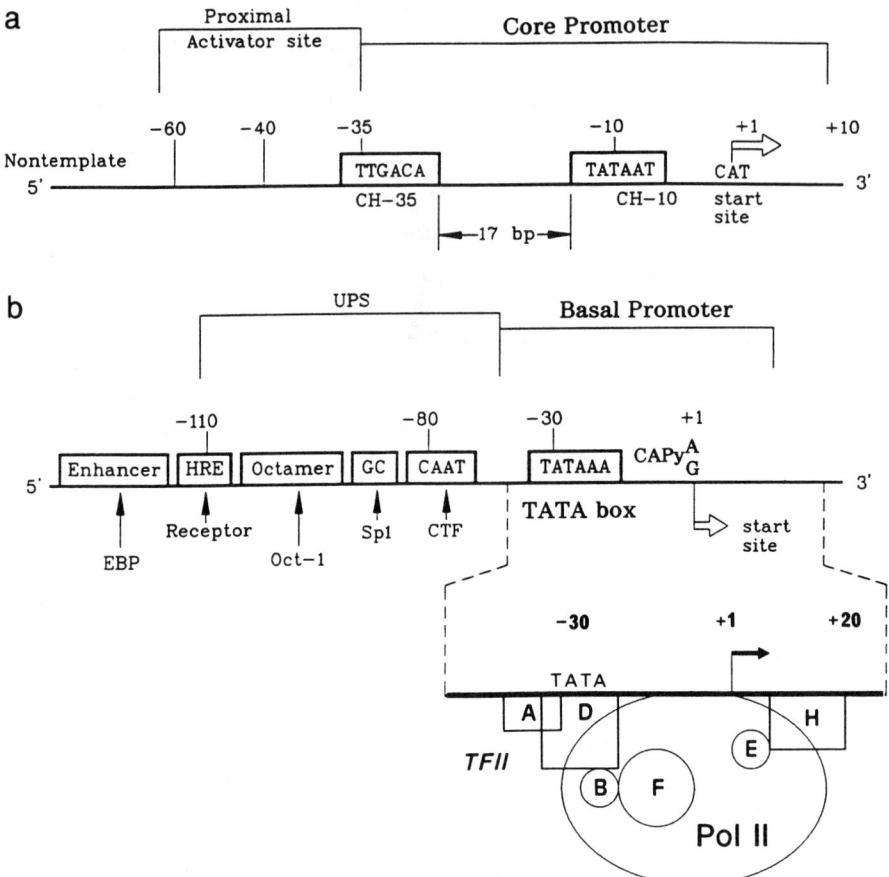

FIGURE 7.1 General schemes for promoter organizations in (a) prokaryotes and (b) eukaryotes. (a) The *E. coli* σ^{70} promoters consist of two regions: a *core promoter* and *upstream activator sites*. The core promoter contains -10 and -35 consensus hexamers (CH). The upstream activator sites can also be divided into "proximal" and "distal" activator sites to which regulatory factors bind. (b) Eukaryotic class II (RNA Pol II) promoters generally consist of a *basal (or core) promoter* and *upstream promoter sequences or sites (UPS)* to which "general" and "specific" transcription factors bind, respectively. The expanded basal promoter region illustrates the interrelationship between RNA Pol II and the general transcription factors (TFII A, B, D, E, F, and H) which make a chain of physical contacts to form a *preinitiation complex*. Abbreviations (see Table 7.1): CTF, CAAT-binding transcription factor; EBP, enhancer-binding protein; HRE, hormone-responsive element; and Py, pyrimidine nucleotide C or T.

and ranges from 14 to 26 bp. Depending on contexts, however, transcriptional efficiency may vary more than 10-fold and the start site may scatter over a distance of 30 bp.

(f) A third recognition element for some *E. coli* promoters is located at the -40 to -60 region, thus called the *upstream (UP) element*. The core promoter consisting of CH-10 and CH-35 is sufficient for specific basal initiation, but the presence of the UP element can stimulate transcription by a factor of 30 in the absence of protein factors other than RNA Pol. The UP element, which is AT rich but has no obvious consensus sequence, is directly recognized by the α subunit (C-terminal region) of RNA Pol (23). Depending on promoter types, the UP and further upstream regions (up to −240) may contain sequence elements recognized by bacterial transcription activators, e.g., FIS (factor for inversion stimulation) and CRP or CAP (for catabolite activator protein). In contrast, repressor (as opposed to the activator) sequence elements (e.g., operator) can occur throughout the promoter, including upstream and downstream regions, and negatively affect transcription efficiencies.

(g) The relative transcriptional efficiency of various promoters, often referred to as *promoter strength*, is determined mainly by two parameters: (i) the binding affinity (10^{12}–10^{14} M^{-1}) of the promoter to RNA Pol and (ii) the rate of "open" complex formation. There is a correlation, although not perfect, between the fit to consensus of the CH-10 and CH-35 and the two parameters. The strength of a promoter thus depends on the positional variation of the consensus sequence, the length of the spacer between CH-10 and CH-35, and the thermodynamic stability of the promoter region, particularly from −9 to −1. In fact, both the consensus and flanking regions are characterized by lower helix stability and higher helix flexibility compared with random sequences. The region (from +1 to +20) downstream of the core promoter also plays a role in promoter strength by influencing the stage(s) of idling (or recycling) and/or promoter clearance.

(h) In recombinant DNA technology, the genes cloned in prokaryotic expression vectors are usually expressed in *E. coli* under the control of strong promoters such as *lacUV5*, *trp*, and λ P_L. An important feature of all these promoters is that their transcriptional activity is regulatable by the presence of the appropriate repressor molecule. A hybrid promoter *tac*, constructed by fusion of the P_{trp} −35 and P_{lac} (UV5)-10 regions, turns out to be more efficient than the parent *lacUV5* and *trp* promoters by 11- and 3-fold, respectively. The *tac* promoter is also regulated by the *lac* repressor and is functional in a broader host range.

(i) Activator proteins stimulate transcription by making dual contacts with regulatory DNA elements and the polymerase, thereby providing additional forces for the polymerase to overcome any of the "defects" in promoter binding, isomerization, or promoter clearance.

As a rule, activators function by bending the DNA of their target sequence or by exacerbating an intrinsic curvature present within or near the binding site. For instance, CRP introduces two discontinuous kinks of ∼45° in each recognition site (TGTGA) of inverted sequence within CRP-cAMP-DNA complexes, resulting in a total bending angle of ∼90° (79). Protein–protein interactions of RNA Pol with one or more activators lead to DNA looping or a cage-like structure with built-in strains which is relieved by transcription initiation.

ii. Operators. Some bacterial polycistronic genes are transcribed as a unit under a regulatory system called an *operon*, e.g., lactose operon, histidine operon, and tryptophan operon. The key regulatory element of transcription initiation in operons is called the *operator,* a specific DNA sequence which binds a regulatory protein called a *repressor.* Since the operator is usually located downstream of the promoter and in fact partially overlaps the promoter, prior or juxtaposed binding of the repressor may physically prevent RNA Pol from binding to the promoter and/or forming a stable initiation complex. Repressors reduce transcription drastically, e.g., ~70-fold in *trp* operon or several hundredfold in *lac* operon.

Typically, operators consist of 16 to 18 bp containing an inverted repeat. For instance, the *lac* operator consists of 24 bp and some 20 bp are related by a twofold axis of symmetry (refer to Fig. 8.2). The operator contains 21 bp that are the template for the first 21 nt of the lactose operon mRNA. Thus the *lac* repressor (LacI, a tetramer, 360 amino acids/subunit) covers the template just where the catalytic center of the RNA Pol would be positioned. The symmetrical shape of the operator allows the repressor to bind and function usually as a single dimer or a dimer of a dimer. The 19-bp *trp* operator (TGT<u>ACTAGT</u> TA <u>ACTAGT</u>AC) features a tandem repeat of 6-bp (underlined) binding sites within an overall context of inverted repeat. The dimeric Trp repressor (108 aa/monomer) can tandemly bind one operator half-site in a 2 : 1 complex (24).

Repressors, typically those belonging to the *lac* repressor family, contain two functional domains: a N-terminal DNA-binding domain (~60 aa) featuring a *helix–turn–helix* (HTH) structural motif and a larger C-terminal domain (~280 amino acids) which assumes the functions of effector binding and oligomerization. Repressors themselves are also strictly controlled both in quantity at the level of repressor synthesis and in quality by interaction with effector molecules. Binding of β-galactosides, its analog IPTG, or other molecules generally known as *inducers* to LacI renders the repressor "inactive" by interfering with the DNA binding. In contrast, some repressors require the binding of small molecules, called *corepressors,* to function as "active" repressors, for example, Trp for *trp* repressor and hypoxanthine and guanine for purine repressor (PurR).

Like other members of the LacI family, the λ repressor (cI, dimer of a dimer), which controls the transcriptions at the phage λ P_L and P_R promoters, binds to its target sites O_L and O_R, respectively, via a HTH motif (25). (*Note:* The cI857, a temperature-sensitive mutant repressor often employed in recombinant DNA technology, allows the expression of cloned genes by temperature shifting, for example, from 32° to 42°C.) The critical helices in the dimer are separated by 34 Å, exactly the length of one helical turn of B-DNA, and thus fits snugly into the major groove. For the λ repressor, the presence of one bound dimer increases the binding of another dimer by a factor of ~10 (26). Although small, this cooperativity is regarded as crucial for the regulation of λ gene expression.

b. Features of eukaryotic regulatory elements

i. Eukaryotic promoters. Compared with prokaryotic promoters, eukaryotic promoters are less well defined in both sequence identity and position (27). Nevertheless, the comparative analysis of different eukaryotic promoters has revealed a common pattern of organization (Fig. 7.1b). The features of the eukaryotic promoters specific for RNA Pol II, the polymerase responsible for the synthesis of mRNA, can be summarized as follows.

(a) The transcription start point is usually Pu (A or G) following the sequence context of CAPy. In the proximal promoter region, a consensus sequence called the *TATA box* (TATAAA) or *Hogness box* is located at positions around -25 to -30. This TATA box is similar to the *E. coli* CH-10 and plays a key role in correctly anchoring the RNA Pol II. This fortuitous homology is believed to permit some eukaryotic genes to be expressed in *E. coli*. The TATA box forms a specific, initial complex with a "general" transcription factor, TFIID (20). The base frequencies in various positions in the promoter region, the TATA box, and transcriptional start sites of human genes show small but significant differences from those of other eukaryotic species and viruses (28).

(b) RNA Pol II promoters do not have the sequence element comparable to the CH-35 of bacterial promoters. Instead, recognizable consensus sequences, e.g., GC box and CCAAT box, are found at variable positions (from -110 to -40) farther upstream from the start site, thus called *UPS* (for upstream promoter sequences or sites) or *UAS* (for upstream activator sequences or sites). These UPS are recognized in a tissue-specific manner by specific transcription factors (activators) in response to various extracellular and intracellular stimuli (Table 7.1). These stimuli include hormones (e.g., steroid, thyroid), metals (e.g., Ca^{2+}, Fe^{2+}), retinoic acid, vitamin D, phorbol esters, interferons, second messengers (e.g., cAMP), and oncogenic proteins (Jun, Fos). The responsive elements are generally regarded as part of enhancers (see below). The UPS activators include receptors and other ligand-activated enhancer-binding proteins as well as the typical nonligand-associated transcription factors.

(c) As opposed to the UPS and the "specific" transcription factors that recognize them, the proximal region of the promoter surrounding the TATA box may be functionally regarded as a "basal" promoter. The transcription factors that participate in the basal transcription complex are classified as *general* transcription factors.

(d) Despite some substantial differences between eukaryotic and prokaryotic promoter sequences upstream from the CH-10, many eukaryotic (and viral) promoters have been shown to drive gene expression in *E. coli* and at the expected sites, albeit at varying strengths, e.g., RSV LTR > rat β-actin > rat insulin 1 > hepatitis B virus precore > human insulin > mouse metallothionein 1 (29).

ii. **Enhancers.** In addition to the consensus sequences that are located within the typical promoter region, certain "activating" sequences known as *enhancers* can drastically influence the level of transcription by RNA Pol II. In general, enhancers are cell type specific and are largely responsible for tissue-specific gene expression. Some features of the enhancers are as follows.

(a) Enhancers are relatively large elements, often including several hundred base pairs. They contain at least two distinct domains which exhibit very little enhancing activity on their own. However, duplication (or tandem repeat) of

TABLE 7.1 Some Typical Eukaryotic Transcription Factors (TFs) and Their Consensus Recognition Sequences

DNA-binding factor		Recognition sequence[a]	Functional characteristics	Ref[b]
Mnemonic	Name/synonym			
AP-1	c-Jun, c-Fos	GTGA(G/C)T(C/A)A	TRE [phorbol ester (TPA) response element] Differs from CRE by a single base deletion	1 2
AP-2	Retinoic acid-inducible mammalian factor	CC(C/G)C(A/G)GGC	Binding site on cAMP- and phorbol ester-inducible genes	3
CAP	Catabolite gene activator protein. CRP (cAMP receptor protein)	TGTGA	Pleitropic bacterial activator	4
C/EBP	CCAAT/enhancer binding protein	CCAAT	CAT box	5
	CP1, CTF or CBF	TGTGG(A/T)(A/T)(A/T)G TATGGTGTCAAAGGTCAAACT	Second binding site	
COUP-TF	Chicken ovalbumin upstream promoter TF	CCAGGGGTCAGGGGGGGGTGCTT	Member of steroid receptor superfamily RIPE for rat insulin II gene promoter element	6
CREB/ATF	CRE (cAMP-response element) binding protein Activating TF	GTGACGTC(A/C)(A/G)	CRE/ATF site Similar to AP-1 site Activates Adenovirus E1A- and cellular cAMP-inducible genes	7 8

E12	kE2-binding proteins	GCAGGTG	E2A human TF E12 is active only as heterodimer with MyoD	9
E47	Rat homologs are Pan1 (E47) and Pan2 (E12)		E47 is crucial in early stages of B cell development	10
GGABP	GGA-binding protein	(C/A)GGA(A/T)	VP16-mediated activation of HSV immediate early genes	11
GAF	(IFN) γ-activated factor	AGTTTCATATTACTCTAAATC	GAS (γ-activated site) IFN-γ induces Tyr-phosphorylation of GAF	
GAL4	A yeast activator	CGGA(C/G)GAC(A/T)GTC(G/C)TCCG	Activation of genes involved in galactose catabolism	12
GATA-1	GATA-1 factor	(T/A)GATA(A/G)	Regulation in hematopoietic cells	13
GCN4	A yeast activator	ATGA(C/G)TCAT	Activation of general amino acid biosynthetic genes Contains AP-1 site	14
GR	Glucocorticoid receptor	GGTACAnnnTGTTCT	GRE (glucocorticoid response element)	15
HSF	Heat-shock TF	nGAAnnTTCnnGAAn	HSE (heat-shock element) Activated form of HSF is a trimer	
Myb	c-Myb nuclear TF (oncoprotein)	CCGTTA	MRE (Myb response element) Activates genes of hematopoietic cells	16, 17
NF-κB	Nuclear TF	GGG(A/G)N(T/A)TYCC	κB-binding element Ig κ light chain enhancer (κB) of B cells Pleiotropic activation of inducible (cytokines) and tissue-specific genes	18

(continued)

TABLE 7.1 (continued)

Mnemonic	DNA-binding factor Name/synonym	Recognition sequence[a]	Functional characteristics	Ref[b]
Oct-1	Octamer-binding protein. OTF-1, NF-A1, or NFIII	ATG(T/C)AAAT	Octamer Activates a wide variety of genes	19 20
Oct-2	Octamer-binding protein	AAGAATAAATTAGA AT(T/G)T(G/C)CAT	Second recognition site Ig gene activator	21 22, 23 24
Oct-3	Octamer-binding protein	ATT(T/A)GCAT TTAAAATTCA	Important in early mammalian development	25
p53	Tumor suppressor protein	CTTGCCT	Transcriptional activator/repressor, dimer of dimer Binds other factors (ex. CBF, TBP, SV40 T Ag)	
SP1	GC box-binding protein	GGGCGG	GC box SRE (Serum response element) SRF is a MADS domain protein Role in cell proliferation and differentiation	26, 27
TBP	TATA box-binding protein	TATAA	TATA box A component of TFIID	28 29
E1a	Adenovirus early gene product (13S)		No direct binding to sequence elements Binds to diverse DNA-binding domains of several cellular TFs and stimulates, via activation regions of both factors, specific (adenovirus early genes) and promiscuous (cellular genes)	30

VP16	HSV major late phosphoprotein. α-TIF or Vmw65	No direct binding to sequence elements C-terminal acidic activation domain interacts with TFs such as TFIIB and Oct-1, and results in pleiotropic activation of HSV immediate early genes and various cellular genes	31

[a] Degenerate nucleotides are denoted by (N/N) or n.

[b] Key to references: (1) Angel, P., Imagawa, M., Chiu, R., Stein, B., Imbra, R. J., Rahmsdorf, H. J., Jonat, C., Herrlich, P., and Karin, M. (1987). *Cell* **49**, 729–739; (2) Lee, W., Mitchell, P., and Tjian, R. (1987). *Cell* **49**, 741–752; (3) Mitchell, P. J., Wang, C., and Tjian, R. (1987). *Cell* **50**, 847–861; (4) de Crombugghe, B., Busby, S., and Buc, H. (1984). *Science* **224**, 831–838; (5) Johnson, P. F., Landschulz, W. H., Graves, B. J., and McKnight, S. L. (1987). *Genes Dev.* **1**, 133–146; (6) Hwung, Y.-P., Wang, L.-H., Tsai, S. Y., and Tsai, M.-J. (1988). *JBC* **263**, 13470–13474; (7) Hai, T., Liu, F., Allegretto, E. A., Karin, M., and Green, M. R. (1988). *Genes Dev.* **2**, 1216–1226; (8) Ziff, E. B. (1990). *TIG* **6**, 69–72; (9) Sun, X.-H., and Baltimore, D. (1991). *Cell* **64**, 459–470; (10) Wasylyk, B., Hahn, S. L, and Giovane, A. (1993). *EJB* **211**, 7–18; (11) Shuai, K., Schindler, C., Prezioso, V. R., and Darnell, J. E., Jr. (1992). *Science* **258**, 1808–1812; (12) Giniger, E., Varnum, S. M., and Ptashne, M. (1985). *Cell* **40**, 767–774; (13) Omichinski, J. G., Clore, G. M., Schaad, O., Felsenfeld, G., Trainor, C., Appella, E., Stahl, S. J., and Gronenborn, A. M. (1993). *Science* **261**, 438–446; (14) Hope, I. A., and Struhl, K. (1987). *EMBO J.* **6**, 2781–2784; (15) Evans, R. M. (1988). *Science* **240**, 889–895; (16) Luscher, B., Christenson, E., Litchfeld, D. W., Krebs, E. G., and Eisenman, R. N. (1990). *Nature* **344**, 517–522; (17) Gabrielsen, O. S., Sentenac, A., and Fromageot, P. (1991). *Science* **253**, 1140–1143; (18) Lenardo, M. J., and Baltimore, D. (1989). *Cell* **58**, 227–229; (19) Puijn, G. J. M., van Driel, W., and van der Vliet, P. C. (1986). *Nature* **322**, 656–659; (20) Fletcher, C., Heintz, N., and Roeder, R. G. (1987). *Cell* **51**, 773–781; (21) Ponce, E., Lloyd, J. A., Pierani, A., Roeder, R. G., and Lingrel, J. B. (1991). *Biochem.* **30**, 2961–2967; (22) Ko, H.-S., Fast, P., McBride, W., and Staudt, L. M. (1988). *Cell* **55**, 135–144; (23) Muller, M. M., Ruppert, S., Schaffner, W., and Matthias, P. (1988). *Nature* **336**, 544–551; (24) Scheidereit, C., Cromlish, J. A., Gerster, T., Kawakami, K., Balmaceda, C-G., Currie, R. A., and Roeder, R. G. (1988). *Nature* **336**, 551–557; (25) Okamoto, K., Okazawa, H., Okuda, A., Sakai, M., Muramatsu, M., and Hamada, H. (1990). *Cell* **60**, 461–472; (26) Briggs, M. R., Kadonaga, J. T., Bell, S. P., and Tjian, R. (1986). *Science* **234**, 47–52; (27) Dynan, W. S., and Tjian, R. (1983). *Cell* **35**, 79–87; (28) Peterson, M. G., Tanese, N., Pugh, B. F., and Tjian, R. (1990). *Science* **248**, 1625–1646; (29) Kao, C. C., Lieberman, P. M., Schmidt, M. C., Zhou, Q., Pei, R., and Berk, A. J. (1990). *Science* **248**, 1646–1650; (30) Liu, F, and Green, M. R. (1994). *Nature* **368**, 520–525; (31) Greaves, R., and O'Hare, P. (1989). *J. Virol.* **63**, 1641–1650.

either domain, for example, the 72-bp repeat enhancer of SV40 DNA, leads to a dramatic increase (400-fold) in transcription (30). An enhancer may actually consist of small, multiple elements called *enhansons* which function synergistically in a tissue-specific manner.

(b) Enhancers can act over considerable distances in a nonpolar fashion, sometimes up to 5 kb or more upstream from the transcriptional start sites.

(c) Enhancers are position independent and function in either orientation. They can be located upstream from the promoter (e.g., albumin gene), downstream of the transcribed region (e.g., β-globin and T-cell receptor genes), or in the middle of a gene (e.g., within an intron of immunoglobulin gene).

(d) Enhancers are both cis- and trans-acting; the element can stimulate the transcription of a gene on the same DNA molecule or on other DNA molecules as long as they are in close proximity of the promoter (31). Enhancers have the capacity to augment the activity of a wide variety of heterologous promoters.

(e) Enhancers are cell or tissue specific, partly explaining the host range of some animal viruses and the differential gene expressions during development.

(f) Enhancers are specifically recognized by *enhancer-binding proteins* (EBP). An EBP expressed from a cloned human gene has a Cys_2His_2 zinc-finger motif, binds a specific sequence (e.g., GGGGATTCCCC), and activates a number of genes including HIV-1 LTR, human IFN-β gene, MHC class I gene (H-2K), and Ig κ gene (32). There is no clear distinction between promoter- and enhancer-binding factors in terms of their ability to mediate transcriptional activation.

c. Regulatory proteins: transcription factors

i. Functional roles of transcription factors. In both prokaryotes and eukaryotes, the initiation of transcription involves a network of RNA Pol and transcription factors that interact with proximal and/or distal regulatory elements. Transcription factors comprise all UPS-binding proteins, including enhancer-binding proteins and certain hormone and secondary messenger receptors as well as the factors generally required for basal transcription. Transcription factors can be divided into two groups: general and specific. "General" transcription factors help RNA Pol form the *basal transcription complex,* also known as *preinitiation complex,* around the core promoter element (TATA box). The preinitiation complex can perform a low, unregulated level of transcription or *basal transcription.* "Specific" transcription factors, also called activators, bind regulatory elements either proximal or distal to the TATA box and regulate basal transcription.

Transcription of eukaryotic genes *in vivo* presents an added level of complexity because regulatory proteins and RNA Pol have to get to their regulatory and/or promoter sites located in the "blocking" environment of chromatin. Transcriptional control at the nucleosomal level is only beginning to be understood. The GAGA transcription factor (of *Drosophila*), which binds to GA/CT-rich sites, and SWI/SNF proteins of yeast are among the available examples of the factors that can induce in an ATP-dependent manner local disruption of the nucleosome structure (a compact bundle of DNA wrapping around an octameric histone core) and stimulate transcription.

All known transcription factors are proteins except for a unique transcription factor (TFIIIR), for RNA Pol III in *Bombyx mori,* which is an isoleucine tRNA (33). This section provides an overview of the structural and functional characteristics of eukaryotic class II transcription factors which act on RNA Pol II and class II promoters. [*Note*: For more extensive information on transcription factors and their properties, readers should consult the *Transcription Factors Database* (TFD) compiled by the National Center for Biotechnology Information, National Library of Medicine, National Institutes of Health, Bethesda, Maryland.]

(a) General transcription factors. General transcription factors constitute, together with RNA Pol II, the basal transcription complex which spans ~80 nt around the transcription start site. The general transcription factors comprise at least six distinct species: TFII A, B, D, E, F, and H (see Fig. 7.1b).

TFIID (300–750 kDa) is a multiprotein complex composed of a TATA (box)-binding protein (TBP) and up to 13 TBP-associated factors (TAFs). TBP (~36 kDa) is probably the most highly conserved eukaryotic transcription factor, with its multifunctional (DNA-binding and protein-interacting) C-terminal domain (180 amino acids) showing >75% sequence identity in a wide variety of species. The N-terminal region of TBP from different organisms differs significantly in length and sequence. TBP makes direct contacts with RNA Pol (C-terminal domain of the large subunit) and other transcription factors like TFIIA and TFIIB, assuming the role of a "commitment factor" whose binding to the TATA box is the prerequisite for the ordered assembly of the preinitiation complex (20). Binding of TBP in a "saddle" shape on a TATA box-containing element (K_d of ~3 × 10^{-9} M) results in sharp (90°) kinks at the end of the TATA sequence, severe unwinding, and a compensating superhelical twist in the double helix. TBP also provides a binding site for certain proximal activators like octamer-binding proteins (Oct-1 and Oct-2) which stimulate transcription at the level of stable preinitiation complex formation. In fact, TBP is a universal factor required for the initiation of transcription by all three classes of nuclear RNA polymerases. Native TFIID mediates both basal and activator-dependent transcription, whereas TBP is competent for only basal transcription. Thus TAFs are essential cofactors for transmitting regulatory signals and distinguishing RNA Pol species.

TFIIA (110 kDa tetramer) stabilizes the binding of TBP to the TATA box.

TFIIB, a ~30-kDa factor, provides multiple points of interaction with TFIID, TFIIF, Pol II, and also with DNA. The stable entry of TFIIB after the TBP binding is presumed to be a limiting step in the recruitment of RNA Pol II, which is stimulated by acidic activators (e.g., VP16). Following the initiaion of transcription, TFIIB dissociates from the promoter. The protein-binding sites are apparently exposed by a conformational change on interaction with VP16 (36). In the absence of an activator, only a small fraction of TFIIB is presumed to be in the alternative "active" conformation in which the TFIIF- and RNA Pol II-binding sites are available. TFIIB also interacts with CBP (CREB-binding protein) which, as a "coactivator," mediates stimulation by signal-targeted transactivators such as c-Jun and CREB (cAMP-responsive element-binding protein).

TFIIE binds to RNA Pol II and also interacts with TFIIH. The interaction with TFIIH apparently stimulates the enzymatic activities associated with the TFIIH factor.

TFIIF, a ~100-kDa heterodimer, binds to both RNA Pol II and TFIIB.

TFIIH (or BTF2), a multisubunit complex, has DNA-dependent ATPase, protein kinase, and ATP-dependent DNA helicase activities which are stimulated by TFIIE. TFIIH interacts with the "acidic" activation domain of transcriptional activators such as VP16 and p53. One subunit of TFIIH is encoded by a gene previously linked to DNA repair, which provides a support to the notion that transcription and DNA repair are tightly coupled.

(b) Specific transcription factors. Specific transcription factors bind regulatory elements either proximal or distal to the TATA box (Fig. 7.1b and Table 7.1) and stimulate or repress the basal transcription (37). As described later in greater detail, transcription factors are built around common structural motifs that are quite well conserved throughout evolution, even among highly divergent species. Therefore, activators from one organism (e.g., yeast) can often function as activators in other organisms (e.g., mammalian cells).

ii. Structural features of transcription factors: domains and motifs. Activators are typically modular proteins having two functionally distinct domains: one "DNA-binding" domain which recognizes a specific DNA sequence and the other "activating" domain which is involved in protein–protein interactions.

Both DNA-binding and activating domains are characterized by highly conserved sequence and/or structural motifs. The ordered conformation may be preexistent or induced on interactions between activator subunits or between the activator and other molecules. By virtue of the features associated with the modular structure and independent function, certain DNA-binding motifs (e.g., zinc fingers) can be assembled to create chimeric transcription factors or, more generally, DNA-binding proteins with novel specificities and customized biological functions. Likewise, activating domains or modules can be functionally interchanged between distantly related species. Examples of the "domain swap" include the GAL4–VP16 and TetR–VP16 hybrid activators in which the activating domain of VP16 is respectively linked to the DNA-binding domain of yeast GAL4 and to the Tet repressor (80).

(a) Single-domain activators. Activators may possess only a single domain in the sense that some activators have only the sequence-specific DNA-binding domain (e.g., Oct-1), whereas others have only the activating domain(s). These single-domain activators function as *coactivators* (or *adapters* or *intermediary factors*) since they stimulate transcription only by their ability to interact with and complement other transcription factors. For some apparent bifunctional activators (e.g., E1a and VP16), the promoter-binding function is actually mediated by interaction with a DNA-binding protein. Single-domain factors generally consti-

tute tissue-specific nuclear factors that regulate gene expression only in one kind of tissue or in certain specific phases of cell differentiation and/or cycle.

(b) DNA-binding domains. The DNA-binding surfaces usually span 60–100 amino acids. Some are highly specific for their DNA-binding site ($K_d \geq 10^{-10}$–10^{-12} M), whereas others show only a moderate degree of specificity and affinity ($K_d \geq 10^{-9}$ M). Note that the affinity (or promoter strength) can be graded depending on the identity of promoter consensus sequence. The factors with modest affinities may achieve a high degree of specificity via cooperative protein–protein interactions with another factor(s) forming a transcription complex, as has been shown by an oncogenic protein Myc and its partner Max (38). Indeed the "loose" binding is presumed to be an inherent property that serves as a medium for combinatorial use of multiple factors and regulation of their activities.

Some activators exhibit equally avid (puzzling) binding to sites on DNA that share only minimal nucleotide sequence similarity. In other cases, two activators from entirely different origins, e.g., MyoD (muscle-specific gene expression) and E2A adenovirus protein, recognize a common consensus sequence CANNTG and yet activate *in vivo* different tissue-restricted sets of genes. In fact there is a family of transcription factors that gather at the cAMP response site (39). The capacity of similar factors to discriminate subtle differences in binding sites is attributed, at least in part, to the fact that flanking and internal position nucleotides contribute to the recognition in a half-site mode (40). In addition, another factor(s) may intervene to strengthen or weaken the binding of the first activator. Bipartite sequence recognition seems to be an important theme of activator functions as has been observed, for example, with Oct-1, NF-κB (41), and nuclear hormone receptors (42).

The structural motifs used by transcription factors for DNA binding fall into three major types (43–45): helix–turn (or loop)–helix, zinc finger, and leucine zipper. Within a major motif, variations exist and a transcription factor may possess more than one motif or combination of motifs. For a more general perspective of protein–nucleic acid interactions, refer to Ref. (46).

(i) Helix–turn (or loop)–helix. HTH or HLH is the DNA-binding motif used in prokaryotic regulatory proteins such as Cro, CAP, and λ repressor and in many eukaryotic activators such as Myc, MyoD, E12, E47, and AP-4. In prokaryotic regulatory proteins, the HTH motif is a tightly packed 20 amino acid structure consisting of two helices connected by a 3 amino acid linker (25). The HTH motif of the protein directly interacts with DNA, the second α helix (the "recognition helix") binding in the major groove of the DNA.

The eukaryotic homolog of the HTH motif consists of two apparent amphipathic helices (60–90 amino acids) connected by a short loop of 10–25 amino acids, thus more appropriately called HLH (47–49). The eukaryotic HLH is physically larger than the prokaryotic counterpart HTH motif. Most significantly, unlike HTH proteins, the actual DNA-binding motif of the HLH proteins is a stretch of about 13 "basic" amino acids adjacent to the HLH whose main role is in protein dimerization. The entire motif comprising the DNA-binding basic region and the HLH is termed *bHLH* (50). The two amphipathic helices, particularly the highly conserved hydrophobic amino acids, are thought to mediate dimerization, which is a prerequisite to DNA binding.

A variation of the HLH motif is the *homeodomain,* about 60 highly conserved basic and hydrophobic amino acid residues that form a HLH of approximately 21 amino acids. The homeodomain is usually a part of the bipartite DNA recognition module known as the *POU domain.* For example, the POU domain of Oct-1 consists of the C-terminal 60 amino acid POU-homeodomain (POU_h) and the N-terminal 75–80 amino acid POU-specific domain (POU_s) connected by a short nonconserved linker (52). The POU_s domain also has a HTH motif which is almost superimposable to the DNA-binding domain of the bacteriophage λ repressor. Oct-1 recognizes the left half of its binding site (the octamer ATGCAAAT) by using POU_s and the right half by using POU_h (51). The POU_h, which interacts with TBP, binds DNA by itself with a low affinity.

(ii) Zinc-finger motif. The zinc finger is one of the most important DNA-binding motifs found first in TFIIIA and subsequently in various other proteins including transcription factors such as SP1 (GC-box binding protein), GLI (human glioblastoma protein), Zif268, and a subfamily of nuclear hormone receptors. A zinc-finger domain is characterized by the consensus sequence X_3-Cys-X_{2-4}-Cys-X_{12}-His-X_{3-5}-His-X_4 (where X is any amino acid residue), which forms a compact globular structure that contains a β sheet and an α helix held together by a central Zn ion. The Zn ion is tetrahedrally coordinated to the Cys_2His_2 or Cys_4 motif, and a zinc-finger protein may contain two or more tandemly repeated or contiguous zinc-finger motifs which are used in a flexible manner to recognize a diverse set of DNA sequences. TFIIIA contains nine tandemly repeated motifs, and different sets of zinc fingers are used for high affinity interactions with RNA and with DNA. In the Zif268–DNA complex, two α helices out of the three zinc fingers wind around the major groove for almost a complete turn of the B helix, making equivalent contacts to one of the two DNA strands (53).

(iii) Leucine-zipper motif. The leucine zipper is another common motif found in many DNA-binding proteins, including transcription factors such as C/EBP, Jun, Fos, GCN4, and HSF. The actual DNA binding is carried out by a 16–25 amino acid, Arg-rich "basic" region adjacent to the leucine zipper, thus more appropriately referred to as a *basic-leucine-zipper* (bZIP) (54). The leucine zipper is an amphipathic α helix containing heptad repeats of Leu residues on one face of the helix and serves as a dimerization module. On dimerization, the leucine-zipper α helices form a parallel-coiled coil based on hydrophobic interfacial side-chain packing (55). The dimerization brings the DNA-binding surface to the positions appropriate for contacting DNA in a scissor-grip mode (56) or in an induced helical fork mode (57). In GCN4 bound to a target site (ATGACTCAT), the two N-terminal α-helical basic regions fit into the major groove of half sites on opposite sides of the DNA double helix. Notably, the basic region of the bZIP

motif appears to assume a stable α-helical structure only when bound to a specific DNA recognition sequence (58–60).

(iv) Role of oligomerization. Although some activators (e.g., TFIID and Myb) bind DNA as monomers, many activators (and repressors as well) exert gene activation (or repression) through oligomerization, i.e., predominantly a dimer and, in some cases, a tetramer (dimer of a dimer). A dimer may consist of either identical (homodimer) or nonidentical (heterodimer) subunits. Members of each family are capable of forming intrafamily and sometimes interfamily heterodimers in a highly selective manner. For instance, nuclear hormone receptors which belong to a superfamily of ligand-activated transcription factors form homodimers as well as intrafamily heterodimers. Fos forms heterodimers with Jun-related proteins but does not form homodimers, whereas Jun forms homodimers as well as heterodimers with all Jun- and Fos-related proteins (61). A heterodimer may display a DNA-binding specificity distinct from each other and from their parental homodimers, providing an important mechanism of gene regulation.

Protein dimerization is most frequently mediated by HLH, leucine zipper, and ankyrin motifs. The ankyrin motif, which is present in a heterodimeric transcription factor, GGABP (for GGA-binding protein), consists of 33 amino acid repeats (62).

(c) Activating domains. Many transcription factors possess, in addition to the DNA-binding domain, one or more separate activating domains which are used for protein–protein interactions. Single-domain activators functioning as coactivators have uniquely protein interaction domains. Depending on the characteristics of constituent amino acids, activating domains largely fall into three types: an acidic domain having a high content of negatively charged residues (e.g., in GAL4, GCN4, and VP16), a proline domain (e.g., in CTF and AP-2), and a glutamine domain (e.g., in SP1 and CP1).

VP16 has an acidic activation domain in the C terminus and binds TFIIB and TFIID. Good correlation has been observed between transactivation activities *in vivo* and the binding of VP16 to TFIID *in vitro* (63). Although the exact roles of acidic residues and secondary structures remain to be clarified, VP16 apparently requires both the negatively charged region and some specific hydrophobic residues for its full activity (64).

(d) Latent forms of activators. Activators with strong activating regions, e.g., steroid hormone-responsive activators (receptors), are controlled so that these regions are exposed in the nucleus in their fully functional form only when a cognate hormone is present. The activation of latent activators can occur in two ways. First, hormone-induced chemical modification can convert a latent activator into an active one. For instance, IFN-γ-induced activation of GAF (γ-activated factor) is mediated by cytoplasmic phosphorylation on Tyr. Second, specific and high-affinity binding of the ligand modifies the receptor so that the ligand–receptor complexes dimerize (or oligomerize) and bind to the cognate palindromic responsive elements resulting in modulation of gene transcription. Otherwise, the recep-

tors stay in the cytoplasm as inactive monomers complexed with an inhibitor protein(s). Among the typical examples are steroid (glucocorticoid) hormone receptors which form complexes with heat-shock proteins Hsp70 and Hsp90 (65,66) and an almost ubiquitous NF-κB which is bound by inhibitor-κB (IκB) (67,68). The activation of latent HSF (heat-shock factor) is induced by heat shock which converts the Hsp70-bound, monomeric non-DNA-binding HSF to active trimeric HSF, leading to their accumulation in the nucleus.

d. Mechanisms of transcriptional activation and repression

i. Models of activator function. Eukaryotic activators stimulate transcription by increasing the assembly of the preinitiation complex. In this assembly, the recruitment of the TFIIB factor is considered to be the rate-limiting step, and activators (e.g., VP16) function during at least two stages: first to recruit the TFIIB and then after TFIIB entry (69). How transactivators that recognize specific UPS elements interact with the preinitiation complex and stimulate transcription remains a key question. According to a *looping model,* a set of activators is thought to juxtapose the otherwise distant regulatory region with the basal transcriptional region by a mechanism involving long-range protein–protein interactions (44,70). The interactions can be a self-association of transactivators (e.g., SP1) bound to multiple sites (71,72) or a heteromeric association of several different activators [e.g., hormone receptors and *Drosophila* homeobox proteins (73–77)] resulting in looping of the intervening DNA. Alternatively, transactivators bound to UPS may interact with the basal initiation complex by a mechanism involving direct bending of DNA by the activator(s) (*wrapping model*). Indeed DNA bending has been shown to be an intrinsic property of many DNA-binding proteins including CAP, TBP, Fos, and Jun (78,79). For instance, the binding of CAP (cAMP-dependent catabolite gene activator protein) induces DNA bending by about 90° via two ~40° kinks.

ii. Regulation of activation or repression. Transcription in eukaryotes (and in prokaryotes as well) can be positively or negatively regulated. Although some factors function uniquely as activators or repressors, other transcription factors exercise both positive and negative controls depending on contexts. Activation ensues when activators interact with coactivators and/or a preinitiation complex in a right combination like in a "jigsaw" puzzle, whereas repression results from unfitting arrangements.

Activators can function in two ways. One is the function of activators as *antirepressors*. This is analogous to the antagonizing function of IPTG on the *lac* repressor which would, in the absence of IPTG, bind the *lac* operator and exert a transcriptional inhibition. The "GAGA factor," originally identified in *Drosophila,* provides an example of this type of activator which enhances transcription by countering the transcriptional repression exerted by a general repressor, histone H1 (81). The other type of activating interactions may be characterized as the function of bona fide activators. This type of activator (e.g., SP1 and GAL4-VP16) enhances transcription in the absence of the repressor H1.

In a manner opposing the modes of activation, repression can be envisaged to arise mainly from three types of interactions. One type occurs when a repressor binds to an overlapping or adjacent site, displacing the true activator in a concentration-dependent, competitive manner (82). A second mode of repression is that of repressors which bind to a site independent from the activator-binding site and yet interfere with activator functions, possibly by bending the DNA such that the preinitiation complex is destabilized. The Engrailed homeodomain protein (En) of *Drosophila* and the wild-type p53 (a tumor suppressor) are among the examples of such active or bona fide repressors (83). The third type occurs when a ligand or repressor interacts directly with an activator(s), forming a tight complex and rendering the activator incapable of further interacting with a coactivator(s) or regulatory sequence. Basically a given species of repressors can titrate out its cognate activator. This "sequestering" or "squelching" type of repression is operational for many activators, including AP-2 (by SV40 T antigen), glucocorticoid receptor (by Hsp90 and calreticulin), retinoid receptor (by SMRT, a nuclear protein), Myc (by Rb-related p107 protein), NF-κB (by IκB), p53 (by MDM2, a human oncoprotein), and Yin-Yang-1 (by c-Myc).

iii. Regulation of regulators and differential gene expression.

The mechanisms by which transactivators are regulated are not clearly known, but hold the key to differential often autoregulatory gene expressions. Temporal and spatial expression of a gene(s) in differentiating cells depends, in addition to the activating (or repressing) mode of interactions of the activators, on the availability of transcription factors. Regulators are affected by at least three modes of regulation. The first mode relates to the production of "active" transactivators, the number of which is subject to a similar complex network of regulations (84). One apparent function of such nested regulatory networks is that the abundance of activator molecules can be regulated in a cooperative manner in rapid response to extracellular signals. The second mode involves posttranscriptional regulation in which different forms and activities of transactivators are created, for example, by differential splicing (85). The third and perhaps the most common form of regulation is the modulation of activities of transactivators by posttranslational modifications either by covalent modifications (e.g., phosphorylation of CREB, AP-1, NF-κB, and Oct-1) or by noncovalent modifications (e.g., protein–protein complex formation). The formation of such transactivator–modulator complexes may either block or enhance the transcriptional activity.

4. TRANSCRIPTION ELONGATION

Once the RNA polymerase overcomes the abortive initiation phase by successfully incorporating 8–12 nt residues, it enters the elongation mode. In *E. coli*, the transition from initiation to elongation modes is marked by the release of the σ factor from the core polymerase. The elongating RNA Pol covers roughly 40 bp of DNA which harbors a "transcription bubble," an 18 ± 1-bp region of unwound DNA containing an RNA·DNA hybrid of up to 12 bp (22). The stability of the elongation complex is determined by the terms of free energy mostly contributed by (a) local denaturation of the 17-bp DNA "bubble" and (b) the binding interaction of RNA Pol with a segment (from 2 to 12 bases) of nascent RNA. Apparently, the RNA · DNA hybrid is transiently formed and peeled apart without much consequence on the stability of the ternary complex, RNA Pol/DNA/RNA. Progression of the elongation complex is influenced by the energy barrier associated with local DNA sequences and is thereby subject to stalling, pausing, or termination. For instance, a strand-specific hybridization of an antisense *peptide nucleic acid* (PNA) with an AT-rich target dsDNA provides a more thermostable duplex structure that inhibits transcription by RNA Pol II (87).

How does an RNA polymerase move along the DNA within a transcription unit? Evidence suggests that the RNA polymerase has two distinct DNA-binding sites and that each site, located at the front and back of the molecule, interacts with about 10 nt. The polymerase moves along the DNA in a discontinuous fashion like an "inchworm," alternately locking at and sliding along two attachment points of DNA, while the polymerase catalytic site located at the back part of the molecule adds successive ribonucleotides (88). The 3'-terminal region of RNA is also presumed to bind the polymerase at two sites. These features of the elongation complex are consistent with the observation that a replicating DNA polymerase can pass the RNA polymerase moving along the same DNA strand in the same direction without displacing the nascent transcript and the RNA polymerase (89).

The elongation phase of transcription is affected by the presence or absence of specific protein factors, called *elongation factors*, such as GreA and GreB for *E. coli* RNA Pol and TFIIS for eukaryotic RNA Pol II. The stimulation of transcriptional elongation is attributed, at least in part, to the elongation factor-associated $3' \rightarrow 5'$ hydrolytic activity which enables the stalled RNA Pol complex to retract from the transcription barriers (e.g., pause sites) and to repeat the forward synthesis, eventually overcoming the barriers.

The fidelity of transcription may be assured by a conformation-dependent mechanism in which the RNA Pol assumes an alternative conformation depending on correct or incorrect nucleotides and placing the incorrect nucleotides away from the terminal 3' OH group of the growing RNA chain (86).

5. TRANSCRIPTION TERMINATION

When the RNA polymerase locked in the elongation mode encounters a termination signal, transcription is brought to a halt. A termination signal can be apparent or cryptic. It can be induced by interactions with certain specific signal recognizing factors. Therefore, the termination of transcription is dependent on whether the RNA Pol perceives the signal under given circumstances. If the termination signal is functional, the polymerase core detaches from the DNA template. If not, as in the case of the lytic growth of phage λ, the RNA Pol reads through multiple transcription terminators. The termination of transcription is thus an important target in the regulation of gene expression (90).

In *E. coli*, at least three distinct transcription termination systems are known to operate: (a) *attenuation* controls in some amino acid operons, (b) hairpin

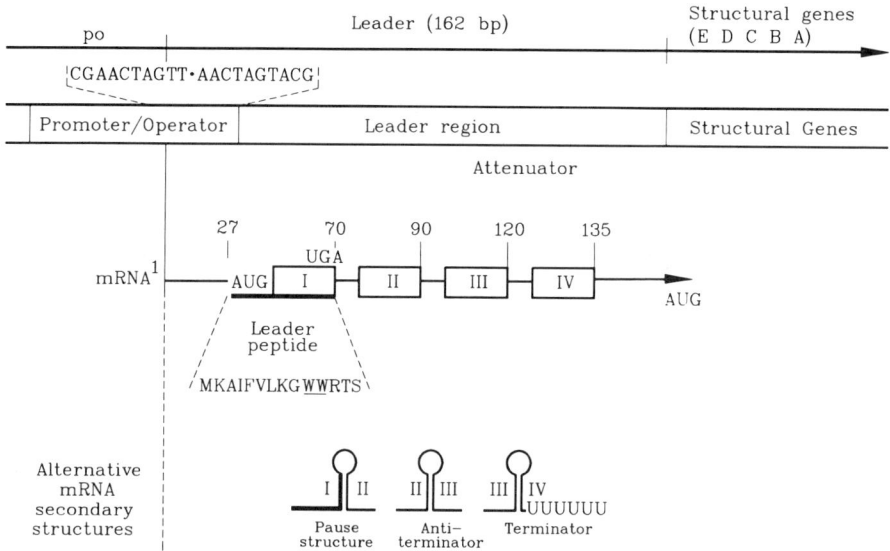

FIGURE 7.2 A model of transcription attenuation in the *E. coli* tryptophan operon. The operator sequence is meant to show only the central axis of symmetry. The 14 amino acids including two Trp (underlined) in the leader peptide are given following one-letter notation (refer to Table 1.3 in Chapter 1). The nucleotide positions and the size of mutually exclusive hairpin structures are not to scale.

structures in *factor-* or *rho-independent termination*, and (c) stalling or pausing sites in cooperation with the termination factor *rho* (*rho-dependent termination*). Phage λ utilizes yet an additional protein factor (N protein) called an *antiterminator* which regulates both rho-dependent and rho-independent termination (22).

The mechanism of eukaryotic transcription termination is significantly different from that of prokaryotes. Eukaryotic transcription termination is coupled to 3′ polyadenylation of the transcripts, thereby involving the poly(A) signal and poly(A) polymerase as well as various signal- and polymerase-associated protein factors.

a. Attenuation control of transcription

Attenuation is a termination mechanism that was initially discovered in polycistronic prokaryotic genes coding for a series of enzymes involved in the biosynthesis of amino acids such as Trp, His, Leu, Thr, Ile, Val, and Phe (91). The unique feature of this termination mechanism is that it is coupled with translation. Expression of the structural genes of the aforementioned operons is regulated by the availability of the product amino acid. Transcription termination occurs right at the 5′ leader region of the transcript and involves the mutually exclusive formation of terminator and antiterminator hairpins (Fig. 7.2). The alternative conformations are determined by a set of amino acid codons which sense the movement of translating ribosomes as a function of specific, charged tRNAs available. For instance, the tryptophan operon contains 2 tandem Trp codons within a total of 14 amino acid codons in its leader sequence as *sensors,* whereas the histidine operon has 7 His codons in a row out of 16 total amino acid codons. The transcription terminator, called *attenuator,* is a region (~80 nt in the *trp* operon) located near the end of the leader sequence and before the first structural gene of the operon. It contains four sequence motifs which form alternative sets of hairpins: I–II, II–III, or III–IV. When the product amino acid (Trp) is scarce, ribosomes stall at the "sensor" codons in region I and allow the formation of hairpin II–III (antiterminator) in the mRNA, which in turn lets the RNA polymerase continue transcription into the structural genes. When Trp is abundant, ribosomes swiftly proceed to the leader peptide termination codon and into region II, allowing the terminator hairpin III–IV to form. This leads to the termination of transcription, as in the case in which there is no accompanying translation. The attenuation mechanism can reduce transcription by a factor of 10.

Transcription attenuation can also occur by other mechanisms, involving RNA-binding proteins in particular. For instance, the attenuation of the *trp* operon in *B. subtilis* is mediated by TRAP (for *trp* RNA-binding attenuation protein, a toroid-shaped molecule) which acts by binding to the (G/U) AG-repeat segment of the nascent transcript (92). The binding prevents formation of an antiterminator structure, and allows terminator formation and transcription termination. The "active" conformation of TRAP is induced by Trp binding, which renders the attenuation mechanism responsive to the availability of the product amino acid.

b. Factor-independent termination

This type of termination signal, called an *operonic* or *intrinsic terminator,* usually defines the end of most prokaryotic transcriptional units and also exists at control points between genes and in most attenuators. The typical feature of *E. coli* ρ-independent terminators is an RNA hairpin which has the following characteristics. (i) A GC-rich segment capable of forming a hairpin with a stem length of 7 ± 3 bp is followed by a series of ~7 U residues. (ii) The size of the loops presents a narrow distribution of tetranucleotides (*tetra-loop*), most abundantly UUCG and GAAA. (iii) Most loops are closed by a CG base pair (93). The release of RNA Pol occurs when the transcript has extended to 8 to 9 bases beyond the end of the stable RNA stem.

Termination positions are defined by the thermodynamic stability of the polymerase complex, whereas the efficiencies of termination are determined by a kinetic competition between RNA elongation and release at each position (94). The termination efficiency is affected mostly by the strength of the stem–loop, the presence of the U-tract, and also by the concentration of NTP substrates. The efficiency of transcription termination also depends on other factors such as the size and shape of the hairpin, not just its stability, most likely involving interactions with the polymerase. Different RNA hairpins exert varying effects on transcription termination, while a cruciform structure may have no or some attenuating effect on transcription. Furthermore, certain changes in the stem sequence designed to maintain or raise the stability of the stem ($\Delta G = -18$ kcal/mol) lead to a decrease of the terminator activity of the λ tR2 terminator (95). Note that a terminator with a strong hairpin structure ($\Delta G = -30$ kcal/mol) can also serve as a positive element stabilizing mRNA and thus augmenting the level of gene expression (96).

c. ρ-Dependent termination

ρ-Dependent termination occurs at the ends of some bacterial transcriptional units, at regulatory points before or between genes, and sometimes within genes. The intragenic terminators are latent and function only when translation becomes uncoupled from transcription either by a mutational change in the gene or by a metabolic stress such as amino acid starvation (97).

ρ protein is a hexamer of 46-kDa subunits which form a symmetric hexagon organized as a trimer of dimers. Each subunit contains separate binding sites for RNA (cofactor) at the N-terminal region and ATP (substrate) at the C-terminal region. ρ-Dependent termination involves dynamic interactions of the ρ protein with the nascent RNA strand. The sequence of events includes initial binding of ρ at a specific site called *rut* and ATP hydrolysis-dependent displacement of the RNA polymerase and DNA template at the 3' end of the RNA. Transcription stop points are just downstream from the *rut* site. However, they can be spread over a fairly large number of specific points depending on sequences in and around the termination sites. The *rut* sites show C > G compositional bias, but no obvious consensus sequences have been found. In fact, the *rut* site could be a cytidine-binding site because the ATPase activity of ρ can be activated by the presence of unpaired C residues in the RNA cofactor, be it poly(C) or poly(U,C).

Although the exact mechanism of ρ-mediated transcription termination is not known, it is likely that ρ uses the *rut* site as an anchor point, making more stable secondary contacts with the 3' segments of the RNA and pulling the RNA away from the RNA Pol (97).

d. Termination of eukaryotic gene transcription

In contrast to the termination of transcription in prokaryotes which takes place precisely at the 3' end of the mRNA, that of eukaryotes occurs usually beyond the normal 3' end of mRNA. The 3' ends of nonhistone eukaryotic mRNAs are generated by an endonucleolytic cleavage with accompanying addition of a poly(A) tail (see Section II, this chapter). This suggests that the termination of eukaryotic transcription involves mechanisms different from those of prokaryotic transcription. Furthermore, the mechanism of termination is presumed to play a role in eukaryotic gene regulation (98).

One unique feature of most, but not all, eukaryotic mRNAs is that the 3'-end poly(A) addition site is preceded by a consensus sequence known as the poly(A) signal, AAUAAA. The actual termination of transcription by RNA Pol II occurs beyond the poly(A) site, often heterogeneously over one or more kilobase of sequence. One attractive model for termination incorporates this unique feature: the 3' processing mechanism plays an active role in the selection of termination sites (98). Evidence suggests that the termination of transcription and 3' processing are indeed not independent but coupled.

For efficient 3' end processing, two separate sequence elements are required: the AATAAA sequence and a second GT-rich element immediately downstream of the cleavage site at which poly(A) is added. Although 3' processing may initiate the termination process, actual termination is believed to occur at the RNA Pol II pause site(s). The pause site may be a hairpin structure and/or a sequence element that binds a specific protein factor. One such termination signal has been identified to be the CCAAT box sequence which functions as an orientation-dependent, position-independent terminator (99). Most probably the termination signal is mediated by the specific-binding protein CP1, similar to the role supposedly played by an AATAAA-binding factor(s).

References

1. Siebenlist, U., Simpson, R. B., and Gilbert, W. (1980). *Cell* **20**, 269–281.
2. Schafer, D. A., Gelles, J., Sheetz, M. P., and Landick, R. (1991). *Nature* **352**, 444–448.
3. Erie, D. A., Hajiseyedjavadi, O., Young, M. C., and von Hippel, P. H. (1993). *Science* **262**, 867–873.
4. Igarashi, K., and Ishihama, A. (1991). *Cell* **65**, 1015–1022.
5. Bowser, C. A., and Hanna, M. M. (1991). *JMB* **220**, 227–239.
6. Helmann, J. D., and Chamberlin, M. J. (1988). *Ann. Rev. Biochem.* **57**, 839–872.
7. Lonetto, M., Gribskov, M., and Gross, C. A. (1992). *J. Bacteriol.* **174**, 3843–3849.
8. Horwitz, M. S. Z., and Loeb, L. A. (1990). *PNARMB* **38**, 137–164.
9. Popham, D. L., Szeto, D., Keener, J., and Kustu, S. (1989). *Science* **243**, 629–635.
10. Collado-Vides, J., Magasanik, B., and Gralla, J. D. (1991). *Microbiol. Rev.* **55**, 371–394.
11. Sentenac, A. (1985). *Crit. Rev. Biochem.* **18**, 31–90.
12. Saltzman, A. G., and Weinmann, R. (1989). *FASEB J.* **3**, 1723–1733.
13. Woychik, N. A., and Young, R. A. (1990). *TIBS* **15**, 347–351.
14. Cornelissen, A. W. C. A., Evers, R., and Kock, J. (1988). In "Oxford Surveys on Eukaryotic Genes" (N. MacLean, Ed.), Vol. 5, pp. 91–131. Oxford University Press, Oxford.
15. Ishihama, A., and Nagata, K. (1988). *Crit. Rev. Biochem.* **23**, 27–76.
16. Sweetser, D., Nonet, M., and Young, R. A. (1987). *PNAS* **84**, 1192–1196.
17. Jeon, Y. H., Negishi, T., Shirakawa, M., Yamazaki, T., Fujita, N., Ishihama, A., and Kyogoku, Y. (1995). *Science* **270**, 1495–1497.
18. Nonet, M., Sweetser, D., and Young, R. A. (1987). *Cell* **50**, 909–915.
19. Suzuki, M. (1990). *Nature* **344**, 562–565.
20. Buratowski, S., Hahn, S., Guarente, L., and Sharp, P. A. (1989). *Cell* **56**, 549–561.
21. Jorgensen, E. D., Durbin, R. K., Risman, S. S., and McAllister, W. T. (1991). *JBC* **266**, 645–651.
22. Perez-Martin, J., Rojo, F., and de Lorenzo, V. (1994). *Microbiol. Rev.* **58**, 268–290.
23. Ross, W., Gosink, K. K., Salomon, J., Igarashi, K., Zou, C., Ishihara, A., Severinov, K., and Gourse, R. L. (1993). *Science* **262**, 1407–1413.
24. Lawson, C. L., and Carey, J. (1993). *Nature* **366**, 178–182.
25. Ptashne, M. (1992). "A Genetic Switch: Phage Lambda and Higher Organisms." Cell Press and Blackwall Scientific, Cambridge, MA.
26. Brenowitz, M., Pickar, A., and Jamison, E. (1991). *Biochem.* **30**, 5986–5998.
27. Nussinov, R. (1990). *Crit. Rev. B. MB.* **25**, 185–224.
28. Penotti, F. E. (1990). *JMB* **213**, 37–52.
29. Antonucci, T. K., Wen, P., and Rutter, W. J. (1989). *JBC* **264**, 17656–17659.
30. Zenke, M., Grundstrom, T., Matthes, H., Wintzerith, M., Schatz, C., Wildeman, A., and Chambon, P. (1986). *EMBO J.* **5**, 387–397.
31. Mueller, H.-P., and Schaffner, W. (1990). *TIG* **6**, 300–304.
32. Fan, C.-M., and Maniatis, T. (1990). *Genes Dev.* **4**, 29–42.

33. Young, L. S., Dunstan, H. M., Witte, P. R., Smith, T. P., Ottonello, S., and Sprague, K. U. (1991). *Science* **252**, 542–546.
34. Pugh, B. F., and Tjian, R. (1992). *JBC* **267**, 679–682.
35. Zwilling, J., Annweiler, A., and Wirth, T. (1994). *NAR* **22**, 1655–1662.
36. Roberts, S. G. E., and Green, M. R. (1994). *Nature* **371**, 717–720.
37. Ptashne, M. (1988). *Nature* **335**, 683–689.
38. Blackwood, E. M., and Eisenman, R. N. (1991). *Science* **251**, 1211–1217.
39. Ziff, E. B. (1990). *TIG* **6**, 69–72.
40. Blackwell, T. K., and Weintraub, H. (1990). *Science* **250**, 1104–1110.
41. Zabel, U., Schreck, R., and Baeuerle, P. A. (1991). *JBC* **266**, 252–260.
42. Forman, B. M., and Samuels, H. H. (1990). *Mol. Endocrinol.* **4**, 1293–1301.
43. Johnson, P. F., and McKnight, S. L. (1989). *Ann. Rev. Biochem.* **58**, 799–839.
44. Mitchell, P. J., and Tjian, R. (1989). *Science* **245**, 371–378.
45. Struhl, K. (1989). *TIBS* **14**, 137–140.
46. Steitz, T. A. (1990). *Q. Rev. Biophys.* **23**, 205–280.
47. Murre, C., McCaw, P. S., and Baltimore, D. (1989). *Cell* **56**, 777–783.
48. Harrison, S. C., and Aggarwal, A. K. (1990). *Ann. Rev. Biochem.* **59**, 933–969.
49. Busch, S. J., and Sassone-Corsi, P. (1990). *TIG* **6**, 36–40.
50. Sun, X.-H., and Baltimore, D. (1991). *Cell* **64**, 459–470.
51. Verrijzer, C. P., Karl, A. J., and van der Vliet, P. C. (1990). *Genes Dev.* **4**, 1964–1974.
52. Rosenfeld, M. G. (1991). *Genes Dev.* **5**, 897–907.
53. Pavletich, N. P., and Pabo, C. O. (1991). *Science* **252**, 809–817.
54. Landschulz, W. H., Johnson, P. F., and McKnight, S. L. (1988). *Science* **240**, 1759–1764.
55. O'Shea, E. K., Klemm, J. D., Kim, P. S., and Alber, T. (1991). *Science* **254**, 539–544.
56. Vinson, C. R., Sigler, P. B., and McKnight, S. L. (1989). *Science* **246**, 911–916.
57. O'Neil, K. T., Hoess, R. H., and DeGrado, W. F. (1990). *Science* **249**, 774–778.
58. Patel, L., Abate, C., and Curran, T. (1990). *Nature* **347**, 572–575.
59. Weiss, M. A., Ellenberger, T., Wobbe, C. R., Lee, J. P., Harrison, S. C., and Struhl, K. (1990). *Nature* **347**, 575–578.
60. O'Neil, K. T., Shuman, J. D., Ampe, C., and DeGrado, W. F. (1991). *Biochem.* **30**, 9030–9034.
61. Hai, T., and Curran, T. (1991). *PNAS* **88**, 3720–3724.
62. Thompson, C. C., Brown, T. A., and McKnight, S. L. (1991). *Science* **253**, 762–789.
63. Ingles, C. J., Shales, M., Cress, W. D., Triezenberg, S. J., and Greenblatt, J. (1991). *Nature* **351**, 588–590.
64. Cress, W. D., and Triezenberg, S. J. (1991). *Science* **251**, 87–90.
65. Allan, G. F., Tsai, S. Y., O'Malley, B. W., and Tsai, M.-J. (1991). *BioEssays* **13**, 73–78.
66. Cadepond, F., Schweizer-Groyer, G., Seard-Maurel, I., Jibard, N., Hollenberg, S. M., Giguere, V., Evans, R. M., and Baulieu, E.-E. (1991). *JBC* **266**, 5834–5841.
67. Lenardo, M. J., and Baltimore, D. (1989). *Cell* **58**, 227–229.
68. Urban, M. B., and Baeuerle, P. A. (1990). *Genes Dev.* **4**, 1975–1984.
69. Choy, B., and Green, M. R. (1993). *Nature* **366**, 531–536.
70. Ptashne, M., and Gann, A. A. F. (1990). *Nature* **346**, 329–331.
71. Mastrangelo, I. A., Courey, A. J., Wall, J. S., Jackson, S. P., and Hough, P. V. C. (1991). *PNAS* **88**, 5670–5674.
72. Pascal, E., and Tjian, R. (1991). *Genes Dev.* **5**, 1646–1656.
73. Brueggemeier, U., Kalff, M., Franke, S., Scheidereit, C., and Beato, M. (1991). *Cell* **64**, 565–572.
74. Schüle, R., Muller, M., Kaltschmidt, C., and Renkawitz, R. (1988). *Science* **242**, 1418–1420.
75. Alan, G. F., Ing, N. H., Tsai, S. Y., Srinivasan, G., Weigel, N. L., Thompson, E. B., Tsai, M.-J., and O'Malley, B. W. (1991). *JBC* **266**, 5905–5910.
76. Han, K., Levine, M. S., and Manley, J. L. (1989). *Cell* **56**, 573–583.
77. Lin, Y.-S., Carey, M., Ptashne, M., and Green, M. R. (1990). *Nature* **345**, 359–361.
78. Kerppola, T. K., and Curran, T. (1991). *Science* **254**, 1210–1214.
79. Schultz, S. C., Shields, G. C., and Steitz, T. A. (1991). *Science* **253**, 1001–1007.

80. Gossen, M., Freundlieb, S., Bender, G., Müller, G., Hillen, W., and Bujard, H. (1995). *Science* **268**, 1766–1769.
81. Croston, G. E., Kerrigan, L. A., Lira, L. M., Marshak, D. R., and Kadonaga, J. T. (1991). *Science* **251**, 643–649.
82. Barberis, A., Superti-Furga, G., and Busslinger, M. (1987). *Cell* **50**, 347–359.
83. Jaynes, J. B., and O'Farrell, P. H. (1991). *EMBO J.* **10**, 1427–1433.
84. Falvey, E., and Schibler, U. (1991). *FASEB J.* **5**, 309–314.
85. Roman, C., Cohn, L., and Calame, K. (1991). *Science* **254**, 94–97.
86. Eichhorn, G. L., Chuknyisky, P. P., Butzow, J. J., Beal, R. B., Garland, C., Janzen, C. P., Clark, P., and Tarien, E. (1994). *PNAS* **91**, 7613–7617.
87. Hanvey, J. C., and 14 coauthors (1992). *Science* **258**, 1481–1485.
88. Chamberlin, M. J. (1994). "The Harvey Lectures," Series 88, 1–21.
89. Liu, B., Wong, M. L., Tinker, R. L., Geiduscheck, E. P., and Alberts, B. M. (1993). *Nature* **366**, 33–39.
90. Platt, T. (1986). *Ann. Rev. Biochem.* **55**, 339–372.
91. Yanofsky, C. (1987). *TIG* **3**, 356–360.
92. Babitzke, P., Bear, D. G., and Yanofsky, C. (1995). *PNAS* **92**, 7916–7920.
93. d'Aubenton-Carafa, Y., Brody, E., and Thermes, C. (1990). *JMB* **216**, 835–858.
94. Wilson, K. S., and von Hippel, P. H. (1994). *JMB* **244**, 36–51.
95. Cheng, S.-W. C., Lynch, E. C., Leason, K. R., Court, D. L., Shapiro, B. A., and Friedman, D. I. (1991). *Science* **254**, 1205–1207.
96. Wong, H. C., and Chang, S. (1986). *PNAS* **83**, 3233–3237.
97. Richardson, J. P. (1991). *Cell* **64**, 1047–1049.
98. Proudfoot, N. J. (1989). *TIBS* **14**, 105–110.
99. Connelly, S., and Manley, J. L. (1989). *Mol. Cell. Biol.* **9**, 5254–5259.

B. T7 RNA Polymerase
[EC 2.7.7.6]

T7 RNA polymerase is a single 100-kDa polypeptide consisting of 883 amino acid residues. Despite its structural simplicity compared with the multimeric prokaryotic and eukaryotic RNA polymerases, T7 RNA polymerase (Pol) performs a highly specific promoter recognition and all the other steps in the transcription process in the absence of any additional protein factors. These and other properties make the T7 (and similarly SP6 and T3) RNA polymerase particularly well suited for studies of the mechanism of transcription. In recombinant DNA technology, T7 RNA polymerase is used as a powerful tool in driving high level expressions of cloned genes and/or in preparing *in vitro* a large amount of defined RNA transcripts. This section describes T7 RNA Pol as the representative of phage (SP6 and T3) RNA polymerases.

1. FUNCTIONS

a. Reaction specificity

Like SP6 and T3 RNA polymerases, T7 RNA Pol displays a stringent specificity for its own promoter to start transcription (1–3). The consensus promoter sequence derived from all 17 T7 promoters is a highly conserved sequence of 23 continuous base pairs (Fig. 7.3). The nucleotide positions from -7 to $+1$ are highly conserved in all three phage promoters. The promoter sequences differ significantly in the region from -8 to -12, suggesting that the base pairs in

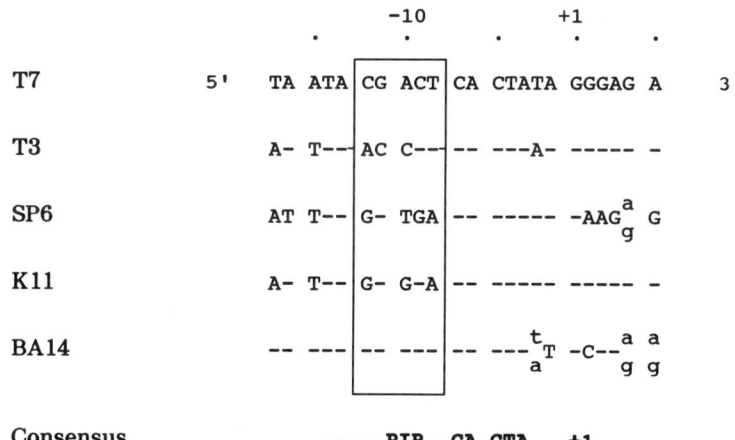

FIGURE 7.3 Consensus promoter sequences for phage RNA polymerases. The promoters for RNA polymerases of T7, T3, SP6, K11, and BA14 phages share a highly conserved structure composed of 23 bp (112). Using the 23-bp consensus sequence of 17 natural T7 promoters as a reference, other consensus phage promoters are aligned using dashes (-) for conserved positions and letters for divergent positions. Lowercase letters designate alternative base pairs that occur approximately 50% of the natural promoters within respective consensus sequences. By convention, the sequence of the nontemplate strand is shown; the transcription starts at position +1. The promoter identity region (PIR) is boxed.

the region, called the promoter identity region (PIR), are important for specific promoter recognition. The features of T7 and other phage promoters are quite distinct from those of *E. coli*. Whereas the phage promoters serve simultaneously as a recognition motif and a transcriptional start site, the bacterial (*E. coli*) promoters extend over 40 bp and exhibit two separate, highly conserved regions, −10 and −35 hexamers, suggesting more complex polymerase interactions (see Section I, this chapter). Despite the similarities of promoter sequences, T7 RNA Pol recognizes, like other phage RNA polymerases, its own promoter and carries out efficient and selective transcriptions of almost any DNA on which the promoter is located.

b. Reaction conditions

i. *Optimal reaction conditions.* T7 RNA Pol is optimally active at pH ~8.0 in Tris–Cl buffer (4,5). At pH 7.0 and 9.0, the activity falls off to 70 and 85% of the maximum, respectively. Although the optimal conditions for a particular reaction may vary slightly depending on templates and the amount and application of the transcripts, the following conditions are generally recommended for *in vitro* transcription reactions (6,7): 40 mM Tris–Cl (pH 8.0), 6 mM $MgCl_2$, 5 mM DTT, 1 mM spermidine, 50 μg/ml BSA, 40 μg/ml DNA template containing a phage promoter, 0.5 mM NTPs, 200–300 U/ml of T7 RNA Pol, and incubation at 37°C for 30 min. For high-yield transcriptions, the concentration of each NTP may be increased to 8.5 mM and the incubation time to 2 hr.

To protect RNA transcripts from possible degradation by contaminating RNases, the RNase inhibitor (RNasin) may be added at 1 U/μl final concentration. For the synthesis of capped transcripts, the GTP concentration is reduced to one tenth that of the noncapping reaction while G(5')ppp(5')G or its methylated cap analog is added at a cap analog/GTP ratio of 4. The reaction is stopped by adding EDTA to 50 mM or by heating to 65°C for 10 min.

For the preparation of labeled transcripts having high specific activities, the concentration of corresponding nonlabeled NTP should be reduced to 10 μM. The transcripts obtained under this or other limiting conditions may contain the products of premature termination and thus have nonuniform lengths. Incubation at a lower temperature, e.g., 30°, 22°, or even 4°C, results in a higher relative yield of full-length transcripts, although the total yield of transcripts may drop in some cases by as much as 50%.

For plasmid DNA templates, the yield can be over 30 μg RNA transcripts per μg of template DNA under the recommended conditions and as high as 200 μg RNA under some high-yield conditions. The DNA template usually does not interfere with subsequent applications of the RNA transcripts. If the removal of the DNA is desired, the reaction mixture can be digested with 20 μg/ml RNase-free DNase for 15 min at 37°C and the DNase removed by phenol/chloroform extraction. Alternatively, the reaction mixture is passed through a Mono Q column to purify RNA transcripts as well as to recycle the DNA template (8).

ii. **Kinetic parameters.** For the production of small transcripts, e.g., tetranucleotide (GACU) or pentanucleotide (GGACU), from the oligonucleotide-based consensus promoter, the K_m for DNA (22 bp) and k_{cat} at pH 7.8 and 37°C are 2 nM and 30 min^{-1}, respectively (9,10,105). The K_m for initiating GTP is ~0.60 mM. The apparent K_m values for each NTP are as follows: 47 μM (ATP), 60 μM (UTP), 81 μM (CTP), and 160 μM (GTP) (4). For large RNA transcripts, the elongation rate is 230 nt/sec at 37°C (11) or 60 nt/sec at 25°C (12), a rate approximately five times faster than that of E. coli RNA Pol.

iii. **Cofactor requirements**

(a) *Divalent cations.* T7 RNA Pol requires Mg^{2+} as a divalent cation. The enzyme has an optimal activity at a Mg^{2+} concentration of ~20 mM (4). [*Note:* The optimal binding of the enzyme to a T7 promoter occurs at 2–5 mM $MgCl_2$ (13).] Neither Mn^{2+} nor Co^{2+} can replace Mg^{2+}. The addition of 1 mM Mn^{2+}, Co^{2+}, or Ca^{2+} to the standard assay mixture containing 20 mM Mg^{2+} reduces the rate of T7 RNA synthesis by 68, 66, and 43%, respectively. Zn^{2+} is strongly inhibitory at concentrations above 10 μM (14). Unlike bacterial and eukaryotic RNA polymerases, T7 RNA Pol is not a Zn-dependent enzyme.

(b) *Monovalent salts.* NaCl and other monovalent ions in the reaction mixture reduce the overall yield of transcription (15,16). In the presence of KCl, the activity of T7 RNA Pol steadily decreases and there is essentially no activity above 0.2 M KCl (4). A similar inhibitory profile is observed with NH_4Cl or NaCl; at

~50 mM NaCl, the activity of T7 (and T3 and SP6 as well) RNA Pol is inhibited by more than 90%. Nevertheless, the polymerase reaction at ~50 mM NaCl may help suppress aberrant transcriptions from templates having 3′-protruding ends from 3–5% to 0.1–0.5%.

Chloride ion is presumed to bind, as does the phosphate ion, to the anion-binding site of the enzyme in competition with the backbone phosphates of substrate DNA. In fact, the apparent K_m, but not k_{cat}, is strongly dependent on the concentration of NaCl; the K_m increases over 20-fold when [NaCl] increases from 10 to 100 mM (105). In contrast to NaCl or KCl, T7 RNA Pol well tolerates potassium glutamate (KGlu), at least up to 100 mM.

iv. Activators, inhibitors, and inactivators.

Chemicals and biologicals: DTT or 2-MSH significantly stabilizes some preparations of T7 RNA Pol. (*Note*: For long-term storage of the enzyme, the addition of DTT is recommended every 6 months.) The enzyme is inactivated by SH reagents such as IAA, iodoacetamide, and PCMB and also by air oxidation.

T7 RNA Pol is stabilized by "carriers" such as BSA, spermidine, or *N,N*-dimethylated casein (18). Tween 20, at 0.05% (w/v), can fully replace the carrier substances in their stabilizing role, suggesting that the improvement of polymerase activity is most likely due to stabilization against nonspecific adsorptions (enzyme aggregations and/or adsorption to container walls) (105).

PEG at a 2.5% final concentration is 20–40% inhibitory to phage RNA polymerases. The presence of ethanol (5%) has a minimal effect on the polymerases.

The RNA synthesis catalyzed by T7 RNA Pol is inhibited by actinomycin D, elsamicin A, heparin, elevated concentrations of salt, and poly(rU). Heparin causes a rapid, quantitative release of RNA transcripts from the Pol–DNA–RNA ternary complex (19). Unlike *E. coli* RNA Pol, T7 RNA Pol is not inhibited by rifampicin or the Tagetin RNA polymerase inhibitor (Epicentre). The rifampicin resistance of T7 RNA Pol has been exploited for the exclusive overexpression in gram-negative bacteria of foreign genes cloned downstream of the T7 promoter.

Nucleic acids: T7 RNA Pol activity is not significantly affected by preincubation of the enzyme with poly(rC), poly(rI), or a variety of heterologous DNA. However, UV-irradiated T7 DNA interferes with T7 RNA Pol in the RNA synthesis reaction due to its competitive binding to the enzyme.

Triple helix-forming oligonucleotides: Oligonucleotides capable of forming triple helices with a purine-rich, double-stranded target (~45 bp) can inhibit RNA synthesis by blocking either transcription initiation (106) or elongation (107) depending on whether the triplex region overlaps the promoter sequence or is located downstream of it. The sequence-specific, triplex-dependent inhibition is observed only when the triplex formation is antiparallel to the most purine-rich strand and stable ($K_d \approx 10^{-9}\ M$), for instance, in the absence of Na$^+$ or K$^+$ salts before the transcription reaction.

T7 lysozyme: T7 phage lysozyme, a zinc amidase (118 amino acids) which cuts the amide bond between *N*-acetylmuramic acid (MurNAc) and L-Ala rather than the glycosidic bond between MurNAc and *N*-acetylglucosamine (GlcNAc), is an inhibitor of T7 RNA Pol. The N terminus of T7 lysozyme is presumed to be a part of the binding site for the polymerase (108). The specific interaction between T7 lysozyme and T7 RNA Pol plays a mutually regulatory role in viral replication. T7 lysozyme thus serves as a useful tool for the temporal control of gene expression in T7 RNA Pol-based expression systems.

c. Activity assay and unit definition

A typical assay mixture (100 μl) contains 40 mM Tris–Cl (pH 8.0), 20 mM $MgCl_2$, 5 mM DTT, 0.4 mM NTPs (including [^{3}H]UTP, 30 cpm/pmol), 60 μM T7 DNA, 50 μg/ml BSA, and T7 RNA Pol. The mixture is incubated at 37°C for 10 min, and the amount of acid-insoluble products is determined. Due to extensive recycling properties of the enzyme, the reproducibility of the assay drops rapidly as the incubation time goes beyond 10 min (20).

Unit definition: One unit is defined as the amount of enzyme that catalyzes the incorporation of 1 nmol of [^{3}H]UMP (or [^{3}H]AMP) into acid-insoluble products per hour. (*Note*: Some unit definitions in the literature are based on a 30-min incubation time.)

d. Substrate specificities

i. Promoter characteristics

(a) Promoter length. The promoters recognized by T7 RNA Pol can be represented by a highly conserved consensus sequence of 23 contiguous base pairs which includes the site of transcription initiation (+1) and extends from -17 to $+6$ (see Fig. 7.3). The -17 to $+6$ promoter defines the minimal length having full promoter strength (k_{cat} = 300 sec^{-1}, K_m = 0.02 μM) (10). The -14 promoter exhibits a substantially increased K_m (0.3 μM) but slightly reduced k_{cat} (180 sec^{-1}). The -12 promoter displays an extremely low level of specific transcription. The binding affinity for the consensus T7 promoter in oligonucleotides is $\sim 10^6$ M^{-1}.

Variants in the $+1$ to $+6$ region of the promoter are transcribed with reduced efficiencies (15,16). The degree of abortive initiation and thus the product yield vary considerably (see Table 7.2). The effects appear to be position independent but additive. The $+1$ to $+6$ sequence variants are, however, useful in preparing a variety of RNA transcripts.

(b) Template and nontemplate strands. Extension of the -12 promoter to the -17 position only in the template strand results in a significant improvement of promoter efficiency, whereas the extension of only the nontemplate strand results in no significant improvement. The nontemplate strand downstream of the position +3 from the transcription start site is not required for transcriptional activity. Partially ssDNA templates which are base paired only in the promoter region (-17 to $+1$) are as active in transcription as linear dsDNA (15).

ii. Templates

(a) Natural or modified DNA. T7 RNA Pol can utilize a wide variety of ss- or dsDNA as template as long as the duplex promoter structure is provided. Among the natural dsDNA, T7 DNA, T3 DNA, and salmon sperm DNA have

TABLE 7.2 Relative Effects of Sequence Variations in +1 to +6 Region on Yield of Transcription[a]

Transcript:	+1 pppG	G	G	A	G	+6 A ...

	Position		
Nucleotide	+1	+2	≥+3
G	[1][b]	[1][b]	
C	0.1	0.5	Smaller effects
A	0.2	0.5	

[a] Adapted from Milligan and Uhlenbeck (1989), *Meth. Enzy.* **180**, 51–62.
[b] Values in brackets are reference values.

appreciable template activities for T7 RNA Pol. The T3 DNA contains a single promoter site that is recognized by T7 RNA Pol, although the transcription from this promoter is relatively weak (at a level of 0.5–1%). *E. coli* DNA contains no apparent T7 promoters and thus is not a substrate for T7 RNA Pol.

The transcription on dsDNA template is arrested when T7 RNA Pol encounters modified bases, e.g., psoralen adduct or cyclobutane pyrimidine dimer, on the template strand but not on the nontemplate strand (19).

The presence of bulky adducts, e.g., benzo[*a*]pyrene, in the template DNA inhibits the transcription by causing premature termination at the adduct site (21).

T7 RNA Pol can transcribe DNA containing 1–24 nt gaps in the template strand, resulting in products that contain corresponding base deletions (17).

(b) Synthetic templates. Transcription of oligonucleotide templates is strongly influenced by template integrity and thus by the length and sequence. Templates prepared by ligating shorter oligodeoxynucleotides give a higher yield of transcripts than the templates synthesized as a long single block (109).

Of the double-stranded homopolymeric templates tested, only $d(I)_n \cdot d(C)_n$ and $d(G)_n \cdot d(C)_n$ are active as templates, both supporting the synthesis of poly(rG) (4). T7 RNA Pol is appreciably active with single-stranded polydeoxynucleotides but only with poly(dT) and poly(dC).

Templates containing stretches of A or U less than 8 bases can be readily transcribed. When a region of greater than 8 A or U is encountered, the polymerase begins to slip, creating a series of molecules with additional N, $N + 1$, $N + 2$, etc., residues of A or U (7). Although rare, the phenomenon of slippage is also observed with stretches of G or C. T7 RNA Pol can efficiently utilize a template with a CG stretch in the left-handed Z-DNA conformation (22).

(c) Template structures. T7 RNA Pol transcribes both linear and supercoiled plasmid DNA templates with almost equal efficiencies. For linearized DNA, the templates should have preferably blunt or 5′-protruding ends. Templates having

3'-protruding ends (e.g., after *Pst*I or *Sac*I digestion) generate a significant amount of by-products in addition to the expected products (23). Among the by-products are long, template-sized RNAs which are copied from the nontemplate strand and contain sequences from the vector DNA upstream of the promoter. DNA templates with blunt or 5'-protruding ends give rise to few (less than 1%) extraneous RNA products.

Similar to SP6 RNA Pol (24), T7 RNA Pol does not use nicks in the DNA as adventitious sites for transcription initiation or for termination.

(d) **RNA templates.** T7 RNA Pol can efficiently and specifically replicate certain RNA templates, whether natural or itself transcribed *in vitro*, in the absence of known promoter-like sequences. The RNA template-directed synthesis occurs with RNA (and not DNA) capable of forming either tRNA-like secondary structures (intramolecular primed templates) via palindromic sequences (25) or intermolecular primed templates. The minimally required structure, which is apparently further stabilized by the polymerase, is presumed to be two Watson-Crick base pairs at the primer terminus (27). The propensity for the RNA templated/primed extension by the T7 RNA Pol is the major cause of dsRNA by-products often observed in *in vitro* transcriptions.

iii. Transcript profile

(a) **Distribution of products.** In addition to the full-length products, the transcription of synthetic DNA yields a large amount (about 10-fold greater level than the full-length products) of smaller oligoribonucleotides 2 to 6 nt long (15). These by-products arise from the abortive cycling of RNA Pol.

Depending on the termini structure of the DNA template and the 3-end structure of the RNA, the runoff transcriptions may produce significant levels of dsRNA by-products (23,27). The dsRNA products behave as the "slow migrating component" in a nondenaturing PAGE, and can thus be readily distinguished from the ssRNA transcripts. Other by-products may also be generated from cryptic promoters located on the plasmid DNA.

The transcripts copied by T7 RNA Pol usually contain one nucleotide more or less than the template (15,24,28). (*Note*: A product containing one extra nucleotide is more prevalent with SP6 RNA Pol.) The full-length transcripts may thus appear as a doublet or occasionally as a multiplet in a gel electrophoresis.

(b) **Products under nonstandard reaction conditions.** Transcription under certain limiting conditions with a template whose message begins with GGGA results in a ladder of poly(rG) transcripts due to slippage of the RNA messages along the DNA template (16). The ladder tends to be limited to a length of 14 bases. This phenomenon is observed when the reaction is carried out in the presence of either GTP as the sole nucleotide or GTP plus CTP, UTP, and 3'-O-methyl-ATP.

iv. Nucleotides

(a) **5'-End nucleotides.** Under the standard reaction conditions containing GTP, the first nucleotide of the transcripts is always pppG. The nucleotide is subject to pyrophosphate exchange with [^{32}P]PP$_i$. The regular initiator GTP can be replaced with other mononucleotides (GDP, GMP, and guanosine) and dinucleotides [NpG, G(5')ppp(5')G, and m^{7}G(5')ppp(5')G] (26,30). In fact, short oligonucleotides up to hexamers having guanosine at the 3' end can be utilized by T7 RNA Pol as the RNA chain initiator, albeit at varying efficiencies (110). Only the 3'-terminal guanosine of these oligonucleotides is encoded by the template DNA at the transcription start point. The oligonucleotides can contain mixtures of deoxyribo-, ribo-, and modified nucleotides such as 2'-O-methylated and biotinylated nucleotides, providing versatility in purification, nonisotopic labeling, and RNA sequencing.

When the capping dinucleotide, m^{7}G(5')ppp(5')G or unmethylated G(5')ppp(5')G, is supplied at a cap analog/GTP ratio of 4 to the *in vitro* reaction, T7 RNA Pol produces capped RNA transcripts with >50% capping efficiency.

(b) 2'-Deoxyribonucleotides. T7 RNA Pol can incorporate a limited number of dNTPs into oligoribonucleotides. The efficiency of incorporation is dependent on the template sequence and the dNMP incorporated; compared to the corresponding rNMP, 70% for dA, 85% for dT or dU, 50% for dC, and 25% for dA plus dT (29). During the incorporation of dNMP, rNMP misincorporations may occur at a frequency of less than 5%.

The substitution of MnCl$_2$ for MgCl$_2$ results in a relaxation of the specificity of T7 RNA Pol, allowing incorporation of deoxynucleotides to 90–100% efficiency. However, the overall yield is lowered by a factor of 30 when compared with that obtained with ribonucleotides only.

A mutant (Y639F) T7 RNA Pol exhibits markedly reduced discrimination between dNTP and rNTP and, as if a DNA Pol, incorporates dNMP as efficiently as rNMP (33).

2'-Fluoro- and 2'-amino-2'-deoxynucleoside 5'-triphosphates are also substrates for T7 RNA Pol. Compared with CTP and UTP, the 2'-modified analogs (2'-F-CTP and 2'-F-UTP) exhibit approximately 500- and 200-fold lower efficiencies (k_{cat}/K_m), respectively, whereas 2'-NH$_2$-UTP is incorporated at a 20-fold lower efficiency (111). Runoff transcription of a 2500-base message with 2'-F-NTPs gives rise to full-length products but generates a considerable amount of shorter fragments as well. The 2'-NH$_2$-UTP gives a cleaner product than the 2'-F-UTP. The transcription of the tRNAs and tRNA-sized genes using any of the analogs is less susceptible to the problem of premature termination.

(c) 3'-Deoxyribonucleotide chain terminators. T7 RNA Pol can incorporate the chain-terminating 3'-deoxyribonucleotides with 23–60% reduced efficiencies as compared to the corresponding ribonucleotides (30): 3'-deoxy-ATP (23%), 3'-deoxy-UTP (30%), 3'-deoxy-CTP (31%), and 3'-deoxy-GTP (59%).

T7 (or SP6) RNA Pol is 100–1000 times more sensitive to 3'-deoxy-NTPs than *E. coli* RNA Pol or Qβ replicase. In a reaction containing 100 μM CTP, only 80 nM 3'-deoxy-CTP will inhibit the T7 RNA Pol activity by 50%.

(d) α-Thiophosphate nucleosides. T7 RNA Pol incorporates 5'-O-(1-thiotriphosphates), NTPαS, into RNA with an apparent K_m (for ATPαS) of ~15 μM,

similar to that for ATP (31). The Sp diastereoisomer is incorporated with an inversion of configuration to Rp. The Rp diastereoisomer is neither a substrate nor a competitive inhibitor. The phosphorothioate-containing RNAs in which a single nucleotide is substituted by the corresponding α-thio-NTP show an increased efficiency in protein synthesis (especially proteins of larger sizes) by stabilizing the mRNA in a prokaryotic *in vitro* translation system (32).

(e) Other nucleotide analogs. Like T3 or SP6 RNA Pol, T7 RNA Pol is capable of incorporating various nucleotide analogs such as digoxigenin-11-UTP (Dig-UTP) and biotin-UTP. These analogs can be used as nonradioactive labels, providing convenient and nonhazardous RNA probe products for detecting and isolating specific DNA and RNA sequences, and assaying polymerase activities.

Dig-UTP (BMB) is a UTP analog containing a steroid hapten isolated from *Foxglove* plants. In the Dig-UTP molecule, digoxigenin is coupled to the C-5 of uracil via an 11-atom spacer arm. Because of the bulky size of the Dig molecule, Dig-UTP labeling of some transcripts is more efficiently performed at a Dig-UTP/UTP ratio of 0.15/0.85 instead of the usual 0.35/0.65. Dig-labeled probes/products can be detected or quantitated *in situ* or on membranes by means of an anti-Dig antibody coupled with a reporter enzyme (e.g., alkaline phosphatase) and of the enzymatic reaction with chromogenic or (chemi)luminescent substrates.

Biotin-UTP is a biotinylated analog of UTP containing a biotin molecule covalently bound to the C-5 position of the uracil ring via various linkers, e.g., allylamine linker arm, C_{11} linker (BRL), and C_4 or 12-atom linker containing a cleavable disulfide bond (34). Compared with the unmodified UTP, the biotin-C_4-UTP is incorporated with an almost equal efficiency, while the biotin-C_{11}-UTP is incorporated with ~50% efficiency. As few as 10 biotin-C_4-UMP residues incorporated per 486-nt transcript result in >90% binding efficiency on a streptavidin/biotin-cellulose affinity column (34).

T7 RNA Pol also incorporates 5-Br-UTP (at 70–80% efficiencies) and 6-aza-UTP or 4-thio-UTP (at ~50% efficiency) to RNA transcripts. These nucleotide analogs are useful for cross-linking the RNA to associated proteins by UV (>300 nm) irradiation (7,35).

v. Processivity. The transcription by T7 RNA Pol is rapid and processive. Transcription of a T7 promoter-containing plasmid DNA (e.g., in pBR322) results in transcripts several times the plasmid length (36). Although pBR322 possesses its own specific set of transcription terminators, they are inefficient in stopping the transcription by T7 RNA Pol.

T7 RNA Pol requires 10–15 sec to initiate a RNA chain (4). The enzyme enters a processive mode (i.e., a relatively stable ternary complex of enzyme–DNA–RNA) after incorporation of at least eight bases (16).

vi. Transcription terminations. The typical termination signal (Tϕ) for phage T7 RNA Pol encodes, similarly to the ρ-independent terminators of *E. coli* RNA

Pol, a RNA sequence that can form a stable, GC-rich stem–loop followed by a run of six U residues. T7 RNA Pol can thus recognize some of the bacterial termination signals such as the *rrnB* T1 terminator at the end of the ribosomal operator, the *rrnC* terminator (32% efficiency), the pBR322 P4 terminator (54% efficiency), and the threonine (*thr*) attenuator. At the *thr* attenuator, which has a 14-bp stem plus a 6-base loop, the termination efficiencies of *E. coli* and T7 RNA polymerases are ~90 and ~40%, respectively (37). In general, the termination signals recognized by T7 RNA Pol may be characterized by a more stable loop, a more extended stem, and/or a longer run of U than those of average signals recognized by the bacterial RNA Pol. The relatively weak efficiencies of non-T7 terminators for the phage enzyme are not unusual when the cognate "strong" T7-Tϕ terminator is only 66% efficient. In fact, the inefficient termination and read-through transcriptions are thought to be required for phage survival, as a region of the genome that lies downstream from the terminator and encodes essential functions is transcribed only by the polymerases that fail to terminate (3).

Although a stem–loop and a 3' U-tract constitute the structural backbone for T7 terminators, the efficiency of termination is influenced by multiple factors such as sequence, stability, and length of the stem–loop, length and context of the U-tracts, promoter type, and distance between the promoter and termination site. Sequences upstream from the canonical stem–loop structure also exert marked effects on the position and efficiency of termination (130). Stem–loop structures that lack a 3' U-tract are not functional as a terminator for the phage enzyme. In contrast, U-tracts having certain 5' sequences that lack an apparent stem–loop structure can function as terminators (131).

Because T7 RNA Pol is relatively unaffected by the presence of non-T7 terminators usually found on cloning/expression plasmid vectors, the polymerase produces complete transcripts from almost any DNA linked to a T7 promoter. For this reason, the termination of *in vitro* transcriptions is usually made to occur at the end of the linear or linearized dsDNA templates, generating so-called *runoff* transcripts. Note that some T7 (or SP6) Pol molecules may not actually "run off" the linearized DNA templates, especially when the DNA templates carry 3'-end overhangs. With such templates, the polymerase may swing around and continue transcription on the opposite DNA strand, generating dsRNA by-products.

The runoff transcription termination strategy has been extended into a more general system which employs an oligonucleotide called "terminator oligo" (38). In this system, the oligonucleotide is annealed to a ssDNA template (e.g., M13 or its derivatives) and is then extended with Pol Ik converting the T7 promoter into an active double-stranded form (see Applications). When the ssDNA is designed to contain the target sequence in a noncoding strand, the 5' end of the oligonucleotide becomes the physical end of the template DNA in transcription.

Another approach to terminate a transcription without relying on a natural terminator sequence is to use a stable DNA–protein complex, e.g., *lac* operator–repressor. This system is efficient in blocking RNA polymerases (e.g., T7, T3, and *E. coli*) and results in the termination of transcriptions (39,40). A similar termination strategy based on a hexamer (GAATTC) binding protein (the Gln-111 mutant of EcoRI endonuclease) was shown to let T7 or SP6 RNA Pol carry out substantial levels of read-through transcription beyond the protein block (41).

e. Mechanism of RNA polymerase action

T7 RNA Pol is the best studied of a family of single-subunit RNA polymerases that alone are capable of performing highly specific promoter recognition and initiation of transcription from the relatively short, well-defined promoter sequences (see Fig. 7.3). In contrast, E. coli and eukaryotic RNA polymerases are multisubunit enzymes and require auxiliary factors to recognize much larger segmented promoter sequences and to initiate transcription. The simplicity of the enzyme and promoter structures and the apparent lack of regulation make T7 RNA Pol an ideal system to study fundamental aspects of promoter recognition and the initiation of transcription.

i. Bimodal recognition of promoters. A variety of biochemical and genetic experiments support the notion that phage promoters are recognized by phage (T7, T3, and SP6) RNA polymerases in a bimodal fashion: a *binding region* that extends from −17 to −8 and an *initiation region* that extends from −7 to +6, with initiation at +1 on the nontemplate strand (42,44,112). Base pairs at the positions −12 to −8, a subregion known as "promoter identity region (PIR)," are responsible for phage-specific contact with the cognate polymerase. The specific binding occurs on the promoter which is duplex upstream of about position −6 and melted downstream through the start site. In the duplex domain, both template and nontemplate bases contribute to recognition. Within the melted initiation region, the polymerase interacts mainly with the template strand of the promoter DNA (105).

For T7 and SP6 promoters, the two base pairs at −9 and −8 positions are the primary determinants of specificity and display the following hierarchy of importance: −8, −9 ≫ −10 > −12 (113). The double substitution at −9 and −8 of T7 promoter to SP6 sequences (−9G, −8A) yields significant (10%) SP6 promoter activities compared with the consensus SP6 promoter and virutally no T7 promoter activity. On the other hand, the SP6 promoter variants with both T7-specific −9C and −8T show good (40%) T7 promoter activities and some (20%) SP6 promoter activities. The sequences outside the PIR also contribute to the contacts with the polymerase but in a less significant manner.

For T7 and T3 promoters, discriminating elements involve primarily −11 position and secondarily −10 followed by −12 positions. Changing the −11G of T7 promoter to −11C of the T3 sequence prevents recognition by T7 RNA Pol but allows recognition by T3 RNA Pol (43). Substitution of −10A by −10C of the T3 promoter results in a 10-fold increase in the K_m of T7 RNA Pol but has little effect on the k_{cat}. The differential recognition is largely mediated by Asn-748 in T7 RNA Pol or Asp-749 in T3 RNA Pol. Changing the Asn-748 to Asp in T7 RNA Pol (T7-N748D) or Asp-749 to Asn in T3 RNA Pol (T3-D749N) switches the specificities of the phage polymerases at position −11 and possibly at -10 as well to the heterologous promoter sequences (114). T3 and T7 RNA Pol recognize the respective functional groups (amino, imino, and/or carbonyl) on -10 and -11 bases along the nontemplate strand wall of the major groove.

In contrast to the phage-dependent variation of sequences in the binding region, the initiation region sequences, especially from -7 to $+1$, are virtually identical for all phage promoters. Binding of a polymerase to its promoter results in melting of the dsDNA in the initiation region, as evidenced by a hypochromic shift and hypersensitivity of the nontemplate strand in this region to attack by single-strand-specific endonucleases. Compared with fully duplex promoters, the promoters lacking portions of the nontemplate strand downstream of position -5 through the start site are recognized with minor changes (within 2-fold) in K_m and permit specific initiation at a high rate (105). Deletion of the nontemplate sequences at position -6 and downstream or substitution of $-6T$ to $-6dU$ results in a >10-fold increase in K_m but no significant changes in k_{cat}. These data suggest that the formation of an "open" complex at the promoter initiation region involves major contacts between the polymerase and the template strand, while the polymerase begins to establish important contacts with the nontemplate bases from about position -6 and upstream.

X-ray crystallographic structures of T7 RNA Pol show that residue Asn-748 is located within the DNA-binding cleft, at a position approximately one helical turn (35 Å) of dsDNA away from the putative catalytic site (86). This corroborates the biochemical evidence that the polymerase binds to the promoter at one face of fully duplex DNA. Asn-748 is also found near a region (residues 144–159) presumed to be important for promoter binding. Incidentally, this region bears a significant sequence homology to region 2.4 of the *E. coli* σ factor which is believed to interact with the bases in the -10 region of *E. coli* promoters.

ii. **Dynamic interactions of RNA polymerase with substrates.** The DNA region protected by T7 RNA Pol in footprinting assays shows evidence of dynamic interactions between the polymerase and the promoter DNA during the course of transcription (13,45). In the absence of nucleotides, T7 RNA Pol binds the promoter DNA (e.g., T7 phage $\phi10$ promoter) weakly ($K_d \approx 25$ nM) and protects 19 bases in the region from -21 to -3. (*Note*: The polymerase binds nonpromoter DNA with a K_d of 48 mM). The binding of an initiating nucleotide (e.g., GTP) does not enhance the promoter binding or induce any conformational change of the enzyme (42,46). However, synthesis of a trinucleotide, r(GGG), in the presence of GTP alone, expands the length of the protected region to 29 bases from -21 to $+8$ positions ($K_d \approx 10$ nM). Transcription of a hexanucleotide, r(GGGAGA), further extends the protected region to 32 bases (-21 to $+11$). Finally, synthesis of a 15-nt message leads to a translocation of the protected region to 24 bases (-4 to $+20$). This sequential transition suggests stepwise conformational changes of the enzyme as it progresses from the initial "closed" complex with the promoter to a transcriptionally active "open" complex. Note that *E. coli* RNA Pol demonstrates a similar sequence of structural changes and nonmonotonic translocations during the initiation and elongation phases of transcription. For *E. coli* RNA Pol, the commitment to elongation occurs after the incorporation of 10 nt (47).

iii. **Incorporation of the first nucleotide.** The initiation of transcription is unique in two respects: (a) the 5′-triphosphate of the first nucleotide incorporated is not utilized in a bond-formation step and (b) the formation of the first bond in the elongation of RNA consists of a single NTP (the first base in the message). In fact, T7 RNA Pol makes no specific interaction with the 5′-phosphate of the initiating nucleotide (9). This is consistent with previous observations that T7 RNA Pol is able to incorporate, albeit at rates roughly corresponding to the strength of Watson–Crick base pairing, various nucleotide analogs as the first base, e.g., GMP, guanosine, ITP, and m^{7}GpppG. A single active site has thus been evolutionarily adapted to perform a dual function, i.e., to accept a 5′-triphosphate for the initiation and/or a 5′-monophosphate ester (RNA) for the elongation. This may also be part of the reason for the observed high rate of abortive fall-off of the initiation complex (Pol–DNA–RNA) even at saturating levels of all four NTPs (10).

iv. **Catalytic mechanism.** A dsDNA template binds the T7 RNA Pol molecule in the cleft formed by "thumb–palm–fingers" subdomains (85,86). As the promoter is functionally divided into two regions, so is the binding contact of the polymerase: a subsite exhibiting binding specificity to promoter/templates and a catalytic "pocket" recognizing the initiation region of the promoter. The template-binding specificity is apparently determined by the structural element defined by Asn-748 and other surrounding residues of the fingers region which is remarkably similar to that of Klenow polymerase and not to that of HIV-1 RTase. The catalytic pocket contains two catalytic residues (Asp-537 and Asp-812), Lys-631 which lies in close proximity to the initiating nucleotide triphosphate, and Phe-882 which provides the binding site to the elongating rNTP. The catalytic efficiency as well as the processivity of T7 RNA Pol is severely affected when the C-terminal Phe and Ala residues are mutated, enzymatically cleaved, or insertionally disrupted.

A 3-D model can be built for the ternary complex (polymerase-DNA template-RNA) in which the 3′ terminus of a small (three base) nascent RNA is placed near the two catalytic residues, while the elongating NTP is positioned, via a base contact with Phe-882, at a distance proper for the next polymerization. In the presence of Mg^{2+}, which is most likely coordinated with the two Asp residues, the phosphodiester bond is presumed to occur by a catalytic mechanism similar to that of Pol Ik.

The N-terminal domain of the polymerase interacts with the nascent RNA (or RNA–DNA hybrid) in the region 4–10 bases from the 3′ end of the RNA, thus providing an essential structural support for polymerase processivity. A truncated form (80 kDa) of the enzyme, obtained by partial digestion with trypsin of the N-terminal 20-kDa domain, carries out only "abortive transcription," i.e., synthesis of short oligonucleotides up to ~8 bases long (88). As the initial "abortive" ternary complex is converted to the "processive" ternary complex, T7 RNA Pol undergoes a conformational change which apparently involves wrapping of the thumb subdomain around the template. The "processive" complex is characterized by a higher processivity, a higher catalytic efficiency, and increased protection of the template DNA from DNase I digestion.

v. **Abortive cycling and chain elongation.** The transcription by the RNA polymerase is characterized by a high rate of *abortive cycling,* a fundamental aspect of early transcription during which dissociation of the enzyme–DNA–RNA ternary complex competes with elongation. The abortive cycling is an intrinsic property of all RNA polymerases.

Under normal transcription conditions, T7 RNA Pol produces many small transcripts (less than eight bases long) for each full-length transcript made. Whether the polymerase enters the abortive cycling or locks into an elongation mode is determined by several factors (16). The primary factor is the sequence of the transcript (or template). Dissociation is favored immediately following incorporation of UMP but less likely after GMP. The ease of melting and reannealing of the DNA duplex in the promoter region does not contribute to abortive cycling, nor do the strong interactions of the enzyme with the promoter. The conversion to a highly processive ternary complex takes place after the incorporation of eight bases. The transition to an elongation mode ensues from increased contacts between the nascent RNA and the DNA template and between RNA and enzyme. The RNA·DNA hybrid in the elongation complex is ~ 7 bp (19). Once in the elongation mode, the ternary complex is fairly stable and chain growth continues until the polymerase encounters a terminator. The elongation complex may lose its RNA component or encounter 1–24 nt gaps in the template strand, but T7 RNA Pol can remain attached to the DNA template for a long time or bypass the gap and continue template-directed transcription. The nascent RNA transcript can be stabilized against release by the copresence of plasmid DNA bearing both a T7 promoter and a T7 terminator but not just one (48).

2. APPLICATIONS

T7 RNA Pol, which is considered to be the representative of phage RNA polymerases, is an invaluable tool used to direct a high level expression of cloned genes *in vivo* under the control of the T7 promoter and to prepare milligram quantities of almost any RNA transcripts *in vitro*. A number of plasmid vectors that feature built-in single or double phage promoters are available for specialized as well as general gene cloning and expression. These transcription vectors make it much easier and versatile to use phage RNA polymerases in a wide variety of recombinant DNA techniques in conjunction with gene cloning. *In vivo* use of such phage RNA polymerase/promoter expression systems provides a means of gene therapy, vaccination, and phenotype alterations. *In vitro*-transcribed RNA can be used in diverse applications, for example, as hybridization probes in Northern (RNA) and Southern (DNA) blots, as substrates for RNA processing and structure–function studies, as substrates in exon–intron mapping of genomic DNA, as templates for reverse-transcription dideoxynucleotide sequencing, as templates for *in vitro* translations, as capture reagents for RNA-binding proteins, and as antisense probes and ribozymes.

Some of the features that make the T7 RNA Pol particularly useful for such applications include a stringent promoter selectivity, a high rate of synthesis, and an ability to produce complete transcripts from almost any DNA that is linked to a T7 promoter. Phage (T7, T3, and SP6) promoters are not recognized by bacterial RNA polymerases, but they can be recognized by some mammalian nuclear factors and be transcribed in mammalian cells even in the absence of phage RNA polymerases. Among various transcription systems containing T7, T3, SP6, or a bacterial synthetic consensus promoter, the T7 system is superior to other systems in several respects. The following examples illustrate the basic principles underlying wide applications of phage RNA polymerases and cognate promoter systems.

a. High level expressions of fusion genes

A relatively small amount of T7 RNA Pol originating from a cloned copy of the Pol gene is usually sufficient to direct a high level transcription from a T7 promoter in a multicopy plasmid in *E. coli*. Both the rate of synthesis and the accumulation of RNAs directed by T7 RNA Pol can reach levels comparable to those of rRNAs in a normal cell (49). Since the RNAs transcribed by T7 RNA Pol can rapidly saturate the translational machinery of *E. coli*, the rate of protein synthesis from such mRNAs depends critically on the efficiency of their translation. In some cases, such as the transcription of the *lacZ* gene in yeast, substantial amounts of intact mRNA may accumulate without an accompanying translation of the target mRNAs (50). In *E. coli,* the desynchronization between extremely efficient transcription and lagging translation may lead to a lower overall polypeptide yield with a T7 RNA Pol/promoter expression system than with an *E. coli* RNA Pol/*lacUV5* promoter system (115). When the mRNA is efficiently translated, the target protein can accumulate to over 50% of the total cell protein in 3 hr or less.

One potential problem associated with a highly efficient gene expression occurs when the gene product is toxic to the host. In those cases, a leaky premature expression of the phage Pol gene in the presence of a T7 promoter–target gene is likely to render the cells unviable before reaching the stage of full expression. This problem could be resolved in principle if the T7 RNA Pol is expressed under a "stringent" control of an inducible promoter, although in practice a small basal level of promoter activity may not be avoided. One interesting approach to circumvent this problem is to supply T7 lysozyme, a natural inhibitor of T7 RNA Pol, from a vector by coinfecting the cells expressing T7 RNA Pol (51,52). In the expression of the recombinant somatotropin gene in *E. coli* (53), a low level synthesis of T7 lysozyme has been shown to be crucial for high level (40%) gene expression. To acquire a better control of the T7 RNA Pol gene expression, various other strategies have been developed (see below).

I. DNA-Directed RNA Polymerases   535

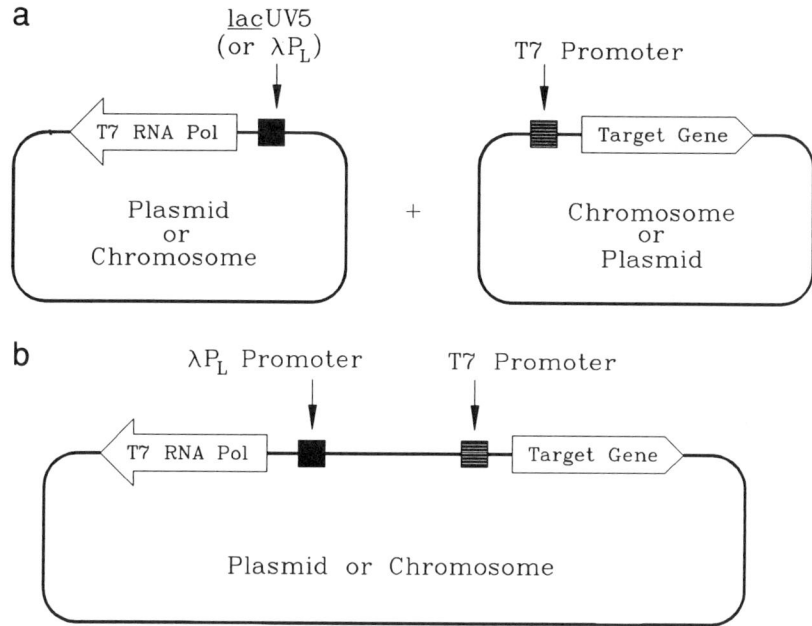

FIGURE 7.4 Two general strategies used in T7 RNA polymerase/T7 promoter–fusion gene expression systems. (a) A coupled T7 expression system and (b) a conjugated T7 expression system. In both systems, the expression of T7 RNA polymerase is controlled under an inducible promoter.

Because T7 RNA Pol is so active, strategies to prevent host or other DNA transcriptions, e.g., by the use of rifampicin, are virtually unnecessary (18). The position of T7 promoter *vis-à-vis* a target gene in a plasmid is usually not a matter of serious concern either, although the T7 promoter is conventionally placed upstream of the target gene. Unless an efficient T7 terminator is incorporated into the vector, the transcription proceeds several times around the plasmid without terminating. Introduction in the expression vector of a cleavage signal (e.g., T7 R1.1 signal) for dsRNA-specific *E. coli* RNase III results in the production of discrete RNAs of plasmid length. The T7 cleavage signal is an approximately 60-nt, irregular RNA hairpin consisting of two dsRNA segments that are separated by an asymmetric 9-nt internal loop (116).

In the following, various expression systems are described which differ in the strategies on how to deliver active T7 RNA Pol to a T7 promoter-linked target gene (Fig. 7.4).

i. Coupled expression systems. Perhaps the most common strategy for a high level expression of a cloned gene is to provide the RNA polymerase *in trans* from a polymerase gene expression system in parallel with the target gene expression system. Note that molecules of T7 RNA Pol can be codelivered with the target gene, thereby allowing a transient expression of the target gene in the cytoplasm.

Coupled expression systems can be designed in various formats: the polymerase and the target gene can be expressed separately or in tandem in the cytosol or chromosome.

*(a) **Chromosomal expression systems.*** A phage RNA Pol gene can be stably integrated into the E. coli chromosome via a λ lysogen and expressed under the IPTG-inducible control of the *lacUV5* promoter. A timely expression of the chromosomal T7 RNA Pol gene would then drive the expression of the T7 promoter-linked foreign gene cloned in a separate vector.

When eukaryotic cells are used as hosts, a similar chromosomal expression of the phage Pol gene can be engineered by transformation of the cells with appropriate DNA. Alternatively, the RNA polymerase expression in the cytoplasm can be targeted to the nucleus to drive the expression of a chromosome-located foreign gene. In such reverse chromosomal expression systems, the phage RNA Pol gene carried on a plasmid vector (54) or vaccinia virus vector (55) is expressed in eukaryotic cells (e.g., mouse L and NIH 3T3 cells) from the mouse metallothionein promoter or vaccinia promoter, respectively. The RNA Pol is then transported to the nucleus of yeast or mammalian cells by the polymerase-linked nuclear localization signal (Pro–Lys–Lys–Lys–Arg–Lys–Val) of the SV40 large T antigen (50,54,56). Unmodified T7 RNA Pol is mainly cytoplasmic and is unable to direct a detectable level of T7 promoter-linked gene expression in the nucleus. When the polymerase is overexpressed to a level of 0.5–1.5% total cell protein, however, some T7 RNA Pol may enter the nucleus passively. The expression from such T7 constructs has been shown to be at least sixfold more efficient than an analogous expression directed from the RSV-LTR promoter which is regarded as one of the strongest known in mammalian cells.

Note that a number of mammalian cell lines exhibit a considerable level of expression from the T7 promoter, even in the absence of T7 RNA Pol. This background expression, which actually interferes with the transcription performed by T7 RNA Pol, is due to the binding of some nuclear factors to the T7 promoter, thereby permitting transcription by RNA Pol II. Use of mutant T7 promoters, for instance, with simultaneous alterations at -13 (A to T) and $+1$ (G to C), substantially decreases the nuclear factor binding and consequently increases the specificity of T7 RNA Pol-dependent transcription in mammalian cells as much as 10-fold (117).

*(b) **Cytosolic expression systems.*** In contrast to the chromosomal expression of either phage RNA Pol or the phage promoter-linked target gene, cytosolic expression can also be obtained in eukaryotic as well as prokaryotic cells. For example, T7 RNA Pol can be directly codelivered with the plasmid containing the target gene into mammalian cells via various means (e.g., liposome) of transfection and induce transient cytoplasmic expression of the gene (118). Alternatively and for more sustained or controlled gene expression, the T7 RNA Pol gene can be expressed under the control of the λ P_L promoter or T7 promoter. The E. coli or eukaryotic cells harboring the phage Pol gene are coinfected with a second plasmid carrying a T7 promoter–target gene (49,57–59). Timely induction of

the RNA Pol drives an amplified expression of the target gene. In fact, the order of coinfection is arbitrary and, when the gene expression involves a toxic protein, a later infection of the vector carrying the phage Pol gene would be preferred. The *coinfection* strategy, initially developed for the expression in *E. coli* of a T7 promoter-fused gene carried on a recombinant plasmid by coinfection with the phage T7 (36), also works in yeast (60), insect cells, and mammalian cells. In a recombinant vaccinia virus gene expression system, the T7 RNA Pol gene is ligated to a vaccinia virus promoter and integrated into the viral genome (61). The T7 promoter–target fusion gene is constructed in a plasmid and is transfected into the cell for cytosolic gene expression. The hybrid vaccinia virus–T7 Pol system has proven to be highly efficient in transiently producing various proteins (antigens) in mammalian cells (62).

ii. **Conjugated expression systems.** Another strategy for delivering RNA polymerase is to use a conjugated system in which both the T7 Pol gene and a T7 promoter–fusion gene of interest are positioned in tandem on the same vector. This construct obviously necessitates tight control of the inducible expression, since a small amount of polymerase from a leaky transcription may not allow the cells to support the presence of a multicopy plasmid. Introduction of a single-copy gene bearing a T7 promoter into the *E. coli* chromosome which also contains the T7 Pol gene downstream of the *lacUV5* promoter has been shown to affect neither cell viability nor growth rate and yet allows the synthesis of a large amount of mRNAs (63).

Another expression system potentially more useful for strict control of gene expression in *E. coli* is the dual control system in which the expressions of both T7 RNA Pol and the T7 promoter–target gene are independently regulated. For example, the *lac* operator–repressor complex can be utilized to interrupt the transcription by T7 RNA Pol efficiently (39). The basic construct of this system consists of a *lacO* sequence (25 bp) centered 15 bp downstream from the transcription start point (40,64). The repression can be relieved by administering IPTG (0.5 mM), which induces a $>10^5$-fold expression of the target gene within 30 min after induction. A similar *lacO*–repressor system conjugated to a T3 promoter has been used for highly regulated expressions of foreign genes in mammalian cells which constitutively express the phage RNA polymerase and *lac* repressor (55,65).

iii. **Expression systems with provisions for high level translations.** One major block to the high level expression of a gene in T7 expression systems is the limited capacity of translation by host cells on overexpressed transcripts. A system (pET and pRSET series transcription–translation vectors) designed to overcome this limitation utilizes an efficient RBS [T7 gene 10 leader (g10-L) sequence] incorporated into the DNA upstream from a target gene (52,58,119). The T7 gene 10 encodes the capsid protein, the most highly expressed protein following a T7 infection (3), and the unusually efficient RBS (g10-L) allows expression of various foreign genes at a level 40-fold higher than that from the "consensus" RBS (66). Due to an extremely high level of expression, T7 gene 10 fusion proteins tend to form insoluble inclusion bodies in *E. coli*, a property that can be exploited in some cases for their rapid purification from other soluble cellular proteins.

Another provision for increasing the level of translation of uncapped transcripts in eukaryotic cells is the insertion of the "internal ribosome entry site" of the EMC virus into the 5' UTR of the target gene in the configuration of the T7 promoter-EMC-gene (120).

The level of gene expression in mammalian cells may critically depend on the stability of mRNA transcripts due to the lack of coupling between the relatively low 5' capping capacity of the cells and the high speed of T7 transcription. The stability of nascent RNA can be significantly increased by introducing stem–loop structures into the 5' UTR of a target gene (67).

b. Preparation of labeled or unlabeled RNA transcripts in vitro

In vitro transcription is an important technique for the preparation of RNA that can be used for various purposes. At present there are no efficient simple techniques available for the synthesis of RNA oligonucleotides comparable to those of current automated DNA synthesis. *In vitro* transcription is also the basis for efficient isothermal gene amplification techniques. *In vitro* transcription can be performed in one of the two formats (Fig. 7.5): by the vector-based approach in which a cloned target (template) DNA is abutted to the T7 promoter in a linearized plasmid (or phagemid) or by the synthetic oligonucleotide-based approach in which the template DNA is linked to a mobile promoter.

i. Transcription from vector-borne promoter(s). One of the earliest examples of this "standard" approach is the synthesis of infectious poliovirus RNA (68). In this approach, the DNA to be transcribed is first cloned into the polylinker of a transcription vector which contains a promoter(s) for T7 (and/or T3, SP6) RNA Pol adjacent to the linker. Unless a transcription terminator is present at or near the 3' terminus of the desired transcript, the vector DNA is usually linearized at a site (preferably avoiding 3'-end overhangs) appropriate for the promoter to be utilized, and then subject to runoff transcription.

To facilitate the cloning and *in vitro* transcription, a number of dedicated transcription vectors have been developed with single or dual phage promoters. The single-promoter vectors include a T7 promoter-containing pTZ(18/19,R/U) (Pharmacia and USB) (69), a SP6 promoter-containing pSP(64/65) (Promega) (24), and the pRSET family of vectors which contain a T7 terminator in addition to the T7 promoter (119). Among the typical dual-promoter vectors containing T7/T3 promoters are pT7/T3α(18/19) (BRL), pT7T3 18U (Pharmacia), and pBluescript II series phagemid and pWE cosmid vectors (Stratagene). Typical dual-promoter vectors containing T7/SP6 promoters include pGEM and pSP70 series vectors (Promega), pSPORT (1/2) (BRL), and pSPT(18/19) (BMB). Compared with single-promoter vectors, dual- and triple-promoter vectors offer the advantage of transcribing a cloned gene from either orientation simply by choosing an appropriate promoter-specific RNA polymerase.

In case no suitable restriction sites are available for plasmid linearization and runoff transcription, a *Fok*I recognition site can be engineered into the vector at

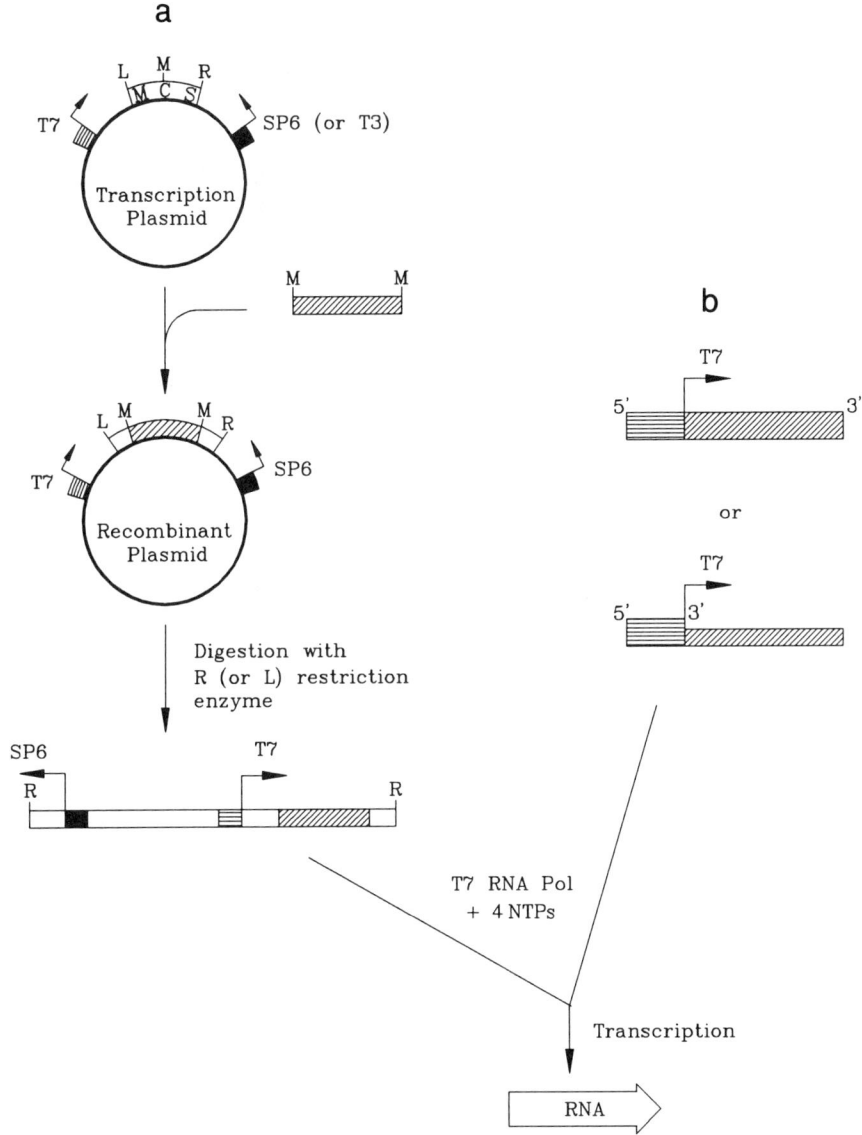

FIGURE 7.5 General schemes of *in vitro* runoff transcriptions. (a) The transcription from a dual promoter-carrying plasmid and (b) the transcription from a synthetic oligonucleotide template (ds or ss) linked to a mobile promoter. L, M, and R represent left, middle, and right restriction sites in the multiple cloning site (MCS) which is flanked by T7 and SP6 (or T3) promoter sequences.

the desired position flanking the expected 3' terminus of the transcript (121). Alternatively, a synthetic oligonucleotide can be used as an external transcription terminator based on a "terminator-oligonucleotide" strategy (28,38) (Fig. 7.6).

Another state-of-the-art technique introduced in *in vitro* (and *in vivo* as well) transcriptions is the use of an RNA transcript-trimming plasmid which incorporates a ribozyme technology (70,71). A typical vector construct has the configura-

540   *Chapter 7   RNA Polymerases*

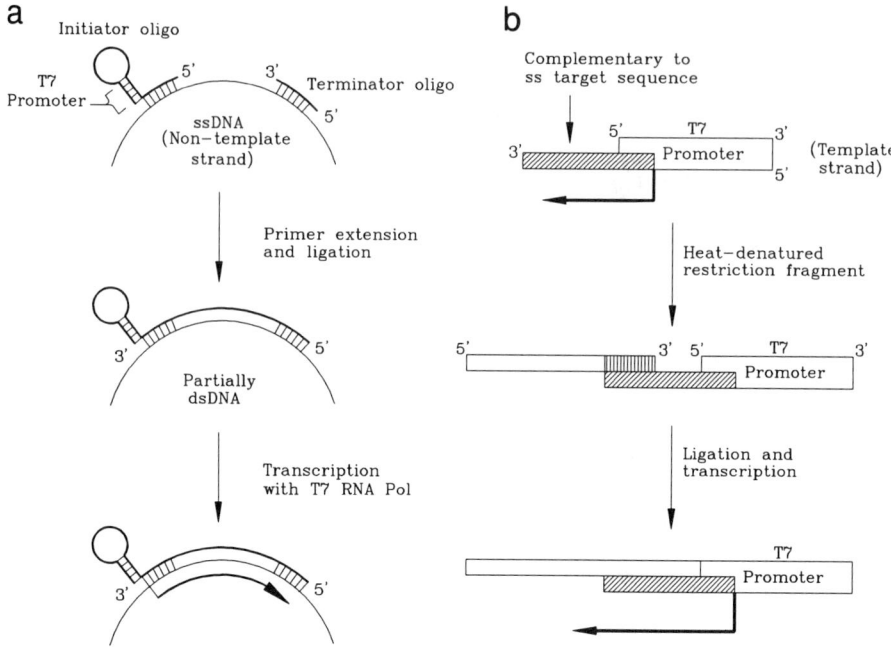

FIGURE 7.6 Two variations of the portable/mobile promoter transcription system. Two typical examples illustrate the strategies for providing a functional promoter to ssDNA templates. A single or two synthetic oligonucleotides are employed to form a partially double-stranded structure by either snap-back or straight annealing. The double-stranded region contains a phage promoter, while the protruding single-stranded region carries complementary sequences to the target DNA. (a) A portable promoter system involving initiator and terminator oligonucleotides (76): The method is particularly adapted for providing a portable promoter to ssDNA prepared from M13 and its derivative phages or phagemids. As a unique feature, the template strand of defined length is synthesized by a DNA polymerase using the "terminator" oligonucleotide as the primer. This versatile transcription system results in runoff transcripts with exact 3′ termini. (b) Another mobile promoter transcription system particularly useful for restriction–fragment templates (73): Single-stranded DNA (template strand), which is typically prepared from a restriction fragment, is annealed and ligated to the complementary part of the bipartite-sequence promoter oligo(s), thereby acquiring a requisite configuration for run-off transcriptions.

tion of a promoter (T7 or other phages)–5′ processing ribozyme–cloned DNA to be transcribed–3′ processing ribozyme (Fig. 7.7). This hammerhead ribozyme-based transcript-trimming system produces uniform RNAs with defined 5′ and 3′ ends from circular templates and with a higher transcriptional efficiency than the conventional runoff transcription (71). Furthermore, the ribozyme system offers other advantages: (a) it can replace the G-free RNA synthesis which was designed for specific transcription assays (72), and (b) it is versatile and can be applied to *in vivo* transcriptions with the possibility of generating multiple transcripts of *trans*-acting ribozymes.

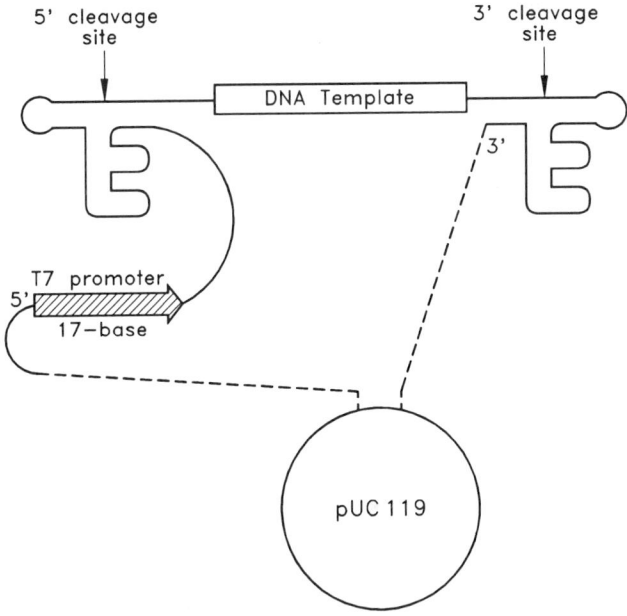

FIGURE 7.7 General configuration of a RNA transcript-trimming vector. A T7 promoter/transcription vector is constructed such that DNA templates are inserted between two *cis*-acting ribozyme sequences in a general order of T7 promoter–5' processing ribozyme–DNA template–3' processing ribozyme (71). The ribozyme sequence incorporated here is a hammerhead ribozyme consisting of a 24-nt active site and short flanking 5' and 3' regions that function as a substrate-binding site. Transcription from circular templates produces linear RNA transcripts with defined 5' and 3' ends derived by autocatalytic cleavages.

ii. Transcription from mobile (or portable) promoter carried on synthetic oligonucleotide. This is perhaps the most convenient method of *in vitro* transcriptions when a large amount of small, defined-length RNAs is desired (15,73). DNA templates provided with a T7 promoter are transcribed by T7 RNA Pol to generate runoff transcripts that initiate at a unique predictable position. Under optimal conditions, milligram quantities of virtually any RNA from 12 to 100 nt in length can be synthesized. Depending on how the promoter complex is assembled and how the template strand is provided, the oligonucleotide-based *in vitro* transcription strategy takes various formats.

(a) Standard format. The most commonly used format consists of two oligonucleotides: one for the universal priming structure and the other for individual templates. This format essentially requires a single template DNA to be synthesized for each type of RNA transcripts desired. The universal priming structure corresponds to the 17-nt fragment (positions -17 to -1) of the T7 consensus promoter. The priming oligonucleotide is annealed to the promoter portion of the synthetic template, reconstituting the double-stranded promoter region.

(b) **Variant formats.** Compared with the mobile/portable promoter strategy applied to synthetic templates, *in vitro* transcriptions are also widely performed with "natural" DNA templates that are prepared enzymatically.

(i) Templates for gene amplification. DNA templates provided with a functional double-stranded promoter(s) can be readily obtained by PCR using bracketing primers containing T7 or SP6 (or T3) promoter sequences at the 5' termini (74,75). When starting with an RNA, it can be converted first to cDNA using a RTase (AMV or MoLV) and a T7-promoter primer. The DNA templates are then transcribed by T7 RNA Pol into multiple copy RNAs which are used by RTase to generate new DNA templates. The combined use of RTase and T7 RNA Pol in a new gene amplification technique, called "3SR" (87) or "NASBA" (90), leads to self-sustained exponential replication of the target sequence under isothermal conditions.

(ii) Terminator-oligo-containing templates. In this portable promoter strategy (Fig. 7.6a), two synthetic oligonucleotides, "initiator-oligo" and "terminator-oligo," are annealed to a ssDNA template and extended by a DNA polymerase (e.g., Pol Ik) to generate the transcription coding strand having a defined terminus (28). The "initiator-oligo" is designed to be a bipartite sequence consisting of a double-stranded T7 promoter in a hairpin structure and a single-strand 5' extension which anneals to the ssDNA template (76).

(iii) Restriction fragment templates. A synthetic mobile promoter containing a template-complementary 3' extension is ligated to a target sequence, forming a mobile promoter complex (Fig. 7.6b). The single-stranded template DNA is prepared by heat denaturation of appropriate restriction fragments.

(c) **Comparison with use of transcription vectors.** Transcription of synthetic DNAs offers several advantages: (i) the unwanted 5'-flanking sequences that are often introduced as part of the cloning procedure can be eliminated, (ii) the sequence of the 3' end of the transcript is not limited by the need for a restriction site to linearize the template, and (iii) the time-consuming cloning, sequencing, and plasmid DNA preparation can be entirely avoided. Note that the RNA transcripts often contain the first (G-rich) 4-nt sequence as a part of the high efficiency promoter. The G residues tend to make the transcripts aggregate and render the RNA samples unsuitable for certain experiments such as NMR (77). This can be avoided by making the G residues fully base paired intramolecularly.

iii. Preparation of labeled or modified RNAs

(a) **Labeling of RNA transcripts.** RNAs can be labeled during the transcription *in vitro* by introducing either radiolabeled NTP (e.g., ^{32}P, ^{3}H, or ^{35}S) or nonisotopically labeled NTP (e.g., biotin-UTP or Dig-UTP). Alternatively, RNA can be radiolabeled posttranscriptionally using T4 polynucleotide kinase (see Section II, Chapter 5) or T4 RNA ligase (see Section II, Chapter 2).

(b) **5'-End capping and in vitro translation of transcripts.** Diverse applications of unlabeled mRNAs include *in vitro* and *in vivo* translations (78). *In vitro* translation is particularly useful in obtaining unstable or toxic proteins, and several eukaryotic and prokaryotic systems have been developed. For *in vitro* translations using wheat germ lysates (WGL) in particular, 5' end capping of transcripts is essential. In contrast, rabbit reticulocyte lysates (RRL) support the translation of uncapped RNA, albeit at a ~70% reduced rate (79). The 5' capping and polyadenylation are essential for the stability of mRNA when the expression of an RNA transcript is studied *in vivo*, for example, following injection into *Xenopus* oocytes (80,81). Polyadenylation of mRNA is not necessary for *in vitro* translation. In fact, polyadenylation has a drastic negative effect (60% reduction) on *in vitro* translation in both WGL and RRL systems.

A cell-free extract (S30) from *E. coli* K12 has been developed as an efficient coupled transcription/translation system which performs protein synthesis *in vitro* from the genes cloned in plasmids under the T7, T3, or SP6 promoter. Capping of the mRNA for eukaryotic proteins is apparently unnecessary with the S30 system. The coupled transcription/translation system, which was originally developed as the S30 translation system (122), can be used in a batchwise or continuous-flow mode (123,124). Use of the ribosome fraction collected from the S30 extracts is reported to improve the yield and efficacy further and more advantageously with nonlinearized plasmids than with linearized plasmids (125).

c. RNA polymerase-based RNA sequencing

RNA sequencing can be directly performed from dsDNA templates cloned downstream of a phage promoter, T7, T3, or SP6 (30,82). Transcription reactions are performed in the presence of a chain terminator, 3'-deoxynucleoside 5'-triphosphate, in a way similar to the dideoxynucleotide sequencing of DNA. The difference is that the DNA sequencing by RNA transcription obviates the need to prepare ssDNA templates or to use sequencing primers.

3. STRUCTURE

   i. **Primary structure.** T7 RNA Pol is a monomeric enzyme consisting of 883 deduced amino acid residues (M_r 98,856) (83). It has 12 Cys (all as free sulfhydryls) and an approximately equal number of acidic and basic amino acid residues. A single SH group ($pK_a \approx 7.8$) has been shown to be essential for catalytic activity (5). The N- and C-terminal residues are Met and Ala, respectively. The initiating Met is apparently retained.

   T7 RNA Pol is fully active as N-terminal fusion proteins; for example, the first two amino acids can be replaced by Val–Pro which is encoded by a connecting polylinker (84). Fusion proteins having foreign codons ahead of the codon 11 of the T7 RNA Pol gene maintain transcriptional activities, but insertion of a foreign sequence at four codons from the C terminus abolishes the polymerase activity (56). In fact, deletions of as many as 20 amino acids from the N terminus have minimal effects on polymerase activity.

   T7 RNA Pol has an overall 82% sequence identity with T3 RNA Pol. With other RNA/DNA polymerases, T7 RNA Pol shares limited but highly significant amino acid sequence conservation centered in three motifs A, B, and C (126):

motif A (amino acids 532–555) contains an active site residue (Asp-537) which corresponds to Asp-705 of Pol IK and is invariant in all polymerase sequences, *motif B* (amino acids 625–652) contains presumed active site residues (Lys-631 and Tyr-639) which respectively align with Lys-758 and Tyr-766 of Pol Ik, and *motif C* (amino acids 805–818) contains another active site residue (Asp-812) corresponding to Asp-882 of Pol Ik and invariant in all polymerase sequences.

ii. Tertiary structure

(a) Overall structure. The crystal structure of T7 RNA Pol, determined at a 3.3-Å resolution (85,86), shows that the polymerase is highly α-helical with overall dimensions of 75 × 75 × 65 Å. The C-terminal portion (~45%) of the molecule displays extensive structural homology to the polymerase domain (~400 residues) of the Klenow fragment and more limited homology to HIV-1 RTase.

The T7 RNA Pol molecule is organized around a deep cleft which could accommodate a dsDNA template. The DNA-binding cleft, which is formed by thumb–palm–fingers subdomains, is of a size that can cover almost two full turns of dsDNA. This is consistent with the observation that T7 RNA Pol protects ~20 bp of DNA from DNase I digestion. The C terminus of the enzyme lies within the cleft and adjacent to the putative catalytic pocket. In contrast, the N terminus lies away from the cleft. Much of the N-terminal portion of the molecule has no counterpart in the Klenow fragment.

(b) Active site. The N-terminal domain (residues from 1 to ~307) contains the subsites for binding nascent RNA and initiating NTP. The Glu-222–Lys mutation enables the polymerase to recognize an expanded range of T7 promoter-like sequences with variations at the -6 to -9 positions (127).

The putative catalytic pocket contains five residues: two catalytic residues (Asp-537 and Asp-812), which are invariant in all polymerases, and a cluster of three residues (Lys-631, Tyr-639, and Gly-640), which are invariant in all DNA-directed polymerases (126). Lys-631 interacts with the phosphates of NTP substrates. The Y639F mutation results in markedly reduced discrimination against dNTP utilization, without affecting promoter specificity or overall activity of the polymerase (33). Asp-569 and Tyr-571 are also functionally important residues.

The C terminus of the enzyme lies close to and immediately above the catalytic pocket and plays a role in catalysis and processivity. Evidence suggests that the penultimate residue (Phe-882) provides a base contact to the elongating NTP (89,92). Mutations of Phe-882 to nonaromatic residues result in an increase in K_m for all NTPs, but this is much greater for purines than for pyrimidines. Mutation of Tyr-571 to Ser results in an enzyme that is catalytically competent but unable to bind to the promoter (128). The promoter specificity is determined by a region of ~80 residues (amino acids 674–752) (91). In fact, Asn-748 of T7 RNA Pol has been identified as the determinant of promoter specificity recognizing the T7 promoter at the −10 region (114,129).

iii. **Physical parameters.** T7 RNA Pol has an ε_{280} of 1.4×10^5 M^{-1} cm^{-1} (14). The A_{280} of a 1% solution is 7.4 (93). The $s^{\circ}_{20,w}$ is 5.9–6.3 S (93), a value consistent with its existence as a monomer of ~100 kDa.

4. GENETICS

i. Primary structure and gene locus. T7 RNA Pol is the product of the T7 gene 1 which is one of the "early" (or "class I") genes of the virulent *E. coli* phage T7. The genome of the phage is a linear dsDNA having 39,936 bp (3). The T7 gene 1 has been cloned (94) and sequenced (83). The structural gene for T7 RNA Pol occupies nucleotide positions 3171–5822 (883 amino acid codons) on the phage DNA. The gene 1 mRNA, a product of the "early" primary transcript processed by a double-strand specific *E. coli* enzyme RNase III (95), corresponds to nucleotides 3139–5887 of T7 DNA.

ii. Regulatory region of the T7 gene 1 and regulation of polymerase activity. A weak *E. coli* RNA Pol promoter, promoter C, is located in the region 5' proximal to the T7 gene 1 (96). Only 9 nt downstream of the T7 gene 1 termination codon (UAA) are two T7 RNA Pol promoters, ϕ1.1A (starting nt 5858) and ϕ1.1B (starting nt 5923) (3). *E. coli* DNA does not have promoters for T7 RNA Pol. In the absence of any exogenous T7 promoter, the growth rate of *E. coli* is unaffected even when >2% of the cellular protein is T7 RNA Pol.

Immediately following the phage infection in *E. coli*, several new (class I or early) proteins are synthesized, probably within the first 5 min, as the result of transcription of some of the T7 genes by *E. coli* RNA Pol. Among the new proteins is the T7 gene 1 protein (T7 RNA Pol), which takes over the transcription of the remaining (80%) T7 genes which fall into two groups: "middle" genes having class II promoters and "late" genes having stronger class III promoters (97). The class II and III promoter sequences are only slightly different, and yet the transcription from the class II promoters is preferentially inhibited by conditions (e.g., ionic strength, temperature, and Me_2SO) that increase the helix stability (98).

T7 RNA Pol activity is regulated (or inhibited) by a direct interaction with the T7 lysozyme (gene 3.5 product) which forms a specific complex with the polymerase (99). Under the conditions of amino acid starvation, for example, in medium lacking some critical amino acids, the *in vivo* activity of T7 RNA Pol is subject to a stringent control similar to that on *E. coli* RNA Pol (100).

iii. RNA polymerase-associated activities. In addition to the role of transcribing the majority of phage T7 genes, T7 RNA Pol functions as a primase, introducing replication bubbles during the initiation of T7 DNA replication (101). T7 RNA Pol also functions as an efficient RNA-directed RNA Pol on a short (~70 nt) RNA template containing a specific T7 promoter.

5. SOURCES

T7 RNA Pol is present in relatively low amounts in T7-infected cells, constituting ~0.1% of the total cellular protein. The polymerase can now be conveniently prepared from overexpression systems in *E. coli* (strain BL21) (102,103). When the gene was cloned into pKC30, a pBR322-derived vector that contains the temperature-inducible λ P_L promoter, T7 RNA Pol could be overexpressed 200-fold (or 20% of the total soluble protein) (57). T7 RNA Pol has also been overexpressed from pAR1219, another pBR322-derived plasmid, under the control of the *lacUV5* promoter and purified to a specific activity of 300–400 units/μg (14,94).

Another potentially rich source of the phage polymerase is an "autogene" system in which the expression of the T7 Pol gene is induced from its own promoter. To keep the activity of the polymerase below the toxic level until the gene is fully induced, a configuration has been designed that allows both inhibition of T7 RNA Pol by T7 lysozyme and repression of T7-*lac* promoter activity by *lac* repressor (104). Neither type of inhibition by itself is sufficient to control the autogene. On derepression of the T7 promoter by adding IPTG in the presence of a low level of T7 lysozyme, T7 RNA Pol can be produced autocatalytically until the cells die due to toxic effects.

During the course of purification, T7 RNA Pol may be proteolytically cleaved at a specific site, most prevalently at Tyr^{178}/Lys^{179}, dependent on the *E. coli* host strain (57). The two peptides (80 and 20 kDa) remain tightly associated throughout the purification. The nicking results in a loss of ~70% polymerase activity (12,88), and the nicked enzyme terminates less efficiently than the intact polymerase at the T7-Tϕ terminator. Note that the T7 RNA Pol is not cleaved in intact cells. To minimize the artifactual cleavage of the polymerase during purification, a protease inhibitor(s) such as leupeptin (103) or bacitracin and benzamidine (102) should be included in the buffer. In fact, the best strategy is to express the T7 gene 1 in an *E. coli* strain lacking the protease activity.

The major source of this endoprotease activity is OmpT, an outer membrane protein of *E. coli* (100). All *E. coli* K strains, with the exception of *ompT* deletion mutants (UT4400 or UT5600), express the protease, although DH1 does so at a greatly reduced level. Two *E. coli* B strains (B834 and BL21) and one C strain (C1757) apparently lack active OmpT protease on the cell surface. *E. coli* BL21 is also naturally deficient in the *lon* protease and grows well in minimal media, making the strain well suited for the expression of cloned genes whose products are sensitive to proteolysis during purification.

References

1. Rosa, M. D. (1979). *Cell* **16**, 815–825.
2. Panayotatos, N., and Wells, R. D. (1979). *Nature* **280**, 35–39.
3. Dunn, J. J., and Studier, F. W. (1983). *JMB* **166**, 477–535.
4. Chamberlin, M., and Ring, J. (1973). *JBC* **248**, 2235–2244.
5. Oakley, J. L., Strothkamp, R. E., Sarris, A. H., and Coleman, J. E. (1979). *Biochem.* **18**, 528–537.
6. Yisraeli, J. K., and Melton, D. A. (1989). *Meth. Enzy.* **180**, 42–50.
7. Milligan, J. F., and Uhlenbeck, O. C. (1989). *Meth. Enzy.* **180**, 51–62.
8. Jahn, M. J., Jahn, D., Kumar, A. M., and Soll, D. (1991). *NAR* **19**, 2786.
9. Martin, C. T., and Coleman, J. E. (1989). *Biochem.* **28**, 2760–2762.
10. Martin, C. T., and Coleman, J. E. (1987). *Biochem.* **26**, 2690–2696.
11. Golomb, M., and Chamberlin, M. (1974). *JBC* **249**, 2858–2863.
12. Ikeda, R. A., and Richardson, C. C. (1987). *JBC* **262**, 3790–3799.
13. Gunderson, S. I., Chapman, K. A., and Burgess, R. R. (1987). *Biochem.* **26**, 1539–1546.
14. King, G. C., Martin, C. T., Pham, T. T., and Coleman, J. E. (1986). *Biochem.* **25**, 36–40.

15. Milligan, J. F., Groebe, D. R., Witherell, G. W., and Uhlenbeck, O. C. (1987). *NAR* **15**, 8783–8798.
16. Martin, C. T., Muller, D. K., and Coleman, J. E. (1988). *Biochem.* **27**, 3966–3974.
17. Zhou, W., Reines, D., and Doetsch, P. W. (1995). *Cell* **82**, 577–585.
18. Chamberlin, M., and Ring, J. (1973). *JBC* **248**, 2245–2250.
19. Sastry, S. S., and Hearst, J. E. (1991). *JMB* **221**, 1091–1110.
20. Chamberlin, M., and Ryan, T. (1982). "The Enzymes," 3rd ed., Vol. 15, pp. 87–108.
21. Nath, S. T., and Romana, L. J. (1991). *Carcinogenesis* **12**, 973–976.
22. Droge, P., and Pohl, F. M. (1991). *NAR* **19**, 5301–5306.
23. Schenborn, E. T., and Mierendorf, R. C., Jr. (1985). *NAR* **13**, 6223–6236.
24. Melton, D. A., Krieg, P. A., Rebagliati, M. R., Maniatis, T., Zinn, K., and Green, M. R. (1984). *NAR* **12**, 7035–7056.
25. Konarska, M. M., and Sharp, P. A. (1990). *Cell* **63**, 609–618.
26. Sampson, J. R., and Uhlenbeck, O. C. (1988). *PNAS* **85**, 1033–1037.
27. Cazenave, C., and Uhlenbeck, O. C. (1994). *PNAS* **91**, 6972–6976.
28. Krupp, G. (1988). *Gene* **72**, 75–89.
29. Wyatt, J. R., and Walker, G. T. (1989). *NAR* **17**, 7833–7842.
30. Axelrod, V. D., and Kramer, F. R. (1985). *Biochem.* **24**, 5716–5723.
31. Griffiths, A. D., Potter, B. V. L., and Eperon, I. C. (1987). *NAR* **15**, 4145–4162.
32. Ueda, T., Tohda, H., Chikazumi, N., Eckstein, F., and Watanabe, K. (1991). *NAR* **19**, 547–552.
33. Sousa, R., and Padilla, R. (1995). *EMBO J.* **14**, 4609–4621.
34. Fenn, B. J., and Herman, T. M. (1990). *Anal. Biochem.* **190**, 78–83.
35. Tanner, N. K., Hanna, M. M., and Abelson, J. (1988). *Biochem.* **27**, 8852–8861.
36. McAllister, W. T., Morris, C., Rosenberg, A. H., and Studier, F. W. (1981). *JMB* **153**, 527–544.
37. Jeng, S.-T., Gardner, J. F., and Gumport, R. I. (1990). *JBC* **265**, 3823–3830.
38. Eperon, I. C. (1986). *NAR* **14**, 2830.
39. Giordano, T. J., Deuschle, U., Bujard, H., and McAllister, W. T. (1989). *Gene* **84**, 209–219.
40. Deuschle, U., Gentz, R., and Bujard, H. (1986). *PNAS* **83**, 4134–4137.
41. Pavco, P. A., and Steege, D. A. (1991). *NAR* **19**, 4639–4646.
42. Muller, D. K., Martin, C. T., and Coleman, J. E. (1989). *Biochem.* **28**, 3306–3313.
43. Klement, J. F., Moorefield, M. B., Jorgensen, E., Brown, J. E., Risman, S., and McAllister, W. T. (1990). *JMB* **215**, 21–29.
44. Jorgensen, E. D., Durbin, R. K., Risman, S. S., and McAllister, W. T. (1991). *JBC* **266**, 645–651.
45. Ikeda, R. A., and Richardson, C. C. (1986). *PNAS* **83**, 3614–3618.
46. Basu, S., and Maitra, U. (1986). *JMB* **190**, 425–437.
47. Krummel, B., and Chamberlin, M. J. (1989). *Biochem.* **28**, 7829–7842.
48. Sastry, S. S., and Hearst, J. E. (1991). *JMB* **221**, 1111–1125.
49. Studier, F. W., and Moffatt, B. A. (1986). *JMB* **189**, 113–130.
50. Benton, B. M., Eng, W. K., Dunn, J. J., Studier, F. W., Sternglanz, R., and Fisher, P. A. (1990). *Mol. Cell. Biol.* **10**, 353–360.
51. Studier, F. W. (1991). *JMB* **219**, 37–44.
52. Studier, F. W., Rosenberg, A. H., Dunn, J. J., and Dubendorff, J. W. (1990). *Meth. Enzy.* **185**, 60–89.
53. O'Mahony, D. J., Wang, H. Y., McConnell, D. J., Jia, F., Zhou, S. W., Yin, D., and Qi, S. (1990). *Gene* **91**, 275–279.
54. Lieber, A., Kiessling, U., and Strauss, M. (1989). *NAR* **17**, 8485–8493.
55. Rodriguez, D., Zhou, Y., Rodriguez, J.-R., Durbin, R. K., Jimenez, V., McAllister, W. J., and Esteban, M. (1990). *J. Virol.* **64**, 4851–4857.
56. Dunn, J. J., Krippl, B., Bernstein, K. E., Westphal, H., and Studier, F. W. (1988). *Gene* **68**, 259–266.
57. Tabor, S., and Richardson, C. C. (1985). *PNAS* **82**, 1074–1078.
58. Rosenberg, A. H., Lade, B. N., Chui, D., Lin, S.-W., Dunn, J. J., and Studier, F. W. (1987). *Gene* **56**, 125–135.
59. Elroy-Stein, O., and Moss, B. (1990). *PNAS* **87**, 6743–6747.

60. Chen, W., Tabor, S., and Struhl, K. (1987). *Cell* **50**, 1047–1055.
61. Fuerst, T. R., Niles, E. G., Studier, F. W., and Moss, B. (1986). *PNAS* **83**, 8122–8126.
62. Fuerst, T. R., Earl, P. L., and Moss, B. (1987). *Mol. Cell. Biol.* **7**, 2538–2544.
63. Chevrier-Miller, M., Jacques, N., Raibaud, O., and Dreyfus, M. (1990). *NAR* **18**, 5787–5792.
64. Dubendorff, J. W., and Studier, F. W. (1991). *JMB* **219**, 45–59.
65. Deuschle, U., Pepperkok, R., Wang, F., Giordano, T. J., McAllister, W. T., Ansorge, W., and Bujard, H. (1989). *PNAS* **86**, 5400–5404.
66. Olins, P. O., Devine, C. S., Rangwala, S. H., and Kavka, K. S. (1988). *Gene* **73**, 227–235.
67. Fuerst, T. R., and Moss, B. (1989). *JMB* **206**, 333–348.
68. van der Werf, S., Bradley, J., Wimmer, E., Studier, F. W., and Dunn, J. J. (1986). *PNAS* **83**, 2330–2334.
69. Mead, D. A., Szczesna-Skorupa, E., and Kemper, B. (1986). *Prot. Eng.* **1**, 67–74.
70. Dzianott, A. M., and Bujarski, J. J. (1989). *PNAS* **86**, 4823–4827.
71. Taira, K., Nakagawa, K., Nishikawa, S., and Furukawa, K. (1991). *NAR* **19**, 5125–5130.
72. Sawadogo, M., and Roeder, R. G. (1985). *PNAS* **82**, 4394–4398.
73. Loewy, Z. G., Leary, S. L., and Baum, H. J. (1989). *Gene* **83**, 367–370.
74. Kovach, J. S., McGovern, R. M., Cassady, J. D., Swanson, S. K., Wold, L. E., Vogelstein, B., and Sommer, S. S. (1991). *J. Natl. Cancer Inst. (USA)* **83**, 1004–1009.
75. Young, I. D., Ailles, L., Deugau, K., and Kisilevsky, R. (1991). *Lab. Invest.* **64**, 709–712.
76. Krupp, G., and Soll, D. (1987). *FEBS Lett.* **212**, 271–275.
77. Szewczak, A. A., White, S. A., Gewirth, D. T., and Moore, P. B. (1990). *NAR* **18**, 4139–4142.
78. Krieg, P. A., and Melton, D. A. (1984). *NAR* **12**, 7057–7070.
79. Roitsch, T., and Lehle, L. (1989). *BBA* **1009**, 19–26.
80. Drummond, D. R., Armstrong, J., and Colman, A. (1985). *NAR* **13**, 7375–7394.
81. Green, M. R., Maniatis, T., and Melton, D. A. (1983). *Cell* **32**, 681–694.
82. Parvin, J. D., Smith, F. I., and Palese, P. (1986). *DNA* **5**, 167–171.
83. Moffatt, B. A., Dunn, J. J., and Studier, F. W. (1984). *JMB* **173**, 265–269.
84. Ostrander, E. A., Benedetti, P., and Wang, J. C. (1990). *Science* **249**, 1261–1265.
85. Sousa, R., Rose, J. P., Chung, Y. J., Lafer, E. M., and Wang, B.-C. (1989). *Proteins (S.F.G)* **5**, 266–270.
86. Sousa, R., Chung, Y. J., Rose, J. P., and Wang, B.-C. (1993). *Nature* **364**, 593–599.
87. Guatelli, J. C., Whitfield, K. M., Kwoh, D. Y., Barringer, K. J., Richman, D. D., and Gingeras, T. R. (1990). *PNAS* **87**, 1874–1878.
88. Muller, D. K., Martin, C. T., and Coleman, J. E. (1988). *Biochem.* **27**, 5763–5771.
89. Mookhtiar, K. A., Peluso, P. S., Muller, D. K., Dunn, J. J., and Coleman, J. E. (1991). *Biochem.* **30**, 6306–6313.
90. Sooknanan, R., and Malek, L. T. (1994). *Bio/Technology* **13**, 563–564.
91. Joho, K. E., Gross, L. B., McGraw, N. J., Raskin, C., and McAllister, W. T. (1990). *JMB* **215**, 31–39.
92. Patra, D., Lafer, E. M., and Sousa, R. (1992). *JMB* **224**, 307–318.
93. Niles, E. G., Conlon, S. W., and Summers, W. C. (1974). *Biochem.* **13**, 3904–3912.
94. Davanloo, P., Rosenberg, A. H., Dunn, J. J., and Studier, F. W. (1984). *PNAS* **81**, 2035–2039.
95. Bardwell, J. C., Regnier, P., Chen, S.-M., Nakamura, Y., Grunberg-Manago, M., and Court, D. L. (1989). *EMBO J.* **8**, 3401–3407.
96. McConnell, D. J. (1979). *NAR* **6**, 525–544.
97. Studier, F. W., and Dunn, J. J. (1982). *CSHSQB* **47**, 999–1007.
98. McAllister, W. T., and Carter, A. D. (1980). *NAR* **8**, 4821–4837.
99. Moffatt, B. A., and Studier, F. W. (1987). *Cell* **49**, 221–227.
100. Yamagishi, M., Cole, J. R., Nomura, M., Studier, F. W., and Dunn, J. J. (1987). *JBC* **262**, 3940–3943.
101. Fuller, C. W., Beauchamp, B. B., Engler, M. J., Lechner, R. L., Matson, S. W., Tabor, S., White, J. H., and Richardson, C. C. (1982). *CSHSQB* **47**, 669–679.
102. Grodberg, J., and Dunn, J. J. (1988). *J. Bacteriol.* **170**, 1245–1253.
103. Zawadzki, V., and Gross, H. J. (1991). *NAR* **19**, 1948.

104. Dubendorff, J. W., and Studier, F. W. (1991). *JMB* **219**, 61–68.
105. Maslak, M., and Martin, C. T. (1994). *Biochem.* **33**, 6918–6924.
106. Maher, L. J., III (1992). *Biochem.* **31**, 7587–7594.
107. Rando, R. F., DePaolis, L., Durland, R. H., Jayaraman, K., Kessler, D., and Hogan, M. E. (1994). *NAR* **22**, 678–685.
108. Cheng, X., Zhang, X., Pflugrath, J. W., and Studier, F. W. (1994). *PNAS* **91**, 4034–4038.
109. Kohli, V., and Temsamani, J. (1993). *Anal. Biochem.* **208**, 223–227.
110. Pitulle, C., Kleineidam, R. G., Sproat, B., and Krupp, G. (1992). *Gene* **112**, 101–105.
111. Aurup, H., Williams, D. M., and Eckstein, F. (1992). *Biochem.* **31**, 9636–9641.
112. McAllister, W. T. (1993). *Cell. Mol. Biol. Res.* **39**, 385–391.
113. Lee, S. S., and Kang, C. (1993). *JBC* **268**, 19299–19304.
114. Raskin, C. A., Diaz, G., Joho, K., and McAllister, W. T. (1992). *JMB* **228**, 506–515.
115. Lopez, P. J., Iost, I., and Dreyfus, M. (1994). *NAR* **22**, 1186–1193.
116. Schweisguth, D. C., Chelladurai, B. S., Nicholson, A. W., and Moore, P. B. (1994). *NAR* **22**, 604–612.
117. Lieber, A., Sandig, V., and Strauss, M. (1993). *EJB* **217**, 387–394.
118. Gao, X., and Huang, L. (1993). *NAR* **21**, 2867–2872.
119. Schoepfer, R. (1993). *Gene* **124**, 83–85.
120. Elroy-Stein, O., Fuerst, T. R., and Moss, B. (1989). *PNAS* **86**, 6126–6130.
121. Liu, M., and Horowitz, J. (1993). *BioTechniques* **15**, 264–266.
122. Zubay, G. (1973). *Ann. Rev. Genet.* **7**, 267–287.
123. Spirin, A. S., Baranov, V. I., Ryabova, L. A., Ovodov, S. Y., and Alakhov, Y. B. (1988). *Science* **242**, 1162–1164.
124. Baranov, V. I., Morozov, I. Y., Ortlepp, S. A., and Spirin, A. S. (1989). *Gene* **84**, 463–466.
125. Kudlicki, W., Kramer, G., and Hardesty, B. (1992). *Anal. Biochem.* **206**, 389–393.
126. Delarue, M., Poch, O., Tordo, N., Moras, D., and Argos, P. (1990). *Prot. Eng.* **3**, 461–467.
127. Ikeda, R. A., Chang, L. L., and Warshamana, G. S. (1993). *Biochem.* **32**, 9115–9124.
128. Rechinsky, V. O., Tunitskaya, V. L., Dragan, S. M., Kostyuk, D. A., and Kochetkov, S. N. (1993). *FEBS Lett.* **320**, 9–12.
129. Raskin, C. A., Diaz, G., and McAllister, W. T. (1993). *PNAS* **90**, 3147–3151.
130. Macdonald, L. E., Durbin, R. K., Dunn, J. J., and McAllister, W. T. (1994). *JMB* **238**, 145–158.
131. Macdonald, L. E., Zhou, Y., and McAllister, W. T. (1993). *JMB* **232**, 1030–1047.

C. SP6 RNA Polymerase
[EC 2.7.7.6]

The RNA polymerase from bacteriophage SP6, like other phage (T7 and T3) RNA polymerases, is an extremely useful enzyme for *in vitro* transcription. SP6 RNA Pol possesses a stringent promoter specificity similar to, but distinct from, that of T7 or T3 RNA Pol making it possible to transcribe bidirectionally *in vitro* a gene or DNA cloned in double-promoter plasmid vectors. In addition, use of SP6 RNA Pol extends the range of possible 5′ sequences of RNA products because the preferred SP6 start site is 5′-GAAGA while T7 RNA Pol prefers 5′-GGGAG. The SP6 start site can be particularly advantageous in large-scale synthesis where RNA with high contents of G residues may self aggregate due to guanine-tetrad formation. Most of the enzymatic properties, principles, and other areas of applications of SP6 RNA Pol are similar to those described for T7 RNA Pol. Therefore the following descriptions on SP6 RNA Pol will focus on the features distinguishing the SP6 RNA Pol from other RNA polymerases.

1. FUNCTIONS

a. Reaction conditions

i. Recommended reaction conditions. The optimum pH for SP6 RNA Pol is 7.5 in Tris–Cl buffer. The pH optimum in K-phosphate buffer is unchanged, but the activity is reduced by 40% (1). The enzyme has a sharp temperature optimum at 40°C, with 30% higher activity than at 37°C (2).

A typical reaction mixture (50 μl) consists of 40 mM Tris–Cl (pH 8.0 at 25°C), 6 mM MgCl$_2$, 2 mM spermidine, 10 mM DTT, 0.5 mM each NTP (optional [α-^{32}P]UTP, 1–10 μCi, 400 Ci/mmol), 50 μg/ml BSA, 50 μg/ml or 100 nM DNA template containing a SP6 promoter, and 10 U/μl SP6 RNA Pol. The addition of 1 U/μl pancreatic RNase inhibitor (RNasin) is optional. The mixture is incubated at 40°C for 2 hr. For short duplex DNA templates, yields are ~100 mol RNA/mol DNA. Higher yields (~300 mol RNA/mol DNA) can be obtained in slightly modified conditions: each NTP is increased to 5 mM, keeping the MgCl$_2$ at a concentration 6 mM in excess of the total NTP concentration, and with the addition of PEG (PEG-8000, 80 mg/ml) and Triton X-100 (0.01%) (10). The typical reaction conditions for SP6 RNA Pol work reasonably well for T7 RNA Pol (3).

ii. Kinetic parameters. The apparent K_m values of SP6 RNA Pol for nucleotide substrates are 30 μM (CTP), 31 μM (UTP), and 67 μM (ATP) (1). For GTP, the double-reciprocal plot is nonlinear with the half-maximal rate at ~50 μM GTP.

iii. Metal ion requirements. SP6 RNA Pol requires the presence of Mg^{2+} with an optimum at around 5 mM (in the presence of 4 mM spermidine) (1). Mn^{2+} cannot substitute for Mg^{2+} at any concentration. The polymerase activity is strongly inhibited (but not inactivated) by monovalent cations such as K$^+$ and NH$_4^+$ even at low concentrations.

iv. Activators and inactivators. Spermidine (≥2 mM) and BSA (50 μg/ml) stimulate polymerase activity (1,2). In the presence of optimal spermidine concentrations, BSA does not give additional stimulation. Sulfhydryl reagents (e.g., NEM, PCMB) inactivate the SP6 RNA Pol (1).

Like T7 RNA Pol but unlike *E. coli* RNA Pol, SP6 RNA Pol is not inhibited by rifampicin (4).

b. Activity assay and unit definition

One *unit* is usually defined as the amount of the enzyme that catalyzes the incorporation of 1 nmol of NMP (e.g., [α-^{32}P]CTP at 5–50 cpm/pmol) into acid-insoluble products in 1 hr under the typical reaction conditions described earlier and using 1 μg SP6 promoter-containing plasmid as the template in a 10-min assay.

c. Promoter specificities

SP6 RNA Pol shows a high degree of specificity for its promoters (5), the consensus sequence of which is a 23-bp contiguous sequence (see Fig. 7.3). When the promoter is modified to start the transcription with the AAAUUG--- sequence (e.g., the 5' end of 16S RNA), SP6 RNA Pol shows a fivefold higher transcription efficiency that T7 RNA Pol with the variant promoter (6). However, 40–50% of the SP6 transcript molecules contain extra adenosine residues in the -1, -2, -3, -4, and -5 positions, apparently as a result of "stuttering" during transcription initiation.

d. Substrate specificities

i. Templates. Like other phage RNA polymerases, SP6 RNA Pol uses as templates a wide variety of DNAs on which the SP6 promoter is found.

(a) Natural dsDNA templates. SP6 RNA Pol is naturally active on SP6 DNA. DNAs from other bacteriophages, including T3 and T7, are inert. Transcription *in vitro* of linear dsDNA can produce transcripts with wide ranging lengths, e.g., from 380 to 5700 bases (2).

(b) Synthetic ssDNA and dsDNA templates. Transcription of short DNA templates (<40 nt) by SP6 RNA Pol should be performed with fully double-stranded templates (10). The efficiency of transcription may be reduced 20-fold with partially double-stranded templates in which only the promoter region is double stranded. The observed difference in transcription efficiency between full and partial duplex DNA templates inversely correlates with the amounts of abortive products (2- through 10-mer) generated with the two templates.

One peculiar property of SP6 RNA Pol is that it binds preferentially to ssDNA in a mixture of ssDNA and dsDNA, and initiates promoter-independent, apparently end-to-end transcriptions (7). The 5'-nucleotide of the transcripts corresponds to the 3'-terminal nucleotide of the ssDNA template. Using up to 65-mer synthetic DNA templates, full-length transcripts can be obtained as single or major products.

SP6 RNA Pol is highly active in the synthesis of poly(G) with poly[(dI) · (dC)] as a template (1). This reaction is quite distinct from the promoter-initiated transcription since the cleavage of the enzyme with trypsin, which completely eliminates the SP6 promoter-transcribing activity, has little effect on the poly(G) synthesis.

(c) Homopolymeric templates. SP6 RNA Pol can transcribe other homopolymeric templates, e.g., 65 T residues, to produce a poly(A) tail (2). It can also transcribe the poly(C), poly(G), and poly(A) stretches of ~30 bases each without difficulty.

ii. Nucleotide requirements. Transcription by SP6 RNA Pol is selectively sensitive to limiting concentrations of ATP. The ATP dependence, which manifests with a linearized or relaxed circular plasmid template but not with supercoiled DNA, is relieved when the ATP concentration is higher than the other three nucleotides (8). The ATP dependence is not due to ATPase activity or to the premature termination of transcription at a low ATP concentration, but is apparently related to promoter topology.

iii. Product profiles. These properties are similar to those of T7 RNA Pol (see Section I,B, this chapter).

2. APPLICATIONS

The principles of SP6 RNA Pol applications are virtually identical to those illustrated for T7 RNA Pol.

3. STRUCTURE

SP6 RNA Pol is a single polypeptide. The predicted sequence contains 874 amino acid residues (M_r 98,561) (4). The N- and C-terminal residues are Met and Ala, respectively.

Comparison of the amino acid sequence with that of T7 RNA Pol shows that there are regions of partial homology along the sequence.

The enzyme has a $s^{\circ}_{20,w}$ of 7.1S in glycerol gradients.

4. GENETICS

SP6 RNA Pol is the product of gene 1 of phage SP6, a small dsDNA-containing, female-specific phage of *S. typhimurium* LT2 (1). The polymerase gene has been cloned into pBR322, expressed in *E. coli* under the inducible control of the *lac*UV5 promoter, and its entire nucleotide sequence has been determined (4). The Pol structural gene contains 2622 bp (= 874 amino acid codons). The ORF is terminated by tandem termination codons, UAA UAG. A Shine–Dalgarno-like sequence, AGGA (−10 to −7 position), is located in the upstream region of the AUG codon.

Immediately following the phage infection, the SP6 RNA Pol gene is transcribed by bacterial RNA polymerase. The SP6 gene 1 can also be transcribed selectively *in vitro* by *E. coli* RNA Pol. The Pol gene is located in the 3.3- to 6.0-kb map region of the SP6 genome (43.5 kbp), similar to that of T7 RNA Pol. The SP6 genome has no gross homologies with the T7 phage at the nucleotide level (9).

5. SOURCES

SP6 RNA Pol has been conventionally purified from phage SP6-infected *S. typhimurium* LT2. Compared with other phage (T3 and T7) polymerases, SP6 RNA Pol is significantly more stable and easier to purify. SP6 RNA Pol is now produced from the gene cloned in a pBR322 vector (pSR3) and overexpressed in *E. coli* BL21.

SP6 RNA Pol can be stored at a concentration as low as 60 μg/ml at −20° (in 50% glycerol) for 1 year with no loss of activity (1).

References

1. Butler, E. T., and Chamberlin, M. J. (1982). *JBC* **257**, 5772–5778.
2. Melton, D. A., Krieg, P. A., Rebagliati, M. R., Maniatis, T., Zinn, K., and Green, M. R. (1984). *NAR* **12**, 7035–7056.
3. Yisraeli, J. K., and Melton, D. A. (1989). *Meth. Enzy.* **180**, 42–50.

4. Kotani, H., Ishizaki, Y., Hiraoka, N., and Obayashi, A. (1987). NAR **15**, 2653–2664.
5. Brown, J. E., Klement, J. F., and McAllister, W. T. (1986). NAR **14**, 3521–3526.
6. Cunningham, P. R., Weitzmann, C. J., and Ofengand, J. (1991). NAR **19**, 4669–4673.
7. Sharmeen, L., and Taylor, J. (1987). NAR **15**, 6705–6711.
8. Taylor, D. R., and Mathews, M. B. (1993). NAR **21**, 1927–1933.
9. Kassavetis, G. A., Butler, E. T., Roulland, D., and Chamberlin, M. J. (1982). JBC **257**, 5779–5788.
10. Stump, W. T., and Hall, K. B. (1993). NAR **23**, 5480–5484.

D. T3 RNA Polymerase
[EC 2.7.7.6]

The RNA polymerase from bacteriophage T3 is structurally and functionally very similar to T7 RNA Pol, yet the T3 RNA Pol exhibits a distinct, nearly exclusive promoter specificity (1–3).

The consensus promoter sequence derived from eleven T3 promoter sites on the T3 genome is a 23-bp consecutive sequence (see Fig. 7.3).

1. FUNCTIONS

a. Reaction conditions

i. Optimal reaction conditions. The pH optimum for T3 RNA Pol is pH 7.8 in 50 mM Tris–Cl (4). A typical reaction mixture (50 μl) consists of 40 mM Tris–Cl (pH 8.0), 25 mM NaCl, 8 mM $MgCl_2$, 2 mM spermidine, 5 mM DTT, 50 μg/ml BSA, 0.5 mM each NTP, and 20–50 μg/ml DNA (containing a T3 promoter). The mixture is incubated at 37°C.

ii. Kinetic parameters. The kinetics of a polymerase reaction catalyzed by T3 RNA Pol follows typical Michaelis–Menten kinetics. The apparent K_m values for NTPs are: ATP (100 μM), CTP (80 μM), and UTP (75 μM) (4,5). The kinetics of GTP does not fit the Michaelis–Menten equation, but the concentration at the half-maximal rate is 120–180 μM. The rate of chain elongation is ~170 nt/sec at 37°C (in the presence of 50 mM KCl), which is approximately five times faster than that of *E. coli* RNA Pol. For the production of small transcripts, e.g., pentamer (GGGAA), from the oligonucleotide-based consensus promoter, the K_m (for DNA) and k_{cat} values are 1.0 nM and 40 min^{-1}, respectively (7).

iii. Metal ion requirements. T3 RNA Pol absolutely requires Mg^{2+}; the optimal concentration is 20 mM. Mn^{2+} cannot replace Mg^{2+} (4).

iv. Activators and inhibitors. DTT is required for sustained enzyme activity. T3 RNA Pol is inactivated by sulfhydryl reagents such as NEM and PHMB (4).

Spermine stimulates polymerase activity. Heparin does not inhibit T3 RNA Pol when added during polymerization (at a minimum polymer chain length of tetranucleotides), but almost completely inhibits the polymerase when added before the transcription initiation (5).

T3 RNA Pol is inhibited by SDS, phenol, and EDTA. KCl at concentrations higher than 100 mM reduces the rate of RNA chain elongation to ~50%, but does not affect the chain length of the RNA transcripts (5).

b. Activity assay and unit definition

A typical activity assay (50 μl) is similar to the optimal reaction conditions described earlier except that [α-^{32}P]UTP (80 mCi/mmol) may be included as a radioactive label. The mixture is incubated at 37°C for 20 min and the incorporation of [α-^{32}P]UMP into TCA-precipitable products is measured.

Unit definition: One unit is conventionally defined as the amount of enzyme that catalyzed the incorporation of 1 nmol of NTP into acid-insoluble products in 1 hr at 37°C.

c. Substrate specificities

The substrate specificities of T3 RNA Pol are essentially similar to those of T7 RNA Pol (refer to Section I,B, this chapter).

2. APPLICATIONS

Refer to Section I,B,2 on T7 RNA Pol.

3. STRUCTURE

T3 RNA Pol is a single polypeptide consisting of 884 amino acids (M_r 98,804) (1). The polymerase has 82% sequence identity with T7 RNA Pol. The predicted amino acid sequence contains nine Cys residues. The N- and C-terminal residues are Met and Ala, respectively.

4. GENETICS

The gene for T3 RNA Pol is T3 gene 1 located at the gene 1.0 region. T3 RNA Pol is encoded by an ORF of 2652 bp (nucleotide positions 143–2794, 884 amino acid codons) (1). With a few exceptions, the genes of T3 phage are highly homologous to those at the corresponding genomic regions of phage T7 (3). At the nucleotide level, the ORF for T3 RNA Pol has over 76% sequence identity with the counterpart of the T7 gene 1 (3).

Just upstream from the start codon, there are a number of closely packed regulatory elements which resemble those of T7 Pol gene. These include an RBS (or S-D sequence, GAGG), a promoter for *E. coli* RNA Pol (−10 sequence TAGAT and −35 sequence TTGACA), and a sequence that is closely related to the RNase III cleavage site and located at the corresponding position in T7 DNA. At the 3' end of the Pol gene, a promoter for T3 RNA Pol is located just 12 bp downstream from the stop codon, UAA.

5. SOURCES

The T3 RNA Pol gene (gene 1) has been cloned into a pBR322 derivative vector and expressed under the inducible control of the *lacUV5* promoter (6). Expression of the cloned T3 gene 1 (pCM56) in *E. coli* BL21 yields a large amount of T3 RNA Pol with little or no cleavage of the enzyme. Despite the nearly identical molecular size, T3 RNA Pol moves slightly faster than T7 RNA Pol in SDS–PAGE.

References

1. McGraw, N. J., Bailey, J. N., Cleaves, G. R., Dembinski, D. R., Gocke, C. R., Joliffe, L. K., MacWright, R. S., and McAllister, W. T. (1985). *NAR* **13**, 6753–6766.
2. Basu, S., Sarkar, P., Adhya, S., and Maitra, U. (1984). *JBC* **259**, 1993–1998.
3. Beck, P. J., Gonzalez, S., Ward, C. L., and Molineux, I. J. (1989). *JMB* **210**, 687–701.
4. Chakraborty, P. R., Sarkar, P., Huang, H. H., and Maitra, U. (1973). *JBC* **248**, 6637–6646.
5. McAllister, W. T., Kupper, H., and Bautz, E. K. F. (1973). *EJB* **34**, 489–501.
6. Morris, C. E., Klement, J. F., and McAllister, W. T. (1986). *Gene* **41**, 193–200.
7. Schick, C., and Martin, C. T. (1993). *Biochem.* **32**, 4275–4280.

II. DNA-INDEPENDENT RNA POLYMERASE: POLY(A) POLYMERASE

[Polynucleotide Adenylyltransferase, EC 2.7.7.19]

Polyadenylate polymerases are the enzymes that synthesize poly(A) tails usually found at the 3′ end of nonhistone eukaryotic mRNAs. Like the poly(A) tailing which occurs posttranscriptionally in the nucleus, the enzyme performs template-independent polyadenylation *in vitro* at the 3′ terminus of a wide variety of RNAs. Its specificity for ATP as an almost exclusive nucleotide substrate, virtual nonspecificity (in the absence of a specificity factor) for acceptor (or initiator) RNA substrates, and distributive mode of poly(A) elongation make the enzyme a useful tool in recombinant DNA technology. The poly(A) polymerase is particularly useful in converting poly(A)$^-$ RNAs to poly(A)$^+$ RNAs which can then be reverse transcribed to cDNAs using oligo(dT) as the primer.

Poly(A) polymerases are found in a wide variety of organisms from higher animals and plants to yeast, bacteria, and certain viruses (1–3). In eukaryotes, the enzymes are presumed to play a key role in the polyadenylation of pre-mRNAs in the nucleus. However, in prokaryotes such as *E. coli* where polyadenylated mRNAs are minor species, the function of poly(A) polymerases remains a matter of speculation. In most species, poly(A) polymerases are single polypeptide (~60 kDa) enzymes and catalyze a linear polymerization reaction as follows (Scheme 7.1).

$$\text{RNA} + n\text{ATP} \xrightarrow{\text{Mg}^{2+}} \text{RNA--(pA)}_n + n\text{PP}_i$$

SCHEME 7.1

Eukaryotic poly(A) polymerases have been the subject of intensive studies on the mechanism and biological roles of polyadenylation.

For current recombinant DNA work, the *E. coli* poly(A) polymerase is the most commonly used enzyme, which is the focus of this section. Most of the enzymatic properties of the *E. coli* poly(A) polymerase are believed to be similar to those of eukaryotic enzymes, although some properties are significantly different. Therefore, some available information on enzymological properties and molecular biological characteristics will also be described for bovine and yeast poly(A) polymerases.

1. FUNCTIONS

a. Reaction conditions

i. Optimal reaction conditions. A typical reaction mixture (50 μl) consists of 50 mM Tris–Cl (pH 8.0), 10 mM $MgCl_2$, 2.5 mM $MnCl_2$, 1 mM DTT, 50 μg/ml BSA, 0.25–1.0 mM ATP, initiator RNA, and 1–5 U of enzyme. The reaction with the *E. coli* poly(A) polymerase further requires 0.3 M NaCl. The reaction is carried out at 37°C and is stopped by adding EDTA (20 mM) or heating to 75°C for 10 min.

Poly(A) polymerase activities are sensitive to pH and ionic strength as follows.

(a) pH. Most poly(A) polymerases are optimally active at pH ~8 (4,5). The activity becomes half-maximal at pH 7.0 and 8.8.

(b) Ionic strength. The *E. coli* poly(A) polymerase is optimally active at ~300 mM concentrations of NaCl (6). The calf thymus poly(A) polymerase is activated at low salt concentrations (~50 mM KCl or NaCl), but is inhibited at concentrations above 150 mM. The inhibition is apparently due to dissociation of the enzyme–initiator complexes at high ionic strengths (7).

ii. Kinetic parameters. With the *E. coli* poly(A) polymerase, the K_m for a tRNA initiator is 0.2 μM and the K_m for ATP is 6 μM. The k_{cat} for initiation and elongation are 13 and 60 min^{-1}, respectively (5).

For $p(A)_3$ as the initiator substrate, the Mn^{2+}-activated calf thymus enzyme shows a K_m of 50 μM (4). The K_m increases to 200 μM when Mg^{2+} (4 mM) is substituted for Mn^{2+} (0.5 mM). The k_{cat} is ~1800 min^{-1}, regardless of the species of metal ions.

iii. Cofactor requirements. The poly(A) polymerase requires Mg^{2+}, Mn^{2+}, or both, depending on several factors such as the source of enzyme, substrate ATP, and initiator species and concentrations. The *E. coli* poly(A) polymerase has a higher activity with Mn^{2+} (2.5 mM) than with Mg^{2+} (10 mM). The enzyme shows the highest activity in the presence of these two metal ions (6).

For cytoplasmic calf thymus enzyme, Mn^{2+} rather than Mg^{2+} is the preferred cofactor (4). The higher activity observed in the presence of Mn^{2+} is ascribed to a 100-fold higher affinity of the polymerase for the initiator terminus (8). Note that some nuclear poly(A) polymerases are more active with Mg^{2+} than with Mn^{2+}.

iv. Inhibitors and inactivators

(a) Antibiotics and nucleic acid intercalators. Certain rifamycin derivatives (e.g., AF/013) inhibit poly(A) polymerases with a 50% inhibition at ~12 μM and in a manner noncompetitive with respect to ATP (9). The poly(A) polymerase is completely inhibited by 10 mM aurintricarboxylic acid, an inhibitor of both RNA polymerase and RNases. Poly(A) polymerases are insensitive to rifampicin and streptolydigin at concentrations that completely inhibit certain RNA polymerases. High doses of actinomycin D, EtBr, and α-amanitin have little effect on poly(A) polymerase activities (10,11).

(b) Metal chelators and substrate analogs. Because of differences in the nature of nuclear and cytosolic enzymes, eukaryotic poly(A) polymerases may exhibit widely different responses for certain inhibitors. The nuclear enzyme is particularly sensitive to the Zn-chelator o-phenanthroline (12) and the substrate analogs cordycepin triphosphate (3'-deoxy-ATP, $K_i \approx 1.3$ μM) and Ara-ATP ($K_i \approx 4$ μM).

(c) Phosphates. Pyrophosphate, which is a product of the polyadenylation reaction, is a noncompetitive inhibitor with respect to both ATP [$K_i \approx 0.2$ mM with E. coli enzyme (5)] and the tRNA inhibitor. Inorganic phosphate is also inhibitory: 50% inhibition occurs at 0.35 mM (6).

(d) Polyamines. Spermidine blocks polyadenylation by interacting with initiator substrates (2).

(e) Chemical reagents. NEM inactivates poly(A) polymerase (calf thymus, human lymphocyte) (11). VRC is a general inhibitor of poly(A) polymerases (13).

b. Activity assays and unit definition

The polymerase activity is assayed by radiolabeled AMP incorporation into an RNA initiator, e.g., tRNAs or poly(A)$_{50}$. A typical assay mixture (50 μl) contains 50 mM Tris–Cl (pH 8.0), 10 mM MgCl$_2$, 2.5 mM MnCl$_2$, 0.25 mM [α-^{32}P]ATP (or [^{3}H]ATP), RNA initiator (tRNA, 0.25 mg/ml), 4 mM DTT, and enzyme (0.1–1 mg/ml). The mixture is incubated at 37°C for 10–30 min, and the reaction is terminated by adding SDS and TCA. The reaction products are precipitated with 10% TCA on a glass fiber filter, washed, and counted.

This assay is most suitable for relatively pure enzymes. For crude or partially purified extracts where other polymerizing activities are present, it may be necessary to isolate first the oligo(A) or poly(A) reaction products by hybridization on oligo(dT)-cellulose and then quantitate the incorporation of ATP (1).

Activity can also be assayed by measuring the release of PP$_i$, which usually coincides with the AMP incorporation (4,5).

Unit definition: One unit is defined as the amount of enzyme that incorporates 1 nmol of AMP per 10 min (or per hour depending on assay systems).

c. Substrate specificities

i. Nucleotides. ATP is by far the preferred nucleotide substrate. Both CTP and UTP are polymerized at less than 5% of the rate obtained with ATP (6). The enzyme does not show any activity with GTP, ADP, and dATP. Cordycepin triphosphate (3'-deoxy-ATP) serves as a chain-terminating substrate.

Eukaryotic poly(A) polymerases can utilize NTPs other than ATP, either singly or in the presence of ATP, at less than 1% of the rate for ATP (3). The same is true for rADP and dATP.

ii. RNA initiators. A wide variety of natural and synthetic oligo- or polyribonucleotides with free 3'-OH termini can serve as initiator substrates. Poly(A) polymerases are essentially nonspecific for the type of initiators and the 3'-terminal nucleotides: poly(A)$^+$ and poly(A)$^-$ RNAs are equally effective as substrates. The length of the polyadenylate synthesized *in vitro* may vary considerably, for example, from 20 to 200 residues depending on reaction conditions. The polymerase activities are also dependent on initiator composition, size, and shape. For instance, tRNA is used as an initiator ~10 times more efficiently than poly(A). Double-stranded RNA is a very poor initiator. Polydeoxynucleotides are not used as initiators.

Of the synthetic polyribonucleotides, poly(A) is the best initiator substrate. The calf thymus enzyme, when activated by Mg^{2+}, uses poly(UG), poly(C), and short poly(A) (e.g., di- and trinucleotides) much less effectively than poly(A) having ~40 nt (10,14). In contrast, the Mn^{2+}-activated enzyme functions equally well with short or long poly(A) (4). In the presence of an "AAUAAA-specificity factor," also called "cleavage and polyadenylation specificity factor (CPSF)," eukaryotic poly(A) polymerases acquire an almost exclusive specificity for RNAs containing the AAUAAA sequence motif (15,16).

d. Reaction mechanism

i. Kinetics of polymerization. The *E. coli* poly(A) polymerase catalyzes polymerization following an ordered Bi–Bi kinetic mechanism. The two substrates, ATP and RNA initiator, bind the enzyme sequentially: the RNA initiator binds first and then the ATP; PPi is released first followed by the polyadenylated product (5). The kinetic mechanism of calf thymus or other eukaryotic poly(A) polymerases is not clearly known, although their similarity in many respects with the *E. coli* enzyme makes it likely that the kinetic mechanisms are also similar. In contrast, eukarytoic poly(A) polymerases and their interaction with and regulation by other factors have been extensively studied, providing as yet incomplete but rich information on the mechanism of eukaryotic polyadenylation.

ii. Mechanism of poly(A) synthesis. Eukaryotic polyadenylation *in vivo* is an event coupled to an endonucleolytic cleavage reaction. *In vitro*, however, polyadenylation can be uncoupled from the cleavage and can be carried out separately with synthetic RNA substrates that end at the cleavage site. The *in* *vitro* polymerization proceeds linearly without any lag phase and in a "distributive" mode (8).

In the presence of nuclear extracts, the polyadenylation reaction is distinctly biphasic: the *initiation* phase during which the first few AMPs are added to the initiator and the *elongation* phase during which the oligo(A) product of the first phase is subject to elongation. Evidence suggests that the two reaction phases are essentially independent of each other.

Initiation of polyadenylation is dependent on the AAUAAA recognition sequence (see below) and slowly reaches a plateau of nine adenosine residues. When the growing poly(A) tail reaches a length of 10–12 nt, the slow oligoadenylation phase locks into a rapid elongation mode. The phase transition to *elongation* mode is presumed to occur on binding of the nuclear poly(A)-binding protein II (PABII) to the oligoadenylates. A minimum length of 10 nt is necessary to relieve the requirements for AAUAAA and CPSF (17). Nevertheless, the elongation is stimulated by the presence of CPSF (18) and much more so when both CPSF and PABII are present. PABII stimulates 30- to 50-fold the extension of the simple poly(A) primer in the absence of CPSF.

The poly(A) polymerase by itself is entirely "distributive," dissociating from its primer after every polymerization step. In the presence of either PABII or CPSF, the processivity increases slightly to less than 10 nt. Only in the presence of both stimulatory factors can poly(A) polymerase synthesize a full-length poly(A) tail of 200 nt in a single processive event (19).

Termination of *in vivo* polyadenylation or the *in vitro* reaction in nuclear extracts occurs over a defined range of lengths (e.g., 60–80 nt in yeast mRNAs, 200–300 nt in mammalian mRNAs) rather than at a precise number of nucleotides. What causes the termination of polyadenylation is not known. It is not an intrinsic property of poly(A) polymerase, although the *in vitro* reaction in the absence of nuclear extracts also results in a defined size distribution (5).

iii. Recognition of substrates

(a) Recognition sequence motifs. The most noticeable structural feature of mRNAs that may serve as the probable recognition site for polyadenylation has been identified as a highly conserved hexanucleotide with a canonical sequence of 5'-AAUAAA-3' (22–24). This consensus sequence is typically located 10 to 30 nt upstream of the poly(A) addition site of mRNAs. Every position in the canonical hexanucleotide is required. Depending on the position and the base introduced, a base substitution may reduce both cleavage and polyadenylation efficiencies (20). It is to be noted, however, that not all AAUAAA in the 3'-noncoding region are utilized for directing mRNA cleavage and polyadenylation. For many mRNAs, not only is AAUAAA (or some of its natural variant sequences) required for polyadenylation, but it is virtually sufficient for inducing polyadenylation. The 2'-OH of the U residue in AAUAAA is required for the binding of the specificity factor and hence for poly(A) addition (21). The cleavage site is a dinucleotide which shows a clear preference for the sequence CA. Some genes

have a third element, a U- or GU-rich sequence, located within 100 nt downstream of the poly(A) site.

A synthetic poly(A) site composed of AAUAAA and a GU/U-rich sequence with a spacing of 22–23 nt between them is fully functional as a poly(A) signal (25). When appropriately placed downstream from the regular poly(A) site, for example, of a globin gene, the synthetic poly(A) site elicits a more efficient polyadenylation than the regular poly(A) sites. The number of nucleotides between the AAUAAA and the 3' end is critical for efficient polyadenylation. Polyadenylation of short synthetic substrates is optimal when the spacer length is 8 or more nucleotides (21). The sequence identity of the spacer hardly affects the polyadenylation efficiency. However, different spacer sequences have been shown to promote slight variations in the position at which poly(A) is added (24,25).

Note that some poly(A)$^+$ mRNAs, e.g., histone mRNA of yeast, do not carry the AAUAAA sequence in their 3' ends. In these mRNAs, two different classes of polyadenylation sites have been recognized, i.e., a unidirectional site with the sequence UUUUUAU and a bidirectional site with a tripartite sequence UAG···UA(U)GU···UUU (26). For other mRNAs that have no apparent recognition sequence motifs in their 3' UTR and yet carry poly(A) tails, other additional or alternative structural features may be present such as further internal sequences (5' to the apparent polyadenylation signal) or secondary structures that are required for proper and/or differential polyadenylation.

(b) Influence of AAUAAA specificity factor (CPSF). The selectivity of polyadenylation imparted by the specificity factor for AAUAAA-containing RNAs is thought to be the result of a direct contact between the specificity factor bound to AAUAAA and poly(A) polymerase, a situation analogous to the interactions between bacterial transcription factor σ and RNA polymerase (21,24). This contact probably decreases the effective dissociation constant of the polymerase for AAUAAA-containing substrates. At sufficiently high substrate or enzyme concentrations, as in classical assays, the poly(A) polymerase adds poly(A) nonselectively to any RNA. At high dilutions and in the absence of the specificity factor, the enzyme is almost completely inactive and indifferent to the presence or absence of the AAUAAA sequence in the substrate RNA. When complemented with CPSF, the enzyme acts only on RNAs carrying the AAUAAA signal. A CPSF may also decrease the nonspecific activity observed at high polymerase concentrations. The CPSF of HeLa cells is a protein complex ($\sim$290 kDa) composed of at least four subunits (16,27). The calf thymus poly(A) polymerase has been shown to be fully responsive to the CPSF of HeLa cells.

(c) Endonucleolytic cleavage and polyadenylation. Most eukaryotic transcriptions proceed beyond the polyadenylation sites. Therefore, polyadenylation *in vivo* is normally coupled to the endonucleolytic cleavage which exposes a 3'-OH terminus within the pre-mRNA. The specific cleavage requires CPSF and one or more apparently dispensable factors. One is known as the cleavage stimulation factor (CStF) or cleavage factor 1 (CFI), which increases the efficiency and specificity of cleavage. CStF/CFI is not involved in polyadenylation. The poly(A) polymer-

ase is also required for cleavage, suggesting that the polymerase is most likely associated with an endonuclease, CPSF, PAB, and other processing factors in a functional ribonucleoprotein (21,27,28).

2. APPLICATIONS

i. Production of poly(A)-tailed RNAs. Poly(A) tails of RNAs have multiple uses *in vitro* in addition to the presumed roles of biological importance. For instance, poly(A) tails provide the basis for the most common method of poly(A)$^+$ RNA purification using oligo(dT) affinity chromatography. The conversion of poly(A)$^-$ RNAs, e.g., ribosomal RNAs (29,30), tRNAs (5), and viral RNAs (31–33), into poly(A)$^+$ RNAs permits molecular cloning of these RNAs via conventional cDNA synthesis with oligo(dT) primers (34). [*Note*: Alternatively, the cDNAs of poly(A)$^-$ RNAs can be prepared by direct reverse transcriptions of poly(A)$^-$ RNA using random hexamer as primers.] Such cDNAs can also be used as hybridization probes. Furthermore, poly(A) tailing allows functional studies on the role of poly(A) tails on mRNAs.

Although the chain growth may be nonsynchronous, it is possible to obtain poly(A) products in defined size ranges by choosing proper reaction conditions. At low enzyme concentrations, the rate of elongation remains relatively constant regardless of the initiator concentration, whereas the initiation rate increases with increasing initiator concentrations. Therefore shorter poly(A) chains can be obtained at low concentrations of enzymes and high concentrations of initiator. Using *E. coli* poly(A) polymerase at low concentrations, a 40-fold increase in initiator concentrations has been shown to reduce the average length of the poly(A) population from >300 to ~50 nt (5).

ii. 3′-End labeling of RNA. RNA molecules can be radiolabeled at the 3′ end by the addition of short oligo(A) tails using [α-^{32}P]ATP (35) or, preferably, by the single addition of the chain terminating 3′-dA using 3′-[α-^{32}P]dATP (36). Such 3′-labeled RNAs are suitable substrates for direct RNA sequencing.

iii. Determination of poly(A)$^+$ RNA contents. A poly(A) polymerase-based method has been developed to quantify the purity of poly(A)$^+$ mRNAs from crude or partially purified mRNA pools in which the contaminating RNAs are largely poly(A)$^-$ rRNAs (13). The method relies on the polymerase to attach nontemplated poly(A) tails on both poly(A)$^+$ and poly(A)$^-$ RNAs. In the presence of cordycepin 5′-triphosphate as the chain terminator and [α-^{32}P]ATP as the radiolabel, only oligo(A) tails having four residues or less in length are synthesized. Subsequently, these products are digested with RNase H in the presence of oligo(dT)$_{12-18}$. This reaction produces radioactive poly(A) tails only from the poly(A)$^+$ RNA, leaving the short oligo(A) tails on poly(A)$^-$ RNA intact. Quantification of the radioactive tag provides an estimate of the mole fraction of genuine poly(A)$^+$ RNA in mixed RNA samples.

3. STRUCTURE

The *E. coli* poly(A) polymerase is a monomeric enzyme of ~55 kDa. A highly purified preparation of the enzyme from *E. coli* K12 apparently exhibits doublet bands having a 3-kDa difference in SDS–PAGE. The N termini of the doublet components are identical and start with Lys. Presumably, poly(A) polymerase is synthesized as a precursor consisting of 472 amino acid residues (calc. 54 kDa), and the N-terminal 17 amino acid peptide is cleaved at the Arg–Lys junction (37). The *E. coli* poly(A) polymerase shows no obvious sequence similarities with the corresponding enzymes from calf thymus (38), vaccinia virus (39), and yeast (40).

The poly(A) polymerase highly purified from calf thymus is a 55-kDa polypeptide (38). However, the active poly(A) polymerase obtained by *in vitro* translation of a bovine cDNA clone has been shown to be an 85-kDa protein.

The purified poly(A) polymerase from yeast is a 63-kDa polypeptide. However, the deduced amino acid sequence from a genomic clone suggests that the polymerase gene encodes a 568 residue polypeptide (PAPI, M_r 64,551) (40). The yeast PAPI does not reveal long stretches of sequence similarity to other known proteins, but the N-terminal 400 residues have a 47% similarity with bovine poly(A) polymerase. In both yeast and bovine enzymes, the N-terminal region contains the RNA-binding domain and polymerase active site.

The vaccinia virus-encoded poly(A) polymerase is contained in an 80-kDa dimer which consists of a 55-kDa subunit (VP55) possessing the polymerase activity and a 39-kDa subunit (VP39) functioning as the CPSF (39).

4. GENETICS

i. Gene structure and organization. The *E. coli* poly(A) polymerase is encoded by *pcnB* (37), a gene that had originally been identified to be responsible for controlling plasmid copy number (41) and was subsequently cloned and sequenced (42). The polymerase ORF is presumed to contain 427 codons, starting at the UUG codon and terminating by UAG. The start codon is found 17 codons upstream from the observed N terminus of the purified enzyme. A putative ribosome-binding site (TGAT) is located 6 nt upstream from the start site. No obvious transcription control elements are located within the 250-nt sequence upstream of the putative translation start site. Deletion of the N-terminal 58 amino acid residues results in a complete loss of polymerase activity.

Bovine poly(A) polymerases are apparently encoded by two alternatively spliced mRNAs. One class of cDNAs (PAP I) consists of 3431 bp and contains a single 2067-bp ORF that is capable of encoding a 689 amino acid protein (calc. 77 or 85 kDa in the gel) (38). The 160-nt 5' UTR of PAP I is GC rich (69%), whereas the ~1200-nt 3' UTR is AU rich (67%). The other class (PAP II), which has a slightly shorter 3' UTR (~900 nt, 60% AU rich), diverges from PAP I at codon 663 and is capable of encoding a 740 amino acid protein (calc. 83 kDa).

ii. *In vivo* role of poly(A) polymerase and polyadenylation. Although the physiological functions of poly(A) polymerases and poly(A) tails of certain mRNAs in prokaryotes remain obscure, the poly(A) found at the 3' terminus of most eukaryotic mRNAs is known to play an important role in gene expression by way of enhancing the initiation of translation and stabilizing mRNAs against potential cellular nucleases. The eukaryotic polyadenylation, which occurs post-transcriptionally in the nucleus, is catalyzed by poly(A) polymerases in the context of various processing factors such as poly(A)-binding proteins 1 (cytosolic) and II (nuclear) and CPSF.

mRNAs are stabilized mostly by two structural elements, i.e., the 5'-cap structure and the 3' poly(A) tail. Once transported to the cytoplasm, mRNA begins to lose its tail. Only after the poly(A) tail has been removed does the mRNA begin to degrade. However, the mRNA need not be completely deadenylated before the next phase of the degradation reaction takes place. Shortening the tail to a length of ~10 nt renders the tail incapable of binding to the cytoplasmic factor (PABI) with a high affinity and is sufficient to induce the next step of degradation.

The role of the poly(A) tail in mRNA translation has not been as obvious as its role in the protection of mRNA from degradation. It is now generally accepted that mRNA translation is most efficient with, and sometimes even completely dependent on, the poly(A) tail (43). In maturing oocytes, the adenylation status of mRNA is directly correlated with its translatability. Furthermore, the translation of previously dormant RNAs occurs after cytoplasmic extension of their poly(A) tail, given the appropriate context of having both a sequence related to UUUUUUAU which is considered to be the cytoplasmic polyadenylation element (CPE) and the AAUAAA signal.

The poly(A) polymerase is also presumed to play a regulatory role in development because the polymerase activity itself is developmentally regulated and/or modulated by various physiological stimuli including hormones (e.g., glucocorticoids), drugs (e.g., phenobarbital), and immunostimulants (e.g., phytohemagglutinin). For instance, certain viral genes of adenovirus (44) and HBV (45) make differential use of poly(A) signals at different stages of infection. Many important cellular genes, e.g., for tropomyosin (46) and IgM (47), contain multiple poly(A) signals whose differential recognition leads to the production of multiple mRNAs that encode distinct protein isoforms. During the early development of frog oocytes, an extensive array of poly(A) addition and removal occurs in the cytoplasm as a part of translational regulation (48).

5. SOURCES

Poly(A) polymerases have been purified to near homogeneity from many sources including *E. coli* (5,6), yeast (49), and calf thymus (1,4,10) using various conventional chromatographies and ATP or poly(A) affinity chromatography. The *E. coli* enzyme, whether purified from natural sources (e.g., strain MRE 600 for BRL enzyme) or from a recombinant expression system (37), tends to aggregate during or after purification at low ionic strengths (e.g., below 0.4 M NaCl) (6).

Expression in *E. coli* of the cloned polymerase gene, even at the uninduced basal level, is apparently highly toxic to the host. Overexpression (~150-fold) of the gene is possible only in a tightly controlled inducible expression system consisting of a phage λ P_L promoter vector (pRE1) and *E. coli* MZ1 which is a λ lysogen containing a temperature-sensitive mutant repressor (cI857).

A recombinant yeast poly(A) polymerase (available from USB) is overexpressed in *E. coli* BL21 (DE3) from a T7 expression vector (49).

Mammalian poly(A) polymerases are commonly observed in multiple forms even from a single type of cells, possibly due to the presence of compartmentally distinct isozymes. Multiple enzyme forms may also arise from different degrees of phosphorylation (50). The cDNA cloning of the gene(s) for bovine poly(A) polymerase has shown that the bovine enzymes are apparently derived from two alternatively spliced mRNAs.

References

1. Edmonds, M. (1990). *Meth. Enzy.* **181**, 161–170.
2. Jacob, S. T., and Rose, K. M. (1983). *In* "Enzymes of Nucleic Acid Synthesis and Modification" (S. T. Jacob, Ed.), Vol. 2, pp. 135–157. CRC Press, Florida.
3. Edmonds, M. (1982). "The Enzymes," Vol. 15, pp. 217–244.
4. Tsiapalis, C. M., Dorson, J. W., and Bollum, F. J. (1975). *JBC* **250**, 4486–4496.
5. Sano, H., and Feix, G. (1976). *EJB* **71**, 577–583.
6. Sippel, A. E. (1973). *EJB* **37**, 31–40.
7. Sethi, V. S. (1975). *FEBS Lett.* **59**, 3–7.
8. Wahle, E. (1991). *JBC* **266**, 3131–3139.
9. Nutter, R. C., and Glazer, R. I. (1979). *Biochem. Pharmacol.* **28**, 2503–2509.
10. Winters, M. A., and Edmonds, M. (1973). *JBC* **248**, 4756–4762.
11. Coleman, M. S., Hutton, J. J., and Bollum, F. J. (1974). *Nature* **248**, 407–409.
12. Rose, K. M., Allen, M. S., Crawford, I. L., and Jacob, S. T. (1978). *EJB* **88**, 29–36.
13. Krug, M. S., and Berger, S. L. (1987). *Meth. Enzy.* **152**, 262–266.
14. Winters, M. A., and Edmonds, M. (1973). *JBC* **248**, 4763–4768.
15. Takagaki, Y., Ryner, L. C., and Manley, J. L. (1988). *Cell* **52**, 731–742.
16. Bardwell, V. J., Zarkower, D., Edmonds, M., and Wickens, M. (1990). *Mol. Cell. Biol.* **10**, 846–849.
17. Sachs, A., and Wahle, E. (1993). *JBC* **268**, 22955–22958.
18. Gilmartin, G. M., and Nevins, J. R. (1989). *Genes Dev.* **3**, 2180–2189.
19. Bienroth, S., Keller, W., and Wahle, E. (1992). *EMBO J.* **12**, 585–595.
20. Sheets, M. D., Ogg, S. C., and Wickens, M. P. (1990). *NAR* **18**, 5799–5805.
21. Wigley, P. L., Sheets, M. D., Zarkower, D. A., Whitmer, M. E., and Wickens, M. (1990). *Mol. Cell. Biol.* **10**, 1705–1713.
22. Proudfoot, N. J., and Brownlee, G. G. (1974). *Nature* **252**, 359–362.
23. Fitzgerald, M., and Shenk, T. (1981). *Cell* **24**, 251–260.
24. Wickens, M. (1990). *TIBS* **15**, 277–281.
25. Levitt, N., Briggs, D., Gil, A., and Proudfoot, N. J. (1989). *Genes Dev.* **3**, 1019–1025.
26. Irniger, S., Egli, C. M., and Braus, G. H. (1991). *Mol. Cell. Biol.* **11**, 3060–3069.
27. Takagaki, Y., Ryner, L. C., and Manley, J. L. (1989). *Genes Dev.* **3**, 1711–1724.
28. Terns, M. P., and Jacob, S. T. (1989). *Mol. Cell. Biol.* **9**, 1435–1444.
29. Hell, A., Young, B. D., and Birnie, G. D. (1976). *BBA* **442**, 37–49.
30. Ackerman, S., Keshgegian, A. A., Henner, D., and Furth, J. J. (1979). *Biochem.* **18**, 3232–3242.
31. Emtage, J. S., Catlin, G. H., and Carey, N. H. (1979). *NAR* **6**, 1221–1240.
32. Gething, M.-J., Bye, J., Skehel, J., and Waterfield, M. (1980). *Nature* **287**, 301–306.
33. Cashdollar, L. W., Esparza, J., Hudson, G. R., Chmelo, R., Lee, P. W. K., and Joklik, W. K. (1982). *PNAS* **79**, 7644–7648.
34. Getz, M. J., Birnie, G. D., and Paul, J. (1974). *Biochem.* **13**, 2235–2240.

35. Winter, G., and Brownlee, G. G. (1978). *NAR* **5**, 3129–3139.
36. Beltz, W. R., and Ashton, S. H. (1982). *Fed. Proc.* **41**, 1450 (#6896).
37. Cao, G.-J., and Sarkar, N. (1992). *PNAS* **89**, 10380–10384.
38. Raabe, T., Bollum, F. J., and Manley, J. L. (1991). *Nature* **353**, 229–234.
39. Gershon, P. D., Ahn, B.-Y., Garfield, M., and Moss, B. (1991). *Cell* **66**, 1269–1278.
40. Linger, J., Kellerman, J., and Keller, W. (1991). *Nature* **354**, 496–498.
41. Lopilato, J., Bortner, S., and Beckwith, J. (1986). *MGG* **205**, 285–290.
42. Liu, J., and Parkinson, J. S. (1989). *J. Bacteriol.* **171**, 1254–1261.
43. Jackson, R., and Standart, N. (1990). *Cell* **62**, 15–24.
44. Falck-Pedersen, E., and Logan, J. (1989). *J. Virol.* **63**, 532–541.
45. Russnak, R., and Ganem, D. (1990). *Genes Dev.* **4**, 764–776.
46. Helfman, D. M., Cheley, S., Kuismanen, E., Finn, L. A., and Yamawaki-Kataoka, Y. (1986). *Mol. Cell. Biol.* **6**, 3582–3595.
47. Galli, G., Guise, J., Tucker, P. W., and Nevins, J. R. (1988). *PNAS* **85**, 2439–2443.
48. Wickens, M. (1990). *TIBS* **15**, 320–324.
49. Lingner, J., and Keller, W. (1993). *NAR* **21**, 2917–2920.
50. Jacob, S. T., and Rose, K. M. (1984). *Adv. Enzyme Regulation* **22**, 485–497.

8
Marker/Reporter Enzymes

In gene cloning, screening, and expression, a selectable marker or reporter is an invaluable tool. When a (cloned) gene of interest possesses no distinguishable genetic or biochemical traits, the presence of a selectable marker on the vector becomes indispensable. A marker/reporter gene fused to a target gene also provides a powerful tool for the study of gene expression and regulation. Cloning and/or expression vectors generally carry at least one marker or reporter gene to facilitate selection or screening of recombinant clones.

The genes commonly used as selection markers and/or reporters are those that code for enzymes (or proteins). They generate readily distinguishable colors when chromogenic substrates are introduced or provide resistance (or sensitivity) to antibiotics such as ampicillin, tetracycline, kanamycin, and chloramphenicol. It should be noted that, although rare, there are extremely sensitive *nonenzymatic reporters* such as the human growth hormone gene system, (green) fluorescent proteins, and the ice nucleation protein (and its gene *inaZ*).

The term "marker" will be used here mainly to designate those genes (or enzymes) which provide resistance either to antibiotics or inhibitory substrate analogs. The term "reporter" will be used for those genes (or enzymes) which generate readily identifiable or assayable traits such as colors and light. Because the distinction between marker and reporter is arbitrary and rests on the kind of substrates used for assays, a gene system may, in some cases, be a marker and reporter at the same time.

An ideal marker/reporter gene is presumed to meet the following criteria. (i) The genetic features of the gene should be known in sufficient detail to be amenable to easy manipulations. (ii) The enzyme (or reporter protein) should be stable either alone or as a fusion protein and should function in diverse heterologous environments. (iii) The (enzymatic) activity encoded by the reporter gene must be readily distinguishable from any similar activity in the host. (iv) There should be no interference or competition from other host-dependent enzymatic activities. (v) The assay for the reporter enzyme should be rapid, sensitive, reproducible, and convenient.

Obviously, there is no single marker/reporter system that satisfies all the criteria enumerated above. Besides, enzymatic properties are intrinsically subject to the nature, environment, and conditions of assay. Nevertheless, there are several marker/reporter systems available that seem to meet fairly closely the ideal characteristics of a marker/reporter.

This chapter focuses on a few selected marker/reporter gene systems to illustrate their general utility in molecular biology and recombinant DNA technology. Among other marker/reporter genes which are particularly useful for gene expression in eukaryotic cells, and which will not be discussed here, are the bacterial dihydrofolate reductase (DHFR) gene that confers resistance to methotrexate; the *E. coli gpt* gene encoding xanthine–guanine phosphoribosyltransferase that allows positive selection of transformants in HAT (hypoxanthine–aminopterin–thymidine) medium; the bacterial (Tn5 transposon) *neo* gene encoding aminoglycoside 3'-phosphotransferase, also called neomycin phosphotransferase II (*npt*II), that confers resistance to antibiotics kanamycin, neomycin, or its analog G418; and the herpes simplex virus thymidine kinase (TK) gene that confers sensitivity to ganciclovir.

This chapter covers β-galactosidase (βGal or LacZ) and luciferase (Lux/Luc) gene systems as typical reporter systems. Note that alkaline phosphatase (PhoA) can also be used as an excellent reporter system and has been described in Chapter 5. As selection markers, the ampicillin resistance system (Bla or Ap^r) will be described. The chloramphenicol acetyltransferase (CAT) gene system will also be discussed as a typical marker/reporter with dual functions.

A. β-Galactosidase
[β-D-Galactoside Galactohydrolase, EC 3.2.1.23]

Galactosidases are a group of glycosidic enzymes that specifically catalyze the hydrolysis of galactosidic linkages in various oligosaccharides. There are two types of galactosidases: α-galactosidases and β-galactosidases, depending on the configuration of the anomeric carbon atom of the substrate molecule.

β-Galactosidases hydrolyze the terminal nonreducing β-D-galactose residues in β-D-galactosides, whereas α-galactosidases hydrolyze the α-D-galactosides. The sugar lactose naturally found in milk is a disaccharide composed of β-D-galactopyranosyl-(1,4)-D-glucose, and therefore is digested only by β-galactosidase.

Among the β-galactosidases that are widely present in nature (bacteria, yeast, fungi, and mammalian cells), *E. coli* β-galactosidase (βGal) is the best characterized enzyme in terms of genetics and biochemistry. βGal is a tetrameric enzyme with a subunit MW of 116,000. It has been the prime model for the study of induction and regulation of genes, especially in the development of the landmark concept of the *operon*. The wealth of genetic information, together with the ease of assay and enzymatic stability, makes the βGal (and the gene *lacZ*) the first and one of the most important marker/reporter/carrier systems employed in recombinant DNA technology. The progress made in the development of versatile vectors featuring inducible gene expression and of synthetic β-D-galactoside substrates allows extremely wide applications of βGal or *lacZ* gene with highly sensitive and reliable detection or assay capabilities based on colorimetry, spectrophotometry, fluorimetry, or luminometry.

1. FUNCTIONS

a. Reaction conditions

i. Optimal conditions for β-galactosidase reaction or detection. βGal can be detected or assayed in a number of ways under varying situations, usually using chromogenic substrates. βGal is optimally active at pH ~7 in solution reactions with ONPG as the substrate. The activity is higher at 30° than at 40°C in the absence of salts (NaCl) but, in the presence of 50 mM NaCl, the activity is higher at 40° than at 30°C. Depending on applications, reaction conditions can be easily adapted or modified. The reaction can be terminated by adding 2 mM phenylethyl-β-D-thiogalactopyranoside (PETG), a βGal inhibitor (see below).

(a) Solution reactions. A liquid reaction of βGal is usually performed in 0.1 M HEPES (pH 7.2) containing 3.5 mM ONPG and 1 mM $MgCl_2$ for 20–60 min at 37°C. Other buffers such as PBS and imidazole are equally useful.

To assay the cellular expression of βGal, cells are lysed or made permeable to substrates by adding 0.5% NP-40 (or Triton X-100) or 20% $CHCl_3$ (or toluene) in the assay buffer. This prior treatment significantly improves the sensitivity of the assay not only with bacterial cells but also with yeast and mammalian cells that lack the permease necessary for substrate transport. When the βGal activity in *E. coli* is assayed, the cells can be lysed using phage T4 at a high multiplicity of infection (>100 phages/cell). This alternative means of cell lysis enables the βGal assay to be performed in 96-well microtiter plates in an automated fashion (1).

The βGal reaction following phage-mediated cell lysis provides a sensitive means of coliphage detection, for example, in water samples. Some eukaryotic cell lysates may exhibit a high background βGal activity due to the presence of inherent lysosomal βGal which is active at pH ≤ 7.0. The cellular βGal activity can be more easily discriminated from the transfected *lacZ* activity by performing the assays at pH 8.0 (0.1 M Na-phosphate).

(i) Colorimetry. The colorimetric assay using the "standard" substrate ONPG (also see Section A,1,b, in this chapter) is only moderately sensitive with a detection limit of ~1 ng (1 × 10^9 molecules) of βGal. Using a different chromogenic substrate, CPRG (chlorophenol red–β-D-galactopyranoside), which has a molar absorption coefficient 21-fold higher than that of ONPG, the sensitivity of detection can be increased about 10-fold (2). The βGal-catalyzed hydrolysis of water-soluble CPRG (yellow) gives rise to a distinctly red product (chlorophenol red) that can be monitored at 570 nm.

(ii) Fluorimetry. For a more sensitive detection but at the expense of the convenience offered by colorimetric detection, βGal can be assayed fluorimetrically using MUG. The fluorimetric assay can detect up to 1 pg (10^6 molecules) of βGal. Fluorescein di-β-D-galactopyranoside (FDG) is a substrate that penetrates viable gram-negative bacteria and thereby allows flow cytometry of Lac$^+$ cells.

(iii) Luminometry. Compared with the assays based on chromogenic or fluorigenic substrates, the assays based on (chemi- or bio-)luminogenic substrates provide the highest sensitivity, enabling the detection of βGal at a subpicogram (~10^4 molecules) level which is suitable for the detection of single cells stably expressing *lacZ*.

One typical *chemi*luminescent substrate is AMPGD (e.g., LumiGal of Lumigen, Detroit, IL, or Galacto-Light kit of Tropix, Bedford, MA) which is a 1,2-dioxetane β-D-galactopyranoside derivative (3). [The full name for AMPGD is 3-(4-methoxyspiro[1,2-dioxetane-3,2′-tricyclo[3.2.1.1^{3,7}]decan]-4-yl)-phenyl-β-D-galactopyranoside.] It is structurally similar to AMPPD, the chemiluminescent substrate for alkaline phosphatase (see Section I,A, in Chapter 5 and also Fig. 5.2), except that the P_i group is replaced by β-D-galactoside. Removal of the galactose moiety from AMPGD generates an unstable dioxetane anion which further decomposes with the concurrent emission of the chemiluminescence. The luminescence can be measured with a luminometer or a liquid scintillation counter.

βGal can also be assayed using a *bio*luminogenic substrate, D-luciferin O-β-galactopyranoside in a successive βGal-luciferase assay (105). The substrate, which is stable and water soluble, is first hydrolyzed by βGal (K_m = 2.9 μM, k_{cat} = 256 sec^{-1}) at pH 7.8 (50 mM phosphate buffer) and 37°C, releasing luminometrically active luciferin. The luciferin is then oxidized to oxyluciferin in a reaction catalyzed by firefly luciferase in the presence of ATP and O_2, thereby producing light.

(b) Solid-phase reactions. In the screening of recombinant bacterial clones or in the detection of eukaryotic cells expressing βGal (or βGal-fusion protein) the βGal reactions usually employ a chromogenic substrate either *in situ* or on indicator plates (agar or cell culture) in the presence or absence of the inducer IPTG. Hydrolysis of chromogenic substrates such as X-Gal and Bluo-Gal (5-bromoindolyl-β-O-galactopyranoside) generates indole derivatives which rapidly undergo oxidation/dimerization under alkaline conditions to form "blue" precipitates. The precipitates are insoluble in alcohol or xylene and do not diffuse through the agar. On agar plates containing X-Gal (or Bluo-Gal, 0.04%) and IPTG, recombinant clones obtained by insertional inactivation of the βGal appear as "white or colorless" colonies (with plasmids) or plaques (with phages). Bluo-Gal may be particularly useful in some histochemical applications because the cleavage product precipitates as fine birefringent crystals that emit a strong optical signal (yellow), allowing enhanced resolution of staining patterns under a polarized light microscope (106).

βGal samples dot-blotted on a NC membrane can be stained with high sensitivity using 6-bromo-2-naphthyl-β-D-galactopyranoside as the substrate and by coupling the 6-bromo-2-naphthyl product with an azo dye (diazo-blue B) followed by treatment with 0.01 M sodium carbonate (5).

Following a similar principle but using different chromogenic substrates, recombinants can be screened by their respective colors: (i) lactose (0.4%)–MacConkey agar (Lac$^+$: red, Lac$^-$: white), (ii) lactose–tetrazolium agar (Lac$^+$: white, Lac$^-$: red) (4), and (iii) lactose-EMB (eosin–methylene blue) agar (Lac$^+$: purple, Lac$^-$: white or pink) (6).

(c) Immunological detections. βGal-fusion proteins are often detected by immunostaining either on microtiter plates or on protein (Western) blots using antibodies directed against either βGal or the fusion protein. Highly sensitive and specific monoclonal anti-βGal Ab are commercially available or can be prepared in-house (7). Secondary Ab conjugated with a reporter enzyme (horseradish peroxidase or alkaline phosphatase) are also commercially available.

ii. *Kinetic parameters.* The reactions catalyzed by βGal follow Michaelis–Menten kinetics for all substrates. The k_{cat} values for some commonly used substrates vary within 2–3 orders of magnitude, whereas the K_m values remain virtually the same (8,9) (Table 8.1). In the absence of Mg^{2+} (or after exhaustive dialysis in EDTA media), βGal maintains some residual activity, although the catalytic efficiency (k_{cat}/K_m) is reduced 10- to 100-fold. Compared with a galactoside substrate PNPG (*p*-nitrophenyl-β-D-galactoside), an arabinoside substrate PNPA (*p*-nitrophenyl-α-L-arabinopyranoside) has a 180-fold higher K_m (10).

iii. *Metal ion requirements*

(a) Divalent cations. βGal requires divalent metal ions such as Mg^{2+} and Mn^{2+} for optimal activity (11). One Mg^{2+} ion is bound per subunit of the tetrameric enzyme. Mn^{2+} has a 1000-fold higher affinity than Mg^{2+} at pH 7, but the Mn^{2+}-enzyme has 82% of the activity of the Mg^{2+}-enzyme (12).

(b) Monovalent cations. Monovalent cations have various effects depending on the concentration and substrate species. With lactose as a substrate, 50 mM KCl is three to five times more stimulating than 50 mM NaCl. With ONPG, however, NaCl (50 mM) is more stimulating than KCl. The stimulation may be

TABLE 8.1 Kinetic Parameters of β-Galactosidase[a]

Substrates	Mg^{2+}	K_m (mM)	k_{cat} (sec^{-1})	Ref.
ONPF	+	0.10	800–1100	8, 9
	−	0.70	120	
MUG	+	0.13	[50]	
	−	0.58	[1]	
DNPG	+	0.18	1290	8
	−	0.32	200	
PNPG	+	0.03	90	8, 9
	−	0.53	53	

[a] To make coherent comparisons among the substrates, kinetic parameters were chosen from data obtained at pH 7.0, 25°C, either in the presence (+) or absence (−) of 1 mM Mg^{2+}. [The k_{cat} for MUG is 50-fold higher in the presence of Mg^{2+} than in its absence.] ONPF, o-nitrophenyl-β-D-galacto-pyranoside; MUG, 4-methylumbelliferyl-β-D-galactoside; DNPG, 3,4-DNP-β-D-galactoside; PNPG, p-nitrophenyl-β-D-galactoside.

subject to counterion effects by anions (e.g., Cl$^-$) at high concentrations. Compared with Cl$^-$, P_i is stimulatory and can reverse the inhibitory effect of Cl$^-$.

Na$^+$ (20 mM), K$^+$ (50 mM), or NH$_4^+$ (50 mM), but not Mg^{2+} or Mn^{2+}, confers complete protection to βGal against otherwise slow inactivation on digestion with pancreatic elastase (13). The monovalent cations also provide significant protection to βGal from the digestion with chymotrypsin.

iv. Inhibitors. βGal is inhibited by various substrate analogs. These analogs have proved to be useful in studying the catalytic mechanism of the enzyme (14).

β-D-*Galactosyltrimethylammonium bromide* is a competitive inhibitor with a K_i of 1.4 mM, whereas tetramethylammonium bromide is not an inhibitor ($K_i >$ 0.2 M) (15).

In contrast to β-galactosylpyridinium salts, which are substrates, N-substituted D-*galactosylamines* are strong competitive inhibitors. The inhibitors with basic nitrogen, e.g., β-D-galactosyl-NH-C$_7$H$_{15}$ ($K_i \approx$ 0.6 nM) and β-D-galactosyl-NH-CH$_2$-C$_6$H$_5$ ($K_i \approx$ 9.5 nM), are bound to Mg^{2+}-enzyme several orders of magnitude more tightly than their weakly basic or nonbasic analogs, e.g., β-D-galactosyl-NH-C$_6$H$_4$-Me ($K_i \approx$ 0.76 mM) and β-D-galactosyl-NH-CO-CH$_2$Br ($K_i \approx$ 1.1 mM) (16). The tight binding of basic inhibitors is observed only in the presence of Mg^{2+}. The substitution of sulfur for basic nitrogen results in a weakening of binding by 2–5 orders of magnitude.

Phenylethyl-β-D-thiogalactoside (PETG) is a very poor substrate or, more appropriately, it acts as an inhibitor.

Furanoses and *lactones* are generally strong inhibitors: L-ribose in the furanose form has a K_i of 0.2 mM at pH 7.0 and D-galactonolactone has a K_i of <0.25 mM (17).

Tris, a common cationic buffer substance, is a fairly strong competitive inhibitor, which can be antagonized by K^+ (18).

b. Activity assay and unit definition

The assay for βGal generally relies on the substrates that produce chromogenic (e.g., 2-nitrophenol and 2,4-DNP) or fluorigenic (e.g., 4-methyl-umbelliferone) products (19).

A typical assay mixture contains 100 mM Na-P (pH 7.0), 5 mM KCl, 1 mM MgCl$_2$, 10 mM 2-MSH, and 2–4 mM ONPG as the substrate. The mixture is incubated at 25° or 37°C. The reaction is stopped by adding Na$_2$CO$_3$ (1 M stock) to shift the pH. At pH 11, βGal is inactive. The hydrolysis of colorless ONPG is measured by the absorption of *o*-nitrophenol which has a sensitive yellow color (ε_{420} = 4700 M^{-1} cm^{-1}). [The A_{420} value at pH 7.0 (before the pH shift) is lower by a factor of ~2.]

Unit definition: One *unit* is defined as the amount of enzyme that hydrolyzes 1 μmol of substrate (ONPG) per min under the assay conditions.

c. Substrate specificities

E. coli βGal shows a wide range of activities depending on the property of the leaving group (also called *aglycon*) (20). βGal also displays strict requirements for the structure of the *glycon* part of the substrate molecule. The natural substrate for βGal is a disaccharide, lactose (1,4-*O*-β-D-galactopyranosyl-D-glucose).

i. Hydroxyl groups of the glycon. The OH groups at positions C-3 and C-4 are critical. Methylation of one or more of the OH groups on C-3 and C-4 renders the modified substrates nonhydrolyzable. The hydroxymethyl group on C-5 can be replaced by a methyl group or a hydrogen atom without any effect. β-D-Galactofuranoside, β-D-glucopyranoside, and α-D-galactopyranoside (e.g., melibiose) are not substrates. β-D-Fucopyranoside is a substrate; this is consistent with the observation that the OH at C-6 is not as important for productive binding as the OH groups at C-3 and C-4 (21).

ii. Influence of aglycon structures. βGal can catalyze the hydrolysis of lactose analogs in which the glucose part is substituted by an alkyl or aryl group, e.g., methyl, ethyl, phenyl, and *o*-nitrophenyl.

Phenyl-β-D-galactopyranoside is an acceptable carbon source for *E. coli*. It is not an inducer per se but, like lactose, it can be converted into an inducer by βGal, albeit much more slowly and inefficiently. Since the growth on phenyl-β-D-galactoside is autocatalytic and depends on the initial level of βGal, the use of phenyl-β-D-galactoside as the carbon source provides a means for enriching *lacO*^{c} (operator-constitutive) mutants (22).

iii. Thiogalactosides. Replacement of the D-galactosidic anomeric oxygen by a sulfur atom renders the analogs extremely poor ($<10^{-5}$) substrates. Indeed, thio analogs such as IPTG, ONP-β-D-thiogalactoside (ONP-S-G), and phenylethyl-β-D-thiogalactoside may better be classified as inhibitors.

IPTG is an inducer of βGal synthesis and/or *lacZ*-fusion gene expression under the control of the *lac* promoter–operator. ONP-S-G is also an inducer and a useful analog for the selection of *lac*⁺ mutants, especially *lacY* or *lacP* (23). In the presence of succinate or acetate (but not glucose) as the sole carbon source, the thiogalactoside analogs can be used to inhibit the basal level activity of βGal arising from *lac* promoter fusions. This reduces lactose utilization to the point where appropriate *lac*⁺ clones can be selected.

iv. Galactosylpyridinium salts. These substrate analogs are "well-behaved" substrates, although the rate of hydrolysis can vary by 2–3 orders of magnitude. Compared with the rate of spontaneous hydrolysis, the catalytic enhancement can reach a factor of $\sim 10^{10}$. When the value for ONPG is taken as [100], pyridinium derivatives have relative k_{cat} values as follows: 3-Cl-pyridine [110], 3-Br-pyridine [90], pyridine [0.61], and 2-Me-pyridine [0.22]. The K_m values range from 0.5 to 1.9 mM (24). Remarkably, the removal of Mg^{2+} from the βGal reaction slightly increases the k_{cat} (and the K_m as well). In contrast, the removal of Mg^{2+} decreases the rate of hydrolysis of aryl galactosides.

v. D-Galactal. D-Galactal is a substrate with very slow binding and release rates ($k_{cat} \approx 2 \times 10^{-3}$ sec^{-1}); it is presumed to be a transition-state analog. D-Galactal can be hydrolyzed to 2-deoxygalactose or it can be transferred to a nucleophile, e.g., glycerol (25), to form 2-deoxygalactosyl derivatives.

D-Galactal is a reversible, competitive inhibitor with a K_i of 5–14 μM in the presence of Mg^{2+} (26,27). In the absence of Mg^{2+}, the K_i is 10 mM, which is an increase of ~1000-fold compared with the K_i value in the presence of Mg^{2+}. This large change is due to the decrease in the binding constant of D-galactal to Mg^{2+}-free enzyme.

vi. Transgalactosylation. βGal can catalyze transgalactosylation in the presence of sugars and aliphatic alcohols. The acceptor specificity of sugars and alcohols varies widely: e.g., D-glucose [380 sec^{-1}], L-glucose [56 sec^{-1}], 3-O-Me-D-glucose [84 sec^{-1}], D-galactose [380 sec^{-1}], D-mannose [610 sec^{-1}], D-mannitol [3,120 sec^{-1}], 2-propanol [1,660 sec^{-1}], and 1-propanol [10,600 sec^{-1}] (28). Regarding the equilibrium binding, glucose is the preferred acceptor due to its slow rate of dissociation (360 sec^{-1} at pH 7.0 or $K_d = 17$ mM). However, glucose is only moderately reactive. Transgalactosylation for the production of galacto-oligosaccharides (e.g., tri- and tetrasaccharides) occurs more efficiently in water-hydrophobic solvent mixture. For instance, the reaction can be efficiently carried out in a water (2.5–10%, v/v)–cyclohexane mixture at pH 6.0 and 60°C (29). With lactose as a donor–acceptor, βGal catalyzes the formation of an isomer allolactose (1,6-O-β-D-galactosyl-D-glucose), a natural inducer of the *lac* operon

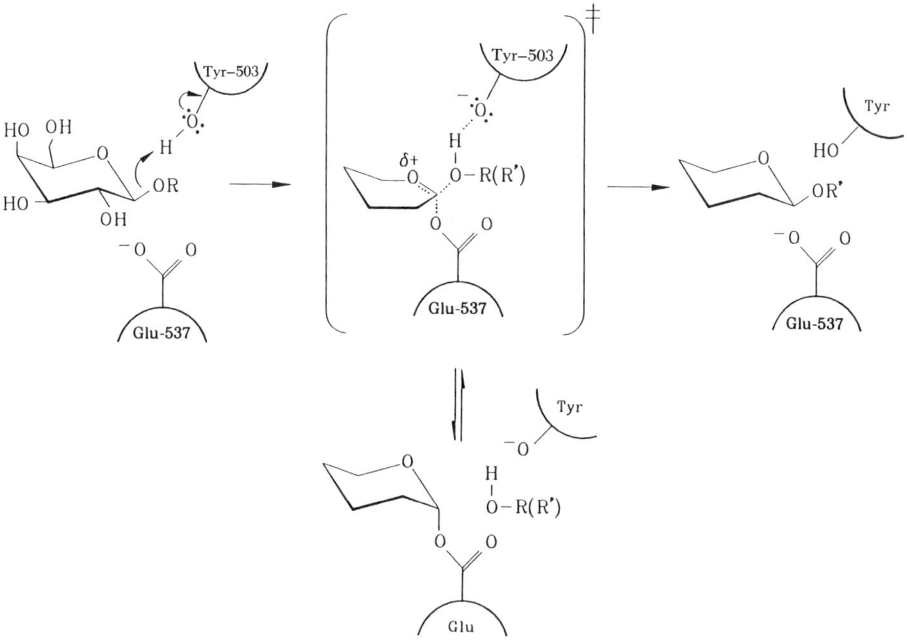

FIGURE 8.1 Mechanism of β-galactosidase catalysis. A double dagger denotes a transition-state intermediate (bracketed). Glu-537 is thought to be the residue that forms the covalent linkage with the intermediate. The electrostatic stabilization of the oxocarbonium ion is presumed to be mediated by the Mg^{2+}-coordinated Glu-461. R' is H for hydrolase activity. When R' is a sugar (or alcohol), the activity of βGal corresponds to transgalactosylation.

(30). The allolactose has a K_d for a *lac* repressor similar in magnitude to that of IPTG. Allolactose itself is a substrate for βGal with a K_m of ~2 mM.

d. Reaction mechanism of β-galactosidase

The hydrolysis (or alcoholysis) of β-D-galactosides by βGal occurs with an overall retention of configuration at the anomeric C-1 carbon (Fig. 8.1). The catalytic pathway thus involves a covalent glycosyl–enzyme intermediate. The initial step is believed to be the acid-catalyzed labilization of the leaving group. An acid catalyst ($pK_a \geq 9$) donates a proton from the β side of the galactose ring. The rate enhancement varies with the nature of the leaving group. An enzyme carboxylate (Glu-537) participates in the covalent bond formation in an axial position at C-1. The formation and breakdown of the covalent intermediate is presumed to occur via an oxocarbonium ion-like transition state which is stabilized by electrostatic interactions with an acid group, presumably the Mg^{2+}-bound Glu-461. Hydrolysis or transgalactosylation occurs via stereo-specific attack of the water or alcohol after the first product (ROH) has diffused.

The kinetics of βGal reaction can be described by a general reaction scheme (Scheme 8.1). Depending on galactoside substrates (G-OR), the RDS can be either the formation (k_2) or the breakdown (k_3) of the intermediate (E-G).

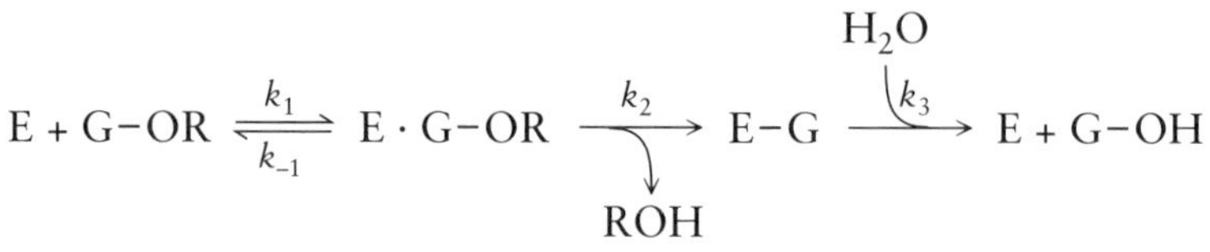

$$E + G{-}OR \underset{k_{-1}}{\overset{k_1}{\rightleftharpoons}} E \cdot G{-}OR \overset{k_2}{\underset{\downarrow}{\longrightarrow}} E{-}G \overset{H_2O}{\overset{k_3}{\longrightarrow}} E + G{-}OH$$
$$ROH$$

SCHEME 8.1

For substrates such as ONPG (9) or DNPG (31), k_2 is larger than k_3, i.e., degalactosylation is the RDS. With Me-β-D-galactoside, which has a relatively poor leaving group, $k_3 \gg k_2$ and the RDS occurs before the degalactosylation step. More elaborate mechanisms have been proposed to incorporate a step for a conformational change(s) of the enzyme subsequent to the formation of the Michaelis complex (16,32).

Acid catalysis requires a contribution by the Mg^{2+} ion, which is thought to induce a conformational change that brings the acidic group into a catalytically competent position. Removal of Mg^{2+} has little effect on the hydrolysis of pyridinium salts but lowers the rate of hydrolysis of O-glycosides. The metal ion, as revealed by experiments with Mn^{2+} as an NMR probe, is not coordinated to the sugar. It is positioned at a distance of ~9 Å from the methyl protons of the Me-S-galactoside bound in the active site (33).

The best evidence that βGal forms a covalent intermediate comes from experiments with D-galactal, a presumed transition-state analog. Using D-[6-^{3}H]galactal, βGal could be covalently labeled and the labeled complex could be separated (34). The covalent enzyme–galactal intermediate spontaneously disintegrates to yield reactivated βGal and 2-deoxy-D-[^{3}H]*lyxo*-hexose. The disintegration proceeds faster at higher pH conditions.

2. STRUCTURE

i. **Overall structure.** *E. coli* βGal is a tetrameric enzyme with identical subunits consisting of 1023 amino acids (M_r 116 353/monomer) (35,36). The X-ray crystallographic structure of tetramer exhibits a dimension of roughly 175 × 135 × 90 Å and a twofold symmetry (107). A monomer contains five compact domains plus ~50 residue N-terminal region corresponding to the "α-complementation" peptide (see below). Each monomer makes two different monomer–monomer contacts. The N-terminal region, which is relatively extended, contributes to the "activating" interface in the sense that it completes the active site of the neighboring monomer.

The N- and C-terminal residues are Thr and Lys, respectively. The enzyme has a high content (39 residues) of Trp. The ε_{280} of βGal is 2.18 $mM^{-1}cm^{-1}$. The secondary structure of the enzyme is estimated to be 35% α helix, 40% β sheet, 12% random coil, and 13% β turn (108).

ii. **N-terminal region.** The removal of N-terminal residues 11 to 41 inactivates βGal, but it does not eliminate the potential to regain the enzyme activity.

A dimeric mutant protein, from a deletion mutant (*lacZM15*), can be fully activated when it is complemented with an N-terminal fragment, e.g., CNBr fragment (residues 3–92) or a smaller tryptic peptide (residues 3–41) (37–39). This phenomenon is known as α-*complementation* (also see Section A,3, in this chapter). The M15 protein also regains enzymatic activity when complexed with an anti-βGal Ab to a maximum level of 10–15% of the activity restored by the α-complementation (40). Another *lacZ* deletion protein (M112), a dimer lacking amino acid residues 23–31, is inactive but shows a similar activation profile as M15 protein (39,41).

A maximum of 41 N-terminal amino acid residues can be substituted by heterologous amino acid sequences without affecting the enzymatic activity significantly (42,43). Nevertheless, such replacements may incur reduced structural stabilities, suggesting that some of the N-terminal residues, especially amino acids 27–31, are involved in dimer–dimer interactions. An exposed loop region (involving amino acids 272 to 287 in domain 2) is tolerant to the insertion of large foreign peptides, thus allowing construction of chimeric proteins which retain βGal activity (44).

iii. C-terminal region.
Most C-terminal alterations result in inactivation of βGal. For example, the βGal with a C-terminal 10 amino acid truncation, as found in the mutant X90, is devoid of enzymatic activity. The X90 protein is a monomer under normal conditions. However, it can be converted to an active enzyme when complemented with an ω *peptide* (40 kDa) or a CNBr fragment containing the C-terminal 32 residues (45), a phenomenon known as ω-*complementation*. The X90 protein cannot be complemented with the exact missing part, the C-terminal 10 residue peptide derived by a tryptic digestion. This suggests that the C-terminal region plays an important role in monomer–monomer subunit interactions. Nevertheless, at least 2 C-terminal amino acids can be removed or substituted with little effect on βGal activity (46).

iv. Active site.
The active site residues of βGal comprise Met-502, Tyr-503, Glu-461, and Glu-537. The four residues, which are strictly conserved in at least five homologous βGal sequences, are found close together and are located around a deep pit (107). Cys residues are not involved in the activity of βGal.

Met-502, because of its active-site location, has some influence on enzyme activity but is not important for catalysis. Modifications of Met-502, either irreversibly (47,48) or reversibly (49), render the enzyme inactive. However, biosynthetic replacement of Met by norleucine gives an enzyme with activities similar to those of the wild-type βGal.

Tyr-503 is the residue apparently involved in general acid/base catalysis. Site-mutagenic replacement of Tyr-503 by Phe reduces the k_{cat} by about 250-fold and the K_m by about 4-fold (50). The pK_a of the group has been estimated to be ≥ 9. Iodination of βGal results in stoichiometric inactivation of the enzyme (51). βGal can be protected from iodination in the presence of IPTG. In these iodination experiments, however, Tyr-253 has been identified to be the most reactive among the Tyr residues in βGal. Tyr-503 is not important in binding Mg^{2+} (10).

Glu-461 is the residue covalently modified by an active site-directed *cis*-epoxide inhibitor (52) and most likely by substrates as well during the course of catalysis. Glu-461–Gln substitution reduces the k_{cat} (with ONPG) by 250-fold and the K_m by 17-fold (53). The same substitution results in a 50-fold decrease in the affinity of the βGal for Mg^{2+} (10). These observations suggest that Glu-461 is most likely a Mg^{2+} ligand. Glu-461 is also presumed to be involved in electrostatic stabilization of a galactosyl cation transition state. Glu-537, and not Glu-461, is covalently modified by 2,4-DNP-2-deoxy-2-fluoro-β-D-galactopyranoside, suggesting its role as the nucleophile (109). Site-specific replacement studies suggest that Glu-537 is important for catalysis (110). In a crystal structure, Glu-537 is located in the immediate proximity to Tyr-503.

The galactose (or glycon)-binding subsite interacts with the OH groups on C-3, C-4, and C-6 of galactose (21). Positions C-3 and C-4 are more critical than C-6. These sites are capable of discriminating against glucosides by a factor of $>5 \times 10^4$ (16). Substrates without OH at C-2 bind the enzyme readily but do not undergo catalysis. The glucose (or aglycon)-binding subsite has a hydrophobic character and shows a high but subtle affinity for hydrophobic aglycone moiety and primary alcohol groups (28).

3. GENETICS

The structural gene for βGal is *lacZ*. It is located, as a part of the *lac* operon, at the 8-min position on the *E. coli* genetic map. *In vivo*, βGal functions as the enzyme responsible for breaking down β-galactosides (e.g., lactose) into component sugars (e.g., galactose + glucose) which can then be utilized as the energy source. In the absence of β-galactoside substrates, cells maintain a basal level (<10 molecules/cell) of βGal. When a substrate (lactose), a suitable substrate analog, or an *inducer* is added to the medium, the amount of enzyme increases within 2–3 min to a maximum of ~5000 molecules per cell (or 5–10% of the total soluble protein). IPTG, a substrate analog not metabolized by βGal, is an efficient "gratuitous" inducer. (ONP-fucopyranoside is a competitive inhibitor of induction.) βGal is also induced by the natural inducer allolactose, which is a product of transgalactosylation catalyzed by βGal. In fact, a more abundant and effective natural inducer is thought to be β-galactosylglycerol which is derived from galactolipids by fatty acid hydrolysis.

[*Note*: If a cell carries a mutation in the *gal* operon, e.g., the *galE* gene that codes for UDP-galactose epimerase, the product galactose can render the cells sensitive to galactose accumulation. Thus the selection for bacteria resistant to lactose or its derivatives in a GalE$^-$ background enables the selection of spontaneous *lac* mutants (54,55).]

    i. Lactose operon. The *lacZ* gene (3069 bp) is the first structural gene of the lactose (*lac*) *operon*, the historical paradigm of gene regulation discovered in the early 1960s by Jacob and Monod (55,56). The *lac* operon consists of regulatory

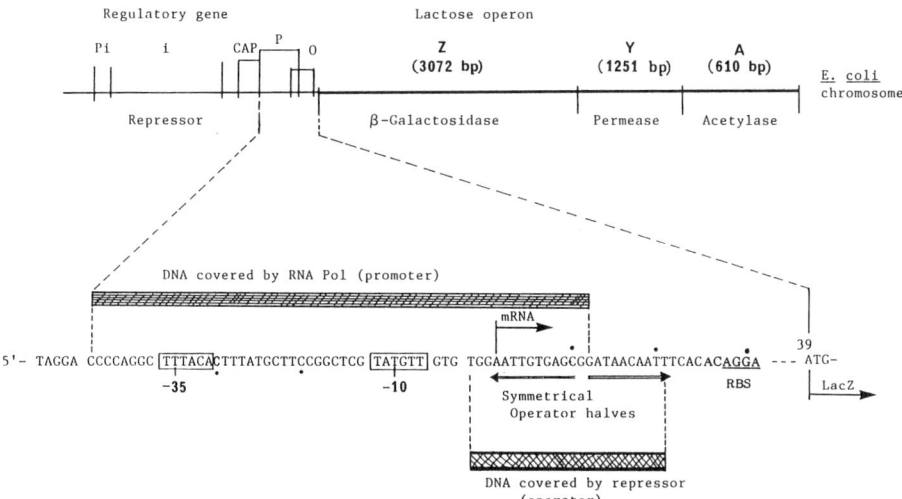

FIGURE 8.2 Organization of lactose operon and features of promoter region. (Top) Schematic gene organization of the *E. coli lac* operon: Pi denotes *lacI* promoter, and CAP denotes the binding site for catabolite gene activator protein. (Bottom) Features of the *lac* promoter with *lac* operator (O_1) sequences displaying an inverted repeat. The ribosome-binding site (RBS) is underlined. A typical hybrid promoter, Ptac, features −35 and −10 regions having consensus hexamer sequences and a 16-nt spacing as follows:

```
    -35                         -10           +1
     |                           |             |
     |                           |             |
   TTGACA ATTAATCATCGGCTCG TATAAT GTGTGG A
```

genes and three structural genes (Fig. 8.2). The regulatory genes comprise *lacI* (*lac* repressor, with its own set of promoter and terminator), *lacP* (promoter, RNA polymerase-binding site), and *lacO* (operator, repressor-binding site). The structural genes are *lacZ, lacY* (1251 bp, β-galactoside permease), and *lacA* (610 bp, β-galactoside transacetylase).

The *lac* repressor-mediated regulation of transcription in the *lac* operon is based on high affinity binding of the repressor protein to the *lac* operator (O_1) and consequent steric hindrance to *E. coli* RNA Pol at the transcription initiation region. The repression on RNA Pol is relieved by prior binding of repressors with inducer molecules which dramatically modulate the affinity of the repressor for operator DNA. As an apparent second level of control, LacI is believed to interfere with transcription elongation by binding to a second site (O_2) that is located within the *lacZ* gene 402 bp downstream from the O_1 (57).

LacI functions as a tetramer of identical subunits. A monomer (38 kDa) consists of 360 amino acid residues (58). The N-terminal 59 amino acid residues constitute the DNA-binding domain (59,60). The C-terminal region, which contains a Leu zipper motif, is implicated in the dimer–tetramer interaction. There are 10–15 tetramer repressor molecules in a wild-type cell. The *lac* repressor binds the inducer IPTG with a K_d of 2 μM (monomer) and the operator with a K_d of 8 × 10^{-11} M (dimer). Binding of IPTG to the repressor keeps the repressor from binding to the operator, thus derepressing the transcription by *E. coli* RNA polymerase.

The *lacP* consists of two important regions, CH-35 (TTTACA) and CH-10 (TATGTT), which are separated by 18 bp in native DNA. Single base deletions which reduce the spacer length to 17 bp increase the strength of the promoter approximately twofold (61,62). A single base substitution (T to G) in the CH-35 to generate consensus sequence TTGACA or substitutions of the CH-10 to a consensus hexamer TATAAT (better known as *lacUV5* promoter) further increases the promoter strength regardless of whether the 17- or 18-bp spacing is maintained. Indeed a family of hybrid promoters derived from *trp* and *lac* promoters, thus called *tac* promoters (see legend to Fig. 8.2), have been constructed by combining the −35 region of the *trp* promoter and the −10 (Pribnow box) downstream region of the *lacUV5* promoter (63,64). Because the *tac* promoters retain the features of *lac* repressor/IPTG-mediated control of gene expression, for example in pBTac vectors (from BM) or pKK223-3 (from Pharmacia), they are frequently employed in the overexpression of recombinant proteins in *E. coli*. (*Note*: The λ P_L promoter, which is also commonly used for controlled gene expression in the presence of the λ repressor cI857, is estimated to be twice as strong as P_{tac}.)

The *E. coli lac* promoter has been widely used for the induced expression not only of the homologous structural genes in the *lac* operon but also of a variety of foreign genes cloned in vectors (see below). Expression of the genes cloned downstream from the *lac* promoter can be conveniently induced by adding IPTG. However, in the absence of an inducer, there is a basal level of expression from this promoter. This inherent "leakiness," which is particularly bothersome when the product of the cloned gene is toxic to the cells, can be virtually eliminated 100% by growing the cells (e.g., transformed with pUC plasmids) in the presence of 0.4% glucose (65). Under this concentration of glucose, even the induction by IPTG is strongly repressed due to the phenomenon of glucose repression and inducer exclusion. As the bacterial culture attains an A_{600} of 0.9, the glucose is sufficiently consumed, and the addition of IPTG (0.4 mM) results in a full expression of the gene from the *lac* promoter. In a large-scale culture (or fermentation), the expression of recombinant genes under the control of the *lac* promoter can be induced most economically using lactose rather than IPTG. The lactose, which is also a carbon source for *E. coli*, is added at a concentration of 2% (w/v) at the late phase of culture in the LB-glucose (0.5%) medium when the glucose concentration drops to below 0.1% (111).

ii. *lacZ* **gene structure.** The complete nucleotide sequence of the *lacZ* (1023 amino acid codons) has been determined (36). Based on the predicted amino acid sequence, the nascent N-terminal Met (AUG) is posttranslationally processed, leaving the second amino acid (Thr) to become the N-terminal residue in the mature protein. The *lacZ* ORF is terminated by three repeats of the UAA termination codon.

The 5' ends of *lac* mRNA are heterogeneous, presenting four species in fully induced cells. The most abundant species (~70%) have 5' A(+1)-A-U-U-G, which is located 38 nt upstream of the AUG translation start (66). A Shine–Dalgarno sequence is present at position 28–31.

The *lacZ* contains a series of latent transcription terminators that are responsible for the polar effects of certain mutations. The terminators are in the vicinity of five positions: 180, 220, 379, 421, and 463 bp downstream of the transcription start point (67). No strong terminators are found further downstream from this region. Termination at all but the 421 position is ρ dependent *in vitro*, with efficiencies ranging from 8 to 56%. Those stop points correspond to 5 of the 11 transcriptional pause sites located within the region.

The RBS has been mapped to a region of ~35 nt (position 25 to 59) (68). It starts from the S-D region and includes a region 7–12 nt distal to the AUG initiation codon.

iii. *lacZ* mutations.

Several nonsense and deletion mutations are located in the early part of the N-terminal region (41). The *lacZ2*, *lacZ2246*, and *lacZU131* are point mutations at the sites corresponding to amino acid residue 23 (Gln), 36 (Trp), and 41 (Glu), respectively. The *lacZM112* produces a dimeric βGal protein containing a deletion of amino acid residues 23–31. The *lacZU118* is an *ochre* (UAA) mutation at the site corresponding to Glu-17 (69). The *lacZM15* is a deletion mutant (amino acid residues 11–41) producing a defective, dimeric βGal (38). The *lacZ90* is a mutation lacking the C-terminal 10 residues (45). A missense mutation at position 794 (Gly to Asp) results in a heat-labile enzyme but with a five to six times higher k_{cat} on lactose than the wild-type βGal (70).

iv. α complementation.

In genetics, the term *complementation* is used to describe a phenomenon in which a biological function that has been lost or altered by a mutation can be restored through mutual compensation by differently altered mutants. The repair of function by complementation occurs at the level of the protein molecule either *in vivo* or *in vitro*, and involves noncovalent interactions of differently altered polypeptide domains. Complementation is thus distinct from the repair of function by genetic *recombination*, which involves the reconstitution of an active gene or a set of genes.

Some mutant βGal proteins can regain enzymatic activity by intracistronic complementation (71). Those that occur in the N-terminal region are called α *complementations*. All mutants carrying operator-proximal segments (α segments) intact are functional α *donors*. In fact, the N-terminal 39 amino acids (residues 3–41, α fragment) are sufficient to promote an α complementation (37,39). All mutants with partial deletions not extending beyond a barrier region (which is situated between the mutations 274 and X64, or at the N-terminal 20–25% region of βGal) are α *acceptors*. The donor restores the activity of the acceptor molecules by converting the dimeric mutant protein to an active tetrameric structure.

Mutations located in the C-terminal region of βGal can also be complemented in a manner similar to the mutations in the N-terminal region, and this is termed ω *complementation*.

v. Evolutionary aspects of β-galactosidases

(a) Homology with dihydrofolate reductase. A segment of C-terminal residues (820–934) of βGal shows a high degree of sequence identity with a segment of DHFR from *E. coli* and chicken. Alignment of the sequences based on the similarity of tertiary structures between the *E. coli* and chicken DHFRs has revealed an interesting feature: the positions of introns 1 and 5 of the chicken DHFR correspond exactly to the start and the end of the DHFR-like domain in the βGal, suggesting the possibility of a common origin (72).

(b) Evolutionary βGal. *E. coli* cells which contain *lacZ* deletions and are selected for growth on lactose produce a new βGal which is encoded by a new gene *ebg*, an acronym for "evolved beta (β)-galactosidase" (73). The *ebg* gene maps at the 68 min on the *E. coli* chromosome. The *ebg* system consists of a repressor encoded by *ebgR* and at least three structural genes: *ebgA*, *ebgC*, and *ebgB*. Mutants with *lacZ* deletions cannot grow on lactose because the enzymatic product of the wild-type βGal gene, *ebg*^{o}, is catalytically too weak to support the growth and because the wild-type repressor is too insensitive to β-galactosides. Through a process of *acquisition evolution* by which simultaneous mutations arise in the *ebgR* gene and the *ebg* structural genes, the cells acquire catalytically more competent, evolved *ebg* βGal (74,75). The evolved forms of the enzyme fall into mutant classes such as *ebg*^{a}, *ebg*^{b}, and *ebg*^{ab}.

The *ebg* βGal is a heterodimer of 120 kDa (76,77), a size very similar to that of *lacZ* βGal. However, the active *ebg* enzyme is an apparent hexamer (not a tetramer) and the two proteins do not cross-react with antisera prepared against each type of protein. Although the two proteins are somewhat different in primary structure, available data suggest that the active-site environment of the two enzymes is similar (78).

4. APPLICATIONS

βGal has wide-ranging applications in biotechnology as well as in recombinant DNA technology. As a biocatalytic tool, βGal is employed in the conversion of sugars producing "lactose-free" milk and galacto-oligosaccharides. βGal is an important reagent used in association with second antibodies (ELISA and dot-blots) or oligonucleotides (hybridization) in enzymatic or immuno(histo)chemical assay/detection. As far as recombinant DNA technology is concerned, βGal (and *lacZ* gene) finds the foremost applications as a reporter/marker/carrier of gene expression in both prokaryotes and eukaryotes. Typical applications include (i) a blue–white selection marker for gene cloning in plasmids, (ii) a marker for the discovery, assessment and optimization of promoter (length and sequence context), (iii) a marker gene to estimate the efficiencies of translation or more generally recombinant gene expression, (iv) a stable carrier in the production of unstable proteins and vaccinal epitope(s), (v) a tracer for gene rearrangement, cellular differentiation, and development, and (vi) an indicator of infectivity of recombinant viruses. General principles and some typical strategies underlying the wide-ranging applications are described below.

a. General principles of reporter gene function and gene fusions

βGal (or *lacZ* gene) is the first and the most important of the marker/carrier enzymes (or genes) used in the molecular cloning of genes and the studies of gene expression and regulation (79). A large body of genetic information on the *lac* operon offers special advantages regarding the manipulation of the gene and the selection and identification of the LacZ$^+$ phenotype.

The *lacZ* gene fusions can be constructed either *in vivo* by means of spontaneous nonhomologous recombination and transposon (phage Mu)-mediated recombination or *in vitro* using recombinant DNA techniques (79,80).

The *in vitro* fusion approach, which is most commonly used, has two obvious advantages: (i) the possibility for direct and precise controls over the process of gene fusion and (ii) the availability of an assortment of general and specialized vectors.

DNA sequences representing either the genes for recombinant proteins or regulatory regions can be fused to the 5'-end region of *lacZ* (N-terminal part of βGal) without adversely affecting the enzymatic activity. Such *lacZ* fusions provide useful tools for the analysis of gene/protein structure and function, as well as for the production of recombinant proteins.

The *lacZ* fusion system is particularly valuable for the expression of recombinant proteins which have not been identified, for which no convenient assay exists, or which are too unstable to be expressed by themselves. Furthermore, inducible (or regulated) expression under the control of the *lac* promoter provides a means of overproducing recombinant proteins that are inherently toxic to the cells or "toxic" simply because of overproduction. Among the earliest examples of foreign structural gene–*lacZ* fusions used to produce recombinant proteins are somatostatin (14-amino acid peptide encoded by a synthetic gene) (81) and insulin A/B chains (82). Recent examples are abundant in the literature, encompassing diverse areas which include: (i) the production of proteins having biological and pharmacological value, (ii) the generation of bifunctional enzymes for the study of protein–protein interactions, and (iii) the study of protein stability and degradation by fused enzymes. The host systems for the expression of *lacZ* fusions have also been substantially expanded, now including bacteria (both gram-positive and gram-negative), fungi, yeast, insect cells, mammalian cells, and plants.

Cloned genes are most commonly linked to the reporter *lacZ* by N-terminal fusions, and most available vectors are constructed in configurations appropriate for such practices. However, the N-terminal fusion is effective only when the insert DNA or gene does not contain a termination codon. Otherwise, considerations should be given to eliminate or mutate the stop codon. Alternatively, a C-terminal fusion can be considered. A number of vectors, e.g., the pUR278-292 series (46), have been developed for such C-terminal fusions consuming up to 2 amino acid residues of βGal. These types of vectors are generally engineered to contain MCS (*Bam*HI, *Sal*I, *Hin*dIII and *Cla*I plus *Xba*I or *Pst*I) in the 3' end of the *lacZ* gene in all three different reading frames.

The large molecular size of the overexpressed βGal fusion proteins tends to drive the fusion proteins into insoluble aggregates called inclusion bodies. This can be taken in some cases as an advantage because the expressed protein is extremely easy to purify, it is stable being largely out of the reach of degradating enzymes, and it is an efficient antigen. In other cases, however, the low solubility of the expressed protein may pose a problem. Also, the cytosolic nature of the unmodified βGal should be taken into account when attempting an extracellular secretion of recombinant proteins.

The work with βGal fusion has been greatly facilitated by the introduction of various restriction sites and additional features to the vector constructs. The improved features include an *Eco*RI site at codon 5 or *Bam*HI, *Eco*RI, and *Sma*I sites adjacent to codon 8, leading to the general use of "polylinkers" or MCS carrying various numbers of unique restriction sites (83). Many of these vector constructs (see below) are designed for *insertional inactivation* of βGal: the parental construct is LacZ$^+$ (blue on X-Gal/IPTG plates) but the insertion of a gene into the MCS results in a LacZ$^-$ phenotype (white). Other vector constructs are built for *insertional activation:* appropriate fusions provide a correction to the out-of-frame parental construct and produce a LacZ$^+$ phenotype (84).

b. Regulatory gene fusions

Construction *in vitro* of *lacZ* (N-terminal) fusions with or without recombinant structural genes has proven extremely valuable. The regulatory DNA–*lacZ* fusions permit a wide variety of applications (79,80), including (i) identification/cloning of regulatory genes or sequences such as promoters, enhancers, and silencers; (ii) optimal positioning of promoters and evaluation of promoter strengths; (iii) characterization of transcriptional and/or translational (RBS and initiator codon selection) control mechanisms in gene expression; (iv) tissue (or cell-type)-specific gene expression and *in situ* analyses of transfected cells or transgenic organisms; and (v) assays for DNA transfection. In essence, these applications enable one to find the combination(s) of regulatory sequences optimal for gene expression and open the way to custom-designing regulatory sequences such as promoter, UPS, operator, enhancer, terminator, RBS, and initiation codons.

The regulatory gene fusions on *lacZ* (or other marker/reporter genes) are usually carried out by "modular" or "cassette" approaches in which "portable" elements, e.g., promoter, RBS, and transcription terminators, are assembled *in vitro* together with a structural gene. One example is the *ORF expression vectors* which employ tripartite construct strategies for *lacZ* fusions. The basic construct contains a promoter, e.g., *lacP* (85) or of *ompF* (86), a RBS, and a coding sequence followed by *lacZ*. The parental hybrid without the coding sequence insert is LacZ$^-$ because *lacZ* is set to be out of frame. When a gene or coding sequence is inserted

TABLE 8.2 Some Commercial Plasmid Vectors Expressing Full-Length *lacZ* Gene

Vector	Promoter	Other features (size, marker)	Source
pEX	λP$_R$	5.8 kb Ap^r C-terminal fusion (λCro-βGal-)	Boehringer Mannheim
pCH110	SV40 early/*E. coli gpt*	7.2 kb Ap^r	Pharmacia
pSV-βGal	SV40 early	6.8 kb Ap^r	Promega
pNEOβGAL	SV40 early	8.6 kb Neo^r	Stratagene
pADβ	Adenovirus major late	7.1 kb Ap^r	Clontech
pCMVβ	CMV immediate-early	7.2 kb Ap^r	Clontech
pTKβ	HSV thymidine kinase	7.5 kb Ap^r	Clontech
pBlueBac	AcMNPV early-to-late	10.3 kb Ap^r (baculotransfer vectors)	Invitrogen

in the ORF at an optimal site, the insertion provides a correction to the *lacZ* reading frame and produces a fusion protein (or *tribrid* in the form of *ompF-ORF-lacZ*) which exhibits βGal activity. Once the configuration for maximal expression is attained, the *lacZ* part may be removed to produce the unfused product of the cloned gene (87).

c. Basic features of lacZ vectors and strategies of vector construction

The *lacZ* gene has been incorporated into a variety of prokaryotic and eukaryotic vectors as an intrinsic marker/reporter of gene cloning/expression. Some expression vectors carry the full-length *lacZ* gene (Table 8.2). Other vectors, known as *multipurpose vectors* because of their versatile use in various gene cloning, sequencing, and expression, generally feature not the full-length *lacZ* but a fragment of it, denoted as *lacZ'* or *lacZα*, capable of encoding N-terminal 145 residues (α-peptide) of βGal. These vectors (single-stranded DNA phages, plasmids, phagemids, and λ phages) carry varying numbers of multiple cloning sites (MCS) within the N- or C-terminal part of *lacZ'*. Insertion of foreign DNA into one of the MCS usually leads to the production of an α peptide defective in α complementation, giving rise to a *lacZ*-minus phenotype. This insertional inactivation of the βGal offers convenience and unambiguity of blue(*lacZ*$^+$)/white(*lacZ*$^-$) selection, a standard technique for screening recombinant clones. The following selected examples illustrate the key features of the designs incorporating the *lacZ'* or *lacZ* gene in common multipurpose vectors (Fig. 8.3).

i. Single-strand phage systems

(a) *M13 vectors.* M13mp series vectors are the prototype ssDNA (~7.2 kb) phage vectors (88). They are high copy number vectors with ~300 molecules/cell of the double-stranded, supercoiled, replicative form (RFI) DNA. These vectors

586   Chapter 8   Marker/Reporter Enzymes

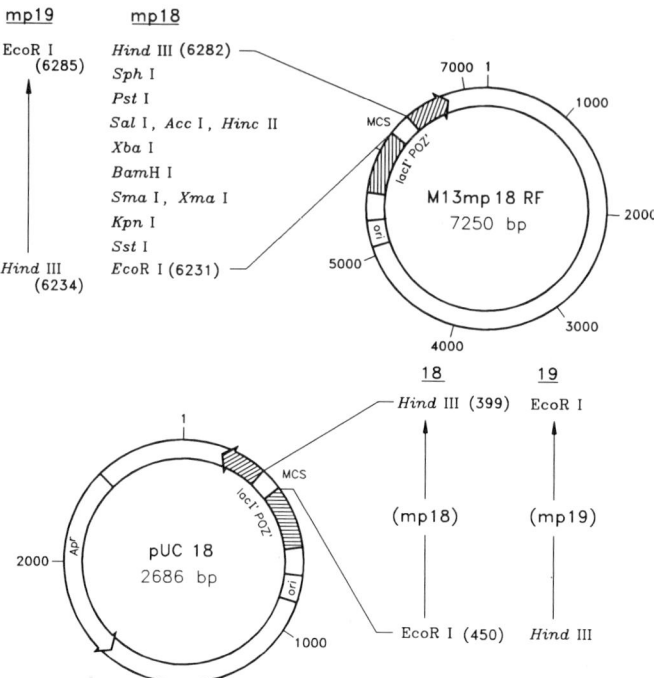

FIGURE 8.3 Basic constructs of the vector systems incorporating the *lacZ* gene as a marker/reporter. The vector systems represent the top line of the vectors often developed in their own series: M13mp18/19, pUC18/19, pUC118/119, pEMBL8, pGEM-7Zf, pBluescript II SK/KS, pTZ(18/19)R, pEX2, λgt22 and λZAP II. The constructs show the general features of the vectors with emphasis on the configuration of the *lacZ'* (or *lacZα*) reporter gene and multiple cloning sites (MCS).

allow simultaneous preparation of ssDNA from the phages recovered from the culture media and of dsDNA plasmids from the cell.

The vectors are constructed out of the phage M13 by inserting a *Hin*dII restriction fragment of *E. coli lac* operon into the nonessential intergenic space (map point 0.083, *Bsu*I site) of the wild type M13 phage genome (89). The inserted fragment contains the *lac* regulatory region [part of the *lacI* (*lacI'*) and the *lacP,O*] and the proximal part of the *lacZ* gene [*lacZ'(α)*]. Within this gene, a single *Eco*RI site was initially introduced by using a methylation mutagenesis technique at the fifth amino acid position changing G (Asp-7) to A (Asn) and giving rise to a new phage vector M13mp2. Despite the introduction of the new *Eco*RI site and, subsequently, a polylinker containing up to 13 unique restriction sites, e.g., in M13mp18/19 (90), the α peptide is active in α complementation. (The notation 18/19 is for alternate orientation of the same polylinker.)

(b) Hosts for M13 vectors. The JM series cells (JM 101–109), which are widely used as the host for M13 vectors, are a derivative of *E. coli* K12 (strain 71-18) in which the *lac pro* region of the chromosome is deleted and is carried

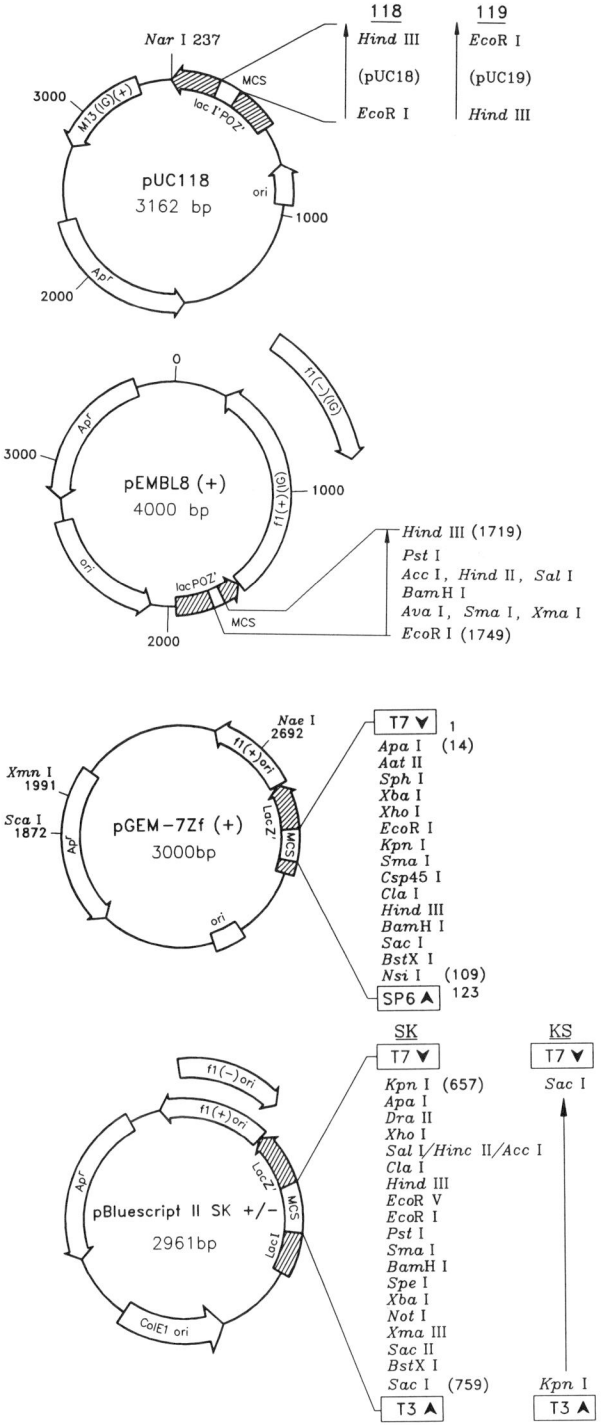

FIGURE 8.3 (*continued*)

Chapter 8 Marker/Reporter Enzymes

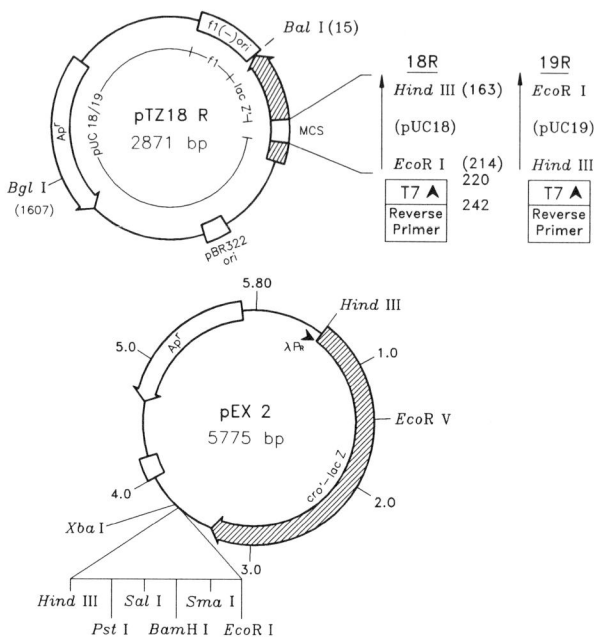

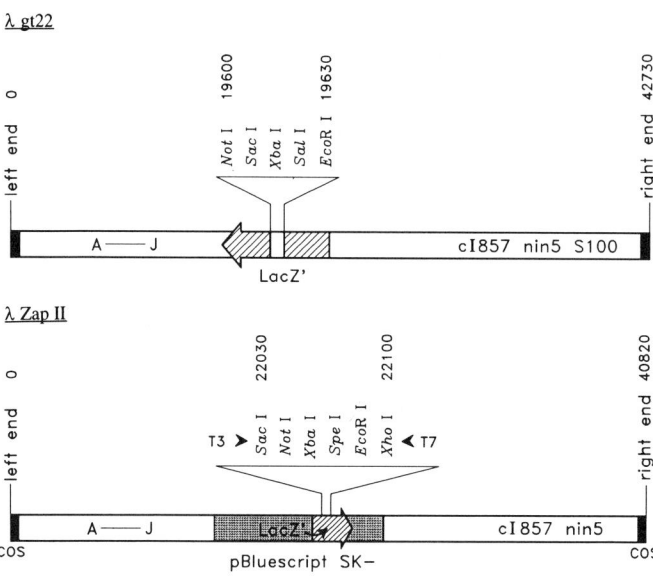

FIGURE 8.3 *(continued)*

on an F episome instead (90). For example, the genotype of JM109 is Δ[lac pro], supE, thi, F'[proAB lacI^q ZΔM15 traD36] recA1, endA1, gyrA96, hsdR17, relA1, λ⁻ (also see Appendix B). The sex factor F encodes F-pili, an adsorption site for male-specific phages for the conjugational transfer of phage genetic material. As the phage-infected cells are rather unstable and have a tendency to lose the F episome, the *pro* marker is used to select for and maintain the F episome in the host cells on minimal media. The *lac* operon contained in the F episome carries two mutations of practical importance: (i) a mutation in the *lac* repressor promoter (*lacI*^{q}) that causes an overproduction of the *lac* repressor, and (ii) a deletion in *lacZ* (ΔM15) corresponding to amino acid residues 11 to 41. The overproduction of the *lac* repressor helps avoid the potential risk of titrating out all the *lac* repressors by the *lac* operators which exist on multiple copies of the intracellular *replicative form* (RF) of the phage. At the same time, it allows the expression of the *lac* operon (and the cloned gene as well) to be regulated by the inducer, IPTG. The defective βGal (M15 protein) of the host serves as the rescue marker which becomes activated by α complementation provided by the LacZ' of the M13 vectors.

A third mutation intentionally carried on the F episome is the *traD36*. This mutation renders the *tra*$^-$ cells incapable of acquiring the F$^+$ phenotype by conjugation. However, the *tra* mutation increases the risk of losing the episome when the cells are cultured in rich broths, the reason for the recommendation to use fresh inocula from the minimal agar plate (+ glucose) each time a new batch of JM cells are to be grown.

Noninfected *E. coli* hosts (JM and related series) are Lac$^-$ since they produce defective βGal (LacZ') from the F' episome. Phage (or plasmid)-infected cells turn into Lac$^+$ due to the α complementation. Cloning into the *Eco*RI site or other sites in the MCS of M13mp18/19 (or derivative vectors) results in the production of inactive α peptides, rendering the infected cells Lac$^-$. Cloning of small sized genes that happen to be in-frame with the adjoining βGal(α) may give rise to true recombinants which appear as Lac$^+$.

Phage particles containing (+)-sense ssDNA are continuously produced by and extruded from the infected cells without cell lysis. Titers of the phage in the cell-free supernatants range up to 5×10^{12} PFU/ml, which is equivalent to 150 μg/ml of phage or ~15 μg of ssDNA (91).

ii. Plasmid and phagemid vector systems

(a) pUC vectors. The foremost example of plasmid vectors that utilize the *lacZ* gene as a marker/reporter is the pUC series vectors. The pUC vector (e.g., pUC18/19, 2.7 kb) is a high copy number plasmid derived from pBR322 by inserting the *lac Hae*II fragment at the pBR322 *Hae*II site (position 2352) (90). This *lacZ* fusion, in the same orientation as the Ap^r gene (*bla*), produces an α fragment of 107 amino acid residues (19 encoded by the polylinker sequences at residue 5). Although smaller than the α peptide produced in M13mp vectors, this peptide is active in α complementation. The parental pBR322 sequence has been modified by removing the *Eco*RI/*Pvu*II fragment containing the Tc^r gene via a

fill-in reaction and blunt-end ligation. The *Eco*RI site has not been regenerated, and the *Pst*I site present in the *bla* gene at position 3610 of pBR322 has been removed by mutagenesis. Other restriction sites have also been removed so that the polylinker contains 13 unique restriction sites for convenient cloning. The bacterial clones harboring pUC vectors can be selected by the Ap^r phenotype, and the recombinant clones can be screened for by the LacZ$^-$ phenotype against the "blue" LacZ$^+$ backgrounds on the X-Gal/IPTG plates.

Since the advent of the pUC vectors, a number of new and/or modified vectors have appeared, greatly increasing the diversity and versatility of the plasmid-based vectors. A new generation of single-strand phage vectors, called *phagemids* (or plages), has been derived from dsDNA plasmids by combining the origin of replication of phage f1 or M13, a genetic trait which allows the production of single-strand phages. For instance, the pUC18/19 plasmids have been upgraded to pUC118/119 by incorporating the M13 *ori* (92). The phagemid has all the best features of both single-strand phages and double-strand plasmids.

Among the new generation vectors described below are the pEMBL series, pTZ series (Pharmacia), pGEM series (Promega), and pBluescript series (Stratagene) vectors, and the pEX series (BMB) expression vectors.

(b) pEMBL vectors. The pEMBL series vectors (4 kb) are derivatives of pUC vectors: a 1300-bp *Eco*RI segment of the f1 genome was inserted at the unique *Nar*I site (93). The DNA segment (IG region) contains the phage f1 origin of replication and thus allows the production of ssDNA phages on superinfection with *helper phages* (phage f1 or M13K07).

(c) pTZ vectors. The pTZ(R/U) series vectors (2.9 kb) are also phagemids, derived from the pUC vectors in a manner similar to pEMBL. In addition, pTZ(R/U) vectors contain phage T7 promoter sequences in conjunction with the MCS (94). The multipurpose pTZ vectors thus facilitate *in vitro* transcription of cloned genes using the T7 RNA polymerase. The R/U notation in pTZ18/19(R/U) indicates the orientation of the f1 origin segment, which determines the sense of ssDNA in progeny phages: R for the ssDNA that hybridizes with the M13 "reverse" sequencing primer, and U for the ssDNA that hybridizes with the M13 "universal" primer. The notation 18/19 corresponds to the alternate orientations of the same MCS as present on M13mp18/19.

(d) pGEM vectors. The pGEM-Zf(+/−) series vectors (3 kb) are also derived from pUC plasmids by incorporating additional features similar to the constructs of pTZ vectors. The pGEM-Zf vectors are gemini systems carrying both T7 and SP6 promoter sequences which respectively flank the MCS. The MCS of pGEM-7Zf(+/−) contains 15 unique restriction sites. These versatile vectors allow a cloned gene to be transcribed *in vitro* from either orientation depending on the use of T7 or SP6 RNA polymerases (see Chapter 7).

(e) pBluescript vectors. The pBluescript vectors, e.g., pBluescript II KS/SK(+/−) (3 kb), are phagemids derived from pUC19 and have essentially similar configurations as pGEM-Zf except that they have T7 and T3 promoters, respectively,

flanking the MCS which is composed of 21 unique restriction sites (95). The K and S denote the two boundary restriction sites *Kpn*I and *Sac*I, respectively, which delimit the polylinker. The vector designated by (+) produces single-strand phages with the *lacZ* in the sense strand.

(f) pEX vectors. The pEX series (5.8 kbp) and pUBEX (6.7 kbp) expression vectors are derived from pBR322 and contain *cro–lacZ* fusion under the control of the phage λP_R promoter and cI857 repressor (96–98). A gene can be cloned, in any reading frame, at a polylinker site in the 3′ end of the *lacZ* gene. The expression in *E. coli* of a cloned gene produces a tripartite fusion protein, Cro(partial)-βGal-foreign protein.

iii. λ Phage systems

(a) λ gt11. The incorporation of the *lacZ* marker into the phage λ DNA has created an efficient expression vector, λgt11 (*lac5 nin5* cI857 S100) useful for cDNA and/or genomic DNA cloning (99). In this vector, a foreign DNA is inserted in a unique natural *Eco*RI site (at Glu-1006 codon) located within the *lacZ*, 53 bp upstream of the βGal termination codon. Following *in vitro* packaging and infection, recombinants are initially amplified in *E. coli* Y1088, which has a genotype of hsd R$^-$ M$^+$, and then are screened in *E. coli* Y1090, which carries a lon^- (protease-minus) mutation (also see Appendix B). Compared with the plasmid-based vector systems described earlier, λ vectors offer two notable advantages: the size of the insert DNA can be substantially larger (7–20 kbp) and the efficiency of infection is usually higher than the efficiency of transformation with plasmids.

Recombinant λ phages generate fusion proteins with an inactive βGal, whereas nonrecombinants (parental phages) present LacZ$^+$ phenotype. The phage vector produces a temperature-sensitive repressor (cI857), which its inactive at 42°C, and contains an amber mutation (S100) that renders it lysis defective in host cells which lack the amber suppressor *supF*. Consequently, the λ phages, either as lysogen or more practically as plaque-yielding lytic phages, can be induced by a temperature shift to 42°C to express large quantities of recombinant products in the absence of lysis. The βGal fusion protein expressed from the λgt11 can be screened using antibody probes. Optionally, recombinants can be screened by hybridization with specific oligonucleotide probes; this is the method of choice when a closely related, non-βGal vector λgt12 is used.

Once the initial cloning and screening stages are over, the DNA insert can be subcloned, following standard procedures (100), into more easily manageable plasmid/phagemid vectors for subsequent purposes, e.g., large-scale production of the gene product or analysis of the cloned gene. An alternative, neat subcloning strategy uses the endogenous plasmid pMC9 normally harbored in the *E. coli* hosts (Y1088 and Y1090) (101). The pMC9 is a plasmid with the 1.7-kbp *lacZ* and *lacI* genes inserted into the *Eco*RI site of pBR322 (102). For subcloning, the phage DNA preparation (containing both recombinant phage DNA and pMC9) is first digested with *Eco*RI, ligated by using T4 DNA ligase, and then used for transformation and screening (of LacZ$^-$ colonies) under the ampicillin selection.

(b) **Advanced forms of λ vectors.** A variety of *lacZ*–λ derivative vectors have been developed with an assortment of improved features. For example, λ*gt22(/23)* vectors have several unique cloning sites in the *lacZ* (103) (see Fig. 8.3). The λ*SWAJ* vector represents a totally different vector configuration; it has a whole plasmid (pGEM2) incorporated into the λgt12 environment and relies on the plasmid for MCS as well as other useful features (104). The λBlueSTAR (Novagen) and λTriplEx (Clontech) vectors are novel-concept, multi-purpose λ replacement vectors featuring, among others, *lacZ*-based blue/white screening of recombinant clones and Cre-*loxP*-mediated autosubcloning. When appropriate *E. coli* strains (e.g., BM25.8 and BNN132) are transduced with the λ vectors and plated in the presence of ampicillin, plasmid or phagemid subclones can be generated *in vivo* by a Cre recombinase-mediated excision. Note that the plasmid or phagemid subclones can also be obtained *in vitro* by treatment of the λ DNA with purified Cre recombinase.

Along the same line, λ*ZAP* vectors contain the whole pBluescript SK plasmid in the environment of λgt12 (95). One additional outstanding feature of the λZAP system is that the phagemid (pBluescript SK) sequences can be excised from the phage λ DNA by the *in vivo* action of gene II protein provided by the f1 (or M13) helper phage. The capabilities for *in vivo* excision from the λ vector as well as for autonomous reconstitution of the f1 origin of replication in the plasmid or phagemid make the advanced λ (βGal-reporter) vectors an extremely useful tool for gene cloning, sequencing, transcription, and various other downstream manipulations and analyses of the cloned gene.

5. Sources

As an enzymatic reagent, β-galactosidase can be produced and purified in a number of classical or modern ways. For example, βGal (and βGal fusion proteins as well) overexpressed from a constitutive mutant (*lacO^c*) strain of *E. coli* or from recombinant gene expression systems can be purified by one-step affinity chromatography on APTG (*p*-aminophenyl-β-D-thiogalactoside)-Sepharose. The method consists of affinity binding of βGal at a high salt concentration (1.6 *M* NaCl, pH 7.4) and of subsequent elution (0.1 *M* sodium borate, pH 10), which gives an overall yield and purity of 85–95% (112). Another high-yield preparation method involves the temperature-inducible expression of the *lacZ* gene under the control of the λ P_L promoter in an *E. coli* host (e.g., MZ-1 and K12 N4830-1) which harbors a chromosomal λ repressor, cI857. The expression of *lacZ* from a large-scale culture can be most efficiently performed under the following set of optimized conditions (113): (i) cell growth in LB-rich medium at 32°C and pH 7.2, to mid-exponential phase, (ii) gene induction for 5 hr by shifting the temperature to 42°C and the pH to 5.5, and (iii) extended induction and degradation control by shifting the temperature to 38°C and the pH to 5.7.

An α-complementing peptide (residues 6–44) has been obtained from the expression of the peptide-encoding DNA using a GST (glutathione *S*-transferase) expression system provided with a thrombin recognition site (114). Treatment of the expressed GST fusion protein with thrombin liberates functionally active α peptides which can be purified by gel filtration or an immunoaffinity column.

Pure βGal has a specific activity of ~500 U/mg protein. A commercially available enzyme (specific activity >300 U/mg) is supplied as white powder obtained by lyophilization in 5 mM sodium phosphate, pH 7.2.

References

1. Arvidson, D. N., Youderian, P., Schneider, T. D., and Stormo, G. D. (1991). *BioTechniques* **11**, 733–738.
2. Eustice, D. C., Feldman, P. A., Colberg-Poley, A. M., Buckery, R. M., and Neubauer, R. H. (1991). *BioTechniques* **11**, 739–743.
3. Jain, V. K., and Magrath, I. T. (1991). *Anal. Biochem.* **199**, 119–124.
4. Lin, E. C. C., Lerner, S. A., and Jorgensen, S. E. (1962). *BBA* **60**, 422–424.
5. Wan, L. G., and Vanhuystee, R. B. (1994). *J. Agricult. Food Chem.* **42**, 2499–2501.
6. Ebright, R. H., Cossart, P., Gicquel-Sanzey, B., and Beckwith, J. (1984). *Nature* **311**, 232–235.
7. Broker, M., and Harthus, H.-P. (1989). *FEBS Lett.* **257**, 118–207.
8. Sinnott, M. L., Withers, S. G., and Viratelle, O. M. (1978). *BJ* **175**, 539–546.
9. Tenu, J.-P., Viratelle, O. M., Garnier, J., and Yon, J. (1971). *EJB* **20**, 363–370.
10. Edwards, R. A., Cupples, C. G., and Huber, R. E. (1990). *BBRC* **171**, 33–37.
11. Wallenfels, K., and Weil, R. (1972). "The Enzymes," Vol. 7, pp. 617–663.
12. Yon, J., and Tenu, J.-P. (1973). *In* "Dynamic Aspects of Conformation Changes in Biological Macromolecules" (C. Sadron, Ed.), pp. 447–458. D. Reidel, Dordrecht.
13. Edwards, L. A., Tian, M. R., Huber, R. E., and Fowler, A. V. (1988). *JBC* **263**, 1848–1854.
14. Lalegerie, P., Legler, G., and Yon, J. M. (1982). *Biochimie* **64**, 977–1000.
15. Case, G. S., Sinnott, M. L., and Tenu, J.-P. (1973). *BJ* **133**, 99–104.
16. Legler, G., and Herrchen, M. (1983). *Carbohyd. Res.* **116**, 95–103.
17. Huber, R. E., and Brockbank, R. L. (1987). *Biochem.* **26**, 1526–1531.
18. Starka, J. (1964). *Folia Microbiol.* **9**, 24–27.
19. Miller, J. H. (1972). "Experiments in Molecular Genetics." CSHL Press, Cold Spring Harbor, New York.
20. Wallenfels, K., and Malhotra, O. P. (1961). *Adv. Carbohyd. Chem.* **16**, 239–298.
21. Huber, R. E., and Gaunt, M. T. (1983). *ABB* **220**, 263–271.
22. Smith, T. F., and Sadler, J. R. (1971). *JMB* **59**, 273–305.
23. Hopkins, J. D. (1974). *JMB* **87**, 715–724.
24. Sinnott, M. L., and Withers, S. G. (1974). *BJ* **143**, 751–762.
25. Lehmann, J., and Ziegler, B. (1977). *Carbohyd. Res.* **58**, 73–78.
26. Wentworth, D. F., and Wolfenden, R. (1974). *Biochem.* **13**, 4715–4720.
27. Viratelle, O. M., and Yon, J. (1980). *Biochem.* **19**, 4143–4149.
28. Huber, R. E., Gaunt, M. T., and Hurlburt, K. L. (1984). *ABB* **234**, 151–160.
29. Shin, H. J., and Yang, J. W. (1994). *Biotech. Lett.* **16**, 1157–1162.
30. Jobe, A., and Bourgeois, S. (1972). *JMB* **69**, 397–408.
31. Sinnott, M. L., and Viratelle, O. M. (1973). *BJ* **133**, 81–87.
32. Sinnott, M. L., and Souchard, I. J. L. (1973). *BJ* **133**, 89–98.
33. Loeffler, R. S. T., Sinnott, M. L., Sykes, B. D., and Withers, S. G. (1979). *BJ* **177**, 145–152.
34. Kurz, G., Lehmann, J., and Vorberg, E. (1981). *Carbohyd. Res.* **93**, C14–C20.
35. Fowler, A. V., and Zabin, I. (1978). *JBC* **253**, 5521–5525.
36. Kalnins, A., Otto, K., Ruther, U., and Muller-Hill, B. (1983). *EMBO J.* **2**, 593–597.
37. Zabin, I. (1982). *Mol. Cell. Biochem.* **49**, 87–96.
38. Langley, K. E., Villarejo, M. R., Fowler, A. V., Zamenhof, P. J., and Zabin, I. (1975). *PNAS* **72**, 1254–1257.

39. Welply, J. K., Fowler, A. V., and Zabin, I. (1981). *JBC* **256**, 6804–6810.
40. Accolla, R. S., and Celada, F. (1976). *FEBS Lett.* **67**, 299–302.
41. Welply, J. K., Fowler, A. V., Beckwith, J. R., and Zabin, I. (1980). *J. Bacteriol.* **142**, 732–734.
42. Brickman, E., Silhavy, T. J., Bassford, P. J., Jr., Shuman, H. A., and Beckwith, J. R. (1979). *J. Bacteriol.* **139**, 13–18.
43. Fowler, A. V., and Zabin, I. (1983). *JBC* **258**, 14354–14358.
44. Benito, A., Mateu, M. G., and Villaverde, A. (1995). *Bio/Technology* **13**, 801–804.
45. Mandecki, W., Fowler, A. V., and Zabin, I. (1981). *J. Bacteriol.* **147**, 694–697.
46. Ruther, U., and Muller-Hill, B. (1983). *EMBO J.* **2**, 1791–1794.
47. Sinnott, M. L., and Smith, P. J. (1978). *BJ* **175**, 525–538.
48. Fowler, A. V., Zabin, I., Sinnott, M. L., and Smith, P. J. (1978). *JBC* **253**, 5283–5285.
49. Naider, F., Bohak, Z., and Yariv, J. (1972). *Biochem.* **11**, 3202–3208.
50. Ring, M., Bader, D. E., and Huber, R. E. (1988). *BBRC* **152**, 1050–1055.
51. Huber, R. E., Fowler, A. V., and Zabin, I. (1982). *Biochem.* **21**, 5052–5055.
52. Herrchen, M., and Legler, G. (1984). *EJB* **138**, 527–531.
53. Bader, D. E., Ring, M., and Huber, R. E. (1988). *BBRC* **153**, 301–306.
54. Malamy, M. H. (1967). *CSHSQB* **31**, 189–201.
55. Miller, J. H., and Reznikoff, W. S. (Eds.) (1980). "The Operon." CSHL Press, Cold Spring Harbor, New York.
56. Jacob, F., and Monod, J. (1961). *JMB* **3**, 318–356.
57. Flashner, Y., and Gralla, J. D. (1988). *PNAS* **85**, 8968–8972.
58. Farabaugh, P. J. (1978). *Nature* **274**, 765–769.
59. Gordon, A. J. E., Burns, P. A., Fix, D. F., Yatagai, F., Allen, F. L., Horsfall, M. J., Halliday, J. A., Gray, J., Bernelot-Moens, C., and Glickman, B. W. (1988). *JMB* **200**, 239–251.
60. Kleina, L. G., and Miller, J. H. (1990). *JMB* **212**, 295–318.
61. Stefano, J. E., and Gralla, J. D. (1982). *PNAS* **79**, 1069–1072.
62. Markoff, A. J., and Oxer, M. D. (1991). *NAR* **19**, 2417–2421.
63. De Boer, H. A., Comstock, L. J., and Vasser, M. (1983). *PNAS* **80**, 21–25.
64. Amann, E., Brosius, J., and Ptashne, M. (1983). *Gene* **25**, 167–178.
65. De Bellis, D., and Schwartz, I. (1990). *NAR* **18**, 1311.
66. Cannistraro, V. J., and Kennell, D. (1985). *JMB* **182**, 241–248.
67. Ruteshouser, E. C., and Richardson, J. P. (1989). *JMB* **208**, 23–43.
68. Murakawa, G. J., and Nierlich, D. P. (1989). *Biochem.* **28**, 8067–8072.
69. Zabin, I., Fowler, A. V., and Beckwith, J. R. (1978). *J. Bacteriol.* **133**, 437–438.
70. Martinez-Bilbao, M., Holdsworth, R. E., Edwards, L. A., and Huber, R. E. (1991). *JBC* **266**, 4979–4986.
71. Ullmann, A., and Perrin, D. (1970). In "The Lactose Operon" (J. R. Beckwith and D. Zipser, Eds.), pp. 143–172. CSHL Press, Cold Spring Harbor, New York.
72. Kuchinke, W. (1989). *J. Mol. Evol.* **29**, 95–97.
73. Hall, B. G., and Hartl, D. L. (1975). *Genetics* **81**, 427–435.
74. Hall, B. G. (1982). *Evol. Biol.* **15**, 85–150.
75. Li, B. F. L., Holdup, D., Morton, C. A. J., and Sinnott, M. L. (1989). *BJ* **260**, 109–114.
76. Hall, B. G. (1976). *JMB* **107**, 71–84.
77. Stokes, H. W., Betts, P. W., and Hall, B. G. (1985). *Mol. Biol. Evol.* **2**, 469–477.
78. Fowler, A. V., and Smith, P. J. (1983). *JBC* **258**, 10204–10207.
79. Silhavy, T. J., and Beckwith, J. R. (1985). *Microbiol. Rev.* **49**, 398–418.
80. Casadaban, M. J., Martinez-Arias, A., Shapira, S. K., and Chou, J. (1983). *Meth. Enzy.* **100**, 293–308.
81. Itakura, K., Hirose, T., Crea, R., and Riggs, A. D. (1977). *Science* **198**, 1056–1063.
82. Goeddel, D. V., Kleid, D. G., Bolivar, F., Heyneker, H. L., Yansura, D. G., Crea, R., Hirose, T., Kraszewski, A., Itakura, K., and Riggs, A. D. (1979). *PNAS* **76**, 106–110.
83. Casadaban, M. J., Chou, J., and Cohen, S. N. (1980). *J. Bacteriol.* **143**, 971–980.
84. Jain, C. (1993). *Gene* **133**, 99–102.
85. Gray, M. R., Colot, H. V., Guarente, L., and Rosbash, M. (1982). *PNAS* **79**, 6598–6602.

86. Weinstock, G. M., Ap Rhys, C., Berman, M. L., Hampar, B., Jackson, D., Silhavy, T. J., Weisemann, J., and Zweig, M. (1983). *PNAS* **80**, 4432–4436.
87. Guarente, L., Lauer, G., Roberts, T. M., and Ptashne, M. (1980). *Cell* **20**, 543–553.
88. Messing, J. (1991). *Gene* **100**, 3–12.
89. Messing, J., Gronenborn, B., Muller-Hill, B., and Hofschneider, P. H. (1977). *PNAS* **74**, 3642–3646.
90. Yanisch-Perron, C., Vieira, J., and Messing, J. (1985). *Gene* **33**, 103–119.
91. Marvin, D. A., and Hohn, B. (1969). *Bacteriol. Rev.* **33**, 172–209.
92. Vieira, J., and Messing, J. (1987). *Meth. Enzy.* **153**, 3–11.
93. Dente, L., Sollazzo, M., Baldari, C., Cesareni, G., and Cortese, R. (1985). *In* "DNA Cloning" (D. M. Glover, Ed.), Vol. I, pp. 49–78. IRL Press, Oxford.
94. Mead, D. A., Szczesna-Skorupa, E., and Kemper, B. (1986). *Prot. Eng.* **1**, 67–74.
95. Short, J. M., Fernandez, J. M., Sorge, J. A., and Huse, W. D. (1988). *NAR* **16**, 7583–7600.
96. Stanley, K. K., and Luzio, J. P. (1984). *EMBO J.* **3**, 1429–1434.
97. Kusters, J. G., and Jager, E. J., and van der Zeijst, B. A. M. (1989). *NAR* **17**, 8007.
98. Banting, G., Luzio, J. P., Braghetta, P., Brake, B., and Stanley, K. K. (1991). *Gene* **107**, 127–132.
99. Huynh, T. V., Young, R. A., and Davis, R. W. (1985). *In* "DNA Cloning." (D. M. Glover, Ed.), Vol. I, pp. 49–78. IRL Press, Oxford.
100. Sambrook, J., Fritsch, E. F., and Maniatis, T. (1989). "Molecular Cloning." A Laboratory Manual," 2nd ed. CSHL Press, Cold Spring Harbor, New York.
101. Chiu, I.-M., and Lehtoma, K. (1990). *Genet. Anal. Techn. Appl.* **7**, 18–23.
102. Coles, M. P., Lebkowski, J. S., and Botchan, M. R. (1983). *PNAS* **80**, 3015–3019.
103. Han, J. H., and Rutter, W. J. (1987). *NAR* **15**, 6304.
104. Palazzolo, M. J., and Meyerowitz, E. M. (1987). *Gene* **52**, 197–206.
105. Geiger, R., Schneider, E., Wallenfels, K., and Miska, W. (1992). *Biol. Chem. Hoppe-Seyler.* **373**, 1187–1191.
106. Aguzzi, A., and Theuring, F. (1994). *Histochem.* **102**, 477–481.
107. Jacobson, R. H., Zhang, X.-J., DuBose, R. F., and Matthews, B. W. (1994), *Nature* **369**, 761–766.
108. Arrondo, J. L. R., Muga, A., Castresana, J., Bernabeu, C., and Goni, F. M. (1989). *FEBS Lett.* **252**, 118–120.
109. Gebler, J. C., Aebersold, R., and Withers, S. G. (1992). *JBC* **267**, 11126–11130.
110. Huber, R. E., Gupta, M. N., and Khare, S. K. (1994). *Int. J. Biochem.* **26**, 309–318.
111. Neubauer, P., and Hofmann, K. (1994). *FEMS Microbiol. Rev.* **14**, 99–102.
112. Ullmann, A. (1984). *Gene* **29**, 27–31.
113. Horiuchi, J.-I., Kamasawa, M., Miyakawa, H., Kishimoto, M., and Momose, H. (1993). *J. Ferment. Bioeng.* **76**, 382–387.
114. Gallagher, C. N., Roth, N. J., and Huber, R. E. (1994). *Prep. Biochem.* **24**, 297–304.

B. β-Lactamase
[Penicillinase/Cephalosporinase, EC 3.5.2.6]

1. INTRODUCTION

a. General properties of β-lactamases

β-Lactamases are a large family of hydrolases that catalyze the hydrolysis of the amide bond in the β-lactam ring of penicillins and cephalosporins (Fig. 8.4). The hydrolysis product, penicilloic acid or cephalosporoic acid, is biologically inactive.

β-Lactamase-catalyzed conversion of the penicillins, the well-known antibiotics discovered by Fleming in 1929, to penicilloic acid is the most common type of resistance determinants in β-lactam-resistant bacteria. Note that bacteria operate various other mechanisms of antibiotic resistance, which function either singly or more often in combination. They include (i) a decrease in membrane permeability (or reduced uptake), (ii) modification of a drug-sensitive site(s), (iii) synthesis of an additional drug-resistant enzyme(s) or overproduction of the drug-sensitive enzyme, (iv) derivatization of antibiotics, (v) sequestration of the drug by protein binding, and (vi) an active efflux of drugs from cells.

β-Lactam antibiotics exert their antibiotic activity mainly by blocking the synthesis of peptidoglycan, a network of glycan chains and interlinking peptides that provide structural integrity to bacterial cell walls. Bacterial cells contain several *penicillin-binding proteins* (PBPs, e.g., about seven species in *E. coli*); the functions of only some of these proteins are understood. The PBPs include the most important penicillin target enzymes, D-*Ala*-D-*Ala-carboxypeptidase/transpeptidase* (DD-CPase/TPase), which catalyze the transpeptidation reaction that cross-links the pentapeptide precursor units during the final step of peptidoglycan biosynthesis. β-Lactam antibiotics, by virtue of their structural similarity to the D-Ala–D-Ala end group of the peptidoglycan precursor, covalently modify the enzymes, inactivate them (at least temporarily), and render the bacterial cells vulnerable to osmotic shocks or other damages.

As the major defensive response against β-lactam antibiotics, both gram-positive and gram-negative bacteria produce varying amounts of β-lactamases, which are hypothesized to be an evolutionary form of a penicillin-sensitive enzyme involved in cell wall synthesis. To cope with the evolving bacterial resistance and to increase the antibacterial activity, especially for gram-negative bacteria, numerous semisynthetic β-lactam derivatives have been synthesized from 6-aminopenicillanic acid (6-APA) or 7-aminocephalosporanic acid (7-ACA) (see Fig. 8.4). This effort has met and will continue to have only partial success because of the continuing molecular evolution of β-lactamases toward extended spectrum enzymes active against newer β-lactam antibiotics (1).

Primarily due to their clinical importance, β-lactamases have been extensively studied with respect to enzyme structure, function, induction, secretion, and transfer of genetic elements. The wealth of biological, biochemical, and genetic information of the β-lactamases and their genes (e.g., *bla*) has led to the use of the *bla* system as one of the most important selection marker/reporters in recombinant DNA technology.

b. Classifications of β-lactamases

β-Lactamases can be classified according to enzymatic, physicochemical, and/or genetic criteria. It has been recognized, however, that the wide diversity of β-lactamases does not make them amenable to any strict scheme of classification.

i. Classification by genetic origins. β-Lactamases can be classified by their genetic origins: (a) enzymes encoded by chromosomal genes and (b) enzymes encoded by the genes carried on autonomously replicating transferable elements called resistance (R) plasmids or *R factors*. The chromosome-encoded β-lactamases are inducible by β-lactam substrates in many bacterial genera, whereas the expression of most of the plasmid-determined β-lactamases are constitutive. The R factor-determined β-lactamases are widely distributed in bacteria (2) due to the fact that the β-lactamase gene is carried on a *transposon*, the genetic element that transfers autonomously to and from plasmids and chromosomes. A single bacterium may produce multiple, distinct β-lactamases of both origins.

ii. **Classification by activity profiles.** β-Lactamases can be divided into several classes based on their activity/inhibition profiles toward penicillins (*penams*), cephalosporins (*cephems*), or some particular members of the antibiotics (3). No known β-lactamases, however, are exclusively specific on penams or cephems. Most chromosomal β-lactamases are active against both penams and cephems, but are in general considerably more active against cephems, and thus belong to the *cephalosporinases*. β-Lactamases having higher activities on penems are called *penicillinases*. TEM (or RTEM) β-lactamase, an R factor-determined enzyme produced by *E. coli* TEM, is the prototype β-lactamase that belongs to the penicillinases. Nevertheless, it has wide-ranging activities against penams and cephems, and there are significant specificity variations within each category of substrates.

iii. **Classification by molecular structures.** The third and perhaps most systematic classification scheme is the one that is based on accumulated knowledge regarding catalytic properties and molecular structures. In this scheme, β-lactamases are divided into A, B, C, and D classes (4). Class A, C, and D β-lactamases are active-site serine enzymes.

(a) *Class A β-lactamases.* These enzymes have an average MW of 29 kDa, are "serine hydrolases" due to the catalytic role of Ser-70, and show a preference for penicillins as substrates. The members of class A β-lactamases show discernible patterns of regional sequence conservation (5). Class A enzymes are mostly R factor encoded and thus are widely distributed among gram-negative and gram-positive bacteria. Class A β-lactamases from gram-negative bacteria (e.g., *E. coli* TEM enzymes) remain associated with the cell in the periplasmic space, whereas the enzymes from gram-positive bacteria (*S. aureus* enzyme, *B. cereus* β-lactamase I) are largely extracellular and are secreted into the surrounding environment. Some class A enzymes such as the β-lactamases from *B. licheniformis* are chromosomally encoded.

The amino acid numbering for class A β-lactamases conventionally follow the numbering scheme (class A β-lactamase or ABL scheme) which is based on the sequence alignment of 20 class A β-lactamases (6).

(b) *Class B β-lactamases.* Class B β-lactamases are metalloenzymes and are rare among clinical isolates. There are two members known at present: *B. cereus* β-lactamase II (26-kDa monomer size) and *Pseudomonas maltophilia* L1 (32-kDa monomer size). They require a metal cofactor (Zn^{2+}) for enzyme activities.

(c) *Class C β-lactamases.* Class C enzymes are mostly chromosome-encoded cephalosporinases with an average MW of 39 kDa. They are serine (Ser-80)

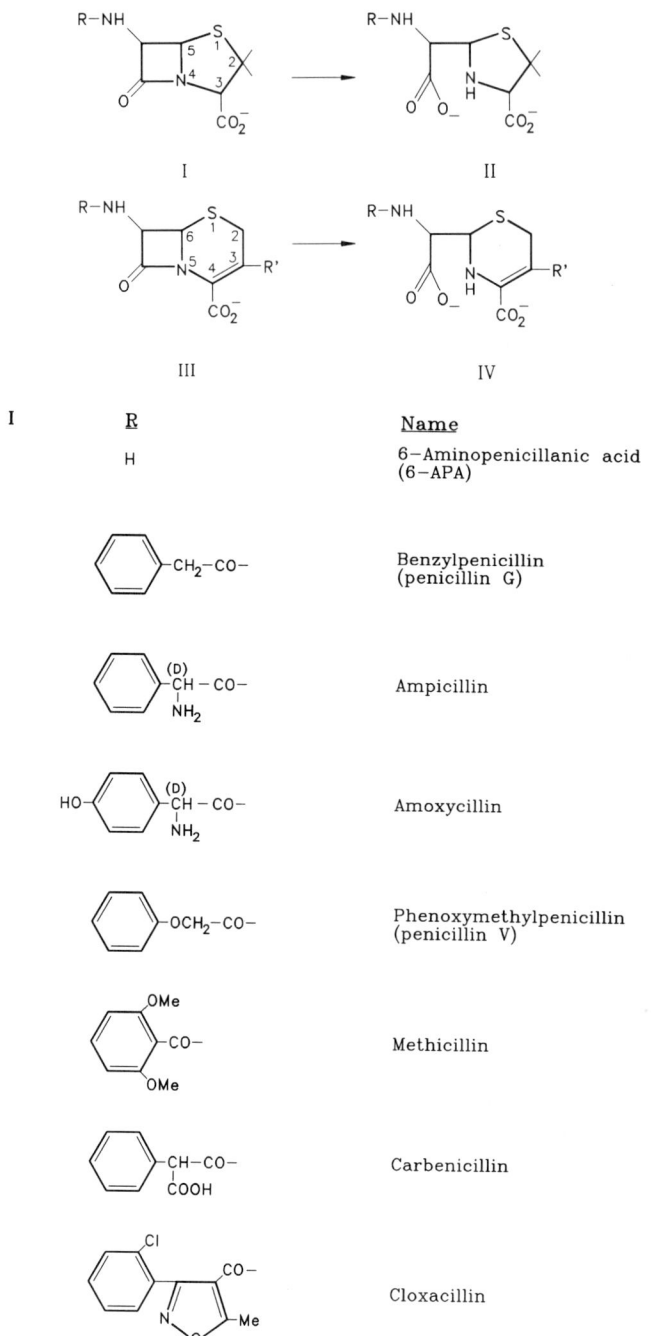

FIGURE 8.4 General reaction of β-lactamases and structures of typical β-lactam substrates. (I) and (II): penicillin(s) and penicilloic acid pair. (III) and (IV): cephalosporin(s) and cephalosporoic acid pair. R denotes the side chain on 6-APA or 7-ACA, and R' denotes the side chain on the C-3 position of cephalosporins.

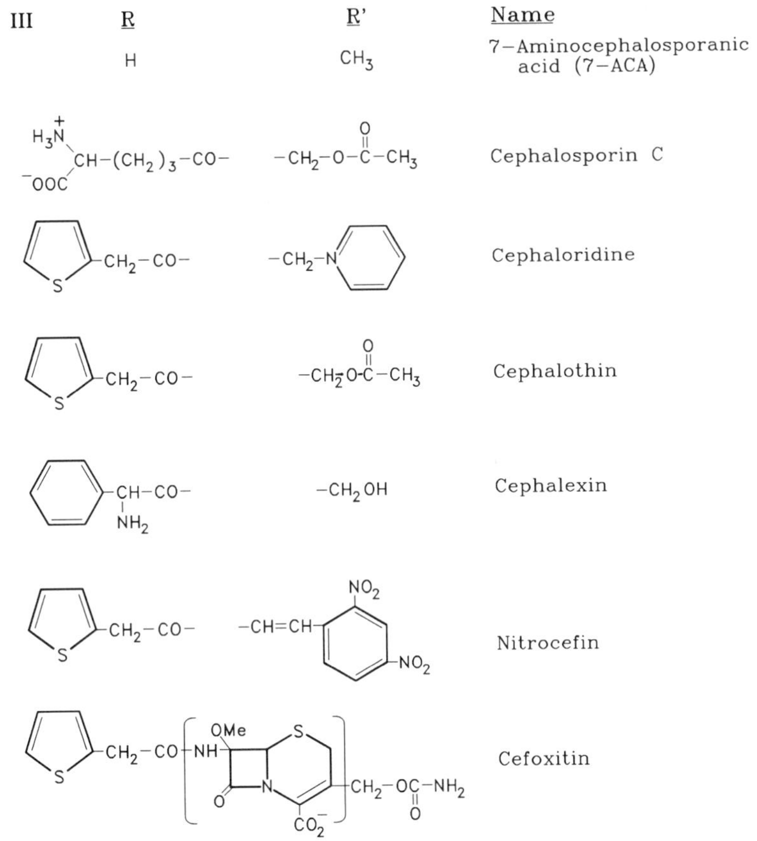

III	R	R'	Name
	H	CH_3	7-Aminocephalosporanic acid (7-ACA)
	$\underset{^-OOC}{\overset{H_3\overset{+}{N}}{>}}CH-(CH_2)_3-CO-$	$-CH_2-O-\underset{\underset{O}{\|\|}}{C}-CH_3$	Cephalosporin C
	thienyl-CH_2-CO-	$-CH_2-N^+$(pyridinium)	Cephaloridine
	thienyl-CH_2-CO-	$-CH_2-O-\underset{\underset{O}{\|\|}}{C}-CH_3$	Cephalothin
	phenyl-$\underset{NH_2}{CH}-CO-$	$-CH_2OH$	Cephalexin
	thienyl-CH_2-CO-	$-CH=CH-$(2,4-dinitrophenyl)	Nitrocefin
	thienyl-$CH_2-CO-NH-$[OMe, full cephem structure]$-CH_2-O\underset{\underset{O}{\|\|}}{C}-NH_2$		Cefoxitin

FIGURE 8.4 (*continued*)

hydrolases like class A enzymes. Class C β-lactamases have no significant sequence homologies with the class A enzymes but, within the same class C, they share extensive sequence identities among bacterial species. Virtually all gram-negative bacteria produce a class C β-lactamase. Some organisms like *E. coli* and *Shigella* species express the enzyme at a low constitutive level, while others, like *Citrobacter freundii, Enterobacter cloacae,* and *P. aeruginosa,* produce β-lactamases that are inducible by various antibiotic or chemical compounds.

(d) Class D β-lactamases. Class D enzymes, also called oxacillinases (OXA β-lactamases), have a monomer size of 32 kDa, and consistently hydrolyze oxacillin and cloxacillin faster than benzylpenicillin.

This section focuses on *E. coli* TEM-1 β-lactamase, although relevant enzymatic properties of other β-lactamases are described for the sake of comparison. TEM-1 β-lactamase is the prototype of class A, gram-negative β-lactamases and is one of the most thoroughly characterized β-lactamases. The TEM β-lactamase and its gene *bla* [or conventionally Ap^r (ampicillin-resistance) gene] are the most widely used selection marker on pBR322 and its derivative plasmids.

2. FUNCTIONS

a. Reaction conditions

i. **Optimal conditions for β-lactamase reaction, detection, or assays**

(a) Optimum pH and temperature. The pH optimum for TEM β-lactamase is between pH 6 and 7 in 0.1 M K-phosphate (7,8). The enzyme is 50% active at pH 9.

TEM β-lactamase is stable at 50°C. At 60°C, the half-life is 1.5 min (8). The β-lactamase reaction can be carried out at any temperature between 20° and 37°C, although the optimal temperature may be considered 30°C.

(b) Choice of assay methods. There are several β-lactamase assays available for general as well as analytical purposes (9). The standard substrate for β-lactamase assays is, unless specified otherwise, benzylpenicillin in conventional methods which include: (i) an acidimetric method in which the carboxyl group newly formed on hydrolysis of the β-lactam bond is titrated, (ii) an iodometric titration which is based on the reducing potential of the hydrolysis product, and (iii) a colorimetric determination of the hydroxamate formed by residual substrates with hydroxylamine.

Modern β-lactamase assays generally rely on spectrophotometry using chromogenic substrates such as nitrocefin (formerly cephalosporin 87/312) and *N*-(2-furyl)acryloylpenicillin.

The β-lactamase assay most commonly used in recombinant DNA techniques is a microbiological assay, which monitors the effect (sensitivity or resistance) on bacterial cultures of biologically active substrates. Although the choice of an assay method depends on a particular application, the development of chromogenic substrates has made it possible to combine the advantages of colorimetry and/or spectrophotometry (convenience and sensitivity) with those of conventional microbiological assays. For example, nitrocefin can be used *in situ* to detect β-lactamase on a gel after isoelectrofocusing (10). Filter papers (or discs) saturated with the chromophore nitrocefin provide a convenient means for rapidly and unambiguously determining sensitive (Bla$^-$, yellow) and resistant (Bla$^+$, red) colonies. Similar filter paper tests (available under the trade name Ampscreen, from BRL) permit rapid selections of Ap^s clones (blue) from Ap^r clones (yellow), thus eliminating the efforts involved in conventional replica plating.

(c) Spectrophotometry of β-lactamase. Spectrophotometric assays are typically carried out in 50 mM Na-phosphate (pH 7.0) and 50 μg/ml BSA with the chromogenic substrate nitrocefin (0.2 mM).

For the assay of β-lactamase in samples from liquid culture, bacterial cells (1 ml, harvested at cell density of $0.6 A_{650}$) are lysed with 40 μl of lysis buffer [0.1 M EDTA, 0.1 M DTT, 50 mM Tris–Cl (pH 8.0), and 100 μg/ml chloramphenicol] and 20 μl toluene. The mixture is vigorously agitated on a vortex for

TABLE 8.3 Kinetic Parameters of TEM β-Lactamase for Some Typical β-Lactam Substrates[a]

Substrate	K_m (mM)	k_{cat} (sec^{-1})
Benzylpenicillin	20–50 (μM)	2000–2200
Ampicillin	40–65 (μM)	1700–2000
Cephaloridine	1.0	1700
Nitrocefin	0.11	900
Cephalothin	0.21	130
Cephalosporin C	0.69	57
Cefoxitin	0.65	0.004

[a] Reactions are at pH 7.0 (phosphate buffer) (13–15). For structures of the substrates, see Fig. 8.4.

15 sec and placed on ice. Toluene is evaporated by a stream of N_2 gas before the activity assay.

Nitrocefin is a convenient substrate because of its high absorption coefficient (ε_{386} = 17,000; $\Delta\varepsilon_{486}$ = 16,000 M^{-1} cm^{-1}) (8,11). To solubilize nitrocefin, a stock solution is made in Me$_2$SO and is used at a final concentration of ≤0.5% in the assay mixture. The following are spectrophotometric parameters for some other useful substrates (6,12): benzylpenicillin (ε_{232} = 1700; $\Delta\varepsilon$ = 940 M^{-1} cm^{-1}), ampicillin (ε_{235} = 1850; $\Delta\varepsilon$ = 1200 M^{-1} cm^{-1}), cephaloridine (ε_{295} = 1300; $\Delta\varepsilon$ = 1000 M^{-1} cm^{-1}), and N-(2-furyl)acryloylpenicillin ($\Delta\varepsilon_{285}$ = 1900 and $\Delta\varepsilon_{330}$ = −3000 M^{-1} cm^{-1}).

Unit definition: One *lactamase unit* is defined as the amount of enzyme that hydrolyzes 1 μmol of substrate (benzylpenicillin) per min under the assay conditions.

ii. Kinetic parameters. The steady-state kinetics of TEM β-lactamase (and other β-lactamases as well) follow typical Michaelis–Menten kinetics with a wide range of K_m and k_{cat} values (13–15) (Table 8.3).

iii. Inhibitors. A variety of β-lactamase inhibitors have been found in nature, have been chemically synthesized, or have been obtained semisynthetically (16). Although the following descriptions apply to β-lactamases in general and to TEM-1 (or -2) β-lactamase in particular, the rate and mode of inhibition or inactivation may vary significantly depending on β-lactamase classes and within members of the classes.

(a) *Product inhibition.* The hydrolysis products (penicilloic acids) or the decarboxylation products (penilloic acids) are reversible inhibitors. For example, the K_i (competitive) value for benzylpenicilloic acid is 40 mM, while the K_i for benzylpenilloic acid is 16 mM (17).

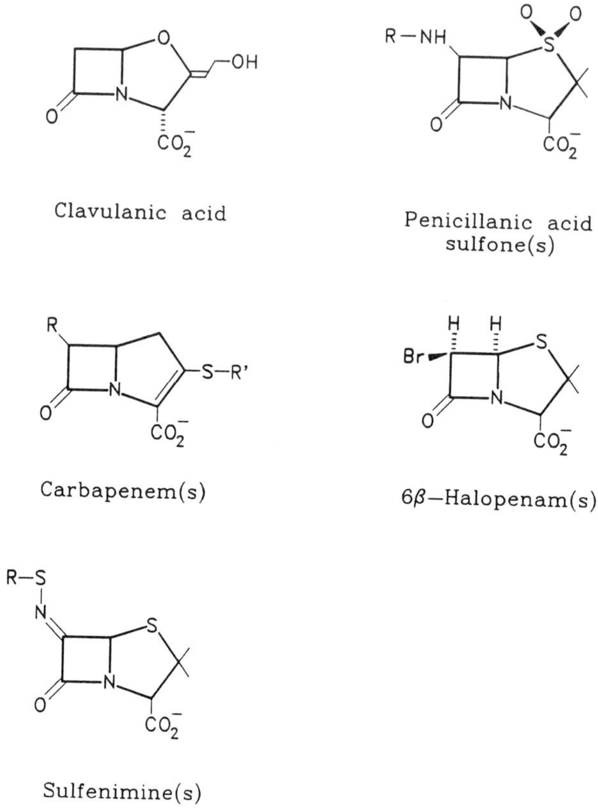

Clavulanic acid

Penicillanic acid sulfone(s)

Carbapenem(s)

6β—Halopenam(s)

Sulfenimine(s)

FIGURE 8.5 Structures of β-lactamase inhibitors: clavulanic acid, penicillanic acid sulfone(s), carbapenem(s), 6β-halopenam(s), and sulfenimine(s).

(b) Clavulanic acid. Clavulanic acid is a naturally occurring β-lactam isolated from *Streptomyces clavuligerus* (Fig. 8.5). It has little antibacterial activity on its own but displays potent inhibitory properties on class A (but not class C) β-lactamases with a K_i (competitive) of 0.8 μM (18). Clavulanic acid is in fact a substrate with low K_m and V values (3.8×10^{-3} sec^{-1} with TEM-2 β-lactamase) and, more appropriately, can be considered a "suicide" or "mechanism-based" inhibitor (19). The rate of conversion from a reversible Michaelis complex to an activated clavulanic acid–enzyme complex (inactive) is 0.027 sec^{-1} ($t_{1/2}$ = 26 sec) at pH 7 and 37°C (18). TEM-1 and TEM-2 β-lactamases are inactivated similarly by clavulanic acid. The TEM-2 enzyme turns over 115 times before complete enzyme inactivation occurs (20). Mutations (M69L, R241C, and/or N276D) in TEM-1 render the enzyme highly resistant to clavulanic acid.

(c) Substrate analogs. The following substrate analogs belong to mechanism-based inhibitors (see Fig. 8.5 for structures): sulbactam (or penicillanic acid sulfones) (21), tazobactam, 6β-halopenams (and not 6α-halopenams) (22,23),

carbapenems (e.g., olivanic acids) (24,25), and sulfenimines (26). For TEM-1 β-lactamase, clavulanic acid is 60 times more inhibitory than sulbactam.

Phosphonate monoesters with a substituted phenol are inhibitors of class A and C β-lactamases. The inhibition is due to phosphorylation of the primary nucleophile (Ser-70) of the β-lactamase active site, resulting in a "stable" phosphonate intermediate with a structure resembling the tetrahedral intermediate of substrates (27).

Boric acid is a rather strong competitive inhibitor with a K_i of 1 mM (8,17).

b. Substrate specificities

i. Structural features of β-lactam substrates. The bicyclic ring system in penicillins and cephalosporins forces the β-lactam amide group to be nonplanar and suppresses the amide resonance in —CO⁻—N⁺H—. β-Lactam antibiotics are therefore activated amides that are more easily hydrolyzed than regular amide bonds. Nevertheless, the substrate efficiency can be drastically modulated by the nature of side chain derivatives (R and/or R′) that are linked to the basal structure, 6-APA or 7-ACA (Fig. 8.4).

Compared with the rate of hydrolysis (k_{cat}) of benzylpenicillin [100], TEM-1 β-lactamase exhibits the following specificities: ampicillin [94], 6-APA [85], carbenicillin [11], oxacillin [4.0], cloxacillin [1.1], methicillin [1], cephaloridine [65], cephalothin [10], and cephalosporin C [0.8] (10,28) (also see Table 8.2).

ii. Relationship between substrate efficiency and antibiotic activity. Bacteria present several resistance mechanisms against antibiotics, as briefly described in the *Introduction*. Depending on the bacterial species, β-lactam antibiotics display a wide range of antibiotic activities in a manner related to the principal resistance mechanism operating in the cells. For example, penicillins such as penicillin G are extremely effective against gram-positive bacteria such as streptococci and staphylococci and gram-negative gonococci and meningococci. These penicillins are, however, much less effective against the more typical gram-negative bacilli, requiring a ~1000 times higher concentration than for gram-positive organisms. Ampicillin, one of the most widely used semisynthetic penicillins, has a drastically increased antibiotic effect on gram-negative bacilli.

A priori, the poorest substrates for β-lactamase may be considered as the most effective β-lactam antibiotics. When subjected to antibiotic-selection pressure, however, the bacteria rapidly develop resistance via high frequency mutations either altering the structure and function of the enzyme toward broader (or extended) substrate specificities or overproducing the wild-type β-lactamases (up-promoter mutations).

c. Mechanism of β-lactamase action

i. Kinetics of acyl-enzyme catalysis. β-Lactamase converts penicillin (or cephalosporin) [S] to the biologically inactive penicilloic (or cephalosporoic) acid [P] via an acyl-enzyme intermediate (E-acyl) (Scheme 8.2).

$$E + S \underset{k_{-1}}{\overset{k_1}{\rightleftharpoons}} ES \xrightarrow{k_2} E\text{-acyl} \xrightarrow{k_3} E + P$$

SCHEME 8.2

The reaction is partly diffusion controlled and the rate constants are such that there is no single RDS for β-lactamase action (Table 8.4), a sign of fully efficient enzymes with good substrates. Another class A enzyme, β-lactamase I of B. cereus, has rate constants for benzylpenicillin similar to those of TEM β-lactamase and S. aureus β-lactamase PC1 (29,30). With cephalosporin C, which is a poor substrate, k_3 is substantially larger than k_2, indicating that the ring-opening step is RDS (31).

With good substrates such as benzylpenicillin, the acyl-enzyme intermediate does not accumulate in TEM β-lactamase nor in other β-lactamases such as β-lactamase I of B. cereus. However, the accumulation of benzylpenicilloyl–enzyme intermediate occurs in β-lactamase PC1. Compared with the other two class A β-lactamases, β-lactamase PC1 exhibits poorer acylation (k_2) and deacylation (k_3) (32).

With poor substrates such as cefoxitin, the acyl-enzyme intermediate can also be demonstrated for TEM β-lactamase (7). At subzero temperatures, the acyl-enzyme intermediate can be trapped and isolated, as has been shown with B. cereus and S. aureus β-lactamases (33).

ii. Catalytic mechanism. The covalent intermediate arises from the attack of the enzymic nucleophile (Ser-70) at the β-lactam carbonyl carbon and subsequent formation of a tetrahedral intermediate (Fig. 8.6). The Ser-70 is believed to be activated by the ε-NH$_2$ of Lys-73 which, acting as a general base, abstracts the proton of Ser-OH. The proton is then shuttled to the adjacent Ser-130 which in turn donates its proton to the thiazolidine ring nitrogen of the leaving group. The

TABLE 8.4 Rate Constants of Benzylpenicillin Hydrolysis Catalyzed by TEM β-Lactamase[a]

Rate constant	Value	Unit
k_1	123	$\mu M^{-1} \text{sec}^{-1}$
k_{-1}	11,800	sec^{-1}
k_2	2,800	sec^{-1}
k_3	1,500	sec^{-1}
K_s (= k_{-1}/k_1)	96	μM
K_m	42	μM
k_{cat}	980	sec^{-1}
k_{cat}/K_m	2.4×10^7	$M^{-1} \text{sec}^{-1}$

[a] Kinetic data at pH 7.0 and 20°C fitted the acyl-enzyme mechanism (Scheme 8.2) (29).

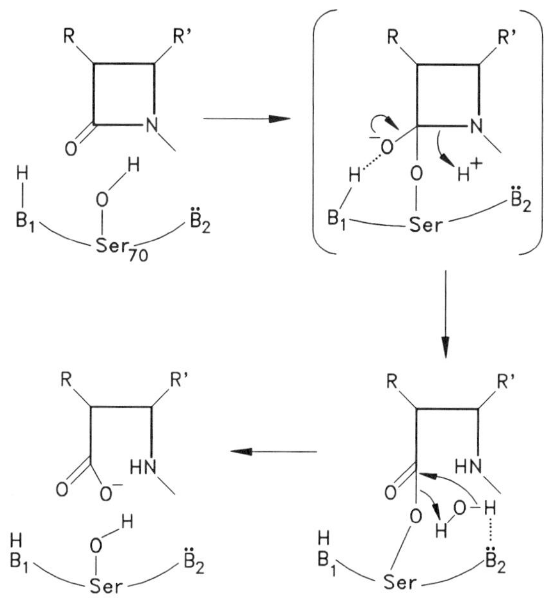

FIGURE 8.6 Schematic mechanism of β-lactamase catalysis.

The acyl-enzyme intermediate is hydrolyzed via a second tetrahedral intermediate by the nucleophilic attack on the penicilloyl carbonyl carbon by the OH$^-$ group of a water molecule. The water molecule is activated by the carboxylate group of Glu-166 which functions as a general base. The proton abstracted from the deacylating water molecule is presumed to be transferred back to Ser-70 to regenerate the Ser-OH, an event concurrent with the breakdown of the second tetrahedral intermediate.

This catalytic mechanism is consistent with data from X-ray crystallography, site-specific mutagenesis, and various biochemical studies. In crystal structures of E. coli TEM-1 β-lactamase (34) and other related class A β-lactamases (35–37), the side chains of the conserved residues (Ser-70, Lys-73, Ser-130, and Glu-166) are, together with a water molecule, clustered around the bound substrate (penicillin G). The presence of a methoxy or hydroxymethyl group on the α face of the β-lactam ring, e.g., in cefoxitin or 6α-(hydroxymethyl)penicillanic acid, results in a displacement of the water (−712), thereby allowing the mechanism-based inhibitors to form rather stable acyl-enzyme intermediates (38). Substitution of either Lys-73 by Arg or of Ser-130 by Ala or Gly impairs the acylation step. Substitution of Glu-166 by Asn or Ala drastically reduces the deacylation step, leading to the accumulation of the acyl-enzyme intermediate (39,40).

3. STRUCTURE

i. Primary structure. The TEM-1 β-lactamase of E. coli plasmid pBR322 consists of 286 amino acid residues (M_r 28,500) (41,42). (Note: The class A β-lactamases have an average size of 25–30 kDa, are active as monomers, and contain no carbohydrate or cofactors.) The first 23 amino acid region (residues 3–25) is a signal peptide and does not appear in the mature enzyme. The mature enzyme has His and Trp as the N-terminal (ABL 3) and C-terminal (ABL 290) residues, respectively.

Mature TEM β-lactamase contains two Cys residues (C77, C123). They form a disulfide bridge which is unique among class A enzymes. The enzyme is not inhibited by PCMB or IAA. Elimination of the disulfide bond by a site-mutagenic replacement (Cys-77–Ser) does not affect the catalytic activity nor the stability of the protein at temperatures below 40°C (11). The substitution of both Cys residues by Ala (Cys–Ala double mutant) does not affect the folding process nor the activity: the kinetic parameters have been found to be similar to those of the wild-type enzyme (43). For the wild-type β-lactamase, however, optimal folding occurs under very oxidizing conditions, e.g., <0.4 mM glutathione (GSH) and 4–8 mM GSSG (oxidized GSH), which contrasts the eukaryotic proteins (e.g., bovine pancreatic RNase) that require a reducing environment for optimal folding, i.e., 1 mM GSH and 0.2 m GSSG (44).

The TEM β-lactamase encoded by the *bla* gene on pBR322 has Gln at position 39. This enzyme is referred to as TEM-1 β-lactamase and has a pI of 5.4 (28,45). TEM-2 β-lactamase is identical to the TEM-1 enzyme except that it has Lys-39 and a pI value of 5.6 (46).

There are now over 15 members in the TEM family of β-lactamases that differ from the prototype TEM-1 in one or more amino acids. The residues susceptible for β-lactamase divergence include Gln-39 (Q39K), Glu-104 (E104K), Arg-164 (R164S), Gln-205 (Q205L), Ala-237 (A237T), Gly-238 (G238S), and Glu-240 (E240K) (47). These TEM-1 variants, referred to as *extended-spectrum β-lactamases*, display varying degrees of altered substrate specificities characterized by reduced activities toward classic β-lactams (e.g., benzylpenicillin) and increased activities toward α-methoxy- and oxyimino-β-lactams. Site-specific substitutions of the aforementioned residues generally confirm the alteration of substrate profiles toward new generation β-lactam antibiotics.

ii. Secondary structure and physical parameters. The secondary structure of the TEM β-lactamase is predicted to have 40% α helix and 20% β sheet (8). The ε_{281} is 29,400 M^{-1} cm^{-1} (7).

iii. Tertiary structure

(a) General features. The crystallographic structure of TEM-1 β-lactamase shows a two-domain structure with an overall topology similar to those of other class A β-lactamases, e.g., from *S. aureus* (32), *B. licheniformis* (35), and *Streptomyces albus* G (37). In particular, the regions of catalytic cleft, which are located at the two-domain interface, have very similar conformations. In class A β-lactamases, at least five residues (Ser-70, Lys-73, Ser-130, Glu-166, and Lys-234) are believed to play a catalytic role, while many other residues are implicated in substrate binding via ionic and/or hydrophobic interactions.

(b) Active-site projections. Ser-70: This is the active site residue that forms the acyl-enzyme intermediate and is a part of the tetrad, Ser–Thr–X–Lys, which is conserved in class A β-lactamases. Replacement of Ser-70 by another nucleophile Cys, called "thiol β-lactamase," is active with a substrate specificity distinct from that of wild-type enzyme (8): the K_m values (for benzylpenicillin and ampicillin) are similar to the wild-type values, whereas the k_{cat} values are ~1% of those of the wild-type enzyme. With nitrocefin as substrate, however, the K_m of thiol β-lactamase is ≥10-fold greater than that of the wild-type enzyme, whereas the k_{cat} is similar to that of the wild-type enzyme. Surprisingly, the Ser-70–Ala mutant of *S. albus* G β-lactamase exhibits a weak (less than 0.1% of wild-type k_{cat}/K_m) but measurable activity for benzylpenicillin and ampicillin (48).

Thr-71: Thr-71 is not essential for catalysis, but is important for the structural integrity of the enzyme. Although Thr-71 is conserved in class A β-lactamases, the enzyme is fairly tolerant to single amino acid substitutions at this position; 14 out of 19 cases yield active enzymes (49). Single mutations at Thr-71 to Gly, Asn, Glu, Gln, and Phe provide significant Ap resistance *in vivo*, whereas double mutants with each of these amino acids at position 71, in addition to the C77S mutation, are unable to confer resistance to Ap above the background level. Thus the destabilizing effect of some mutations at Thr-71 is dramatically amplified in an environment where the disulfide bridge is eliminated (11). Site-directed mutagenesis of the Ser–Thr dyad to Thr–Ser results in an inactive enzyme (50).

The tetrad region (Ser^{70}–X–X–Lys) does not appear to be involved in substrate recognition; this region is conserved in natural β-lactamases that exhibit differences in substrate specificity (51). Available evidence suggests that this region is required for enzyme stability. Mutations within the region from Pro^{66} to Ser^{70} are at least partly responsible for alterations in substrate specificity (52).

Lys-73: Lys-73 apparently plays a significant role in the enhancement of catalytic function. The residue is required for electrostatic stabilization of the anionic tetrahedral intermediate. The Lys-73–Arg mutation in the β-lactamase I of *B. cereus* 569/H results in an enzyme with 2–3% of the k_{cat} of the wild-type enzyme (53). A similar site-specific mutation in the class C cephalosporinase of *C. freundii* GN346 results in an enzyme that retains 35% of the k_{cat} of the wild-type enzyme, whereas nonconservative replacement of the Lys to Thr or Glu results in a nearly complete loss of activity (54).

Ser-130: This residue is located close to Ser-70 and Lys-73. In the wild-type TEM-1, Lys-73 is linked to Ser-70 (2.7 Å), but not to Ser-130 (3.7 Å), by a H-bond (34). In the acyl-enzyme complex, however, slight conformational changes bring Ser-130 to a H-bonding distance (2.8 Å) with Lys-73 while Ser-70 is distanced (3.3 Å). Mutation of Ser-130 to Ala, Gly, or other residues impairs the catalytic function of the enzyme, in particular in the acylation step, and alters the substrate-binding specificities as well (55,56).

Glu-166: This residue is a strong candidate for a general base catalyst of β-lactamase. Substitution of Glu-166, which is conserved in class A β-lactamases, by Gln in the β-lactamase I of *B. cereus* 569/H (53) or the β-lactamase of *B. licheniformis* 749C (39) results in a nearly complete loss of catalytic activity.

Lys-234: This residue is conserved in class A β-lactamases and forms a strong H-bond to Ser-130. Lys-234 plays an important role in optimizing the electrostatic potential of the transition state during acylation (34,57). Substitution of Lys-234 by Ala, Glu, or Thr has minor effects on substrate binding but strongly reduces the k_{cat}. The conservative substitution of Lys-234 by Arg, one of the residues characteristically found in carbenicillin-hydrolyzing β-lactamases (CARB/PSE enzymes), yields an enzyme with virtually unaltered k_{cat} toward cephalosporins and most penicillins (with 10-fold higher K_m) but with 6- and 10-fold higher k_{cat} and K_m, respectively, toward carbenicillins (58).

Residues 237–241: These residues are located at the edge of the substrate-binding pocket of the active site. Replacement of one or more of these residues alters the substrate profile of the TEM-1 enzyme toward extended spectrum cephalosporins such as cefotaxime and ceftazidime. For instance, the mutants selected for 100-fold greater ceftazidime resistance display G238S, E240K, and R241G substitutions (59). The A237N mutant has a 4-fold higher activity on cephems (cephalothin and cephalosporin C) than the wild-type enzyme, whereas the activity toward penams is reduced to ~10% that of the wild-type enzyme (9). A237V exhibits only 2% of the wild-type catalytic activity on carbenicillin, whereas it retains 30–60% of the hydrolytic activity toward other penicillins and cephalosporins (60).

iv. Precursor β-lactamase. TEM pre-β-lactamase carries a 23 amino acid hydrophobic leader sequence: MSIQH FRVAL IPFFA AFCLP VFA-. The pre-β-lactamase is catalytically "inactive," i.e., the k_{cat} is ~10 times lower than that of the mature enzyme, while the K_m is only slightly higher. Compared with the mature enzyme, the precursor molecule folds ≥15 times more slowly (43).

This retarded folding property is uniquely attributed to the presence of the leader sequence and is deemed to be an essential and perhaps general feature to secure a transport competent conformation. The translocation of the precursor β-lactamase across the (inner) membrane is a posttranslational event (as opposed to cotranslational process) and does not require the presence of the C terminus or the processing of the signal peptide (61).

Deletion mutants, in which the *bla* signal sequence is almost completely removed, e.g., Met-Ser-(263 amino acid β-lactamase) or Met-Arg-Ser-(263 amino acid β-lactamase), produce but do not secrete the β-lactamase (62). (Note: The extra 2 or 3 amino acids at the N terminus do not affect the catalytic activity of the enzyme.) The fusion protein expressed from the *bla* signal codons fused to an intracellular protein (e.g., triose-phosphate isomerase) gene is neither secreted into the periplasm of *E. coli* nor proteolytically processed.

v. Homologies and evolution. There is ample evidence that class A (and C as well) β-lactamases are functionally related to DD-CPase/TPase, the presumed target for β-lactam antibiotics (63,64). A strong 3-D structural similarity also exists between them (65,66), although the overall amino acid sequence homology is low. Note that, among the class A β-lactamases, there is an overall 20% amino acid sequence homology, although there are a number of highly conserved regional sequences (5). Sequence comparisons of known β-lactamases and PBPs which include DD-CPase/TPase have revealed highly conserved regions of sequence identity such as the active site Ser (Ser^{70}–X–X–Lys), KTG motif (Lys^{234}–Thr–Gly), and SDN motif (Ser^{130}–Asp–Asn) (65,67). These data support the hypothesis of a common, yet distant, ancestral origin for β-lactamases and the PBPs, particularly DD-CPase/TPase.

4. GENETICS

i. Identity of *bla* gene. The complete nucleotide sequence of the TEM-1 β-lactamase gene (*bla*) carried on pBR322 has been determined (42). The deduced amino acid sequence coincides with that of the TEM-2 β-lactamase determined by amino acid sequencing except at position 39, i.e., Gln (CAG) for TEM-1 and Lys (AAG) for TEM-2 (41).

The TEM-1 *bla* gene originates from the transposon Tn3, which derives from the plasmid R1 (68), the R factor (originally R$_{7268}$) isolated in 1963 in London from wild strains of *Salmonella paratyphi* B (69). The Ap^r gene was translocated to pBR313 from the Ap transposon (TnA) of pRSF2124 [a ColE1-Ap^r plasmid (70)] via pMB9 (71). In this and subsequent vector constructs, the *bla* gene has been no longer transposable. Molecular arrangements of pBR313 gave rise to pBR322 (72).

The TEM-2 *bla* gene derives from the plasmid R6K (formerly RTEM) which was isolated from *E. coli* in 1963 in Athens (69,73).

ii. Gene organization. In pBR322 (4363 bp), the *bla* gene encodes TEM-1 β-lactamase in an 858-bp (286 amino acids) ORF from positions 3298 to 4155 (74). (Position 1 starts from the unique *Eco*RI site.) The plasmid contains a second selection marker, the Tc^r gene.

The *bla* gene is transcribed by at least two promoters, the *bla* promoter and the *tetR* (Tc repressor gene) promoter. The *bla* promoter, designated P3, is located at position 4190 (35 bp from the ATG codon). (P1 and P2 designations have been assigned to *tet* gene promoters in pBR322.) The *bla* gene is also transcribed in the sense orientation by the *tetR* promoter (P1) which is located in the region of positions 1–80 and in the antisense orientation by promoter P4 mapped to the region 40–80. The transcriptional termination of the *bla* gene has been located at three sites at approximate positions 3210, 3110, and 3050.

The 5′ UTR of the *bla* gene contains sequence AATATTG AAA AAGGA AGAGT ATG AG, which includes a Pribnow box-like sequence (AATATTG), the Shine-Dalgarno sequence (AAGGA), and the initiator codon (ATG). A continuous stretch of 286 amino acid codons is terminated by an ochre codon (TAA). The 5′ UTR segment contains the minimal region responsible for the translational repression induced by heat shock (temperature shift from 30° to 42°C) (75). The synthesis of β-lactamase mRNA is not repressed under the heat-shock conditions.

Note that the structural gene for β-lactamase is called *blaZ* in some other species (e.g., *S. aureus*) where induction is regulated by a classical repressor, BlaI, and positive activators, BlaR1, BlaR2, etc. (76).

iii. Expression of *bla* gene. In *E. coli*, the expression of TEM β-lactamase is constitutive. The resistance level toward β-lactam antibiotics is directly related to the number of copies of the TEM *bla* gene and consequently to the amount of enzyme synthesized. The copy number of plasmid R1 is higher at slow growth rates and at anaerobic growth with a corresponding *gene dosage* (R factor copies per genome) effect for TEM β-lactamase. The specific activity of β-lactamase increases linearly with the gene dosage over a 10-fold range (77). Therefore the specific activity of the TEM-1 β-lactamase can be used, unlike other plasmid-encoded antibiotic marker enzymes, to determine the plasmid copy number (78).

iv. Chromosome-encoded *E. coli* β-lactamase. Wild-type *E. coli* cells constitutively express a low-level β-lactamase (cephalosporinase) encoded by a chromosomal gene *ampC*, which is mapped at the 94-min position on the *E. coli* K12 genetic map. The regulatory region of the *ampC* gene is designated *ampA*. A locus *ampB*, most likely a second regulatory region, is distinct from *ampA*.

AmpC β-lactamase of *E. coli* K12 consists of 358 amino acids (M_r 39,600). Its precursor form contains a 19 amino acid signal peptide (79). AmpC β-lactamase shows no recognizable homology in DNA and amino acid sequences with the TEM or other type A β-lactamases, although the AmpC enzyme apparently follows a similar catalytic mechanism involving Ser (Ser^{80}–Val–Ser–Lys) in the active site.

5. APPLICATIONS

The β-lactamase gene (often called the Ap^r gene) is the most widely used antibiotic selection marker carried on cloning/expression vectors, typically on pBR322 (71,73) and other pBR322-derived advanced vectors such as pUC18/19, pTZ18/19, pGEM, and pBluescript (see Section A, this chapter, and also Fig. 8.3).

The principles underlying the use of antibiotic marker genes are essentially similar to those of reporter genes described earlier, although the circumstances for full expression or insertional inactivation of the *bla* gene appear to be less sophisticated. In addition to the major use of the β-lactamase (and *bla* gene) as an antibiotic selection marker, the signal sequence of the *bla* has been used as an efficient vehicle to achieve extracellular secretions of recombinant proteins.

In recombinant DNA work involving gene cloning, cDNA library construction, site-specific mutagenesis, and amplification/expression of cloned genes in *E. coli*, ampicillin is the antibiotic of prime importance due to the high sensitivity of the gram-negative bacteria to this particular antibiotic. The application of ampicillin (at 50–100 μg/ml) in liquid cultures or Petri plates allows the growth of only the cells harboring the plasmid which carries the functional *bla* gene.

Ampicillin is rather unstable and is hydrolyzed when incubated for an extended period of time even at ambient temperatures. Long incubations with Ap-resistant cells may also consume all the active ampicillin molecules due to overexpression of the β-lactamase. Note that β-lactam antibiotics do not kill bacterial cells unless they are actively dividing under the exposure conditions. Therefore, the potential always exists for the opportunistic growth of cells that are originally Ap sensitive and are present as a contaminant or used as lawn cells during the gene cloning procedures. This type of growth can be readily observed as satellite colonies on Ap plates.

A typical gene cloning in pBR322 uses either the unique *Pst*I site (position 3612) in the Ap^r gene or the unique *Bam*HI site (position 375) in the Tc^r gene. The insertion of a DNA fragment at the *Pst*I site inactivates the β-lactamase, whereas the insertion at the *Bam*HI site inactivates the *tet* gene product (tetracycline efflux protein). The insertional inactivation of one of the two marker genes renders the cells sensitive to Ap or Tc, and the recombinant clones have either a (Ap^s, Tc^r) or (Ap^r, Tc^s) phenotype.

6. Sources

TEM β-lactamase is normally secreted into the periplasmic space of *E. coli* cells. Because the natural constitutive β-lactamase promoter is weak, overexpression of the enzyme is best achieved by the use of the λP$_L$ promoter or *tac* promoter (43). *E. coli* β-lactamase has also been expressed in *S. cerevisiae* as pre-β-lactamases that were correctly processed and secreted into the culture medium. One case is from a gene fusion in which the mature β-lactamase was fused in frame to the 16 amino acid leader sequence of the yeast invertase (SUC2-bla°) under the control of the yeast SUC2 promoter (80). Another case is from the *bla* gene carrying the natural 23 amino acid signal sequence under the control of the yeast ADCI promoter in a *S. cerevisiae* ts1 mutant (strain TY4) (81). Conventional laboratory strains of *S. cerevisiae* are incapable of processing and secreting the pre-β-lactamase that contains the bacterial signal sequence.

The enzyme has been purified by employing diverse combinations of separation techniques (7,82). It can be efficiently purified by immunoaffinity chromatography on a MAb-conjugated Sepharose 4B matrix (83). Elution with either benzylpenicillin or cloxacillin yields highly purified, concentrated, active enzyme preparations.

References

1. Collatz, E., Labia, R., and Gutmann, L. (1990). *Mol. Microbiol.* **4**, 1615–1620.
2. Hedges, R. W., Datta, N., Kontomichalou, P., and Smith, J. T. (1974). *J. Bacteriol.* **117**, 56–62.
3. Richmond, M. H., and Sykes, R. B. (1973). In "Advances in Microbial Physiology" (A. H. Rose and D. W. Tempest, Eds.), pp. 31–88. Academic Press, New York.
4. Ambler, R. P. (1980). *Phil. Trans. R. Soc. London B* **289**, 321–331.
5. Pastor, N., Pinero, D., Valdes, A. M., and Soberon, X. (1990). *Mol. Microbiol.* **4**, 1957–1965.
6. Ambler, R. P., Coulson, A. F. W., Frere, J.-M., Ghuysen, J.-M., Joris, B., Forsman, M., Levesque, R. C., Tiraby, G., and Waley, S. G. (1991). *BJ* **276**, 269–272.
7. Datta, N., and Richmond, M. H. (1966). *BJ* **98**, 204–209.
8. Schultz, S. C., Dalbadie-McFarland, G., Neitzel, J. J., and Richards, J. H. (1987). *Proteins (S.F.G)* **2**, 290–297.
9. Sykes, R. B., and Matthew, M. (1979). In "Beta-Lactamases" (J. M. T. Hamilton-Miller and J. T. Smith, Eds.), pp. 17–49. Academic Press, London.

10. Matthew, M., Harris, A. M., Marshall, M. J., and Ross, G. W. (1975). *J. Gen. Microbiol.* **88**, 169–178.
11. O'Callaghan, C. H., Morris, A., Kirby, S. M., and Shingler, A. H. (1972). *Antimicrob. Agents Chemother.* **1**, 283–288.
12. Waley, S. G. (1974). *BJ* **139**, 789–790.
13. Fisher, J., Belasco, J. G., Khosla, S., and Knowles, J. R. (1980). *Biochem.* **19**, 2895–2901.
14. Sigal, I. S., DeGrado, W. F., Thomas, B. J., and Petteway, S. R., Jr. (1984). *JBC* **259**, 5327–5332.
15. Healey, W. J., Labgold, M. R., and Richards, J. H. (1989). *Proteins (S.F.G)* **6**, 275–283.
16. Cole, M. (1979). *In* "Beta-Lactamases" (J. M. T. Hamilton-Miller and J. T. Smith, Eds.), pp. 205–289. Academic Press, New York.
17. Kiener, P. A., and Waley, S. G. (1978). *BJ* **169**, 197–204.
18. Labia, R., and Peduzzi, J. (1978). *BBA* **526**, 572–579.
19. Knowles, J. R. (1985). *Acc. Chem. Res.* **18**, 97–104.
20. Fisher, J., Charnas, R. L., and Knowles, J. R. (1978). *Biochem.* **17**, 2180–2184.
21. Fisher, J., Charnas, R. L., Bradley, S. M., and Knowles, J. R. (1981). *Biochem.* **20**, 2726–2731.
22. Pratt, R. F., and Loosemore, M. J. (1978). *PNAS* **75**, 4145–4149.
23. Knott-Hunziker, V., Orlek, B. S., Sammes, P. G., and Waley, S. G. (1979). *BJ* **177**, 365–367.
24. Charnas, R. L., and Knowles, J. R. (1981). *Biochem.* **20**, 2732–2737.
25. Easton, C. J., and Knowles, J. R. (1982). *Biochem.* **21**, 2857–2862.
26. Gordon, E. M., Chang, H. W., Cimarusti, C. M., Toeplitz, B., and Gougoutas, J. Z. (1980). *JACS* **102**, 1690–1702.
27. Rahil, J., and Pratt, R. F. (1994). *Biochem.* **33**, 116–125.
28. Labia, R., Barthelemy, M., Fabre, C., Guionie, M., and Peduzzi, J. (1979). *In* "Beta-Lactamases" (J. M. T. Hamilton-Miller and J. T. Smith, Eds.), pp. 429–442. Academic Press, New York.
29. Christensen, H., Martin, M. T., and Waley, S. G. (1990). *BJ* **266**, 853–861.
30. Martin, M. T., and Waley, S. G. (1988). *BJ* **254**, 923–925.
31. Bicknell, R., and Waley, S. G. (1985). *BJ* **231**, 83–88.
32. Pratt, R. F., McConnell, T. S., and Murphy, S. J. (1988). *BJ* **254**, 919–922.
33. Virden, R., Tan, A. K., and Fink, A. L. (1990). *Biochem.* **29**, 145–153.
34. Strynadka, N. C. J., Adachi, H., Jensen, S. E., Johns, K., Sielecki, A., Betzel, C., Sutoh, K., and James, M. N. G. (1992). *Nature* **359**, 700–705.
35. Moews, P. C., Knox, J. R., Dideberg, O., Charlier, P., and Frere, J.-M. (1990). *Proteins (S.F.G.)* **7**, 156–171.
36. Herzberg, O., and Moult, J. (1987). *Science* **236**, 694–701.
37. Dideberg, O., Charlier, P., Wery, J.-P., Dehottay, P., Dusart, J., Erpicum, T., Frere, J.-M., and Ghuysen, J.-M. (1987). *BJ* **245**, 911–913.
38. Matagne, A., Lamotte-Brasseur, J., Dive, G., Knox, J. R., and Frere, J.-M. (1993). *BJ* **293**, 607–611.
39. Escobar, W. A., Tan, A. K., and Fink, A. L. (1991). *Biochem.* **30**, 10783–10787.
40. Adachi, H., Ohta, T., and Matsuzawa, H. (1991). *JBC* **266**, 3186–3191.
41. Ambler, R. P., and Scott, G. K. (1978). *PNAS* **75**, 3732–3736.
42. Sutcliffe, J. G. (1978). *PNAS* **75**, 3737–3741.
43. Laminet, A. A., and Pluckthun, A. (1989). *EMBO J.* **8**, 1469–1477.
44. Walker, K. W., and Gilbert, H.F. (1994). *JBC* **269**, 28487–28493.
45. Matthew, M., Hedges, R. W., and Smith, J. T. (1979). *J. Bacteriol.* **138**, 657–662.
46. Barthelemy, M., Peduzzi, J., and Labia, R. (1985). *Ann. Inst. Pasteur. Microbiol.* **136A**, 311–321.
47. Jacoby, G. A. (1994). *Eur. J. Clin. Microbiol. Inf. Dis.* 13(Suppl. 1), 2–11.
48. Jacob, F., Joris, B., and Frere, J.-M. (1991). *BJ* **277**, 647–652.
49. Schultz, S. C., and Richards, J. H. (1986). *PNAS* **83**, 1588–1592.
50. Dalbadie-McFarland, G., Cohen, L. W., Riggs, A. D., Morin, C., Itakura, K., and Richards, J. H. (1982). *PNAS* **79**, 6409–6413.
51. Dale, J. W., Godwin, D., Mossakowska, D., Stephenson, P., and Wall, S. (1985). *FEBS Lett.* **191**, 39–44.
52. Dube, D. K., and Loeb, L. A. (1989). *Biochem.* **28**, 5703–5707.

53. Madgwick, P. J., and Waley, S. G. (1987). *BJ* **248**, 657–662.
54. Tsukamoto, K., Tachibana, K., Yamazaki, N., Ishii, Y., Ujiie, K., Nishida, N., and Sawai, T. (1990). *EJB* **188**, 15–22.
55. Jacob, F., Joris, B., Lepage, S., Dusart, J., and Frere, J.-M. (1990). *BJ* **271**, 399–406.
56. Juteau, J.-M., Billings, E., Knox, J. R., and Levesque, R. C. (1992). *Prot. Eng.* **5**, 693–701.
57. Ellerby, L. M., Escobar, W. A., Fink, A. L., Mitchinson, C., and Wells, J. A. (1990). *Biochem.* **29**, 5797–5800.
58. Lenfant, F., Petit, A., Labia, R., Maveyraud, L., Samama, J.-P., and Masson, J.-M. (1993). *EJB* **217**, 939–946.
59. Venkatachalam, K. V., Huang, W. Z., Larocco, M., and Palzkill, T. (1994). *JBC* **269**, 23444–23450.
60. Barthelemy, M., Peduzzi, J., Rowlands, D., Paul, G., Moreau, G., and Labia, R. (1994). *FEMS Microbiol. Lett.* **117**, 333–340.
61. Koshland, D., Sauer, R. T., and Botstein, D. (1982). *Cell* **30**, 903–914.
62. Kadonaga, J. T., Gautier, A. E., Straus, D. R., Charles, A. D., Edge, M. D., and Knowles, J. R. (1984). *JBC* **259**, 2149–2154.
63. Granier, B., and 17 coauthors (1994). *Meth. Enzy.* **244**, 249–266.
64. Waxman, D. J., and Strominger, J. L. (1983). *Ann. Rev. Biochem.* **52**, 825–869.
65. Kelly, J. A., Knox, J. R., Moews, P. C., Hite, G. J., Bartolone, J. B., Zhao, H., Joris, B., Frere, J.-M., and Ghuysen, J.-M. (1985). *JBC* **260**, 6449–6458.
66. Samraoui, B., Sutton, B. J., Todd, R. J., Artymiuk, P. J., Waley, S. G., and Phillips, D. C. (1986). *Nature* **320**, 378–380.
67. Spratt, B. G., and Cromie, K. D. (1988). *Rev. Infec. Dis.* **10**, 699–711.
68. Burman, L. G., Nordstrom, K., and Boman, H. G. (1968). *J. Bacteriol.* **94**, 438–446.
69. Datta, N., and Kontomichalou, P. (1965). *Nature* **208**, 239–241.
70. So, M., Gill, R., and Falkow, S. (1975). *MGG* **142**, 239–249.
71. Bolivar, F., Rodriguez, R. L., Betlach, M. C., and Boyer, H. W. (1977). *Gene* **2**, 75–93.
72. Bolivar, F., Rodriguez, R. L., Greene, P. J., Betlach, M. C., Heyneker, H. L., Boyer, H. W., Crosa, J. H., and Falkow, S. (1977). *Gene* **2**, 95–113.
73. Kontomichalou, P., Mitani, M., and Clowes, R. C. (1970). *J. Bacteriol.* **104**, 34–44.
74. Balbas, P., Soberon, X., Bolivar, F., and Rodriguez, R. L. (1988). In "Vectors: A Survey of Molecular Cloning Vectors and Their Uses" (R. L. Rodriguez and D. T. Denhardt, Eds.), pp. 5–41. Butterworths, Boston.
75. Kuriki, Y. (1989). *J. Bacteriol.* **171**, 5452–5457.
76. Wang, P.-Z., Projan, S. J., and Novick, R. P. (1991). *NAR* **19**, 4000.
77. Engberg, B., and Nordstrom, K. (1975). *J. Bacteriol.* **123**, 179–186.
78. Uhlin, B. E., and Nordstrom, K. (1977). *Plasmid* **1**, 1–7.
79. Jaurin, B., and Grundstrom, T. (1981). *PNAS* **78**, 4897–4901.
80. Bielefeld, M., and Hollenberg, C. P. (1992). *Curr. Genet.* **21**, 265–268.
81. Broker, M. (1993). *Res. Microbiol.* **144**, 575–580.
82. Dale, J. W. (1979). In "Beta-Lactamases (J. M. T. Hamilton-Miller and J. T. Smith, Eds.), pp. 73–98. Academic Press, New York.
83. Bibi, E. (1989). *BJ* **263**, 309–311.

C. Chloramphenicol Acetyltransferase
[Chloramphenicol Acetyltransferase, EC 2.3.1.28]

Chloramphenicol (Cm) is a broad spectrum antibiotic that acts by inhibiting the function of bacterial ribosomes. Cm binds the 50S subunit of bacterial (70S) ribosomes and inhibits peptidyltransferase ($K_d \approx 3$ μM), effectively blocking prokaryotic protein synthesis. Because of its broad spectrum activity on ribosomal function, Cm also causes some serious side effects in eukaryotic hosts. (In humans

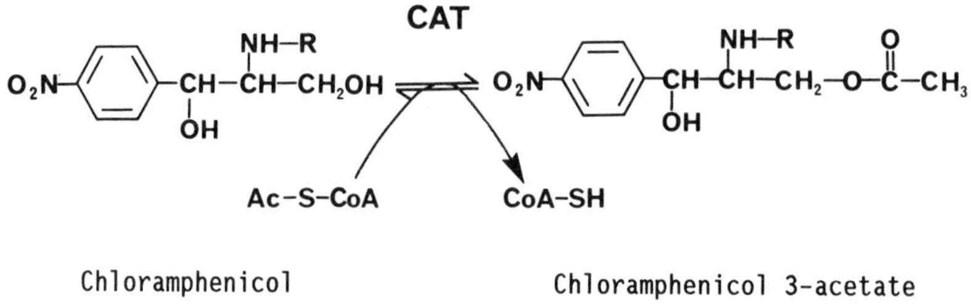

Chloramphenicol → Chloramphenicol 3-acetate

$O_2N-C_6H_4-CH(OH)-CH(NH-R)-CH_2OH + Ac\text{-}S\text{-}CoA \xrightarrow{CAT} O_2N-C_6H_4-CH(OH)-CH(NH-R)-CH_2-O-CO-CH_3 + CoA\text{-}SH$

FIGURE 8.7 Principal reaction catalyzed by chloramphenicol acetyltransferase. R = -COCHCl$_2$ (Cm), -COCH$_2$OH (2-hydroxyacetamido-Cm), -COCH$_3$ (2-acetamido-Cm).

and animals, Cm has been implicated in medullary aplasia, anemia, and cancer, which is the reason for the ban on clinical use.)

The high level resistance of certain bacteria to Cm is due to the enzyme *chloramphenicol acetyltransferase* (CAT) which modifies the Cm to a biologically inactive derivative. CAT is an intracellular, trimeric enzyme with an average monomer size of 25 kDa. CAT catalyzes the transfer of an acetyl group from donor acetyl-CoA to the primary (C-3) hydroxyl of Cm, generating chloramphenicol 3-acetate and CoA-SH as products (Fig. 8.7). The acetylated Cm is incapable of binding to bacterial ribosome and is devoid of antimicrobial activity.

CAT activity has been found in a wide range of Cm-resistant eubacteria. Although some variant CAT genes (e.g., of *Streptomyces* sp.) are of chromosomal origin, most CAT genes are carried on plasmids, often in combination with other genes that confer resistance to other antibiotics such as ampicillin and kanamycin. Naturally occurring variants of CAT from gram-negative bacteria have been classified into three types (type I–III), whereas the variants of CAT from gram-positive bacteria have been divided into several subgroups (A, B, C, and D). These classifications are based on parameters such as electrophoretic mobility, kinetic data, susceptibility to inhibitors, reactivity with antisera, and the ability to form hybrid enzymes with other variants (1).

Type I CAT (or CAT$_I$), which is typically encoded by transposon Tn9, has been widely used both as a selection marker and as a reporter for studying gene expression in eukaryotic cells. The CAT reporter system is a powerful and sensitive tool for such studies since eukaryotic cells frequently used in such experiments do not have endogenous CAT or CAT-like activities. Furthermore, CAT activity can be detected and quantitated with relative ease and at high sensitivities.

In this section, type I CAT (and the *cat* gene) will be described as a typical example of a dual function marker/reporter system, although much of the available biochemical information on CAT comes from the CAT$_{III}$ that is specified by plasmid R387, a non-F enteric plasmid.

1. FUNCTIONS

a. Optimal conditions for reaction, detection, or assays

CAT activity is optimal over a broad pH range with the maximum at about pH 7.6; at pH 6.0, the remaining CAT activity is ~60% of that at pH 7.8 (2). The CAT reaction is usually carried out at 37°C.

CAT activity can be quantitated by measuring either of the two reaction products: (a) acyl-Cm by radiometry, fluorometry, or HPLC, or (b) the free SH group on CoA by spectrophotometry.

Unit definition: One unit is the amount of enzyme that catalyzes the formation of 1 μmol of product per min under the assay conditions.

i. Radiometric assays

(a) Assay principles based on substrate choice. Although many variations exist, CAT activity is most commonly assayed *in vitro* using [^{14}C]- or [^{3}H]Cm as the acceptor substrate and nonradioactive Ac-CoA as the acetyl donor. Typically, the product Ac-Cm and unreacted Cm are first extracted from the reaction mixture using an organic solvent (e.g., ethyl acetate) and are resolved by TLC on silica plates. (Ac-CoA remains in the aqueous phase.) The product can be quantitated by autoradiography followed by densitometric scanning, by scintillation counting of the radiolabeled Ac-Cm recovered from the TLC plate, or more conveniently by phosphorescence imaging (e.g., Molecular Imager from Bio-Rad and PhosphorImager from Molecular Dynamics).

When more hydrophobic *n*-butyryl-CoA and [^{3}H]Cm are used as the donor and acceptor pair of substrates, phase extraction of the products can be more efficiently carried out using xylenes (3). This method is reported to have a 25-fold higher sensitivity and a greater effective range of enzymes by ~3 orders of magnitude at a substantial saving in cost. The assay can be completed in as little as 2–3 hr and can detect 3×10^{-4} units (2 pg) of CAT (3,4).

Alternatively, [^{14}C]Ac-CoA can be used as the donor substrate and nonradioactive Cm as the acceptor (5). With this assay format, the reaction mixture can be added to a toluene-based scintillation fluid and the [^{14}C]Ac-Cm product can be counted directly since [^{14}C]Ac-Cm is far more soluble than [^{14}C]Ac-CoA in toluene (6). More conveniently, the reaction product can be quantitated by liquid scintillation counting immediately after extraction into ethyl acetate (5). An improved version of this two-phase extraction strategy is the two-phase diffusion system in which the radiolabeled product is selectively diffused into the water-immiscible liquid scintillation cocktail (e.g., Econofluor, NEN) which is overlaid on the aqueous reaction mixture (7). The diffusion assay method allows for continuous and convenient data collection within minutes to several hours.

(b) Source of CAT sample. The usual source of CAT to be assayed is the crude cell extract, although an *in vivo* assay can sometimes be carried out with distinct advantages (see below). To prepare a crude cell extract, cells are harvested in 250 mM Tris–Cl (pH 8) and lysed by sonication (or by three freeze-thawing cycles or homogenization). Note that detergents such as Triton X-100, NP-40, SDS, and Na-deoxycholate inhibit CAT activity and should not be employed for cell lysis. A 10-min incubation at 65°C in the presence of 5 mM EDTA serves to inactivate interfering enzymes (e.g., deacetylase, thioesterase, and/or proteases) that may counteract CAT activity (5,8). Under the incubation conditions, CAT_I is not inactivated. The denatured cellular debris is removed by centrifugation at 14,000 *g* for 15 min at 4°C. Many eukaryotic cells such as CV-1 (African green

monkey kidney cell), NIH 3T3 (mouse cell), HeLa, CHO (Chinese hamster ovary), or chicken embryo fibroblast cells contain little, if any, Cm-modifying enzymes, whereas hepatoma and some neuroendocrine cell lines contain a substantial amount of the enzymes that interfere with correct CAT assays (8).

(c) In vivo assay. CAT assays can be performed *in vivo* as simply and efficiently as the *in vitro* (cell extracts) assays because Cm and its acetylated products rapidly equilibrate between the cells and the surrounding tissue culture medium (9) (also see the fluorescent Cm assay below). The *in vivo* assay conditions are essentially identical to those of the *in vitro* assay described below.

(d) "Standard" assay conditions. CAT is typically assayed in a 180-μl reaction mixture containing cell extracts (or intact cells) (10–150 μl), 1–10 μl [^{14}C]Cm (0.2 μCi), and 250 mM Tris–Cl (pH 7.8). The reaction is initiated by adding fresh Ac-CoA to a 1 mM final concentration. After incubation at 37°C for 1–2 hr, [^{14}C]Ac-Cm products can be processed by any of the methods described earlier. For example, the products can be extracted with two successive 0.5-ml aliquots of ethyl acetate [or xylenes for *n*-butyryl-Cm (3)], pooled, evaporated under vacuum, and then redissolved in 30 μl ethyl acetate. To separate the product from the substrate, a 10-μl aliquot is applied to a 25-mm silica TLC plate. Chromatography is carried out in chloroform–methanol (95:5, v/v), and the plate is dried and autoradiographed (10).

ii. **Fluorescent Cm assay.** This assay uses a fluorescent Cm derivative, e.g., BODIPY-Cm (Molecular Probes, Eugene, OR), as substrate. BODIPY (boron dipyrromethene difluoride)-Cm is an effective acetyl group acceptor substrate for CAT with a K_m of 2 μM (compared to 12 μM for C^{14}-labeled Cm) and a V of 120 pmol/min (compared to 180 pmol/min for the radioactive substrate) (11,12). It is yellow in visible light with the absorption maximum at 505 nm (ε = 73.2 mM^{-1} cm^{-1} in methanol) and emits at 512 nm. Like Cm, BODIPY-Cm is taken up by tissue culture cells; this enables the direct measurement of CAT activity in intact cells.

Typically, 60 μl of the cell extract containing 15 μM BODIPY-Cm is incubated at 37°C for 5 min, then 10 μl of 4 mM Ac-CoA is added, and the mixture is incubated for an additional hour (11). Following the incubation period, the products are extracted with ethyl acetate and separated on ascending TLC on silica plates using methanol:chloroform (9:1, v/v) as the solvent. The order of R_f values is 1,3-diacetyl-Cm > 3-Ac-Cm > 1-Ac-Cm > Cm. Potential complications arising from the product isomerization can be alleviated using BODIPY-1-deoxy-Cm (Green and Yellow, Molecular Probes). The movement of the fluorescent (BODIPY) products is visible during the chromatography. Following chromatography, the plate is air-dried and used for quantitation. It can be photographed on a UV transilluminator using Kodak Ektachrome 400 or Polaroid type 667 film. Alternatively, the fluorescent product spot can be recovered, extracted with methanol, and quantitated. This BODIPY fluorescence assay is reported to be as sensitive as the radiometric assay method with all the convenience of working with nonradioactive material.

iii. HPLC assay. This method of product detection/quantitation is based on the ability to unambiguously separate on a HPLC column (e.g., Pharmacia ODS-2) the product 3-Ac-Cm from unreacted Cm or from by-products such as 1-Ac-Cm and 1,3-diAc-Cm (59,60). Following the reaction with nonradioactive Cm and Ac-CoA under "standard" assay conditions, products are extracted with ethyl acetate and dried. Shortly before use, they are dissolved in methanol. HPLC is performed isocratically at 37°C using a solvent system, MeOH:H$_2$O:HAc (111:186:3). The HPLC response monitored at 325 nm is linear, accurate, and reproducible within a range of 0.001 μg to 10 μg 3-Ac-Cm. Furthermore, the separation conditions help suppress the troublesome nonenzymatic conversion of 3-Ac-Cm to 1-Ac-Cm.

iv. Spectrophotometric assay. This assay measures the rate of appearance of DTNB-titratable free thiol on the product CoA-SH. This assay is especially suitable for kinetic measurements of CAT activity. The standard assay mixture consists of 50 mM Tris–Cl (pH 7.5), 100 mM NaCl, 0.1 mM EDTA, 1 mM DTNB, 0.4 mM Ac-CoA, and 0.1 mM Cm (13,14). The reaction is initiated by the addition of the enzyme and is monitored at 412 nm (ε_{TNB} = 13.6 mM^{-1}cm^{-1}) at 25°C.

Note that SH titration with the ionizable modifying reagent DTNB can be carried out in the background of "negligible" reactivity of the Cys-SH groups intrinsically present in the CAT protein. Under certain circumstances in which the protein structure is altered (e.g., by site-specific mutagenesis) such that the reactivity of SH groups is substantially enhanced or a new reactive Cys residue is introduced into the protein, the usefulness of the DTNB method may be compromised.

b. Kinetic parameters

The K_m values at pH 7.8 for Cm and Ac-CoA are 7 and 46 μM, respectively (15). The k_{cat} of CAT$_I$ is about 20% of that of CAT$_{III}$ (1). For comparison, the K_m (Cm) and k_{cat} of CAT$_{III}$ are 12 μM and 600 sec^{-1}, respectively, at pH 7.5 and 25°C (13). The K_m for Ac-CoA is ~90 μM, while the K_m for propionyl-CoA and butyryl-CoA are 170 and 270 μM, respectively.

c. Inhibitors and inactivators

Fusidic acid: Unlike CAT$_{II}$ or CAT$_{III}$, CAT$_I$ binds the steroidal antibiotic fusidic acid (15). It is a competitive inhibitor with respect to Cm ($K_i \approx$ 11.7 μM). Phe-146 located in the Cm binding site of CAT$_I$ provides a critical hydrophobic surface for the interaction.

Sulfhydryl reagents: CAT$_I$ (as well as CAT$_{III}$) is sensitive to inhibition by uncharged SH-modifying reagents such as iodoacetamide, 4,4'-dithiopyridine, and MMTS, whereas it is resistant to inactivation by ionizable SH-modifying reagents such as DTNB and IAA. The presence of Cm protects the enzyme from modification.

Basic triphenylmethane dyes: Crystal violet and ethyl violet are inhibitors of CAT. They are competitive against Cm with K_i values of 300 and 80 nM, respectively. The two dyes are noncompetitive against Ac-CoA with K_i values of 1 and 0.3 μM, respectively (16).

3-Halo-3-deoxychloramphenicols, the derivatives of Cm in which the 3-OH is replaced by a halogen, are competitive inhibitors. The affinity of the inhibitors to the enzyme shows a linear free energy relationship with hydrophobicity (17).

Ethyl-CoA is also an inhibitor which forms nonproductive ternary complexes with CAT in the presence of Cm.

d. Substrate specificities

i. Overall reaction. CAT catalyzes acetylation of Cm at only one (C-3) of the two available hydroxyl groups (Fig. 8.7). The natural donor substrate for CAT is acetyl-CoA (Ac-CoA). The product 3-Ac-Cm is biologically inactive. Of the four diastereoisomers originating from the two chiral centers (C-1 and C-2) of Cm, only the D-*threo* form is active as an antibiotic.

ii. Donor substrate. Normally the acetyl group is donated from Ac-CoA. Propionyl-CoA and butyryl-CoA can also be used as donor substrates, with transfer efficiencies (in k_{cat}/K_m) of 71 and 14% that of Ac-CoA, respectively. Acetylpantetheine is a poor substitute for Ac-CoA. In the absence of the acceptor Cm, CAT hydrolyzes Ac-CoA at a rate $\sim 10^4$-fold lower than that of the transacetylation reaction.

iii. Acceptor substrate. The normal acceptor substrate is Cm. With Ac-CoA as the acyl donor, CAT catalyzes 3-O-acetylation of Cm. 3-Ac-Cm undergoes a nonenzymatic and pH-dependent rearrangement to 1-Ac-Cm, which can then be acetylated a second time to yield the 1,3-diacetyl derivative. The second acetylation is significantly slower than the first. For the purpose of accurate CAT activity assay, the Cm product isomerization is a nuisance. This can be alleviated using 1-deoxy-Cm as a substrate which yields only one product. The k_{cat} and K_m values for 1-deoxy-Cm are twofold lower and twofold higher than those for Cm, respectively (61).

2-Hydroxyacetamido-Cm and 2-acetamido-Cm (see Fig. 8.7) are acetyl acceptor substrates as competent as Cm (18). Isomers of Cm and analogs having various other substitutions on side groups in Cm have widely ranging substrate efficacies (from 0 to 200%) for CAT_I (1).

iv. Product profile. 1-Ac-Cm and 3-Ac-Cm are stable for at least 3 months in ethyl acetate at 4°C (2). The stability of the products is highly pH dependent. The two products are also very stable at pH 6.0 in Tris–malate buffer with negligible interconversion, but rapidly equilibrate at pH values higher than pH 7.8. Significant nonenzymatic conversion of 3-Ac-Cm to 1-Ac-Cm occurs at pH 6.8, and 1,3-diacetyl-Cm is produced within 1–5 hr by type I and III CAT enzymes but not by type II CAT (19).

e. Mechanism of CAT action

The acetylation and inactivation of Cm by CAT is basically reversible *in vitro* (Fig. 8.7). The equilibrium is in favor of the oxyester (Ac-O-Cm) rather than the thioester (Ac-S-CoA), i.e., $K_{eq} \approx 16$ (2,16).

The reaction proceeds via a ternary complex mechanism (CAT · Ac-CoA · Cm $\rightleftharpoons$ CAT · CoA · Ac-Cm) with a random order of addition of substrates (18). The progression from the binary complex to the ternary complex involves subtle changes in the structure of the enzyme and/or the conformations of bound substrates and accompanies a threefold decrease in affinity: the K_d of Cm and Ac-CoA in respective binary complexes are 4 and 30 μM, whereas the corresponding K_m values of the ternary complex are 12 and 90 μM, respectively (20). The RDS is thought to involve both product release and ternary complex interconversion. No evidence exists for the formation of an acetyl-enzyme intermediate.

Data from kinetic and crystallographic studies consistently suggest that the imidazole of His-195 plays a catalytic role as a general base (21,22). The imidazole ring (N-1) of His-195 is conformationally constrained and polarized through a network of interactions involving a H-bond with the carboxylate of Asp-199 which in turn interacts with the side chain of Arg-18. The N-3 atom of the imidazole ring abstracts a proton from the primary (C-3) hydroxyl of Cm (or from water in the absence of Cm), thereby promoting a nucleophilic attack at the carbonyl of the thioester of Ac-S-CoA (Fig. 8.8). This results in a transient formation of the tetrahedral intermediate, which eventually breaks down to the 3-acetoxy ester of Cm and CoA-SH. The oxyanion intermediate is presumed to be stabilized by two H-bonds: one with the hydroxyl group of Ser-148 and the other with a water molecule (water 252) bound to Thr-174. In the CAT–Cm binary complex, water 252 is H-bonded to the hydroxyl group of Thr-174, the 1-hydroxyl of Cm, and to another bound water molecule (water 249). Ser-148–Ala substitution results in a 53-fold decrease in k_{cat}, whereas the K_m is similar to that of wild-type CAT (62). Substitution of Thr-174 by Ala or Ile results in a ~2-fold decrease in k_{cat} with a 5- to 10-fold increase in K_m for Cm. The H-bond between water 252 and the oxyanion of the tetrahedral intermediate is estimated to contribute 0.9 kcal/mol toward transition state stabilization (61). Replacement of His-195 by Ala or Gln results in an apparent decrease in k_{cat} of $>10^5$-fold, whereas K_m values for both substrates (Cm and Ac-CoA) remain unchanged (63). The His-195–Tyr mutation in CAT_I results in a complete loss of activity (23). However, His-195–Glu retains 80% of the wild-type CAT activity, suggesting that Glu can replace His-195 as a general base, albeit inefficiently.

2. Structure

i. Primary structure. CAT is a timer of identical subunits. The complete amino acid sequence has been determined for CAT_I by direct protein sequencing (24) and by deduction from DNA sequences (25). One subunit consists of 219 amino acid residues (M_r 25,668). The N- and C-terminal residues are Met and Ala, respectively.

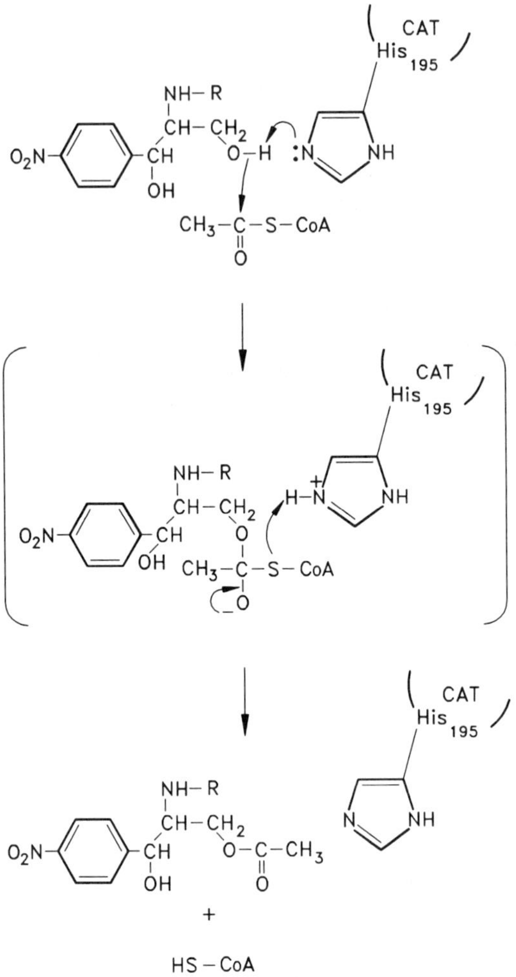

FIGURE 8.8 Catalytic mechanism of chloramphenicol acetyltransferase. The catalytic model features His-195 as the general base initiating the catalysis which involves a tetrahedral intermediate (the structure in parentheses). The oxyanion of the tetrahedral intermediate is thought to be stabilized by two H-bonds: one with the OH group of Ser-148 and the other with a water molecule (water 252) bound to Thr-174.

Note that the amino acid positions used here are those based on the alignment of various CAT sequences that achieves the maximum sequence homology (26). In this numbering system, positions 79 and 142 of CAT_I are vacant.

CAT_I shares an overall conservation of primary structure, the identities ranging from 30 to 70%, with the family members of CAT from various gram-positive and gram-negative bacteria. There are also significant diversions as shown by the fact that only 25 residues are strictly conserved in all known CAT variants (26).

ii. Secondary and tertiary structures. The 3-D structure of CAT has been determined only for the type III variant in the form of CAT_{III}–Cm and CAT_{III}–CoA binary complexes (22,27). CAT_{III} is composed of 213 amino acid residues (25 kDa) and shares a 46% amino acid sequence identity with CAT_I (28). The enzyme monomer is a compact assembly of five α-helical segments and six large and three small β strands. Type I and type III variants can form stable and fully functional heteromeric trimers in *E. coli* (29).

The three subunits associate to form a stable, disk-shaped trimer 60 Å in diameter and 45 Å thick. The symmetrical arrangement of the subunits around the threefold axis presents three subunit interfaces, on which are located three active sites. One subunit provides the catalytic residue (His-195), whereas the opposing subunit contributes the residues involved in substrate binding. This structural arrangement implies that a monomer by itself cannot be catalytically competent.

Lys-137 is implicated in subunit interactions, but it can be substituted by Gln without adverse effects. Pro-136 apparently plays a critical role because its replacement by other amino acids results in significant decreases in CAT activity. Deletion of nine residues from the C terminus results in a folding-defective mutant protein which deposits as cytoplasmic inclusion bodies in *E. coli* and gives rise to a Cm^s phenotype (23). The folding defect can be relieved by fusion to a wide variety of peptides, the smallest being dipeptides.

iii. Active sites. The active site of CAT_{III} (three sites/molecule) features a long tunnel, approximately 25 Å long, formed by the two binding sites at the subunit interfaces (22,27). Cm and Ac-CoA approach the active site from opposite ends of the tunnel to bring the 3-OH of Cm and the sulfur atom of CoA into close proximity. Cm is bound in an extended conformation, which is quite different from the most likely structure of the free antibiotic in solution. The 3-OH of Cm is placed deep in one cleft within H-bonding distance (2.8 Å) from the N-3 of the imidazole of His-195. The apparent ΔG for this H-bonding is estimated to be 1.5 kcal/mol (17). The sulfur of CoA located in the opposite cleft is also within H-bonding distance from the N-3 of the imidazole and from the 3-OH of Cm as well. The active site tunnel is occupied by solvent in the absence of ligands.

CAT_I differs from CAT_{III} by 8 of the 17 residues that align the Cm-binding pocket. The fact that CAT_I can strongly bind a bulky steroidal antibiotic fusidic acid, whereas CAT_{III} cannot, may be at least partly attributed to the differences in the amino acid side chains in the pocket. Presumably the substitutions Leu-29–Ala and Phe-24–Ala open out the pocket. Furthermore, substitutions such as Tyr-25–Phe, Gln-92–Cys, Asn-146–Phe, and Tyr-168–Phe would make the binding site less polar. Nevertheless, these differences have to be reconciled with the practically identical K_m values for Cm between CAT_I and CAT_{III}.

His-195 is the catalytic residue which acts as a general base in deprotonating the 3-OH of Cm (see Fig. 8.8). It can be covalently modified by an active site-directed inhibitor, 3-bromoacetyl-Cm, which results in inactivation of CAT (21).

The alkylation occurs solely at the N-3 position of the imidazole ring. His-195 is absolutely conserved and lies in a sequence of nine residues (192 to 200) that shows a very high degree of sequence conservation. Replacement of His-195 by Tyr in CAT_I results in a complete loss of activity (29).

Ser-148 and, to a lesser extent, Thr-174 are the residues involved in stabilization of the oxyanionic tetrahedral intermediate (refer to Mechanism).

CAT_{III} has conserved Arg-18 and Asp-199 which form a completely buried salt bridge. This ion pair apparently plays some functional as well as structural role because R18V and D199A mutants are active with 9- and 13-fold reduced k_{cat}, respectively, and are significantly more thermolabile than the wild-type enzyme (13). The Asp-199–Asn mutant shows a dramatically reduced k_{cat} (1500-fold) as a consequence of structural changes in the microenvironment rather than the loss of the ion pair.

A single Cys residue (Cys-31) is preferentially reactive with sulfhydryl reagents such as iodoacetamide and NEM (30). The modification results in inactivation of CAT. Protection is conferred in each case by Cm but not by Ac-CoA. Replacement of Cys-31 by Tyr reduces the relative specific activity by 50- to 100-fold (23). Nevertheless, Cys-31 is not an essential residue: it is believed to play a structural role near the active site.

3. GENETICS

i. Origin and primary structure of CAT gene. The *cat* gene that codes for CAT_I has been cloned and sequenced (25,31). The type I *cat* gene originates from an *E. coli* transposon Tn9, which confers resistance to the antibiotic Cm. Tn9 consists of a 1102-bp CAT cistron flanked by two 768-bp IS1 elements. Tn9 has been transferred from an R factor via bacteriophage P1 [P1 CM (32)] to λ phage derivatives (33) and subsequently to the plasmid pBR322, resulting in pBR322–Tn9 constructs (25). Note that the gene for CAT_{III} was orginally isolated from *incK* plasmid R387. CAT_{II} is typically encoded by the *incW* transmissible plasmid pSa.

The CAT_I ORF consists of 219 codons, beginning with the AUG initiator codon at nucleotide position 224 and terminating with the ochre codon (UAA). The deduced amino acid sequence is identical to the sequence obtained by direct amino acid sequencing (24).

ii. Regulatory elements of CAT gene. Two potential RBS are present upstream from the AUG: AGGAG at position 205 and TAAGGA at position 211. The type I *cat* gene has been shown to be controlled at the transcriptional level by cAMP (34,35). Note that other CAT genes from gram-negative bacteria have upstream sequences different from those of type I *cat* and from each other, and are not subject to cAMP control. Consistent with other promotor sequences (e.g., *lac*, *ara*, and *gal* operons) known to be regulated by the cAMP/CAP (cAMP receptor protein) complex, the *cat* promoter region also contains a sequence motif GTGA----TCAC at position 29.

Transcription of the *cat* gene initiates at the G residue at position 165. In addition, the CAP-binding site, identified as TGAGA CGTTG ATCGG <u>CACGTA</u> A (positions 113 to 134), includes or juxtaposes the −35 region (underlined) (34). This promoter region is followed by a Pribnow box (CATAATG) at position 152, fitting to a configuration in which CAP binding enhances the recognition of promoter sequences by RNA polymerase and thus stimulates transcription.

In contrast to the *cat* genes in *E. coli* and other gram-negative bacteria, the *cat* genes in gram-positive bacteria, e.g., of the plasmid-encoded CAT (*cat-86*) in *B. subtilis*, are inducible by Cm. The Cm-induced resistance or the Cm-mediated induction of *cat*-86 expression occurs by a mechanism involving translation attenuation at the leader codons: the drug stalls a ribosome at a specific "induction" site in the *cat* leader mRNA and thereby unmasks the RBS for the *cat*-coding sequence. Selection of the stalling site is apparently correlated with the synthesis of a leader-encoded pentapeptide (MVKTD) which itself binds the ribosome (specifically 23S rRNA) and exhibits an antipeptidyltransferase activity (36).

4. APPLICATIONS

CAT (*cat* gene) is an extremely useful and efficient marker/reporter for the study of eukaryotic gene expression and regulation. The principles underlying the applications of the CAT gene as reporter, selectable marker, and/or fusion protein carrier are similar to those for other marker/reporter systems, namely *lacZ* (βGal) and *bla* (Ap^r) (see Sections A and B, this chapter).

i. Use of CAT gene as reporter. Because CAT is a bacterial enzyme, levels of CAT activity expressed in eukaryotic cells can be readily assayed with little or no background from endogenous CAT or CAT-related activities. Commerically available CAT expression vectors (Table 8.5) fall into two classes. One is the plasmid vectors that express the *cat* gene under the control of a eukaryotic promoter and/or enhancer, thereby serving as a reference for transfection and gene expression. The CAT expression can be made singly or simultaneously with, but independently from, a gene cloned upstream or downstream of the *cat* gene. The other class of CAT vectors is not provided with a promoter and thus cannot express CAT by itself. These vectors are designed to study regulatory sequences (promoters, enhancers, and other UPS) that are inserted in front of the *cat* gene. In addition to the commercially available CAT vectors, several custom-designed CAT expression vectors are useful for the latter applications. They include pCAT series plasmids (10), pBluescript KS-based constructs (37), pLinkCAT (38), and pBRAMScat vectors (39). Other types of vector constructs have also been developed that facilitate the subcloning of the promoterless *cat* gene (gene cassette or cartridge) (40,41,64) or that produce C-terminal peptide (or protein) fusions to CAT (65,66). A note of potential importance for the efficient use of these dedicated CAT vectors concerns the presence of an upstream *Sph*I restriction site (GCATG/C) in the MCS region. This site introduces an ATG triplet that may be out of frame with the initiation codon of the *cat* gene and interferes with the translation

TABLE 8.5 Some Commercial Plasmid Vectors Expressing CAT Reporter Gene

Vector	Promoter	Other features (size, marker)	Source
pMAM$_{neo}$-CAT	MMTV LTR (P)[a] +RSV LTR (E)	9.2 kb Ap^r, Kan^r (control vector)	Clontech
pCAT Basic	—	4.3 kb Ap^r (promoter selection)	Promega
Enhancer	SV40 (E)	4.6 kb Ap^r (promoter selection)	Promega
Promoter	SV40 (P)	4.5 kb Ap^r (enhancer selection)	Promega
Control	SV40 (P + E)	4.7 kb Ap^r (control vector)	Promega
pMSG-CAT	MMTV LTR	8.4 kb Ap^r, Neo^r (control vector)	Pharmacia
pKK232-8	—	5.1 kb Ap^r (promoter selection)	Pharmacia
pCM7	—	4.1 kb Ap^r (*cat* cartridge)	Pharmacia

[a] P and E denote enhancer and promoter, respectively.

of the coding sequence. Deletion of the *Sph*I site from the plasmid increases the CAT activity four- to fivefold (42).

Parameters affecting the (CAT) gene transcription can be evaluated in a wide variety of eukaryotic cells (10,43,44) and plant cells (45) and in transgenic animals as well. The CAT gene can also be used for the study of translational controls, e.g., initiation of protein synthesis in *E. coli* from a termination codon (UAG) in the presence of a complementary anticodon mutant fMet-tRNA$_{CUA}$ (46). When eukaryotic cells are transfected with *cat* gene-fusion vectors, the CAT activity can usually be detected after 2 days, although the exact time of appearance of the CAT activity is subject to many variables such as the cell lines and culture conditions. The half-life of CAT in mammalian cells is estimated to be about 50 hr. Control plasmids containing SV40 promoter and enhancer sequences can be used to monitor the transfection efficiency. They also serve as a convenient internal standard for comparing promoter and enhancer strengths. When the *cat* gene-vector construct contains a *bla* gene under its own promoter, β-lactamase expression can be used as an internal standard to estimate the relative promoter strengths that control the expression of the CAT reporter gene (47).

Some promoterless vectors give a relatively high background of CAT activity to such a level that the study of "weak" promoters may be hampered. This situation arises mostly due to some cryptic promoters and enhancers (cAMP-inducible) located in the vector sequence. Positioning a strong transcription terminator, e.g., the upstream mouse sequence (UMS) derived from the mouse c-*mos* gene, at a site upstream of the CAT gene helps reduce the background up to threefold (48,49). CAT-like activities or CAT inhibitors may also be present in the target eukaryotic cells, especially of hepatic origin (50).

Especially at high concentrations, Cm may display toxic side effects on eukaryotic cells: D-*threo*-Cm (but not L-Cm) specifically inhibits the synthesis of mitochondrial proteins such as cytochrome-c oxidase (51) and induces mutations on the 16S rRNA in mitochondria (52). Cells that harbor CAT vectors show resistance to Cm toxicity. The pharmacological fate of Cm in mammalian cells primarily involves the inactivation mediated by glucuronyltransferase.

ii. Use of CAT gene as selectable marker. Cm resistance (Cm^r) can be used as a phenotypic selection marker in site-directed insertion and deletion mutagenesis. The whole transposon Tn9 or just the CAT gene fragment can be used as a mutating agent and selectable marker (53) either singly or in combination with the Ap^r gene (54). The mutation created *in vitro* of a cloned DNA fragment can be efficiently reintroduced into the original replicon of the gene, i.e., an *E. coli* chromosome or a plasmid, by linearizing the cloned DNA and transforming a *recB recC sbcB* mutant. Vectors such as pUC18CM plasmids that contain a *cat* gene cassette flanked by multiple restriction sites facilitate the manipulation and use of the *cat* gene in diverse situations (55).

In another application using a CAT vector which encodes a C-terminally truncated, folding-defective CAT protein (Cm^s), successful cloning of a target DNA results in CAT fusion proteins which can be selected by Cm^r (23).

iii. Use of Cm as antibiotic for plasmid amplifications. ColE1-type plasmids, such as pBR322 and pBR327, continue to replicate when the protein synthesis and initiation of chromosome replication in the host bacterium are inhibited by the antibiotic Cm (56,57). Consequently, the number of plasmids increases under these conditions in comparison to total cell mass and chromosomal DNA: an effect known as *plasmid amplification*. Conventionally the plasmid amplification has been performed at >150 μg/ml Cm, a high concentration which produces total inhibition of protein synthesis. It has been shown that the replication or amplification of plasmids pBR322 and pBR327 is maximal (5- to 10-fold) during the partial inhibition of protein synthesis by low concentrations (10–20 μg/ml) of Cm in rich medium (LB) (58). This and often higher levels of plasmid amplification are observed only with $relA^+$ *E. coli* strains (e.g., HB101). Although the exact role of *rel* gene in plasmid amplification remains to be clarified, one reasonable explanation is that the replication of plasmids carrying the ColE1 origin requires an unstable *rel*-associated host factor; this factor may continue to be synthesized during the partial inhibition of protein synthesis.

5. Sources

Because the CAT_I expression is known to be subject to cAMP-mediated catabolite repression (35,56), bacterial cells grown on glycerol rather than on glucose produce up to fivefold higher yields of CAT.

CAT can be purified by a variety of procedures involving both conventional protein purification techniques (15) and, preferably, affinity chromatography (2,4). One efficient affinity matrix used for rapid preparation of highly purified enzymes consists of the substrate Cm bound to the carboxyl group of substituted Sepharose via the amino group of 2-amino-Cm.

References

1. Shaw, W. V. (1983). *CRC Crit. Rev. Biochem.* **14**, 1–46.
2. Thibault, G., Guitard, M., and Daigneault, R. (1980). *BBA* **614**, 339–349.
3. Seed, B., and Sheen, J.-Y. (1988). *Gene* **67**, 271–277.
4. Zaidenzaig, Y., and Shaw, W. V. (1976). *FEBS Lett.* **62**, 266–271.
5. Sleigh, M. J. (1986). *Anal. Biochem.* **156**, 251–256.
6. Robison, L. R., Seligsohn, R., and Lerner, S. A. (1978). *Antimicrob. Agents Chemother.* **13**, 25–29.
7. Neumann, J. R., Morency, C. A., and Russian, K. O. (1987). *BioTechniques* **5**, 444–447.
8. Crabb, D. W., Minth, C. D., and Dixon, J. E. (1989). *Meth. Enzy.* **168**, 690–701.
9. Alter, D. C., and Subramanian, K. N. (1988). *BioTechniques* **6**, 526–530.
10. Rosenthal, N. (1987). *Meth. Enzy.* **152**, 704–720.
11. Hruby, D. E., Brinkley, J. M., Kang, H. C., Haugland, R. P., Young, S. L., and Melner, M. H. (1990). *BioTechniques* **8**, 170–171.
12. Young, S. L., Barbera, L., Kaynard, A. H., Haugland, R. P., Kang, H. C., Brinkley, M., and Melner, M. H. (1991). *Anal. Biochem.* **197**, 401–407.
13. Lewendon, A., Murray, I. A., Kleanthous, C., Cullis, P. M., and Shaw, W. V. (1988). *Biochem.* **27**, 7385–7390.
14. Shaw, W. V. (1975). *Meth. Enzy.* **43**, 737–755.
15. Bennett, A. D., and Shaw, W. V. (1983). *BJ* **215**, 29–38.
16. Tanaka, H., Izaki, K., and Takahashi, H. (1974). *J. Biochem.* **76**, 1009–1019.
17. Cullis, P. M., Lewendon, A., Shaw, W. V., and Williams, J. A. (1991). *Biochem.* **30**, 3758–3762.
18. Ellis, J., Bagshaw, C. R., and Shaw, W. V. (1995). *Biochem.* **34**, 16852–16859.
19. Powell, M., Davy, K. W. M., and Livermore, D. M. (1989). *J. Antimicrob. Chemother.* **24**, 897–903.
20. Ellis, J., Bagshaw, C. R., and Shaw, W. V. (1991). *Biochem.* **30**, 10806–10813.
21. Kleanthous, C., Cullis, P. M., and Shaw, W. V. (1985). *Biochem.* **24**, 5307–5313.
22. Leslie, A. G. W., Moody, P. C. E., and Shaw, W. V. (1988). *PNAS* **85**, 4133–4137.
23. Robben, J., Vanderschueren, J., Verhasselt, P., Aert, R., and Volckaert, G. (1995). *Prot. Eng.* **8**, 159–165.
24. Shaw, W. V., Packman, L. C., Burleigh, B. D., Dell, A., Morris, H. R., and Hartley, B. S. (1979). *Nature* **282**, 870–872.
25. Alton, N. K., and Vapnek, D. (1979). *Nature* **282**, 864–869.
26. Shaw, W. V., and Leslie, A. G. W. (1991). *Ann. Rev. B. BC.* **20**, 363–386.
27. Leslie, A. G. W. (1990). *JMB* **213**, 167–186.
28. Murray, I. A., Hawkins, A. R., Keyte, J. W., and Shaw, W. V. (1988). *BJ* **252**, 173–179.
29. Packman, L. C., and Shaw, W. V. (1981). *BJ* **192**, 541–552.
30. Zaidenzaig, Y., and Shaw, W. V. (1978). *EJB* **83**, 553–562.
31. Marcoli, R., Iida, S., and Bickle, T. A. (1980). *FEBS Lett.* **110**, 11–14.
32. Kondo, E., and Mitsuhashi, S. (1964). *J. Bacteriol.* **88**, 1266–1276.
33. Scott, J. R. (1973). *Virol.* **53**, 327–336.
34. Le Grice, S. F. J., and Matzura, H. (1981). *JMB* **150**, 185–196.
35. de Crombrugghe, B., Pastan, I., Shaw, W. V., and Rosner, J. L. (1973). *Nature NB* **241**, 237–239.
36. Gu, Z. P., Harrod, R., Rogers, E. J., and Lovett, P. S. (1994). *J. Bacteriol.* **176**, 6238–6244.
37. Clark, A. R., Boam, D. S. W., and Docherty, K. (1989). *NAR* **17**, 10130.
38. Wu, Y., Tam, S.-P., and Davies, P. L. (1990). *NAR* **18**, 1919.

39. Iyer, R. R. (1991). *Gene* **105**, 97–100.
40. Close, T. J., and Rodriguez, R. L. (1982). *Gene* **20**, 305–316.
41. Dobrowolski, P. (1991). *Gene* **102**, 139–140.
42. Alam, J., Yu, N., Irias, S., Cook, J. L., and Vig, E. (1991). *BioTechniques* **10**, 423–425.
43. Gorman, C. M., Moffat, L. F., and Howard, B. H. (1982). *Mol. Cell. Biol.* **2**, 1044–1051.
44. Mercola, M., Goverman, J., Mirell, C., and Calame, K. (1985). *Science* **227**, 266–270.
45. Fromm, M., Taylor, L. P., and Walbot, V. (1985). *PNAS* **82**, 5824–5828.
46. Varshney, U., and RajBhandary, U. L. (1990). *PNAS* **87**, 1586–1590.
47. Trukhan, M. E., Gorovits, R. L., Lebedeva, M. I., Lapidus, A. L., and Mashko, S. V. (1989). "Molecular Biology," pp. 828–837. [An English translation of Molekulyarnaya Biologiya (1988). **22**, 1033–1044.]
48. Heard, J.-M., Herbomel, P., Ott, M.-O., Mottura-Rollier, A., Weiss, M., and Yaniv, M. (1987). *Mol. Cell. Biol.* **7**, 2425–2434.
49. Piaggio, G., and DeSimmone, V. (1990). *FOCUS (BRL)* **12**, 83–84.
50. De Maio, A., and Buchman, T. G. (1990). *BBA* **1087**, 303–308.
51. Fettes, I. M., Haldar, D., and Freeman, K. B. (1972). *Can. J. Biochem.* **50**, 200–209.
52. Kearsey, S. E., and Craig, I. W. (1981). *Nature* **290**, 607–608.
53. Winans, S. C., Elledge, S. J., Krueger, J. H., and Walker, G. C. (1985). *J. Bacteriol.* **161**, 1219–1221.
54. Marinus, M. G., Carraway, M., Frey, A. Z., Brown, L., and Arraj, J. A. (1983). *MGG* **192**, 288–289.
55. Schweizer, H. P. (1990). *BioTechniques* **8**, 614–616.
56. Clewell, D. B., and Helinski, D. R. (1972). *J. Bacteriol.* **110**, 1135–1146.
57. Clewell, D. B. (1972). *J. Bacteriol.* **110**, 667–676.
58. Frenkel, L., and Bremer, H. (1986). *DNA* **5**, 539–544.
59. Young, S. L., Jackson, A. E., Puett, D., and Melner, M. H. (1985). *DNA* **4**, 469–475.
60. Robins, R. J., Hamill, J. D., Bent, E. G., and Evans, D. M. (1990). *Meth. Mol. Cell. Biol.* **2**, 91–94.
61. Lewendon, A., and Shaw, W. V. (1993). *JBC* **268**, 20997–21001.
62. Lewendon, A., Murray, I. A., Shaw, W. V., Gibbs, M. R., and Leslie, A. G. W. (1990). *Biochem.* **29**, 2075–2080.
63. Lewendon, A., Murray, I. A., Shaw, W. V., Gibbs, M. R., and Leslie, A. G. W. (1994). *Biochem.* **33**, 1944–1950.
64. Mongkolsuk, S., Vattanaviboon, P., Rabibhadana, S., and Kiatpapan, P. (1993). *Gene* **124**, 131–132.
65. Robben, J., Massie, G., Bosmans, E., Welles, B., and Volckaert, G. (1993). *Gene* **126**, 109–113.
66. Dykes, C. W., Bookless, A. B., Coomber, B. A., Noble, S. A., Humber, D. C., and Hobden, A. N. (1988). *EJB* **174**, 411–416.

D. Luciferases

Luciferase is a light-producing enzyme naturally found in insect fireflies and in luminous marine and terrestrial microorganisms. Introduction of the luciferase and other light-emitting proteins (e.g., green fluorescent proteins) as a visualizing marker/reporter has drastically expanded the versatility of reporter gene technology. Expression of the luciferase gene fusion product confers on the host the ability to glow in the dark. The reporter visualizes transcriptional and translational expression of the attached foreign gene, and enables one to localize it to particular domains, cells, or organelles of almost any organism. This can be done noninvasively, with inexpensive, nonisotopic, easily available substrates, and at extremely high sensitivity under no significant endogenous background.

Two types of luciferase genes, cloned from bacteria and firefly, are used as sensitive reporter systems in a wide variety of cells such as bacterial, yeast, insect, animal, and plant cells. Bacterial luciferases are flavoenzymes composed of two subunits each encoded by the *luxA* and *luxB* genes, while the firefly luciferase is a single polypeptide specified by the *luc* gene. The two types of luciferase catalyze

Bacterial luciferase (Lux)

$$FMNH_2 + O_2 + RCHO \longrightarrow FMN + RCOOH + H_2O + light$$

Firefly luciferase (Luc)

$$Luciferin + O_2 + ATP \longrightarrow Oxyluciferin + CO_2 + H_2O + AMP + light$$

$$\begin{bmatrix} E + LH_2 + ATP \longrightarrow E \cdot LH_2-AMP + PP_i \\ E \cdot LH_2-AMP + O_2 \longrightarrow E \cdot P \text{ (oxyluciferin)} + CO_2 + AMP + light \end{bmatrix}$$

SCHEME 8.3

different reactions (Scheme 8.3). The bacterial luciferase oxidizes decanal (and some homologous long-chain aldehydes) with the energy transfer from $FMNH_2$ and produces blue–green light with an absorption maximum at 490 nm. The firefly enzyme couples the oxidation of luciferin with the energy transfer from ATP and produces yellow–green light with a pH-dependent absorption maximum. The bacterial and firefly luciferase systems present respective advantages and disadvantages associated with inherent differences in substrate profiles and enzyme structures. Recent developments and improvements on the two luciferase systems have rendered the two systems almost equally amenable to a wide variety of applications. To utilize the reporter systems to their full capacity, however, one has to make a judicious choice based on an understanding of their respective characteristics.

This section describes the bacterial luciferase (Lux) in the first part and the firefly luciferase (Luc) in the second part.

Bacterial Luciferases
[Alkanal Monooxygenase (FMN-Linked), EC 1.14.14.3]

The luciferases from marine bacteria such as *Vibrio harveyi* (Vh) and *Vibrio fischeri* (Vf) are heterodimeric enzymes (40-kDa α subunit, 36-kDa β subunit). Many of the physicochemical and genetic properties of the two enzymes are practically identical. However, characteristic differences exist and, unless specified, the properties of *V. harveyi* luciferase will be described as the prime example.

1. FUNCTIONS (LUX)

a. Optimal conditions for reaction, detection, and assays

The optimal pH for Vh luciferase is between pH 6.5 and 7.0 in Tris–malate (25 mM) or other buffers (1). [The Vf luciferase has an optimal pH of 6.2 in either Tris–acetate or Na,K-phosphate buffer (2).] The luciferase assay is usually performed at room temperature.

The assay for bacterial luciferase and the detection of light emission can be carried out in a number of ways. The *in vivo* assay has become particularly important and is described below together with the standard *in vitro* assays. Some notable characteristics of bacterial luciferases are as follows. First, the *in vitro* luciferase assay is based on a single turnover of the enzyme. Unless special conditions are provided for the steady supply of $FMNH_2$, light emission rises rapidly to a maximum and then decays slowly. The maximum light emission corresponds to the amount of enzyme in the form of a stable enzyme–flavin–oxygen–aldehyde intermediate. The decay of the intermediate and luminescence follows the first-order kinetics with half-times between 1 and 20 sec at room temperature. Second, $FMNH_2$ is nonenzymatically oxidized within 1 sec at 20°C, a rate faster than the enzymatic reaction. Therefore, most *in vitro* assays involve rapid mixing of the luciferase and the substrates. Third, $FMNH_2$ is not available for the reaction in most eukaryotic cells. Thus unlike bacterial systems, eukaryotic cells need to be disrupted for the detection of luciferase activity. This means that bacterial luciferase is more suitable in bacteria than in eukaryotic cells as *in vivo* reporter, although yeast has been shown to generate low levels of luminescence on the addition of aldehyde.

i. *In vivo* assays

(a) Standard assay conditions and unit definition. Bacterial cells are transferred to and resuspended in ice-cold *lux buffer* [50 mM Na-P (pH 7.0), 50 mM 2-MSH, and 2% BSA] to an A_{595} of 1. Aliquots of 10–100 μl are taken, mixed with 400 μl lux buffer, and placed in a luminometer (3). The decanal substrate, prepared as a sonicated suspension in lux buffer (or in water) at 0.01% (v/v, ~6.4 mM), is injected to a final concentration of 40 μM into the sample in the luminometer. The total light produced is recorded during the first 10 sec after injection of the substrate.

Unit definition: The *in vivo* assay is measured as *light units (LU) per milliliter of cells* at A_{595} of 1.0. For a sensitive instrument, 1 LU should range between 5×10^9 and 10^{10} quanta/sec as referenced to the light standard of Hastings and Weber (4).

(b) Luxdot assay. An alternative assay for the cells from the liquid culture is to deposit the cells as "dots" on a membrane filter (NC or nylon) in a regular dot-blot (or slot-blot) hybridization apparatus. The substrate decanal is supplied by capillary action from underneath the blot filter using a sandwiched, equal-

sized piece of filter paper (e.g., Whatman 3MM) wetted with a 0.5% decanal suspension (5).

(c) Solid culture assay. Bacterial or eukaryotic cells that express luciferase genes on solid culture can be screened in Petri or culture plates. The substrate decanal is conveniently introduced in the form of an aldehyde vapor by smearing a small amount onto the inside lid of the cover plate. After 5–10 min, the plates can be taken to a dark room for photographic detection.

ii. *In vitro* assays

(a) Standard (FMNH$_2$ injection) assay and unit definition. Aliquots (500 µl) of the resuspended cells (as prepared in the *in vivo* assay) are sonicated and the extracts are centrifuged. Aliquots of 1 to 100 µl of the extracts are mixed with 400 µl lux buffer, placed in a luminometer, and the luciferase reaction is started by injecting 500 µl of FMNH$_2$ in tricine buffer (200 mM, pH 7.0) and 10 µl of the diluted decanal (40 µM final concentration). The height of the light peak produced during the first 10 sec is taken as the luciferase activity.

Unit definition: The *in vitro activity* is given as $LU/\mu g$ total protein in the extracts.

(b) Dithionite assay. An alternative form of the *in vitro* assay is based on *in situ* generation of FMNH$_2$ from FMN (25 µM) using sodium dithionite (Na$_2$S$_2$O$_4$) (6). This assay is especially useful when low FMNH$_2$ concentrations (<10 µM) are employed, for example, in K_m determinations or for mutant luciferases having lower aldehyde-binding affinities. Note that high concentrations of aldehyde inhibit the standard FMNH$_2$ injection assay but not the dithionite assay. Since the dithionite also removes the O$_2$ from the solution, the reaction should be initiated by injection of an air-equilibrated sonicated suspension of aldehyde.

iii. **Detection of light emission.** The light emission from bacterial or firefly luciferase assays can be detected and measured at differing levels of sensitivity by photometric, photographic, and/or video imaging methods.

(a) Photometry. The luminometer is the most common instrument used for photometric detection, although the standard liquid scintillation counter may also be used for some assays not requiring a high sensitivity. Commercial luminometers are based on two different types of photon-processing designs: "photon counting" luminometers (e.g., Hamamatsu Argus-50) that count individual photons and "direct current" luminometers (e.g., Turner Designs Model 20) that measure electrical current produced by, and proportional to, the photon flux passing through the photomultiplier tubes. Both designs allow the measurement of ultralow light emission, but the sensitivity, reproducibility, and configuration of the mixing chamber differ from one instrument to another, and this necessitates careful calibration of the instruments with a standard luminating substance. One typical calibration method relies on chemiluminescence produced by the luminol reaction, i.e., the reaction of 3-aminophthalhydrazide with H$_2$O$_2$ to produce light (7).

A luciferase molecule gives rise to about one photon in the luciferase reaction. With a sensitive luminometer, 3×10^5 luciferase molecules (or 0.03 pg) can be detected, a level of expression often obtainable in a single cell. This sensitivity is several orders of magnitude higher than that of any non-light-producing enzyme reaction. Note that an initial burst (or flash) of light can be detected only with apparatus equipped with a sample autoinjector.

(b) Photographic detection and/or video imaging. Light emission can be detected using Polaroid photography or exposure to X-ray films. The photographic detection method is especially useful for keeping visual records. The detection of low level emission requires the use of an image intensifier; otherwise, the photographic detection may take more than 3 days of continuous exposure even with ISO 1600 film.

Perhaps the most advanced method of light detection is video imaging based on a photon-counting image processor and a microscope attached to a video camera. This imaging method allows real-time analysis of the spatially and/or temporally regulated gene expression *in vivo*.

b. Kinetic parameters

Due to the substrate aldehyde inhibition of the luciferase, the enzymatic reactions cannot be adequately described by simple Michaelis–Menten kinetics (8). The K_m for $FMNH_2$ with dodecanal as the acceptor substrate is 0.3 μM at pH 7 (50 mM Na-P) and 23°C (9). The K_d values for $FMNH_2$ with *n*-octanal and *n*-decanal as substrates are 1.6 and 0.9 μM, respectively (10). The K_d for FMN is ~0.1 mM in the binary complex (E·FMN) and ~3 μM is the ternary complex (E·FMN·myristic acid) (11).

The $K_{m,app}$ for *n*-decanal is ~1.5 μM in Bis–Tris (pH 7.0) or phosphate buffer (8). In 20 mM phosphate buffer (pH 7.0) containing 0.2% BSA, the $K_{m,app}$ for decanal increases to 9 μM. Under the same assay conditions, V is about 260 LU.

c. Activators and inhibitors

i. Activators. Divalent anions such as sulfate and phosphate act as activators for luciferase, giving a 10- to 15-fold increase in light emission when neutral flavin analogs are used as the substrate (11). Stimulation is not observed with any flavin that has a negatively charged side chain.

Salts such as NaCl and KCl or monovalent anions such as Cl^- and COO^- have very little effect on luciferase activity.

ii. Inhibitors and inactivators. *Phosphate* is a competitive inhibitor (K_i = 0.22 M) against $FMNH_2$.

Amines: 2,2-Diphenylpropylamine is a competitive inhibitor (K_i = 0.1 mM) with respect to aldehyde (decanal) binding while it enhances the binding of $FMNH_2$ to the enzyme about fourfold (12). 2,2-Diphenylpropylamine binds more tightly to the enzyme · $FMNH_2$ or enzyme · phosphate complex than to the enzyme alone (13).

Anesthetics: Many anesthetic compounds inhibit luciferase by competitively binding to the substrate (aldehyde)-binding site. However, the strength of competitive inhibition has a poor correlation with the strength of general anesthetics (14).

Heat: The $\alpha\beta$ heterodimeric bacterial luciferases display 50% loss of activity within 8 min at 45°C in 0.1 M phosphate buffer (pH 7.0) containing 1 mM EDTA and 0.1% BSA but not substrates (2).

d. Substrate specificities

i. Aldehyde homologs. The R (in R-CHO) is an aliphatic moiety of at least 7 carbon atoms. Although *n*-decanal is the commonly used substrate, tetradecanal (14 carbons) is believed to be the natural substrate *in vivo* (15). The relative *in vitro* activities (in brackets) for aldehydes in 20 mM Bis–Tris (pH 7.1) are in the order of *n*-dodecanal [1] < *n*-octanal [2] < *n*-decanal [15] (1). These substrate differences are partly due to secondary reversible binding of the substrates and consequent substrate inhibition (see below): *n*-dodecanal (K_m = 0.2 μM, K_d = 0.4 μM), *n*-decanal (K_m = 1.1 μM, K_d = 3.3 μM), and *n*-octanal (K_m = 5.0 μM, K_d = 14 μM).

Externally added fatty aldehyde substrates can instantly penetrate living cells and become available for luciferase reaction. This property enables one to follow nondestructively and nonintrusibly the activity of the luciferase *in vivo*.

ii. Substrate inhibition. Bacterial luciferase is subject to aldehyde substrate inhibition, which varies according to the aldehyde chain length, buffer composition, concentration, and other factors (8). In 0.2 M phosphate (pH 7.0), luciferase is much less sensitive to aldehyde inhibition than in low phosphate (~20 mM) or cationic buffers such as Bis–Tris. The addition of BSA protects the enzyme from aldehyde inhibition. The inhibition is due in part to the formation of the E · aldehyde complex that cannot bind $FMNH_2$ until the aldehyde dissociates from the enzyme and in part to the formation of the E · $FMNH_2$ · aldehyde ternary complex which reacts with O_2 at a rate 100 times slower than E · $FMNH_2$ (16).

iii. Flavin specificity. Bacterial luciferase is highly specific for reduced flavin mononucleotide ($FMNH_2$). Alteration of the flavin ring and of side-chain groups or removal of the phosphate affects the stability of the intermediates and gives varying degrees of activity decrease (11). N^5-methyl- and N^5-ethyl-1,5-dihydroflavin mononucleotides show no bioluminescence activity with the enzyme. A negatively charged group, e.g., carboxyl or more preferably phosphate, is required for the highest activity. Inorganic anions can supplement neutral flavins and stimulate the activity, but they do not affect the binding of the neutral flavins.

$FMNH_2$ appears to be sufficiently abundant in most bacteria but not in eukaryotic cells. In both *E. coli* and *Vibrio* species, the oxidized flavin (FMN) is efficiently reduced by NAD(P)H:FMN oxidoreductase [FMN + NAD(P)H → $FMNH_2$ + NAD(P)] and is continuously available to cytoplasmic luciferase.

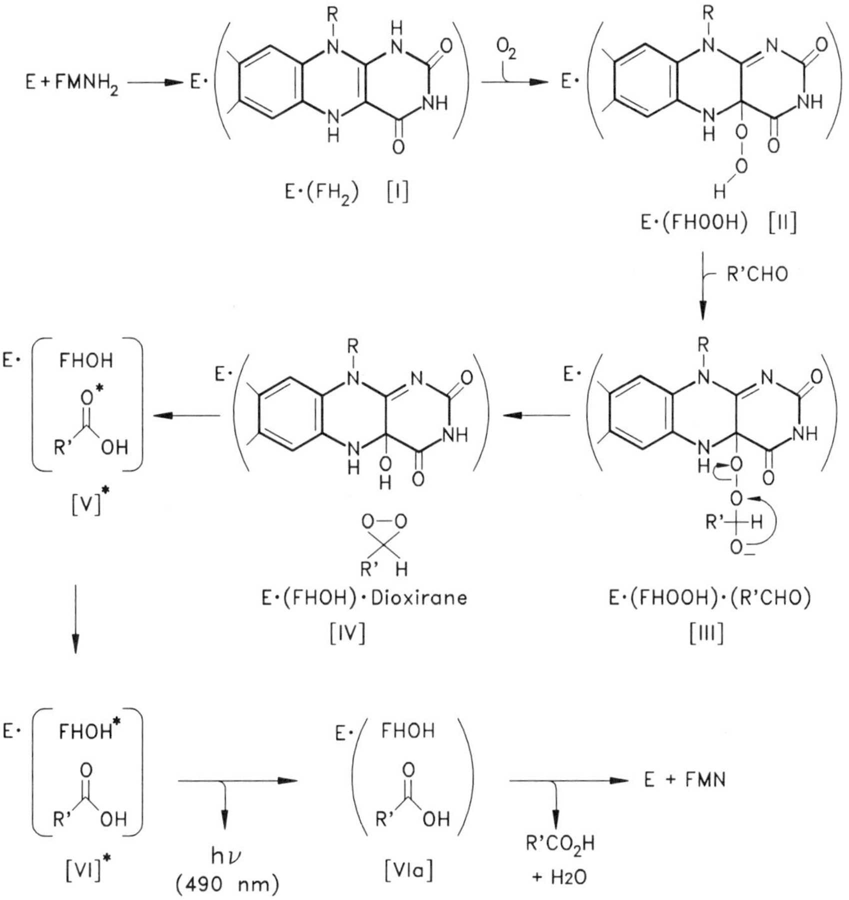

FIGURE 8.9 A scheme of the bacterial luciferase-catalyzed reaction pathway. The FMN part is abbreviated as F in the scheme and R is -$CH_2(CHOH)_3CH_2PO_4^{2-}$. Intermediate II is the C4a-hydroperoxyflavin (FHOOH); FHOH is C4a-hydroxyflavin.

e. Reaction mechanism of bacterial luciferase

Bacterial luciferases catalyze the reaction of $FMNH_2$, O_2, and an aliphatic aldehyde to yield the carboxylic acid, FMN, H_2O, and blue–green light (see Scheme 8.3). According to the mechanism proposed by Raushel and Baldwin (17), the light-producing reaction occurs via multiple steps involving at least five intermediates (Fig. 8.9). In the first step, the enzyme-bound $FMNH_2$ [I] reacts with O_2 to form a 4a-hydroperoxyflavin [II] (18). This intermediate can be isolated by chromatography in a relatively stable form under certain conditions (9,19). Data with site-specifically mutagenized luciferases suggest that two forms of 4a-hydroperoxyflavin, one active and the other essentially inactive in bioluminescence, can be generated by a single luciferase species (20). 4a-Hydroperoxyflavins can emit light nonenzymatically in the presence of aldehydes, albeit with low efficiency (21). Subsequent reaction of the intermediate [II] with an aldehyde gives rise to an unstable tetrahedral intermediate peroxyhemiacetal [III]. These two steps do not differ from those of other mechanisms previously proposed. The novel feature of the present mechanism is that the tetrahedral intermediate is converted to the flavin C4a-hydroxide and dioxirane [IV] rather than undergoing a Baeyer–Villager rearrangement with the removal of the α proton. The dioxirane then undergoes homolytic cleavage to yield the oxygen diradical which rearranges to form the carboxylic acid in either the triplet or singlet state [V]. The high energy intermediate [V] is presumed to be the primary emitter which transfers its energy to a suitable fluorophore such as C4a-hydroxyflavin [VI] and other secondary emitters. The "fluorescent transient" intermediate has a fluorescence quantum yield of ~ 0.3 and a lifetime of ~ 10 nsec (22).

Certain proteins have been known to function as secondary emitters, affecting the wavelength and efficiency of the light emission. For example, a yellow fluorescent protein (YFP) of a *V. fischeri* strain accepts excitation energy from a luciferase-bound intermediate and emits yellow light (23). In contrast, the lumazine protein (21 kDa), a fluorescent protein from *Photobacterium phosphoreum* (24), causes an apparent blue shift of the light produced by the luciferase reaction. The absorption maximum shifts from 496 to 475 nm in the presence of the lumazine protein.

2. STRUCTURE (LUX)

i. Primary and secondary structures. The luciferase from *V. harveyi* is a 76.5-kDa heterodimer composed of α (or LuxA, *luxA* gene product) and β (or LuxB, *luxB* gene product) subunits. The enzyme contains a single active center residing on the α subunit. The α subunit consists of 355 amino acid residues (M_r 40,108) (25). The N-terminal residue is Met encoded by the initiation codon. The C-terminal residue is Gln. The predicted secondary structure of the α subunit is composed of 34% α helix and 12% β sheet.

The β subunit consists of 324 amino acid residues (M_r 36,349) (26). The N- and C-terminal residues are Met and Ser, respectively. The β subunit is predicted to have 47% α helix and 27% β sheet. The $\alpha\beta$-dimer exhibits 28% α helix and 14% β sheet; these values are more in keeping with those of the α subunit. The α and β subunits share 32% sequence identity, including the first four identical amino acid residues.

The LuxA (355 amino acids) and LuxB (327 amino acids) of Vf luciferase share 63 and 52% sequence identities with their respective counterparts of the Vh enzyme (27).

ii. Subunit requirements and activity. The α or β subunit alone is enzymatically inactive and requires the presence of the other subunit for full activity. The α and β subunits can be purified separately by chromatography in the presence of 5 M urea and then renatured on elimination of the denaturing agent. Within

24–48 hr, the two renatured subunits can form a functional enzyme complex having up to 80% of the activity of the native complex. The two subunits simultaneously expressed in E. coli from individually cloned luxA and luxB transcriptional units also form a fully functional enzyme complex (28). Genetically engineered single αβ-fusion proteins are fully active as a monomer and, depending on fusion constructs, may be highly sensitive to temperature.

Experiments with mutant luciferases (29) or complex enzymes whose subunits have been exchanged between the wild-type and chemically modified, inactivated subunits (2,30) suggest that catalytic residues are exclusively located in the α subunit. Truncation of ~10 C-terminal residues of the β subunit has little effect on bioluminescence or stability of the dimeric structure per se, but it gives a temperature-sensitive phenotype (32). However, replacement of βHis-82 to Ala, Asp, or Lys results in a decrease of enzyme efficiency by 2–3 orders of magnitude (33), suggesting that the β subunit is essential for the bioluminescent reaction. The β subunit is required for active conformation and thermal stability of the α subunit. It also contributes to the interaction of the enzyme with $FMNH_2$.

iii. **Tertiary structure.** The Vh luciferase has an overall dimension of ~75 Å × 45 Å × 40 Å (31). The α and β subunits display very similar folding patterns, both assuming a single domain composed of an eight-stranded β/α barrel motif. The two subunits associate through extensive but flat surface contacts. The α subunit contains a "large" internal cavity which is believed to be the active site.

Bacterial luciferases are exquisitely sensitive to inactivation by proteases. The protease labile region is on the α (Vh) subunit, in particular the disordered loop (Phe^{272}–Thr^{288}). The β subunit, which has a shorter loop without the α sequence, is resistant to proteases. Binding of FMN, $FMNH_2$, and/or aldehyde renders the enzyme insensitive to proteolysis. Binding of the substrates also protects the otherwise reactive SH group (αCys-106) from modification by NEM or IAA. The substrate binding apparently induces an altered conformation in which the loop region is shielded from protease attacks.

αHis-44 and αHis-45 are essential residues and are located in the putative active site. αTrp-194 and αTrp-250, which are also located in the active site, directly interact with the bound flavin. The replacement of His-44 and His-45 by Ala, Asp, or Lys results in a 10^4- to 10^7-fold reduction in bioluminescence (20). Mutation of Trp-194 or Trp-250 to Phe results in greatly reduced bioluminescence activity and decreased affinity for flavin. C106V (34) but not C106S or C106A mutation results in a substantial decrease of enzyme activity.

The N-terminal hydrophobic domain of the α subunit is indispensable for luciferase function. The deletion of two N-terminal amino acids reduces luciferase activity by 95%. The removal of four N-terminal amino acids results in a complete loss of activity.

iv. **Physicochemical properties and purification.** Vh luciferase has an ε_{280} of 7.4×10^7 M^{-1} cm^{-1}. Bacterial luciferases are, unlike firefly luciferase, highly soluble enzymes.

Bacterial luciferase can be purified by various conventional chromatographic techniques (6) and by using a series of HPLC columns (7). One of the new purification techniques relies on an affinity matrix composed of a luciferase inhibitor, 2,2-diphenylpropylamine, conjugated to Sepharose (13,35). The affinity chromatography procedure is reported to give a routine yield (60%) of 600–700 mg of over 90% pure luciferase from a 500 g batch of cell paste.

3. GENETICS (*LUX*)

i. Gene structure and organization: *lux* operon. The enzyme luciferase is the key component in the phenomenon of bioluminescence detected in many marine bacteria such as *V. harveyi* (also previously called *Beneckea harveyi*, a planktonic bacterium), *V. fischeri* (previously called *Photobacterium fischeri*, a light organ symbiont of the fish *Monocentris japonicus*) (36,37), and *P. phosphoreum* (38). In these microorganisms, the structural genes coding for the luciferase are found within a cluster of genes forming the *lux* operon (39). In *V. harveyi*, the *lux* operon consists of seven structural genes which are organized as *luxCDABEGH*. It encodes a polycistronic mRNA (>8 kb) containing 5′-noncoding 26 bases. The 3′ end of *luxH* is followed by a GC-rich hairpin, a classical ρ-independent termination signal. The *lux* operon is regulated by a transcriptional activator, LuxR (see below).

Two of the seven structural genes, *luxA* and *luxB*, encode the luciferase α and β subunits, respectively. The other three genes, *luxC* (fatty acid reductase, 55 kDa), *luxD* (acyltransferase, 305 amino acids, 33 kDa), and *luxE* (fatty acid synthetase, 42 kDa) encode the enzymes that, as a large multienzyme *fatty acid reductase complex,* are involved in the biosynthesis of myristic (tetradecanoic) acid and its reduction to tetradecanal, the natural substrate for luciferase (15,39). A dark mutant (M17) of *V. harveyi* can produce a bright luminescence when supplemented with myristic acid. The exact functions of *luxG* capable of encoding a 26-kDa protein (233 amino acids) and of *luxH* capable of encoding a 25.3-kDa protein (230 amino acids) are not known. Based on sequence similarities with other known enzymes, LuxG and LuxH are presumed to be implicated in the production of $FMNH_2$ and in the synthesis of riboflavin, respectively.

The *lux* operon of *V. fischeri* is organized in a manner similar to that of *V. harveyi* except for a few differences. Notably, the Vf *lux* operon (*luxICDABEG*) lacks a gene corresponding to the Vh *luxH*, while Vf has a new structural gene *luxI* which has no counterpart in Vh *lux* operon. The *luxI* gene is essential for autoinducer biosynthesis (see below). The Vf *luxG* is immediately followed by a hairpin loop that functions as a bidirectional transcriptional terminator (40).

ii. *luxA* and *luxB* genes. The *luxA* and *luxB* genes of *V. harveyi* have been cloned, sequenced, and expressed in *E. coli* (25,26,41–43). The luciferase genes of *V. fischeri* (27) and *P. phosphoreum* (38) have also been sequenced.

The *luxA* and *luxB* genes of *V. harveyi* are contained within a 4.0-kbp *Hind*III genomic fragment and are immediately adjacent to each other (44).

The *luxA* gene (355 codons) is contained in the 1.85-kbp *Eco*RI fragment. There are 34 bp separating the 5′ end (AUG) of *luxA* from *luxD*. The 3′ end of *luxA* is followed by a 26-bp intergenic region to the start (AUG) of the *luxB* gene. The *Eco*RI fragment also contains the N-terminal 13 codons of the *luxB* structural gene, while most of the *luxB* gene is found in the 1.30-kbp *Eco*RI/ *Bgl*II fragment.

The *luxB* gene comprises 324 codons and, like the *luxA*, terminates with UAA. The stop codon is followed by a 27-bp spacer before the start of *luxE*. A 20-bp region of a potential stem–loop structure is located 3′ to the *luxB* ORF and that includes the 5′-proximal 9 bases of *luxE* (32).

Thus the *lux* gene map of *V. harveyi* is *lux*CD-(34 nt)-*A*-(26 nt)-*B*-(27 nt)- *E*-(1 nt)-*G*-(22 nt)-*H*.

iii. *luxAB* fusion genes. The *luxA* and *luxB* genes have been converted to a single ORF by deleting the intergenic spacer and site mutagenically converting the *luxA* stop codon TAA to a Leu codon CTG (45). Other full-length (A-B) fusion proteins containing 10 amino acid linkers have been produced in *E. coli* under the transcriptional control of phage T7 promoter (46). The product is fully active both as a monomer and as a dimer at 23°C. However, it is highly sensitive to high temperatures *in vivo*, exhibiting merely 0.002% of the wild-type luciferase activity at 37°C. Note that the heterodimeric forms of bacterial luciferases are stable at 37°C and higher temperatures. An active LuxAB fusion protein has also been obtained by fusing the *luxA* and *luxB* ORFs with an *Xba*I linker encoding only 2 amino acid residues (Ser–Arg) (47). This fused gene has been expressed in *E. coli*, yeast, and plant cells.

iv. Gene regulation. In *V. fischeri*, two regulatory genes (*luxR* and *luxI*) have been found immediately upstream of the *lux* structural genes (48,49). LuxI (193 amino acids) is required for the synthesis of an autoinducer that controls the expression of bacterial luminescence. The gene *luxR*, which is located 218 bp upstream from the *luxI*, encodes a transcriptional activator (LuxR, 250 amino acids) which is itself activated by the binding of the autoinducer. The *luxI* is separated from the adjacent *luxC* gene by 52 bp and forms the same rightward operon (or operon R) *luxICDABEG*, whereas the *luxR* gene forms a separate leftward operon (or operon L).

In *V. harveyi*, a regulatory region, also designated *luxR*, has been identified downstream of the *lux* operon, but it has no sequence identity with Vf *luxR*. Vh LuxR binds to an upstream region of the Vh *lux* operon which is highly AT rich without any extended ORF and functions as a transactivator in an autoinducer-independent manner. Vh genes homologous to the Vf *luxI* have not been identified, despite the similarity of autoinducer molecules (see below) and of the mechanism of *lux* operon induction.

The intensity of light emission in most luminescent marine bacteria and in other bacteria harboring the cloned *luxA* and *luxB* genes depends on a number of parameters, including the cAMP system and, in particular, the density of the cell suspension (43). In fact, it is not the cell density per se that directly influences the luminescence but it is the accumulation of an endogenously synthesized and secreted autoinducer, N-(3-oxohexanoyl)-L-homoserine lactone for *V. fischeri* (50) and N-(3-hydroxybutanoyl)-L-homoserine lactone for *V. harveyi* (51). The maximum luminescence of a bacterial culture is attained at an autoinducer concentration of ~5 μM (or 1 $\mu g/ml$), resulting in a 5000-fold stimulation compared with the basal level luminescence. Homoserine lactones carrying different, strain-specific homologous fatty acids apparently constitute a family of signal transductants that control certain types of gene expression in bacteria (39). The "autoinducer" signal–response system of *V. harveyi* is a two-component system consisting of *luxLMN* and *luxPQ*.

Binding of autoinducer to LuxR (at the N-terminal domain) enables the regulatory protein (C-terminal domain) to interact with the operator region between the *luxICDABEG* and *luxR* operons. In the absence of autoinducer, the N-terminal domain inhibits transcriptional activation by the C-terminal domain. The autoinducer-mediated gene induction results in increased syntheses of LuxI and consequently autoinducer, leading to a positive feedback loop. The same LuxR–autoinducer complex apparently exercises a negative control on the expression of *luxR* most likely at the translational level, thereby modulating (or autoregulating) the otherwise runaway activity of operon R (52). An operator sequence, also called *lux* box, consisting of a 20-bp inverted repeat (ACCTGTAGGA| TCGTACAGGT) is centered at -40 bp from the start of *luxI* transcription.

4. APPLICATIONS

i. **General considerations.** Luciferase (and its gene) is proving to be a powerful molecular reporter system having unparalleled sensitivity. Depending on the applications, the luciferase system offers several obvious advantages over other useful reporter systems: (a) most prokaryotic and eukaryotic cellular systems do not contain any enzymatic activities equivalent to or competing with the luciferase system, thus making the data interpretation straightforward; (b) the sensitivity reaches 0.03 pg (or 3×10^5 luciferase molecules), which amounts to 1000 times higher sensitivity than that of, for example, CAT reporter; (c) in addition to the precision, the luciferase assay can be performed in a broad linear range (3 orders of magnitude); and (d) compared with other reporter systems, the luciferase assay is faster, less expensive, and uses no radioactivity. Furthermore, the luciferase assay provides a noninvasive *in situ* detection of gene expression.

Both firefly luciferase (see below) and bacterial luciferase are sensitive reporters of gene expression suitable for the study of regulatory sequences (promoter, enhancer, and terminator) of cellular differentiation and morphogenesis, and of responses to environmental and developmental changes (53,54). The two reporter systems have been used in bacteria, yeast, fungi, insects, plants, and mammalian cells. As far as *in vitro* assays are concerned, both reporter systems may be considered equally useful. For *in vivo* applications, however, one system may turn out to be more suitable than the other due to the strengths and weaknesses inherent to each reporter system.

(a) *Substrate diffusibility.* Decanal, the substrate for bacterial enzyme, is freely diffusible across cell membranes, whereas luciferin is not. In addition, enzyme-saturating levels of luciferin are somewhat toxic *in vivo*. However, the development of esterified luciferins which can readily cross the cell membranes and be deesterified by cellular esterases has largely removed the disadvantages of insect luciferase. On the other hand, the cosubstrate ($FMNH_2$) for bacterial enzyme is not available for reaction in most eukaryotic cells, a potential handicap for using bacterial luciferases as an *in vivo* reporter in eukaryotes. With an appropriate choice of substrates or reaction conditions, firefly luciferase can be readily used as a reporter in prokaryotes as well as in eukaryotes. [*Note*: The restrictions imposed by the substrate availability are relieved if *green fluorescent protein* (originally isolated from the jellyfish *Aequorea victoria*) or its blue- or red-shifted mutants are used as an *in vivo* reporter (79–81).]

(b) *Enzyme structure and gene fusion.* Firefly luciferase is a single polypeptide, giving advantages for gene manipulations over the two-subunit bacterial enzymes. The creation of bacterial *luxA–luxB* fusion genes expressing a functional single polypeptide should make the bacterial enzymes equally convenient to use. However, currently available *luxA–luxB* fusion constructs tend to produce thermally unstable luciferases, obliging *in vivo* expressions to be performed at below 30°C. Such protein instability may pose no problems for applications in some plants but it would be a serious drawback in mammalian cells. In contrast, two-subunit bacterial enzymes are, like firefly luciferase, stable at 37°C in most circumstances.

The firefly enzyme might have its own problem by the fact that it is localized *in vivo* (yeast, mammalian, and plant cells as in the firefly lantern) in small vesicular structures called peroxisomes (55). This is due to the presence of a *peroxisomal translocation signal* located at the C-terminal domain of the molecule. The peroxisomal localization may present an additional membrane barrier for *in situ* detection of the enzyme, although it offers a particular advantage for studying protein transport or targeting to peroxisomes. Removal of the peroxisomal translocation signal has been shown to provide an alternative system in which the "modified" firefly luciferase is expressed as a cytoplasmic enzyme like bacterial luciferases.

ii. **Specific considerations: Construction of *lux* fusion genes.** The following information, which is largely based on the description of Olsson *et al.* (3), should serve as a useful guide in the construction of *luxA* fusion genes.

(a) Introduction of a short coding sequence upstream of the natural RBS of *luxA* is possible and may even positively affect *luxA* expression due to translational coupling.

(b) The absence of a properly spaced Shine–Dalgarno sequence causes a significant reduction in *luxA* translation.

(c) Short sequences (up to 17 amino acids) can be added to the N-terminal part of the intact *luxA* without altering the luciferase activity. The N-terminal addition may, however, have varying effects on the stability of the fusion proteins.

(d) Deletion of only two N-terminal amino acids (Met–Lys) of LuxA can reduce the luciferase activity by 95% even in the presence of heterologous N-terminal amino acid residues. Removal of four amino acids (Met–Lys–Phe–Gly) results in an almost complete loss of activity.

(e) The addition of 24 amino acids to the C terminus of the α subunit does not alter the enzyme activity. The removal of 9 amino acids from the C terminus of the β subunit has little effect on the bioluminescence reaction or stability of the dimeric structure per se, but it does play an important role in proper folding of the protein (32).

(f) Insertion of a foreign DNA in the *luxB* gene results in an inactive enzyme and loss of the Lux phenotype (56). However, insertion into *luxB* of a synthetic oligonucleotide containing four restriction sites produced an active β subunit (57). A cloning vector that contains the normal *luxA* and the mutated *luxB* allows a gene to be cloned into the MCS of the *luxB* and the recombinants to be screened easily by the Lux$^-$ phenotype.

Firefly Luciferase
[*Photinus*-Luciferin 4-Monooxygenase (ATP-Hydrolyzing), EC 1.13.12.7]

The luciferase from the North American firefly *P. pyralis* is a monomeric enzyme (62 kDa) consisting of 550 amino acid residues. The firefly luciferase produces light by the ATP-dependent oxidation of D(−)-luciferin (LH$_2$) (see Scheme 8.3). The reaction involves an enzyme-bound luciferyl adenylate intermediate. The peak light emission occurs at 562 nm (yellow–green, quantum yield of 0.88) in dianionic form between pH 7 and 8 or at 610 nm (red, quantum yield of 0.2) in monoanionic form at pH values below 7. The red shift also occurs in the presence of Zn^{2+} (2.3 mM) or Cd^{2+} (12 mM) (58).

1. FUNCTIONS (LUC)

a. Firefly luciferase assays and reaction conditions

The pH optimum is 7.8 in various buffer solutions (59). A typical *in vitro* assay consists of 100 μl enzyme solution (cell lysate, see below), 360 μl luciferase buffer [25 mM Gly–Gly (pH 7.8), 15 mM MgSO$_4$, 4 mM EGTA, 15 mM K-P (pH 7.8), and 1 mM DTT], and 2 mM ATP (60). The reaction is initiated by injecting 200 μl of 0.2 mM luciferin into the solution and the light output is measured for 20 sec at 25°C.

The cell lysate is prepared in 1% Triton X-100, 25 mM Gly–Gly (pH 7.8), 15 mM MgSO$_4$, 4 mM EGTA, and 1 mM DTT. Lysis by detergent is superior to lysis by freezing and thawing, which is detrimental to luciferase activity. Cell debris is pelleted by centrifugation in a microfuge for 5 min at 4°C and the supernatants are transferred and used as luciferase solution.

The *in vivo* assay of Luc is not as straightforward as the *in vitro* assay due to the membrane impermeability of the normal substrate luciferin (see below).

The maximum light output and the course of its decay are functions of multiple factors, especially the concentrations of enzyme and ATP, and, in the case of crude extracts, the presence of substrate-modifying activities. A typical firefly luciferase reaction produces an initial flash of light due to a first turnover of the enzyme. Then the light emission gradually decreases at a rate that corresponds to slow dissociation of the enzyme–product complex. The luciferase reaction under modified conditions with coenzyme A or other modulatory compounds (see below) produces light of "constant" intensity which gradually decreases after 30 sec. The light output is linearly proportional to the amount of Luc within 3 orders of magnitude. Detergents (e.g., Tween and Triton X-100) and PEG increase the light emission, apparently by promoting the dissociation of inhibitory products from the enzyme (61).

b. Substrate specificities

(i) Firefly luciferase requires D(−)-luciferin [(S)-4,5-dihydro-2-(6-hydroxy-2-benzothiazolyl)-4-thiazolecarboxylic acid] as substrate. The enzyme is inhibited by dehydroluciferin or L(+)-luciferin which reacts with ATP to liberate PP_i, but does not undergo further light-emitting reactions (58). Luciferin is not readily permeable to lipid membranes under the usual assay conditions of neutral pH where the carboxyl group is ionized. Depending on host cells, the following approaches may help alleviate the substrate limitation *in vivo*. Acidic buffers which keep the luciferin in its unionized form or neutral medium (pH 7.2) supplemented with permeabilizing agents (e.g., 1% Me_2SO) and/or detergents (e.g., 0.01% Tween-20) facilitate substrate entry into cells. Alternatively, and more effectively, luciferin derivatives may be employed as *in vivo* substrates. For instance, esterification with DMNPE [1-(4,5-dimethoxy-2-nitrophenyl)diazoethane] and formation of DMNPE "caged" luciferin allow rapid equilibrium of the luciferin ester in the cell (62). The luciferin ester is then converted to luciferin either by photolysis or by endogenous esterases, thus enabling steady measurement of luciferase activity.

(ii) Firefly luciferase has a specific requirement for ATP. An ATP concentration of 2 mM is optimal for usual assays in which there is a flash of light followed by a decline of light production. The biphasic time course of light production is switched to a monophasic constant light production (for >5 min) in the presence of several cytidine nucleotides, e.g., CTP, CDP, dCTP, dCDP, ddCTP, and periodate-oxidized CTP (oCTP) (63). Periodate-oxidized ATP (oATP) and oADP also give rise to monophasic light production. Note that the light production at low concentrations of ATP (<10 μM) is virtually constant and is not influenced by the modulatory compounds. CMP and cytidine are inhibitory to Luc or are ineffective in changing the flash pattern. CoA and related compounds also modify the time course of light production and enhance the signal up to fourfold. The activating effects of the compounds are attributed to a faster dissociation of the inhibitory product (oxyluciferin) from the enzyme and/or dissociation of ATP from an allosteric binding site of the enzyme. The effects of two activators are not additive.

The specificity for ATP makes it possible to perform Luc assays in the presence of other NTPs and NDPs. Indeed, this specificity allows the firefly luciferase to be used as a sensitive reagent for the ATP assay (64). In an elegant *in vivo* application, the ATP assay is exploited as a rapid method for distinguishing drug-sensitive and drug-resistant mycobacteria following transformation of the bacteria (clinical isolates) with luciferase reporter phages (65).

(iii) Micromolar concentrations of inorganic PP_i stabilize light emission by preventing the formation of the Luc–oxyluciferin complex (66). Higher concentrations of PP_i depress the signal by inhibiting the ATP binding to luciferase.

(iv) Certain dyes and anesthetics act as competitive inhibitors. The principle of substrate competition on the hydrophobic binding site of the enzyme has been exploited as a sensitive assay for amphipathic substances at subnanomolar concentrations (67).

Phenylbenzothiazole, which is a substrate analog, is a competitive inhibitor. When introduced *in vivo*, this compound protects the luciferase from proteolytic degradation and results in more than 10-fold increase in Luc activity (68).

2. Structure (Luc)

Firefly luciferase from *P. pyralis* is a 50-kDa single polypeptide (550 amino acids). A C-terminal tripeptide sequence, –Ser–Lys–Leu, serves as the peroxisomal translocation signal, and its removal abolishes import of the enzyme into peroxisomes. Luc is hardly soluble in water, and its solubilization requires some salt. In solvents of relatively low ionic strengths, the protein aggregates in a rapidly reversible manner as the solubility limit is approached (69).

The enzyme has two binding sites for ATP with K_m values of 0.11 and 0.02 μM (70). The loose ATP-binding site is responsible for the initial flash which is apparently inhibited by the product for further light emission. The tight ATP-binding site catalyzes the continuous low emission of light. Occupation of both sites is required for luciferase activity.

3. Genetics (Luc)

a. Gene structure and organization

The *luc* gene encoding luciferase in *P. pyralis* consists of seven exons. The six introns are all less than 60 bases in length. The *luc* gene cDNA has been cloned in λ vectors and expressed in *E. coli* (71,72).

The *luc* gene consists of 550 amino acid codons starting from nucleotide position 52 (cf. +1 is the start site of mRNA transcript). The *luc* cDNA contains four TATA(-like) boxes between the nucleotide positions −25 and −145. The cDNA also contains a presumed polyadenylation signal (AATAAA) at base 1813, which contributes to a higher level of expression in plant and animal cells.

b. luc gene fusions and expression

A number of *luc* reporter vectors have been constructed, usually with the MCS positioned immediately upstream of the *luc* gene (60,73). Typical examples include a pBluescript SK$^+$-derived pLUC plasmid retaining the advantage of βGal-based blue–white selection (74) and a mammalian expression vector pMAM$_{neo}$–LUC (Clontech). Insertion of a strong transcriptional terminator (e.g., from mouse c-*mos* oncogene) upstream from the *luc* gene substantially reduces the background transcription in NIH 3T3 cells, thereby allowing the study of weak promoters (75). The *luc* gene has been expressed under the control of various viral, bacterial, and eukaryotic promoters such as the phage λ P$_R$ promoter, SV40 early promoter, RSV-LTR, cauliflower mosaic virus (CaMV) 35S promoter, bacterial heat-shock protein 60 (HSP60) and *Drosophila* HSP70 promoters, prolactin promoter, and yeast ADH1 and Gal1-10 promoters.

Several modifications have also been introduced in the *luc* gene itself, resulting in an "improved" *luc* gene, designated as *luc*+ (76). Some notable modifications involve removal of the C-terminal peroxisomal translocation signal, of potential N-glycosylation sites, and of some internal restriction sites (e.g., *Cla*I, *Eco*RI, *Eco*RV, and *Xba*I) as well as nucleotide changes for better codon usage in mammalian cells. Incorporation of the *luc*+ gene in pSP−luc+ and pGL3 family vectors (Promega) enables expression of the luciferase in the cytoplasm. These Luc reporter vectors also offer the versatility of gene fusions and insertion of promoters and other regulatory elements by virtue of the conveniently located MCS.

Several gene fusions have been made successfully at the 5' ends of the *luc* gene and the truncated *luc* gene missing the first 6 amino acids. The fusions at or near the N terminus of Luc range from 7 amino acids of λ Cro protein (71) to a 30-kDa size-modified *S. aureus* protein A (77). Expression in *E. coli* of the Luc linked at its N terminus to an OmpA signal sequence (30 amino acids) produces a Luc (+9 extra amino acids) which is secreted into the periplasmic space (78). Deletions in the 5'-end (11 amino acids) and 3'-end (12 amino acids) coding regions result in a significant loss of Luc activity. A larger fusion (e.g., of CAT gene) at the 5' end of the *luc* gene also results in inactivation of the luciferase. Most C-terminal fusions result in disruption of the luciferase function. Luc has a half-life of 3 hr in mammalian cells, making it highly responsive in stably transfected cell lines.

References

1. Nicoli, M. Z., Meighen, E. A., and Hastings, J. W. (1974). *JBC* **249**, 2385–2392.
2. Meighen, E. A., Nicoli, M. Z., and Hastings, J. W. (1971). *Biochem.* **10**, 4069–4073.
3. Olsson, O., Koncz, C., and Szalay, A. A. (1988). *MGG* **215**, 1–9.
4. Hastings, J. W., and Weber, G. (1963). *J. Opt. Soc. Am.* **53**, 1410–1415.
5. Peabody, D. S., Andrews, C. L., Escudero, K. W., Devine, J. H., Baldwin, T. O., and Bear, D. G. (1989). *Gene* **75**, 289–296.
6. Hasting, J. W., Baldwin, T. O., and Nicoli, M. Z. (1978). *Meth. Enzy.* **57**, 135–152.
7. O'Kane, D. J., Ahmad, M., Matheson, I. B. C., and Lee, J. (1986). *Meth. Enzy.* **133**, 109–128.
8. Holzman, T. F., and Baldwin, T. O. (1983). *Biochem.* **22**, 2838–2846.
9. Tu, S.-C. (1986). *Meth. Enzy.* **133**, 128–139.

10. Chen, L. H., and Baldwin, T. O. (1989). *Biochem.* **28**, 2684–2689.
11. Meighen, E. A., and MacKenzie, R. E. (1973). *Biochem.* **12**, 1482–1491.
12. Holzman, T. F., and Baldwin, T. O. (1981). *Biochem.* **20**, 5524–5528.
13. Holzman, T. F., and Baldwin, T. O. (1982). *Biochem.* **21**, 6194–6201.
14. Curry, S., Lieb, W. R., and Franks, N. P. (1990). *Biochem.* **29**, 4641–4652.
15. Ulitzur, S., and Hastings, J. W. (1979). *PNAS* **76**, 265–267.
16. Francisco, W. A., Abu-Soud, H. M., Baldwin, T. O., and Raushel, F. M. (1993). *JBC* **268**, 24734–24741.
17. Raushel, F. M., and Baldwin, T. O. (1989). *BBRC* **164**, 1137–1142.
18. Ghisla, S., Hastings, J. W., Favaudon, V., and Lhoste, J.-M. (1978). *PNAS* **75**, 5860–5863.
19. Balny, C., and Hastings, J. W. (1975). *Biochem.* **14**, 4719–4723.
20. Xin, X., Xi, L., and Tu, S.-C. (1991). *Biochem.* **30**, 11255–11262.
21. Kemal, C., Chan, T. W., and Bruice, T. C. (1977). *PNAS* **74**, 405–409.
22. Lee, J., Wang, Y., and Gibson, B. G. (1991). *Biochem.* **30**, 6825–6835.
23. Baldwin, T. O., Treat, M. L., and Dauber, S. C. (1990). *Biochem.* **29**, 5509–5515.
24. O'Kane, D. J., Karle, V. A., and Lee, J. (1985). *Biochem.* **24**, 1461–1467.
25. Cohn, D. H., Mileham, A. J., Simon, M. I., Nealson, K. H., Rausch, S. K., Bonam, D., and Baldwin, T. O. (1985). *JBC* **260**, 6139–6146.
26. Johnston, T. C., Thompson, R. B., and Baldwin, T. O. (1986). *JBC* **261**, 4805–4811.
27. Foran, D. R., and Brown, W. M. (1988). *NAR* **16**, 777.
28. Gupta, S. C., O'Brien, D., and Hastings, J. W. (1985). *BBRC* **127**, 1007–1011.
29. Cline, T. W., and Hastings, J. W. (1972). *Biochem.* **11**, 3359–3370.
30. Cousineau, J., and Meighen, E. (1976). *Biochem.* **15**, 4992–5000.
31. Baldwin, T. O., Christopher, J. A., Raushel, F. M., Sinclair, J. F., Ziegler, M. M., Fisher, A. J., and Rayment, I. (1995). *Curr. Opin. Struct. Biol.* **5**, 798–809.
32. Sugihara, J., and Baldwin, T. O. (1988). *Biochem.* **27**, 2872–2880.
33. Xin, X., Xi, L., and Tu, S.-C. (1994). *Biochem.* **33**, 12194–12201.
34. Xi, L., Cho, K.-W., Herndon, M. E., and Tu, S.-C. (1990). *JBC* **265**, 4200–4203.
35. Baldwin, T. O., Holzman, T. F., Holzman, R. B., and Riddle, V. A. (1986). *Meth. Enzy.* **133**, 98–108.
36. Ziegler, M. M., and Baldwin, T. O. (1981). *Curr. Topics Bioenerg.* **12**, 65–113.
37. Nealson, K. H., and Hastings, J. W. (1979). *Microbiol. Rev.* **43**, 496–518.
38. Ferri, S. R., Soly, R. R., Szittner, R. B., and Meighen, E. A. (1991). *BBRC* **176**, 541–548.
39. Meighen, E. A. (1994). *Ann. Rev. Genetics* **28**, 117–139.
40. Swartzman, E., Kapoor, S., Graham, A. F., and Meighen, E. A. (1990). *J. Bacteriol.* **172**, 6797–6802.
41. Engebrecht, J., and Silverman, M. (1986). *Meth. Enzy.* **133**, 83–98.
42. Baldwin, T. O., Berends, T., Bunch, T. A., Holzman, T. F., Rausch, S. K., Shamansky, L., Treat, M. L., and Ziegler, M. M. (1984). *Biochem.* **23**, 3663–3667.
43. Belas, R., Mileham, A., Cohn, D., Hilmen, M., Simon, M., and Silverman, M. (1982). *Science* **218**, 791–793.
44. Engebrecht, J., Nealson, K., and Silverman, M. (1983). *Cell* **32**, 773–781.
45. Olsson, O., Escher, A., Sandberg, G., Schell, J., Koncz, C., and Szalay, A. A. (1989). *Gene* **81**, 335–347.
46. Escher, A., O'Kane, D. J., Lee, J., and Szalay, A. A. (1989). *PNAS* **86**, 6528–6532.
47. Kirchner, G., Roberts, J. L., Gustafson, G. D., and Ingolia, T. D. (1989). *Gene* **81**, 349–354.
48. Engebrecht, J., and Silverman, M. (1987). *NAR* **15**, 10455–10467.
49. Devine, J. H., Shadel, G. S., and Baldwin, T. O. (1989). *PNAS* **86**, 5688–5692.
50. Eberhard, A., Burlingame, A. L., Eberhard, C., Kenyon, G. L., Nealson, K. H., and Oppenheimer, N. J. (1981). *Biochem.* **20**, 2444–2449.
51. Cao, J.-G., and Meighen, E. A. (1989). *JBC* **264**, 21670–21676.
52. Engebrecht, J., and Silverman, M. (1986). *Gene. Eng.* **8**, 31–44.
53. Schauer, A., Ranes, M., Santamaria, R., Guijarro, J., Lawlor, E., Mendez, C., Chater, K., and Losick, R. (1988). *Science* **240**, 768–772.

54. Carmi, O. A., Stewart, G. S. A. B., Ulitzur, S., and Kuhn, J. (1987). *J. Bacteriol.* **169**, 2165–2170.
55. Keller, G.-A., Gould, S., DeLuca, M., and Subramani, S. (1987). *PNAS* **84**, 3264–3268.
56. Legocki, R. P., Legocki, M., Baldwin, T. O., and Szalay, A. A. (1986). *PNAS* **83**, 9080–9084.
57. Sevigny, P., and Gossard, F. (1990). *Gene* **93**, 143–146.
58. Seliger, H. H., and McElroy, W. D. (1964). *PNAS* **52**, 75–81.
59. Leach, F. R., and Webster, J. J. (1986). *Meth. Enzy.* **133**, 51–70.
60. Brasier, A. R., Tate, J. E., and Habener, J. F. (1989). *BioTechniques* **7**, 1116–1122.
61. Kricka, L. J., and DeLuca, M. (1982). *ABB* **217**, 674–681.
62. Yang, J., and Thomason, D. B. (1993). *BioTechniques* **15**, 848–850.
63. Ford, S. R., Hall, M. S., and Leach, F. R. (1992). *Anal. Biochem.* **204**, 283–291.
64. Lundin, A., and Thore, A. (1975). *Anal. Biochem.* **66**, 47–63.
65. Jacobs, W. R., Jr., Barletta, R. G., Udani, R., Chan, J., Kalkut, G., Sosne, G., Kieser, T., Sarkis, G. J., Hatfull, G. F., and Bloom, B. R. (1993). *Science* **260**, 819–822.
66. Gandelman, O., Allue, I., Bowers, K., and Cobbold, P. (1994). *J. Biolum. Chemilum.* **9**, 363–371.
67. Naderi, S., and Melchior, D. L. (1990). *Anal. Biochem.* **190**, 304–308.
68. Thompson, J. F., Hayes, L. S., and Lloyd, D. B. (1991). *Gene* **103**, 171–177.
69. Denburg, J. L., and McElroy, W. D. (1970). *Biochem.* **9**, 4619–4624.
70. DeLuca, M., and McElroy, W. D. (1984). *BBRC* **123**, 764–770.
71. de Wet, J. R., Wood, K. V., Helinski, D. R., and DeLuca, M. (1985). *PNAS* **82**, 7870–7873.
72. de Wet, J. R., Wood, K. V., DeLuca, M., Helinski, D. R., and Subramani, S. (1987). *Mol. Cell. Biol.* **7**, 725–737.
73. Nordeen, S. K. (1988). *BioTechniques* **6**, 454–458.
74. Caricasole, A., and Ward, A. (1993). *Gene* **124**, 139–140.
75. de Martin, R., Strasswimmer, J., and Philipson, L. (1993). *Gene* **124**, 137–138.
76. Sherf, B. A., and Wood, K. V. (1994). *Promega Notes* **49**, 14–21.
77. Subramani, S., and DeLuca, M. (1988). *Gen. Eng.* **10**, 75–89.
78. Lu, J.-R., Yang, J., and Jin, Z.-H. (1993). *Biotech. Lett.* **15**, 1111–1116.
79. Chalfie, M., Tu, Y., Euskirchen, G., Ward, W. W., and Prasher, D. C. (1994). *Science* **263**, 802–805.
80. Heim, R., Prasher, D. C., and Tsien, R. Y. (1994). *PNAS* **91**, 12501–12504.
81. Delagrave, S., Hawtin, R. E., Silva, C. M., Yang, M. M., and Youvan, D. C. (1995). *Bio/Technology* **13**, 151–154.

APPENDIX A
Important Molecular Biological Methods

I. DNA LABELING

Labeled DNAs play key roles in molecular diagnostics as well as in various molecular biological detection and analytical techniques such as gene mapping, screening of recombinant clones, and nucleotide sequencing. DNA can be labeled either with radionucleotides containing, for example, ^{31}P, ^{33}P, or ^{35}S (Table A.1), or with nonradionucleotides carrying ligands such as biotin, dioxigenin, and fluorescent compounds. Together with the development of sensitive chromogenic and in particular chemiluminescent substrates, nonisotopic labeling and nonisotopically labeled nucleotides are playing increasingly important roles in nucleic acid probe technology. Nevertheless, radioisotopic labeling remains the method of choice when sensitivity is the most important criteria.

Isotopic labeling of DNA (and RNA as well) at the 3' or 5' end is typically performed using polynucleotide kinase and [γ-^{32}P]ATP (refer to Section II, Chapter 5). Some 3'-end labeling can also be performed with DNA polymerase or terminal deoxynucleotidyltransferase and [α-^{32}P]dNTP or [α-^{32}P]ddNTP. Labeling by incorporation of isotopically or nonisotopically labeled nucleotides usually relies on DNA (or RNA) polymerases, although nonisotopically labeled oligonucleotides

TABLE A.1 Properties of Some Commonly Used Radioisotopes

Element	Isotope	Half-life ($t_{1/2}$)	Decay mode[b]	Energy (MeV)	Maximum range	Decay product (stable)
Phosphorus ($_{15}$P)	^{32}P	14.3 days	β^-	1.718	6 m (air) 8 mm (water)	$^{32}_{16}$S
	^{33}P	25.4 days	β^-	0.248	46 cm (air)	
Sulfur ($_{16}$S)	^{35}S	87.1 days	β^-	0.167	24 cm (air)	
Hydrogen ($_1$H)	^{3}H (tritium)	12.3 years	β^-	0.0179	4.7 mm (air)	$^3_2$He
Carbon ($_6$C)	^{14}C	5730 years	β^-	0.154	22 cm (air)	$^{14}_7$N
Chromium ($_{24}$Cr)	^{51}Cr	27.8 days	γ K	0.32 0.004		
Iodine ($_{53}$I)	^{125}I	60 days	γ X	0.035 0.027		
	^{131}I	8.1 days	β^- γ	0.606 0.364 0.637	165 cm (air)	$^{131}_{54}$Xe

Emission[a]

[a] One Curie (Ci) is equivalent to the amount of isotope undergoing 3.70×10^{10} DPS (nuclear disintegration per second) or 2.22×10^{12} DPM (disintegration per minute).
1 Ci = 3.70×10^{10} Bq (Becquerel), 1 Bq = 2.70×10^{-11} Ci
(*Note*: Specific activity refers to the amount of radioactivity per unit amount of substance.)
[b] Decay mode: β^-, negative β particles; γ, γ rays; K, K-electron capture (Auger electron); X, X-ray (internal electron conversion).

can be readily prepared by chemical synthesis. Focusing on DNA polymerase-catalyzed labeling for the purpose of preparing hybridization probes, the most commonly used methods include nick translation, random-primer labeling, and PCR labeling.

1. Nick Translation Labeling Method

Nick translation is one of the DNA-labeling techniques conventionally used to prepare hybridization probes. Nick translation utilizes combined activities of DNase I which introduces nicks (or single-strand breaks) and the 5′-(exo)nuclease and polymerase activities of *E. coli* DNA Pol I. While the 5′-nuclease activity of Pol I removes the nucleotides from the 5′-phosphoryl terminus, the polymerase activity carries out the sequential addition of nucleotides to the 3′-hydroxyl terminus, thus translocating the nick. When a highly radioactive nucleotide, e.g., [α-^{32}P]dATP, is included during the reaction, nick translation results in the uniform labeling of duplex molecules with a specific activity $>10^8$ cpm/μg DNA (1). Nick translation produces labeled DNA probes from both strands. Pol Ik, which lacks 5′-nuclease activity, cannot perform the nick translation, but it can carry out a strand displacement synthesis.

Nicks are randomly introduced in dsDNA using pancreatic DNase I. For optimal nick translation, the amount of DNase I (40–80 pg/μl) is adjusted to yield a probe size of ~500 nt and the reaction is carried out at 16°C for ~1 hr (2,3). If the reaction is carried out at higher temperatures, a considerable amount of "snapback" DNA may be generated due to Pol I copying newly synthesized strands. The conditions used for simultaneous enzymatic reactions with DNase I and Pol I are not really optimal. A better way to control the reaction is to separate the action of the two enzymes: first, DNase I and then polymerase (4). This approach is based on the following observations: (i) the DNase activity is about eightfold higher at 37°C than at 14°C and (ii) dilution of the DNase–DNA mixture greatly decreases the activity of the enzyme. Briefly, the DNA is digested with an appropriate amount of DNase I (cf. this needs to be titrated to give the desired fragment length) in NT (nick translation) buffer at 37°C for 15 min (or longer for larger amounts of DNA sample). The NT buffer consists of 50 mM Tris–Cl (pH 7.5), 10 mM MgCl$_2$, 0.1 mM DTT, and 50 μg/ml BSA. Subsequently, dNTPs and Pol I (~12 U) are added, and the mixture is incubated at 14°C until the polymerase reaction is completed, e.g., overnight or until a specific activity of 10^9 dpm/μg DNA is reached. Nonisotopic labeling of DNA can also be carried out by nick translation using, for example, biotin- or digoxigenin-labeled nucleotides (biotin-11-dUTP, biotin-7-dATP, or Dig-11-dUTP). Biotin-labeled probes can be detected using an avidin– or streptavidin–alkaline phosphatase conjugate (e.g., BlueGene system of BRL) or an avidin–peroxidase conjugate (e.g., of Vector Laboratory) to a sensitivity of 0.25 pg target DNA. Despite a number of advantages over radiolabeling, biotin-labeled probes tend to give higher backgrounds than ^{32}P-labeled probes. The size of biotinylated probes giving an optimal signal-to-noise ratio is estimated to be ~200 nt, with 10–30 biotinylated nucleotides per kilobase. Digoxigenin-labeled probes can be detected using anti-Dig Ab conjugated with alkaline phosphatase or peroxidase (e.g., Dig labeling system of BMB). When labeled at ~40 Dig-dUMP per kb, the Dig-labeled probes provide a sensitivity similar to that of the biotin-labeled system.

2. Random-Primer Labeling Method

The random-primer labeling method is an alternative to nick translation for producing uniformly labeled DNA. It is an efficient labeling method that is suitable for smaller sized DNA (down to 100 bp). DNA is first denatured by boiling to separate the double strands and is quickly quenched in ice. Random primers are then added and annealed to template DNA (optimally less than 25 ng/reaction) at room temperature. A DNA polymerase extends the primers at 37°C, generating uniformly labeled second-strand DNAs with a specific activity greater than 10^9 dpm/μg DNA. The technique, in its original form developed by Feinberg and Vogelstein (5,6), employed random hexamers which were prepared by DNase I digestion of calf thymus DNA. Original reactions were carried out at pH 6.6, which is not optimal for polymerase activity, mainly to minimize the deleterious effect of residual $5' \rightarrow 3'$-exonuclease activities present in the "apparent" Pol Ik and also in the hexamer primers. As the cloned Pol Ik became available, the random-primer labeling reaction can now be performed at pH 7.5, closer to the optimum pH of the polymerase (7). (*Note*: Pol Ik is preferred to T7 DNA Pol due to lower backgrounds in hybridization.) This increases the efficiency of radionucleotide incorporation, which now exceeds 65% in a shorter incubation time of 15–30 min. Nonisotopic labeling can also be easily performed by including biotin- or digoxigenin-labeled nucleotides in the reaction mixture (see *Nick translation*). Use of $3' \rightarrow 5'$-exonuclease-deficient (exo-minus) Pol Ik in place of wild-type Pol Ik further contributes to the stability of the labeled probes. Another improvement to the labeling method is the use of longer synthetic oligonucleotides, e.g., random octamers, nonamers, decamers, or tetradecamers. The increase of primer length to 14-mer from the hexamer results in more stable annealing at 25°C, which lowers more than 250-fold the primer mass (e.g., 0.02 A_{260} of 14-mer per ml) used in the reaction. For efficient labeling, superhelical plasmid DNA should be linearized or denatured with alkali prior to labeling reaction.

3. PCR Labeling

DNA labeling by PCR is an increasingly popular technique because of its inherent advantages associated with repetitive cyclic, high-temperature polymerase reactions. PCR labeling allows one to use subnanogram quantities (0.1 to 1 ng) of template DNA as opposed to the larger amounts ($\geq$20 ng) required for nick translation or random-primer labeling. PCR labeling generates labeled products of uniform length rather than the heterogeneously sized products obtained by other methods. Unlike other labeling methods, however, PCR labeling requires a pair of ~30-nt-long oligonucleotide primers that specifically define the two ends of the amplified DNA products. Because the incorporation of nucleotides, especially nonisotopically labeled nucleotides, critically depends on the property of the polymerase, typically *Taq* DNA pol, it is important to optimize the reaction conditions for efficient labeling. Efficiency here implicates the yield and, in the case of isotopic labeling, the specific activity. Considering that the K_m values for dNTPs are 10–15 μM, the radionucleotide concentrations usually employed for the labeling reaction fall far below the K_m values, thereby resulting in a low rate of polymerization on the one hand and a high probability of nucleotide misincorporation and of premature chain termination on the other. Although exact optimum conditions may vary depending on the probe size, the optimal yield and specific activity can be obtained in general with a 2- to 10-fold but not a >20-fold imbalance of each unlabeled dNTP versus labeled dNTP (8).

References

1. Rigby, P. W. J., Dieckmann, M., Rhodes, C., and Berg, P. (1977). *JMB* **113**, 237–251.
2. Sambrook, J., Fritsch, E. F., and Maniatis, T. (1989). "Molecular Cloning: A Laboratory Manual," 2nd ed., pp. 10.6–10.12. CSHL Press, Cold Spring Harbor, New York.

3. Meinkoth, J., and Wahl, G. M. (1987). *Meth. Enzy.* **152**, 91–94.
4. Koch, J., Kolvraa, S., and Bolund, L. (1986). *NAR* **14**, 7132.
5. Feinberg, A. P., and Vogelstein, B. (1983). *Anal. Biochem.* **132**, 6–13.
6. Feinberg, A. P., and Vogelstein, B. (1984). *Anal. Biochem.* **137**, 266–267.
7. Hodgson, C. P., and Fisk, R. Z. (1987). *NAR* **15**, 6295.
8. Mertz, L. M., and Rashtchian, A. (1994). *Focus (BRL)* **16**, 45–48.

II. NUCLEOTIDE SEQUENCING

Nucleotide sequences can be determined by either enzymatic or chemical sequencing methods. The best known enzymatic sequencing method is the *dideoxynucleotide chain termination* method in which the DNA polymerase-catalyzed synthesis of second strands or complementary strands is randomly terminated by incorporation of a 2′,3′-dideoxynucleotide. The reaction generates a nested set of DNA fragments terminated at the specific nucleotides for which the dideoxynucleotide is substituted. In contrast to the enzymatic method, the *chemical sequencing* method relies on direct, base-specific cleavages of DNA using specific sets of chemical reagents.

1. DIDEOXYNUCLEOTIDE CHAIN TERMINATION METHODS

Dideoxynucleotide sequencing methods, originally developed by Sanger *et al.* (1), typically consist of the following steps (2,3): (i) an oligonucleotide primer is annealed to a ssDNA template; (ii) the primer is extended using a polymerase in four separate reaction mixtures, each containing one labeled dNTP, one chain-terminating ddNTP, and a mixture of the other three unlabeled dNTPs; (iii) four sets of reaction products terminating respectively at G,A,T, or C are separated according to the size by electrophoresis in four adjacent lanes on a high-resolution polyacrylamide–urea gel; and (iv) the image of the gel is produced by autoradiography, and the DNA sequences are read from the "sequence ladder." An RNA template can be sequenced, albeit inefficiently, in the same manner using a reverse transcriptase.

To increase the efficiency of labeled nucleotide incorporation, the polymerase reaction is usually carried out in two stages: the first is a labeling reaction during which labeling nucleotides are maximally incorporated in the presence of limiting concentrations of the corresponding unlabeled nucleotides, and the second is an extension/termination reaction during which the primers are further extended and terminated in the presence of an ample supply of nucleotides. The size distribution of fragments can be controlled by adjusting the ratio of ddNTP/dNTP. Note that the polymerase reaction does not need to be divided into two stages when 5′-labeled primers are employed in place of or in addition to the labeled dNTP(s).

The dideoxynucleotide sequencing protocol has undergone considerable refinements in various aspects, including sample DNA (or RNA) preparation, quality and variety of reagents, separation techniques, signal detection and data acquisition, and automation of a large part of the procedures (4). Applying PCR to dideoxynucleotide sequencing as a means of template preparation and/or as a form of direct sequencing reaction has further extended the speed and efficiency of nucleotide sequencing. Nevertheless, each sequencing task, especially with long DNA calls for strategies specifically adapted for generating and subcloning smaller DNA fragments. Furthermore, the choice of DNA polymerase affects the quantity and quality of nucleotide sequences obtainable from sequencing reactions (2,5). The following is a brief account of the features expected of a polymerase frequently employed in dideoxynucleotide sequencing.

a. E. coli DNA Pol Ik

The Klenow fragment of *E. coli* DNA Pol I is the first and most widely used enzyme for nucleotide sequencing by the "standard" dideoxy method with ssDNA templates (2,3). The reaction is generally performed at 37°C, and occasionally at 50°C. Pol Ik has also been used in fluorescence-based automated DNA sequencing which utilizes the primers labeled at the 5' ends with fluorescein dyes (6,7). Although Pol Ik is a reliable enzyme with a high fidelity ($\sim 4 \times 10^{-5}$), its utility is limited to short-range sequencing (≤ 250 nt) due to its thermolability, low rate of polymerization, and low processivity. Modified Pol Ik lacking $3' \rightarrow 5'$-exonuclease activity improves the utilization of nucleotide analogs. Pol Ik lacks the 5'–nuclease activity of Pol I, but has a strand displacement activity. Pol Ik is particularly sensitive to secondary structures and/or high GC content of the template.

b. Modified T7 DNA Pol

This enzyme, better known as Sequenase (USB/Amersham), is a $3' \rightarrow 5'$-exonuclease-deficient T7 DNA Pol possessing excellent qualities as a sequencing polymerase (8). It has a high processivity and a high rate of polymerization, making the enzyme particularly suitable for long-range sequencing. In fact, the utility is only limited by the resolution limit of the separation methods, i.e., ~ 500 nt in the standard urea–PAGE. Because of its high processivity, Sequenase is capable of going through highly structured regions and is the enzyme of choice for sequencing heat- or alkali-denatured dsDNA. Sequenase exhibits relaxed substrate specificities in the presence of Mn^{2+} (2 mM in 15 mM isocitrate), allowing efficient incorporation of nucleotide analogs such as ddNTP, dNTPαS, and c^7dGTP (8,9) and contributing to uniform, readily resolvable sequence ladders. The ability to incorporate fluorescence-tagged ddNTP (dye terminator) efficiently also lends the enzyme to automated DNA sequencing methods based on fluorescence detection systems (10,11). Sequenase is heat labile and is thus unsuitable for use in thermal cycling sequencing (see below). Despite this minor drawback, Sequenase remains fully useful for sequencing PCR-amplified dsDNA as well as the ssDNA templates prepared by digestion with the T7 gene 6 exonuclease (11).

c. Reverse transcriptases

Reverse transcriptases, e.g., from AMV and MoLV, are frequently used for dideoxy sequencing of DNA and are indispensable tools for dideoxy sequencing of RNA. RNA sequencing is typically performed via reverse transcriptase-catalyzed cDNA synthesis in the presence of ddNTPs, similar to standard dideoxy DNA sequencing (12–14). The RNA templates obtained from *in vitro* transcription of PCR-amplified DNA which carries a phage promoter sequence are also suitable for direct dideoxy sequencing. The RNA polymerase-directed amplification of PCR products provides a powerful strategy for the preparation of RNA templates (15).

Reverse transcriptases (AMV and MoLV) can efficiently substitute for Pol Ik in sequencing DNA regions with high GC contents, homopolymer tracts (e.g., as many as 45 G/C bases), or secondary structures. AMV RTase is generally more useful than the thermolabile MoLV RTase for sequencing purposes because AMV RTase reactions can be performed at 42°C and often at 50°C for certain highly structured templates. AMV RTase can utilize fluorescence-tagged (succinylfluorescein dyes) ddNTPs with efficiencies comparable to that of the corresponding unsubstituted ddNTPs and has been used in the automated sequencing of DNA (16) and RNA (9). Perhaps the most widely used form of RNA sequencing techniques is the one based on the initial RTase-catalyzed conversion of the RNA into cDNA and then amplification of the DNA by PCR, a process commonly referred to as *reverse transcription (RT)–PCR*. This technique offers critical advantages: (i) DNA templates can be prepared in sufficient amounts from subnanogram quantities of starting RNA material and (ii) DNA sequencing can be performed by employing more efficient and convenient DNA polymerases. The standard format for RT–PCR consists of a cDNA synthesis reaction, using either AMV or MoLV RTase, that is followed by standard PCR amplification. The PCR amplification and/or sequencing (see below) can be performed with nanogram quantities of cDNA present in a fraction (≤ 0.1) of the original cDNA mixture. The unusual property of thermostable *Tth* DNA Pol exhibiting high RTase activity in the presence of $MnCl_2$ provides a streamlined RT–PCR process. Here cDNA synthesis and PCR amplifications are performed successively in a single tube using the same polymerase simply by adjusting the buffer and other reaction conditions.

d. Thermostable polymerase

Thermostable DNA polymerases (Table A.2) play increasingly important roles in nucleotide sequencing. The polymerases are used in part for the preparation of ssDNA or dsDNA templates via the process of PCR and in part as the direct sequencing enzymes in high temperature dideoxy sequencing. Compared with the conventional methods of nucleotide sequencing performed at 37–50°C, high temperature (65–72°C) sequencing with thermostable polymerases offers unique advantages: (i) due to high temperatures used for primer annealing and sequencing reactions, much of the sequencing artifacts arising from nonspecific priming are eliminated and (ii) regions of high GC content, long homopolymer tracts, and/or secondary structures can be readily sequenced. Furthermore, the thermal stability makes it possible to perform repeated cycles of heat denaturation, primer annealing, and primer extension, a sequencing technique known as *thermal cycle sequencing*. Thermal cycle sequencing amplifies the dideoxynucleotide-terminated DNA fragments in a linear fashion, unlike the exponential amplification in standard

TABLE A.2 Thermostable DNA Polymerases[a]

Polymerase	Source organism	MW (kDa)	Enzymatic activity				Reaction (°C)	Fidelity[b]	Thermostability $t_{1/2}$ (°C)	Commercial sources		Remarks
			5'-3' Pol	3'-5' Exo	5' Nuc	Nuc trans				Trade name	Company[c]	
Bca Dpol	Bacillus caldotenax		Yes	Yes	Removed		65-70			Ladderman Dpol	Takara Biomed	Cloned
Bst Dpol Large fragment	Bacillus stearothermophilus		Yes	Yes	Removed		65			IsoTherm Dpol	Bio-Rad Epicentre	Native Cloned
Pfu Dpol	Pyrococcus furiosus	60	Yes	Yes	No	No	72	12	~4 h (96)		Stratagene	Cloned
Pfu Dpol (exo-)			Yes	Removed	No		72					
Deep Vent Dpol	Pyrococcus (strain GB-D)		Yes	Yes	No		~75	3-5	23 h (95)		NEB	Cloned
Deep Vent(exo-)			Yes	Removed	No		~75	1.5	23 h (95)			
Pwo Dpol	Pyrococcus woesei	90	Yes	Yes	No	Yes	~75		>2 h (100)		BM	Cloned
Taq Dpol	Thermus aquaticus	94	Yes	No	Yes	Yes	72	[1]	10 m (97.5) 1 h (95)	AmpliTaq Dpol	P-E/R [d]	Cloned/Native
Taq Dpol (exo-)		61	Yes Yes	No No	Removed Removed		72 72	~2	1.4 h (95)	Thermo Sequenase KlenTaq1 KlenTaq5	Amersham Ab Peptides Amersham	F667Y mutant Cloned
		56	Yes	No	Removed		72	~2	1.4 h (95)	Stoffel Fragment	P-E/R	
Tbr Dpol	Thermus brockianus		Yes	No	Yes	No	72	2	2.5 h (96)	DynaZyme Thermalase Tbr	Finnzyme Amresco	Native
Tfl Dpol	Thermus flavus	82	Yes	No	No	No	72			Pyrostase	Promega MGR Inc.	
Tli Dpol	Thermococcus litoralis	85	Yes	Yes	No	No	~75	3-5	1.8 h (100)	Vent Dpol	NEB	Cloned
Tli Dpol (exo-)			Yes	Removed	No		~75		6.7 h (95)	Vent (exo-) Dpol	Promega NEB	Cloned
9°Nm Dpol	Thermococcus (strain 9°N-7)		Yes	1-5%	No		~75	1.5	6.7 h (95)	9°Nm Dpol	NEB	Cloned
Tma Dpol	Thermotoga maritima	70	Yes	Yes	Yes	No	72		40 m (97.5)	UlTma Dpol	P-E/R	Cloned
Tth Dpol	Thermus thermophilus	94	Yes	No	Yes	Yes	65(RT) ~75		2 m (97.5) 20 m (95)	TET-z Pol	BM, P-E/R Prom, Pharma Amersham	Cloned/Native RT activity with Mn(II)
Tth Dpol (exo-)			Yes	Removed	No					Tth Dpol	Clontech	Cloned
Tub Dpol	Thermus ubiquitus		Yes	Yes	No		~75	~1		Hot Tub Dpol	Amersham	

[a] Pol, polymerase; Dpol, DNA polymerase; Exo, exonuclease; Nuc, nuclease; Nuc trans, nucleotidyltransferase; and RT, reverse transcriptase.
[b] Fidelity of DNA polymerases relative to the estimated value(s) of ~10^{-5} errors/bp/cycle for Taq DNA polymerase under usual PCR conditions.
[c] BM, Boehringer Mannheim; NEB, New England Biolabs; P-E/R, Perkin-Elmer/Hoffman-La Roche; Pharma, Pharmacia Biotech; Prom, Promega.
[d] Major suppliers of Taq DNA Pol include Perkin-Elmer/Roche, Amersham, Boehringer Mannheim, GIBCO/BRL, Pharmacia Biotech, Promega, and Stratagene.

PCR. As a consequence of the most efficient use of the primer, the sequencing technique produces sequence ladders with highly enhanced signals. This feature allows sequence information to be obtained from an extremely small amount of template DNA (or RNA) recoverable from a single bacterial or phage colony and ancient relics or fossils. Thermal cycle sequencing can be performed with dsDNA or preferably ssDNA. Either ssDNA or dsDNA templates can be prepared by various conventional as well as PCR-based methods (refer to PCR, below).

In keeping with the need for high-volume nucleotide sequencing, 96-well microtiter plates containing prealiquoted, dried reagents have been used as the support for sequencing reactions. For this particular type of applications, *Bst* DNA Pol is an excellent enzyme with the ability to withstand the dried conditions and to generate sequences of consistently high quality (17).

2. CHEMICAL SEQUENCING METHODS

Chemical DNA sequencing methods, generally known as the Maxam–Gilbert sequencing methods (18), are based on limited chemical cleavages at one or two specific bases of an end-labeled DNA. Unlike the dideoxynucleotide sequencing which relies on enzymatic synthesis of a second-strand DNA, chemical sequencing provides sequence information directly from the DNA sample. Chemical sequencing methods are most useful for short-range sequencing (≤250 nt from the labeled terminus); the sequencing efficiency drops significantly for longer DNAs.

In the Maxam–Gilbert method, guanosine is preferentially cleaved by dimethyl sulfate, purines (G + A) are depurinated and cleaved by acid, and pyrimidines (T + C) are cleaved by hydrazine (Table A.3). When the hydrazine reaction is performed at high concentrations of salt, the reaction of thymines with hydrazine is preferentially suppressed, thereby resulting in selective cleavages at cytosines. Additionally, and as an alternative chemical cleavage procedure (19), an adenine-enhanced (A > C) reaction can be performed with NaOH. This reaction results in preferential cleavages at adenines and minor cleavages at cytosines. On the other hand, G residues of dsDNA can be specifically cleaved by reaction (30 min at 22°C) with osmium tetroxide (OsO_4) in a HEPES buffer (pH 7.9) containing 10 mM $CaCl_2$ but not pyridine (20). These base-specific reactions generate a nested set of fragments at controlled reaction conditions in which each molecule of DNA is cleaved on average only once randomly. The fragmented products are denatured and separated by urea–PAGE according to size, much like the corresponding step in the enzymatic sequencing methods described earlier.

Chemical sequencing methods can be applied to ssDNA or dsDNA as long as one labeled end (3' or 5') is available. The 5' ends are usually labeled with ^{32}P or ^{35}S using T4 polynucleotide kinase in the presence of [γ-^{32}P]ATP or [γ-^{35}S]ATP. The 3'-end labeling is typically carried out using terminal deoxynucleotidyltransferase in the presence of ^{32}P- or ^{35}S-labeled ddNTP, 3'-deoxy-ATP (cordycepin triphosphate), or rATP. (*Note*: For most of the procedures involving radiolabeling, nucleotides labeled with ^{33}P should be considered as a preferred substitute.)

TABLE A.3 Base-Specific Chemical Cleavages Used in Maxam–Gilbert Sequencing Method

Cleavage	Base modification	Modified base displacement	Strand scission[e]
G	DMS[a]	Piperidine	Piperidine
G + A	Acid[b]	Acid	Piperidine formate
C + T	Hydrazine[c]	Piperidine	Piperidine
C	Hydrazine[c] (+1.5 M NaCl)	Piperidine	Piperidine
A > C	NaOH[d]	Piperidine	Piperidine

[a] Reaction of 0.5% DMS (dimethyl sulfate) in 50 mM cacodylate buffer (pH 8.0) at 20°C for 10 min. DMS methylates the N-7 of guanine. The extent of methylation and consequently the size distribution of resultant DNA fragments can be controlled by varying reagent concentrations, reaction times, and temperatures.

[b] Depurination reaction with 0.13 M aqueous piperidine formate at pH 2.0, 20°C, and for 60 min. The depurination is followed by strand scission in the presence of piperidine.

[c] Hydrazine initially attacks the pyrimidine ring at C-4 and C-6 of both cytosines and thymines, leading to ring opening and cyclization. In the presence of salt (1.5–2 M NaCl), the reaction of hydrazine (15–18 M) at 20°C is selectively directed to cytosines.

[d] Reaction with 1.2 N NaOH at 90°C for 10 min.

[e] The strand-cleavage reaction is carried out with 1 M piperidine (in H_2O) at 90°C for 30 min. Piperidine is subsequently removed by lyophilization.

With dsDNA as the starting material, the enzymatic labeling produces DNA labeled at both ends. The DNA fragments labeled at a single end are then obtained either by separation of the two strands after denaturation or by cleavage with a suitable restriction enzyme(s).

The application of PCR technology to amplify DNA with 5'-labeled primer and to obtain singly labeled dsDNA directly has drastically increased the utility of chemical sequencing methods (21,22). In an entirely different PCR-based approach, chemically fragmented DNAs are subsequently amplified by linear PCR using a defined 5'-labeled primer and are then analyzed on a sequencing gel (23). This approach, dubbed *genomic sequencing*, incorporates an enzymatic step as part of the procedure, substantially increasing the speed and sensitivity of the chemical sequencing method. In the original form of the genomic sequencing method pioneered by Church and Gilbert (24), restriction enzyme-digested fragments of genomic DNA were chemically cleaved, separated by a sequencing gel, and subsequently hybridized with specific, labeled probes. The versatility of the technique allows it to be used in *genomic footprinting* for the study of the methylation state of DNA or DNA–protein interactions *in vivo*.

PCR technology has also been applied to a chemical sequencing technique based on phosphorothioate-substituted nucleotides. In this sequencing method (25), DNA is amplified by PCR using *Taq* DNA Pol in the presence of dNTPαS. Four separate PCRs in the presence of each different dNTPαS produce a full set of DNA samples in which dNMPαS is randomly and site specifically incorporated. The original version of the method relies on ^{32}P-labeling of the PCR products at the 5' ends, but the product labeling can be done much more easily using a 5'-labeled PCR primer. Chemical degradation of the phosphorothioate-containing DNA fragments is carried out using 2,3-epoxy-1-propanol. This reagent initially alkylates the 3'-thiophosphate and then induces strand scission. Degradation products are separated on a sequencing gel, and the sequence is read off the autoradiogram.

References

1. Sanger, F., Nicklen, S., and Coulson, A. R. (1977). *PNAS* **74**, 5463–5467.
2. Sambrook, J., Fritsch, E. F., and Maniatis, T. (1989). "Molecular Cloning: A Laboratory Manual," 2nd ed., pp. 13.1–13.104. CSHL Press, Cold Spring Harbor, New York.
3. Ausubel, F. M., Brent, R., Kingston, R. E., Moore, D. D., Seidman, J. G., Smith, J. A., Struhl, K., Wang-Iverson, P., and Bonitz, S. G. Eds. (1989). *In* "Short Protocols in Molecular Biology," pp. 201–231. Wiley, New York.
4. Hunkapiller, T., Kaiser, R. J., Koop, B. F., and Hood, L. (1991). *Science* **254**, 59–67.
5. Mardis, E. R., and Roe, B. A. (1989). *BioTechniques* **7**, 840–850.
6. Ansorge, W., Sproat, B., Stegemann, J., Schwager, C., and Zenke, M. (1987). *NAR* **15**, 4593–4602.
7. Connell, C., and 14 coauthors (1987). *BioTechniques* **5**, 342–348.
8. Tabor, S., and Richardson, C. C. (1987). *PNAS* **84**, 4767–4771.
9. Bauer, G. J. (1990). *NAR* **18**, 879–884.
10. Wilson, R. K., Chen, C., and Hood, L. (1990). *BioTechniques* **8**, 184–189.
11. Lee, L. G., Connell, C. R., Woo, S. L., Cheng, R. D., McArdle, B. F., Fuller, C. W., Halloran, N. D., and Wilson, R. K. (1992). *NAR* **20**, 2471–2483.
12. Karanthanasis, S. (1982). *Focus (BRL)* **4**(3), 6–7.
13. Geliebter, J. (1989). *Focus (BRL)* **9**, 5–8.
14. Hahn, C. S., Strauss, E. G., and Strauss, J. H. (1989). *Meth.Enzy.* **180**, 121–130.
15. Sommer, S. S., Sarkar, G., Koeberl, D. D., Bottema, C. D. K., Buerstedde, J.-M., Schowalter, D. B., and Cassady, J. D. (1990). *In* "PCR Protocols: A Guide to Methods and Applications" (M. A. Innis *et al.*, Eds.), pp. 197–205. Academic Press, San Diego.
16. Prober J. M., Trainor, G. L., Dam, R. J., Hobbs, F. W., Robertson, C. W., Zagursky, R. J., Cocuzza, A. J., Jensen, M. A., and Baumeister, K. (1987). *Science* **238**, 336–341.
17. Earley, J. J., Kuivaniemi, H., Prockop, D. J., and Tromp, G. (1993). *DNA Sequence* **4**(2), 79–85.
18. Maxam, A. M., and Gilbert, W. (1980). *Meth.Enzy.* **65**(I), 499–560.
19. Ambrose, B. J. B., and Pless, R. C. (1987). *Meth.Enzy.* **152**, 522–538.
20. Dobi, A. L., Matsumoto, K., Santha, E., and v.Agoston, D. (1994). *NAR* **22**, 4846–4847.
21. Higuchi, R., Krummel, B., and Saiki, R. K. (1988). *NAR* **16**, 7351–7367.
22. Ohara, O., Dorit, R. L., and Gilbert, W. (1989). *PNAS* **86**, 5673–5677.
23. Saluz, H., and Jost, J.-P. (1989). *PNAS* **86**, 2602–2606.
24. Church, G. M., and Gilbert, W. (1984). *PNAS* **81**, 1991–1995.
25. Nakamaye, K. L., Gish, G., Eckstein, F., and Vosberg, H.-P. (1988). *NAR* **16**, 9947–9959.

III. POLYMERASE CHAIN REACTION

Polymerase chain reaction (PCR) is a powerful technique used to amplify enzymatically a segment of DNA through multiple cycles of a primer-directed polymerization reaction. PCR enables one to amplify over one millionfold a picogram amount of starting DNA from rare samples, a single-copy gene, or plasmid. A DNA segment (usually ≤2000 bp) can be amplified exponentially using a pair of oligonucleotide primers which define the boundaries. Under "long PCR" conditions (refer to Section II,D, Chapter 6), DNA fragments with sizes ≥20 kb can be readily amplified.

Basic PCR procedures consist of three steps: (i) thermal denaturation of dsDNA, (ii) annealing of two bracketing primers to their complementary sequences, and (iii) primer extension with a DNA polymerase.

The original PCR techniques, which were developed using Pol Ik (1,2), have undergone a number of improvements and refinements, making the techniques more widely acceptable and applicable. Focusing on the enzyme components among various aspects of PCR technology (3,4), Pol Ik is now replaced by thermoresistant polymerases, typically *Taq* DNA Pol. *Taq* DNA Pol provides two important advantages over thermolabile polymerases such as *E. coli* DNA Pol Ik and T7 DNA Pol: (i) because annealing and polymerization reactions can be performed at elevated temperatures and repetitive cycles, the specificity and yield of extension products are substantially higher, and (ii) PCR protocols are amenable to automation, thereby permitting precision control of the cycling parameters with greater ease, reliability, and reduced danger of contamination.

If PCR technology has been developed for the amplification of DNA, it is because the technology relies on thermostable DNA polymerases. For the amplification of RNA by PCR, the RNA should first be converted to cDNA and then be amplified as DNA, a process known as reverse transcription (RT)–PCR.

Taq DNA Pol has been used successfully for some RT–PCR. *Tth* DNA Pol is now the enzyme of choice because of its ability to reverse transcribe RNA efficiently in the presence of Mn^{2+} and subsequently to amplify DNA in a single tube at a buffer condition adjusted for pH and Mg^{2+} concentration (5). In the coupled RT–PCR, *Tth* DNA Pol is estimated to be over 100-fold more efficient than *Taq* DNA Pol. The task of reverse transcription in RT–PCR can also be assumed efficiently by dedicated reverse transcriptases such as AMV and MoLV RTases. In fact, sequential reactions with RTase and then *Taq* DNA Pol have been shown to give results at least comparable to those of *Tth* DNA Pol protocols.

Although standard PCR techniques are designed to generate dsDNA products, slight modifications regarding the relative concentrations of two bracketing primers result in ssDNA products that are more advantageous than dsDNA as sequencing templates or as labeled probes. In the modified technique known as *asymmetric PCR* (6–9) as opposed to the standard *symmetric PCR*, one of the two primers is employed at a limiting concentration (e.g., a molar ratio of 1:20–100). This reaction condition permits exponential amplification of DNA for 10–15 cycles, essentially until the low concentration primer is depleted. Subsequent amplifications proceed in a linear fashion, generating copies of only one strand.

References

1. Saiki, R. K., Scharf, S., Faloona, F., Mullis, K. B., Horn, G. T., Erlich, H. A., and Arnheim, N. (1985). *Science* **230**, 1350–1354.
2. Mullis, K. B., and Faloona, F. A. (1987). *Meth. Enzy.* **155**, 335–350.

3. Bloch, W. (1991). *Biochem.* **30**, 2735–2747.
4. Erlich, H. A., Gelfand, D., and Sninsky, J. J. (1991). *Science* **252**, 1643–1651.
5. Myers, T. W., and Gelfand, D. H. (1991). *Biochem.* **30**, 7661–7666.
6. Innis, M. A., Myambo, K. B., Gelfand, D. H., and Brow, M. A. D. (1988). *PNAS* **85**, 9436–9640.
7. Gyllensten, U. B., and Erlich, H. A. (1988). *PNAS* **85**, 7652–7656.
8. Mihovilovic, M., and Lee, J. E. (1989). *BioTechniques* **7**, 14–16.
9. Wilson, R. K., Chen, C., and Hood, L. (1990). *BioTechniques* **8**, 184–189.

IV. SITE-SPECIFIC *IN VITRO* MUTAGENESIS

In vitro mutagenesis comprises a group of molecular biological techniques used for structural and functional analyses of nucleic acids and/or proteins by introducing mutations *in vitro* on a segment of DNA in a controlled manner. The mutagenesis studies provide key information required for designing DNA, RNA, or proteins with a novel function, tailored specificity, and altered stability to heat or organic solvents.

Mutagenesis is called "random" if one or more bases are modified throughout the sequence. Mutagenesis is called "site specific" when a base or bases are altered within a restricted target region. Whereas random mutagenesis is useful to identify the location and boundaries of a particular function and is most profitably applied to poorly characterized DNA and proteins, site-specific mutagenesis provides a means to explore the role of specific nucleotides or amino acids in relatively well-characterized DNA and proteins. Random mutagenesis generally relies on chemical mutagens to modify bases, whereas site-specific mutagenesis usually resorts to sequence-specific oligonucleotides.

Oligonucleotide-directed site-specific mutagenesis, which was initially developed by Smith *et al.* (1,2), provides the greatest control in introducing a mutation(s) at a defined target site in a DNA of known sequence (3–5). In addition, oligonucleotide-directed mutagenesis provides unparalleled facility for creating site-specific deletions and insertions in a cloned DNA, without being constrained by the natural availability of suitable restriction sites. The application of PCR in site-specific *in vitro* mutagenesis has further increased the convenience and utility of oligonucleotide-directed mutagenesis, fully exploiting the power of PCR as a versatile means of constructing engineered DNA molecules.

Although a number of variations are available for oligonucleotide-directed mutagenesis, the following examples illustrate the core techniques and strategies that provide the highest mutagenic efficiencies: (i) uracil-based mutagenesis, (ii) phosphorothioate-based mutagenesis, (iii) coupled-priming mutagenesis, and (iv) PCR mutagenesis. With some exceptions, the first three "conventional" methods largely rely on the amplification of mutagenic strands *in vivo* by applying a selective pressure either against the wild-type strand or for the mutant strand. PCR mutagenesis differs from the other methods in that the mutagenic dsDNA fragment is synthesized and amplified *in vitro* using PCR and is put back into a plasmid vector replacing the wild-type DNA segment before the plasmid is intro-

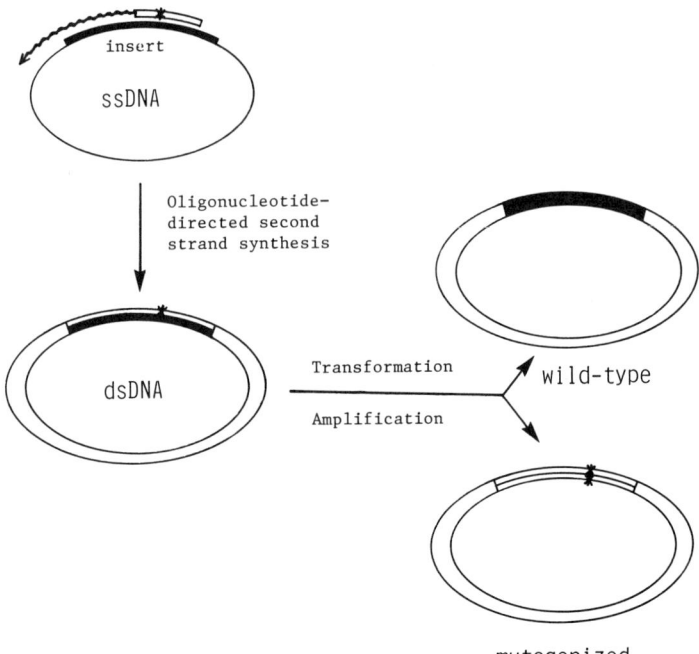

FIGURE A.1 A general scheme illustrating the principle of conventional oligonucleotide-directed site-specific mutagenesis. Following the DNA Pol-catalyzed synthesis of the second (+) strand from a mutagenic oligonucleotide primer, the heteroduplex is introduced into a suitable *E. coli* host. Both the parent (wild-type) and new (mutagenized) strands segregate into homoduplex plasmids usually at unequal efficiencies. The star indicates a mutagenic site.

duced into cells. The PCR mutagenesis is thus an extended form of "cassette" mutagenesis.

Although detailed strategies may vary depending on individual methods, the basic procedures of conventional oligonucleotide-directed mutageneses consist of the following (Fig. A.1): (i) preparation of ssDNA templates from M13 or its derivative phagemid vectors, (ii) annealing to the ssDNA template of a mutagenic oligonucleotide complementary to the target DNA except for a limited sequence(s) to be mutagenized, (iii) extension *in vitro* of the mutagenic primer by the use of a DNA polymerase to generate an entire copy of the gene plus the vector, (iv) transfection of the full or gapped "heteroduplex" DNA into *E. coli* (or other suitable host cells) for amplification, and (v) screening of mutant progeny phages and verification by nucleotide sequencing.

The efficiency of oligonucleotide-directed mutagenesis is critically dependent on several factors most notably involving the primer, DNA polymerase, and amplification strategies.

a. Mutagenic oligonucleotide primers

The mutagenic oligonucleotides used for primer extension with a DNA polymerase at 37°C or below has a general configuration of ~20-mer containing one or two mutagenic nucleotides in the middle of the sequence. The primer extension with a thermostable DNA polymerase at or near 72°C requires longer nonmutagenic bracing arms, e.g., a 30- to 40-mer with one or two mutagenic nucleotides in the middle. Oligonucleotide primers have different hybridization efficiencies related to their length, sequence, and composition. Primers that are GC rich at their 3' ends exhibit high priming efficiency but are also prone to false priming. As a rule, the T_m of a mutagenic primer is lowered by 1–1.5°C for every 1% base mismatch. Primer annealing is usually carried out under conditions that are ~10°C below the T_m.

Polymerase reaction is typically performed with a single mutagenic primer, but some mutagenesis methods employ more than one primer for simultaneous extension. The coupled-priming strategies can be used to introduce either a single mutation (6) or multiple mutations (7). In the coupled-priming mutagenesis, the primers (also called "selection" primers) used for secondary mutations are often targeted to one or more convenient selectable markers. Multiple mutagenesis at multiple sites can then be performed by sequential single-site mutagenesis. For multiple amino acid substitutions at a single site, a highly efficient strategy combines the primary site-specific *in vitro* mutagenesis generating an amber mutation with an "*in vivo amber* suppression" mutagenesis. In this combination mutagenesis strategy, the plasmid containing the *amber* (TAG) stop codon is introduced into selected "suppressor" strains of *E. coli*. The suppressor strains, available from Promega (8), carry specific *amber* suppressor tRNAs which can insert 1 of the 12 amino acids (C, E, F, G, H, K, L, P, Q, R, S, and Y) at the site of the stop codon.

b. DNA polymerases used for primer extension

Because the purpose of the operation is the rapid, faithful extension of a mutagenic primer(s) without displacing them from the template, various DNA polymerases can be used with their respective advantages and disadvantages. For instance, T4 DNA Pol is generally preferred to *E. coli* DNA Pol Ik. In contrast to T4 DNA Pol, *E. coli* Pol Ik has a strand displacement activity which may cause lower mutagenic efficiencies. Native T7 DNA Pol or a modified version (Sequenase) that lacks the 3' → 5'-exonuclease activity may in some cases be more efficient than T4 DNA Pol: (i) polymerization is faster with T7 DNA Pol (usually complete within 30 min rather than 1 hr for T4 DNA Pol) and (ii) T7 DNA Pol does not require the use of a ssDNA-binding protein (e.g., T4 gene 32 protein) to copy efficiently the templates with secondary structures. T7 DNA Pol is particularly useful for extending mutagenic primers on ssDNA templates prepared from dsDNA by heat or alkali denaturation. In contrast to the native T7 DNA Pol, however, the 3' → 5'-exonuclease-lacking Sequenase has a strand displacement activity.

Sequenase is particularly useful for introducing mismatching nucleotides during the oligonucleotide-directed second-strand synthesis in the presence of a biased nucleotide pool (9). This forced misincorporation mutagenesis technique is useful for generating localized random mutants.

c. Amplification strategies

When heteroduplex plasmid DNA constructed *in vitro* is introduced into *E. coli*, one may expect, according to the principle of independent segregation, 50% of the progeny clones to be desired mutants. In reality, the mutation efficiency falls far short of 50% because of (i) incomplete or spurious primer extension *in vitro*, (ii) primer displacement by the DNA polymerase after gap filling or full extension of the mutagenic primer(s), and (iii) *in vivo* mismatch repair mechanisms operating on the unmethylated, newly synthesized DNA strand. Depending on host strains used for transformation, the *in vivo* repair mechanism may be the most critical factor limiting the efficiency of conventional mutagenesis techniques. Use of *E. coli* strains (e.g., *mutL* or *mutS*) that are deficient in point mismatch repair can significantly increase the recovery yields of mutants, although the theoretical maximum of 50% cannot be overcome. (*Note*: Strains that carry the *mutS* allele have 50- to 100-fold higher spontaneous mutation rates.) The *in vivo* genetic restriction mechanism is not always undesirable. In fact, as noted later, certain mutagenesis methods make use of the restriction mechanism as a means of increasing the mutagenesis efficiencies. For successful mutagenesis attempts, therefore, the genetic traits of host bacterial strains should be carefully taken into account (refer to Appendix B).

Apart from the various *in vitro* and *in vivo* factors that contribute to the reduction of mutation efficiency, a mutant clone may not display a readily discernible phenotype, rendering the screening process complicated and labor-intensive. The methods described below illustrate how a variety of techniques and strategies can be used to intervene in or take charge of the mutation-fixing process and improve the efficiency of *in vitro* mutagenesis such that the majority of the resulting clones carry the desired mutations.

1. URACIL-BASED MUTAGENESIS

The uracil-based oligonucleotide-directed mutagenesis, also known as the Kunkel method (10,11), consists of the following steps (Fig. A.2): (i) uracil (U)-containing target ssDNA (U-DNA) is obtained from a phage grown in a U-permissive *E. coli*, (ii) second-strand DNA is synthesized using a mutagenic oligonucleotide as the primer and the U-DNA as the template, and (iii) the duplex is introduced into a U-nonpermissive *E. coli* host, which results in selective degradation of the U-containing parental strand and a highly efficient (typically >70%) production of mutant progenies. (A mutagenesis kit based on the Kunkel method is available from Bio-Rad: Muta-Gene In Vitro Mutagenesis kit.)

Single-stranded template DNA can be prepared using an M13 or phagemid vector, e.g., pTZ18/19 (R/U), pEMBL, or pUC118 (12,13), in an *E. coli* strain lacking two enzymes of DNA metabolism: (a) dUTPase-minus (dut^-) which

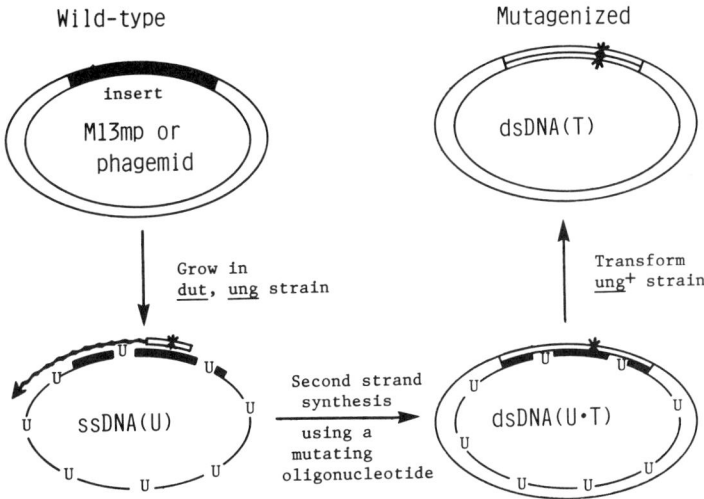

FIGURE A.2 A general scheme for uracil-based mutagenesis. The parent strand containing dU bases at some of the T sites serves as the ssDNA template for the synthesis of a mutagenic T-containing strand. The star indicates a mutagenic site.

results in a high concentration of intracellular dUTP and (b) uracil N-glycosylase-minus (ung^-) which is incapable of removing misincorporated uracils from DNA. All DNAs originating from the $dut^-\ ung^-$ strain (e.g., CJ236) have a small number of U stably incorporated in place of thymines. For *in vitro* synthesis of second-strand DNA using phage T4 or T7 DNA polymerase, the U-DNA template is indistinguishable from normal thymine-containing DNA (T-DNA). The newly synthesized (or negative) strand in the heteroduplex is then ligated using the T4 DNA ligase to form a covalently closed circular DNA. (*Note:* This requires the use of 5'-P primers.) Although the circularization is not absolutely necessary, it generally increases the transformation efficiency. Transfection of the heteroduplex (U-DNA · T-DNA) into a ung^+ *E. coli* strain (e.g., MV1190, DH5α, or TG1) results in excision of the uracils by the active intracellular uracil N-glycosylase (UNG) and the creation of apyrimidinic sites, thereby favoring replication of the mutagenic strand.

Alternatively, the parental U strand can be selectively removed *in vitro* using UNG. Transformation of a ung^+ strain with the UNG-treated DNA gives ~10% higher mutation efficiencies compared with the *in vivo* method. Insertion or deletion mutagenesis can be performed in a manner similar to the substitution mutagenesis with an appropriately designed primer whose annealing would generate loop-in or loop-out intermediates at the mutagenic segment.

Various versions of uracil-based mutagenesis exist that differ in the preparation and use of dsDNA templates. In one variation, gapped duplex molecules are used as the target template for oligonucleotide-directed polymerization (14). The gapped duplex circular templates are generated by mixed annealing between linearized DNA molecules of U-containing plasmid and normal (T-containing) plasmid from which the target fragment has been removed.

Another variation, which is adapted for mutagenesis with double-stranded plasmid DNA, consists of the following steps (15): (i) preparation of U-containing dsDNA, (ii) denaturation at 0.2 N NaOH (and 0.2 mM EDTA, 15 min at 37°C) followed by neutralization, (iii) adsorption to and elution from NC filter (by centrifugation) of the denatured DNA, (iv) annealing and T7 DNA Pol-catalyzed extension of a mutagenic primer, (v) second NC filter centrifugation to remove nonconverted denatured DNA, and (vi) transformation of ung^+ host and screening of mutants.

2. Phosphorothioate-Based Mutagenesis

The phosphorothioate-based oligonucleotide-directed mutagenesis, developed by Eckstein et al. (16–18), relies on two distinct properties of phosphorothioate-substituted nucleotides toward the polymerase and nucleolytic enzymes: (i) DNA polymerases can stably incorporate α-thio-dNTPs, and (ii) the phosphorothioate-containing nucleotides in the (−) strand are resistant to hydrolysis by certain restriction enzymes (e.g., NciI, AvaI, HindII, PstI, and PvuI) which consequently inflict selective nicking in the phosphate-containing (+) DNA strand. Following the polymerase-catalyzed extension of a mutagenic primer in the presence of dNTPαS, the new (or minus) DNA strand is covalently closed using T4 DNA ligase (Fig. A.3). The circular heteroduplex is then treated with an appropriate restriction enzyme (RE1 or RE2) to generate nicked DNA. Depending on the position of the target site with respect to the nick, a $3' \rightarrow 5'$-exonuclease (e.g., Exo III) or a $5' \rightarrow 3'$-exonuclease (e.g., T7 or λ exonuclease) is used to create a gap in the wild-type (+) strand past the mismatch (19). The gap is then repaired by a second-round polymerase reaction. Transfection with such engineered DNA yields desired mutants with an efficiency higher than 70%. (A T7 DNA Pol-based mutagenesis kit is available from Amersham: Sculptor In Vitro Mutagenesis System.)

The phosphorothioate-based methods of mutagenesis are not limited to base substitutions but are applicable with equal efficiencies for deletions and insertions. Duplex DNA can also be used, following denaturation (with 0.2 N NaOH) and ethanol precipitation (in the presence of NH_4Ac, pH 4.5), as an efficient template for the mutagenesis (20).

3. Coupled-Priming Mutagenesis

In contrast to the uracil-based or phosphorothioate-based mutagenesis methods which employ a single mutagenic primer, other efficient mutagenesis methods have been developed that employ two or more primers simultaneously (21–26). The second oligonucleotide primer, also referred to as a selection primer, is targeted to a marker/reporter gene, altering the nucleotide sequence (e.g., restriction site) or inactivating the gene product which normally confers antibiotic resistance. The desired mutation, which segregates together with the phenotypic marker, can thus be conveniently selected. Coupled-priming mutagenesis is espe-

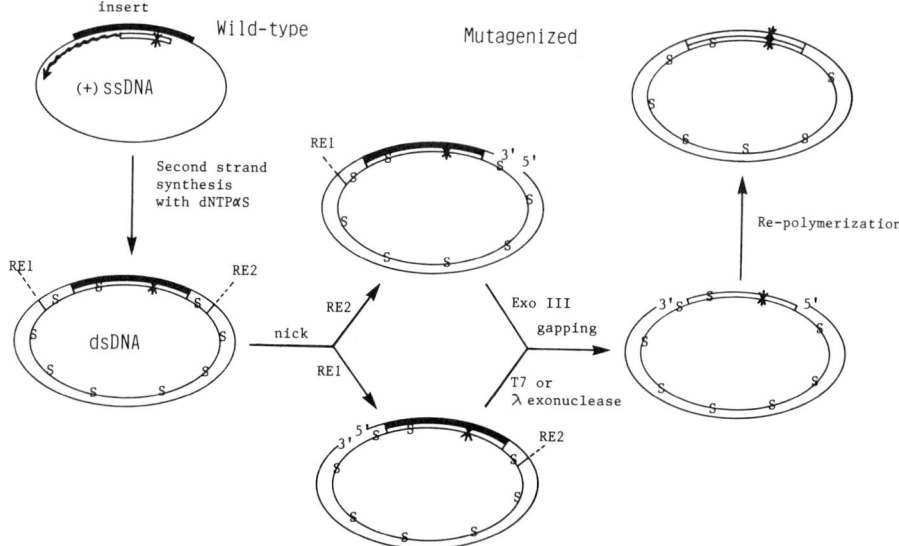

FIGURE A.3 A general scheme illustrating phosphorothioate-based mutagenesis. The mutagenic oligonucleotide primer is extended by DNA Pol in the presence of a dNTPαS. Digestion of the heteroduplex with a restriction enzyme introduces a nick only in the parent strand, rendering it susceptible to subsequent exonuclease digestion. A gap-filling reaction produces a homoduplex containing the desired mutation(s). The asterisk indicates a mutagenic site.

cially suited for multiple, successive rounds of mutagenesis. An extension of the coupled-priming strategy is the multiple-priming mutagenesis in which numerous regions of a DNA can be mutagenized simultaneously.

a. Restriction–selection

In this category, two methods using double-priming strategy will be described. The first method relies on the *EcoB/EcoK* restriction system of *E. coli* to apply selection pressure *in vivo* against the wild-type strand (21,22). The procedure consists of (i) annealing two oligonucleotide primers to a circular ssDNA template and (ii) extending the two primers simultaneously. One (mutant) primer introduces the desired mutation on the target gene, while the other (selection) primer interconverts one restriction site to the other. The two genetic markers, *EcoK* (5'-AAC-N_6-GTGC-3') and *EcoB* (5'-TGA-N_8-TGCT-3'), are interconvertible by a single base change. The heteroduplex DNA containing a hybrid *EcoB* and *EcoK* site is then introduced into an $EcoK^+$ or $EcoB^+$ *E. coli* host so that only the parent strand is subject to restriction. This positive *in vivo* selection for a mutagenic strand results in desired mutants with efficiencies as high as 70%. Alternate use of the two selection primers in conjunction with the $EcoK^+$ and $EcoB^+$ hosts enables one to perform an easy, stepwise construction of multiple mutations.

In the second method which is designed to apply the selection pressure *in vitro* (26), the selection primer is directed to a unique, nonessential restriction site, either eliminating it or converting it to an altered site: for instance, EcoRI (GA<u>AT</u>TC)/EcoRV (GA<u>TA</u>TC), HindIII (A<u>A</u>GC<u>T</u>T)/MluI (A<u>C</u>GC<u>G</u>T), ScaI (AG<u>TA</u>CT)/MluI (A<u>C</u>GC<u>G</u>T), SspI (A<u>AT</u>A<u>T</u>T)/StuI (A<u>GG</u>CC<u>T</u>), and SspI (A<u>AT</u>A<u>TT</u>)/EcoRV (<u>G</u>A<u>T</u>A<u>TC</u>). The two (selection and mutagenic) primers are annealed to circular ssDNA and are extended simultaneously by the use of T4 DNA Pol. The resulting dsDNA is transformed into a mismatch repair-defective E. coli strain (e.g., BMH 71-18 mutL) in which the mutagenic and parent strands segregate with equal proficiency during the first round of DNA replication. Transformants are selected en masse in liquid medium containing an appropriate antibiotic. Plasmid DNAs are then prepared and digested with the enzyme that recognizes the unique restriction site. This processing enriches the mutant plasmids that are resistant to digestion, remain circular, and transform bacteria more efficiently than the linearized DNAs in a second-round transformation. This mutagenesis method yields desired mutants with ~80% efficiencies. [Several mutagenesis kits based on this restriction–selection method are commercially available: for example, Transformer site-directed mutagenesis kit (Clontech), U.S.E. (for unique site elimination) mutagenesis kit (Pharmacia Biotech), and Chameleon mutagenesis kit (Stratagene).] In an alternative case where mutant plasmids are designed to be sensitive to restriction enzyme digestion while parent plasmids are resistant, the restriction–selection method applies in a slightly different but equally efficient format: the linearized plasmids are purified, for example, by gel electrophoresis, recircularized, and then used to transform competent bacteria.

b. Antibiotic selection

The strategy of double-priming mutagenesis can also be applied on antibiotic markers, thereby exerting a formidable selection pressure in favor of the mutagenic strand which restores the antibiotic resistance phenotype (23). Application of the strategy on two alternately convertible antibiotic selection markers carried on the same vector provides an efficient means of performing multiple rounds of mutagenesis (24). In this triple-priming strategy (Fig. A.4), a segment of or a whole gene is inserted first into a suitable vector which disposes two selectable markers such as Ap/Tet, Ap/Cm, or Tet/Cm. [The pALTER family of phagemid vectors is available commercially as a part of the "Altered Sites Mutagenesis system" (Promega).] The mutagenic primer annealed on ssDNA templates is extended using T4 or T7 DNA Pol; so are a *knockout* primer (e.g., $Cm^r \rightarrow Cm^s$) and a *repair* primer (e.g., $Ap^s \rightarrow Ap^r$). Transformation of competent E. coli cells with the heteroduplex and culture on Ap plates produce desired mutants with ~80% efficiencies.

4. PCR MUTAGENESIS

The ease and versatility of amplifying a mutagenic DNA fragment by using a mutagenic primer have made the PCR mutagenesis an important technique in site-specific mutagenesis. Mutagenic dsDNA can be reinserted into the gene or plasmid via appropriate restriction sites, adapter sequences (refer to Fig. 6.3), or overlap annealing. PCR mutagenesis is similar to and extends the utility of *cassette*

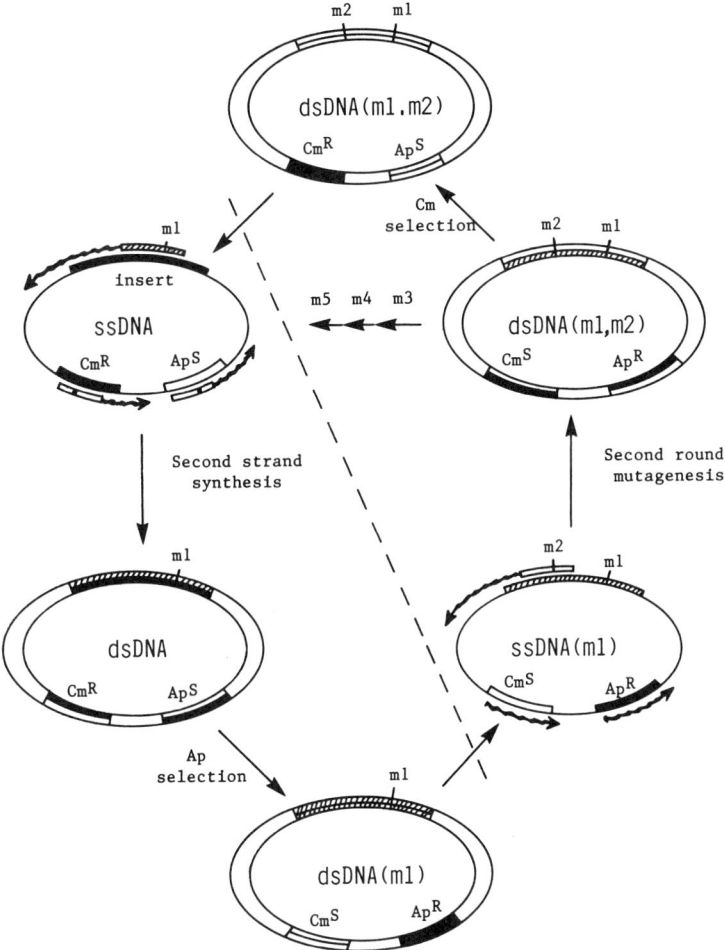

FIGURE A.4 A general scheme for triple-priming mutagenesis and consecutive introduction of mutations (m1, m2, m3, etc.). The method employs, in addition to the mutagenic primer, two selection primers called "knockout" and "repair" primers, respectively. On simultaneous extension, the selection primers alternatively convert two antibiotic markers [ampicillin (Ap) and chloramphenicol (Cm)] between resistance (R) and sensitive (S) phenotypes.

mutagenesis in which the wild-type gene or DNA segment is wholly replaced by a new or mutagenic DNA fragment (27). PCR mutagenesis is distinct from other conventional mutageneses by the fact that mutagenic dsDNA is amplified *in vitro,* thus obviating the potential bias applied against the mutagenic strands if heteroduplexes were introduced into the cells. In addition to the convenience of using dsDNA as the starting material, PCR mutagenesis is particularly suited for deletion mutagenesis because the two sequences flanking the deletion site can be readily incorporated into dsDNA and linked to each other via a sequence overlap.

Likewise, PCR mutagenesis is amenable to insertion mutagenesis because new or altered sequences can be readily incorporated in the overlapping region. PCR mutagenesis can be divided into three major groups of strategies: (a) overlap extension of mutagenized DNA cassettes, (b) nonoverlap extension followed by ligation, and (c) overlap circle PCR.

a. Overlap-extension PCR mutagenesis

A number of methods developed for PCR mutagenesis (substitution, insertion, and deletion) are based on overlap-extension strategies (28–31). The prototype consists of the following steps (Fig. A.5): First, two sets of DNA fragments are amplified separately using two (inner and outer) pairs of oligonucleotide primers;

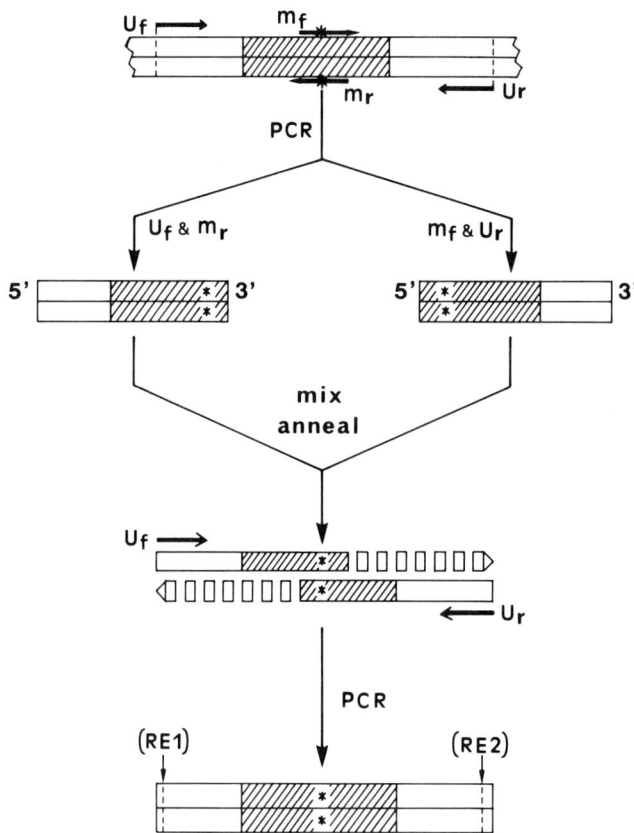

FIGURE A.5 A general scheme for standard PCR mutagenesis which employs two mutagenic primers (forward and reverse, m_f and m_r, respectively) and two universal (or amplification) primers (u_f and u_r). The two PCR products are joined by annealing at the overlap sequences and are further extended and PCR amplified. The overlap-extension strategy produces a homoduplex mutant cassette which is then inserted into a suitable expression vector. RE1 and RE2 indicate two available (natural or engineered) restriction sites.

the inner (mutagenic) pair overlaps the mutagenic region. Second, the two PCR products are mixed, denatured, and partially annealed via the overlapping region. Third, single-stranded regions extending from the double-stranded overlap region are converted by polymerase-catalyzed reaction into an elongated duplex, which is further amplified by PCR using the outer (universal) pair of primers. The PCR-generated mutated fragments are then inserted into an expression vector at appropriate restriction sites.

A number of variations have been developed that directly improve the efficiency of mutagenesis and/or streamline the mutagenesis and cloning procedures. For example, the use of biotinylated primers facilitates the isolation of PCR products (32). Another method introduces a λ exonuclease digestion step after the first PCR (using 5'-P mutagenic primers) to generate ssDNA, thus increasing the efficiency of the next hybridization step (33). When template DNA is prepared from dam^+ E. coli, the carryover of the parent DNA can be substantially reduced by treatment with DpnI which specifically restricts the DNA at G^mATC sites. PCR products are unmethylated and so are resistant to DpnI digestion.

In another elegant variation featuring ligation-free cloning (34), both mutagenic and amplification primers are synthesized using dU in place of the T base(s) at their 5' ends. The two amplified products are then digested *in vitro* with uracil N-glycosylase. This treatment generates 3' complementary overhangs which permit efficient annealing between the two PCR fragments and also between the PCR products and the complementary-ended vectors.

Some variations in the use of primers employ one mutagenic and two flanking (universal) primers instead of the standard two–two combination. In these methods (35–38), the first mutagenic fragment (e.g., m_f/u_r-PCR, see Fig. A.5) serves as a "long" primer in the second PCR on wild-type templates to generate full-length mutant products (e.g., u_f/u_r-PCR). If an asymmetric PCR is performed in the presence of limiting mutagenic primers, the first PCR product can be obtained as a single-stranded mutagenic fragment which offers the obvious advantage in the subsequent step of overlap annealing (38). Still another variation (Fig. A.6) employs one mutagenic and three universal primers to efficiently carry out the mutagenesis (39).

b. Nonoverlap PCR mutagenesis

In contrast to the methods based on overlap-extension strategies, nonoverlap PCR mutagenesis employs at least one set of tail-to-tail primers for PCR amplification and blunt-end ligation of the products. In one such method (40), two small nonoverlapping DNA fragments, one of which contains the desired mutation, are generated by separate PCR. The two tail-to-tail fragments, i.e., mutagenic and nonmutagenic "cassettes," are digested with restriction enzymes to open respective cohesive termini and then are ligated to a similarly treated plasmid. The DNA with free tail-to-tail ends is circularized by a blunt-end ligation.

In other PCR mutagenesis methods that are especially suited for small inserts or plasmid vectors, the entire circular plasmid is amplified using two (mutagenic and amplification) primers in a tail-to-tail configuration. This technique is an

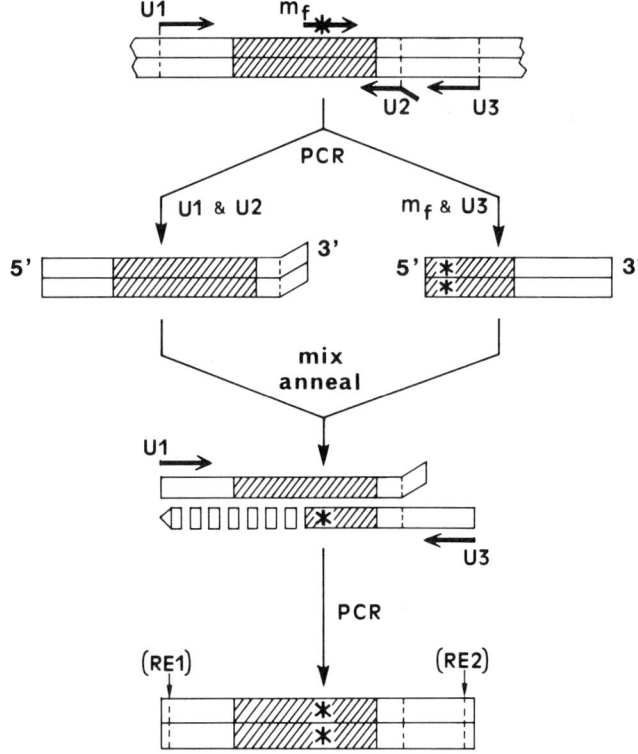

FIGURE A.6 An alternative overlap-extension PCR mutagenesis scheme. The method employs a single mutagenic primer (m_f) and three universal primers (u_1, u_2, and u_3). Note that the 5' end of primer u_2 contains mismatches (8 nt) designed to prevent the termini extension in the second PCR.

adaptation of a PCR strategy known as *inverse PCR*, which was developed to amplify and clone unknown sequences flanking a region of known sequences (41,42). The linear plasmid molecules are then isolated, recircularized by blunt-end ligation, and used for transformation (43). This method easily lends to deletion mutagenesis when the primer pair is designed to leave a gap spanning the target region between the 5'-end sequences of the primers (44).

The "long PCR" should make the nonoverlap mutagenesis as well as the following overlap circle mutagenesis methods particularly efficient.

c. Overlap Circle PCR Mutagenesis

This method, as yet another variation of the inverse PCR, makes use of two or more overlapping fragments generated by separate PCR. When these fragments are mixed, denatured, and annealed together, they give rise to recombinant circular plasmids. For some small-sized plasmids, two pairs of primers may be enough to generate two overlapping linear fragments and ultimately to obtain gapped duplex recombinant circles (45). The recombinant circles of DNA can then be directly used to transform competent cells, obviating the inefficient blunt-end ligation step involved in nonoverlap PCR mutagenesis. More generally, and for larger plasmids, the whole plasmid containing an insert to be mutagenized can be amplified by PCR in three or more overlapping fragments. Complete duplex plasmids can then be reconstituted by mixed annealing and used for bacterial transformation (46,47). This PCR mutagenesis method is rapid, flexible, and amenable to the introduction of two or more mutations simultaneously.

References

1. Smith, M. (1985). *Ann. Rev. Genet.* **19**, 423–462.
2. Hutchison, C. A., III, Phillips, S., Edgell, M. H., Gillam, S., Jahnke, P., and Smith, M. (1978). *JBC* **253**, 6551–6560.
3. Botstein, D., and Shortle, D. (1985). *Science* **229**, 1193–1201.
4. Zoller, M. J., and Smith, M. (1983). *Meth. Enzy.* **100**, 468–500.
5. Osinga, K. A., Van der Bliek, A. M., Van der Horst, G., Groot Koerkamp, M. J. A., Tabak, H. F., Veeneman, G. H., and Van Boom, J. H. (1983). *NAR* **11**, 8595–8608.
6. Zoller, M. J., and Smith, M. (1987). *Meth. Enzy.* **154**, 329–350.
7. Perlak, F. J. (1990). *NAR* **18**, 7457–7458.
8. Lesley, S. A., Ziegelbauer, J., and Maffitt, M. (1994). *Promega Notes* **46**, 2–5.
9. Liao, X., and Wise, J. A. (1990). *Gene* **88**, 107–111.
10. Kunkel, T. A. (1985). *PNAS* **82**, 488–492.
11. Kunkel, T. A., Roberts, J. D., and Zakour, R. A. (1987). *Meth. Enzy.* **154**, 367–382.
12. McClary, J. A., Witney, F., and Geisselsoder, J. (1989). *BioTechniques* **7**, 282–289.
13. Wang, L.-M., Geihl, D. K., Choudhury, G. G., Minter, A., Martinez, L., Weber, D. K., and Sakaguchi, A. Y. (1989). *BioTechniques* **7**, 1000–1010.
14. Hofer, B., and Kuhlein, B. (1989). *Gene* **84**, 153–157.
15. Jung, R., Scott, M. P., Oliveira, L. O., and Nielsen, N. C. (1992). *Gene* **121**, 17–24.
16. Nakamaye, K. L., and Eckstein, F. (1986). *NAR* **14**, 9679–9698.
17. Sayers, J. R., Schmidt, W., Wendler, A., and Eckstein, F. (1988). *NAR* **16**, 803–814.
18. Sayers, J. R., and Eckstein, F. (1988). *Gen. Eng.* **10**, 109–122.
19. Sayers, J. R., and Eckstein, F. (1991). In "Directed mutagenesis: A Practical Approach" (M. J. McPherson, Ed.), pp. 49–69. IRL Press, Oxford.
20. Sugimoto, M., Esaki, N., Tanaka, H., and Soda, K. (1989). *Anal. Biochem.* **179**, 309–311.
21. Carter, P., Bedouelle, H., and Winter, G. (1985), *NAR* **13**, 4431–4443.
22. Carter, P. (1987). *Meth. Enzy.* **154**, 382–403.
23. Lewis, M. K., and Thompson, D. V. (1990). *NAR* **18**, 3439–3943.
24. Stanssens, P., Opsomer, C., McKeown, Y. M., Kramer, W., Zabeau, M., and Fritz, H.-J. (1989). *NAR* **17**, 4441–4454.
25. Perlak, F. J. (1990). *NAR* **18**, 7457–7458.
26. Deng, W. P., and Nickoloff, J. A. (1992). *Anal. Biochem.* **200**, 81–88.
27. Wells, J. A., Vasser, M., and Powers, D. B. (1985). *Gene* **34**, 315–323.
28. Higuchi, R., Krummel, B., and Saiki, R. K. (1988). *NAR* **16**, 7351–7367.
29. Ho, S. N., Hunt, H. D., Horton, R. M., Pullen, J. K., and Pease, L. R. (1989). *Gene* **77**, 51–59.
30. Yon, J., and Fried, M. (1989). *NAR* **17**, 4895.
31. Horton, R. M., Hunt, H. D., Ho, S. N., Pullen, J. K., and Pease, L. R. (1989). *Gene* **77**, 61–68.
32. Hall, L., and Emery, D. C. (1991). *Prot. Eng.* **4**, 601.
33. Shyamala, V., and Ferro-Luzzi Ames, G. (1991). *Gene* **97**, 1–6.
34. Owen, J. L., Hay, C., Schuster, D. M., and Rashtchian, A. (1994). *Focus (BRL)* **16**(2), 39–44.
35. Kammann, M., Laufs, J., Schell, J., and Gronenborn, B. (1989). *NAR* **17**, 5404.
36. Landt, O., Grunert, H.-P., and Hahn, U. (1990). *Gene* **96**, 125–128.

37. Nelson, R. M., and Long, G. L. (1989). *Anal. Biochem.* **180**, 147–151.
38. Perrin, S., and Gilliland, G. (1990). *NAR* **18**, 7433–7438.
39. Mikaelian, I., and Sergeant, A. (1992). *NAR* **20**, 376.
40. Kadowaki, H., Kadowaki, T., Wondisford, F. E., and Taylor, S. I. (1989). *Gene* **76**, 161–166.
41. Triglia, T., Petersen, M. G., and Kemp, D. J. (1988). *NAR* **16**, 8186.
42. Ochman, H., Medhora, M. M., Garza, D., and Hartl, D. L. (1990). *In* "PCR Protocols: A Guide to Methods and Applications" (M. A. Innis *et al*. Eds.), pp. 219–227. Academic Press, San Diego.
43. Hemsley, A., Arnheim, N., Toney, M. D., Cortopassi, G., and Galas, D. J. (1989). *NAR* **17**, 6545–6551.
44. Imai, Y., Matsushima, Y., Sugimura, T., and Terada, M. (1991). *NAR* **19**, 2785.
45. Jones, D. H., and Howard, B. H. (1990). *BioTechniques* **8**, 178–183.
46. Jones, D. H., and Winistorfer, S. C. (1992). *BioTechniques* **12**, 528–535.
47. Watkins, B. A., Davis, A. E., Cocchi, F., and Reitz, M. S., Jr. (1993). *BioTechniques* **15**, 700–704.

APPENDIX B
Genotypes of *Escherichia coli* Strains

A *genotype* indicates the genetic trait which is at least partly responsible for the observed behavior (*phenotype*) of an organism. *E. coli* genotypes list only genes that are defective. If a gene is not mentioned, then it is assumed to be nonmutated. Exceptionally, genes listed on the F' episome indicate wild-type alleles unless specified otherwise. Proper notation for genotype omits, in contrast to that for phenotype, superscript + or −, but these are sometimes used redundantly for clarity. Deletion mutations are noted as Δ, followed by the names of deleted genes in parentheses, e.g., Δ(lac-pro). Specific mutations are given allele numbers that are usually italic arabic numerals, e.g., *hsdR17*. Strains are λ⁻ unless specified otherwise.

The following lists for (i) *genotypes of some E. coli strains* and (ii) *useful genetic markers and phenotypic traits* have been compiled based on the following publications: NEB catalog, GIBCO/BRL catalog, Promega catalog, Stratagene catalog, Sambrook *et al.* ["Molecular Cloning: A Laboratory Manual," 2nd ed., pp. A.9–A.13 (1989) Cold Spring Harbor Laboratory Press], *E. coli* linkage map by Bachmann, B. J. ["*Escherichia coli* and *Salmonella typhimurium:* Cellular and Molecular Biology" (F. C. Neidhardt *et al.*, eds.), Vol. 2, pp. 807–876 (1987)

American Society of Microbiology, Washington, D.C.], and "Selectable Phenotypes of E. coli and S. typhimurium" by R. T. Vinopal (ibid. pp. 990–1015).

I. GENOTYPES OF SOME FREQUENTLY USED E. COLI STRAINS

BL21 (DE3) F^- $hsdS(r_B^- m_B^-)$ gal ompT (λ cI857 ind-1 nin-5 Sam-7 lacUV5-T7 gene 1). An E. coli B strain with a λ prophage (DE3) carrying the T7 RNA Pol gene under the lacUV5 promoter. The strain is also deficient in Lon protease

BNN93 F^- $hsdR(r_K^- m_K^+)$ e14$^-$(McrA$^-$) mcrBC fhuA21 lacY1 leuB6 supE44 thi-1 thr-1. A general host for phage λ and recombinant λ vectors (e.g., λgt10)

BNN97   BNN93(λgt11). A λgt11 lysogen producing λgt11 phage at 42°C.

C600 F^- e14$^-$(McrA$^-$) fhuA21 lac Y1 leuB6 rfbD1 supE44 thi-1 thr-1. A general host for phage λ and recombinant λ vectors (e.g., λgt10). Some of the strains circulating as C600 are actually BNN93. The original C600 is EcoK(r^+m^+) $mcrBC^+$.

CC118 F^- Δ(ara-leu)7679 appR araD139 argE(am) galE ΔlacX74 ΔphoA20 recA1 thi Spc^r Rif^r

CJ236 F^- dut-1 ung-1 thi-1 relA1 pCJ105(Cm^r F'). A strain used for mutagenesis based on uracil-substituted DNA.

DH1 F^- $hsdR17(r_K^- m_K^+)$ endA1 gyrA96(Nal^r) recA1 relA1 supE44 thi-1. A recombination-deficient suppressing strain generally useful for plasmid amplification.

DH5α DH1 Δ(lacZYA-argF)U169 deoR ϕ80dlacZΔM15. This strain has a higher transformation efficiency than DH1 and permits α complementation (blue–white) selection of recombinants for M13 and derivative vectors.

DH5αF'   DH5α plus F'. This strain permits superinfection by M13 and derivative phages.

DH5αMCR   DH5α mcrA Δ(mrr-hsdRMS-mcrBC). A strain suitable for construction of genomic and cDNA libraries with methylated DNAs.

DH10B F^- mcrA Δ(mrr-hsdRMS-mcrBC) araD139 Δ(ara-leu)7697 deoR endA1 galK galU nupG recA1 rpsL ΔlacX74 Φ80dlacZΔM15. A high efficiency host for genomic library construction with pUC vectors containing methylated bases.

DH11S   NM522 mcrA Δ(mrr-hsdRMS-mcrBC) Δ(recA)1398 deoR rpsL srl. A good general host for cDNA library construction or for amplification of M13-type plasmid and phagemid

I. Genotypes of Some Frequently Used E. coli Strains 675

	vectors. Compared with JM109, DH11S allows for the preparation of highly purified ssDNA from phagemid vectors due to the presence of the wild-type *endA* gene.
DH20	DH1 F'(*lacI*^{q}*Z proAB*)
DM1	*hsdR2*($r_K^- m_K^+$) *mcrB dam13*::Tn9(Cm^r) *dcm ara gal-1 gal-2 lac leu su° thr ton tsx*. A subcloning efficiency host for pUC-type vectors. Plasmid DNAs prepared from the host are unmethylated at *dam* (GATC) and *dcm* [CC(A/T)GG] sites.
HB101	F$^-$ *hsdS20*($r_B^- m_B^-$) Δ(*mcrBC-mrr*) Δ(*gpt-proA*)62 *ara-14 galK2 lacY1 leuB6 mtl-1 recA13 rpsL20*(Str^r) *supE44 xyl-5*. An *E. coli* K12 × *E. coli* B hybrid. A good general host for plasmid DNA; produces relatively stable, unmethylated DNA.
JM83	F$^-$ *ara* Δ(*lac-proAB*) *rpsL*(Str^r) φ80d*lacZ*ΔM15. An *E. coli* K12 host providing the facility for LacZ α complementation selection.
JM101	F'[*proAB-lacI*^{q}*Z*ΔM15-*traD36*] Δ(*lac-proAB*) *supE44 thi-1*. A prototype *E. coli* K12 strain developed by J. Messing for use as the host for M13 and derivative vectors. This and other JM series strains permit blue–white selection of recombinants based on *lacZ* α complementation.
JM105	JM101 *hsdR4* ($r_K^- m_K^+$) *endA1 rpsL*(Str^r) *sbcBC*
JM107	JM101 *hsdR17* ($r_K^- m_K^+$) e14$^-$(McrA$^-$) *endA1 gyrA96* (Nal^r) *relA1*
JM109	JM107 *recA1*. This strain is defective for cell wall synthesis and forms mucoid colonies on minimal media.
JM109 (DE3)	JM109 λ(DE3) [see BL21(DE3)]
JM110	JM101 *hsdR17*($r_K^- m_K^+$) *dam dcm ara fhuA galK galT lacY leu rpsL*(Str^r) *thr tsx*
JS5	MC1061 *rpsL*(Str^r) *thi recA1*/F'::Tn*10*(Tet^r)*proAB lacI*^{q}*-Z*ΔM15. A derivative of MC1061 for *lacZ* blue–white selection cloning. A good M13 host.
K802	F$^-$ *hsdR2* ($r_K^- m_K^+$) e14$^-$(McrA$^-$) *galK2 galT22 lacY1* or Δ(*lac*)6 *mcrB1 metB1 rfbD1 supE44*. A suppressing strain used to propagate phage λ vectors. Note that K802 is identical to WA802.
LE392	F$^-$ *hsdR514*($r_K^- m_K^+$) *galK2 galT22 lacY1 metB1 supE44 supF58 trpR55*. A suppressing strain used to propagate phage λ vectors. An alternative host to Y1090; LE392 does not allow color selection of recombinants or IPTG induction of expression.
MC1061	*hsdR2*($r_K^- m_K^+$) *mcrA mcrB1 araD139* Δ(*ara-leu*)7697 *galK galU* Δ(*lac*)*X74 strA*. A *recA*$^+$, high electrotransformation efficiency strain.

Appendix B Genotypes of Escherichia coli Strains

MV1190   F'[proAB-lacI^{q}ZΔM15-traD36] Δ(lac-proAB) supE thi Δ(srl-recA)306::Tn10(Tet^r)

MV1193   F'[proAB-lacI^{q}ZΔM15-traD36] hsdR4 endA Δ(lac-proAB) rpsL(Str^r) spcB15 thi Δ(srl-recA)306::Tn10(Tet^r). A strain used for enrichment of mutant DNA; adapted for lacZ blue–white selection cloning.

MZ-1   attLΔBamN$_7$N$_{53}$bio cI857 ΔH1 galKΔ8 his ilv N$^+$. A lysogenic strain producing temperature-sensitive λ repressor (cIts); permits heat-inducible expression of the genes cloned downstream of the λP$_L$ promoter. The strain also lacks O-polysaccharide in LPS, displaying "rough" phenotype.

MZ9   F$^-$ Δ(lac)X74 phoA20 phoB$^+$ phoR$^-$ trp(am) strA

NM522   F'[proAB-lacI^{q}ZΔM15] Δ(hsdMS-mcrBC)5($r_K^- m_K^-$) Δ(lac-proAB) supE thi. A general host for M13 and derivative vectors. To maintain the F', the strain should be grown on minimal (M-9) plates.

PMC103   mcrA Δ(mcrBC-hsdRMS-mrr)102 recD sbcC. A general host for plasmid and λ phages. An instability problem may occur with plasmids containing direct repeat or certain retroviral sequences.

RR1   HB101 recA$^+$. A general host for high efficiency transformation with plasmids.

SRB   F'[proAB-lacI^{q}ZΔM15] e14$^-$(McrA$^-$) Δ(mrr-hsdRMS-mcrBC)171 endA1 gyrA96 lac recJ relA1 sbcC supE44 thi-1 uvrC umuC::Tn5(Kan^r)

STBL2   JM109 Δ(mcrBC-hsdRMS-mrr). A derivative of JM109/J5. A good general host (from GIBCO/BRL) for the cloning of plasmids and phagemids. No sequence-dependent plasmid instability has been observed.

SURE   SRB recB Tn10(Tet^r). A general host (from Stratagene) for plasmid and phagemid cloning/sequencing vectors; enables blue–white selection of recombinants based on lacZ α complementation. Instability problem may arise with plasmids containing direct repeat or certain retroviral sequences.

TB1   JM83 hsdR($r_K^- m_K^+$)

TG1   JM101 Δ(hsd)5. An EcoK$^-$ derivative of JM101.

TG2   TG1 Δ(srl-recA)306::Tn10(Tet^r). A recombination-deficient derivative of TG1.

WA802   see K802

WM1100   MC1061 Δ[(srl-recA)306::Tn10]. A recA derivative of MC1061; a useful strain for general cloning.

χ1776   F$^-$ hsdR2($r_K^- m_K^+$) Δ(bioH-asd)29 Δ(gal-uvrB)40 cycA1 cycB2 dapD8 fhuA53 gyrA25(Nal^r) metC65 minA1 minB2 rfb-2 oms-1 (tte-1) oms-2 supE44 thyA142

XL1-Blue   F'[*proAB-lacI*^{q}*Z*Δ*M15*-Tn*10*(Tet^r)] *hsdR17*($r_K^-m_K^+$) *endA1 gyrA96*(Nal^r) *lac recA1 relA1 supE44 thi-1*. A recombination-deficient, α complementation competent strain. A good general host for plasmid, phagemid, and λZAPII recombinants but not λZAP which contains Sam100 mutation.

XL1-Blue MRF'   XL1-Blue Δ(*mcrA*)*183* Δ(*mcrBC-hsdMRS-mrr*)*173*. A good general host for plasmid and phagemid cloning/sequencing vectors.

Y1088   F$^-$ Δ(*lac*)*U169 supE supF hsdR*($r_K^-m_K^+$) *metB trpR fhuA21 proC*::Tn*5* pMC9(Tet^r Ap^r). A strain used for IPTG-induced protein production from λgt11; suitable for initial antibody screening of recombinants. The *supF* marker suppresses Sam100 to allow cell lysis. Note: pMC9 is pBR322 with *lacI*^{q} inserted.

Y1089   F$^-$ Δ(*lac*)*U169 lon-100 araD139 strA hflA150*::Tn*10* pMC9. See Y1088. A strain used to generate (recombinant) phage λ lysogen.

Y1090   F$^-$ Δ(*lac*)*U169 lon-100 araD139 rpsL*(Str^r) *mcrA supF trpC22*::Tn*10* pMC9. A strain for phage λ (e.g., λgt11) used for immunological screening of the proteins expressed from the *lac* promoter by IPTG induction. The *supF* marker suppresses Sam100 of λ vectors to allow cell lysis. The *lon* protease deficiency ensures a high yield production of proteins. For initial recombinant phage screening, λgt11 should not be plated directly on Y1090, but should be first amplified through Y1088 which is *hsdR*$^-$.

II. USEFUL GENETIC MARKERS AND PHENOTYPIC TRAITS

am   *amber* (UAG) mutation
araA   Mutation in L-arabinose isomerase. The strain is unable to utilize arabinose
araC   Mutation in the regulatory gene (activator/repressor protein)
cyc   Mutants are resistant to D-cycloserine
dam   Structural gene for DNA adenosine methylase. Mutation blocks methylation at the recognition sequences G^{m}ATC (methylated on A). Useful in preparing unmethylated DNA which is susceptible to cleavages by such restriction enzymes as *Bcl*I and *Cla*I. Dam mutants exhibit lower transformation efficiencies than dam$^+$ cells

dcm   Structural gene for DNA cytosine methylase. Mutation blocks methylation at the recognition sequences (C^mC(A/T)GG). Useful in preparing unmethylated DNA which is susceptible to cleavages by some restriction enzymes such as *Ava*II, *Eco*RII, and *Stu*I

DE3   Strains contain a chromosomal copy of the gene for T7 RNA polymerase. Suitable for a high level expression of a cloned gene in a T7 promoter vector

deoR   Mutation in the regulatory gene for *deo* (deoxyribose) operon. It allows more efficient uptake of large DNA fragments. It also allows growth with inosine as the sole carbon source

dut   dUTPase activity abolished. In combination with *ung*, it allows incorporation of uracil into DNA

e14   Revised designation for previous *mcrA1* genotype. The loss of the excisable prophage-like element e14, which carries the *mcrA* gene, is responsible for the McrA⁻ character of many strains. Other *mcrA* alleles may also result from excisions of e14

endA1   A mutation inactivating nonspecific DNA endonuclease I. It improves the yield and quality of isolated plasmid DNA

F (or F′)   F(fertility) or F′ episome which produces sex pili. F′ factor carries portions of the *E. coli* chromosome, most notably the *lac* operon on F′(*lac-proAB*). F′ factor is necessary for infection by M13 phage or helper phages (M13K07, R408). Useful for production of single-stranded phage (or phagemid) DNA

fhuA   Mnemonic for ferric hydroxamate uptake. Synonym with *tonA*.

galK   Synonym with *galA*, structural gene for galactokinase. Mutants are unable to utilize galactose. The mutant strains permit use of galactose-linked expression vectors

galT   Synonym with *galB*, structural gene for galactose-1-phosphate uridylyltransferase. Mutant strains permit use of galactose-linked expression vectors

gyrA   Structural gene for the subunit A of DNA gyrase. The *gyrA96* mutation results in resistance to nalidixic acid

hflA   A high frequency lysogeny mutation. The mutation increases phage λ lysogeny by inactivating a specific protease that would otherwise degrade the repressor (cI)

hsdM   DNA methylase of *Eco*K restriction–modification system. Mutation blocks specific adenine methylation in the sequence A^mACN6GTGC or GC^mACN6GTT. DNA isolated from a HsdM⁻ strain will be restricted by a HsdR⁺ host

hsdR   The mutation abolishes *Eco*K restriction, but *Eco*K protective methylation is retained. The mutation permits the introduction

	of foreign DNA without the danger of being cleaved by type I restriction enzymes
hsdS	Both *Eco*K restriction and methylation are abolished
cI857	A temperature-sensitive λ repressor (cIts). Gene expression under the control of λP_L or λP_R promoter can be induced by a temperature shift to 42°C
lacI^q	Mutation in the *i* (gene for repressor) promoter which leads to a 10-fold overproduction of *lac* repressor. Useful for turning off expression from the *lac* promoter
lacY	Galactose permease (M protein) activity abolished, blocking lactose utilization
lacZ	β-Galactosidase activity abolished, permitting use of vectors which carry a *lacZ* reporter system. *lacZ*ΔM15 expresses a C-terminal fragment that complements the Lac α fragment encoded by many vectors. These vectors yield a blue color on X-Gal only if the host carries ΔM15. ΔU169, ΔX111, and ΔX74 all delete the entire *lac* operon from the chromosome, in addition to varying amounts of flanking DNA. ΔX111 deletes *proAB* as well, so that the cell requires proline for growth on minimal medium, unless it also carries F' *lac proAB*
lon	Structural gene for Lon protease (ATP-dependent). The mutation results in increased stability for certain foreign proteins. *E. coli* B and its derivatives are naturally Lon⁻. Colonies of the *lon*⁻ strains appear mucoidal
malA	A mutation that renders the strain incapable of utilizing maltose
mcr	Mutation that affects either or both of two methylcytosine-specific restriction systems, McrA and McrB. The *mcrA* gene, located at 25 min on the *E. coli* K12 chromosome (on the e14 excisable element), codes for a methylation restriction activity with specificity for C^{m}CGG. The *mcrB* region, located at 99 min on the chromosome, contains two genes, *mcrB* and *mcrC*, the products of which form a heterodimeric GTP-dependent nuclease. For a third 5-methylcytosine restriction activity, called *mcrF*, refer to *mrr*. Note: Dcm-modified DNA is not restricted by Mcr⁺
min	Mutation leads to formation of minute cells containing no DNA
mrr	As part of the methylation-dependent restriction system (MDRS), the gene confers restriction of DNAs containing methylated adenines. The *mrr* gene has a restriction activity against some sequences (e.g., CpG residues) containing methylated cytosines. This cytosine restriction activity is referred to as *mcrF*. Note: Dam, Dcm, or *Eco*K-modified DNA is not restricted by Mrr⁺ or any other *E. coli* K12 restriction system

mutD	(or *dnaQ*) Structural gene for DNA polymerase III subunit. The mutation results in an increased frequency of spontaneous mutation. Only useful for mutagenesis work
proA	γ-Glutamylphosphate reductase activity abolished, leading to a deficiency in proline biosynthesis. Mutant requires Pro for growth in minimal media
proB	γ-Glutamate kinase activity abolished, leading to a deficiency in proline biosynthesis. Mutant requires Pro for growth in minimal media
recA1	Mutation abolishes homologous recombination and increases the stability of certain cloned DNA sequences. Strains grow more slowly and transform at lower efficiencies
recBC	Structural genes for the subunits of exonuclease V. Mutant strains show deficiency in RecBCD enzyme-dependent recombination and DNA repair. Useful for stabilizing certain cloned DNA sequences, especially inverted repeat sequences. Strains grow slowly and are difficult to transform
recJ	Mutation in exonuclease involved in a plasmid-by-plasmid homologous recombination
relA	Mutation associated with a relaxed phenotype. It permits RNA synthesis in the absence of protein synthesis
rfbD	Mutation in TDP-rhamnose synthase. It confers penicillin resistance
rpsL	Mutation in 30S ribosomal subunit protein S12. It confers resistance to streptomycin. Also written *strA*
Sam7	Amber mutation which prevents lysis of the infected or expression-induced cells that lack the *amber* suppressor *supF*
Sam100	Also denoted as S100. See Sam7
sbcB	Structural gene for exonuclease I. The mutation, which is usually found with *recBC sbcC*, results in a deficiency in the RecF-dependent recombination pathway. The quadruple mutants (*recBC sbcBC*) are recombination proficient
srl	Mutation affecting sorbitol metabolism. Mutants are resistant to sorbitol plus xylitol
str	Streptomycin resistance gene specifying an enzyme that modifies the antibiotic and inhibits its binding to the ribosome. Mutant strains become streptomycin sensitive
supE	Suppressor of *amber* (UAG) mutation by inserting Gln. Strains carry the gene (*glnV*) for glutamine tRNA2. Required for growth of some phages
supF	Suppressor of *amber* (UAG) mutation by inserting Tyr. Strains carry the gene (*tyrT*) for tyrosine tRNA1. Required for lytic growth of S7 or S100 λ phage (e.g., λgt11)
thi (or *thiA*)	Mutation in the synthesis of thiazole. Mutants require an

	exogenous supply of thiamin (vitamin B_1) for growth in minimal media
Tn5	A transposon that normally codes for kanamycin resistance (Kan^r)
Tn10	A transposon that normally codes for tetracycline resistance (Tet^r)
tonA	Synonym with fhuA. A mutation (T-one) in the outer membrane protein which serves as the receptor for ferrichrome, colicin M, and phages T1, T5, and ϕ80
traA	A defect in the gene encoding pilin
traD	An F factor gene involved in conjugal DNA transfer. The mutation severely reduces the self-transmissibility of the F factor and plasmid DNA as well
tsx	Mutation in an outer membrane protein, which serves as a receptor for phage T6 and colicin K. Mutant is defective in the active transport of nucleosides
umuC	The mutation blocks SOS repair pathway, rendering the cells UV sensitive
uvrC	Mutant loses the ability to repair UV-damaged DNA
uvrD	Mutation in the gene for DNA helicase II. Repair of UV-damaged DNA affected
ung	Uracil N-glycosylase activity abolished. Mutant permits incorporation of uracil residues into DNA. In ung$^+$ host, the U residues in U-DNA are removed, leaving abasic sites
xylA	Mutation in D-xylose isomerase, rendering the strain unable to utilize xylose
(ϕ80)	A lambdoid prophage. A defective ϕ80 prophage carrying the lacM15 deletion is present in some strains
(λ^-)	Strain is not lysogenic for λ phage.

APPENDIX C

Practical Guide for Enzyme Handling*

1. For the best stability, enzymes should be stored in their original commercial form, undiluted, and at the appropriate temperature as specified on the label.
2. For enzyme solutions and assay/reaction buffers, use the highest purity H_2O available. Although glass distilled water may be acceptable, a higher purity water (>18 Megohms·cm resistance, e.g., from Milli-Q system of Millipore) is preferred. The "water" below designates the highest available purity water free of microbial or RNase contamination.

    In case of doubt, autoclave the water for 15 min at 120°C (or 15 lb/sq.in. on liquid cycle). This will eliminate microbial contamination but will not sufficiently inactivate RNases. Water can be rendered RNase free by treatment with diethyl pyrocarbonate (DEPC) at 0.1%. Residual DEPC, which is reactive with amines (and is a suspected carcinogen), can either be left overnight at room temperature to decompose to ethanol and carbon dioxide or preferably be destroyed by autoclaving.

* Adapted from the Probe (Boehringer-Mannheim) Feb. 10, 1990 issue with permission.

3. Enzymes should be handled in the cold (0° to 4°C). Dilute for use with ice-cold buffer or water as appropriate for each enzyme. While using the enzyme solution or suspension at the bench, keep it in the ice bath or ice bucket.
4. Diluted enzymes are generally unstable. The amount of enzyme required for the experiment should be diluted within 1–2 hr of use. Enzymes should not be diluted for long-term storage.
5. Do not shake crystalline suspensions (e.g., ammonium sulfate suspension) as oxygen tends to denature the enzyme. The material should be resuspended with gentle swirling or by rolling the bottle.
6. Do not freeze crystalline enzyme suspensions. Freezing and thawing in the presence of high salt cause denaturation and loss of activity.
7. Vials containing lyophilized enzymes or cofactors (e.g., NADH) should be warmed to room temperature before opening. This prevents condensation of moisture on the powder, which can cause loss of activity or degradation. If the reagent is hygroscopic, one such mishandling may ruin the entire vial.
8. Avoid repeated freezing and thawing of diluted enzymes. Store in small aliquots. Thaw one portion at a time and store that portion, once thawed, at 4°C.
9. Detergents and preservatives should be used with caution as they may affect enzyme activity as well as stability. Sodium azide, for example, inhibits many enzymes that contain heme groups (e.g., peroxidase). Detergents added at a concentration above their critical micellar concentration form micelles that may entrap and denature the enzyme. Commercial enzymes of molecular biology grade are often supplied in a buffer containing 50% (v/v) glycerol as a stabilizer. To avoid potential reaction artifacts, these enzymes should be used at a final glycerol concentration of <10%.
10. Enzymes should be handled with care to avoid contamination. Use a fresh pipette (or tip) for each aliquot that is removed from the parental vial. Never return unused material to the parental vial. Wear gloves to prevent contaminating the enzyme with proteases, DNases, RNases, and inhibitors often found on fingertips. Never pipette by mouth.
11. Choose a buffer that does not interact with the enzyme (or other reactants as well) or interfere in the enzyme activity. Choose a buffer whose pK_a ranges within ±1 unit from the desired pH. Lists of common buffers are available from a number of sources, for example, (i) Sambrook *et al.* (1989). "Molecular Cloning: A Laboratory Manual," 2nd ed., pp. B20–B28. Cold Spring Harbor Laboratory Press; (ii) J. S. Blanchard, (1984). *Meth. Enzy.* **104**, 404–414; and (iii) I. H. Segel (1976). "Biochemical Calculations," 2nd ed., pp. 403–406, Wiley, New York.
12. Adjust the pH of the buffer at the closest temperature at which the enzyme will be used. Many common buffers (e.g., Tris, glycylglycine,

Tes, Bicine, Hepes) change pH rapidly as the temperature changes. For instance, Tris buffer suffers a 0.3 pH unit decrease for every 10°C rise. A solution (50 mM) of Tris, adjusted to pH 7.5 at 25°C, will have a pH of 8.1 at 4°C or 7.2 at 37°C. The change in pH per 10°C temperature rise for other buffers is: Aces, −0.20; Bes, −0.16; Bicine, −0.18; glycylglycine, −0.28; HEPES, −0.14; and Tes, −0.20 [N. E. Good et al. (1966). Biochem. 5, 467].

13. The absorbance of 280 nm, widely used to quickly determine the protein concentration of an enzyme solution, actually is due to the presence of Tyr and Trp in the protein. If an enzyme has a low content of these two amino acids, it will not absorb significantly at 280 nm.

Index

Abzyme, catalysis, 2
Actinomycin D, effect on cDNA synthesis, 437
Alkaline phosphatase, *see also* Bacterial alkaline phosphatase; Calf intestinal alkaline phosphatase
  biological role, 308
  human enzymes
    genes, 330
    homology, 330
    intestine, 331–332
    liver/bone/kidney, 332
    placenta, 331
  pH optima, 307
Alpha helix, structure, 19–20
Amino acid
  helix-forming propensity, 20
  optical absorption, 24
  physiochemical properties, 19
  pK_a, 4
  side chains
    hydropathy scale, 19
    structure, 4
  substitution and protein folding, 25
Aminoacyl-tRNA synthetase
  classes, 102
  tRNA interactions, 101–103

AMV, *see* Avian myeloblastosis virus
AMV reverse transcriptase, *see also* Reverse transcriptase
  activity assays
    polymerase, 454
    ribonuclease H, 454
  applications
    cDNA labeling, 451
    first-strand cDNA synthesis, 450
    nucleotide sequencing, 652–653
    second-strand DNA synthesis, 450–451
  domains, 461
  fidelity, 458
  inhibition, 452–454
  ionic strength effects, 452
  kinetic parameters, 451
  metal requirements, 451
  physical properties, 461
  processivity, 458
  pyrophosphate exchange, 458–459
  reaction conditions, 450–451
  sources, 462
  structure, 449, 460–461
  substrate specificity
    endonuclease, 459–460
    polymerase

AMV reverse transcriptase (*continued*)
    nucleotides, 457–458
    primers, 456–457
    templates, 454–456
  ribonuclease H, 459
Antibiotic
  chloramphenicol mechanism, 613
  β-lactam mechanism, 596
  resistance mechanisms in bacteria, 596
Anticodon, wobbling, 5
Avian myeloblastosis virus
  phenotypes, 461–462
  reverse transcriptase, *see* AMV reverse transcriptase

Bacterial alkaline phosphatase
  activators, 310
  active site, 321–322
  applications
    detection tool, 318
    exporter signal, 319
    phosphomonoesterase, 317–318
    reporter of fusion genes, 318–319
  assays
    bioluminescence-enhanced assay, 312
    chemiluminescent assay, 312–314
    colorimetric dye precipitation assay, 311–312
    fluorometric assay, 311
    radioactive assay, 311
    spectrophotometric assay, 311
  catalytic process, 315–316
  gene structure, 322–323
  inhibitors, 310
  isozymes, 320–321
  kinetic parameters, 309
  metal ion
    binding site, 321
    reaction role, 316
    requirements, 309–310
  optimal reaction conditions, 309
  physical properties, 322
  precursor, 320
  rate-determining steps, 316–317
  reaction, 308
  sources, 323–324
  stability, 310
  structure, 320–322
  substrate specificity
    nicks and gaps, 314
    phosphate ester analogs, 315
    phosphomonoesters, 314
    phosphoric acid anhydrides, 314

Bacterial luciferase
  activators, 631
  active site, 635
  assay
    dithionite assay, 630
    in vitro assays, 630
    in vivo assays, 629–630
    luminometry, 630–631
    luxdot assay, 629–630
    optimal reaction conditions, 629
    photographic detection, 631
  gene
    fusion gene construction, 637, 639–640
    lux operon, 636
    regulation, 637–638
    structure, 636–637
  inhibition, 631–632
  kinetic parameters, 631
  physical properties, 635
  purification, 636
  reaction, 628
  reaction mechanism, 633–634
  as reporter, sensitivity, 638
  structure, 628, 634–635
  substrate diffusibility, 639
  substrate specificity, 632
Bal31, *see* Nuclease Bal31
BAP, *see* Bacterial alkaline phosphatase
Beer–Lambert law, 24
Beta sheet, structure, 20
Bi–bi mechanism
  ordered, 43
  ping-pong, 44
  random, 43
Branch point sequence, intron splicing role, 87
Buckle, nucleic acid conformation, 69, 71
Buffer, selection for enzyme solutions, 684–685

Calf intestinal alkaline phosphatase
  active site, 329–330
  activity assay, 327
  applications, 328
  homology with other phosphatases, 329
  inhibition, 327
  isozymes, 329
  kinetic parameters, 327
  metal requirements, 327
  optimal reaction conditions, 326–327
  physical properties, 330
  sources, 330
  structure, 329–330
  substrate specificity, 327–328

Catalysis constant, range for enzymes, 37
Catalytic triad, consensus motif, 27
cDNA, *see* Complementary DNA
Chargaff's rule, 53
Chemical nuclease, types, 146
Chloramphenicol acetyltransferase
  active sites, 621–622
  applications
    plasmid amplification, 625
    reporter gene, 623–625
    selectable marker gene, 625
  assay
    cell extract, 615–616
    fluorescent assay, 616–617
    high performance liquid chromatography assay, 617
    in vivo assay, 616
    optimal reaction conditions, 614–615
    radiometric assay, 615–616
    spectrophotometric assay, 617
  classification, 614
  genetics, 622–623
  inhibition, 617–618
  kinetic parameters, 617
  mechanism of action, 619
  reaction, 614
  sources, 625–626
  structure, 619–622
  substrate specificity, 618–619
CIAP, *see* Calf intestinal alkaline phosphatase
Codon
  degeneracy, 2, 5
  genetic code, 3
  missense mutation, 15
  stop sequences, 8
Competitive inhibition, 39
Complementary DNA
  labeling, 451
  synthesis using reverse transcriptase
    actinomycin D effects, 437
    detergent effects, 437
    efficiency, 430–431, 433–437
    free primer, 431, 433–435
    glycerol effects, 437
    linker/adapter primer, 435–436
    product characteristics, 429–430
    RNA template, 430–431
    spermidine effects, 437
    transcriptase requirements, 436–437
    vector primer, 436

Dam methyltransferase
  assay, 301

biological role, 300
effect on restriction enzymes, 301–302
genetics, 304–305
inhibitors, 301
kinetic parameters, 300–301
mechanism, 302–303
optimal reaction conditions, 300
sources, 305
structure, 303–304
substrate specificity, 301–302
Dcm methyltransferase
  biological role, 300
  genetics, 304–305
  mechanism, 303
  sources, 305
  structure, 304
  substrate specificity, 302
Deoxyribonuclease I
  active site, 157
  activity assays, 149
  applications
    DNA fragment generation for dideoxy sequencing, 155
    DNA labeling by nick translation, 154
    footprinting, 155–156
    nonspecific degradation, 154
    random mutagenesis, 154–155
  components, 147, 156
  genetics, 157–158
  hypersensitivity in presence of actinomycin D, 151
  inhibition, 148–149
  kinetic parameters, 147
  mechanism
    cleavage by nucleophilic catalysis, 153–154
    DNA binding mode, 152
    double-strand cleavage pattern, 152–153
    metal ion role, 151
  metal requirements, 148
  optical reaction conditions, 146–147
  sources, 158
  structure, 156–157
  substrate specificity
    base specificity, 150
    DNA conformation, 150
    minimal size, 149–150
    nicking activity, 150–151
    ssDNA, 149
    synthetic substrates, 150
DNA, *see also* Nucleic acid
  base bias, 56
  counterions, 58
  density and sedimentation properties, 57–58

DNA (continued)
  melting temperature
    calculation, 63
    factors affecting, 62–63
    –protein interaction, 72
  solvents, 58
  structure
    A-DNA, 73–74
    B-DNA, 73–74
    circular structures, 80
    cruciform, 79–80
    descriptive parameters, 69–71, 82
    G-DNA, 80
    heterogeneity, 71–72
    parallel-stranded double helix, 75–76
    supercoil, 81–82
    Z-DNA, 74–75
  triple helix
    biological significance, 78–79
    detection, 78
    purine–purine–pyrimidine triplex, 77–78
    pyrimidine–purine–pyrimidine triplex, 76–77
DNA exonuclease, see Exonuclease III; Nuclease Bal31
DNA labeling
  nick translation, 648–649
  polymerase chain reaction, 650
  radioactive labeling, 647–648
  random-primer labeling, 649–650
DNA ligase, see also Escherichia coli DNA ligase; T4 DNA ligase
  activity assays
    ATP-PP$_i$ exchange, 111
    circularization, 110–111
    λ HindIII ligation, 111
    oligonucleotide ligation, 111
  applications
    circularization of dsDNA, 116
    DNA structural analysis, 118–119
    end group identification, 118
    ligation amplification reaction, 117
    mismatched nucleotide detection, 117
    nick activity assay of other enzymes, 118
    nick measurement, 118
    RNA synthesis, 117–118
    termini structure identification, 118
  cofactors, 109, 120–121, 126
  macromolecular stimulants, 113, 126–127
  mechanism, 113, 115
  reaction rate, factors affecting, 115–116
  substrate specificity
    blunt ends, 113, 115, 121–122, 130
    heteroduplex substrates, 112
    nick sealing, 111–112
    nucleotide composition, 112
    nucleotide length, 112
    RNA substrates, 112
DNA methylation, see also Methyltransferase
  biological roles, 285
  effect on DNA–protein interactions, 285–286
  effect on restriction enzymes, 288
  locating methylated bases, 287
DNA polymerase, see also Reverse transcriptase; T4 DNA polymerase; T7 DNA polymerase; Taq DNA polymerase; Terminal deoxynucleotidyltransferase
  biological roles, 345–346
  exonuclease activity, 349
  families, 349
  fidelity, 14, 346–349
  primer, 345–346
  processivity, 346–347
  template, 346
  template-based classification, 349–350
DNA polymerase I, see also Klenow fragment
  active sites
    exonuclease, 371
    5'-nuclease, 371
    nucleotide binding site, 370–371
    polymerase, 370
  activity assay, 356
  applications
    amplification reactions, 367
    blunt-end formation of duplex DNA, 367–368
    duplex DNA labeling, 368
    hybridization probe preparation, 367
    mutagenesis, 368
    nucleotide sequencing, 366
    second-strand synthesis, 366
    time-point restriction mapping, 369
  3'→5' exonuclease, substrate specificity, 360–361
  fidelity, 360, 365–366
  gene
    mutation studies, 372–373
    structure, 372
  inhibitors, 355–356
  kinetic parameters, 353–354
    exonuclease, 361
    polymerase, 360, 362
  mechanism
    exonuclease, 364–365
    fidelity, 365–366
    polymerase, 362, 364
    substrate shuttling, 365

metal requirements, 354
5'-nuclease
  domain, 362
  substrate specificity, 361–362
pH dependence
  exonuclease, 352
  polymerase, 352
  pyrophosphorolysis, 353
physical properties, 371
polymerase substrate specificity
  discontinuous template, 358
  DNA–RNA hybrid, 358
  nicked DNA, 357
  nucleotides, 359–360
  RNA template, 358
  single-stranded DNA, 357
  template structure, 356
primer
  length, 358
  mismatch, 359
processivity, 360
reaction conditions, 352–353
reaction specificity, 351–352
relationship to other *Escherichia coli* DNA
  polymerases, 373
sources, 373–374
structure, 369
DNase I, *see* Deoxyribonuclease I
Domain
  linkage, 20
  structure determination, 21

EcoB1, *see* Type I restriction endonuclease
EcoKI, *see* Type I restriction endonuclease
EcoP1, *see* Type III restriction endonuclease
EcoRI, *see also* Type II restriction endonuclease
  active site residues, 263–264
  genetics, 264–265
  homology with RsrI, 264
  kinetic parameters, 251–252
  mechanism
    catalysis, 261–262
    conformational change, 258–259
    DNA binding, 257, 260
    recognition specificity, 257–258, 260
    role of DNA conformation, 259
  metal requirements, 252
  optimal reaction conditions, 251
  physical properties, 262
  site specificity, modulation
    heparin, 256
    intercalators, 255–256
    polyphosphate, 256

sources, 265
structure, 262–264
substrate specificity
  base analog effects, 255
  chain-length dependence, 254
  methylation effects, 255
  polarity, 254–255
  sequence, 252–253
  site specificity, 253–254
  star activity, 254
  triplex DNA, 257
  Z-DNA, 256
*Eco*RI methyltransferase
  active site, 298
  genetics, 299
  inhibitors, 294–295
  kinetic parameters, 294, 296–297
  mechanism, 296–297
  optimal reaction conditions, 294
  sources, 299
  structure, 297–298
  substrate specificity
    DNA structure, 296
    effect of base substitution, 295–296
    methylation sensitivity, 296
    relaxation, 296
    sequence, 294–295
    triplex DNA, 296
Enzyme
  active site
    residue determination, 29–30
    structure, 25–26
  allosteric enzymes, 44–45
  average density, 23
  coenzyme, 30
  cofactor, 30
  denaturation, 22–23
  efficiency, 1
  factors affecting catalysis
    ionic strength, 47
    pH, 46–47
    solvent polarity, 47–48
    temperature, 47
  family classification, 27–28
  handling, practical guide, 683–685
  inhibition, 38–40, 42
  isozymes, 28
  kinetic mechanisms, 35, 42–44
  mutation studies, 16, 27
  prosthetic group, 30
  rate-determining step, 35
  reaction specificity, 28–29
  specific activity, 33
  specificity constant, 37–38

Enzyme (*continued*)
  stabilizing agents, 23
  steady-state kinetics, 34–38, 42
  stereospecificity, 26–27
  substrate binding forces
    electrostatic forces, 31
    hydrogen bonding, 31
    hydrophobic forces, 31
    steric repulsion, 31
  substrate specificity, 1, 26–27
  transition-state theory and catalysis, 31–32
Enzyme Commission code, 28
Equilibrium constant
  calculation, 33
  Gibbs free energy relationship, 34
Escape mutant, 15
Escherichia coli
  genetic markers, 677–681
  genotypes of frequently used strains, 674–677
  phenotypic traits, 677–681
  RNA polymerase, *see* RNA polymerase
Escherichia coli DNA ligase, *see also* DNA ligase
  activators, 121
  active site, 122
  applications, 122
  cofactors, 120–121
  gene
    locus, 123
    mutation analysis, 123–124
    regulatory regions, 123
  kinetic parameters, 120
  optical absorption, 122–123
  optimal reaction conditions, 120
  sources, 124
  structure, 122
  substrate specificity
    blunt ends, 121–122
    single strands, 122
Ethidium bromide, DNA intercalation and fluorescence, 255–256
Exonuclease III
  activity assays, 217–218
  applications
    DNA mapping, 223
    DNA sequencing, 222–223
    polymerase repair synthesis, 223
  gene
    mutants, 224
    organization, 224
    sources, 224
  inhibition, 217
  metal requirements, 217

  processivity, 221
  reaction conditions, 216
  structure, 215, 223–224
  substrate recognition, 221
  substrate specificity
    endonuclease, 220
    phosphatase, 220
    RNase H activity, 221

Fidelity
  components of process, 348
  DNA polymerase, 14, 346–349
  factors affecting, 348–349
Firefly luciferase
  assay, 640–641
  gene
    fusion, 643
    structure, 642
  peroxisomal translocation signal, 639
  reaction, 628, 640
  structure, 642
  substrate specificity, 641–642
*Fok*I
  applications
    gene cloning, 270
    gene synthesis, 269–270
    gene trimming, 272
    linker mutagenesis, 272
    methylated base localization, 272
    oligonucleotide amplification, 272
    universal restriction enzyme, 269
  genetics, 273–274
  metal requirements, 268
  optimal reaction conditions, 268
  sources, 274
  stability, 268
  structure
    endonuclease, 273
    methyltransferase, 273
  substrate specificity
    methylase specificity, 269
    methylation effects, 268–269
    sequence, 267
Formaldehyde, effect on nucleic acid melting temperature, 68
Formamide, effect on nucleic acid melting temperature, 67

β-Galactosidase
  active site, 577–578
  activity assay, 573

applications, 582–583
detection
  colorimetry, 570
  fluorimetry, 570
  immunoassay, 571
  luminometry, 570
  solid-phase reaction, 570–571
  solution reaction, 569–570
dihydrofolate reductase homology, 582
gene
  α complementation, 581–582
  evolution, 582
  fusion
    approaches, 583–584
    inclusion body formation, 584
    regulatory genes, 584–585
    restriction sites, 584
  lactose operon structure, 578–580
  mutations, 581
  structure, 580–581
  vector incorporation
    λ phage, 591–592
    M13, 585–586, 589
    pBluescript, 590–591
    pEMBL, 590
    pEX, 591
    pGEM, 590
    pTZ, 590
    pUC, 589–590
induction, 578
inhibitors, 572–573
kinetic parameters, 571
mechanism of reaction, 575–576
metal requirements, 571–572
reaction, 568–569
sources, 592–593
structure, 569, 576–577
substrate specificity
  aglycon structure influence, 573
  D-galactal, 574
  galactosylpyridinium salt, 574
  glycon hydroxyl group, 573
  thiogalactosides, 573–574
  transgalactosylation, 574–575
Glycerol, effect on cDNA synthesis, 437
Guanidinium thiocyanate, effect on nucleic acid melting temperature, 67

Heat shock proteins, protein folding role, 10
HinfIII, see Type III restriction endonuclease
Homopolymeric tailing
  advantages in cloning, 484
  chimeric DNA formation, 484

nucleotide selection, 483–484
strategies, 483
Hoogsteen base pairing, 56

Inclination, nucleic acid conformation, 69, 71

Klenow fragment, see also DNA polymerase I
  exonuclease-deficient Klenow, 371
  kinetic parameters, 353–354, 360
  metal requirements, 354
  nucleotide sequencing, 652
  reaction conditions, 352
  reaction specificity, 351–352
  sources, 374
  structure, 369–370
Kozak sequence, translation initiation role, 6–7

β-Lactamase
  active site, 607–608
  antibiotic resistance role, 596
  applications, 610–611
  assay, 600–601
  classification, 596–597, 599–600
  evolution, 608–609
  gene
    expression, 610
    identification, 609
    locus, 610
    organization, 609–610
  inhibitors
    clavulinic acid, 602
    product inhibition, 601
    substrate analogs, 602–603
  kinetic parameters, 601
  mechanism
    catalytic mechanism, 604–605
    kinetic acyl-enzyme mechanism, 603–604
    optimum reaction conditions, 600
  precursor, 608
  reaction, 595
  sources, 611
  structure, 605–608
  substrate specificity, 603
Lactose operon
  genes, 578–579
  promoter, 580
  repressor protein, 579–580
Lambda phage vector, lacZ incorporation, 590–591
Ligase, see DNA ligase; T4 RNA ligase

Linking number, calculation, 82
Luciferase, see Bacterial luciferase; Firefly luciferase

M13 vector
  hosts, 586, 589
  lacZ incorporation, 585–586
Marker
  defined, 567–568
  ideal criteria, 568
  types, 568
Maxam–Gilbert method, see Nucleotide sequencing
Messenger RNA
  3'-adenylation, 86–87
  5'-capping, 86
  editing, 89–91
  intron splicing, 87–89
  polycistronic mRNA, 17, 86
  recoding signals, 7–8
  secondary structure and function, 91–92
  stop signals, 8
Methyltransferase, see also Dam methyltransferase; Dcm methyltransferase; EcoRI methyltransferase; FokI
  applications
    cloning, 289, 292
    enhancement of endonuclease site specificity, 289–290
    genomic footprinting, 290–292
    partial DNA cleavage, 290
  assays
    methyl transfer, 287–288
    protection assay, 288
  donor substrate, 234, 284, 286
  gene cloning, 292–293
  homology, 286
  nomenclature
    enzymes, 236, 238
    genes, 238
  types, 234–235, 286–287
Michaelis constant
  determination, 37
  range of values, 37
Michaelis–Menten equation
  assumptions, 35–37
  Briggs–Haldane mechanism, 37
  graphic representations, 36, 38
  kinetic parameters, 37–38
  limitations, 42
Micrococcal nuclease, see Staphylococcal nuclease
Missense mutation, 15

Mixed-type inhibition, 40
Molecular mass, determination, 23–24
Molecular weight, calculation, 23
Moloney murine leukemia virus
  genome organization, 473
  phenotype, 472–473
  reverse transcriptase, see MoLV reverse transcriptase
MoLV reverse transcriptase, see also Reverse transcriptase
  active sites
    polymerase, 472
    ribonuclease, 472
  applications
    first-strand cDNA synthesis, 465
    nucleotide sequencing, 652–653
    second-strand synthesis, 466
  assays
    polymerase, 468
    ribonuclease H, 468
  domains, 471–472
  fidelity, 469
  gene expression, 474
  inhibition, 467–468
  kinetic parameters, 466
  metal requirements, 466–467
  processivity, 469–470
  reaction conditions, 465–466
  ribonuclease H-negative enzyme, 470–471, 474
  sources, 474–475
  structure, 471–472
  substrate specificity
    polymerase, 468–469
    ribonuclease, 469–470
mRNA, see Messenger RNA
Multienzyme complex, substrate channeling, 22
Mung bean nuclease
  activity assay, 213, 215
  inactivators, 212
  metal requirements, 212
  optimal reaction conditions, 212
  sources, 215
  structure, 214
  substrate specificity
    cleavage preference, 213
    mismatched sites, 213
    phosphatase, 214
    product profile, 213
    single-strand specificity, 213
Mutagenesis, site-specific, see Site-specific mutagenesis
Mutations, 14–16

Nick translation, DNA labeling, 648–649
Noncompetitive inhibition, 39
Nonsense mutation, 15
Nuclease Bal31
  activity assay, 227
  applications, 230–231
  genetics, 231
  inhibitors, 227
  kinetic parameters, 227
  mechanism, 229–230
  metal requirements, 227
  processivity, 228
  reaction conditions, 226–227
  sources, 231
  structure, 231
  substrate specificity, 228–229
Nuclease S1
  activity assay, 207
  applications
    cleavage of termini, 209
    hairpin loop opening, 209
    RNA structural analysis, 209–210
  inhibitors, 206–207
  metal requirements, 206
  phosphomonoesterase activity, 208–209
  reaction conditions, 205–206
  sources, 210–211
  structure, 210
  substrate specificity
    circular DNA, 208
    cleavage preferences, 207
    homopolymers, 208
    phosphorothioates, 208
    single-strand specificity, 207–208
Nucleic acid, see also DNA; RNA
  base
    chemical reactivity, 58–59
    pairing, 53, 55–56
    stacking, 68–69
  cellular dry weight, 50
  components, 51–52
  denaturation
    factors affecting, 62–63
    monitoring methods, 61
  hybridization, 59
    factors affecting
      base composition, 66
      base mismatch, 66
      denaturing agents, 67–68
      ionic strength, 66–67
      length, 66
      pH, 67
      temperature, 68
      viscosity, 67
    second-order reassociation kinetics, 63–65
  structural parameters, 69–71
  sugar pucker, 69, 72
Nucleotide
  distribution in DNA, 56
  physiochemical properties, 56–57
  structure, 51–52
Nucleotide sequencing
  chemical reactivity of bases, 58–59
  dideoxynucleotide chain termination method
    efficiency, 651–652
    polymerases
      Klenow fragment, 652
      reverse transcriptase, 652–653
      Sequenase, 652
      thermostable polymerase, 653–655
    steps, 651
  Maxam–Gilbert method
    genomic sequencing, 636
    hydrazine reaction, 655
    polymerase chain reaction, 656–657
Null mutation, 15

Okazaki fragment, synthesis, 346
Opening, nucleic acid conformation, 69, 71
Operator
  lactose operator, 578–580
  lux operon, 636
  repressor binding, 500
  structure, 500

pBluescript vector, lacZ incorporation, 590–591
PCR, see Polymerase chain reaction
pEMBL vector, lacZ incorporation, 590
Peptidyl prolyl isomerase, protein folding role, 10
pEX vector, lacZ incorporation, 591
pGEM vector, lacZ incorporation, 590
Phage λ vector, lacZ incorporation, 591–592
phoA gene
  phosphate regulation, 323
  structure, 322–323
Phosphatase, see Alkaline phosphatase
Point mutation, reversion, 15
Pol I, see DNA polymerase I
Poly(A) polymerase
  activity assays, 557–558
  applications
    3′-end labeling, 561
    poly(A) tailing of RNA, 561
    quantitation of mRNA, 561–562

Poly(A) polymerase (*continued*)
  biological role, 555, 563
  genetics, 562–563
  inhibitors, 557
  kinetic mechanism, 558
  kinetic parameters, 556
  mechanism, 559
  metal requirements, 556–557
  optimal reaction conditions, 556
  reaction, 556
  sources, 563–564
  structure, 562
  substrate recognition
    effect of AAUAAA specificity factor, 560
    endonucleolytic cleavage, 560–561
    sequence motifs, 559–560
  substrate specificity, 558
Polymerase chain reaction
  annealing temperature, 416
  denaturation, 409, 658
  DNA labeling, 650
  efficiency, 414–415
  hot-start technique, 416–417
  Maxam–Gilbert sequencing, 656–657
  nested primer technique, 417
  nucleotide sequencing, 653, 655
  primer annealing, 409, 416, 658
  primer extension, 409, 658
  ramping time, 409–410
  reverse transcription technique, 420, 445, 658
  site-specific mutagenesis
    nonoverlap mutagenesis, 669–670
    overlap circle mutagenesis, 670–671
    overlap-extension mutagenesis, 668–669
    principle, 659–660, 666–668
    specificity and temperature, 415–417
  thermostable polymerases, 653–655, 658
Polynucleotide kinase, *see* T4 polynucleotide kinase
Posttranslational modification
  methionine formylation, 8
  polyprotein processing, 13
  types, 8–9
Processivity
  DNA polymerase, 346–347
  estimation, 347
Promoter
  activator proteins, 499, 501
  enhancers, 501, 506
  eukaryotic structure, 500–501
  Pribnow box, 497
  prokaryotic structure, 497–499
  strength, 499
  TATA box, 501

Propeller twist, nucleic acid conformation, 69, 71
Protein concentration, assays, 24, 685
Protein disulfide-isomerase, protein folding role, 10
Protein folding
  cooperativity, 24
  intermediates, 10
  packing effects, 25
  protein cofactors, 10
  recombinant protein aggregation during refolding, 10
  stability of folded state, 25
  stabilizing forces, 25
Protein sequencing, methods, 19
Protein translocation, mechanism, 9
Protein turnover
  C-terminal rule, 14
  N-end rule, 13–14
  targeting, 13–14
pTZ vector, *lacZ* incorporation, 590
pUC vector, *lacZ* incorporation, 589–590

Random coil, structure, 20
Rational drug design, 21
Reporter
  defined, 567–568
  ideal criteria, 568
  types, 568
Restriction endonuclease, *see also* *Eco*RI; *Fok*I; Type I restriction endonuclease; Type II restriction endonuclease; Type III restriction endonuclease
  discovery, 233
  methyltransferase activity, 233–235, 276
  nomenclature
    enzymes, 236, 238
    genes, 238
  types, 234–235, 238–239, 267, 275
Restriction–modification system
  enzyme activities, 234
  independent systems, 235–236
  role in bacteria, 233
  type II system, properties, 234, 236
  types, 234–235, 267
Reverse transcriptase, *see also* AMV reverse transcriptase
  applications
    nucleotide sequencing, 445–446
    reverse transcription–PCR, 445
    site-specific forced misincorporation mutagenesis, 446
  biological roles, 443, 445

complementary DNA synthesis
 actinomycin D effects, 437
 detergent effects, 437
 efficiency, 430–431, 433–437
 free primer, 431, 433–435
 glycerol effects, 437
 linker/adapter primer, 435–436
 product characteristics, 429–430
 RNA template, 430–431
 spermidine effects, 437
 transcriptase requirements, 436–437
 vector primer, 436
domains
 endonuclease, 443
 polymerase, 442
 ribonuclease H, 442–443
fidelity, 437–438
kinetic mechanism, 440–441
precursor processing, 441–442
primer requirements, 429
processivity, 438
reaction specificities, 427–428
ribonuclease H activity
 metal requirements, 438
 processivity, 440
 reaction specificity, 439
 temporal relationship with polymerase activity, 439–440
sources, 427
structure, 441–443
template specificity, 428
template switching, 428–429
Ribonuclease A
 active site, 180–181
 activity assays
  cyclic nucleotide spectrophotometric assay, 173
  Kunitz spectrophotometric assay, 171–172
  precipitation assay, 172–173
  titrimetric assay of cyclic phosphate hydrolysis, 173
 applications
  helix destabilization, 178
  protein structural studies, 179
  removal of unhybridized regions of RNA, 178
  RNA degradation in DNA preparations, 178
  RNA sequencing, 178
 binding of substrates and inhibitors, 176–177
 catalytic mechanism, 177–178
 genetics, 181
 glycosylation, 170

 inhibitors
  aurintricarboxylic acid, 175
  inactivators, 176
  nucleotide analogs, 174
  placental RNase inhibitor, 175–176
  ribonucleoside–vanadyl complexes, 174–175
 kinetic parameters, 171
 optimal reaction conditions, 170
 proteolysis and RNase S, 180
 sources, 181–182
 structure, 179–181
 substrate specificity, 173–174
Ribonuclease H
 active site, 192
 activity assays, 186
 activity in polymerases
  AMV reverse transcriptase, 454, 459
  exonuclease III, 221
  MoLV reverse transcriptase, 468, 472
 applications
  cDNA synthesis, 189
  hybrid-arrested translation by antisense oligonucleotides, 190–191
  poly(A) tail
   quantitation, 189–190
   removal, 189
  RNA detection in DNA, 191
  specific fragmentation, 190
 gene
  isozymes, 193
  mutagenesis, 193
  structure, 192–193
 inhibitors, 185
 kinetic parameters, 185
 mechanism of action, 188–189
 metal ion requirements, 185
 optimal reaction conditions, 185
 sources, 194
 species distribution, 184
 structure, 191–192
 substrate specificity
  cleavage at RNA–DNA junction, 188
  cleavage specificity, 186–187
  methyl phosphonate bond, 187
  product profile, 186
  size, 186
  sugar configuration, 187–188
Ribonuclease P, see Ribozyme
Ribonuclease S, see Ribonuclease A
Ribonuclease T1
 active site, 199–202
 activity assays, 198
 applications, 200–201

698 Index

Ribonuclease T1 (continued)
  genetics, 202
  inhibitors, 197–198
  kinetic parameters, 197
  mechanism, 199–200
  reaction conditions, 196–197
  sources, 202–203
  structure, 201–202
  substrate specificity, 198–199
Ribosomal RNA
  eukaryotes, 93
  prokaryotes, 92–93
Ribosome
  hopping, 18
  polysome formation, 7
  slipping, 17–18
  tRNA-binding sites, 7
Ribozyme
  catalysis, 1–2, 93, 146
  hammerhead ribozyme
    activity, 93–94
    mechanism, 95
    structure, 93–94
  ribonuclease P
    active site, 98–99
    catalytic properties, 98
    eukaryotes, 99
    structure, 98
  self-splicing introns, 93
  Tetrahymena ribozyme
    activity, 95
    catalytic properties
      aminoacyl esterase, 97–98
      DNA cleavage, 97
      polymerization, 97
      transesterification, 95–97
    structure, 95
Rise, nucleic acid conformation, 69–70
RNA, see also Messenger RNA; Nucleic acid; Ribosomal RNA; Ribozyme; Transfer RNA
  alkali hydrolysis, 58
  functional types, 83
  reverse transcription, 2–3
  secondary structure
    analysis, 84
    mRNA function, 91–92
    prediction, 84
    pseudoknot, 85–86
    stem–loop features, 84–85
RNA polymerase, see also Poly(A) polymerase; SP6 RNA polymerase; T3 RNA polymerase; T7 RNA polymerase
  elongation mechanism, 513–514

  eukaryote, 495–496
  fidelity, 494
  phage, 496
  prokaryote, 494–495
  σ factor, 494–495
  subunit functions, 494–495
RNA polymerase II
  promoter, 501
  subunit function in yeast, 495–496
RNase A, see Ribonuclease A
RNase H, see Ribonuclease H
RNase T1, see Ribonuclease T1
Roll, nucleic acid conformation, 69, 71

Sequencing, see Nucleotide sequencing; Protein sequencing
Shear, nucleic acid conformation, 69–70
Shift, nucleic acid conformation, 69–70
Shine–Dalgarno sequence, translation initiation role, 5–6
Signal peptide
  cleavage, 11–13
  domains, 11–12
  export signals, 11–12
  membrane attachment signal, 12
  protein conformation role, 11–12
Signal recognition particle, 11
Site-specific mutagenesis
  amplification strategies, 662–671
  coupled-priming mutagenesis
    antibiotic selection, 666
    principle, 664–665
    restriction–selection, 665–666
  phosphorothioate-based mutagenesis, 664
  polymerase chain reaction mutagenesis
    nonoverlap mutagenesis, 669–670
    overlap circle mutagenesis, 670–671
    overlap-extension mutagenesis, 668–669
    principle, 659–660, 666–668
    polymerase selection, 661–662
  primer design, 661
  steps in conventional method, 660
  uracil-based mutagenesis, 662–664
Slide, nucleic acid conformation, 69–70
SNase, see Staphylococcal nuclease
SP6 RNA polymerase
  assay, 551
  genetics, 552
  kinetic parameters, 550
  metal requirements, 550
  promoter specificity, 551
  reaction conditions, 550
  sources, 552–553

spermidine activation, 550
structure, 552
substrate specificity, 551–552
Spermidine
effect on cDNA synthesis, 437
effect on type II restriction endonuclease site specificity, 242
Spliceosome, intron splicing role, 88
Stagger, nucleic acid conformation, 69–70
Staphylococcal nuclease
activity assays, 162
applications, 165–166
genetics, 167
inhibitors, 161
kinetic parameters, 161
mechanism of catalysis, 163–165
metal ion requirements, 161
optimal reaction conditions, 160
site-specific mutagenesis, 164–165
sources, 167
structure, 166–167
substrate specificity
cleavage preferences, 162–163
dinucleotides, 163
mononucleotide esters, 163
polynucleotides, 162
Stretch, nucleic acid conformation, 69–70
StyLTI, see Type III restriction endonuclease
Subunit
allosterism role, 22
channeling, 22
characteristics in proteins, 21–22
Suicide inhibition, 40, 42
Superhelical density, calculation, 82
Suppressor mutation, 16

T3 RNA polymerase
activators, 554
genetics, 554–555
inhibition, 554
kinetic parameters, 553
metal requirements, 554
reaction conditions, 553
sources, 555
structure, 554
substrate specificity, 554
T4 DNA ligase, see also DNA ligase
applications, 130
cofactors, 126
gene
mutant analysis, 131
primary structure, 130
regulatory regions, 130

inhibition, 127–128
kinetic parameters, 126
macromolecular stimulants, 113, 126–127
optimal reaction conditions, 125, 128
sources, 131–132
structure, 130
substrate specificity
blunt ends, 130
gapped duplex DNA, 129
mispaired bases, 129–130
partially double-stranded substrates, 129
temperature of reactions, 128
T4 DNA polymerase
accessory proteins, 390–391
activity assays
exonuclease, 379
polymerase, 378–379
applications
blunt-end formation of duplex DNA, 383
deletion subclone generation for sequencing, 386
directional cloning of amplified products, 383–384
3'-end labeling, 386–387
gene trimming, 384–385
site-specific mutagenesis, 387–388
stable DNA lesion detection, 388
biological role, 390
domains, 388–389
$3' \rightarrow 5'$ exonuclease substrate specificity, 380–381
fidelity, 381–382
gene
mutant phenotypes, 391–392
regulation, 389
structure, 389
inhibition, 378
kinetic parameters, 377, 380
mechanism, 381–382
metal requirements, 377
physical properties, 388
polymerase substrate specificity
effect of DNA binding proteins, 379
nucleotides, 379–380
template structure, 379
processivity, 379
reaction conditions, 377
replication complex, 390
sources, 392
structure, 388–389
T4 polynucleotide kinase
activators, 335
applications
endonuclease activity, 341

T4 polynucleotide kinase (*continued*)
   3'-phosphatase, 341
   5' radiolabeling, 340–341
   assay, 336
   cofactors, 334–335
   dephosphorylation substrate specificity, 338
   genetics, 342–343
   inhibitors, 335
   kinetic parameters, 334
   mechanism, 339–340
   optimal reaction conditions, 333–334
   3'-phosphatase mutant, 341
   5'-phosphate exchange
      optimal reaction conditions, 338–339
      substrate specificity, 338–339
   phosphorylation substrate specificity
      acceptor substrates, 336
      DNA termini structure, 336
      donor substrates, 336
      modified nucleotides, 337
      nonnucleotides, 337–338
      terminal nucleotide preference, 336–337
   reaction, 333–334
   sources, 343
   structure, 341–342
T4 RNA ligase
   activators, 134–135
   active site, 140–141
   activity assays
      AMP complex formation assay, 135–136
      ATP–PP$_i$ exchange, 135
      ligation–phosphatase digestion, 135
   applications
      internal nucleotide modification, 140
      3'-labeling, 139
      production of elongated nucleotides, 139–140
      T4 DNA ligase stimulation, 140
      5'-tailing of DNA with RNA, 139
   cofactors, 133–134
   gene
      phage functions, 141–142
      structure, 141
   inhibitors, 135
   kinetic parameters, 133
   mechanism, 138–139
   optimal reaction conditions, 133
   physical properties, 141
   sources, 142
   structure, 140
   substrate specificity
      acceptor substrates, 137–138
      circularization reaction, 138
      donor substrates, 136–137
      duplex structures, 138
      β-substituted ADP derivatives, 136
T7 DNA polymerase
   activity assays
      exonuclease, 396
      polymerase, 395–396
   applications
      blunt-end formation, 400
      dideoxynucleotide sequencing, 398–399, 652
      3'-end labeling, 400
      probe synthesis, 400
      site-specific mutagenesis, 399
   exonuclease-deficient enzyme, 402
   5'-exonuclease gene, 403–404
   inhibition, 395
   kinetic parameters, 395
   mechanism of catalysis, 397–398
   metal requirements, 395
   phage gene 5 protein subunit
      gene, 403
      structure, 400–401
   reaction conditions, 394–395
   sources, 405
   substrate specificity
      exonuclease, 397
      polymerase, 396
   subunit assembly, 400
   thioredoxin subunit
      functions, 404
      gene organization, 404
      structure, 401–402
T7 RNA polymerase
   activators, 523
   active site, 544–545
   assay, 524
   cofactor requirements, 522–523
   genetics, 545–546
   high-level expression of fusion genes
      chromosomal expression systems, 536
      conjugated expression systems, 537
      cytosolic expression system, 536–537
      host toxicity, 534–535
      translation efficiency, 534, 537–538
   inhibitors, 523
   kinetic parameters, 522
   mechanism
      abortive cycling and chain elongation, 533
      catalysis, 532–533
      dynamic interactions with substrates, 531–532
      first nucleotide incorporation, 532
      promoter binding, 530–531
   optimal reaction conditions, 521–522

physical properties, 545
primase activity, 545–546
processivity, 528–529
promoter recognition, 520–521, 530–531
reaction specificity, 520–521
RNA sequencing application, 543
RNA transcript preparation
  5'-end capping, 543
  labeling, 543
  mobile promoters, 541–542
  PCR-amplified templates, 542
  restriction fragment templates, 542
  terminator-oligo-containing templates, 542
  vector-borne promoters, 538–541
sources, 546
structure, 520, 543–544
substrate specificity
  nucleotides, 527–528
  promoter, 524
  templates, 524–526
  transcript profile, 526
termination of transcription, 529–530
Taq DNA polymerase
active site, 423
applications
  DNA quantitation, 422–423
  3'-end labeling, 422
  molecular cloning, 421–422
  nucleotide sequencing, 420–421
  polymerase chain reaction
    cycling parameters, 409–410
    efficiency, 414–415
    specificity and temperature, 415–417
    primer-directed cleavage, 423
assay, 410
5'→3' exonuclease
  reaction conditions, 418
  substrate specificity, 418–419
fidelity, 417–418
genetics, 424
inhibition, 407–408
kinetic parameters, 407
Klentaq, 424
metal requirements, 407
polymerase substrate specificity
  degenerate primers, 411–412
  matching primers, 411
  mutagenic primers, 412
  nucleotides, 412–413
  product profile, 413–414
  template structure, 410
processivity, 413
reaction conditions, 406–407
sources, 424–425

stabilizers, 408
Stoffel fragment, 424
structure, 423
TdTase, *see* Terminal deoxynucleotidyltransferase
Temperature sensitive mutation, 15
Terminal deoxynucleotidyltransferase
applications
  3'-end labeling, 484–485
  homopolymeric tailing, 483–484
  misincorporation mutagenesis, 485
assay, 479
biological role, 487
genetics, 486–487
inhibition, 478–479
kinetic mechanism, 482–483
kinetic parameters, 478
metal requirements, 478
product chain length
  determination, 482
  distribution, 482
reaction catalyzed, 477
reaction conditions, 477
sources, 487
structure, 486
substrate specificity
  initiators, 479–480
  nucleotides, 480–481
  product profiles, 481
Thioredoxin, T7 DNA polymerase subunit, 401–402, 404
Tilt, nucleic acid conformation, 69, 71
Tip, nucleic acid conformation, 69, 71
Transcription, *see also* Operator; Promoter; Transcription factors
elongation, 513–514
initiation steps, 496–497
machinery, 491–493
regulation, 492–493, 512–513
template, 493
termination
  attenuation control, 515–516
  eukaryotes, 517–518
  factor-independent termination, 516–517
  ρ-dependent termination, 517
  T7 RNA polymerase, 529–530
Transcription factors
function, 506–507
general factors, 507–508
recognition sequences, 502–505
regulation, 513
specific factors, 508
structure
  activating domains, 511

Transcription factors (continued)
  DNA-binding domains, 509–511
    helix–turn–helix motif, 509–510
    latent forms of activators, 511–512
    leucine-zipper motif, 510
    oligomerization role, 511
    single-domain activators, 508
    zinc-finger motif, 510
  transcriptional activation
    models, 512
    regulation, 512–513
  types in eukaryotes, 496, 502–505
Transfer RNA
  aminoacyl-tRNA synthetase interactions, 101–103
  base modification, 99
  processing, 99–100
  secondary structure, 100
  Tertiary structure, 99, 101
  wobble, 2, 5
Transition state
  analog, 26
  theory of enzyme catalysis, 31–32
Translation
  elongation, 7–8, 18
  errors, 14
  initiation
    eukaryotes, 6–7
    nonstandard modes, 17–18
    prokaryotes, 5–6
  internal initiation, 18
  termination, 8, 18
Tryptophan, optical absorption, 24, 685
Turnover number, see Catalysis constant
Twist, nucleic acid conformation, 69, 71, 82
Type I restriction endonuclease
  activity assay
    ATPase, 276
    methyltransferase, 276–277
    restriction endonuclease, 276
  applications, 278
  genetics, 279–280
  mechanism, 278
  reaction specificity
    ATPase, 276
    methyltransferase, 276
    restriction endonuclease, 275–276
  sources, 280
  substrate specificity
    methyltransferase, 277
    restriction endonuclease, 277
    sequence, 277
  subunit structure, 278–279
  types, 275

Type II restriction endonuclease, see also EcoRI
  activity assay, 246
  base analogs and inhibition, 244–245
  characterization of cleavage sites
    computer-assisted approach, 246
    direct sequencing, 246
  double digestion strategies
    difficulties, 247–248
    direct digestion, 248
    spacer-carrying vectors, 248
  site specificity, factors affecting
    activator DNA, 242–243
    DNA-binding proteins, 243–244
    DNA methylation, 243
    intercalators, 243
    oligonucleotide analogs, 244
    spermidine, 242
    triplex-forming oligonucleotides, 244
  genetics, 249
  number of discovered enzymes, 239
  optimal reaction conditions, 247
  sources, 249
  structure, 249
  substrate specificity
    DNA–RNA heteroduplex, 240
    isoschizomers, 240–241
    secondary activity, 242
    sequence, 239, 241
    single base mismatch tolerance, 241
    site structure, 239–240
    star activity, 241–242
    termini structure of products, 239
Type III restriction endonuclease
  endonuclease
    reaction conditions, 281
    substrate specificity, 281–282
  genetics, 283
  methyltransferase
    mechanism, 282
    reaction conditions, 282
    substrate specificity, 282
  reaction specificity, 281
  sources, 283
  structure, 282–283
Tyrosine, optical absorption, 24, 685

Uncompetitive inhibition, 39–40
Urea, effect on nucleic acid melting temperature, 67

Water, quality for enzyme solutions, 683
Watson–Crick base pairing, 53, 56
Wobble base pair, 55
Writhe, nucleic acid conformation, 82